Vertebrate Zoology

Fatik Baran Mandal
Bankura Christian College
Bankura, West Bengal, India

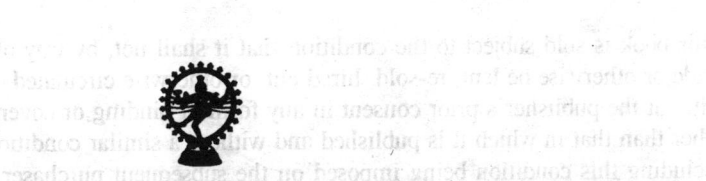

OXFORD & IBH PUBLISHING CO. PVT. LTD.
New Delhi

Oxford & IBH Publishing Company Pvt. Ltd.
113-B Shahpur Jat,
Asian Games Village Side
New Delhi 110 049, India

Fax: (011) 4151 7559
Email: oxford@oxford-ibh.in

ISBN 978-81-204-1769-4

Printed at Chaman Enterprises, New Delhi.

Preface

We are familiar with the fascinating world of vertebrates. We, the *Homo sapiens sapiens*, are also vertebrate. Vertebrate's history extends over 500 million years from the late Cambrian up to now. The first vertebrate had no jaw. Jawed vertebrates appeared 100 million years later in the Silurian period. Vertebrates show relationship with other chordates, with their own unique characters.

Animals have diverse features. Morphological features were the only key for identifying animals. Now anatomical-, physiological-, and cytological characters make clear distinction between the animals. External structure, size and shape, divisions of body, and nature of superficial area like scales, feathers, skin, and hair are the morphological features. Anatomy of nervous -, respiratory-, digestive-, circulatory- , reproductive and muscular system is important in studying evolution. Physiological processes, including contractility, support, sensitivity, digestion, and circulation, exhibit gradual complexity from simple organism to complex organism. Knowledge about animals beginning from the protozoa to mammal, their variability, and importance are documented but yet to be completed. The issues and challenges involved in the biology of animal are becoming more vital in the context of anthropogenic causes of biodiversity loss, particularly of higher vertebrates and obviously a matter of concern for our bio future. The characteristics, anatomy, evolutionary trends, classification, and importance of animals constitute the foundation of Zoology. The present book will help the student to grasp the key concepts and understanding chordate biology. Other activities involving problem solving would help them to acquire some skills.

Students may enjoy extra benefit in studying ahead of their classes. Students should use the textbook regularly. Zoological terms are often a barrier to the students. Students may consult the dictionary of Zoology to overcome this problem. This book has been written consulting the contemporary literature to help the students as well as to enhance their interest in the subject. Sufficient illustration are used in the book to supplement the text.

However, I do believe that still there is room for improvement. I shall appreciate constructive criticism and suggestion for further improvement of the book in the future.

May, 2013   **Fatik Baran Mandal**

Acknowledgements

Publication of this book has been made with the support and encouragement of my many well-wishers, teachers, friends, colleagues and family members. My teacher, Professor Samen kumar Maitra, Visva-Bharati, Santiniketan, West Bengal was always a source of inspiration to me. I would like to express my sincere thanks to Dr. N. C. Nandi, Former Additional Joint Director, Zoological Survey of India, Kolkata for his encouragement. I am also thankful to my few students for their help. I am also thankful to Mr. Barun Mitra, for typing the whole manuscript. Lastly, I also wish to thank my wife, Amita for her encouragement during this project.

Contents

vi

viii

x

1

Introduction

Chordates have considerable uniform general body plan. They show self-maintenance, occupy diverse habitats and show four features, some only during the embryonic stage. The four features are the presence of notochord, presence of dorsal hollow nerve chord, presence of pharyngeal gill slits, and presence of muscular post-anal tail. Chordates, arthropods, and pulmonate molluscs now dominate the terrestrial life. Successful aquatic and amphibious types including frogs and toads are worldwide in distribution. Bony fish in fresh water shows much variety. Notochord, a flexible rod between the digestive tract and nerve cord, is made of fluid-filled cells and remains surrounded by stiff, fibrous tissue to provide skeletal support. Muscles work against it for movement.

Chordates, belong with the Deuterostomia, are bilaterally symmetrical. They are divided into Subphylum Cephalochordata, Urochordata, and Vertebrata. Vertebrates arose as marine species. Appearance of limbs help them to colonize in the land. About 52,000 vertebrate species constitute about 5% of animal diversity. In vertebrates, jointed vertebrae develop around the notochord. In humans, notochord persists as disks in the spine. In chordates, the nerve cord develops from ectoderm during the embryonic stage and is located dorsal to the notochord. It finally develops into central nervous system along with brain and spinal cord. The pharyngeal slits develop below the mouth of embryo and function as suspension feeding apparatus, allow water to enter mouth and exit without passing through digestive system for gas exchange. Except the tetrapods, the pharyngeal slits develop into gills.

In chordate embryo, tail containing skeletal and muscle elements is used in swimming or balance. Development of head, brain, and brain box has resulted in complex, coordinated movement and behaviours. Hoxgenes differentiate the lanceolates and tunicates from the craniates or vertebrates. Neural crest cells produce various structures. Craniates have digestive tract with muscle for digestion and metabolism. Circulatory system includes at least a two-chambered heart, rbc with haemoglobin, and blood vessels. Kidneys remove waste products.

Hagfishes with cartilaginous skull lack the jaws .They use segmented muscles against their rigid notochord to swim. A small brain, eyes, ears, nasal opening, and tooth-like formations are found in them. Of 30 described living hagfish species, most live as bottom dwelling scavengers. The slime glands produce slime, a deterrent for predators which suffocate fish, when they attempt to swallow a hagfish. Jawless agnathans and jawed gnathostomes share some traits. Vertebrates differ with other chordates. In adult vertebrates, the notochord is either rudimentary, or absent and is replaced by a cartilaginous vertebral column. Outgrowths of the vertebral column extend dorsally to protect the nerve cord. Vertebrates may be provided with, or without jaws.

The jawless lamprey is mostly parasitic, marine, or freshwater species with a toothed, sucking mouth. Lamprey larvae are suspension feeders, but adults parasitize fish by attaching to their sides with sucking mouth. Lampreys have large eyes, one nostril on the top of head, and seven gills on each side, but they lack paired fins.

1.1 Vertebrate Ancestry

As annelids and vertebrates possess metamerism, annelid ancestry of vertebrates is suggested. Muscular system, coelomic pouches, and nephridia are similar in both vertebrates and annelids. Nephridial ducts in both transport the sex cells. However, the evolution of continuous pronephric duct of vertebrates from separate ducts of annelid is difficult to explain. Ventral nervous chain of annelid is comparable with spinal cord of vertebrates including dorsal roots of nerves with spinal ganglia. Transverse blood vessels of annelids are similar to the aortic arches and inter costal vessels of vertebrates. Brain of annelid is comparable with vertebrate forebrain. A sub intestinal tube of some annelid has been doubtfully compared with notochord, but structures homologous with gill slits in annelids are the unknown. Metamerism, relation of their nephridia to coelom, and relationships of vascular and nervous systems to digestive tract in annelids resemble the conditions of lower vertebrates and vertebrate embryos. Metamerism, the marked resemblance between vertebrates and annelids may have arisen independently and may be the result of convergent evolution. By inverting the position of body of an annelid, the fundamental systems can be brought into the arrangement as in vertebrates. Such a shift in body orientation is common in animal groups. Notochord, a distinctive feature of chordates, has a counterpart in supporting fibers accompanying the annelid nerve chain (Figure 1.1).

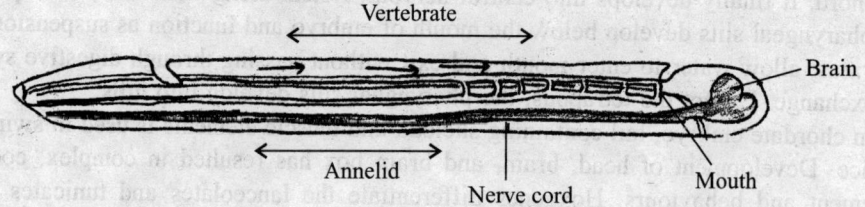

Figure 1.1 Possible transformation of an annelid into a vertebrate.

Hubrecht advocated nemertine stand in direct line of vertebrate ancestry, the arguments being the homology between the proboscis sheath of nemertine and notochord of chordates. Lateral nerve cords of nemertine could assume a dorsal position like chordates through a change in body orientation. Chordates are related to echinoderms and their allies. Adoral ciliary band of auricularia probably carry food into the mouth for which it is turned into the floor of pharynx. Garstang suggests that endostyle has derived from loop of adoral band. Pharyngeal method of food-collection, thus, has been replaced by tentacles in adult. The whole endostyle and atrium become developed for protecting the gills. However, the similarity of arrangement of pharynx in tunicates, *Amphioxus*, and cyclostomes is remarkable.

1.2 Trends in Vertebrate Evolution

Evolution is the central unifying theme in biology. Understanding evolution gives a full insight of the variety, relationships, and functioning of living organisms. It also helps in the significant study of anatomy, physiology, behavior, interactions among organisms, and the history of species. Theodosius Dobzhansky (Figure 1.2) wrote an article entitled, "Nothing in Biology Makes Sense Except in the Light of Evolution". So, to understand biology, one should develop understanding about

- Systems, order, and organization,
- Evidence and explanation,
- Constancy and change,
- Form and function.

Figure 1.2 Theodosius Dobzhansky.

Biological evolution with a historical dimension explains the unity and diversity of species, the history and diversity of life on earth. The diversity of life has evolved over time by mutation, selection, and genetic change. The evolution drives the diversity and unity of life. Life on earth evolved over billions of years and every gene, individual and species is the result of a historical process. Lineages have evolved only once by the unique combination of genes, environments, and chance. Describing the tree of life completely is difficult, but can be continuously explored. Evolution involves trade offs between costs and benefits. Big brain has advantages, but makes birth more difficult. Standing upright has advantages along with chance of back pain. Organisms living today have homologies where similar structures, or functions are inherited from a common ancestor.

Biological systems use free energy to grow, reproduce, and maintain dynamic homeostasis. Living systems store, retrieve, transmit, and respond to information. Interactions in biological systems are the complex. Kinship between vertebrate and invertebrate stand as a proof of vertebrate origin from invertebrate and demonstrates undisputed interrelationships between chordates and non-chordates. Chordate life began probably at sea as ciliated larvae. Ciliary action of pharynx and gill slits, play role in respiration and food collection. Oxygen was carried by colourless blood. Circulatory system perhaps involved an irregular set of intercellular spaces. Distinct contractile vessels, blood with distinct composition appeared later (Figure 1.3). Excretion perhaps involved no specialized cells, but carried over by the whole body surface. Epidermis later became thickened dorsally in walls of neural tube. Special receptor organs were simple and present in skin, or within tube, with little density at the front end without brain. The stimulus from the surrounding world activates the receptor organ. Large masses of nervous tissue were perhaps absent. Specialized glands for internal secretion were absent. Reproduction was perhaps sexual with development followed by radial cleavage, and gastrulation by invagination. Young was provided with yolk for development, but not cared otherwise by their parents.

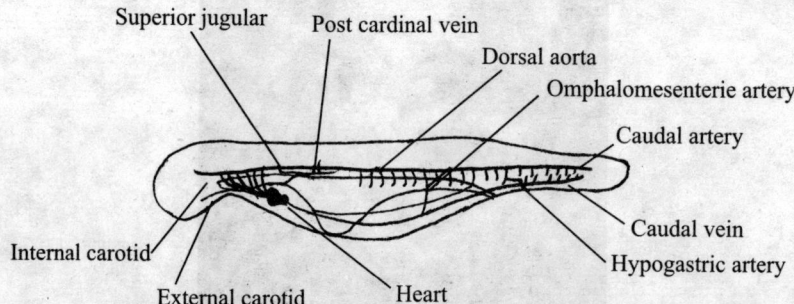

Figure 1.3 Circulation in an early stage of a small-yolked vertebrate (amphibian).

Evolution of chordates shows that since the Cambrian their organization faced situations different from the sea surface. The earlier changes include the transfer from sea surface to fresh waters and to sea bottom. Entrance of animals into fresh water required adjustment. Development of jaws about 350 million years ago added a further quantum to the evolution. Heavy armour of the early animal was given up. Body form was improved with development of air bladder in fish. Some fish left the water in the Devonian and gradually developed devices to meet the new conditions. Modern amphibian, reptiles, birds, and mammals have been produced through this process. They now live in the varied situations on the land and water. The rate of evolutionary change is now measurable. Changes may be considered on a percentage rather than an absolute basis by considering the time required for a unit increase in natural logarithm of a variant, or one standard deviation.

1.3 General Features of Vertebrate

Vertebrata, a subphylum of Chordata, has the basic chordate features along with others as well. Vertebrate's history goes back to more than 500 million years. In vertebrates, the front end of nervous system is differentiated into brain associated with special receptors, nose, eye, and ear. Through receptors they respond to environment. Skull is developed as a skeletal thickening around the brain, mainly for protection, but later for attachment of elaborate muscle systems. Some can discriminate between visual shapes and colours and in auditory field between patterns of tones, and between hosts of chemical substances. Motor organization aids in movements to suit situations. The swimming by passage of waves down the body brings about changes in shape of fish. Besides median fins lateral paired ones, serve at first for stabilizing and steering and then converted into organs of locomotion on ground, or air to suit the individual into changing environment. The earliest vertebrates collected food

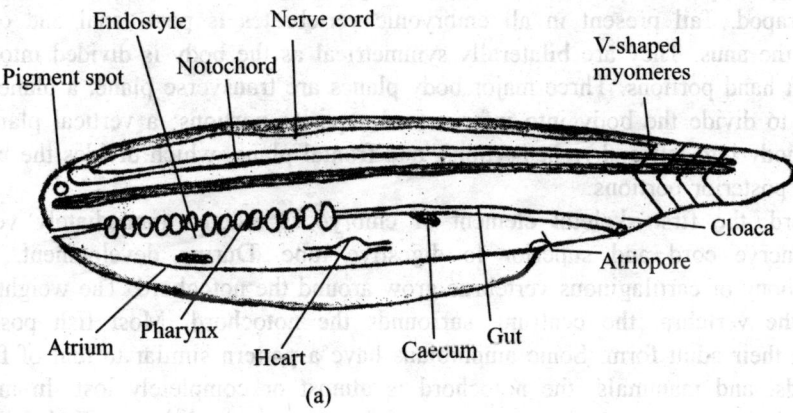

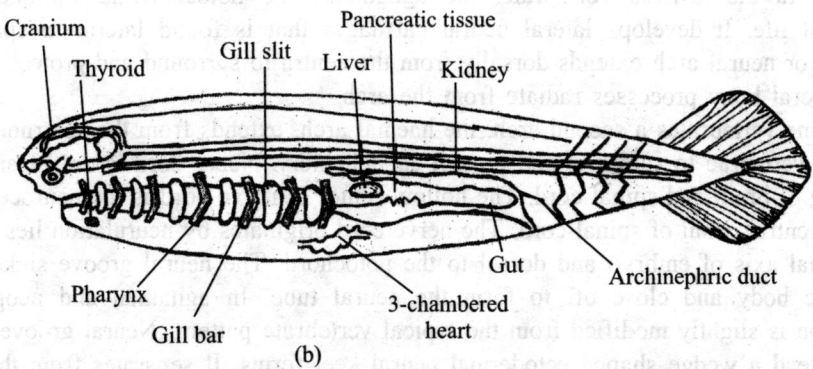

Figure 1.4 (a) Generelized non-vertebrate chordate design.
(b) Primitive hypothetical vertebrate.

by ciliary action, which has long been left except rare cases. Pharynx of most vertebrates is small. There are relatively few respiratory gill-slits. Generally, the anterior arches between gills are modified to form jaws, to seize and hold food and to manipulate the environment. Efficiency of circulatory system aids in dominance of vertebrates. In air- breathing birds and mammals, the respiratory and circulatory systems allow the huge amount of energy expenditure per unit mass of animal, so that extravagant device can be used for survival under conditions that would otherwise not support life. Excretory system consists of mesodermal funnels, leading primarily from coelom to exterior. Elaborate devices for regulation of osmotic pressure are developed and mesodermal kidneys play a role in such regulation (Figure 1.4).

1.4   Organ Systems

Typical vertebrate body consists of head, trunk, and tail. Head develops through cranialization and contains the brain, sense organs, often jaws, and gill in fish .Trunk contains the coelom that houses the visceral organs. Body wall surrounds the coelom .Trunks mostly have paired pelvic and pectoral appendages. Neck connects the head with trunk in tetrapod. Tail present in all embryonic vertebrates is post- anal and originates posterior to the anus. They are bilaterally symmetrical as the body is divided into 2 equal right and left hand portions. Three major body planes are transverse plane, a plane running horizontally to divide the body into inferior and superior portions; a vertical plane which divides the body into left and right portions, and frontal plane, which divides the body into anterior and posterior portions.

Notochord, the first skeletal element of embryo is located immediately ventral to developing nerve cord and superior to digestive tube. During development, in most vertebrates, bony or cartilaginous vertebrae grow around the notochord. The weight bearing portion of the vertebra, the centrum, surrounds the notochord. Most fish possess the notochord in their adult form. Some amphibians have a pattern similar to that of fishes. In reptiles, birds, and mammals, the notochord is almost or completely lost. In mammals, remnants of the notochord remain as portions of the intervertebral discs called the Nucleus Pulposus. In the jawless vertebrate, the agnathans, the notochord grows and remains throughout life. It develops lateral neural cartilages that is found lateral to spinal cord. Vertebral or neural arch extends dorsally from the centra to surround and protect the spinal cord. Several bony processes radiate from the arch.

In some vertebrates a second arch, the haemal arch, extends from the centrum ventrally in caudal vertebrae to surround and protect caudal artery. Nerve cord is dorsal and hollow consisting of brain and spinal cord. The hollow center of nerve cord is the neurocoel, which includes central canal of spinal cord. The nerve cord originates by neurulation lies along the longitudinal axis of embryo and dorsal to the notochord. The neural groove sinks into the embryonic body and close off to form the neural tube. In agnathan and neopterygian, neurulation is slightly modified from the typical vertebrate pattern. Neural groove does not form; instead a wedge shaped ectodermal neural keel forms. It separates from the surface ectoderm, form a cavity to become a nerve cord. Nerve cord expands anteriorly to form brain.

Cranial and spinal nerves develop and radiate out from the nerve cord. These nerves communicate between the CNS and rest of the body. Many organ systems comprise the body of advanced animals. Level of organization in animals ranges from atoms to organ system. Eleven major organ systems are generally present within animals. Head contains four senses and a brain. Muscular system aids in movement, produces heat, maintains posture, and supports the body.

Integument is composed of skin and hypodermis. Skin, the protective layer prevents water loss and invasion of pathogen into the body. The epidermis is the outer, thinner waterproof layer of skin because of keratin. Dermis, the layer of fibrous connective tissue, contains sweat glands, hair follicles and oil glands. The subcutaneous layer is made up of loose connective tissue. In some animals, it produces dermal bone and aids in the production of scales. Adipose tissue of this layer serves for insulation.

Pharynx shows the relationship between vertebrates and other chordates. It produces gills in fish, and lungs, some endocrine glands, and the middle ear in tetrapods. It is a source of stem immune cells in human foetus. Vertebrate embryos exhibit a basic pharyngeal architecture. Pharyngeal pouches arise as outpocketings of gut endoderm and migrate towards the surface. Ectodermal grooves grow towards the developing pouches. Finally, the two structures grow close together and remain separate by a thin membrane, the branchial plate. Branchial plate ruptures to produce the pharyngeal slit. In fish, pharyngeal slits are found throughout life. In tetrapod, pharyngeal slits are temporary structures.

The columns of tissue called the pharyngeal arches are found between the adjacent pharyngeal pouches and gill slits. Each arch has 4 components namely supportive skeletal elements also termed the pharyngeal or visceral skeleton, skeletal muscles called the branchniomeric musculature due to its relation with gills to operate the arch, cranial nerve including both sensory and motor branches to innervate the arch, an aortic arch to connect the dorsal and ventral aorta. These 4 components are found anterior to first pharyngeal arch and often posterior to last pharyngeal arch. The anterior most arch or the first pharyngeal arch is the mandibular arch as it contains the upper and lower jaws and related structures. The second arch is the hyoid arch. The remaining arches are numbered 3 through 7 sometimes called branchial arches 1 through 5 as they resemble unmodified gill arches. Cranial nerves that send branches into pharyngeal arches are CN 5, 7, 9, & 10.

The space between the body wall and digestive tact is the coelom. The coelom in fish, amphibian, and reptile is divided into pericardial cavity which houses heart and the pleuroperitoneal cavity which houses the rest of viscera. In tetrapod, the lungs are located in this cavity. A fibrous partition, the transverse septum in tetrapod, separates the pleuroperitoneal cavity from pericardial cavity. The coelom of birds and mammals is divided into thoracic cavity having a pericardial cavity for heart and a pair of pleural cavities for lungs. In some, there is a mediastinal cavity. The abdominopelvic cavity has an abdominal/peritoneal cavity and a pelvic cavity.

Respiratory system exchanges gas between the lungs and external environment, maintains blood pH, and facilitates exchange of CO_2, and O_2. Digestive system digests and absorbs food, eliminates wastes and recycles water. Most vertebrates conduct external respiration by well vascularized membranes which are located in the pharyngeal pouches, or are derived from the pharyngeal floor. Digestive tube and several accessory organs constitute the digestive system. Digestive tube is a long tube running from the mouth to cloaca. It has several regional specializations along its length. These specialized regions include the oral cavity, pharynx, oesophagus, stomach, small intestine, and large intestine. Accessory organs including the liver, pancreas, and gall bladder located outside the digestive tube, release their products into tube by ducts. Circulatory system including heart, capillaries, arteries, and veins transports O_2, CO_2, nutrients, waste products, immune components, and hormones. The lymphatic system transports excess fluids to and from the circulatory system. The urinary and reproductive systems are often considered as single system, the urogenital system due to their shared origins and structures. Kidneys and gonads develop close together on the coelomic roof. Both the male reproductive and urinary systems use the urethra in mammals. The urinary system removes wastes from the body to maintain homeostasis. Vertebrates have a closed circulatory system where muscular heart pumps blood through vessels. Immune system protects the body from infection. Nervous system coordinates and controls actions of organs and body systems. Memory, learning, and conscious thought are the functions of the nervous system. This system operates heart beat, breathing, and control of involuntary muscle actions. Endocrine system works in coordination with nervous system to control internal organs and long-range response to external stimuli, regulate body metabolism, growth, and reproduction. Reproductive system works under the influence of the endocrine system for survival and perpetuation of species.

Skeleton is internal and composed of cartilage/bone, forms a framework for body shape, the anchoring of muscles for movement, and protection of fragile organs. Skeleton is divided into two portions. Axial skeleton is composed of the skull, vertebrae, and rib cage. Appendicular skeleton is composed of limb bones and girdles. Skeletal system, besides providing rigid framework for movement and protection, serve as attachment sites for muscles.

1.5 Homeostasis

Homeostasis maintains stable internal environment and functioning of its components. Enzymes function best within a range of temperature and pH. The internal environment is maintained for maximum efficiency. The ultimate control of homeostasis is governed by the nervous and endocrine system. Often this control works as negative feedback loops. Heat control, a major homeostatic function involves the integration of skin, muscular, nervous, and circulatory systems. Negative feedback turns off the stimulus that caused it in the first place. Positive feedback amplifies the stimulus by reaction. Negative feedback causes a reverse response. Blood levels of TSH serve as feedback for TSH production. Positive feedback control during uterine contractions produces oxytocin that increases the frequency and strength of contractions which in turn causes further production of oxytocin. Homeostasis depends on action and interaction of body systems to maintain conditions within which the body can perform the best.

1.6 Classification

An outline classification of the Phylum Chordata is given below:

Subphylum UROCHORDATA
Subphylum CEPHALOCHORDATA
Subphylum CRANIATA
Superclass Agnatha
Superclass Gnahostomata
Pisces
Class AMPHIBIA
Class REPTILIA
Class AVES
Class MAMMALIA

Urochordates

Urochordates consist of three classes namely Ascidacea, Thaliacea and larvacea. Ascidians are tiny, short living animals, found in colonies, or solitary. The larval form consists of trunk and muscular tail. Tail is locomotory and possesses a notochord. Running along the notochord, there are 36 to 1,600 uninucleated, striated muscle fibers. In larva, the viscera are reduced as they do not feed. Nervous system in larva consists of a dorsal, tubular hollow nerve cord, several ganglia, and nerves. The primitive brain includes few ganglia with other sensory structures. Ocellus, a receptor for light sensitivity and otolith, a receptor for stereoreception are also found. In respiration, water is drawn into mouth, passes through pharynx over the gills, and then into a chamber called the atrium. Oxygen depleted water exit the atrium via the atriopore. When the larva metamorphoses into adult, three adhesive papillae form to attach the sessile adult to a substrate. Notochord is reabsorbed and used as nutrition source to supply the energy requirement of metamorphosis, and the viscera changes. Mouth becomes an incurrent siphon. The atriopore transforms into an excurrent siphon for filter feeding. A ciliated, glandular endostyle develops on floor of pharynx, which traps food and passes it to oesophagus by ciliary beating. The ascidian also develops a stomach and intestine. Wastes are dumped into the atrium and are released along with deoxygenated water by atriopore. Larvaceans occur as free-floating plankton in waters of less than 100m depth and below the light penetrance level. Their small body measures about 8 mm in length and notochord supports a long tail. Thaliaceans are free living or colonial, alter between free living and colonial forms over succeeding generations, and resemble adult ascidians. They lack a tail. They propel themselves by squirting water out of the atriopore.

Cephalochordates

Branchiostoma is also called the lanceolets due to their cylindrical bodies. This is small, marine, found in sandy beaches leaving only their heads exposed to filter feed. Body consists mostly of a trunk with muscle termed myomeres for locomotion. Myomeres are composed of uninucleated, striated muscle fibers. Adjacent myomeres are separated by a myoseptum, or myocomma. It has a semitransparent skin composed of 2 layers. Epidermis is composed of a single layer of cells. Thin dermis facilitates cutaneous respiration. Integument also has thin hypodermis. They possess pharyngeal gill slits which do not open directly to exterior instead open into the atrium, a fluid-filled cavity that surrounds the pharynx. Food cleansed waters are pumped from the pharynx into atrium and expelled out of body by atriopore. The slits facilitate filter feeding. Number of pharyngeal slits is up to 60. The pharyngeal bars reinforce them. Notochord runs from rostrum to tip of the tail. Notochord consists of arranged muscular discs. Muscular contractions increase the stiffness of notochord for digging. CNS consists of brain, spinal cord and is covered by vascular membrane. Brain has two subdivisions. Special senses are rudimentary. Chemoreceptor are scattered over the body surface and are particularly concentrated in anterior pharyngeal region and mouth. Light sensitive pigmented ocelli remain embedded along the length of spinal cord. Anterior to the mouth, there is vestibule surrounded laterally by the oral hood to collect seawater. Caudally, the mouth perforates the velum. Mouth opens into pharynx. Cilia on pharyngeal bars pull water through vestibule, through the mouth, and into the pharynx. Within the vestibule, there is a wheel organ with stubby projections to trap food missed by the mouth. The opening of vestibule is lined by buccal cirri to strain partially incoming water. Buccal cirri have chemoreceptors to monitor incoming water. Food is processed in pharynx which requires some specialized features. The pharynx has a ventral invagination lined with cilia and mucus called the endostyle or hypobranchial groove. A dorsal invagination is called the epibranchial groove. Pharyngeal bands connecting the 2 grooves cover the pharyngeal bars. Mucus secreted by these structures trap food. Mucus and food become sticky mucus strands that are propelled by ciliary beating into epibranchial groove and then pass caudally into gut. In the gut, digestive enzymes are added to food strand. Ciliary beating pushes the food-enzyme-mucus mixture into intestine. Some mixture is passed to caecum for incorporation of more enzymes. Intestine ends at anus. Coelom is reduced in adult, becomes compressed as the larvae elongate and more pharyngeal bars are added. Circulation differs from that of vertebrates in absence of heart and formed elements. Blood is colorless serum. The venous pathways are similar those of embryonic vertebrates. They lack an organized kidney. Cyrtopodocytes filter blood which is intermediary between the protonepheridia and podocytes. In adult, the gonads are prominent and bulge into atrium. Gonads release sperm or eggs into atrium by atriopore. *Branchiostoma* is dioecious as they develop only ovaries, or testes. It resembles the larval form of the lamprey called the ammocoete.

Agnathans

The vertebrates can be divided into 2 super classes based on presence, or absence of jaws. The jawed vertebrates belong to the Gnathostomata. The vertebrates lacking jaws are the Agnathostomata. Agnathans are the most primitive vertebrates. Today only two extant groups– lampreys and hagfish exist. Several extinct agnathans called the ostracoderms were armored fish-like organisms of the late Cambrian through the early Devonian. Ostracoderms are the oldest known vertebrates with fossils dating back to Ordovician. They were perhaps existed before late Cambrian. Ostracoderms were small marine species, 2 to 3 cms in length; some grew up to 1 m in length, lacked jaws and paired fins, but with a body of overlapping dermal plates. The plates were largest on head forming a bony shield and smaller, more tile-like bones on trunk and tail. Ostracoderms had 3 eyes. A pair of upwardly oriented orbital eyes was present beside a 3rd singular eye called the pineal eye. They had a single naris which opened into a nasopharyngeal duct connecting to an olfactory sac. They had a small, jawless mouth and rows of external gill slits. Along with dermal bone, they also had an endoskeleton composed of bone and cartilage.

Extant Agnathans

Hagfishes and lampreys are not closely related and belong to 2 separate classes. Hagfishes belong to class Mixini. Lampreys belong to class Cephalospidomorphi. Extant agnathans share some common characteristics. Their prominent notochord forms the axial skeleton. They lack a bony skeleton, scale, dermal bony plates, bony jaws, and vertebral column. These missing features perhaps evolved to maximize the life span as a parasite. Agnathans have few semicircular canals than gnathans. Lampreys have 2 semicircular canals. Hagfish has only one. Buccal funnel surrounds the mouth. Hagfish is the marine, eel-like, bottom dwelling scavengers with shallow, buccal funnel surrounded by finger-like papillae resembling short tentacles and vestigial eyes covered by a thin, opaque membrane. Lampreys have better developed eyes. Number of gill slits varies between species. They resemble hagfish being eel-like aquatic organisms. Most are marine, but some spend part of their lives in freshwater. Lampreys have changed little from the Carboniferous. They are parasitic with buccal funnel and tooth-like structures that anchor them to its prey. They consume the blood of fishes. Muscular tongue coated with tooth-like denticles scrap away the flesh of prey. All lampreys have 7 pairs of gill slits.

Pisces

Placoderms named for their bony dermal plates are the earliest jawed vertebrates existing during Paleozoic era and during Devonian Period. They were evolved from the ostracoderms and are now extinct. They ranged in size from few inches in length to 20 ft and had paired pelvic and pectoral fins. Chondrichthyes is an ancient class with 2 extant subclasses; Elasmobranchii and Holocephali. Elasmobranches have a cartilaginous skeleton. The only bony component is found in teeth and placoid scales. Male elasmobranchs have claspers as intromittent organs. Elasmobranchs consists of sharks and rays. They first arose in late Silurian period and exist today. They have exposed gill slits, typically 5 pairs, and 2 spiracles. Holocephalians are called the chimera. They differ from elasmobranchs in operculum that covers the gill slits and bony plates replaced the teeth.

Acanthodians are named for their hollow spines. Along with spines, bony plates also cover the body. They had an operculum. Skeleton had both cartilaginous and bony elements. They became extinct in Carboniferous. Osteichthyes are ray-finned fishes including both ancient and modern bony fish and is divided into Actinopterygii and Sacropterygii. They first arose between the early Devonian and late Silurian. They have membranous fins supported by bony rays, an operculum and scales. Chondrostei, a primitive group of osteichthyes have few extant species with ganoid scales, now represented only by the paddlefish and sturgeons. Endoskeleton composed mostly but not exclusively of cartilage. Unlike the chondrosteins, their skeleton is mostly ossified. Teleosts are the modern successful group of vertebrates which make up 96% of all extant fish species and have three suborders. Lobe-finned fishes have a fleshy lobe to their paired fins. Sarcopterygians gave rise to tetrapods. They have internal nares that open internally into pharyngeal cavity. They retain a gas filled air sac. Sarcopterygians have two orders: Actinistians, the coelacanths; Rhipidistians, the lungfishes.

Amphibia

Amphibians are the first tetrapod appeared in the late Devonian and began to radiate in Carboniferous. The class is divided into Labyrinthodontia and Lissamphibia. Labyrinthodonts are a diverse group with similar pattern on apical surfaces of their teeth. They share many features with rhidipstian fishes such as dermal scales, a fish-like tail supported by rays, rhidipstian-like skulls, and the lateral line at least on skull. One group of labyrinthodont was the *Anthracosaur* which has perhaps given rise to amniotes generally and to reptiles particularly. They include the Seymouria and Temnospondyls. Temnospondylian skulls display features of modern amphibians and are the ancestor of Lissamphibia. Lissamphibians are the modern amphibians and include Anura/Salienta, Urodela/Caudata and Apoda/Gymnophiona

Reptilia

The reptiles, the first amniotes lay shell-covered eggs. Reptiles developed from labyrinthodonts most likely an *Anthracosaur* during the early Carboniferous show terrestrial adaptiveness.Two reptilian lineages are suggested based on skull structure. Synapsids lineage gave rise to mammals and has perhaps evolved separately from other reptiles. The skull possesses one temporal fossa on each side. Reptilia are divided into 2 subgroups based on skull structure. Anapsid skulls lack temporal fossa as found in basal reptiles. Today, the only extant representative is the turtle. Anapsids gave rise to the diapsids. Diapsid skulls have 2 temporal fossa on each side. Diapsids are all the extant reptiles and many extinct species include squamates, snakes and lizards, archosaurs, dinosaurs, and crocodilians. The diapsid reptiles gave rise to euryapsids that have lost one pair of temporal fossa. Euryapsids are the plesiosaurs and ichthyosaurs. Reptiles are ectothermic, covered by scales composed of cornified epidermal scales to decrease dehydration. They have a neck with better developed limb girdles and claws on their digits. They possess the metanepheric kidney; display the partial or complete division of heart into right and left chambers.

Aves

Aves are bipedal, endothermic amniotes, covered with insulating feathers and are evolved for flight. Origin of birds is still controversial. They have perhaps evolved from small theropod dinosaurs or share a common ancestor with them. Both the groups share some anatomical features. Lightweight bones many of which possess air sacs, loss of teeth, modified forelimbs are traits for flight. Of two generally recognized subclasses, Archaeornithes are the extinct, ancient birds while the Neornithes are the modern birds.

Mammalia

Mammals are endothermic with insulating hair .Females produce milk for their young. Several features found in modern mammals were present in therapsids such as 2 occipital condyles, a secondary palate, and heterodontic teeth. Mammals first appeared in the Triassic. Modern mammals display a synapsid skull. Mammalian features are 3 auditory ossicles, muscular diaphragm, and absence of cloaca in adult in all mammals except the monotremes. Majority possesses sweat glands, heterodontic teeth. Homodontic teeth of cetaceans are a derived characteristic. Loss of the fourth aortic arch, a single dentary bones on each side of lower jaw that articulates with squamosal and presence of pinna to gather sound waves, specialized larynx, development of cerebral cortex are mentionable. Prototheria are egg laying mammals.Metatheria give rise to live young. Metatheria gives birth to almost larval young that migrate to, and finish development in a pouch. They lack a placenta. Yolk sac serves the role of placenta.

Summary

1. Chordates having considerable uniform general body plan shows 4 prominent features: presence of notochord, dorsal hollow nerve chord, pharyngeal gill slits, and muscular post-anal tail.

2. Chordate life began probably at sea as ciliated larvae. Ciliary action of pharynx and gill slits serves for respiration and food collection. Nerve cord, located dorsal to notochord, develops from ectoderm during embryonic stage .It finally develops into central nervous system along with brain and spinal cord. Pharyngeal slits develop below the mouth of embryo for suspension feeding, allow water to enter mouth and exit for gas exchange, without passing through digestive system

5. In urochordates, the pharynx has some specialized features for food processing. The pharynx has a characteristic ventral invagination lined with cilia and mucus, called the endostyle or hypobranchial groove.

3. Of 30 described living hagfish species, most are bottom dwellers. The jawless lamprey is mostly parasitic, marine, or freshwater species. They have large eyes, one nostril on the top of head, and 7 gills on each side, but they lack paired fins. Lamprey larvae are suspension feeders.

4. In the jawless vertebrate, the agnathans, the notochord persists throughout life. The jawed vertebrates belong to the Gnathostomata. The vertebrates lacking jaws are called the Agnathostomata. Agnathans are the most primitive vertebrates.

4. Vertebrates arose as marine species .Appearance of limbs help them to colonize the land .About 52,000 vertebrate species constitute about 5% of animal diversity. They have jointed vertebrae around the notochord. The front end of nervous system is differentiated into brain with special receptors. Typical vertebrate body consists of head, trunk, and tail. Head contains the brain, sense organs, often jaws, and gills in fish .Trunk mostly have paired pelvic and pectoral appendages. It contains the coelom that houses the visceral organs. Body wall surrounds the coelom.

5. Metamerism in annelids and vertebrates advocates the annelid ancestry of vertebrates. Muscular system, coelomic pouches, and nephridia are similar in both. Nephridial ducts transports the sex cells in both. Ventral nervous chain of annelid is comparable with vertebrate's spinal cord including dorsal roots of nerves with spinal ganglia .Lateral nerve cords of nemertine could assume a dorsal position like chordates through change in body orientation.

6. Bony fish in fresh water exhibits high diversity. Chondrichthyes, an ancient class has 2 extant subclasses: Elasmobranchii and Holocephali. Elasmobranches have cartilaginous skeleton. Bony component is found in their teeth and placoid scale. Osteichthyes, the ray-finned fish including both ancient and modern bony fish is divided into Actinopterygii and Sacropterygii.

7. Modern amphibian, reptiles, birds, and mammals are the products of evolution. Amphibian, the first tetrapod, appeared in the late Devonian. Reptiles developed from labyrinthodonts during the early Carboniferous. Aves are bipedal, endothermic amniotes, covered with feathers. In reptiles, birds, and mammals, the notochord is almost or completely lost. Mammals are endothermic with insulating hair .Females produce milk for their young .Remnants of notochord persist in them as portion of intervertebral discs called Nucleus Pulposus.

Review Questions

Short Answer Questions

1. Mention four features of the chordates.
2. What is notochord?
3. Mention the differences between chordates and vertebrates.
4. What is coelom?
5. What is "Nucleus Pulposus"?
6. What is neurocoel?
7. Mention two functions of skin.
8. What is epidermis?
9. What are pharyngeal arches?
10. Mention two functions of nervous system.
11. Name the components of axial skeleton.
12. Name the parts of appendicular skeleton.
13. What do you understand by negative feedback?
14. What do you understand by positive feedback?
15. Name three classes of Urochordata.
16. What is endostyle?
17. What do you understand by the term dioecious?
18. Mention one feature of Gnathostomata.
19. Mention the differences between elasmobranches and holocephalians.
20. Name two orders of Sarcopyerygia.
21. Name two living orders of the Class Amphibia.
22. What do you understand by anapsid skull?
23. What do you understand by diapsid skull?

24. Mention two primary characters of birds.
25. Give two salient features of Mammalia.

Long Answer Questions

1. Write a note on the ancestry of vertebrates with suitable diagram.
2. Give ten features of vertebrates.
3. Describe the evolution of chordates in light of recent study.

INTRODUCTION

24. Mention two primary characters of birds.
25. Give two salient features of Mammalia.

Long Answer Questions

1. Write a note on the diversity of vertebrates with suitable diagram.
2. Give the features of vertebrates.
3. Discuss the evolution of chordates in brief. recent study.

CHAPTER

2

Evolution, Taxonomy, and Zoogeography

Chemical processes on the primitive earth produced the precursor molecules for origin of life about 3.5 billion years ago. In 1953, Miller and Urey (Figures 2.1, 2.2) synthesized amino acids from the abiotic substances like those existed on the primitive earth. Organic soup of carbohydrate, amino acids, and lipid accumulated in the oceans with energy from lightning and UV radiation of the sun. Amino acids formed polypeptide during its drying. Heterotrophic protocells with cell membrane were appeared next. After becoming reproducible, the protocell was transformed into a true cell to start the evolution. These ancient cells probably had RNA as genetic code and later cells had DNA.

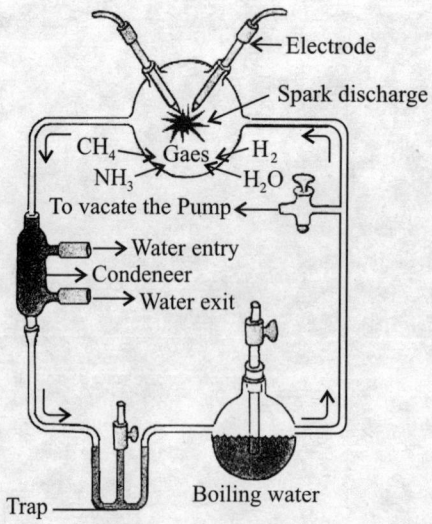

Figure 2.1 Experiment of Muller and Urey on the origin of life.

Figure 2.2 Harold Urey (Left), Stanley Miller (Right)

Life first evolved in conditions like the deep sea thermal vents where chemoautotrophic bacteria are found. Rock layers deep below the continents and ocean floors previously believed to be nutrient poor to sustain life, now found to contain microorganisms 2 miles below the surface at 75°C. Such microorganisms derive their nutrients from carbon, hydrogen, iron and sulphur. Deep subterranean communities could have evolved underground, or on surface and become buried or transported down into subsurface rock strata. Old subterranean bacteria may have shaped many geological processes.

The earliest evidence for the cyanobacteria dated back between 3.5 and 2.75 billion years ago. Firstly, the photosynthetic organisms released oxygen into atmosphere. Before this, the atmosphere was composed of CO_2, N_2, CO, CH_4, H_2 and sulphur. It took over 2 billion years from the initiation of photosynthesis to reach the present level of atmospheric oxygen concentration. As oxygen level increased, some early anaerobic species became extinct; others became restricted to habitats free of oxygen. Anaerobic cells might, initially, have been incorporated into aerobic cells when aerobes engulfed them as food. Anaerobes perhaps invaded the aerobic hosts and parasitized them. An intimate symbiotic relationship between aerobic- and anaerobic cells made the survival of each cell dependent on other cell.

Symbiotic relationship shaped the complex characteristic of eukaryotes. The earliest single-celled eukaryotes are at least 2.7 billion years old. Over 2.7 billion years, some genes of the invading anaerobe have been lost, or transferred to nucleus of host, the aerobe cell. The genomes of ancestral invader and ancestral host thus have mingled and the two entities are now considered as one from a genetic standpoint.

Understanding the evolution is the major achievement in science. Evolution explains the diversity of life, constitutes the foundation of modern biology and modern research for developing technologies to combat diseases, and to promote human well-being. Scientists observe nature, construct explanations of evolution based on evidence, and refine their ideas. Explanations are altered, or rejected based on contradictory evidence. Well established scientific explanations become scientific theories, which refer to comprehensive explanation of an important feature of nature supported by facts. Evolution stands on foundation of observation, experiment, and evidence.

Biological traits pass from parents to offspring. Some traits change between generations. When a new trait does better in its natural surroundings and produce more offspring, the trait becomes widespread. If the new trait makes the offspring to leave fewer offspring, the trait tends to extinct. Some traits succeed and others fail through natural selection. Many unsuccessful types became extinct. The basic facts of evolution are unquestionable. Fossil in rock supports the evolutionary concept based on which the evolution of species has been proposed.

2.1   Fossils

The fossils are the preserved remains of animals, plants, or their parts in various strata of earth, may be an entire organism, or a part which got buried, a mould, or cast, footprints, or imprints .Fossils form by several methods. Petrifaction is the common method in which dead and buried organisms turn into stones due to formation of sedimentary rocks under water. The soft parts disappear due to decaying; hard parts get preserved due to mineralization. Preservation happens in original strata. Sometimes muscles and other soft organs were mineralized to form rocky fossils. Moving animals on soft mud leave footprints which if left undisturbed, hardened to form rocky fossils. Imprints provide clues of body form and characters of extinct animals.

Fossilized moulds are found in volcanic ashes. Several invertebrate moulds provide details of the physical features of animal. Insects were entangled in resin. The dried material was fossilized which even show the colour of organism. Entire animal can get frozen and preserved in ice. In such fossils, the body parts remain intact. Fossils of woolly mammoth from Siberia are classical example of this nature. The age of fossils can be assessed accurately, using radioisotope method. Fossils tell the story of evolution and give clues about climatic conditions of prehistoric periods. Study of fossils also simplifies the phylogenetic discussions.

2.2   Extinct Animals and Mass Extinction

Extinction is the termination of a lineage without issue, or abrupt disappearance of specific groups of organisms without leaving descendents. Extinction may be true extinction, or pseudo extinction. In true extinctions, a particular lineage totally disappears. In Pseudo extinction or phyletic extinction; a group may disappear leaving descendents with evolutionary changes. In horse evolution, the ancestor *Eohippus* became extinct, but its descendent survived to produce *Equus*. The extinction of dinosaurs as a group is the true extinction. Similar extinctions happened in trilobites.

Fossil record reveals few patterns of extinction. Major group of herbivorous vertebrates are susceptible than carnivorous vertebrates for extinction. Larger organisms easily became extinct. VanValen recorded constancy in extinction rate in several groups. He explained this using 'MacArthur's law', which states that every new adaptation encourages the survival of a possessor and decreases the fitness of other related species of that area. Extinctions are regular events in earth and are due to (1) a mass extinction as a result drastic changes in environment; (2) any adaptive advance in one species decreases the fitness of other species, (3) over specialization, (4) epidemic disease without control, (5) increase in population strength of herbivores causes rapid food shortage and extinction for interrelated groups, (6) cosmic radiation causes the death of large organisms, (7) dust storm due to falling of meteorite is a cause for disappearance of dinosaurs. In the history of earth; extinctions of major group of organisms were due to natural causes. By the end of Permian, nearly 60 percent of the varieties then existed, became extinct. Similar large scale extinctions occurred by the end of Mesozoic and during Cenozoic. Now extinction of animals and plants is mostly due to human interference.

Natural selection is a driving force behind such changes. Molecular biology and genetics explain evolution at the level of molecules. DNA studies also support findings. Human and chimp DNA sequences for the gene encoding the leptin reveals only 5 differences in 250 nucleotides. Where the human and chimpanzee sequences differ, the corresponding nucleotide in gorilla is used to derive the nucleotide that likely existed in common ancestor of humans, chimpanzees, and gorillas. In 2 cases, the gorilla and human nucleotides match, while in 3 cases, the gorilla and chimpanzee sequences are same. The common ancestor of gorilla, chimpanzee, and human probably have had the nucleotide that is the same in 2 of the 3 modern-day organisms as it would require 1 DNA change rather than 2.

Many structures and behaviors among species are common. Comparing fossils in structure and age shows the evolutionary success of new species from an ancestral species. For any 2 living species, evolutionary lines can be traced back in time until the 2 lines intersect in a common ancestor and to predict the type of creature existed in particular geologic period.

Diversity of species, ecosystems, and landscapes are the product of 3.7 billion to 3.85 billion years of evolution. Eukaryotes constitute 3 groups - the Viridiplantae or green plants, Fungi, and Metazoa. Basal eukaryotes are extremely diverse and evolutionarily ancient. The Rhodophyta (red algae) includes fossils dating from Precambrian, 1025 billion years ago. Stramenopiles includes small, single-celled organisms like diatoms, fungus-like species of water moulds, and very large, multicellular brown sea weeds like kelps.

The earliest green plants, the green algae dating from Cambrian are at least 500 million years old. By the end of Devonian, plants were diverse including members similar to modern plants. Green plants shaped the environment. Fueled by sunlight, they produce carbohydrates. Evolution and ecology of pollinating insects influenced the evolution of (angiosperms) flowering plants, since the Jurassic and Cretaceous periods. Fungi, which date back to Precambrian times, shaped the biodiversity by recycling nutrients. Flowering plants are amazingly diverse, constitute the foundation of terrestrial ecosystems and are essential to human survival. They provide food and medicine. *Archaefructus liaoningensis* is a 125-million-year-old fossil plant from China with distinct angiosperm traits. Angiosperms occupy every habitat except the highest mountain tops, the deepest oceans, and some polar regions.

Metazoa, which dates back over 500 million years have shaped many ecosystems, from the specialized tubeworms of deep sea hydrothermal vent communities to birds living in high altitudes. Many parasites affect the behavior and life cycles of their hosts. Evolution of earth has physically and biologically shaped the contemporary environment. Many existing landscapes are based on remains of earlier life forms. Some existing large rock formations are the remains of ancient reefs formed 360 to 440 million years ago by algae and invertebrates.

2.3 Fundamental Concept of Evolution

The concept of evolution is found in the literature of Greek philosophers. Lamarck (Figure 2.3) proposed a theory of evolution based on acquired characters. Darwin's theory of evolution is based on natural selection. Natural selection, mutation, isolation of populations, and mass extinctions are mechanisms of evolution. Production of more offspring by the individual animal results competition among the offsprings. Individuals with favourable trait, get the better chance of survival, called the survival of the fittest; the central idea of natural selection. Separation of a small population in a geographically isolated area from the main population tends to result in separate species. Fossil evidence and modern view suggest that speciation is most rapid after a mass extinction. After the extinction of dinosaurs, mammals and birds were evolved rapidly. Some species underwent little change over millions of years as they are well suited to their habitat. Evolutionary evidences come from artificial selection, comparative anatomy and embryology, vestigial structures, fossil records and from comparisons of DNA.

Figure 2.3   Jean Baptiste Lamarck (1744-1829)

Ernst von Baer treatise set the stage for studying ontogeny, the development of individual through a single life cycle to phylogeny, the relatedness of species through descent from a common ancestor. Von Baer studied the mammalian egg and the vertebrate animals during early stages of their development with a common design, where the adult forms showed difference. Arm buds of different species are indistinguishable when they first appear on the embryo, although they may develop into a wing, an arm, or a flipper.

In the early stages of growth, the ontogeny of all vertebrates is very similar. As the fertilized egg transforms into adult, the general vertebrate plan is modified during growth as each species acquires its adult pattern. The late fetal/newborn/adult stages reflect the emergence of species-specific body plans due to differential growth. A comparison of developmental stages in vertebrates led Ernst Haeckel (Figure 2.4) to propose the principle, "ontogeny recapitulates phylogeny." Haeckel claimed that development of an individual (ontogeny) reflects the stages through which individual's species pass during its evolution (phylogeny). All vertebrate embryos follow a common developmental plan for having a set of genes that gives the same instructions for development. As each organism grows, it diverges according to its way of life.

Figure 2.4   Ernst Haeckel (1834-1919)

Breeds of dog have evolved through artificial selection from a wolf ancestor. Fossil evidence for transitional forms like *Arcaeopteryx* is strong. Similar limb bones of amphibians, reptiles, birds and mammals suggest common ancestry. All vertebrates are similar in their early embryonic development having a notochord, gill pouches, and tail. Chimpanzees and humans share 99% common DNA.

The appendix and embryonic tail are vestigial in humans. The first fish were jawless but today, the only surviving jawless fishes are the lamprey. The early jawed fish evolved into cartilaginous and bony fish. Some early lobe-finned fish evolved into the first amphibians. Some early amphibians evolved into the first reptiles. Some early reptiles evolved into the first birds and mammals. Over the past 530 million years, the vertebrate lineage branched out from fish ancestor to evolve into various fish, birds, reptiles, amphibians, and mammals.

2.4   Heart of Evolution

Heart of evolution is located in ecosystems. That no 2 individual in a population are exactly alike, lead Darwin to advocate the theory of natural selection. Natural variation exists in population. Darwin calculated that a single pair of elephants would produce 19 million elephants after 750 years, if a pair produces 6 offspring over 60 years of fertility. Such geometric explosion does not happen due to some factors.

Darwin opined that organism best adapted with environmental limiting factors survives, reproduce and consequently pass their traits to next generation. If traits for success are changed by natural selection and pass to next generations, evolution would not stop. Thus environmental change is important. Natural selection modifies adaptations to adjust with the present conditions. Species are seldom modified consistently unidirectional, which is long enough for accumulation of significant evolutionary changes.

Local populations modify the entire species in a way which is difficult by natural selection. In the fixed environment, natural selection hones adaptations to perfect them further. In changing environment, the adaptations gradually modify the organism to adjust into new conditions. Species may move to regions where their present set of adaptation suits. The outcome of most environmental change is insignificant. Ecosystems and species are fixed entities. Of a succession of 13 Paleozoic communities, each community lasted 5, or 6 million years. Of many marine invertebrates known from each interval, few show any appreciable evolutionary change. An average of 20 % species survives in the next interval. Drastic changes in ecosystem are important. The extinction and appearance of new species occurs suddenly. Mammals were not diversified appreciably till the extinction of dinosaur .In disrupted ecosystems; fragmentation of habitats causes isolation of populations leading to rapid evolution of new species. The driving force is biological in mass extinction.

By disrupting ecosystems, we are driving the species towards extinction. Regulatory elements that orchestrate the gene activity helped in identification of 3 periods of evolutionary innovation in gene regulation, which caused change in gene frequency during vertebrate evolution. In the first period, the regulatory innovations affected genes which were involved in embryonic development.

Important evolutionary changes results from the gain, loss, or modification of gene regulatory elements, rather than from new protein-coding genes. Millions of these regulatory mechanisms are still the same in species which evolved separately over long periods. Natural selection conserves these sequences for their important function. Mutations that change them would be harmful to species.

Conserved sequences outside the known genes are perhaps the gene regulatory elements. By comparing the genomes of species whose evolutionary lineages diverged at different times, one can determine the first appearance of particular conserved sequence. The results provide insight into gene regulation mechanism and evolution at molecular level.

The first indication of distinct phases in vertebrate molecular evolution dominating during different periods can be traced. In the first period of evolutionary innovation, besides changes affecting developmental genes, enrichment in transcription factors, which bind to DNA and regulate other genes, are recorded. New regulatory elements affecting transcription factors reached a peak in early vertebrate ancestors, and then declined to background levels by the time of evolution of mammals. The next trend affected genes involved in cell-to-cell communication. The increase in regulatory innovations near these genes occurred. A third trend in placental mammals during the past 100 million years is known. Such changes tweaked the complex cross-talk between molecules that coordinates all cellular activities.

2.5   Early Evolutionary History of Chordates

The first chordates probably evolved before the Cambrian explosion. The first true vertebrate, the fish, first appeared in the Cambrian and exploded in terms of diversity in Devonian. The first amphibian occurred in Devonian. Terrestrial environments were attractive to live with plenty of yummy plants and insects, and abundant water. The land was free of predator, with plenty of food but without animals on land. Amphibians arose in Devonian from the

Osteichthyes, the bony fish that dominate today's oceans and freshwater bodies. Some bony fish have relatively flexible fins that lack muscles. They are called the ray-fin fish.

A second group of bony fish has more muscular, segmented bone support network within their fines called the lobe-fin fish, which perhaps lacked mobility and the ability to change swimming directions .They could crawl along the sea floor and swim. Lobe-fin fish possessed a pair of nostrils to draw air into mouth. In the Devonian, many fish had lungs and gills .Some lobe-fin fish could stick their noses into air for breathing fresh air before they gone back underwater. One group of Devonian fish evolved into lungfish. They are capable of switching their mode of breathing from gills to lungs. Australian lungfish could hide in the mud until it rains again and then switched back to gill breathing. The group of lobe-fin fish called dipnoan is double breathers.

Another group of Devonian lobe-fin fish the crossopterygians evolved into amphibians which are non-amniotic tetrapods because they laid naked eggs and had 4 legs. The crossopterygians include the coelacanth, the living fossils which were once thought as extinct, but now inhabit deep water between Madagascar and Indonesia. So, amphibians were evolved from a group of lobe-fin bony fish that moved into terrestrial suburbs of Devonian world. Like their fish ancestors, their eggs were non-amniotic. To be fertilized and for the embryos to develop, the eggs had to remain wet. There are still plenty of living members of amphibian including frogs and salamanders which generally possess thin porous skin that allows contaminants in the water to pass directly into their tissue.

Amphibians to Reptiles

Amphibians began to evolve into specialized individuals. Some were plant eaters, others were carnivores. All needed water to reproduce, a significant problem beginning in Permian and culminating in Triassic because of assembly of Pangaea. This super continent that formed from collision of Laurasia, Gondwanna, Baltica, Siberia and other cratonic fragments produced a huge arid landmass. In the egg, the evolutionary change was the addition of a shell to keep embryo wet and protected while permitting gas exchange with atmosphere. These amniotic eggs allowed some tetrapods to take water with them. In addition, changes in skull structures and epidermal composition finally lead to reptiles.

The first reptiles appeared a short time after the first amphibians, about 22 million years later in Mississippian. The first known reptile was *Westothiana*. In Mesozoic, reptiles were huge, dominated the land, much of the water, and even dominated the airways for almost 200 million years. The first reptiles (stem reptiles) had a major evolutionary advantage as they laid amniotic eggs. During Permian and into Triassic, they began to radiate and replace the amphibians. Bipedal reptiles include a group, the thecodonts that evolved tooth-in-socket jaws and produced reptiles like crocodiles. Quadripeds without fins called therapsids were mammal-like reptiles and are likely candidates for stem mammal. Reptiles are all cold-blooded animals. Some pterosaurs had no legs. The pterosaurs along with crocodiles and dinosaurs comprise the Archosaurian clade of reptiles. The stem Archaosaur was a thecodont. Archosaurs, the ruling lizards were the largest and probably the most dangerous animals.

The term dinosaur literally means "terrible lizards". Dinosaurs were the first vertebrates with a fully upright posture and distinguished bone structure. Their pelvis bones became specialized. Dinosaurs are classified into many groups. Saurishians are characterized by triradial reptilian-hipped pelvises. Sauropods were herbivores and theropods were carnivores. In the Ornithischians, the structure of the pelvis was bird-like. The birds, the descendants of reptiles, branched off from the stem *Archaosaur* in Jurassic, or Triassic. The first "bird" was *Archaeopteryx*.

Reptiles to Mammals

The end of Cretaceous came about with a big bang. A large bolide impacted the earth on the Yucatan Peninsula near Chicxulub Mexico. Coupled with stressful environment, the impact resulted in the third great extinction. The K-T boundary marks the end of Mesozoic and the start of Cenozoic. It marks the end of dinosaurs, plesiosaurs, mesosaurs, ichthyosaur and ammonite and the emergence of mammal. The first mammal was tiny and appeared in the early Jurassic, probably from the therapsid reptiles. Mammals evolve and now occupy most niches. Three distinct varieties of mammals based primarily on reproductive strategy are:

1. Monotremes, the egg-laying mammals were the first of the class to appear in the fossil record. Now *Echidna* and *Platypus* found only in Australia represents them.
2. Marsupials have live births. The young is little more than embryos. After escaping the uterus, they crawl sightless into a pouch for protection and nutrition. In kangaroos, female produce 2 types of milk from separate teats for 2 different generations of young - one for a nearly self-sufficient joey and one for a day old sightless lump of flesh. She might also be pregnant at the same time. Advantage of such reproductive strategy is 2-fold. Female kangaroos remain pregnant constantly to grow the population relatively fast .Secondly, by not laying eggs; they don't have to hang around protecting the eggs.
3. Placental mammals with exception of many primates, gives birth to young that are about small adults. The young moves very soon, after their birth. The transition that mammal made to invade from land back into sea is interesting. Environmental changes have driven mammals to radiate.

2.6 Convergent and Parallel Evolution

Animals that feed on fishes acquired long jaws and numerous teeth as in fish, crocodiles, phytosaurs, ichthyosaurs, plesiosaurs, birds, and mammals. The duck-bills sift small invertebrates from mud. Other examples of the same sort, the large mouth of insect eating animals: the frogs, swallow, swifts, and bats. The five types of ant-eater is an example of convergence; all have an elongated snout, long sticky tongue, large salivary glands, and other features. The hands of moles show how a similar result may be arrived at by various slightly differing means. Elasmobranchs maintain equilibrium in the horizontal plane by a heterocercal tail, driving the head downwards, and horizontal pectoral fins and flattened front end of body, having the opposite effect. From this type of organization, creatures with fin have several times developed, flattened dorsoventrally and obtaining their propulsive thrust from the pectoral fins. Conversely in teleosts, where the compression is in a transverse plane, the bottom-living types are flattened laterally, as in the sole or plaice. Where the swim bladder has been lost, there may be dorsoventral flattening. The 'tree-frog' and 'burrowing' conditions have been evolved both from true frogs and from toads, probably several times in each case. The habit of burrowing underground with loss of the limbs has appeared in several squamate reptiles; the slow worms, amphisbaenas, and subterranean skinks with distinct lines, each contains more than one. The snakes are probably derived from another group that went underground.

2.7 Conservative and Radical Influence in Evolution

The evidence about the history of vertebrates shows us the following facts:

1. In all or nearly all populations the organization of life-processes changes often slow.
2. The later types of organization usually replace the earlier in any one environment.
3. Evolutionary change is not always associated with environmental change.
4. When different populations adopt the same habit of life, they often develop similar organizations. Probably no population is stable in genetic, or phenotypic characteristics. The number of animals in a population is rarely constant. Probably the genetic and phenotypic characters of the members also vary with time, perhaps in correlation with fluctuations of number. Feature of population lies in its ability to produce later populations, which is influenced by frequency of reproduction, number of offspring produced, and viability of these in face of climatic and biotic factors, availability of food, and persistence of predators. These factors may lead to various distinct organizations.

A complex relationship between such factors as frequency of reproduction and numbers of young, rate of growth, time of maturity, size, and likelihood of death from predators and pathogens, capacity to adapt and to store information in nervous system is advocated. The characteristics of population depend on its surroundings. Many variables act with an intensity depending on activity of adult organism and the energy and ingenuity with which these find situations suitable for their life and their offspring. Adult activity depends on hereditary- and environmental factors. Thus the productivity and increase of a population depend on a system of interconnected variables.

Tendency to reduce evolutionary change presumably include

1. Copying or reproductive tendency that makes offsprings alike.
2. Anything that reduces the number of offspring.
3. Prevalence of predators.
4. Absence of geographical barriers.
5. Existence of stable external climatic and physical factors.
6. Characteristics within the population tending to keep the animals in their existing conditions, rather than to seek new ones.

Possible reasons for very slowly evolving populations may be the

1. Low mutation rate or low variability
2. It is often implied that some special habit help in survival, such as being nocturnal, or abyssal, but others with the same habits evolve fast.

3. Survival sometimes seems to be helped by isolation, but there is no evidence that this is necessarily a factor.

4. Low rate of evolutionary change is not a function of 'primitive' organization.

5. Long survival must depend on some special relationship between the genetical and information- carrying powers of the species, the risks imposed by the environment, and the stability of the latter.

6. If the adaptive zone is a narrow one, it must be stable and persist.

7. Long survival is perhaps more to be expected in a broad adaptive zone such as the ocean or shore, lowland rivers, or forest belts, especially in the tropics.

8. Bradytelic populations must be genetically so integrated that any deviation is the subject to counter-selection.

9. Simpson concludes that these bradytelic organisms 'have run the whole repertory of baffles and . . . persist indefinitely'.

2.8 Molecular Clock

In the early 1960s, Zuckerkandl and Pauling sequenced the amino acids of hemoglobin of several different species and counted the differences. They insightfully observed that number of amino acid differences among these species corresponded roughly in portions to geological time, when they juxtaposed these two lines of evidence. Primate (human), horse and mouse lineages were thought to have originated about 70 mya, birds about 270 mya, frogs (amphibian origins) about 350 mya, and sharks (cartilaginous fish origins) about 450 mya. Hemoglobin was acting like a molecular clock. Thus clearly biomolecules changes at a regular rate over hundreds of millions years.

Sarich and Wilson found that the albumins of human, chimpanzee, and gorilla differed by 1% .Each of them differed form albumins of Old World monkeys by 6%. Sarich and Wilson picked an event form fossil record to calibrate their molecular findings. The event they chose was the divergence of hominoids and Old World monkey lineage and the estimated time, 30 mya. Therefore, 1% differences among African hominoids represents one sixth of 30, or 5. Sarich and Wilson were then able to deduce that human's chimpanzees and gorillas had a common ancestor about 5 mya. 14 million years old *Ramapithecus* could not have been a hominid. These finding were first reported in 1967as controversial. Many paleontologists insisted that *Ramapithecus* dental morphology proved it to be hominid denying the validity of molecular clocks and rejected a 5 million year ape human divergence as being too recent. Subsequently more molecular data supported the idea of a 5 million year divergence.

Fossil bones indicated that *Ramapithecus* was a tree living ape, not a upright walking human. Hominid fossils 3 to 4 million years old are very like chimpanzees. The 5 million year estimate for common ancestor of apes and humans fits with current hominid fossil record. Fossils tell us when and where ancient ancestors lived and they might have looked like. Molecular data provide quantitative information on species relationships and estimates of when in the past the lineages diverged. These two kinds of information, paleontological and molecular are complementary, not contradictory and both are essential for reconstructing evolutionary history.

2.9 Geological History

The Cambrian period lasted 100 million years and included perhaps 3 inundations. The Ordovician lasted for 60 million years and included 3 floods in North America. Powerful earth movements at the end of this period are called the Tactonian revolution. The Silurian, lasting for 40 million years, included a single cycle of inundation, ending in an elevation of land, which was marked in Europe as Caledonian revolution, producing the range of mountains. Throughout early Palaeozoic periods, the fossils were entirely those of aquatic animals, with traces of land plants and arthropods at the end of Silurian. The oldest remains of vertebrates are the fish scales from the Ordovician.

Devonian is considered by some to include a single main period, about 50 million years long, with one flood at the middle and more arid conditions at the end, but other authorities divide it into several periods. The first forests appeared at this time, and first signs of vertebrate terrestrial life in form of fossil lung-fishes and amphibians are found. Carboniferous in Europe includes 2 major periods of about 40 million years each in America, the Mississippian and Pennsylvanian. In the early Mississippian, there were many swamps in North America. In the northern hemisphere, the Pennsylvanian was a time of warm, moist conditions. The Permian constitutes a single 45-million-year period, with active orogenesis, leading to a more arid climate with deserts in some parts of world and glaciations in others. These conditions continued into the Triassic.

The reptiles, first found in Permian, developed throughout the Triassic and flourished in Jurassic. The Cretaceous period probably lasted for more than 60 million years, including 2 major cycles of inundation. The lower Cretaceous certainly included extensive periods of flooding. Then later, towards the end of the upper Cretaceous, there were extensive orogenic movements, the Laramide revolution. The temperature was warm until near the end of the Cretaceous. Some groups of dinosaurian reptiles seem to have died out suddenly, but not all disappeared at the same time. In North America, Paleocene period was long. Probably the whole time since the end of Cretaceous has been about 70 million years. During early part of Tertiary, the climate was cold, but as erosion of mountains that had been produced at the end of Cretaceous proceeded the conditions became warmer, and throughout Eocene and Oligocene, there were large forests and humid conditions. Then during Miocene, there were marked earth movements, leading to elevation of land accompanied by more arid conditions. Weather became gradually colder through Pliocene culminating in ice ages of Pleistocene. Geological evidence shows many changes in climate and geography, some of them proceedings at very slow rates in comparison with rhythms of individual animal lives. It is uncertain whether evolutionary changes follow these slow geological changes, or are a result of instability imposed on living things by climatic rhythms with shorter periods, such as those of days, years, and sunspot cycles.

Paleozoic Era: This era is known as the cradle of ancient life.

Cambrian Period: Cambrian started with the plants and animals that were successful.

Precambrian Period: Among plants, thallophytes were well established and diversified into various groups. The aquatic arthropods and echinoderms came to prominence.

Ordovician Period: Coral, rocks, molluscs and echinoderms marked this period. The semi terrestrial bryophytes were getting established. This period saw the origin of first vertebrates which was the major event that happened in evolution of animals. The trilobites were more prominent during this period.

Silurian Period: The oldest land plant originated which possessed conducting tissues and colonized the land. Among invertebrates, except for insects all others flourished. The corals diversified. Jawed fish originated. Origin of paired fins and jaws is considered as major events in chordate evolution.

Devonian Period: Land living plants were more successful. The forests were filled with varieties of ferns and cycads. Fish became dominant. They diversified by adapting themselves to live in various aquatic ecosystems. This period is called as the "Age of fishes."

Mississippian Period: Several changes happened to land structure including massive upraising of land in several places which resulted in formation of several mountain ranges. Huge water bodies were broken into smaller lakes. Lung fish evolved

Pennsylvanian Period: The land living forms became more successful during this period. There were huge forests of ferns and cycads. Due to geotectonic changes, several forests got buried under the soil. The Pennsylvanian and the earlier Mississippian were collectively known as Carboniferous period.

Permian Period: It was the last period in the Paleozoic era. This period was marked by extinctions of several older groups of animals and plants. Nearly 60% of the organism that survived at that time became extinct. Some amphibians dramatically laid the land eggs.

Mesozoic Era: The land living forms dominated the middle period. The reptiles became more dominant, increased in size and number. This era is named as the "Golden age of reptiles."

Triassic Period: Fossils of turtles, crocodiles, and dinosaurs have been obtained from this period. Aquatic and flying reptiles thrived during this time. The mammal originated from reptiles.

Jurassic Period: Dinosaur diversified into carnivorous and herbivorous forms. The first birds originated from the reptiles. The modern bony fish were diversified into several groups.

Cretaceous Period: The larger marine molluscs became extinct. The dinosaurs of the Mesozoic era abruptly became extinct during this period. Fossils of dinosaurs were not obtained from later periods.

Cenozoic Era: All modern animals and plants were represented in these fossils. This era is subdivided into Tertiary and Quarternary periods. Further this era contains seven epochs. Through fossils, we can trace the origin and evolution of independent groups of animals, camels and man.

Paleocene Epoch: Modern placental mammals originated.

Eocene Epoch: Ungulates originated. The ancestral form of modern horses lived.

Oligocene Epoch: Modern mammalian families were established. The apes originated during this epoch.

Miocene Epoch: Large Priaries were formed. The evolution of fast *Triceratops*, a horned dinosaur occurred. Carnivorous mammals came to prominence.

Pliocene Epoch: The rodents became more successful. Mammals increased in number.

Pleistocene Epoch: Several glaciations happened. This epoch is popularly called the 'Ice age'. The evolution of horses and man reached the final stages. The melting of ice that happened 1,500 years ago is considered as the last stage of this epoch (Table 2.1).

Table 2.1   Geological Time scale and history of life (age years × 10^6)

Era	Period	Epoch	Aage	Animal group	Geological condition	Dominant Life
Caenozoic (Cenos, recent)	Quaternary	Recent (Holocene)	0.01	Dominance of Humans	End of last Ice Age, Climate warmer.	Age of man
		Glacial (Pleistocene)	2	Orgin of Humans.	Periodic glaciations.	
	Tertiary	Pliocene	7	Hominidae	Continued rise of mountains of western north America, Volcanic activity.	Aage of mammals.
		Miocene	26	Adaptive, radiation of mammals.	Sierra & Cascade mountains formed, cooler climate.	
		Oligocene	38	Dogs and bear appeared.	Lands lower, climate warmer.	
		Eocene	54	Apes and Pig appeared.	Mountains eroded, climate warmer.	
		Palaeocene	65	Horse, cattle and elephants.	Folding in Rockies mountains.	
Mesozoic (mesos, middle)	Cretaceous		135	Extinction of ammounites & dinosaurs; origin of modern fish & placental mammals.	Earlier inland seas and swamps; chalk, shale deposited.	Age of reptiles (dinosaurs').
	Jurassic		195	Dinosaurs' dominant; origin of birds and mammals; insects abundant.	Marine transgression in S. India, continents fairly high.	
	Triassic		225	Dinosaurs' appeared; adaptive radiation of reptiles	Marine transgression in Kutch. Widespread desert conditions.	
	Permian		280	Adaptive radiation of reptiles, beetless appeared, extinction of trilobities.	Increasing glaciations and aridity.	Age of amphibia
	Carboniferous		350	origin of reptiles and insects; adaptive radiation of amphibia.	Great coal swamps	
Palaeozoic	Devonian		400	Origin of amphibia and ammonites; spiders appeared; adaptive radiation of fish.	land higher; more arid, glaciations	Age of fishes

Era	Period	Epoch	Aage	Animal group	Geological condition	Dominant Life
	Silurian		440	Origin of jawed fish; earliest coral reefs	Low lands increasingly arid as land rose	Age of higher shelled invertebrates
	Ordovician		500	Origin of vertebrates, Jawless fish, trilobites, molluscs, and crustacea	Great submergence of land; warm climates even in arctic	
	Cambrian		570	Origin of non-vertcbrate phyla	Climate mild; earliest rocks with abudant fossils, volcanic activity, extemne erosion eroson	
Archaeo-zoic	Precambrain		1000	Selected organism		
			2000	Primitive enkaryotes		
			3000	Bacteria		
			3500	Orisin of life? Origion of earth ?		

2.10 Classification

Aristotle, after observing insects, fishes, birds and whales, emphasized that animal could be classified according to their living, actions, habits and body parts. Insect orders like Coleoptera, Diptera were created by him. He is considered as the 'father of biological classification'. The first work for modern taxonomy was due to John Ray. His work 'Synopsis Methodica Animalium Quadrupedum et Serpentini Generis' was published in 1693. He divided animals into those with blood and those without blood and classified animals based on gills, lungs, claws, teeth and other structures. He defined the species as 'producing unit'.

The Swedish naturalist Linnaeus (Figure 2.5) called the father of taxonomy published *Systema Nature* in 1758. He first introduced the hierarchic system, both in animal and plant kingdoms. He followed four categories namely class, order, genus, species for animal world and introduced binomial nomenclature. Michael Adamson stressed many characters in classification. His concept helped to develop numerical taxonomy. Lamarck attempted to improve Linnaen system and published seven volumes of his 'Histoire Naturelle des Animauxsans Vertebres' and arranged animals according to evolution. He displayed the animal group as a branching tree. Cuvier (Figure 2.6) divided animals into 4 branches. They are Vertebrata-fishes to mammals; Mollusca-mollusca and barnacles; Articulata-annelids, crustaceans, insects and spiders; and Radiata echinoderms, nematodes and coelenterates.

Figure 2.5 Carolus Linnaeus   **Figure 2.6** Georges Cuvier

Darwin in 1859 published 'Origin of species'. Several species were described. The development of modern taxonomy started during 1930s based on population studies. E. Mayr (Figure 2.7) considered species as "groups of interbreeding natural populations". His book 'New Systematics' is a landmark in taxonomy.Morphological characters were studied along with behaviour, sound, ecology, genetics, zoogeography, physiology and biochemistry. Thus taxonomy was transformed into 'biological taxonomy.

Figure 2.7 Ernst Mayr

Many aspects of our traditional classification are based on general similarities of organisms and may include both grades and clades. Cladistics is a philosophy of classification and phylogeny reconstruction based on the recognition of shared-derived characters. The father of cladistic method of classification was Willi Hennig.It is a more objective approach and does not include general similarities but only branches defined by synapomorphic characters. Taxon is the taxonomically recognized group of organisms .Sister taxons are the closest relative to a taxon. Plesiomorphy indicates a primitive character. Synapomorphy means shared derived character. Autapomorphy is derived character unique to single taxon.

Monophyletic is a group that includes all the descendants of a single common ancestor. Paraphyletic is a group that does not include all the descendants of a single common ancestor. Out-group method is used to determine whether characters are derived .Structural grade refers to different organisms from different evolutionary lineages that share broad structural or functional similarities. Cladogram is a branching diagram that represents hypothesized evolutionary relationships.

Phylogenetic Classification Methods

Traditional classification includes the Order Crocodilia in the Class Reptilia because of general similarities shared by crocodilians and other living reptiles. Cladistic classification includes Crocodilia as part of a clade with Class Aves due to shared derived characters. Monotypic species possesses traits or characters that are uniform over its entire range. Polytypic species possesses different traits throughout its range, and shows distinct geographic variation relative to specific traits.

2.11   Taxa and Species

The term taxonomy has been derived from taxis means arrangement, and nomos means law. Taxonomy is the science of "theory and practice of classifying organisms". The term systematics originates from the Greek word systema meaning 'placing together' means classifying living things following their natural relationships. Simpson (1961) defines systematics as "… the scientific study of the kinds and diversity of organisms and of any and all relationships among them". The classification simply means the work of classification. "Zoological classification is the ordering of animals into groups based on relationships". Based on specific characteristics, animals are grouped into categories called taxa. "A taxon is a taxonomic group of any rank that is sufficiently distinct to be worthy of being assigned to a definite category". Several taxa are the Phylum, Class, Order, Family, Genus and Species. The arrangement from Phylum to Species is called hierarchic system. Each taxon, a natural assemblage is based on specific characters of a group of organisms. The taxon 'Phylum' is the largest group. Several Phyla constitute the Kingdom. Members of a Phylum have distinctive features. Each Family contains several genera. Each Genus is subdivided into species. The species, the most important taxon is a reality and the fundamental unit in taxonomy which represents a natural unit. All other taxa remain arbitrary. Evolution operates at the species level.

Concept of Species

Initially the species was considered as group of organism with similar characters. Modern workers have identified 3 main concepts regarding species. Typological species concept began from the essentialism concept of Aristotle. According to this concept, a species is recognized by its essential characters expressed in morphology. In nominalistic species concept, species are the man made ideas. Nature produces individuals and not species and a species is a mental creation. In biological species concept "Species are the groups of interbreeding natural populations that are reproductively isolated from other such groups". This concept is mostly accepted by present day taxonomists.

Phenetic Method or Numerical Taxonomy

This method involves clustering or grouping of individuals of a taxon or several taxa. Based on similarity, identifications are made. The desired size of clusters is called operational taxonomic unit. Identification involves measurement using a scale of 0 to 1 of taxon to taxon similarity, or dissimilarity. In this method, vast amount of data are collected for related groups. Analyses are made, using statistical tools and computers.

Cytotaxonomy

The characterization and identification of complete chromosome set of a cell is called karyotyping, the first step in using chromosomes in taxonomy. Karyotypes within interbreeding population of species are usually constant. Variation in chromosome number and size occurs between the species. Final stages of chromosomal aberrations such as inversions and translocations give clues regarding intermediary stages.

Chemotaxonomy

This refers to use of information about small molecules produced by enzymatic action. Protein fractions isoelectrophoretic techniques, identification of amino acids in chromatography, prevalence of isoenzymes in tissue materials are employed in chemotaxonomy. The occurrence of specific pheromones, colour pigments, and toxins help as keys in taxonomy.

Palaeotaxonomy

It depends on identification and dating of fossils. A complete fossil provides better chance for identification. Several sections of fossils have provided the identification features. The fossils are generally studied along with other accompanying fossils, its geographic location and other factors,even though it is possible to assign a fossil to a genus, or other higher level.

2.12 Nomenclature Methods

Nomenclature is an integral part of taxonomy. Linnaeus specified the species by the combination of its specific and generic names. As it requires two names, it is called binomial system. Now International Commissions are responsible for naming each major group of organisms. Several such commissions authorize the usage of scientific names. Naming is monitored by International Code of Zoological Nomenclature. The rules are set out in the 'codes'. Science congresses modified the codes.

Principles of Nomenclature

1. Providing stability in naming and classifying with one correct name for a species.
2. If two or more names are in use, the correct name will be the one that was published earlier. This system is called the law of priority.
3. If two or more workers at a time describe the same organism using different names, it results in synonyms. In such case, the validity is provided to the senior synonym.
4. When names referring to two separate taxa of the same nomenclatural level are spelt the same, the two names are called homonyms. This condition is called homonymy; the junior name is invalid and a new replacement has to be proposed.
5. A material on which an original description is based gets a special status. This idea is called the type concept. The concept of genus and species are fixed by type genus, or type species.
6. Names that were used before those included by Linnaeus in the "Systema Naturae", tenth edition, 1758 are not recognized.
7. Scientific names must be either Latin or Latinized and should be mentioned in italics.
8. The genus name should be a single word beginning with a capital letter.
9. The species name should be a single, or compound word beginning with a small letter.

The zoological classification system dates from 1757 when Linnaeus published the 10th edition of Systema Naturae. Linnaeus introduced the system of binomial nomenclature. The scientific name of a species is a binomen .The primary Linnaean categories are Kingdom, Phylum, Class, Order, Family, Genus, Species .Other Linnaean categories are Subphylum, Subclass, Super order, Suborder, Super family, Subfamily, Tribe, Subgenus, and Subspecies.

Identification Keys

Identification is an integral part of taxonomy and could be made through literature, keys, pictures and comparison with type specimens. The commonly used method is the use of keys which is printed information, or a computer software package. A good key is dichotomous and not having more than 2 alternatives. The language of a key is telegraphic. The key may be either bracketed or indented. In a bracketed key, alternative contrastive characters are used. The number on the right side shows the next alternative character for consideration. In an indented key, a series of choices are provided for identifying a taxon.

Importance of Taxonomy

Diverse kinds of living organisms flourish in various environments and occupy the earth. The world is estimated to have 5 to 30 million species of living organisms. Now about 2.5 million species of living organisms are identified. Over 1.5 million of them are animal species and out of which 750,000 are insect species. About 350,000 species of plants include algae, fungi, mosses and other higher forms. The existence of different forms of species or genus and diverse adaptations for varied surroundings are called biodiversity.

The survival of such a vast range of living beings depends on their habitats. The term biosphere highlights the interdependence of living and non-living world and represents a stable environment of various physical and biological factors. The organic continuity of the system rests on a delicate network of interdependent relationships. The air, water, animals, plants, microbes and human beings are all interlinked in a life sustaining system, called the environment.

Safeguarding the entire biosphere with all its intricacies has prime importance. The nations of the world have convened several conferences and adopted important resolutions for safeguarding the mother earth. In this background, the United Nation's 'Environmental Agency' organized the "International Conference on Human Environment" at Stockholm in 1972. This conference adopted the motto 'Only one earth'. In 1982, a UN conference on Environment was held at Nairobi. The UN again convened "Earth summit" at Rio de Janeiro highlighting "Our Common Future", in 1992. Once again a world summit on sustainable development was organized in Johannesberg in 2002. One of the agenda commonly placed and accepted in all these meets was the significance of biodiversity and its conservation to ensure sustainable earth.

2.13   Animal Distribution and Evolution of Continents

According to Friday and Ingram(1985), during Cambrian and Ordovician, four continental masses viz. North America, Europe, Asia and Gondwanaland (containing southern continents of Africa, South America, India, Antarctica and Australia) appear to be existed. In Silurian, North American and European continents collided in the line of Appalachian, Caledonian and Scandinavian mountain to form Euramerica. By Carboniferous, three continental masses were approaching one another and South America possibly was in contact with Europe.The fossil reptiles illustrates the paleogeography of time, mostly confined to Euramerica.

During Permian; all the continental masses got united to form Pangaea surrounded by a single world ocean. Fauna of Pangaea are known to be remarkably uniform. Extensive reduction in shelf area was possibly responsible for a major wave of extinction in marine invertebrates. Pangaea persisted throughout the succeeding Triassic period and animals like dinosaurs and early mammals got the chance to spread to all areas. In Jurassic, Pangaea began to break-up. This resulted in partial fragmentation of land areas and evolutionary divergence among land animals, particularly the dinosaurs. During Cretaceous, Africa and India separated from Gondwanaland and drifted northwards. Fragmentation of land areas, diversification of terrestrial organisms particularly the vertebrates is reported. Northern continental mass called Laurasia broke into three portions viz. Europe and Eastern North America, Western North America and Asia. This situation is reflected in dinosaur faunas of three areas, as for example Ceratopsian dinosaurs confined to western North America in upper Cretaceous.

During Paleocene, the North America and Europe remain connected as shown by strong similarity of their mammalian fauna. South America and Australia remained connected via Antarctica, which explains why marsupials nowadays occur in these two continents and nowhere else. In Paleocene, there were partial exchanges of land mammals between Africa and Eurasia showing that barrier prevented free migration of most animals. During Oligocene and Miocene, extensive glaciations of the continent were probably the cause of extinction of its entire terrestrial fauna. Comparison of Miocene mammal fauna suggests that some interchange occurred between Eurasia and North America across the region of Bering straits, but in general mammals of two regions underwent separate radiations.

During Pliocene, North and Central America were reunited with South America resulting in a spectacular interchange of fauna, but immigration was mainly from North to South. During Pleistocene, extensive glaciations resulted in falls of sea level of up to 100 meters or more. This connected chain of Islands (which are now separated) joined the British Isles to the European continent and Asia to North America across Bering straits, allowing faunal and floral dispersal. Australasia however, remained an "Isolated continent" as it is to the present day with many unique faunal elements. The continental drift theory (Figure 2.8) could explain migration pattern of eels. Eels travel to North Atlantic spawning grounds from both Americas and European rivers. If North America and Europe actually were part of the Holarctic landmass (Laurasia) as postulated, ancestral eels could originally have spawned in a common Central Holaractic river system. Their present descendents still might travel from both West and East to same ancestral spawning area which now however happens to lie in the middle of an ocean as recorded elsewhere.

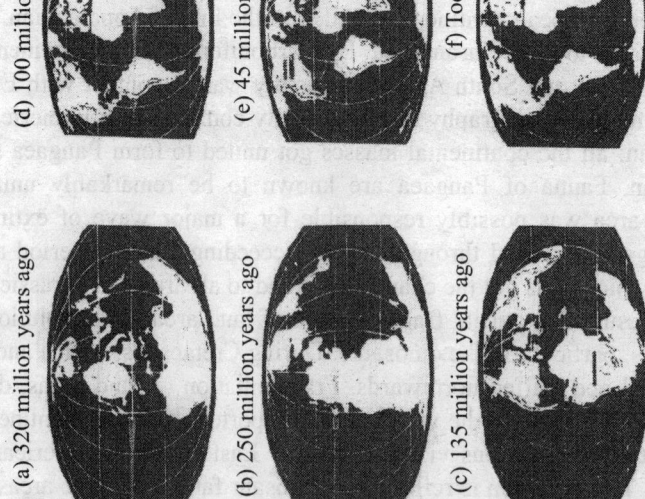

Figure 2.8   Continental drift (a) 320 mya, (b) 250 mya, (c) 135 mya, (d) 100 mya, (e) 45 mya, and (f) Today.

As per Young (1981), throughout Devonian period, there were two separate continental masses. North mass is known as Laurasia and the Southern mass is called Gondwanaland. The sea of uncertain extent between the two continents is called the Hercynian Ocean. Two continents then converged and met during the middle Carboniferous to make a single super continent, horseshoe like in shape known as Pangaea. In subsequent years, various changes occurred and now the whole outer surface of the Earth is divided into 8 major lithospheric plates and some minor ones.

2.14 Zoogeographical Realms

Distribution of animals over the earth surface is worldwide but not uniform. Physical, climatic and biological barriers influence the distribution of animals. Each species has definite range of distribution where it can thrive best. Scalter (1858) divided earth surface into six faunal regions or Zoogeographical realms. A.R. Wallace (1876) further divided each realm into sub realms. Six Zoogeographical realms (Figures 2.9, 2.10) are Nearctic realm, Neotropical realm, Palearctic realm, Ethiopian realm, Oriental realm, and Australian realm. Udvardy in 1975 classified biogeographical realms of the world as Ocenian, Nearctic, Neotropical, Palearctic, Afro tropical, Indimalayan, Australian and Antarctic realms.

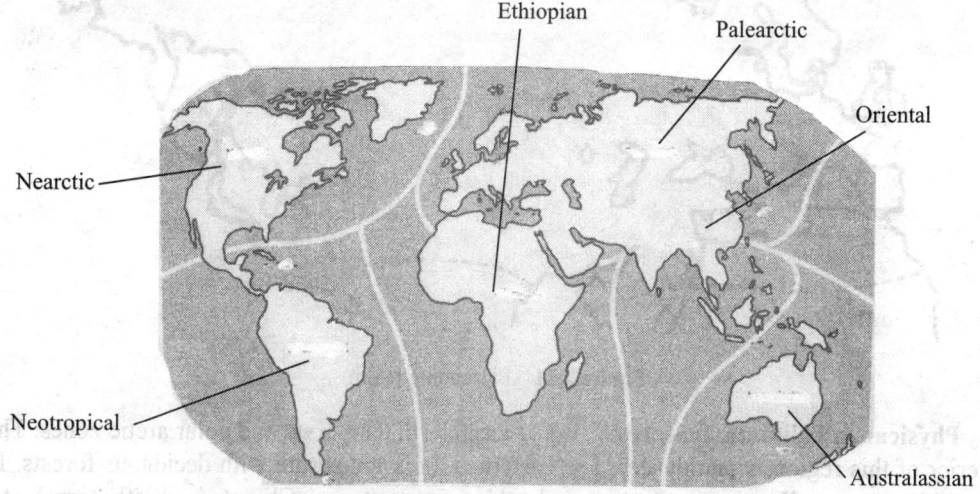

Figure 2.9 Zoogeographical realms

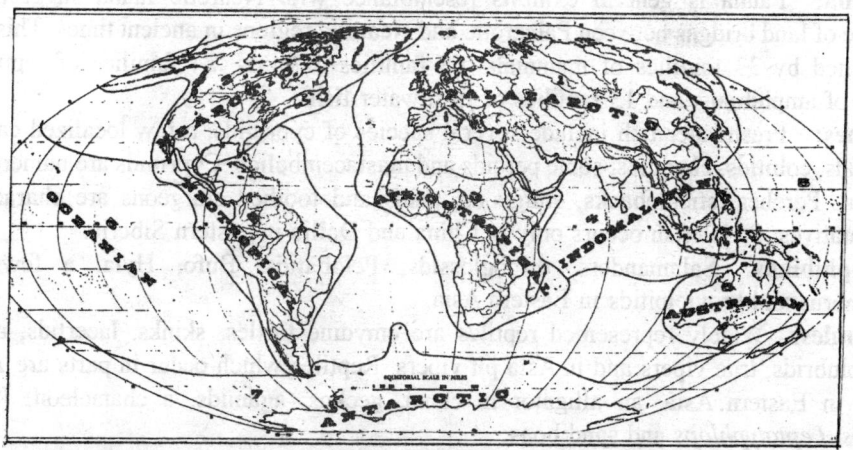

Figure 2.10 Terrestrial biogeographical realms of the World. Modified after (M. D. F Udvardy, 1975). A classification of the biogeographical Proviness of the world IUCN occasional Paper No. 18. 1UCN. Switzerland.

Palearctic Realm

Geographic range: This is the largest realm (Figure 2.11) with an approximate area of 14,000,000 square miles and extends over whole of Europe, China, Japan, Africa, North Sahara, Siberia, Mediterranean and Manchuria, Asia North of Himalayas and North of Arabia. This realm is continuous with its neighboring Ethiopian and Oriental realms.

Figure 2.11   Palaearctic realm

Physical and climatic features: It includes North temperate and polar arctic zones. The interior of this region is mainly dry. The eastern Asia is temperate with deciduous forests. In the northern zone, there is temperate grassland known as steppe. Climate is chiefly temperate, although there is wide range of temperature fluctuation. Rainfall varies greatly.

Fauna: Fauna is general exhibits resemblance with Nearctic fauna suggesting the existence of land bridges between Palaerctic and Nearctic regions in ancient times. This region is inhabited by 33 families of mammals, 68 families of birds, 24 families of reptiles, 10 families of amphibians and 13 families of freshwater fish.

Fishes: Freshwater fish include several species of cyprinids, a few localized catfishes, anabantids, cobitids, channids, some percids and mastacembelids. Cyprinids are numerous and dominant. Perches, sticklebacks, sturgeons, pikes and toothed sturgeons are characteristic representatives. Paddlefish occurs only in China and Dallia in eastern Siberia.

Amphibians: Salamanders, discoglossids, Pelobatids, Bufo, Hyla, a few Rana, Rhacophorus and brevicipitids in Eastern Asia.

Reptiles: Widely represented reptiles are emydine turtles, skinks, lacertids, anguids, many colubrids, true vipers and in Asia pit vipers. Reptiles which occur in parts are *Testudo*, *Trionyx* in Eastern Asia, an alligator in china, geckos, agamids, a chameleon, *Varanus*, *Typhlops*, *Leptotyphlops* and sand boas.

Aves: They include hawks, herons, storks, ducks, cuckoos, rails, plovers, pigeons, owls, kingfishers, swifts, woodpeckers, larks, swallows, thrushes etc. Avian fauna bears resemblance with the neighboring region.

Mammals: The exclusive families are Spalacidae (*Spalax*- mole rat) and Seleviniidae (*Selevinia*). The principal mammals are hedgehogs, shrews, moles bears, pandas, pigs, deers, squirrels, flying squirrels, jumping mice and bats. Palaearctic mammals are partly endemic, partly shared with the old world tropics and partly shared with the North America. The North American relationships become very strong in the North.

Sub-region: This region has been divided into four sub-regions:

European sub-region: It includes Northern and central Europe, Black sea and Caucasus. A mammal namely Myogale is peculiar. Birds like tits, thrushes, wagtails and mammals like hedgehogs, shrews and moles are very common.

Mediterranean sub-region: It includes remaining parts of Europe, all the African and Arabian portions, Asia Minor, Persia, Afghanistan and Baluchistan. The region is supposed to be the richest part of Palaearctic region comprising of about 120 families of vertebrates. Elephant shrews, hyenas, hyrax are the characteristic mammals.

Siberian sub-region: Northern Asia-North to Himalayas represents it and is characterized by unsettled and extremes of climatic conditions. The families of yak, musk deer and moles are almost exclusively confined to this region. Phoca sibirica (fresh water seal found in Baikal Lake) is another characteristic form of this sub-region.

Manchurian sub-region: Japan, Mongolia, Korea and Manchuria belong to this region. Among mammals Tibetan langur, great panda, Chinese water deer, tufted deer are peculiar to this region only.

Oriental Realm

Geographic range: This region includes India, Indo-China, Ceylon, Burma, Malaya, Sumatra, Java, Borneo, Formosa and Philippines. It is bounded by Himalayas in the North. There is no physical boundary in the Southeast corner (Figure 2.12).

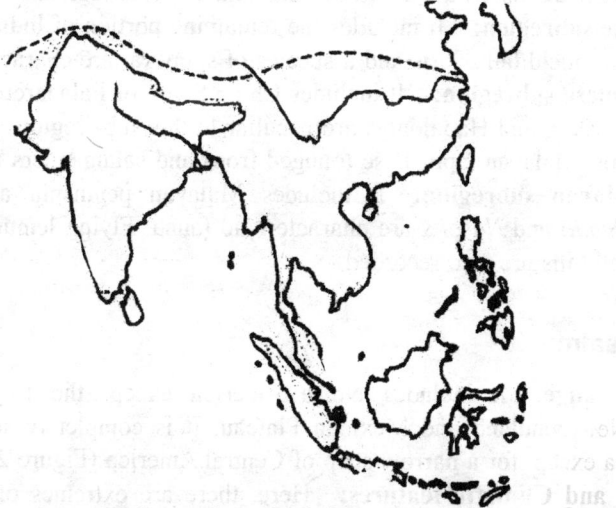

Figure 2.12 Oriental realm

Physical and climatic features: Mainly tropical but Northern part of India is temperate. Eastern part including Burma, Indo-China and North East Asia has rain forests. Indian sub-region has luxuriant tropical forests. There are thick forests in Ceylon, indo-china and Malaya.

Fauna: Fauna exhibits considerable resemblance with that of Ethiopian region. Resemblances are so marked that some zoogeographers prefer to club these two regions under a single realm, the Palaeotropical region.

Pisces: Carps and catfishes are dominant. Notopterids, anabantids, osteoglossids, cypriniforms, cobitids are also found. Homalopteridae, Pristolepidae are exclusive families.

Amphibians: Tailed amphibians like salamanders and caecilians are few. Tailless amphibians like bufonids, ranids and rhacophorids are numerous.

Reptiles: Emydine turtles, trionychids are oriental turtles. Crocodiles, *Gavialis*, geckos, skinks, agamids, chameleon, *Varanus*, *Python*, viper, pit viper, *Typhlops*, *Leptotyphlops* are widespread. The exclusive forms are Uropeltidae and Xenopeltidae.

Aves: Of 66 families of oriental birds, 53 are widely distributed elsewhere; 3 are shared mainly with Africa; 4 have special geographical relationships; 5 are shared chiefly with the Australian region. Exclusive oriental family is the Trenidae. The fairy blue birds and leaf birds with 4 genera, and 14 species are recorded. Parrots, shrikes, hoopes, honey guides are few but pigeons and pheasants are numerous. Peacock is the typical Indian bird.

Mammals: Out of 30 families, 4 are endemic. There are shrews, rabbits, squirrels, murid mice, cats, dogs and bovids. It shares hedgehogs, porcupines, civets, and hyenas with Palaerctic and Ethiopian realms. The loris, apes, pangolins, elephants, rhinoceros etc. are shared with the Ethiopian region. Exclusive animals are tree shrews, tarsiers, cynocephalus, spiny dormouse, flying lemurs and Indian bison.

Sub-region: Oriental realm is divided into following sub-regions;

Indian subregion: It includes Central and Northern India from river Indus and foot of Himalayas Southwards up to Goa and Mysore. Colubrine snakes, antelopes, Indian bears, peacocks and Indian bisons are exclusive mammals to this region.

Ceylonese subregion: It includes the remaining portion of Indian peninsula and Island of Ceylon. The shield tails, loris and a species of spiny rat are characteristic.

Indo-Chinese subregion: It includes China South of Palaearctic boundary, Burma and Siam. Panda, Takin, and Hapalomys are peculiar to this sub-region. Gibbons, flying lemurs, Java rhinoceros, Malayan tapir, Disc tounged frogs and salamanders are also found.

Indo-Malayan subregion: It includes Malayan peninsula and Islands of Malaya Archipelago.*Simia* and *Nasalis* are characteristic fauna. Flying lemurs, rhinoceros, Malayan tapir and broad bills are also recorded.

Nearctic Realm

Geographic range: It includes North America except the tropical part of Mexico, Greenland, Newfoundland and Mexican Plateau. It is completely separated from all other regions by sea except for a narrow strip of Central America (Figure 2.13).

Physical and Climatic features: Here, there are extremes of temperature and great variation in climatic conditions. Extensive mountain ranges exist in the West running from North of South. Arctic belts of Greenland with ice of unknown thickness are located in the

Figure 2.13 Nearctic realm

North. In eastern part of North America, there are deciduous and mixed forest, extensive grassland in the Central part and deserts in the South West of North America.

Fauna: Fauna bears resemblance with the Palaearctic fauna. A total of 26 families of mammals, 49 families of birds, 21 families of reptiles, 14 families of amphibians and 24 families of fishes are recorded.

Pisces: Characteristic representatives are the paddlefish, bowfin, moon eyes, suckers, cyprinids, catfishes and percids. The Eastern of Mississippi is rich in fish.

Amphibians: Dominating tailed amphibians are salamanders, *Amphiuma*, *Ambystoma*. Characteristic tailless amphibians are *Bufo*, *Hyla*, and *Rana*. A primitive frog, *Ascaphus* is found.

Reptiles: Common reptiles are *Trionyx*, musk turtle, emydines, alligators and crocodiles. Lizards like Geckos, Ophisaurus, and *Phrynosoma* are found in Texas only while *Heloderma* only on South- West Central America. Common snakes are pit- vipers, coral snakes and colubirds. *Pituophis* and *Chilomeniscus* are exclusive snakes.

Aves: Common birds are grebes, loons, pelicans, hawks, herons, ducks, quails, rails, cranes, cuckoos, plovers, larks, sandpipers, gulls and owls. Peculiar ones are *Centronyx*, *Catherpes*, *Hylotomus*, *Cupidonia*, *Auriparus*.

Mammals: The only marsupials, Virginian opossum (*Didelphys virginiana*) is found in North America. Shrews, rabbits, squirrels, moles, cats, bats, deers, and bovids are commonly found. Star nosed moles, Canadian porcupine, long legged bats and leaf nosed bats are exclusive forms.

Subregions: This region is divided into following sub regions:

Californian sub region: It embraces a narrow strip of North America between Sierra, Nevada, and Cascade Range extending from Vancouver Island and part of British Columbia.

Fauna includes 86 families of terrestrial vertebrates. Vampires and free tailed bats are the characteristic.

Rocky mountain subregion: It is situated in the East of California and includes the dry and elevated mountain covered area. A total of 107 families of terrestrial vertebrates are recorded. *Antilocapra*, *Haplocerus*, *Heloderma*, American bison and *Cynomys* are characteristic animals.

Alleghanian sub region: It includes Eastern part of U.S.A. i.e. East of Rocky Mountains sub region and South of great lakes. Mammals like opossums, star nosed moles, vampire bats and birds like passenger pigeon, Carolina parrot and turkeys are peculiar fauna.

Canadian sub region: It includes the remaining part of North America and Greenland. Bison, sheep, polar bears and arctic fox are peculiar forms.

Neotropical Realm

It includes south and Central America, West Indies and Southern Mexico (Figure 2.14).

Figure 2.14   Diagrammatic sketch of neotropical realm

Physical and Climatic features: Tropical conditions exist except Southern part of South America. Huge tract of rain forest in Amazon Valley and separate, smaller tracts elsewhere in South and Central America are seen. Extensive areas of Savanna and grassland are found in tropics. Desert and sub desert areas are seen especially in western South America but widely scattered elsewhere.

Fauna: Rich faunal make-up comprises of 155 families of terrestrial vertebrates, of which 39 are exclusive. The rest shows affinity with Nearctic and other tropical regions.

Pisces: Characins, gymnotid eels and catfishes are dominating forms. Lungfish, osteoglossids, mandids, gar pikes, cichlids are also found.

Amphibians: Among 14 anuran families, *Pipa*, *Hyla*, *Bufo*, and *Rana* are common. Of these, 4 families namely Rhinophrynidae, Hylaplesidae, Plectomantidae and pipidae are peculiar. Caecilians are represented by *Siphonopsis* and *Rhinotrema*. Peculiar urodel is *Spelerpes*

Reptiles: Reptilian fauna exhibits affinities with Nearctic, Ethiopian and Oriental region. Few species of crocodiles, alligators, turtles and tortoises are recorded. Out of 15 lizard families, 5 families namely Helodermidae, Anadiadae, Chirocolidae, Cercosauridae and Iphisiadae are peculiar. Typholps, Leptotyphlops, colubirds, coral snakes and pit viper are common. Characteristic animals are *Dromicus*, *Epicrates*, *Elaps* and *Ungalia*.

Aves: Avian fauna is so striking and diverse that South America is called the "Bird Continent". Out of 67 families, 23 are endemic and 2 orders are exclusive. Rhea and Tinamus are endemic fightless birds. Todies, puff birds, oil birds, horned screamers, ant thrushes, sugar birds, plant cutters are peculiar. Herons, storks, ducks, plovers, pigeons and kingfishers are common.

Mammals: Opossums, caenolestid marsupials, shrews, monkeys, anteaters, sloths, armadillos, mustelids, cats, tapirs, camels, rabbits, squirrels and pocket mice are found in Central America. Out of 32 families of mammals, 10 are endemic. Endemic mammals are chinchillas, sloths, cavies, and didelphids.

Sub region: It is divided into following sub regions:

Chilean sub region: It includes Western coast of South America, summits Andes of Peru and Bolivia. Chinchillas, oilbird and rhea are common.

Brazilian sub region: It includes tropical forest region of South America, terminating Northwards at Isthmus of Panama. American monkeys, vampire bats, American porcupines, sloths, armadillos, tapirs, cavies and spiny mice are recorded.

Mexican subregion: Neotropical region north to Isthmus of Mexican sub region represents panama. A total of 24 families of mammals, 67 families of birds, 26 families of reptiles, and 10 families of amphibians are recorded. Tapirs, Anguidae and Plethodontidae are characteristic animals.

Antelian sub region: It comprises of Islands of West Indies except Tobago and Trinidad. This sub region is wholly made up of islands and vertebrate fauna is poor.

Australian Realm

Geographic range: It includes Australia, New Zealand, Tasmania, Moluccas and neighboring Islands. It has no land connection with any other region (Figure 2.15).

Physical and climatic features: This region is partly tropical and partly temperate. New Guinea is tropical and is mostly covered with rain forests. Northern part of Australia is tropical but most of the interior is arid. Tasmania is cool and temperate.

Fauna: Vertebrate fauna is composed of 134 families of terrestrial vertebrates, of which 30 are peculiar. A total 8 families of mammals, 17 of birds, 3 of reptiles and 2 of amphibians belong to peculiar category. Monotremes and marsupials are restricted to this region only.

Pisces: Freshwater fish, osteoglossids and *Neoceratodus* are restricted in distribution.

Amphibians: Amphibians include frogs, hylids and a few ranids. Ceratobatrachidae and Genyophrynidae are exclusive.

Reptiles: Pygopodidae, Hatteridae and Crettochelydidae are peculiar families. Crocodiles, agamids, varanus, Typhlops, pythons and elapids are common in New Guinea and parts of Australia.

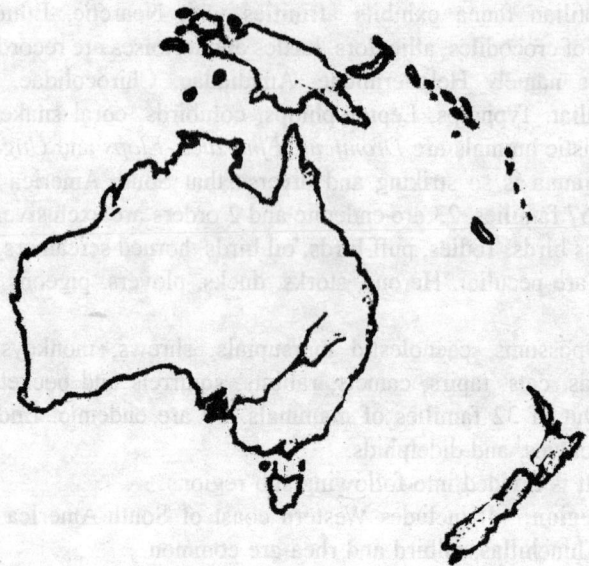

Figure 2.15   Australian realm

Aves: Out of 58 families, 44 are worldwide in distribution. Twelve exclusive bird families include cassowaries, emus, honey suckers, bower birds, legendary birds of paradise, frogmouths, scrub birds etc. Trogon, kingfishers, hawks, cuckoos, pigeons and parrots are common birds.

Mammals: Complete absence of higher eutherian mammals although monotremes and marsupials are exclusive. Monotremes belong to 2 families and 2 genera (*Ornithorhynchus* and *Tachyglossus*). Six families comprising of 52 genera of marsupials include wombats, oppossum, kangaroos, phalangers, marsupial mole etc. Insectivores and frugivorus bats are also recorded. Mice, Australian dog and European rabbit have also been introduced.

Subregion: This region is divided into following 4 subregions:

Austro - Malayan subregion: It comprises of all Islands of Malaya archipelago together with New Guinea, Moluccas and Solomon islands. Peculiar families include Ceratobatrachidae and Genyophyrnidae besides Fly-River turtle family and crowned pigeon family. In New Guinea, Peculiar marsupials are *Dendrolagus*, Dasyuridae and true flying phalangers.

Australian subregion: This subregion comprises of whole of Australia and Tasmania. Wombats, marsupials, moles, duck bills, scrub bill sand emus are exclusive. This subregion is known as home of marsupials. Characteristic fauna includes bandicoots, kangaroos, honeyeaters, swallows, shrikes, cobras and lizards.

Polynesia subregion: It comprises of Polynesia and Sandwich islands. Three families of bats, 37 families of birds, 9 families of reptiles and 2 families of amphibians are recorded. Tooth billed pigeons are Confined only to Samoa islands, Kagu in New Caledonia.

New Zealand subregion: It includes New Zealand, Norfolk Island, Auckland, Campbell and Macquarie islands. Three families of mammals, 27 families of birds, 3 families of reptiles and only 1 family of amphibians are recorded. Most interesting flightless birds of New Zealand are the kiwi. Living fossil, *Sphenodon Punctatum* is found.

Ethiopian Realm

Geographical range: This region includes whole of Africa (South Tropic of cancer) Southern Arabia and Madagascar. Darlington did not include Madagascar in this region (Figure 2.16).

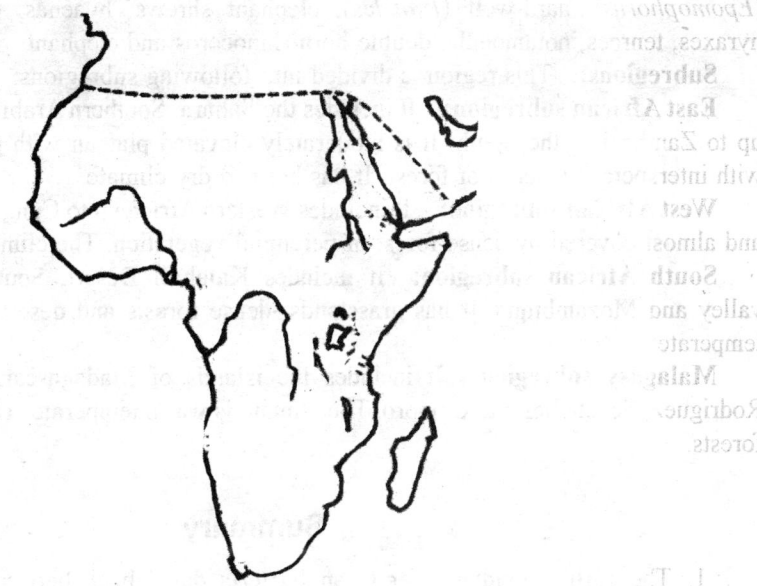

Figure 2.16 Ethiopian realm

Climate: This region has tropical climate in the Northern part and warm climate in Southern region and is characterized by exceptionally large rivers, deserts (e.g. Great Sahara Desert) grasslands, luxuriant tropical and temperate forests and mountains. Its southern part has temperate climate.

Fauna: The fauna is rich, varied and well marked.

Fresh water fishes: Nile-lung fish (*Protopterus*), *Polypterus*, *Notopterus*, *Clarias* are peculiar.

Amphibians: Clawed - toads, frogs, tree frogs and certain caecilians are present. *Rana* (frog) and *Bufo* (toad) are absent.

Reptiles: Snakes, lizards, crocodiles, turtles and tortoises are present. Snakes belong to Colubridae (*Leptorhynchus*, *Rhamnophis*, *Herpeterthiops*, *Grayia*,) Dendrophidae (*Hopsidrophis*, *Bucephalus*). Dryophidae (*Langalia*), Dipsadidae (*Pythonodipsas*), Lycodontidae (*Boedon*, *Lycophidion*) Pytonidae, Elapedae and Viperidae (*Atheris*).

Birds: A rich avian fauna include weaver birds, sun-birds, fruit thrushes, fly-catchers, herons, cuckoos, goat suckers, larks, thrushes, swallows, bee-eaters, ostriches, collies, and hoopoes, helmet birds, secretory birds, barbets, crows, shrikes, hammer headed birds, mouse birds, plantain eaters, hawk-like birds, starlings, rollers, white eyes, hornbills, vultures and eagles. *Serpentarius* is most characteristic hawk-like bird of this region.

Mammals: Lemurs, aye-aye (*Chiromys*), chimpanzee (*Troglodytes*), gorilla, monkeys like thumbless monkey (*Colobus*), long- tailed Cereopithecus, baboons (*Theropithecus* and *Cynocephalus*), antelopes (gemsbok, eland, okapi), gnu , giraffe, hippopotamus, civet cats, hunting dogs, aard vark (*Orycteropus*), insectivores (golden mole), fruit - eating bats (*Epomophorus*), aard-wolf (*Proteles*), elephant shrews, hyaenas, elephant, horse, zebra, hyraxes, tenrecs, potamogale, double-horn rhinoceros and elephant

Subregions: This region is divided into following subregions:

East African subregion: It includes the Sahara, Southern Arabia, North East Africa and up to Zambezi in the South. It is moderately elevated plateau with grassy vegetation along with interspersed patches of forest. It has hot and dry climate.

West African subregion: It includes Western Africa up to Congo South of river Gambia and almost covered by dense forest of perennial vegetation. The climate is hot and moist.

South African subregion: It includes Kalahari Desert, Southern African Limpopo valley and Mozambique. It has grasslands, dense forests and deserts. The climate is warm temperate.

Malagasy subregion: It includes the islands of Madagascar, Bourbon, Maurititius, Rodriguez, Scychelles and Comoro. The climate is warm temperate. The Sub region has dense forests.

Summary

1. The earliest evidence for Cyanobacteria dated back between 3.5 and 2.75 billion years. The earliest single-celled eukaryotes are about 2.7 billion years old. Diversity of species, ecosystems, and landscapes are the product of 3.7 billion to 3.85 billion years of evolution. Natural selection, mutation, isolation of populations, and mass extinctions are mechanisms of evolution.

2. Regulatory elements helped in identification of 3 periods of evolutionary innovation in gene regulation which changed gene frequency necessary for vertebrate evolution. The first chordate probably evolved before the Cambrian explosion. All vertebrates have notochord, gill pouches, and tail in their embryonic stage. Over the past 530 million years, the vertebrate lineage branched out from the fish ancestor to evolve into various groups.

3. Fossils are the preserved remains of animals, plants or their parts in strata of earth. Fossil records reveal few patterns of extinction. Larger organisms easily become extinct. Fossil evidences for transitional forms between two animal groups like *Archaeopteryx* exist. Similar limb bones of amphibians, reptiles, birds and mammals suggest common ancestry. Geological evidence shows many changes in climate and geography in the history of the earth.

4. Number of animals in population is rarely constant. Probably the genetic and phenotypic characters of individuals in a population vary with time. Primate (human), horse and mouse lineages are thought to have originated about 70 mya, birds about 270 mya, frogs (amphibian origins) about 350 mya, and sharks (cartilaginous fish origins) about 450 mya.

5. Comparison of vertebrate developmental stages led Haeckel to propose the principle, "ontogeny recapitulates phylogeny." Amphibians arose in Devonian. The first reptiles (stem reptiles) laid amniotic eggs. During Permian and into Triassic, reptiles began to radiate replacing the amphibians. The K-T boundary marks the end of dinosaur and the emergence of mammal. The first tiny mammal was appeared in the early Jurassic, probably from the therapsid reptiles.

6. Species are "groups of interbreeding natural populations". Cladistics recognizes shared-derived characters. Based on specific characteristic, animals are grouped into categories called taxa, a taxonomic group of any rank that is sufficiently distinct to be assigned to definite category.

7. The zoological classification system dates from 1757, when Linnaeus published the 10th edition of "Systema Naturae". The arrangement from Phylum to Species is called hierarchic system. The taxon 'Phylum' is the largest group. Several Phyla constitute the Kingdom. Members of a Phylum have distinctive features. Each family contains several genera. Each genus is subdivided into species. The species is the fundamental unit in taxonomy.

8. The continental drift theory could explain migration pattern of eels as well as distribution of animals. If North America and Europe actually were part of the Laurasia, ancestral eels could originally have spawned in a common Central Holaractic river system.

9. Scatler divided the earth surface into 6 faunal regions or zoogeographical realms. A.R. Wallace further divided each realm into sub realms. Six zoogeographical realms are Nearctic realm, Neotropical realm, Palaerctic realm, Ethiopian realm, Oriental realm, and Australian realm.

Review Questions

Short answer Questions

1. Why study of evolution is important in understanding biology?
2. Define petrifaction.
3. What are moulds?
4. What do you understand by extinction?
5. Name three groups of eukaryotes.
6. What was the basis of Lamarck's theory of evolution?
7. What was the basis of Darwin's theory of evolution?
8. Explain the Hackel's theory of recapitulation.
9. Name the period in which amphibians arose.
10. What is Pangaea?
11. Name the first known reptile.

12. What do you understand by K-T boundary?
13. What is convergent evolution?
14. What do you understand by the term" molecular clock"
15. What is geological history?
16. Define geologic period.
17. Define Cambrian period.
18. Who is known as the "father of biological classification?"
19. Who is called the father of taxonomy?
20. What do you understand by traditional classification?
21. What is meant by synapomorphy?
22. What is meant by autapomorphy?
23. Define monophyletic group.
24. Define polyphyletic group.
25. What is cytotaxonomy?
26. Define zoological classification.
27. What is zoogeography?

Long Answer Questions

1. Write a note on the importance of taxonomy
2. Give six reasons for extinction of species. Describe the continental drift theory in relation to animal distribution
3. Write the Principles of Zoological nomenclature
4. Describe the geographical range and vertebrate fauna of Oriental realm
5. Describe the geographical range and vertebrate fauna of Ethiopian realm.

3

Subphylum Urochordata

Urochordates or tunicates are small, common, marine animals called the squirts .Their chordate features in young include tadpole-shaped body, central dorsal nervous system, notochord in caudal region, and gill slits on the pharyngeal region. They are sessile, benthic or planktonic, solitary or colonial. In adult stage, they degenerate, attach to rock, or other object. Except the Copelatae, the tail becomes absorbed and notochord being lost. Body becomes twisted. Both gill slits and vent empty into a common atrial chamber. Precambrian fossil *Yarnemia* is an urochordate. Complete body fossil is rare

3.1 External Features

They are mostly sac-like. Tunic, with calcareous secretions of various shapes, covers the fixed body .Tunic is composed mainly of the tunicin which closely related to cellulose and with which glycoprotein combines. The name tunicata refers to the pigment in tunicin. Pigment may be derived from the blood. Colour changes over period. Muscle-fibres in mantle run longitudinally in various directions to draw the animal together with production of jet of water.

Epidermis secretes it with special cells from the mesoderm. Mantle is covered by a single-layered epidermis and lines the tunic. Water passes through an opening at the top, and passes out through a lateral opening. Spicules are rarely described .About 1250 species are divided into Ascidiacea (benthic sea squirts; *Aplousobranchia*, *Phlebobranchia*, and *Stolidobranchia*), Thaliacea (pelagic salps, pyrosomes, and doliolids), and Appendicularia (pelagic larvaceans). Ascidiaceans are often brightly coloured.

3.2 Internal Features

Pharynx, the major body part forms a sac which is attached to mantle along one side and is surrounded dorsally and laterally by a cavity, the atrium. Pharynx collects food by ciliary action .Its stigmata pierce wall. Vertical crack formed by subdivision of 3 original gill slits with cilia produces food current that enters through mouth and leaves the atriopore. A ring of tentacles guards the entrance of pharynx. Tongue bars divide each slit to form horizontal synapticulae and produce numerous holes in pharyngeal wall. Papillae with muscles and cilia are found in stigmata. Endostyle has 3 rows of mucus cells on each side, separated by rows of ciliated cells and with a single median set of cells with very long cilia.

Mucus secreted by endostyle is caught up on papillae. Muscles of papillae move to spread mucus inside the pharynx. Food particles are entangled in mucus, which moves upwards and then passes back to oesophagus by cilia of dorsal lamina or of languets. Iodination occurs in some cells lying above the glandular tracts of endostyle. Iodine is found in tunic also. Ciliated surface of pharyngeal wall and muscular contractions aids in passage of water inward. Pressure of exhalant current drive the water. Oesophagus leads to stomach with a folded wall containing gland cells, which produce enzymes like amylase, invertase, lipase, and protease. Pyloric gland opens into lower end of stomach. A short intestine leads upwards from stomach to open in atriopore. In many ascidians, contractions of siphons and body musculature occur in rotation with a frequency of 8-27 per hour. Contractions move more water than by ciliary current.

Heart remains surrounded by pericardium, lies below the pharynx, and communicates with blood spaces. Mesoderm becomes grooved and folded to form the heart and pericardium. Haemocoelomic spaces around the pharynx and elsewhere consist of mesenchyme. A pair of outpushings form pharynx. The epicardia ends blindly on either side of heart. The biggest blood vessel, the hypobranchial vessel lies below the endostyle, from which branches pass to pharynx. A large visceral vessel springs from the opposite end of heart and others pass to dorsal side of pharynx, tunic, and body wall. Heart beat proceeds in either direction. After passing blood into the hypobranchial vessel and gills for a few beats, its direction changes for passing the blood to viscera. Two pacemaker centres cause reversal, each initiate rhythmic contraction one at either end of heart. Warming and cooling controls the reversal of beat.

The capillary is absent .Blood cavity is haemocoel. Blood-plasma is colourless with corpuscles. Blood contains orange, green, or blue pigment. The green and other pigments contain vanadium.Vanadocytes contain sulphuric acid. Haemovanadin reduces cytochrome. Some individuals contain niobium. Blood is isotonic with sea water. Ascidians perhaps cannot regulate their osmotic pressure. Epicardia may be compared with coelomic cavities. *Ciona* allow sea water to circulate about the heart and help in excretion. In other ascidians, the epicardium loses its connection with the pharynx.

The closed sac serves as excretory organ in some. Tubular excretory organs are absent. Ninety five percent of nitrogen is excreted as ammonia. Nephrocytes found in blood and elsewhere may be stored in excretory sac.

The central nervous system consists of a round, solid ganglion, lying above the front end of pharynx. A cell layer covers the ganglion outside. From the ganglion nerves proceed to siphon, other parts of mantle, muscles, and viscera. Receptor cells with nerve-fibres end around the base, especially in siphons. Stronger stimuli cause closure of both siphons. Strong stimulation contracts the whole body and ejects the water in pharynx and atrium. Body surface is light sensitive. Nerve cells are found in body wall. The ocelli are cup-like collections of orange- pigmented cells around siphons. The neuromuscular system serves as a reflex apparatus for movements. Neural gland is sac, lying beneath the ganglion and open by a ciliated funnel on roof of pharynx. It arises from ectoderm of larval nervous system partly from pharynx and comparable with infundibulum and hypophysis of vertebrates. Sub neural gland is similar to pituitary as it controls the release of gametes. Nervous system shows little sign of spontaneous behaviour. Rhythmical activities are sometimes started within the animal. Food collection in pharyngeal wall involves rhythmic movement of papillae.

3.3 Development

When eggs, or sperms of the same species are present in water, signals from neural gland cause discharge from gonad. Pathway of signals is partly hormonal, partly nervous. Further similarities with pituitary are due to presence of vasopressor and oxytocic substances in sub neural gland. Tunicates are the hermaphrodite. Ovary and testis lie close to intestine and open by ducts near atriopore. Fertilization is external in solitary forms, but internal in colonial forms. The ocellus, situated in posterior wall of cerebral vesicle, contains 3 parts, a lens cell, a pigment cell, and a retina. Lens cell contains 3 lens vesicles. Pigment cell contains melanin granules to protect the photoreceptor from stray light. Processes from retinal cells penetrate the pigment cell. Development results in a fish-like ascidian tadpole.

Cleavage is total and produces a blastula with few cells. Gastrulation occurs by invagination. Tadpole possesses oval head and long tail. Forty cells make up the entire rod, becomes vacuolated and elongated by swelling. Three rows of muscle cells, 18 on each side run on either side of notochord. Other cells of this tissue migrate to form the pericardium, heart, and mesenchyme. Muscle cells contain cross-striated myofibrils at the periphery. Nervous system makes a hollow, dorsal tube extending into tail and enlarged in front into a cerebral vesicle containing ocellus and otolith. Nerve fibres proceed to front end of muscle rows and rest of cord lack nerve cells.

Larva takes no food and the gut is not well developed. In larva, pharynx with single pair of gill-slits, open into an atrium. Suckers are formed below or around the mouth. Whole development completes in 1 or 2 days. Larva set free which is positively phototropic and negatively geotropic and proceeds to sea surface. Within a day, or two, its tropisms reverse and it passes to bottom, turns to any dark place and finds a suitable surface. It attaches by suckers, loses its tail, develops a large pharynx, and transforms into an adult. Sexual reproduction; regeneration and budding are found. Bud consists of outer epicardial, mesenchymal, pharyngeal or atrial tissue. Epidermis develops only more tissue like it. All other tissues are formed from inner mass by folding to make a central cavity. The nervous system, intestine, and pericardium are formed by further folding. Members of Class Larvacea and Thaliacea are pelagic and specialized to live in plankton (Figure 3.1).

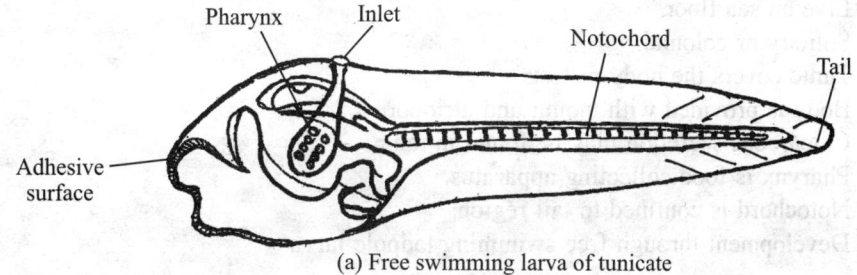

(a) Free swimming larva of tunicate

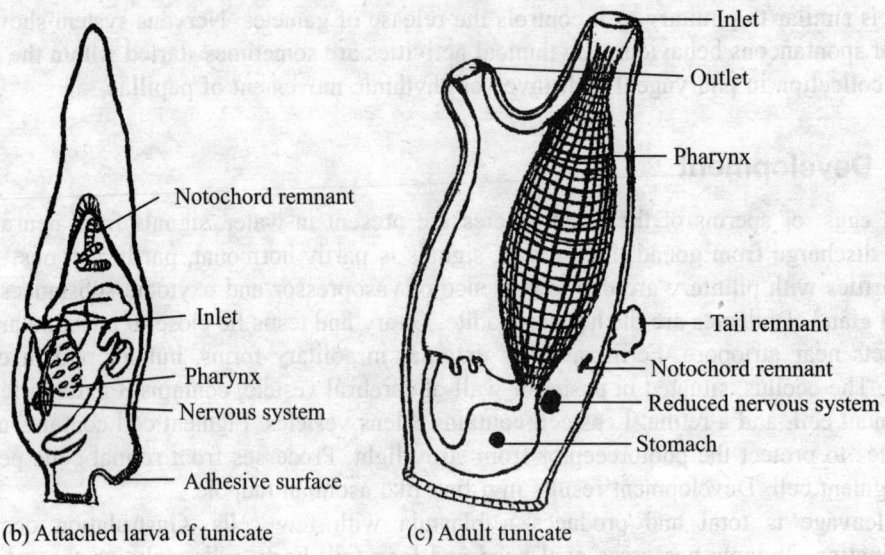

(b) Attached larva of tunicate   (c) Adult tunicate

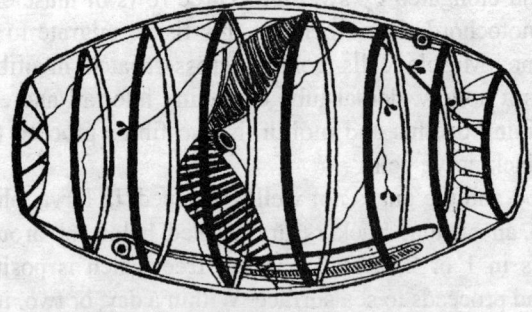

(d) *Doliolum* sp

Figure 3.1   Development in tunicate.

3.4   Characters of Urochodates

1. Bottom-living and filter feeders.
2. Sac-like creatures.
3. Live on sea floor.
4. Solitary or colonial.
5. Tunic covers the body or test.
6. Body is provided with mouth and atriopore.
7. Calcareous secretion may be found in tunic.
8. Pharynx is food collecting apparatus.
9. Notochord is confined to tail regions
10. Development through free swimming tadpole larva.

3.5 Class Ascidacea

They are typically sessile, found in all seas in different forms, as solitary or colony. Colonial forms produced by budding consist of several neighbouring individuals, or of gelatinous test in which individuals are embedded. Body shapes relate to substratum.They inhabit diverse habitats, mostly the littoral zone. Few like *Hypobythius calycodes* inhabit deep-sea. Many live for a short time, mature in the first year. Some live over a second winter, reduce in size; grow again in the following spring. Ascidians are the largest and most diverse group of Tunicata comprising about 3,000 species. Adults bear little resemblance to chordates. Their larvae clearly exhibit dorsal tubular nerve cord, notochord, pharyngeal gill slits and a post-anal tail. The larvae have a sensory vesicle containing 2 darkly-pigmented sense organs - a light sensitive ocellus and a statolith. Following settlement; the lecitotrophic larvae undergo retrograde metamorphosis and lose all characteristics except for the pharyngeal gill slits. Tunic forms a flexible skeleton. Various proteins, some blood cells and spicules occur in the tunic.

Ascidians are hermaphrodites, but avoid self fertilization by developing only egg, or sperm at a time. Most solitary forms release eggs and sperm into water for external fertilization .Under natural conditions, ascidian larvae do not disperse very far. Majority filters food from the water column via oral siphon that brings water into pharyngeal chamber. Cloacal siphon then expels the water. Particles suspended in the current are trapped in mucous on gill slits. The net pores allow filtering of small particulate matter. Substances caught in mucous are transported to stomach for digestion. Larvae and juveniles are preyed by gastropods, oysters, polychaetes, crabs and fish during their early life. Adults are preyed on by prosobranchs and nudibranchs. *Halocynthia roretzi* and *Microcosmous sabatieri* provide a human food source in Japan, Korea, Europe, and Chile.

During the last 2 decades progress has been achieved in the development, evolution, immunology, natural products and ecology of ascidians. The study of self/non-self recognition in ascidians provides information regarding the evolutionary origin of vertebrate immune system. Some species survive in highly polluted areas, accumulate arsenic, cadmium, chromium, cobalt, copper, iron, lead, mercury, selenium, tin and zinc and may be used in bioassays for pollutants and as bioindicators. Many species avoid predation or fouling by producing noxious metabolites. Some ascidians are potential source of anti-cancer compounds. Ecteinascidin 743, a potent antitumor agent isolated from extracts of *Ecteinascidia turbinata* is now in clinical trials. Antimalarial compounds have been isolated from the *Microcosmus goanus, Ascidia sydneiensis* and *Phallusia nigra*.

The classification of Lahille divides the class into three orders based on structure of adult branchial sac: Aplousobranchia, Phlebobranchia and Stolidobranchia. Ascidian's genera are small in number relative to number of species. More than half of known species belong to genera *Aplidium, Didemnum, Molgula, Polycarpa, Ascidia, Styela, Eudistoma, Pyura, Cnemidocarpa,* and *Synoicum.*

Coral Reef Ascidians

Ascidians, a minor benthic component on exposed surfaces of the coral reefs are often found in cryptic environments like caves, crevices and the sides of rocks and corals. In exposed sites, solitary species protect themselves better than colonial species from the predation, abrasion and physical damage. Their rigid tunic covered by epibionts provides camouflage and physical protection. Some colonial Didemnidae host the photosynthetic prokaryote symbiont *Prochloron* and thrive on surfaces exposed to high irradiance on reef flat. Rapid spread of several species in tropical regions is reported.

Surveys along the Red Sea coast of Israel have shown that all ascidian species recorded on artificial substrates are found in the natural environment. *Botryllus eilatensis* found on dead coral skeletons colonize artificial substrates in the coral reefs of Eilat. Increase human activity resulting in unfavorable condition to corals has a competitive advantage to ascidians. Coral reefs are facing increased eutrophication which creates favorable conditions for filter-feeders like ascidians and sponges. In deteriorating environment, dead portions of corals may be rapidly colonized by ascidians and others. *Herdmania nomus* is a common solitary ascidian found along the Mediterranean and Red Sea coasts of Israel. *H. nomus* in the Mediterranean restricted to artificial substrates is found at greater depths and has a limited reproduction season.

3.6 Class Thaliacea

The chief characters of the class are:

1. Found in all seas.
2. Solitary or colonial.
3. Atrium opens through atriopore.
4. Brain, in the from of solid nerve ganglion, is present.

Example: *Enterogona, Pleurogona*.

They are pelagic; barrel shaped filter feeders, and live in warm water. About 70 species are known. They spend their lives swimming slowly in seas. Class is divided into 3 orders. Pyrosomida are the colonial. Salpida and Doliolida are solitary. Circular muscle bands enable them to shoot through the water by jet propulsion. They feed by drawing water current by beating cilia through inhalant siphon and pass out the same through exhalant siphon at opposite end of the body. Between 2 siphons water passes through pores of pharynx. Mucous secreted by special cells aids in collection of plankton.

Colonial *Pyrosoma* consisting of individuals form barrel-shaped colony. Mouths open outwards and atria inwards into a single cavity with terminal outlet. Budding from epicardium and other features show affinity with *Doliolum* and *Salpa*. *Pyrosoma* resembles ascidians as its zooids are sexual and produce bud. Yolky eggs develop in parent without larva. Large masses of *Pyrosoma* illuminate the sea. Light is produced on stimulation by waves of rough sea. Sudden flashes of light protect against enemies by flight reaction. Each zooid within common test draws water in from outside and passes it inside or lumen of communal from. Water is driven into centre of barrel by individual. Water leaves by open end of barrel. This constant flow pushes community ever forward through sea. Individual is generally small communal but they can be large up to several metres long.

Pyrosomids are filter feeders. Doliolids are barrel-shaped, transparent zooids with 8-9 circumferential muscle bands in body wall. Life cycle includes a solitary sexual - and colonial asexual generation. Gonozooid, a solitary, sexually reproducing generation, develops asexually from phorozooid. During planktonic life, it releases gametes. Fertilization leads to zygote which develops into tadpole larva. Tadpole matures into colonial oozooid or nurse. Oozooid gives rise, through asexual budding, to blastozooids which mature to diverse zooids including original oozooid, trophozooids and phorozooids. Trophozooids are feeding colony. Phorozooids finally leave colony, produce gonozooids to complete life cycle.

Body is covered by thin, elastic tunic. Ducal siphon lies at the anterior end. Margins of its aperture are scalloped with 10 buccal lobes which bear sensory receptors. Buccal siphon opens into pharynx. Gill slits perforate pharynx. Endostyle is a ciliated groove .Posteriorly pharynx narrows to form short esophagus which dilates to form stomach. Intestine leads from stomach to anus. Anus opens into the spacious atrium. Lateral cilia on gill slits generate feeding current that enters buccal siphon and passes posteriorly into pharynx. Long lateral cilia protrude into gill slits. They extend almost half way across the gill slit. From pharynx, water passes to side through gill slits and into surrounding atrium. Endostyle produce mucous. Mucus acts as a filter to remove fine food particles from water. Mucus and entangled food particles move posteriorly by cilia into oesophagus and stomach. Atrial siphon is a large opening at posterior end. It is ringed by 12 atrial lobes similar to buccal lobes at anterior end. Body is encircled with 8 swimming muscle bands whose contractions force a water jet out the atrial siphon, resulting in forward motion. Hermaphroditic gonad, an elongate sac beside posterior pharynx opens into atrium.

Nervous system includes cerebral ganglion and single statocyst. Neural gland, a ciliated duct arises in anterior dorsal pharyngeal wall ventral to ganglion. In *Doliolum*, the muscle bands pass right round the body. They are small, only 1cm long in average with a spur. Body shape is cylindrical. Reproductive cycle is complicated. After attaining a certain size, they produce 3 types of buds to form temporary colony.

Salp

Salp is common in equatorial, temperate, and cold seas, abundant in Southern Ocean, occur singly or in long colonies. Complex life cycle exhibits obligatory alternation. Salp is barrel-shaped, free-floating, 1.5 to 19 cms long, and normally colorless. They live most of their lives as asexual zooid. Portions of life cycle exit in seas. Solitary life phase called oozed reproduces asexually by producing chain of hundreds of individuals which are released from parent. Chain of salps is the total portion of life cycle. Aggregate individuals called blastzooids remain attached while swimming and feeding. Each individual grows in size. Each blastozooid reproduces sexually, with a growing embryo oozoid attached to body wall of parent. Growing oozoids released from parent blastozooids grow as solitary asexual phase, thus closing the life cycle. Alternation of generations allows for fast generation time, with both solitary individuals and total chains live and feed together in sea.

Salps move by contracting which pumps water through its body. It consumes phytoplankton that is strained from water. Salps in Southern Ocean are often abundant than krill. Since the early 1900s, krill populations have been declining and salp population is increasing. Salp with complex life cycle may play role in climate changes. Fecal matter and dead salp bodies sink to the ocean floor. In regions, abundant salp population would have major effect on the ocean's biological pump. Changes in distribution of salp in large numbers may alter the carbon cycle of the ocean. Salps are similar to jellyfish in their simple form and free-floating life. The tiny groups of nerves found in salps may be the first examples of primitive nervous system which would have evolved into central nervous systems of true vertebrates.

3.7   Class Larvacea

The chief characters of the class are:

1. Free swimming, planktonic.
2. Buccal and atrial siphons are at opposite ends of the body.
3. In most, the ciliary feeding current is also responsible for locomotion.
4. Pelagic.
5. Circular band of muscles present.
6. Mouth and atriopore are at opposite ends of body
7. Adult lacks notochord and tail.
8. Swimming, and planktonic.
9. Buccal and atrial siphons are located at opposite ends of body.

Example: *Pyrosoma, Salpa, Doliolum, Dolichinia.*

Larvacea are minute neotenous live in plankton. Each individual builds house by secretion from the oikoplastic epithelium of skin. Tail is a broad, held at an angle to rest of body. Its movement produces a current in which food is carried and caught by filter. Water enters the house by a pair of filtering windows and is passed through filter pipes in part of house in front of mouth. Very minute flagellates are stopped by these pipes and sucked. Pharynx has 2 gill-slits, an endostyle and peripharyngeal bands. General organization is like a typical ascidian tadpole. They differ in many ways from ascidian tadpoles. Tail is highly developed for locomotion. A continuous fin is supported by a notochord of 20 cells. Bands of 10 large striped muscle cells extend down each side, giving an appearance like metameric segmentation. Nerve cord is hollow tube with ganglionic thickenings, each containing 1 to 4 nerve cells. From these cells, fibres proceed to muscle and skin in series of roots that usually remain separate, the motor being more dorsal.

Example: *Oikopleura* (Figure 3.2)

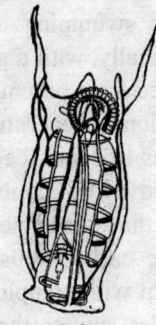

(a) *Herdmania*   (b) Ascidia   (c) *Salpa*

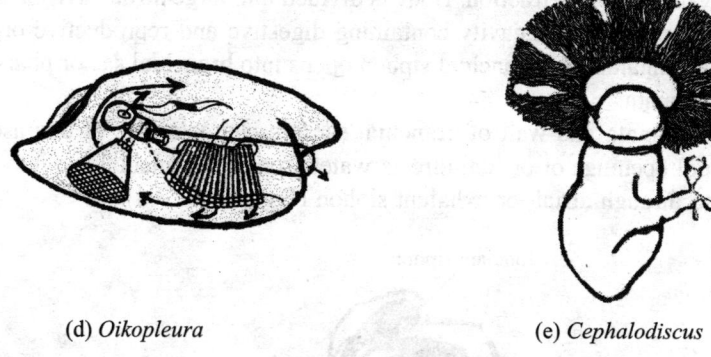

(d) *Oikopleura*   (e) *Cephalodiscus*

Figure 3.2 Few tunicates

3.8 *Ciona Intestinalis*

The solitary ascidian, *Ciona intestinalis* commonly known as Sea Vase tunicate, is a temperate species, cosmopolitan in distribution. *C. intestinalis* has perhaps originated in the Northeast Atlantic. Linnaeus in 1767 mentioned the "European oceans" as the Type Locality for the species. Major natural populations are found in shallow protected inlets in Denmark, along the west coast of Sweden and extending northwards along the coast of Norway. Coldwater or sub-Arctic records also exist for the Faeroe Islands, the east coast of Greenland and as far north as Spits Bergen and Bear Island. Its population outbreaks cause biofouling problems for aquaculture in South Africa, New Zealand, Chile, and Scotland.

In Canadian waters, this cryptogenic species has been observed in high densities recently. Its invasive potential bears ecological implications. Lahille's 1886 classification scheme includes this species in the Order Phlebobranchia. Linnaeus first described in 1767 *C. intestinalis* (*Ciona tenella*) as *Ascidia testinalis*. Adult measures 15 cm in length and 3 cm in diameter.

Body is cylindrical, soft, and translucent; vary in colour from pale greenish/yellow to orange. In old individuals, the tunic becomes leathery .Two openings, or siphons are located at one end of the body. The long inhalent siphon has 8 lobes and the small exhalent siphon has 6 lobes, both with yellow margins and sometimes with orange/red pigment spots. When disturbed, the organism retracts siphons using longitudinal muscles located beneath tunic. *C. intestinalis* is a sessile filter feeder, remains attached to hard substrates by projections of the tunic. It may occur in dense aggregations. *C. intestinalis* is distinguished from *C. savignyi* by the presence of red rather than white spot at the end of the sperm duct. *C. intestinalis* from Atlantic coast populations lack this red spot.

Structure

Cylindrical body includes 2 siphons and a gelatinous tunic composed of tunicin, a polysaccharide chemically similar to cellulose. Inside the tunic, thin sac-like membrane composed of an external epithelium, connective tissue, muscles and blood vessels encloses the internal organs. Contraction of longitudinal muscle bands and circular muscle fibres embedded in this membrane allow rapid retraction. Body is divided into large atrial cavity which contains branchial sac and small visceral cavity containing digestive and reproductive organs. At the upper end, the oral (inhalant) or branchial siphon opens into branchial sac or pharynx which is perforated with stigmata.

Structure of stigmata and wall of branchial sac are characteristic of this ascidian. Cilia located in stigmatal openings or ostium directs water from the branchial sac into atrial cavity where it is ejected through atrial, or exhalant siphon (Figure 3.3, 3.4).

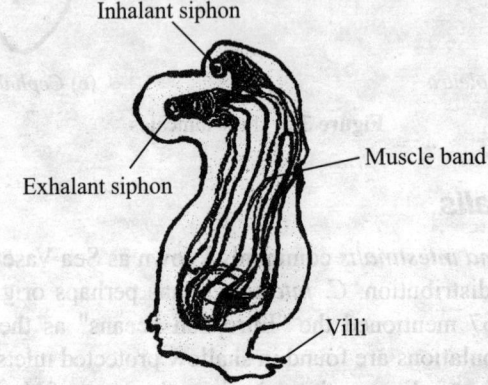

Figure 3.3   External view of *Ciona*.

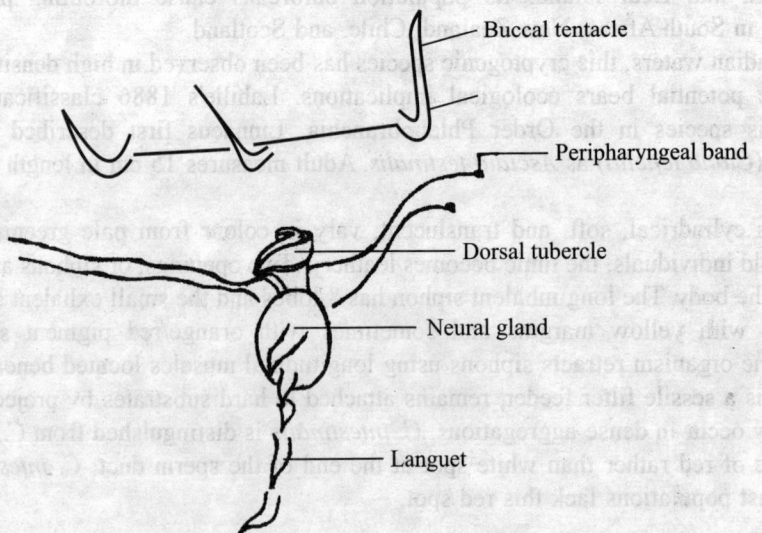

Figure 3.4   Dorsal middle line of anterior end of pharynx in *Ciona*.

This is a mucus feeder. Rod-like endostyle lies along the wall of branchial sac and secretes mucus net. As cilia transport this net across the wall of branchial sac, incoming particles become entrapped in mucus. Mucus-bound particles are gathered into a cord by dorsal lamina or languets and transported to oesophagus and stomach which lie below the branchial sac.

Material passes into intestine which curves upward from the base of stomach to join rectum lying against the wall of atrial cavity. Faeces discharged into the atrial cavity are expelled through the atrial siphon. This has an open blood system driven by a tubular heart which lies adjacent to digestive tract.

Blood from the digestive tissues collects in ventral sinus below the branchial sac and is forced up through vessels lying between the stigmata. After several minutes, the heartbeat slows and then starts beating in opposite direction which moves fluid from the dorsal sinus down into the spaces around the viscera. This primitive circulatory system distributes dissolved organic material and O_2 throughout the body. Blood may contain high level of vanadium and other trace metals which are taken up in the branchial sac (Figures 3.5, 3.6).

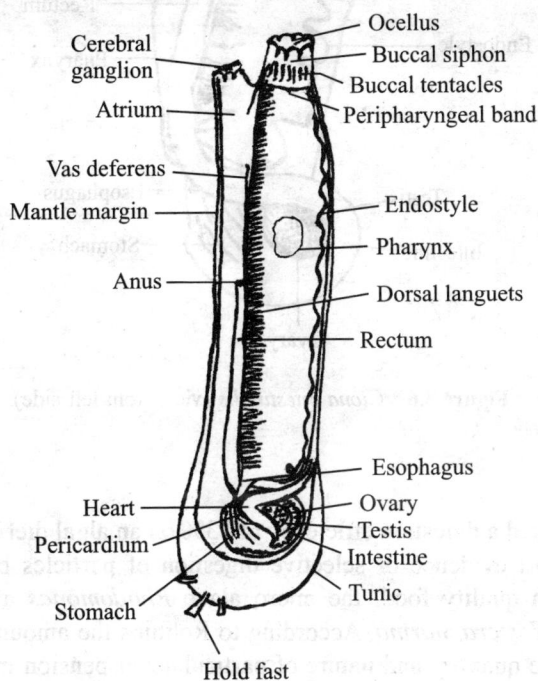

Figure 3.5 *Ciona intestinalis* (view from right side).

Feeding and Respiration

This is an active suspension feeder or filter feeder and removes particles from the surrounding environment. Fiala-Médoni reported that *C. intestinalis* does not exhibit any pumping rhythm. Water flow through the atrial siphon is constant. Filtration rates based on removal of algal cells were also continuous without any noticeable rhythm. Robbins reported declined filtration rate with increasing concentration of inert inorganic particles. They also confirmed that filtration rate varies in response to gut fullness, or gut clearance rate. Sigsgaard et al. confirmed that filtration rates declined and ingestion rates stabilized with increasing algal cell concentration.

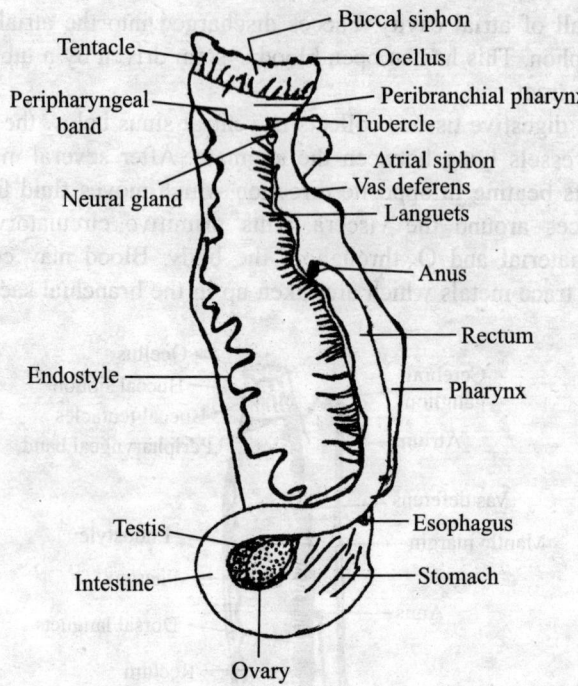

Figure 3.6   *Ciona intestinalis* (view from left side).

Digestion

Fiala-Médioni estimated a digestion efficiency of 83% on an algal diet of *Monochrysis lutheri*. Sigsgaard et al. report evidence of selective digestion of particles by *C. intestinalis* when provided with a high quality food, the micro algae *Rhodomonas* and a low-quality food, particles of eelgrass *Zostera marina*. According to Robbins the amount of faeces produced is dependent on both the quantity and nature of particulate suspension ingested.

Reproduction and Development

This is hermaphrodite. Both male and female reproductive organs are located in lower section of tunic. Ovary lies between the stomach and intestine. Testis is dispersed in diffuse mass over the intestinal surface and part of stomach. In adult, the ovary has a pinkish appearance. The testis is white. Separate gonoducts run parallel to rectum but extend further up into atrial cavity. Gametes discharged into atrial cavity are expelled into environment.

Duration of embryonic development varies with temperature. "Tadpole" larva consists of trunk with a slender muscular tail surrounded by granular test cells. The larva has 6 developed organ systems and 4 rudimentary organ systems including a primordial branchial sac. Notochord extends the length of tail. Trunk has 2 black spots: a large photosensitive ocellus and the small gravity sensitive statolith .Adhesive papillae located at the anterior end of trunk are used to attach to substrate at settlement. Newly hatched larvae may escape from mucus-bound egg strings to disperse in plankton, or may be retained until settlement.

Settlement and Metamorphosis

Towards the end of tadpole phase, the larvae tend to sink or swim downwards and become strongly photonegative with a preference for dark or shaded surfaces in zones with reduced water movement and light intensity. The highest levels of settlement occur in morning. Tadpole larva attaches to substrate with its adhesive papillae, located at anterior end. After settlement, the larval tail is rapidly resorbed into the trunk, reducing the size of early juvenile from 1.3 mm to 350mm.

Human Uses

The larval form has many features resembling vertebrates and provides a model for studying organ development uncomplicated by the variations and elaboration in other organisms whose organs are constructed of many cells. The tadpole larva perhaps represents the closest living form to the ancestral chordate, and provides insights into chordate evolution. This species is also a model for studying the evolution of gene function and regulation as it has an unduplicated compact genome. It shows some promise in development of novel pharmacological or products.

Solitary ascidians possess a wide range of antimicrobial agents in their blood and other tissues. Morula cells of *C. intestinalis* exhibits potent anti-bacterial activity. It is probable that they play role in neutralization of bacteria, and possibly other micro-organisms. Hemocytes of *C. intestinalis* show natural cytotoxic capacity. *C. intestinalis* may be useful as an indicator of marine pollution, as it tends to accumulate and sequester trace elements such as heavy metals vanadium, manganese, zinc and iron. Developmental performance of *C. intestinalis* could be used as an indicator of seawater quality, specifically heavy metal concentrations.

Control

The rock crab was very efficient at cutting open and extracting the body tissues of *C. intestinalis* in laboratory trials. Rock crabs feed on clubbed tunicates attached to mussel sleeves. Another group of natural predators are the small gastropods like *Mitrella lunata* or *Anachis* which feed on recently settled juvenile tunicates. Gastropods like the periwinkle *Littorina littorea* may dislodge juvenile tunicates during normal grazing activities. Unlike crabs, this herbivorous species can be inserted inside bags without risk to the shellfish. Various chemical and/or mechanical methods are suggested for eliminating *C. intestinalis* from aquaculture gear such as cages or nets. However, there are few options for treating sleeved mussels due to the risk of fall-off.

Summary

1. Urochordates or tunicates are small, common, marine animals. Tunic forms their flexible skeleton. Various proteins, some blood cells and spicules occur in the tunic. Pharynx in the form of sac collects food by ciliary action. Ring of tentacles guards the entrance of pharynx. Endostyle has 3 rows of mucus cells on each side, with a single median set of cells with very long cilia.

2. Fertilization is external in solitary forms, but internal in colonial forms. Their youngs have tadpole-shaped body. Larva takes no food and their gut is not well developed. Pharynx with single pair of gill-slits, open into atrium in the larva. Suckers are formed below, or around the mouth.

3. Ascidians are the largest and most diverse group of Tunicata with about 3,000 species. Adults bear little resemblance to chordates. Their larvae have dorsal tubular nerve cord, notochord, pharyngeal gill slits, post-anal tail, and sensory vesicle with 2 darkly-pigmented sense organs, the ocellus and statolith. Larvae undergo retrograde metamorphosis and lose all characteristics except for pharyngeal gill slits.

4. Thalieaceans are pelagic; barrel shaped filter feeders, and lives in warm water. About 70 species are known. Salp is common in equatorial, temperate, and cold seas, abundant in Southern Ocean. Their complex life cycles show obligatory alternation. Salp is free-floating, 1.5 to 19 cms long, and normally colorless.

Review Questions

Short Answer Questions

1. What is tunic?
2. What is the significance of the term "Tunicata"?
3. What are epicardia?
4. What is haemocoel?
5. Mention two differences between Ascidacea and Thaliacea.
6. Give two examples of Ascidacea.
7. Give two examples of Thaliacea.
8. Name the class to which salps belong.
9. What are trophozooids?
10. What are phorozooids?

Long Answer Questions

1. Give six characters of urochordates.
2. Mention six characters of Larvacea.
3. Describe the anatomy of *Ciona intestinalis* with suitable diagram.

4

Subphylum Cephalochordata

Cephalochordates are small fish like marine chordates with persistent notochord which extends forward beyond the brain. Hence they are called cephalochordates signifying the presence of notochord in head. The epidermis is single layered. Paired fins are absent. Muscles, nephridia and gonads are segmentally arranged. The pharynx is large with numerous gills. It is a filter feeder.

Branchiostoma illustrates 3 fundamental chordate characters. It has a dorsal hollow nerve cord and many gill slits in the pharynx. This is world wide in distribution and found on sandy shores at small depths all around the tropical and warm temperate seas. It draws small organisms into the mouth along with the current set up by ciliary apparatus at the anterior end of body. Occasionally, it emerges out of the burrow and swims about by sinuous movements of the body. When it senses some disturbance, it buries itself in sand again either by the head end, or by the tail end.

This is a burrowing animal, but can swim freely. Supposing it a slug, the German zoologist Pallas called it *Limax lanceolatus*. The name *Amphioxus lanceolatus* was given by Yarrell in the year 1836. Costa in 1834 coined the name *Branchiostoma* which now stands as the valid following the rules of priority. Eight *Branchiostoma* spp. are known. In general organization, *Asymmetron* resembles *Branchiostoma*, but they have gonads on right side only.

4.1 Salient Features

This is translucent; 0.5 to 3 cm long with long, laterally compressed and pointed tips at both ends, but lacks a head. Dorsal fin runs along the whole length of mid dorsal line. It expands around the tail as caudal fin or tail fin and then continues ventrally as a ventral fin, extending for about one-third of the body length. Fin rays support the dorsal and ventral fins. The anterior body is triangular in section, because the 2 lateral side project down as flaps called metapleural folds which are continuous anteriorly with the oral hood. The posterior one-third is oval in outline. Below the anterior end of the body, a membrane called oral hood surrounds a median funnel shaped cavity called vestibule or preoral chamber. Twenty two delicate, stiff tentacle-like processes called oral cirri are found in oral hood to guard the opening (Figure 4.1).

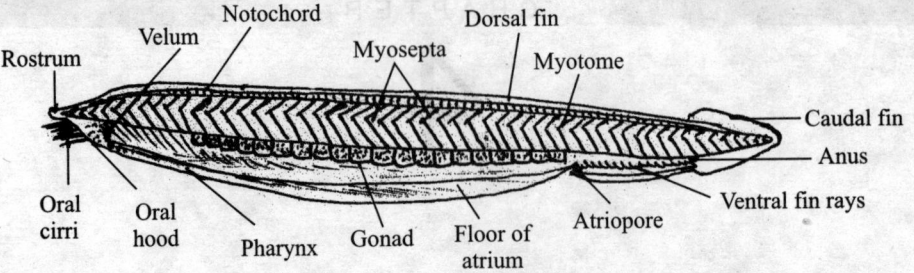

Figure 4.1 External features.

Mouth is situated at the bottom of vestibule and in the middle of velum. Velum is fringed with 12 velar tentacles. Inside the oral hood, the epithelial lining projects into ciliated finger-shaped processes. Movements of these processes cause a flow of water to the mouth resulting in wheel-like movement. The tract is called the wheel organ. A ciliated pit called the Hatschek's pit lies on roof of oral hood. Anus is placed slightly towards left side in front of tail fin.

Atrium: Gill slits open into a special cavity called the atrium. It is a space lined by ectoderm between the pharyngeal wall and body wall, and is enclosed by lateral extensions of body wall. It surrounds the pharynx laterally and ventrally, but not dorsally. The cavity encroaches into body cavity almost completely replacing coelom. Atrium opens outside by a small aperture called atriopore, lying in front of anus and protects gills, preventing them from being choked up with sand.

Body wall: This is soft, thin, translucent and composed of outer epidermis and inner dermis. Epidermis consists of a single layer of columnar epidermal cells, bearing at places sensory hairs and unicellular glands. Dermis consists of an outer layer of connective tissue (cutis) containing fibres and an inner thick layer of matrix (sub cutis) invaded by few fibres, blood channels and nerve endings. Beneath the dermis, muscular layer arranged segmentally as muscle blocks are called the myotomes or myomeres. Number of myotome varies from 60, or more. Muscles are thickened dorsally than ventrally.

Body is divided into head, trunk, and tail. Head located at the anterior end is small and indistinct. Rostrum extends anteriorly. The large mouth lies under the rostrum and opens into a spacious buccal cavity. Mouth is surrounded by a ring of tentacle-like buccal cirri which are probably chemoreceptive and mechanically sorts food particles .Roof and walls of buccal cavity form the oral hood. Trunk extends posteriorly from the head to anus, including large conspicuous pharynx, and musculature (Figure 4.2).

Paired appendages are absent but on either side of trunk a ventro-lateral longitudinal ridge, the metapleural fold is found. These ridges run from oral hood to a position just posterior to gonadal region. Atrium opens to exterior via the atriopore, located on midventral margin at the point where 2 metapleural folds join ventral margin. Further posteriorly, beside a slight dip between the ventral fin and caudal fin, there is anus, located slightly to the left side

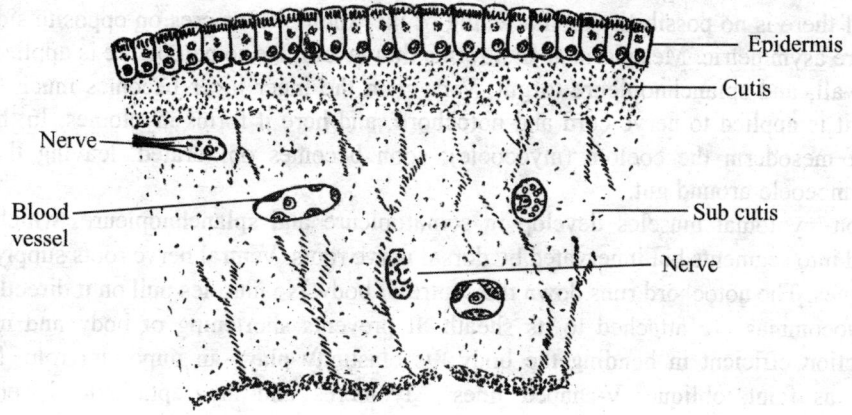

Figure 4.2 V.S. of body wall.

of ventral midline. Anus is the posterior external opening of gut and marks the posterior limit of trunk. Region of the body posterior to the anus is the tail. A posterior caudal fin extends around the dorsal and ventral margins of tail. A long dorsal fin extends along most of dorsal margin of the body. A short ventral fin is located on ventral margin of trunk anterior to caudal fin. It extends from atriopore to anus. Adult *Branchiostoma* is less than 2 in. long with fish-like organization. Elongated body is flattened, pointed at both ends and lacks prominent head. Separate eyes, nose, ears, and jaws are not found.

Skin: Integument includes a monolayered, nonciliated epidermis with a basal lamina, but lacks pigment and extra cellular cuticle. Below the epidermis, the fibrous cutis, and below this, gelatinous material containing fibres, the sub cutis are found. Both these layers are secreted by scattered fibroblast-like cells. Epidermis is very thin, composed of a single layer of cells, ciliated in young, and with outer border slightly cuticularized in adult. Receptor cells are absent in skin.

Skeletal Structures: The skeleton consists of notochordal tissue and gelatinous material containing fibres. Cells are absent within this material, but the cells around the outside secretes it, which retain the epithelial arrangement of mesoderm from which they were derived. This connective tissue continues as a sheath around the nerve cord and above this into fin-ray boxes, which support the median ridge. These are more numerous than the segments and each contains a more rigid material called cartilage. The chief supporting structure is the notochord. The fins and pharynx are connected with the oral hood. Skeletal rods occur in the cirri around the mouth and in gill bars.

Movement: Segmental arrangement of axial musculature is visible. Muscles are arranged in 50-75 V-shaped segmental bundles called myomeres. Successive myomeres are separated from one another by connective tissue partitions called myosepta. The animal is metameric. Contraction of myotome bends body and results in transverse motion of body inclined at varying angles to result in forward propagation. Each myotome contracts after the contraction of the front one to produce an S-bend that moves backwards through water. Serial contraction depends on breaking up of longitudinal muscle into blocks.

Contraction of longitudinally arranged muscle fibres only produce a sharp bending of body, if there is no possibility of shortening of the whole. Myomeres on opposite sides of the body are asymmetric. Mesoderm at first forms thin layers, the somatopleure is applied to outer body wall, and splanchnopleure to gut. Very soon the inner layer becomes much thickened where it is applied to nerve cord and notochord, and here it forms myotomes. In this dorsal part of mesoderm the coelom (myocoele), soon becomes obliterated, leaving the ventral splanchnocoele around gut.

Non-myotomal muscles develop in somatopleure and splanchnopleure, which are not divided into segments but innervated by dorsal nerve roots. Ventral nerve roots supply only the myotomes. The notochord runs down the centre of body. No muscles pull on it directly, though the myocommas are attached to its sheath. It prevents shortening of body and makes the contraction efficient in bending the body. Its elasticity plays an important role. Myosepta appear as faint, oblique, V-shaped lines. Myomeres and myosepta extend ventrally to metapleural folds. Myomeres are derived from segmental coelomic compartments and the muscles are derived from the epitheliomuscular cells of coelomic mesothelium.

Notochord is composed of a series of flattened plates surrounded by a fibrous sheath. Plates are arranged regularly with their flat surfaces in transverse plane of body. Two types of plates are the fibrous and homogeneous, which alternate with each other. Each plate develops as a highly vacuolated cell, the nuclei being later pushed aside to dorsal or ventral edge. This structure is well suited by the turgidity of its cells enclosed in sheath to resist the forces tending to shorten the body. The cord extends from the very tip of head to end of tail, projecting beyond the level of myotomes, a condition perhaps associated with burrowing habit.

Branchiostoma probably does not often swim free in water. The body is not adapted for fast movements. It has no elaborate fins like fishes to ensure static stability, or to allow active control of direction of swimming. A low dorsal ridge continues behind as a small caudal fin. Definite paired fins are absent. Metapleural folds may be compared to lateral fin folds from which vertebrate limbs are probably derived. They are distended with coelomic fluid and with dorsal ridge, probably protect the body during rapid dives by which the creature enters the sand. Habit of swimming with the front end downwards suggests the presence of a gravitational receptor mechanism. Larvae of lampreys swim in a similar way.

Mouth, Pharynx and Feeding: Mouth lies at the bottom of vestibule, leads into a large laterally compressed chamber, the pharynx which occupies the most anterior part of body cavity. Numerous gill slits (up to 180) perforate the pharynx (Figures 4.3a, b) wall on each side. From pharynx, the narrow tubular midgut or intestine extends backwards to anus.

On the ventral side of intestine, a large diverticulum, the liver diverticulum or hepatic caecum emerges. Running along the inner wall of pharynx both on dorsal and ventral middle lines there are 2 grooves lined with cilia. The dorsal groove, the hyperbranchial groove leads into intestine and ventral groove, the endostyle is composed of 4 tracts of mucous glands separated by tracts of ciliated cells. The glands secrete sticky mucous which entangle food materials. Anteriorly 2 ciliated tracts connect the endostyle and hyperbranchial grooves, which encircle the pharynx just behind the mouth. These ciliated tracts are called peripharyngeal bands. Inner wall of the pharynx is lined with cilia.

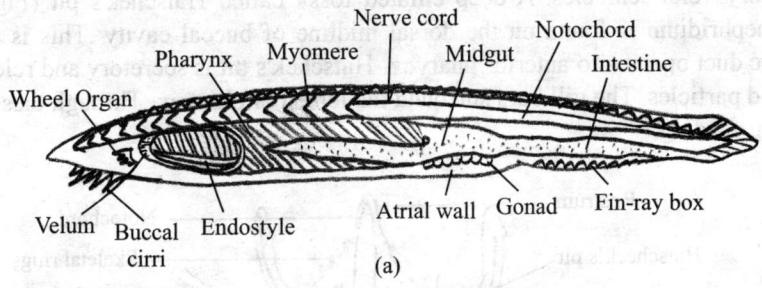

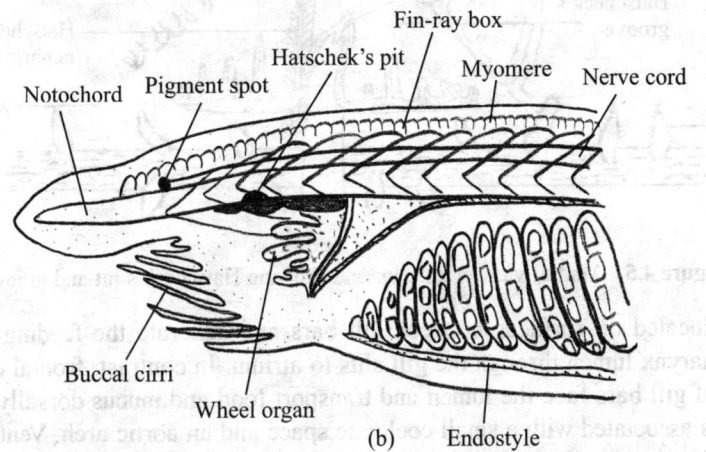

Figure 4.3   (a) L. S. through the entire body showing anatomical features (b) L.S through the anterior end.

Pharynx is the largest, conspicuous, and distinctive region of the gut .It begins just posterior to velum and is a large, intensely red-staining, oval structure with numerous, narrow, oblique gill slits separated by narrow tissue gill bars. Gill bars are supported by a collagenous branchial skeleton probably homologous to visceral (gill) skeleton of vertebrates. Gills function in filter feeding and gas exchange. In larvae, the number of gill slits equals the number of myomeres. The gut lumen within the oral hood is the transverse, muscular velum. An aperture, of adjustable diameter in the center of velum connects buccal cavity with pharynx.

Anterior to velum (Figure 4.4), the walls of buccal cavity bear a series of thick ciliated grooves to form the wheel organ. Cilia in these grooves trap food particles in mucus for digestion further posterior in gut. On the posterior side of velum, mouth is surrounded by

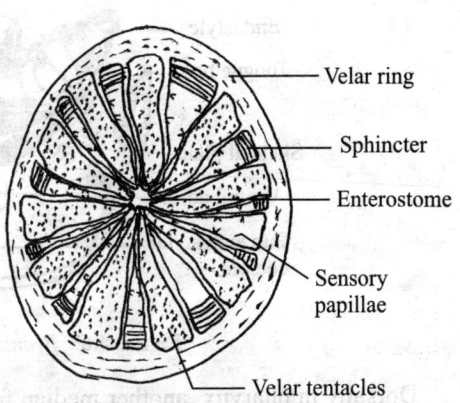

Figure 4.4   Velum.

slender sensory velar tentacles. A deep ciliated fossa called Hatschek's pit (Figure 4.5) or Hatschek's nephridium is found on the dorsal midline of buccal cavity .This is an unpaired kidney whose duct opens into anterior pharynx. Hatschek's pit is secretory and releases mucus to entrap food particles. The gill bars surround the lumen of pharynx. Through these slits water passes during feeding.

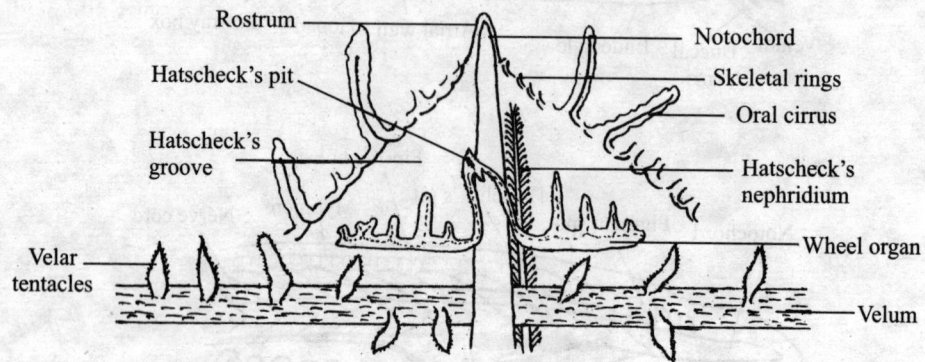

Figure 4.5   Ventral view of anterior end showing Hatscheck's pit and groove.

Cilia are located on surfaces between gill bars and generate the feeding current which passes from pharynx lumen through the gill slits to atrium. In contrast, frontal cilia are on the inner surface of gill bars face the lumen and transport food and mucus dorsally to esophagus. Each gill bar is associated with a small coelomic space and an aortic arch. Ventrally along the pharynx there is a longitudinal, median, ciliated groove, the endostyle, which secretes copious mucus. Mucus contains iodine and endostyle is homologous to vertebrate thyroid gland. Ventral to endostyle (Figures 4.6) there is a small endostylar coelom surrounding an even smaller ventral aorta .Coelom is homologous to the pericardial coelom of vertebrates.

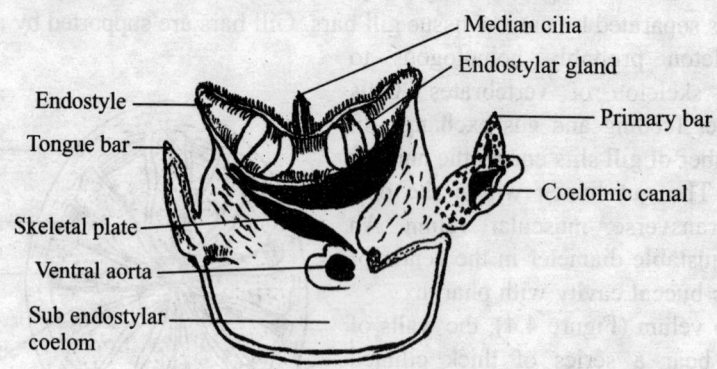

Figure 4.6   *Branchiostoma* : T.S. of endostyle.

Dorsally in pharynx, another median furrow, the epibranchial groove is found. The paired dorsal aortae are located dorsolateral to it. Most space surrounding the pharynx is the atrium, which is enclosed by an ectodermal epithelial lining. During feeding the lateral cilia of gill bars generate water current that enters pharynx through mouth. Water passes laterally through gill slits, passes out of pharynx, and into surrounding atrium. Mucus secreted by endostyle is carried upward over the inner surface of gill bars by ciliary currents generated by frontal cilia. Mucus entangles food particles attempting to pass between the gill bars and transports them to epibranchial groove. Cilia of the groove then move mucus and trapped food posteriorly into esophagus. Food is moved (Figures 4.7a, b) through the gut by its ciliated walls. Epithelium of hepatic cecum secretes digestive enzymes into gut lumen.

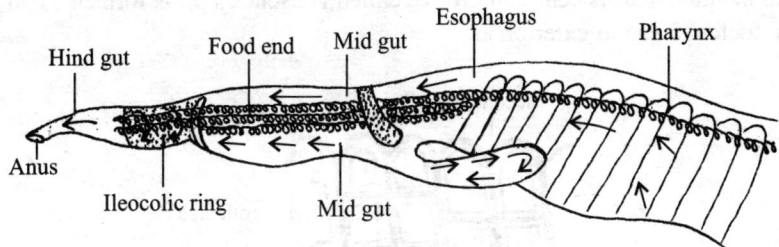

Figure 4.7(a) Path of food current in post-pharyngeal region.

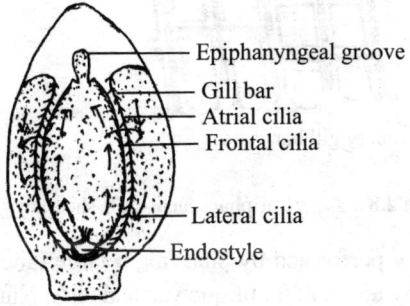

Figure 4.7(b) Diagrammatic transverse section of pharynx showing feeding current.

Digestion occurs extracellularly in stomach and iliocolon and its products are absorbed by epithelium of cecum. Although inconspicuous, coelom occupies its expected position surrounding the gut. Pharynx is surrounded by a large water chamber, an invagination of the surface ectoderm called atrium. Atrium, occupies most space between pharynx and body wall. It is U-shaped in cross section and encloses the pharynx on all sides except dorsally, and opens to exterior through a large atriopore. Posterior to pharynx, the gut narrows to form a short esophagus which connects the pharynx with stomach. Anteriorly, the stomach bears a ventral anteriorly directed diverticulum, the secretory and absorptive hepatic caecum, which projects into atrium on right side of the pharynx. By the structural, positional, and intermediate criteria, the hepatic cecum seems to be homologous to the liver of vertebrates.

Stomach narrow posteriorly to become the darker iliocolon, which bears a ring of ciliated epithelium, that rotates the food mass and mixes it with enzymes. Posterior to ring, the gut continues as intestine, finally ends at the anus, located at the base of tail. Asymmetrically located anus lies just to left of ventral fin. The post anal tail extends posteriorly from the anus. Food is collected by extracting small particles from a stream of water, which is drawn by cilia.

Pharynx and gill bars (Figure 4.8) occupy more than one-half of whole body surfaces. Special arrangements are found for support and protection of this ciliated surface. Wall of pharynx is mostly subdivided and needs protection of an outer layer, the atrium. When feeding, the buccal cirri is curved to form a funnel-like sieve preventing the entry of large particles. Around the mouth, a ring of sensory tentacles, the velum is present. Oral hood contains ciliated tracts, the 'wheel organ' of Muller, which plays a part in sweeping the food particles into mouth. Near its centre, a groove called Hatschek's pit is formed as an opening of the left first coelomic sac to exterior.

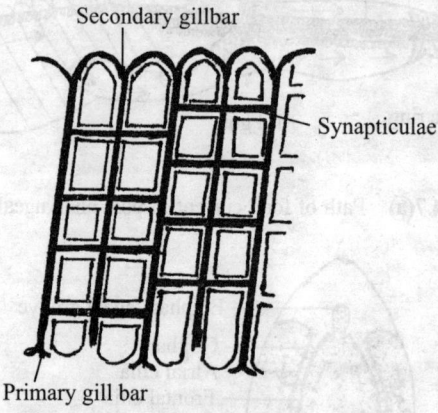

Figure 4.8   A part of pharyngeal-wall showing gill bar.

Food collection is mainly performed by pharynx, a large tube, flattened from side to side, whose walls are perforated by about 200 oblique vertical slits. Number of ventral slit increases as the animal gets older. Slits are separated by bars containing skeletal rods and further subdivision is provided by cross bars. Since the bars slope diagonally many bars are cut in a single transverse section, but they are the vertical portions of main walls of body and pharynx, where these have not been perforated by a gill slit. Such a portion of body wall contains a coelomic space. An increase of ciliary surface is produced by down growth of secondary or tongue bars from the upper margin, dividing each primary slit. These secondary bars lack coelom.

Coelomic spaces in primary bars communicate above and below with continuous longitudinal coelomic cavities. Cilia are found on sides and inner surfaces of gill bars, the lateral ones mainly drive the water outwards through atrium and draw the feeding current of water in at the mouth. In the floor of pharynx lies the endostyle, containing columns of ciliated cells, alternating with mucus secreting cells, which produce sticky threads in which food particles become entangled. Various currents then draw the sticky material along until it reaches the midgut.

Frontal cilia of the gill bars produce an upward current, driving the mucus from endostyle into a median dorsal epipharyngeal groove, in which it is conducted backwards. Cilia of endostyle move mucus along peripharyngeal ciliated tracts, behind the velum, to join the epipharyngeal groove. Radioactive iodine is concentrated by one column of the endostyle and secreted with the mucus. Barrington suggests that these may be the precursors of the thyroid cells, producing iodinated mucoproteins, which are then absorbed further down the gut. Pharynx narrows at its hind end to open dorsally into the mid gut.

A large midgut diverticulum reaches forward from this region on right-hand side of pharynx. From its position this organ is called the liver, but Barrington has considered it as the seat of production of digestive enzymes. Zymogen cells, similar to those of mid gut, are found in its walls. Its strong dorsal and ventral ciliation maintains a circulation of food materials and secretion. Its cells are capable of phagocytosis and secretory activity. *Branchiostoma* combines intracellular- with extra cellular digestion, doubtless about its microphagous habit. Particles placed in diverticulum are swept backwards and join the main food cord that passes through mid gut.

Hind end of mid-gut (Figure 4.9) is marked by a ciliated region, the ileocolon ring, whose cilia rotate the cord of mucus and food. Movement passes to the portion of food cord in midgut and presumably helps in taking up of enzymes that emerge from diverticulum. Extra cellular digestion takes place in mid gut. pH of the contents varies from 6.7 to 7.1. Amylase is present in extracts of diverticulum, midgut, and hindgut, but not in pharynx. Lipase and protease are present in the same regions, the latter having an optimum action about pH 8.0. Protease is not yet recorded. Behind the ileocolon ring, the intestine runs as a straight hind gut to anus.

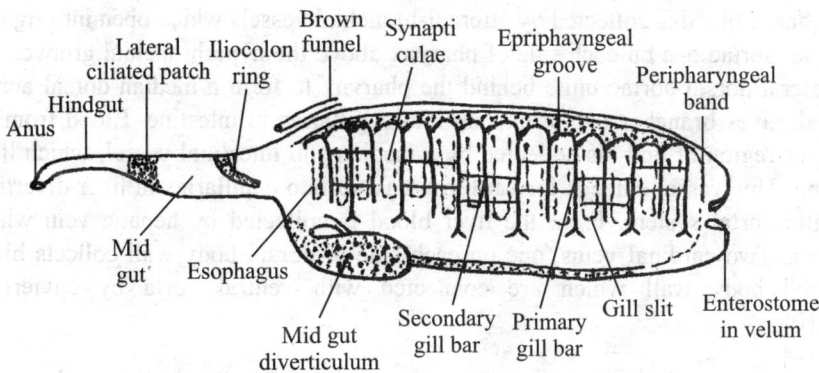

Figure 4.9   Alimentary canal showing midgut and hindgut.

Absorption of food takes place here, and perhaps also in midgut, partly by intracellular digestion. Feeding current is regulated by rate of ciliary beat and degree of contraction of inhalent and exhalent apertures. Atrium walls contain an elaborate system of afferent and efferent nerve fibres. Receptors include a set of large peripheral nerve cell bodies, lying beneath the atrial epithelium and sending axons in by dorsal roots. Motor fibres pass through dorsal roots and run without synapse to cross-striated fibres of pterygial muscle to form the floor of atrium. Stream flowing into pharynx is tested by receptors of velum and atrium. When noxious material is present, the water is expelled by closing atriopore and contracting pterygial muscle.

The system could distinguish between suspensions of food material and inorganic particles. When sufficient food has been taken, collection is suspended until it is digested. Atrial nervous system probably regulates spawning and feeding. It has been compared with sympathetic system of craniates. Nerve cells in it are receptors and peripheral synapse on efferent pathway characteristic of true autonomic system is absent. Atrial system is developed relative to filter feeding and has perhaps been completely lost in higher forms that feed by other methods and have developed new methods.

Branchiostoma feeds on minute organisms found suspended in water. While feeding, a continuous current of water enters the mouth and expelled out through atriopore via pharynx and atrium. Water current is maintained by actions of cilia in the wheel organ and gill bars of pharynx. Velar tentacles behind the mouth fold across and strain off the sand particles. Small organisms are brought into the endostyle and get entangled in mucus. Cilia drive them along the gill bars and peripharyngeal bands into hyper branchial groove and thence to intestine. Digestion occurs mainly in midgut. Digestive enzymes are secreted in midgut diverticulum and poured into the midgut. Digested food is absorbed in midgut.

Respiration and Circulation:　Gaseous exchanges occur in the gill bars when the water passes through the gill slits. The O_2 diffuses into the blood flowing in the channels of gill bars and the CO_2 from the blood diffuses into the running water. Specialized heart and respiratory pigments are absent. Blood is colourless. A median contractile vessel is found in ventral wall of pharynx below the endostyle. It ends anteriorly by branching in snout. Ventral aorta sends paired afferent branchial vessels into the primary gill bars. From gill bars (Primary and secondary bars), blood is collected by efferent-branchial vessels which open into right and left lateral dorsal aortae one on each side of pharynx above the hyperbranchial groove.

The lateral dorsal aortae unite behind the pharynx to form a median dorsal aorta which runs behind, gives branches and breaks up into capillaries in intestine. Blood from intestine and posterior region of body is collected by a median sub intestinal vessel, which lies below the intestine. This vessel extends forward to break up into capillaries in liver diverticulum to form hepatic portal system. From the liver blood is collected by hepatic vein which joins ventral aorta. Two cardinal veins, one on each side of dorsal body wall collects blood from muscles and body wall which are connected with ventral aorta by cuvierian ducts (Figure 4.10).

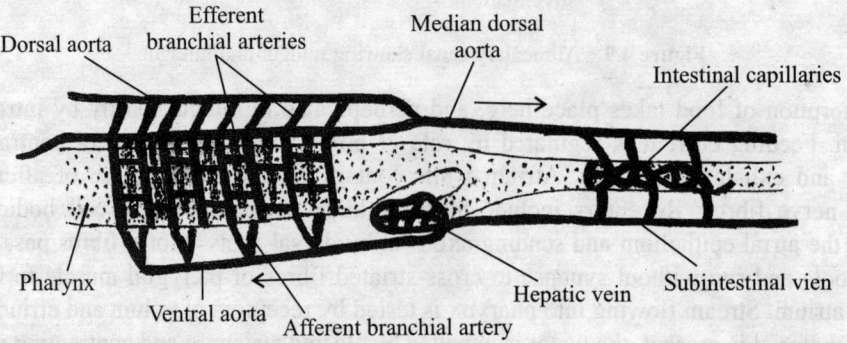

Figure 4.10　Circulatory system.

Slow waves of contraction occur in various separate parts in a way to drive the blood forwards in ventral vessels, backwards in dorsal ones. Below the hind end of pharynx, there is a large sac, the sinus venosus into which blood from all parts of body is collected. From this, there proceeds forwards a large endostylar artery from which spring vessels carrying blood up the branchial arches. At the base of each primary bar, a little bulb functions as a branchial heart. From the gill bars, blood is collected into paired dorsal aortae, which join behind the pharynx. From the paired and median aortae blood is carried to system of lacunae that supplies the tissues.

True capillaries are absent. From the lacunae blood is collected into veins. The important veins are caudals, cardinals, and a plexus on the gut. Cardinals are a pair of vessels in dorsal wall of the coelom. They collect blood from muscles and body wall and lead to sinus venosus by a pair of vessels, ductus Cuvieri, which pass ventrally and across coelom to join the sinus venosus on gut floor. Caudal veins join the plexus on gut, from which blood is collected by a large sub intestinal vein running onto liver; from here another plexus leads to sinus venosus. Contractions arise independently in sinus venosus, branchial bulbs, sub intestinal vein, and elsewhere. Rhythms are very slow, irregular, and apparently not coordinated by any control system. Blood contains no cells. Presumably, the tension of dissolved O_2 is acquired by simple solution which is sufficient for small energy needs of the animal.

Excretion: The paired nephridia (about 90 pairs) of ectodermal origin are segmentally arranged on each side of pharynx in its dorsolateral walls. Each nephridium is a closed, bent tube. Each tube has an anterior vertical and posterior horizontal limb. Vertical limb lies in coelomic canal of primary gill bar. It ends blindly. Horizontal limb projects into dorsal coelom and opens to exterior through nephridiopore. Both limbs of the nephridia are provided with groups of flame cells or solenocytes. A nephridium (Figures 4.11, 4.12) bears about 500 solenocytes.Each solenocyte is a hollow cell with a flagellum hanging down from wall. There are blood vessels passing to nephridia from afferent branchial vessels. Nitrogenous waste is collected by solenocytes from blood and coelomic fluid and is passed to the body of nephridium by the flagella. From this, it is discharged into atrium. It finally goes through atriopore. Flame-cells similar to those are found in platyhelminthes, molluscs, and annelids.

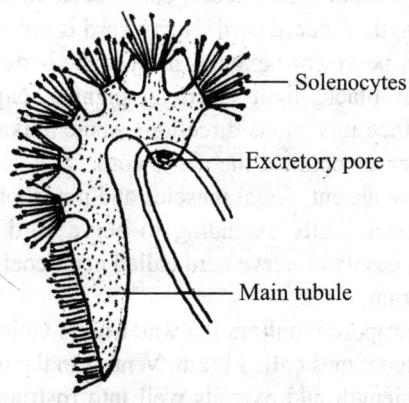

— Solenocytes

— Excretory pore

— Main tubule

Figure 4.11 Entire nephridium.

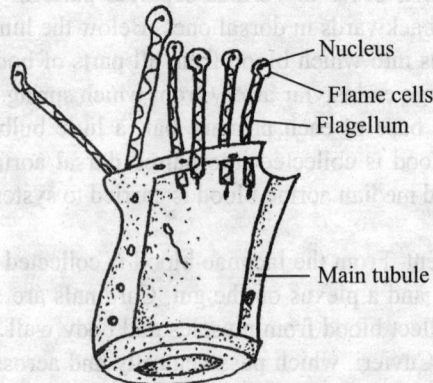

Figure 4.12   A portion of nephridium magnified and tubule out open.

Excretory organs differ from those of other chordates, echinoderms, brachiopods, and polyzoa. Nephridia lie above the pharynx. To each primary gill bar there corresponds a sac, opening by a pore to atrium and studded with numerous elongated flame cells. Flame cells do not open internally, but are in close contact with special blood vessels, whose walls separate the flame cells from coelomic epithelium. Excretion occurs probably by diffusion through flame cell wall, the liquid being driven down the tube by cilia. Coloured particles injected into blood stream are not excreted by nephridia. They are in no ways comparable to pronephros of vertebrates. Brown funnels, the blind sacs at the front of atrium invaginates into epibranchial coelom. They are probably receptor organs. Some parts of atrial wall perform excretion. Masses of cells in atrial floor, the atrial glands, contain granules that may be excretory, but may have been taken up from the food current. In gonads, especially the testes, there are large yellow masses, containing uric acid, which are extruded with the gametes.

Nervous System: The system consists of a median, hollow nerve cord lying above notochord and below the fin rays. It extends anteriorly to about the base of rostrum, where it bears a larger, terminal ocellus called eyespot that extends to posterior end of body into the base of tail. It is contained in a canal called neural canal. A small longitudinal cleft, the dorsal fissure extends throughout length of nerve cord. Nerve cord is not differentiated into brain and spinal cord. The dorsal hollow nerve cord extends most of the body length. It has a distinctive, irregular, longitudinal row of black, light-sensitive, pigment cup ocelli running along its ventral margin. These ocelli face in various directions, some dorsal and some ventral. Paired sensory and visceral motor nerves connect the nerve cord.

Somatic motor neurons are absent. Axial muscles and notochordal cells are innervated by cytoplasmic processes of muscle cells extending to nerve cord. Connective tissue sheath envelopes the nerve cord. The cavity of nerve cord called neurocoel opens to exterior by dorsal neuropore at the base of rostrum.

Chemoreceptor in the neuropore monitors the water in its vicinity. Neurocoel is expanded anteriorly to form a vesicle sometimes called brain. Ventral to the nerve cord is the notochord. It is longer, relative to body length and extends well into rostrum. Nerve-cord is somewhat modified at the front end. Nervous system is connected with periphery by a simple set of nerve-roots, a dorsal and a ventral on each side in each segment. Roots do not join. Ventral roots lie opposite the myotomes, to which they carry motor fibres, and these end on muscle fibres with motor end-plates. Dorsal root runs between myotomes and carries all afferent fibres of segment and motor-fibres for non-myotomal muscles of ventral part of body.

Receptor system is made up of about continuous column of bipolar cells of Retzius, together with smaller cells of various types. These receptor cells are equivalent to dorsal root ganglion cells of vertebrates. Other receptor cell is the giant Rohde cell, which has a large axon and elaborate dendritic system. At least, some of these cells possess a peripheral axon running in dorsal root, i.e. longitudinal connective cell. Visceral motor cells are arranged segmentally one per segment.

Somatic motor cells are found at various levels in cord. Other cells in cord are various types of internuncials. Fibres of peripheral nerves lack thick myelin sheath. An epineurium with connective tissue cells surrounds nerve trunks. Schwann cells accompany the nerve fibres.

At least 3 types of central neuron send fibres that end as free nerve endings in skin. On the head and tail, there are peripheral receptor cells, sending fibres centrally, also complicated encapsulated organs in metapleural folds. There are numerous large multipolar nerve-cells, presumably afferent, just beneath atrial epithelium. Many branched dendrites and an axon runs through a dorsal root to spinal cord. Spinal cord has only a narrow lumen and its elements are arranged as in vertebrates, namely, ependyma close to canal, cell layer, and outer fibrous layer.

Cells are not arranged clearly in horns as they are in vertebrates. The conspicuous cells are the giant cells, which lie dorsally in anterior and posterior parts, but are absent from about the 13th to 39th segments. Each of these cells has many dendrites, branching in region of entry of dorsal root fibres, and a single axon, which runs backwards in front part of body, forwards in hind, passing in each case for whole length of cord. A median giant fibre, which runs ventrally for the length of cord, lies close to visceromotor cells that probably produce the coughing movements of atrium. There are no isolated or local movements; the effect of any stimulus like touch on side of body produce waves of myotomal contraction. These may vary from strong waves going the whole length of body to single rapid twitches. Giant cells participate in spread of these waves.

It seems likely that arrangement ensures that touch on anterior part of body, normally exposed when feeding, and produces backward movement, but touch on hind part causes the reverse movement of emergence and escape. At the front end, central canal is enlarged to form a cerebral vesicle. Whole neural tube is hardly wider here than in region of spinal cord and there is no thickening of the walls, which are indeed mostly formed of a single layer of ciliated epithelial cells. From the region of cerebral vesicle spring the first two dorsal roots, to which there are no corresponding ventrals. These roots carry impulses from the receptors of oral hood and its tentacles.

Cells of infundibular organ beat in opposite direction to those found in the rest of vesicle. From them, fibres run backwards down the cord. Infundibular organ , the site of origin of Reissner's fibre which is a thread of non-cellular material, present at the centre of neural canal. It is secreted at the front end, passed backwards and is collected and absorbed in a sac at the hind end of spinal cord. Infundibular organ of *Branchiostoma* is not exactly similar to that of other vertebrates, although the Reissner's fibres are clearly comparable; an interesting problem in homology. Infundibular organ similar to neurosecretory material found in fibres of hypophysial tract contain material that stains with Gomori method. Organ occupies a central position in control system as a receptor, originator of nerve-fibres, and of two types of secretion.

In young stages, the cerebral vesicle (Figure 4.13) opens through an anterior neuropore, and at the point where the closure takes place there develops a depression of the skin, lined by special epithelium, and known as Kolliker's pit. It lacks special innervation. Cells of the front end of cerebral vesicle contain pigment (Figure 4.14) probably which prevents photic stimulation. Other cells lying in spinal cord are clearly photoreceptors. The front part of body is unprotected, whereas posterior part is pigmented for protection from light. When swimming freely in water, *Branchiostoma* moves spirally about its axis, turning clockwise. A small beam of light produces movements, when it is directed onto region of body or tail, not when it shines on head. Since the animal normally lies with head protruding, it is assumed that pigment spot prevents light that strikes down vertically from stimulating the photoreceptors in cord.

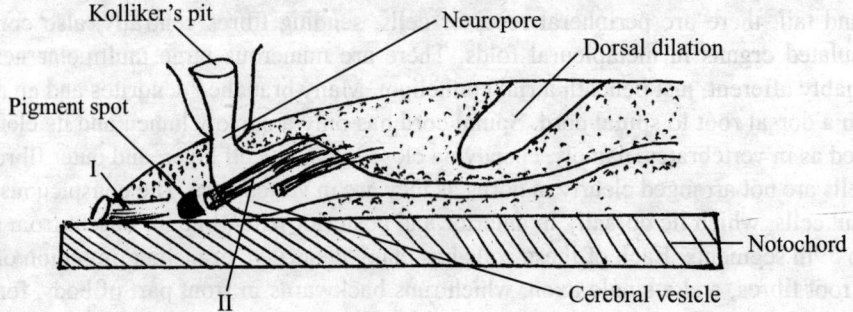

Figure 4.13   Median longitudinal section through brain.

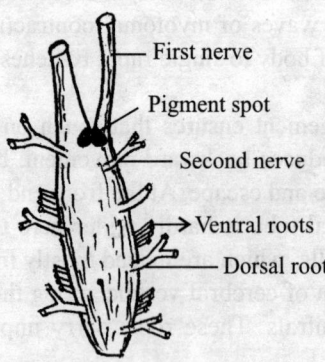

Figure 4.14   *Branchiostoma* : Anterior part of CNS.

Branchiostoma is provided with receptor and motor systems that keep it in its sedentary position, able to collect food from current that it makes by cilia. There are mechanisms that help it to make appropriate movements of escape when it is touched or, when the body is illuminated. Touch receptors of buccal cirri produce rejection of large particles and those of velum are chemoreceptors. Infundibular organ may be form of gravity, or pressure receptor. By these receptor organs and its simple movements of swimming, burrowing, and closing the oral hood, the animal is maintained, probably mainly by trial and error behaviour, an environment suitable for its life.

Gonads and Development: Two sexes are separate. There is no sexual dimorphism. Twenty-six pairs of paired, rectangular swellings along the ventral margins of myomeres are segmental gonads, which are simple hollow sacs and mesodermal in origin. Gonoducts are absent. Gonads (Figure 4.15) bulge atrial cavity. When the germ cells ripe, the sacs burst and genital elements are set free into atrial cavity, from where they pass outside by atriopore. Fertilization and development occur in sea water.

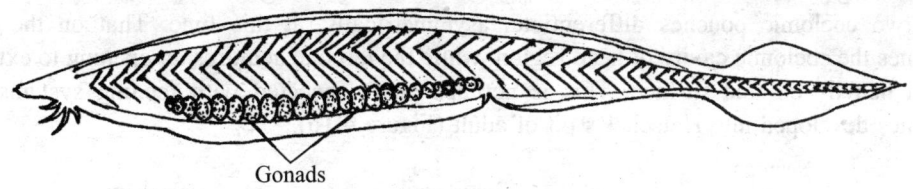

Gonads

Figure 4.15 Diagram showing the location of gonads.

Branchiostoma is gonochoric. Gonads are hollow sacs. Each sac develops from a single cell, at the base of myotomes in branchial region. Genital cells develop on walls. Genital products are shed by dehiscence into atrium. Extrusion of gametes occurs in spring. Fertilization is external. Development occurs in water. Numerous small yolky eggs are produced. Complex flowing movements occur after fertilization. Cleavage is rapid, complete, producing a blastula of a dome of smaller and floor of larger cells. These latter then invaginate to form archenteron, which opens by blastopore and later transforms into anus. At this stage, the gastrula becomes encircled by flagella.

The animal elongates and its dorsal side flattens and finally sinks into form neural tube. At this time, dorsal side of inner layer begins to fold near front end, to make a pair of lateral pouches. Walls of these pouches are future mesoderm and cavity is the coelenterons.

Coelom is continuous at first with archenteron. Roof of archenteron arches up dorsally to form notochord. The gut wall is formed by approximation of edges of rest portion of inner layer. Formation of neural tube, mesoderm and notochord and completion of gut roof involve an upward movement of cells to the mid-dorsal line (Young, 1981). This process of convergence is a marked feature of development. As the animal elongates, further mesodermal pouches are produced, each separating completely from endoderm and from its neighbours.

Cells of each pouch push down ventrally on either side of gut, the outer ones applying themselves to body wall to form somatopleure, the inner to gut wall as splanchnopleure. Inner wall of mesoderm on either side of nerve- cord thickens to form myotome, and a tongue of cells growing up between this and the nerve-cord forms the sheaths of the latter and probably also fin-ray boxes and other mesenchymal tissues. Upper part of coelomic cavity, the myocoele becomes separated from ventral splanchnocoele. Whereas the former becomes almost completely obliterated, the latter expands to form adult coelom, cavities between the adjacent sacs breaking down. While this differentiation of mesoderm has been proceeding, the animal has elongated into a definitely fish-like form.

Neural tube is a small dorsal canal, opening by an anterior neuropore and continuous behind through a neurenteric canal with the gut. Larva hatches when only two segments have been formed and swims at the sea surface by its ciliated epidermis, turning on its axis from right to left as it proceeds with the front end forwards. Mouth now appears as a circular opening and then moves over to left side and becomes very large. From this time onward, the whole development is asymmetrical. First gill-slit also forms near midline but moves up onto right side.

About the same time, the right side of pharyngeal wall develops into a V-shaped thickening, the endostyle. Behind this there forms a tube, the club-shaped gland, joining the pharynx to outside and formed by closure of a groove in the side of pharynx. It is connected with feeding process, which begins at this stage. It has been thought to represent a gill-slit. The first two coelomic pouches differentiate, asymmetrically, at this time. That on the right becomes the coelomic cavity of head region, while the left one acquires an opening to exterior and a heavily ciliated surface. This is perhaps also connected with feeding systems and becomes developed into Hatschek's pit of adult (Figure 4.16).

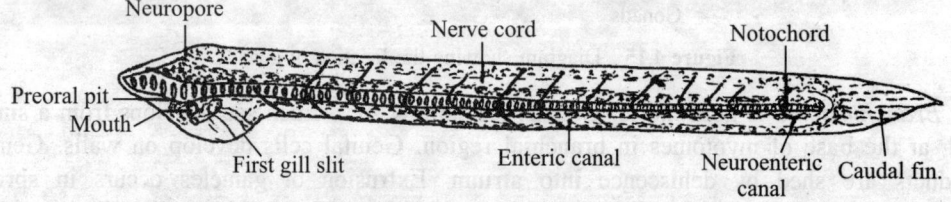

Figure 4.16   Young larva.

First coelomic cavity opens to exterior in other early chordates and in some vertebrates. Further gill-slits develop in mid-ventral line and move over onto right side until fourteen have been so formed. Meanwhile, a further row of 8 slits appears above that already formed. These are definitive slits of right side and presently larva proceeds to become symmetrical by movement of eight of the first row of slits over to left side, the remainder disappearing. At this critical stage with eight pairs of slits, larva pauses for sometime before further changes. This is the time at which it most nearly represents what might have been an ancestral craniate, with 8 branchial arches. Further slits are then gradually added in pairs on both sides.

Each slit becomes subdivided, soon after its formation, by down growth of a tongue bar. Atrium is absent from early larva. Metapleural folds then appear on either side and are united from behind forwards to form a tube below the pharynx. During later stages of development, larva sinks and finally rests on bottom while undergoing the migration of gill-slits that constitutes its metamorphosis. In other species, larva remains longer in plankton, becoming large and even showing quite large gonad rudiments. These were at first thought to be adults of a new genus.

Cleavage, invagination, and mesoderm formation recall those of echinoderms and other forms similar to ancestors of the chordates. There are some special features connected with method of life of larva (Figure 4.16), and especially with its asymmetry. The strange sequence of gill formation, the immense left-sided larval mouth, perhaps the club shaped gland, and Miiller's organ, may show considerable modifications of relatively recent date. The earliest chordates probably fed by cilia and were planktonic.

Segmentation of mesoderm of vertebrates is restricted to the dorsal region. In the lowest chordates, there are 3 coelomic cavities, but it is probable that many segments of vertebrates arose to provide a set of muscles able to contract in a serial manner for swimming. Their segmentation would thus be a relatively late development, not related to segmentation of annelids. Accordingly, the ventral part of vertebrate coelom usually remains unsegmented. But in *Branchiostoma*, it is subdivided from its first appearance and only becomes continuous later.

4.2 Generalized Chordate Features of *Branchiostoma*

Branchiostoma retains the habit of ciliary feeding, and bottom-living habit. There are many specializations. However, the general arrangement of body is simple. Amphioxides larvae show signs of such a change to suspect that *Branchiostoma* is not an ancestral type but a simplified derivative of vertebrates, perhaps a paedomorphic form. Neoteny might explain the regular segmentation, separate dorsal and ventral roots, and other features, but can hardly account for method of obtaining food, for skin condition, or for presence of nephridia. *Branchiostoma* shows the condition of the early fish-like chordates, living in Silurian some 400 million years ago, and it has undergone relatively little change in all the time since.

Summary

1. Cephalochordates are small fish like marine chordates with persistent notochord which extends forward beyond the brain. They are called cephalochordates because of the presence of notochord in head. *Branchiostoma*, a typical representative has a dorsal hollow nerve cord and many gill slits in the large pharynx. Numerous gill slits (up to 180) perforate the pharynx wall on each side.

2. Running along the inner wall of pharynx both on dorsal and ventral middle lines, there are 2 grooves lined with cilia. The dorsal groove, the hyperbranchial groove leads into intestine. The ventral groove, the endostyle is composed of 4 tracts of mucous glands separated by tracts of ciliated cells. The glands secrete sticky mucous which entangle food materials. Anteriorly 2 ciliated tracts connect the endostyle and hyperbranchial grooves. These ciliated tracts are called peripharyngeal bands.

3. Pharynx is the distinctive region of the gut .It begins just posterior to velum and bears oblique gill slits separated by narrow tissue gill bars. Gill bars are supported by a collagenous branchial skeleton probably homologous to visceral (gill) skeleton of vertebrates. Gills function in filter feeding and gas exchange. Nephridia lie above the pharynx. Flame cells do not open internally, but are in close contact with special blood vessels whose walls separate the flame cells from coelomic epithelium.

4. *Branchiostoma* retains the habit of ciliary feeding, and bottom-living habit with many specializations. The general arrangement of their body is simple. Amphioxides larvae show signs of change to suspect that *Branchiostoma* is not an ancestral type but a simplified derivative of vertebrates.

5. Larvacea, the minute neotenous live in plankton. Each individual builds house by secretion from oikoplastic epithelium of skin. The solitary ascidian, *Ciona intestinalis*

has perhaps originated in the Northeast Atlantic. This is cosmopolitan in distribution. Linnaeus in 1767 mentioned the "European oceans" as the type locality for it

Review Questions

Short Answer Questions

1. What are myotomes?
2. What are gill rods?
3. What is oral hood?
4. What is wheel organ?
5. What is "pit of Hatschek"?
6. What is primary gill bar?
7. What is secondary gill bar?
8. What is endostyle?
9. What do you understand by the term "microphagous"?
10. Comment on myocoel.
11. Comment on protonephridia
12. What are solenocytes?

Long Answer Questions

1. Draw and describe the mechanism of feeding and digestion in *Branchiostoma*.
2. Draw and describe the circulatory system of *Branchistoma*.
3. Describe the organs associated with excretion of *Branchistoma* with suitable illustrations.
4. Describe the pharynx of *Branchiostoma* with diagrams.

5

Agnatha

The term Agnatha is derived from the Greek word gnathos meaning jaw, and "a" is the prefix for "without." These jawless fishes, the first vertebrates, date back to 470 million years. The extant groups include lampreys and hagfish.

The name *Petromyzon* signifies a rock sucker, as it is found clinging by its mouth to stones. Agnatha became large by feeding on detritus at the bottom of rivers and lakes. They evolved into various types, mostly heavily armoured and slow-moving forms. Lampreys and hag-fishes have been derived from early Agnatha by evolution of sucking mouth, perhaps with loss of bony skeleton and paired limbs. Living agnathous animals are Cyclostomata, lampreys and hag-fishes. The ostracoderms, the first vertebrates appear in fossil series, found in Silurian and Devonian strata, show agnathous condition. Modern cyclostomes are parasites, or scavengers in adult. Lamprey larvae feed on microscopic material using endostyle and muscular contraction rather than ciliary action, to produce a feeding current. Lampreys begin life as burrowing, freshwater larvae. Mineralization of cartilage seems to have first developed in them and they became scavengers and predators as opposed to teethed filter feeders. Endoskeletons become bony, starting with skull.

5.1 General Features of Vertebrates

1. Skull is developed.
2. Pharynx is mostly small.
3. There are relatively few gill slits.
4. Except the most ancient form, the more anterior arches between the gills form jaws.
5. Heart has at least three chambers.
6. Presence of haemoglobin within corpuscles.
7. Mesodermal kidneys play large part in osmoregulation.
8. The front end of the nervous system is differentiated into an elaborate brain.
9. Brain is associated with special receptors, the nose, eye and ear.

10. Besides the median fins, there develop lateral paired ones. Vertebrata consists of two Superclass Agnatha and Gnathostomata.

5.2 Characters of Agnatha

1. Jaws are absent
2. Presence of round, suctorial mouth.
3. Semi-parasitic on larger, jawed fish.
4. Rasp with the host flesh by a tongue and horny teeth within the mouth cavity.
5. Sexes are separate.
6. Fertilized eggs hatch to form filter-feeding ammocoete larvae.
7. Stream lined shape, and pouched gills allow a tidal flow of water over them.
8. Primitive features include a persistent notochord, absence of paired fins, ducts from the gonads, and seven pairs of gills.

Class: Pteraspidomorphi

1. Extinct.
2. Head was broad, flattened and covered with carapace.
3. Double nosed.

Subclass: Heterostraci

1. Marine, bottom dwellers.
2. They were mostly only a few contimetres long.
3. The bone called isopedin is held by some to be an acellular, primitive type.
4. There were probably seven gills pouches and a posterior branchial aperture.
5. Most had paired eyes and median pineal eyes.

Example: *Psammosteus, Eglonaspis*

Subclass: Thelodonti

1. Small placoid-like scales or spines were present.
2. Existed between lower Silurian and middle Devonian.

Example: *Thelodus, Lanarkia*

Class: Cephalaspidomorphi

1. Single nosed.
2. Extinct.
3. Head was broad, fattened and covered with carapace.

Order: Osteostraci

1. Tail was heterocercal.
2. There were paired eyes and only two semicircular canals.
3. The ventral surface of head was flat and covered with small scales.

Example: *Hemicyclaspis, Cephalaspis*.

Order: Anaspida

1. Rows of bony scales were present.
2. The tail had a long lower lobe than the upper (hypocercal)
3. They possessed a ventral or ventro-lateral fin fold or perhaps a series of them.
4. Large paired eyes were present.

Example: *Birkenia, Jaymoytius*.

Order: Galeaspida

1. The bony covering of the flattened, front part of body was drawn out into lateral spines.
2. Presence of eyes and a nasal or nasohypophyseal canal.
3. Some structures on either side of the body have been called gills.

Example: *Galeaspis*.

Order: Cyclostomata

1. Mouth is circular in outline and suctorial.
2. Skin is naked and glandular.
3. Median fin without true fin-rays is present.
4. Notochord is continuous and enveloped by two sheaths.
5. Alimentary canal is straight and cloaca is absent
6. Respiratory organs consist of gill pouches.
7. The conus arteriosus and renal portal system are absent.
8. Cerebellum is very small.
9. Gonad is unpaired and is without gonduct.

Example: *Petromyzon, Lampetra, Geotria, Myxine*.

5.3 Lampreys

Lampreys are found in temperate zones of both hemispheres. Life-history comprises ammocoete larva and adult. Larva is fresh water form, microphagous, remains buried in mud. Adult is about 30 cm long, black on the back and white below, usually marine with sucking mouth, and feeds on other fish. Of 20 lamprey species most inhabits temperate regions. Lake Lamprey injures fishes while Brook lamprey do not .They are small and do not occur in lake. Two species spawn in identical places. Habitat of larvae and external features are also same.

Brook lamprey spawns 1 to 2 weeks earlier than Lake Lamprey. Head bears a pair of eyes and a round sucker. A single nasal opening is found behind which is a gap in pigment layers of skin through which the third or pineal eye appears as a yellow spot. Seven pairs of round gill openings are found. Paired fins are absent. Tail bears a median fin, which is expanded in front as a dorsal fin. Sex differences exist in shape of dorsal fins of adults. Female bears anal fin. Adult swims in water like a fish. Membranous expansions on its back and on dorsal and ventral sides of its tail serve as fin. Adult does not take food. Larva completes feeding (Figures 5.1, 5.2).

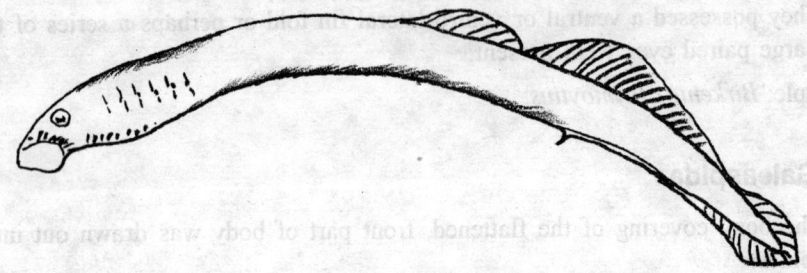

Figure 5.1 American Brook Lamprey, *Lamperta appendix*.

Figure 5.2 River Lamprey.

Skin and Muscles

Surface is smooth with no scales but multilayered. The outermost cells with a striated cuticular border along with many gland cells produce slime. Below the epidermis, the dermis, a layer of bundles of collagen and elastin fibres, run mostly circularly. Dermis is marked off from a layer of subcutaneous tissue containing blood vessels, fat, and connective tissue. Pigment cells are found in dermis and subcutaneous tissue. Chromatophores make the skin dark, or pale especially in larva. Lamprey uses myotomes for mechanically inefficient swimming. Waves pass down the body are of short period relative to length. They rest attaching itself with sucker to stones, or prey. Trunk musculature consists of W-shaped myotomes separated by myocommas. Muscle fibres are striped and run longitudinally.

Skeleton

Skeleton consists of notochord and cartilage, large vacuolated cells which remain separated by a matrix of protein chondrin. In other regions, the fibro cartilage contains more cells and few matrixes. In larva, mucocartilage – an elastic material serves as an antagonist to muscles rather than for attachment. Notochord sheath is continuous with connective tissue layer, which surrounds spinal cord and joins myocommas and subcutaneous connective tissue. Rigidity of notochord depends on turgor of cells. It prevents shortening of body when myotomes contract.

Skull in adult shows basic arrangement found in embryo of higher vertebrates. Skull floor is formed of paired parachordals on either side of notochord and in front of these paired trabeculae a series of incomplete cartilaginous boxes surround brain .Organs of special sense are found. Skull supports sucker and gills. In skull, there is a hole containing pituitary gland. Side walls are strong but roof is composed of a tough membranous fibro cartilage. Auditory capsules are compact boxes surrounding auditory organs. Olfactory capsule, imperfectly paired, is almost detached from cranium. Other ridges of cartilage lie below the eyes and there is a complex support for sucker. Skeleton of branchial region consists of a system of vertical plates between gill-slits.

Alimentary Canal

It has no jaws. Mouth is large circular disk, thickly studded with large, strong, chitinous spines or teeth to grasp its victim. Disk is surrounded by a soft membrane, which fits tightly over surface and helps to adhere firmly to object by suction. Tongue rasps away with saw-like teeth until it has worn through skin or scales of its victim. It attaches to fish, suck blood, often rasping away at its raw flesh, and make the hole deeper until the abdominal wall is perforated. Often the intestines or other organs of fish are attacked .It may fasten itself at another place if its victim has any blood left, or if not it seeks another fish. Lamprey attack catfish, suckers, carp, lake herring, and pickerel, sturgeon, pike, bass, perch, lake trout, wall eyed pike, red horse or mullets, eel, drum, white bass.

Sucker is bounded at edges by sensory lips, which make a tight attachment when lamprey sucks. Arrangement of teeth in sucker varies in different species. Teeth are horny epidermal thickenings, supported by cartilaginous pads. The sharp and large teeth are borne on movable tongue. Annular muscle runs round just above the lips of sucker and narrows the margin to release the fish. Remaining muscles are mostly attached to tongue and base of sucker. Collar of circular fibres around the front end of cardioapical muscle, lock the tendon for suction. Dorsal and ventral to the main tendon, groups of muscles rock the tongue up and down to produce a rasping action. Muscles of sucker are derived from the lateral plate and are innervated from the trigeminal nerve.

At hind end, buccal cavity divides into the oesophagus and respiratory tube leading to gill pouches but is closed behind. At the mouth of respiratory tube, velar tentacles separate mouth and oesophagus from respiratory tube during feeding. Seven branchial sacs are lined by a folded respiratory epithelium and surrounded by muscles, and these together with elastic cartilages and valves pump the water in and out of external openings. In front of the first sac, remains of an 8th pouch without respiratory surface is found. Salivary glands are paired pigmented sacs, within hypobranchial muscles with a folded wall, from which a duct proceeds forward to open below tongue.

The glands produce a secretion to prevent coagulation of fish blood. Secretion rapidly turns black on exposure to air. Esophagus leads into a straight intestine without true stomach. Typhlosole run a spiral course and increase intestinal surface. Liver, gall-bladder, and bile-duct are present. Separate pancreas is absent. In the wall of anterior intestine, large patches of cells resembling those of acini of pancreas of higher forms contain secretory granules. Extracts of this region have a proteolytic power, the enzyme being of tryptic type, with its optimum between pH 7.5 and 7.8. Around the junction of fore-gut and intestine, the groups of follicles do not communicate with intestinal lumen.

Circulatory System

Well-developed heart lies behind gills are a portion of sub intestinal vessel which is folded into S-shape and divided into 3 chambers. Heart is suspended in pericardium with walls supported by cartilage. In larva, heart first appears as a straight tube and may fail to develop in S-shape. In heart, contraction proceeds in chambers from behind forwards. The most posterior chamber is a thin-walled sinus venosus into which veins pour blood. This leads to a thin walled auricle lying above the sinus. Atrium passes blood into ventricle below it. Ventricle provides main force for sending blood round the body. Heart receives nerve fibres from vagus nerve and contains nerve cells. Stimulation of vagus nerve produces acceleration of heart-beat followed by slowing. Acetylcholine also accelerates heart.

Blood leaves ventricle by a large ventral aorta, running forwards between the gill pouches, to which it sends 8 afferent branchial arteries. These break up into capillaries in gills. Efferent branchial arteries collect to a pair of dorsal aortae, running backwards to join and form main dorsal aorta. This passes down the trunk and carries blood to all body parts by segmental arteries and special vessels to gut, gonads, and excretory organs.

Many arteries are provided with valves. Valves are absent where efferent branchials join dorsal aorta, and at points of exit of renal arteries. Valves perhaps reduce pressure in most arteries, while leaving it high in those to kidneys. Venous system consists of a network of sinuses with contractile venous hearts. A large caudal vein divides where it enters abdomen into 2 posterior cardinals. These run forward in dorsal wall of coelom, collect blood from kidneys, gonads, and open into heart by a single ductus Cuvieri on right-hand side. Anterior cardinals collect blood from front part of body. A conspicuous ventral jugular vein drains venous blood from muscles of sucker and gill pouches. There is a large system of venous sinuses especially in head.

Blood, from the gut, passes by a hepatic portal vein through contractile portal heart to liver, from which hepatic veins proceed to heart. Blood contains haemoglobin, enclosed in corpuscles, and here nucleated which increases the oxygen-carrying capacity of blood. Haemopoietic tissue occurs in intestinal wall of larva. In adult, the blood-forming tissue lies below spinal cord and in kidney. White corpuscles resembling lymphocytes and polymorphonuclear cells are produced by lymphoid tissue in kidneys and elsewhere. Distinct system of lymphatic channel is absent.

Urinogenital System

When lamprey remains in fresh water, blood contains a high concentration of salts than surrounding water and they remove water without losing salt. Nephrotome gives rise to kidney lies between dorsal scleromyotome and ventral lateral plate mesoderm. This tissue differentiates during development for making segmental funnels and opens into common archinephric duct. The most anterior funnel open into pericardium, usually there are 4 such canals in freshly hatched larva, opening into a single duct, which reaches back to an aperture near anus. Close to each funnel, tangle of blood vessels develops the glomerulus. Perhaps the osmotic flow of water into body is relieved by pressure of heart-beat forcing water out from glomeruli into coelomic fluid. Anterior funnels constitute the pronephros.

As animal grows, they are replaced by the posterior mesonephros. There is a gap of several segments in which no tubules appear. Pronephric tubules gradually disappear and in adult, organs are mass of lymphoid tissue. Mesonephros develops as larger fold, hang into coelom and contain extensive winding tubules. They do not open to coelom (at least in adult) but to a small sac, the Malpighian corpuscle containing a portion of coelom and glomerulus. This efficient method allows the heart to pump excess water out of blood and down the tubules. The latter have become elongated and make up the main bulk of organ. Mesonephros extends at its hind end until it forms adult kidney, a continuous ridge of tissue reaching back to hind end of coelom. Kidney contains lymphoid tissue and fat, and it plays a part in formation and destruction of red and white corpuscles.

Gonads are unpaired ridges medial to mesonephros. Primordial germ-cells, set aside very early in development, migrate into these ridges and develop into eggs or sperms. Differentiation of gonad occurs relatively late, so that in young ammocoetes the organ is hermaphrodite, containing developing oocytes and spermatocytes. Ripe ovary consists of ova with single-layered follicular epithelium, which ruptures and liberates the egg into coelom, whence it escapes by pores. Testis consists of several follicles containing sperms. When ripe they rupture into coelom, which becomes filled with spermatozoa and these escape, like ova, by pores. Apertures by which the gametes escape are similar in two sexes and consist of short channels, one on each side, leading from coelom to lower end of kidney duct. They normally open only a few weeks before spawning. Fertilization is external, but there are modifications of cloaca in two sexes to assist in fertilization and proper placing of eggs in nest. Lips of cloaca of ripe male form a narrow penis like tube. Cloacal lips of female are enlarged and often red. Anal fin in female is probably used to make a nest.

Nervous System

The front end of spinal cord is enlarged into a complicated brain. Nerves connected with anterior segments are modified to form special cranial nerves. Spinal nerves in dorsal and ventral roots do not join. Ventral roots contain motor fibres passing to myotomes. Dorsal roots consist of sensory fibres with bipolar cell bodies collected into dorsal root ganglia includes proprioceptor fibres from myotomes. In young larva, many afferent fibres are the processes of cells lying in spinal cord. Few types of cells in cord at this time produce simple reflex arcs. Some generalized and some special features are seen in autonomic nervous system. Gut is mainly innervated by vagus, which extends far back along intestine. Numerous fibres run from spinal nerves to rectum, ureters and cloacal region, and numerous postganglionic neurons are found here. Nerve cells are found in intestinal plexuses.

Sympathetic system consists of isolated fibres running in both dorsal and ventral roots. Many of these run directly to their endings without interpolation of neurons. Few postganglionic cells are present which seldom collected into ganglia. Adrenal system is diffuse. Scattered masses of interrenal tissue and large groups of suprarenal cells are found in walls of veins and heart. Suprarenal tissue receives preganglionic fibres from spinal nerves. Its cells sometimes connect with each other by fibres like neurons .They exerts control intermediate between nervous and hormonal. Myelin sheaths are absent in nerve fibres.

Conduction is slow in such non-medullated fibres. Spinal cord is uniformly transparent, grey coloured and is flattened dorsoventrally, to allow access of oxygen, and metabolites. Blood vessels are absent in cord. Nerve cell bodies lie towards the centre, but synaptic contacts are not made in this grey matter but at the periphery. Outer part of cord is made up of a neuropil or nerve felt work, formed of terminations of incoming sensory fibres and dendrites of motor cells. These cells lie in ventral part of cord. Their axons run out to make large fibres of ventral roots and their dendrites pass to all parts of peripheral regions of both the same and opposite sides of cord. They are stimulated directly by impulses in processes of afferent fibres that end in these regions.

Muller's fibres originate from giant cells in reticular formation of brain, whose large dendrites receive fibres from several higher centres, providing an uncrossed common pathway to spinal cord. In the earliest larva, coordination occurs by a pair of giant Mauthner cells. Other nerve cells in more dorsal parts of cord lack long axons and connect the neuropil of various regions. Afferent fibres reaching the cord in dorsal roots give off branches that ascend for a short distance and descend for long distances. Pathways to brain pass through multiple relays. Brain is built on typical vertebrate plan, with thickenings and evaginations corresponding to various organs of special sense. Forebrain is connected with smell, midbrain with sight, hind brain with acoustico-lateral and taste-bud systems. Forebrain and olfactory sense are moderately well developed in adult, as is the visual sense, with its chief centre in midbrain. Auditory and acoustic lateral systems are not well marked, and cerebellum is small. Taste is less developed than in higher fish.

Parts of Brain

These are forebrain (prosencephalon), cerebral hemispheres (telencephalon) between-brain (diencephalon), midbrain (mesencephalon), optic lobes, hind-brain (rhombencephalon), cerebellum (metencephalon) and medulla oblongata (myelencephalon). The upper surface of brain is covered by vascular pad, the choroid plexus or telachoroidea. This extends into ventricles of brain at three points - into third ventricle of diencephalon, into iter leading through midbrain from third to fourth ventricles, and into fourth ventricle itself. Roof of the brain is non-nervous in these regions. Perhaps the vascular membranes of brain are highly developed due to absence of cerebral blood vessels. From the lower part of mid- and hind-brain arise all the cranial nerves except olfactory and optic.

Cranial nerves represent nerves similar to dorsal and ventral nerve roots of trunk, modified due to special development of head. They carry afferent fibres from skin of head and gills and motor fibres for moving eyes, sucker, and branchial apparatus. The largest part of brain, the medulla oblongata is well developed due to extensive sucking apparatus, innervated from trigeminal nerve. Forebrain consists of a pair of large cerebral hemispheres open by foramina of Munro into a median third ventricle, whose walls constitute the diencephalon or between-brain. Diencephalon connects forebrain with midbrain, includes thalamus. Its ventral part, the hypothalamus, is well developed as a central organ controlling visceral activities. Nerve fibres from supraoptic nucleus of hypothalamus proceed to pars nervosa of pituitary and are filled with granules of neurosecretory material, which perhaps controls pituitary action. A simple portal system of blood vessels connects the hypothalamus with pituitary.

Pineal Eyes

The eyes are formed from the diencephalon. The so-called third, epiphysial, or median eye, is developed consisting of an unequally developed pair of sacs, that on the right, the pineal, being larger and placed dorsal to morphologically left parapineal. Sacs form by evagination from brain and connect with dorsal epithalamic or habenular region of between brains by two stalks. Two organs are similar in structure, consisting of irregular flattened sacs with a narrow lumen. Both upper and lower walls of each organ contain receptor cells, with processes that project into lumen and nerve fibres which mostly end within organ, in contact with ganglion cells whose axons run to unequal right and left habenular ganglia. There are supporting and pigment cells in retinas. Retinal cells cause movements, found differently under conditions of illumination and darkness.

The pineal cells are sensitive to light. In the portions of diencephalic wall of pineal organs ciliated cells of ependyma are specialized as photoreceptors. They show the same general plan as paired eyes, but without differentiated dioptic apparatus. Pineal apparatus is concerned with adjustment of internal activities to correspond with change conditions of illumination. Control may be affected by impulses carried in large tract that proceeds from habenular ganglion to hypothalamus, in the floor of diencephalon. The latter is concerned with integration of internal activities of animal.

Pituitary Body and Hypophysial Sac

The lower portion of diencephalon, the hypothalamus, forms a prominent pair of sacs, the lobi inferiores, which contain a partly separated diverticulum of third ventricle and end below in infundibulum. Pituitary gland is pressed against the underside of hypothalamus. Lower wall of brain in this region consists of a single epithelial layer, corresponding to pars nervosa of pituitary of higher forms. Major portion of pituitary gland is a mass of secreting cells in which two parts, the parts anterior and intermedia can be recognized. Pituitary contains oxytocic and water balance hormones and develops by a pocket of buccal ectoderm, whose walls become folded, so that the part in front of lumen becomes the pars anterior, that behind the pars intermedia.

Hypophysial rudiment is continuous with that of olfactory epithelium. The latter then moves dorsally and two remain connected throughout larval life. At metamorphosis, this acquires a lumen and forms a tube extending from nostril below the pituitary and brain. Inside the single nostril, guarded by a valve, are openings into the nasal sacs, which are cavities with folded walls. Some cells of these walls are olfactory receptors and give off the axons that make up olfactory nerves, entering the olfactory bulbs on anterior end of hemisphere. Behind the nasal sacs lie numerous glandular follicles opening into the sac in larva, but completely closed in adult. They may be comparable to Jacobson's organ. Naso-hypophysial tube proceeds back behind the pituitary to a closed sac lying between the first pair of gill pouches.

Lateral Line Organs

Lateral line receptors are little patches of sensory cells found along certain lines on head and trunk. They are all innervated by cranial nerves, those on body and tail being served by a special backward branch of vagus nerve. Receptor cells carry long hairs and are able to detect either movement of water relative to fish or of fish itself. Objects moving nearby set up disturbances that may also be detected. In lamprey the lateral line organs are very simple, and open to exterior and not sunk in a canal as in higher forms. Rows are somewhat irregular, especially those on body.

Vestibular Organs

Labyrinth as a specialized portion of lateral line system is concerned with recording the position of head end and angular accelerations. Labyrinth develops by an inpushing of the wall of head which becomes closed off from exterior. Internal folding divides the sac into chambers. There is a large central vestibule, into which open below several partially separate sacs, provided with patches of sensory hairs. These correspond, from in front backwards, to maculae of utricle, saccule, and lagena of higher forms. Hairs of maculae are loaded with otoliths. Two broad semicircular canals, each with an ampulla, contain a receptor ridge, the crista. Also opening to vestibule are two large sacs, covered with cilia, whose beat produces counter currents in dorsoventral plane. These function as a gyroscope, compensating for absence of a horizontal canal. It is claimed that this has cristae at both ends. Macular system also does not show characteristic subdivisions but is a single macula communis.

Paired Eyes

Paired eyes are formed like pineal eyes by evaginations of diencephalon wall. The so-called optic nerve is the portion of brain. Extrinsic muscles move the eyes. Cornea consists of two distinct layers, separated by a gelatinous substance. Attach to outer cornea is corneal muscle which flattens cornea and pushes lens closer to retina. Iris outlining a round pupil changes little in diameter under different illuminations. Most Lamprey species are diurnal, move towards white objects and probably use both eyes and nose to find their prey. In ammocoete larva, the paired eyes are buried below pigmented skin and make no movements when light falls on this region. Optic tracts of adult end in roof of midbrain which is a differentiated stratified region.

Besides optic fibres, it receives impulses from fibres ascending from spinal cord and others from auditory and lateral line centres. Midbrain is the important parts of brain. Its cells control movements by fibres that run to make connection with dendrites of the large Muller's cells, whose axons pass down the spinal cord; other fibres from tectum opticum reach to various parts of brain, and it is probable that its activities are closely correlated with those of many other regions.

Skin Photoreceptors

Light-sensitive cells are found in skin and eyes and are abundant in tail. Impulses are carried forwards by lateral line nerves. Sensitivity of lateral line organs to light is not found in other fish-like vertebrates. Receptors are not strictly lateral line organs but pigmented epidermal cells. Sensitivity curve shows a sharp peak at 530 m/x, this being the region of the spectrum at which light penetrates farthest into sea water. Pigment is probably a porphyropsin. In hag-fishes, head and cloacal regions are more sensitive to light than is the rest of the body. Impulses from skin are conducted through the spinal nerves.

Habits and Life-history

Life cycle involves metamorphosis. Adult passes about three years in lake, exclusively by sucking blood from fishes. In springtime, they start out independently from lake, each one forsaking its prey and swimming vigorously. They fasten to fish and are carried along up the stream. Male builds nest by removing and placing stones with suctorial mouth. Female joins the male, and together they work until making a basin or a ditch. They than fasten with their mouths to stones at the upper edge of the basin, and their bodies swing downstream and sway in current.

Lampreys are found on beds in the inlet in a single season. In nests the eggs, after being fertilized, sink to bottom and adhere firmly to sand and stones, being covered by the lampreys stirring up the sand with their tails. After some days, the eggs are hatched into young lampreys like small angle-worms which burrow into sand and soon make their way to sand along the banks of stream. Here they remain for two years, or longer, with their rudimentary eyes and valvular mouths. They feed on minute organisms. Adults probably die soon after spawning. Young metamorphoses into adults, find their way down the inlet into lake, and begin the parasitic bloodsucking life like their parents. They belong to class Marsipobranchii, or pouch gill as the gills form a series of pouches .They receive water through many independent gill-openings.

Spawning migrations may take them for hundreds of miles. They perform climbing, leaping from stone to stone and hanging on by their suckers. During migration, some lampreys assume brilliant orange and black colour patterns. Lampreys landlocked in the lakes of New York feed in fresh water and ascends only a few miles up streams to breed. Once in the river, the lampreys do not feed again but live over the winter on reserves accumulated as fat, especially under skin and in muscles. During winter the gonads ripen progressively and secondary sexual characters begin to become apparent only in February. Females then develop a large anal fin, while in male, a penis-like organ appears and the base of dorsal fin becomes thickened. Spawning occurs in spring and is preceded by a form of nest building.

Fertilization is secured by copulation in which the male fixes by sucker onto fore-part of female and the two then become intertwined and undergo rapid contortions, the eggs being squeezed into the water, while sperms are ejected through the penis. Fertilization is therefore external, but sperms must be placed very close to the eggs, as they remain active only for about one minute after entering the fresh water, which provides the stimulus that begins them. Eggs and sperms are not all laid at once; mating is repeated several times until all the products have been shed, after which the animals are exhausted and soon die. Movements of animals stir up the sand in nest ensuring that eggs are covered up as they are carried away by the current.

Ammocoete Larva

Eggs contain yolk, but their cleavage is total. After about 3 weeks, the young hatches as the ammocoete larva, about 7 mm long. At first this is a tiny transparent creature, but during its long larval life, it grows into an opaque eel-like fish up to 170 mm long.

Larval life is spent buried in mud. The animal emerge occasionally to change their feeding-ground. There is no sucker, mouth being surrounded by an oral hood. Muscles and skin cover paired eyes. Head is slightly sensitive to light, but they begin to swim if the tail is illuminated. Photoreceptors in tail are connected with lateral line nerves. In larva, these are main photoreceptors. If several larvae are left in a vessel containing a layer of mud in bottom they rapidly disappear and remain hidden indefinitely. Heads perhaps just visible in small depressions made by rhythmic respiratory movements. When disturbed, they always swim with the head downwards which leads them to burrow rapidly. Nasal and hypophysial sacs are poorly developed, and sense of smell hardly serves this purpose.

Feeding occurs by intake of water through mouth and separation of small food particles from it in pharynx. Mucus secreted by endostyle, a sac below the pharynx, aids in feeding. Endostyle consists of a pair of tubes, on floor of which there are four rows of secretory cells. There is a single opening to pharynx, by a slit about middle of length. As development proceeds, the inner rows of cells at hind part of organ become coiled upwards.

At the end of larval life, the endostyle forms a large mass below pharynx. Probably no enzymes are secreted by endostyle. There is a ciliated groove in floor of pharynx. The velum, a pair of muscular flaps, provides the main current during rest. Branchial basket can expand and contract by muscles. The system enables to feed efficiently on small unicellular algae and bacteria of mud. The muscular feeding-system allows a relatively small pharynx to feed a fish 170 mm long and weighing up to 10 grams. It allowed the animals to escape from limitation of size imposed by ciliary method of feeding.

After the development of jaws to form a more efficient feeding mechanism, the rhythmic movement of branchial apparatus persists for purpose of respiration. Mucus-secreting columns of endostyle shrink and whole organ becomes reduced to a row of closed sacs, lying below pharynx. Each sac lined by an epithelium, contains a structure less colloid substance, and is similar to a thyroid vesicle. Endocrine gland that regulates basal metabolism is derived from the part of feeding-system that in the earliest chordates provide the raw materials of metabolism. The great change in endostyle is only part of complete metamorphosis by which ammocoete larva changes into an adult.

Mouth becomes rounded and its teeth, tongue, and complex musculature develop. Paired eyes appear; the olfactory organ becomes internally folded, and olfactory nerve and tracts much enlarged. Nasohypophysial sac grows backwards to gills. In pharynx, the gills develop into sacs opening to branchial chamber. Changes also take place in intestine. Yellow-brown colour of larva gives place to black with silver underside of adult. The animal often leaves mud and finally migrates to sea to begin its parasitic life.

5.4 *Petromyzon*

Petromyzon, an ectoparasite of fishes, almost worldwide distribution is found in the seas of North America, Europe, Japan, Australia, Tasmania and West Africa. Common marine species are *Petromyzon marinus* and *P. planeri*. Freshwater forms are *Pertomyzon fluviatilis* found in European rivers. It attaches itself to body of hosts with help of buccal funnel and scrapes flesh with help of its rasping tongue which bears teeth, also attached to rock or any other substratum with the help its buccal funnel.

External Features

Body is divisible into head, trunk and tail and without scales. Mouth is situated at bottom of buccal funnel. On dorsal side of head, a single median nasal aperture leads into nasal sac. Paired fins are absent. Median dorsal fin is divided into two unequal parts by a notch. A caudal fin is present. Small anal opening is present on the ventral side. A small urinogenital aperture, borne on a small papilla, is situated just behind anal opening. Eversible penis is found in male. At the anterior end, on ventral aspect of head there is a depression, the buccal funnel (Figure 5.3).

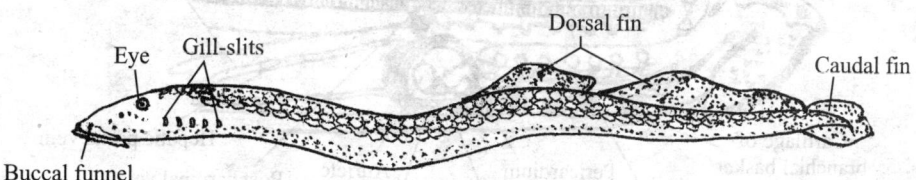

Figure 5.3 External structure of *Petromyzon*.

Body Wall

Skin is soft. Epidermis is composed mostly of unicellular gland cells, granular cells and club cell (Kolbenzelle). Dermis is composed of bundles of collagenous and elastic fibers. Subcutaneous layer between dermis and body wall musculature contains pigment cells, blood vessels and fatty tissue. Pigment cells are found in dermis.

Muscular System

Trunk musculature is well developed. Muscles are arranged to form myomeres i.e. there are series of myotomes separated by myocommas. Myotomes are w-shaped muscle blocks divided into dorsal and ventral parts by horizontal septum. Muscle fibers are striated.

Skeletal Structures

Axial skeleton consists of ill developed skull. Notochord and collection of cartilages are persistent. Notochord is composed of large vacuolated notochordal cells enveloped by a notochordal sheath. A connective tissue layer further surrounds notochordal sheath covers spinal cord and remains connected with myocommas. Connective tissue layer contains cartliaginous element. Cartilage is composed of large cells embedded in a matrix of chondrin. Numerous cartilaginous rods extend dorsally and ventrally to support median fin.

Skull: Skull is primitive. Its floor is formed by basal plate made by union of parachordals and trabeculae. Immediately anterior to notochord, there is a large aperture basicranial fontanelle through which passes naso-hypophysial sac. Side walls of the cranium are strong, but roof is exclusively composed of membranous fibro-cartilage excepting a narrow transverse bar. Two compact auditory capsules are united with the posterior end of basal plate. Olfactory capsule is an unpaired, concavoconvex plate and is pierced by paired apertures for olfactory nerves. On each side of basal plate, there lies the sub ocular arch. A slender styloid process hangs from the subocular plate. Each styloid process is connected with a small cornual cartilage. Cranium is attached with the skeleton that supports gills and buccal funnel (Figure 5.4).

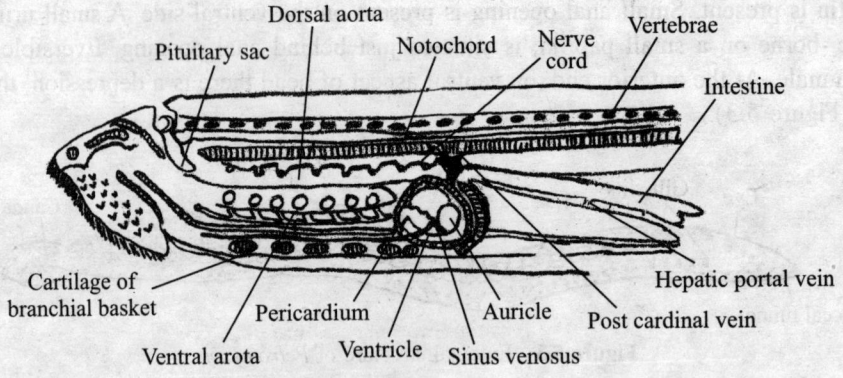

Figure 5.4   Internal anatomy of *Petromyzon*.

Branchial Basket: It is composed of nine irregularly curved ventral bars of cartilage on each side. Bars are united together by four longitudinal rods. First vertical bar is situated just posterior to styloid process, second is in front of first gill slit and rest are just posterior to remaining gill slits. Posterior part of bronchial basket is extended to form pericardial cartilage to accommodate heart. An annular cartilage supports buccal funnel. An elongated lingual cartilage supports tongue (Figure 5.5).

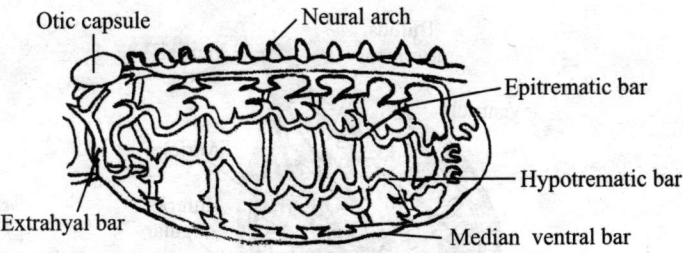

Figure 5.5 Branchial basket of *Petromyzon*.

Digestive System

Mouth is situated inside buccal funnel which acts as a sucker. Numerous teeth present in buccal funnel are laminated horny cones. Teeth present on tongue are elongated and sharply pointed. Vacuum cup is produced inside oral cavity by the action of cardioapicalis muscle. Release of sucker is effected by contraction of an annular muscle running along the margin of buccal funnel.

Mouth leads into the buccal cavity which is communicated behind with two passages. Dorsal passage is called gullet or esophagus, while ventral one is respiratory tube which is closed behind and leads into seven branchial or gill-pouches. The velum guarded the opening of respiratory tube with velar tentacles. Esophagus leads into a straight intestine. Lumen of intestine is crescentic owing to presence of a typhlosole. Posterior end of the intestine is widened to form rectum which ends at the cloaca. Liver is a large bilobed organ. Gall bladder and bile duct are sometimes absent in adult. In both larval and adult lampreys, epithelium aggregations of seemingly secretory cells of two different kinds occur in the gut. There is no pancreas as such.

Respiratory System

There are 7 pairs of gill pouches or branchial pouches or branchial sacs. Seven pairs of external gill pores are visible. Gill pouches open directly into respiratory tube without direct connection with the enteric canal. Gill pouches as biconvex lenses, with numerous gill-lamellae developed on inner surfaces are separated from one another by wide inter-branchial septa.

Circulatory System

Heart enclosed by pericardium, an S - shaped structure situated behind gills and composed of one sinus venosus, one auricle, one ventricle, and one conus anteriosus. Sinus venous is a thin -walled chamber and leads into a thin walled auricle lying above sinus venosus. Auricle opens into thick walled ventricle. Blood is red and hemoglobin remains inside the nucleated and circular R.B.C. Blood forming tissue is present in the spiral valve, kidney and spinal cord (Figure 5.6).

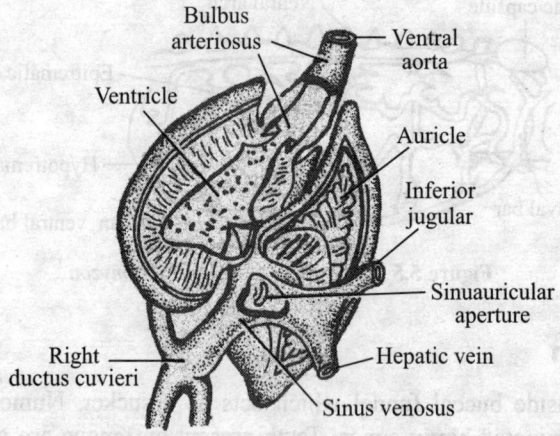

Figure 5.6   Heart of *Petromyzon*.

Arterial System: A large ventral aorta runs anteriorly between gill pouches from ventricle. Ventral aorta gives eight afferent bronchial arteries to gill-pouches. From gills, blood is collected by eight efferent branchial arteries. Efferent branchial arteries open into paired dorsal aortae. These paired aortae run backward and then unite to form a single median dorsal aorta, which in turn gives series of segmental arteries to myotomes. Diffused chromaffin cells are present on segmental arteries. Special arteries from unpaired dorsal aorta supply blood to gut, kidneys and gonads (Figure 5.7).

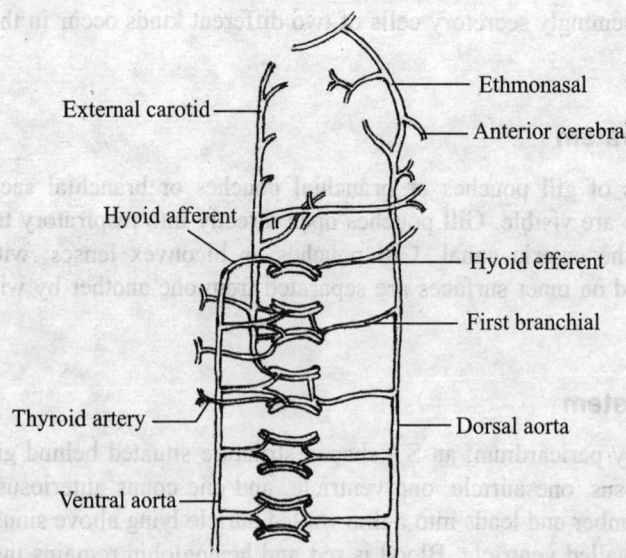

Figure 5.7   Schematic lateral view of the anterior part of the arterial arches of *Petromyzon*.

Venous System: Caudal vein carries blood form the tail region and divides anteriorly into 2 posterior cardinal veins. Cardinals collect blood form kidneys, gonads, myotomes and ultimately open to heart by a single ductus cuvieri. Left ductus cuvieri is absent in adults. Anterior cardinal veins collect blood form the anterior regions. A large median inferior jugular vein drains blood form the musculature of buccal funnel and gill- pouches. Renal portal vein is absent. Hepatic portal vein drains blood from the gut and enters liver though a contractile portal heart. Hepatic veins carry blood to heart form the liver. Branchial sinuses consist of ventral branchial sinus, interior branchial sinus, and superior branchial sinus.

Brain

Brain is very small relative to body weight than in any other vertebrate. Upper surface of brain is covered by an extsnsive vascular pad, choroid plexus or telachoroidea. Neural tube becomes modified anteriorly into a complicated brain, and posterior part transforms into spinal cord. Brain is divided into prosencephalon, mesencephalon and front part consists of cerebral hemispheres and olfactory lobes. Mesencephalon forms optic lobs. Rhombencephalon is divided into metencephalon and myelencephalon. Olfactory bulbs contain small ventricles, and give paired olfactory nerves to nasal capsule.

Cavities of cerebral hemisphere are lateral of first and second ventricle. Lateral ventricles connect with the third ventricle or diocoel through foramen of monro. Diencephalon has a membranous roof which protrudes outwards as a conspicuous saccus dorsalis and includes thalamus which receives a few fibres form optic nerves and hypothalamus. Hypothalamus is well developed and bears inconspicuous inferior lobes and indistinct saccus vasculosus. Supraoptic nucleus of hypothalamus sends nerves to pars nervosa of anterior pituitary lobe. On ventral side of diencephalon the simple crossing of optic nerves do not form chiasma lies. Velum transversum is absent (Figure 5.8).

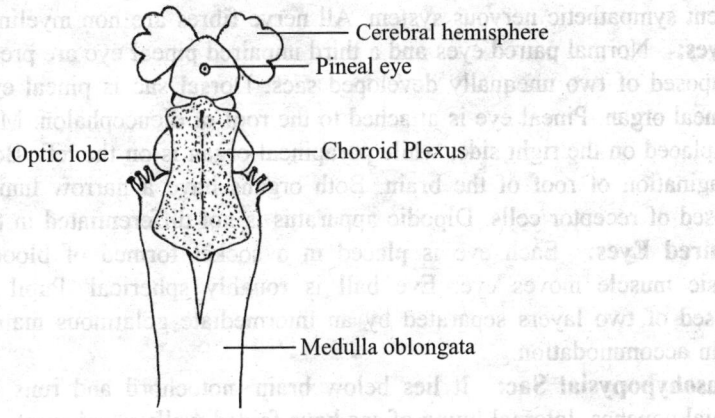

Figure 5.8(a)   Brain of *Petromyzon*, Dorsal view having the choroid plexus intact.

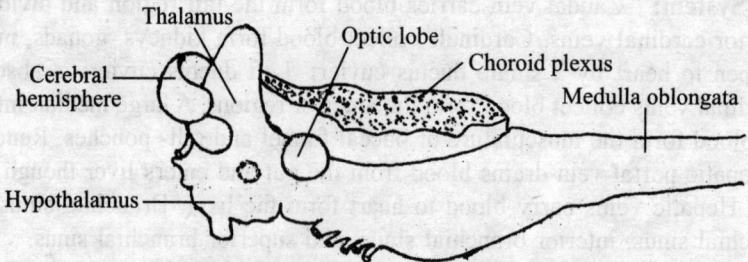

Figure 5.8(a) Brain of *Petromyzon*, Lateral view.

Midbrain is represented by two optic lobes and between optic lobes a membranous choroid plexus fuses with plexus lying over the fourth ventricle. Metencephalon transforms into a longest and well developed cerebellum. Third and fourth ventricles are communicated through a duct passing through midbrain. This duct is called iter or aqueduct of sylvius. Choroid plexus extends into ventricles of brain at three points-into third ventricle of diencephalon, into iter leading through midbrain from third to fourth ventricles, and into fourth ventricle itself.

Spinal Cord: This is dorsoventrally flattened band- like structure without blood vessel.

Cranial Nerves: Except the olfactory and optic nerves, other 8 pairs of cranial nerves emerge out form the mesencephalon and rhombencephalon. Olfactory nerves are composed of many nerve fibres. Optic nerves form the simple crossing without making complicated chaisma formation. Oculomotor nerves lack decussation. Trigeminal and facial nerves are intimately associated. Roots of seventh and eighth cranial nerves are closely placed. Vagus originate as a single outgrowth becomes subsequently branched.

Spinal Nerves: Dorsal and ventral roots remain separate. Ventral root consists mainly of motor fibres that innervate myotomes and dorsal root possesses mostly sensory fibres coming form the myotomes. Isolated nerve fibres running in both roots of spinal nerves represent sympathetic nervous system. All nerve fibres are non myelinated.

Eyes: Normal paired eyes and a third unpaired pineal eye are present. Pineal apparatus is composed of two unequally developed sacs. Dorsal sac is pineal eye and ventral one is parapineal organ. Pineal eye is attached to the roof of diencephalon. Morphologically, pineal eye is placed on the right side, while parapineal organ is on the left. Both organs are formed by evagination of roof of the brain. Both organs have a narrow lumen and the walls are composed of receptor cells. Dipodic apparatus is not differentiated in the pineal apparatus.

Paired Eyes: Each eye is placed in a socket formed of blood and lymph spaces. Extrinsic muscle moves eye. Eye ball is roughly spherical. Pupil is round. Cornea is composed of two layers separated by an intermediate gelatinous matrix. Cornealis muscle helps in accommodation.

Nasohypopysial Sac: It lies below brain, notochord and runs between first pair of branchial pouches. Internal lining of sac have folded walls which contain olfactory receptors.

Ears: Internal ear or membranous labyrinth is subdivided into several chambers by infolding of sac. Two semicircular canals open into a large central sac called vestibule which is subdivided into anterior and posterior utricular chambers. Below the utricular chambers there is a small sacculus and a lagena. Endolymphatic duct is a short closed tube.

Lateral Line Sense Organ

Lateral line receptors consist of small patches of sensory cells and sensory papillae along certain lines. Line receptors are innervated by branches form 5th and 10th cranial nerves. Integumentary photoreceptors are plenty in tail region.

Excretory System

Adult functional kidneys are mesonephric. Mesonephric tubules are elongated. Anterior end of each tubule bears Bowmann's capsule containing glomeruli. Fatty and lymphoid tissues also participate in formation of kidneys. Each kidney remains attached to coelom by peritoneal sheet. Ureters open posteriorly into the urinogenital sinus. First set of kidney in larva is pronephros (Figure 5.9).

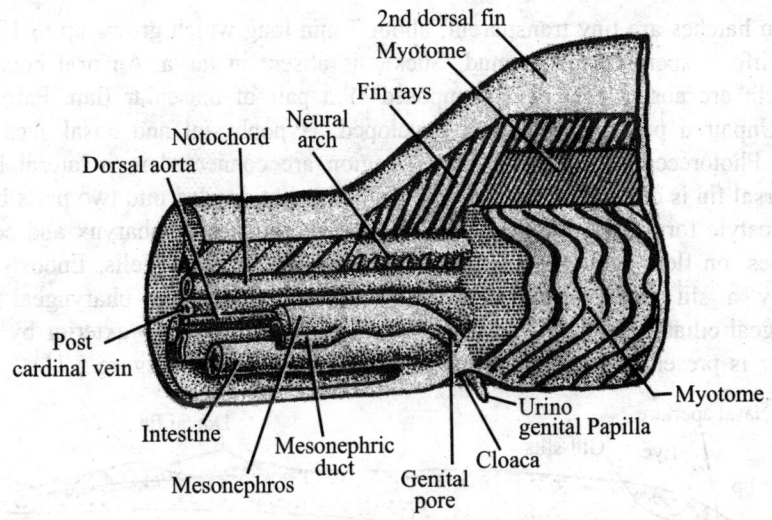

Figure 5.9 Cloacal region of *Petromyzon*.

Reproductive System and Development

Sexes are separate. Gonad occupies a median position to mesencephalon and extends almost to whole of body cavity and lacks the reproductive duct. Abdominal pores are present in wall of urinogenital sinus which opens before spawning. Testes consist of many follicles containing male gametes. Ovary is composed of eggs which are covered by a layer of follicular epithelium. Fertilization is external. Egg is telolecithal. Cleavage is holoblastic but unequal. Gastrulation occurs by invagination. Central nervous system develops as a solid cord or keel which hollows to form lumen of nerve cord .These types of C N.S. formation is called thickened keel method. Existence of ammocoetes larva in life history is significant (Figure 5.10)

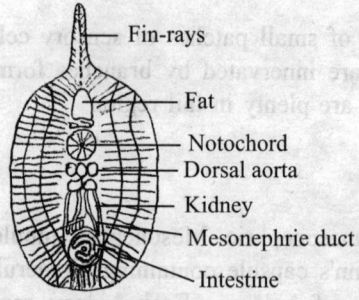

Figure 5.10   Trunk region of *Petromyzon*.

Ammocoetes Larva

Larva when hatches are tiny transparent, about 7 mm long which grows up to 170 mm. The portion of life is spent buried in mud. Sucker is absent in larva. An oral hood surrounds mouth. Teeth are absent. Velum is composed of a pair of muscular flap. Paired eyes are vestigial. Unpaired pineal eye is well developed. Hypophysial and nasal sacs are poorly developed. Photoreceptors abundant in tail region are connected with lateral line nerves. Median dorsal fin is continuous with caudal fin and is not divided into two parts by a median notch. Endostyle forms early in development as a sac below the pharynx and consists of a pair of tubes, on floor of which there are four rows of secretory cells. Endostyle opens to pharynx by a slit like aperture and is connected with hyper pharyngeal groove by peripharyngeal ciliated bands. Seven pairs of branchial sacs open to exterior by gill slits. A gall bladder is present. Pronephros persists as excretory organs (Figure 5.11).

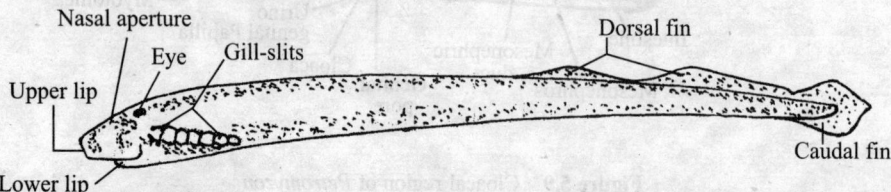

Figure 5.11   Ammocoetes larva of *Petromyzon* in lateral view.

Feeding takes place by intake of water through mouth and separation of small food particles form it in pharynx. Mucus produced by the endostyle entangles food particles which are transported to intestine by action of cilia in pharynx. Velum and muscular system of branchial basket help in maintenance of water current. Animals feed efficiently on small unicellular algae and bacteria of mud. Muscular feeding system of ammocoetes allows a relatively small pharynx to feed an animal of 170 mm long and weighing up to 10 grams. Larval life is long, may continue for 3 to 4 years. In favorable conditions, preferably in winter, the larva metamorphoses into adult.

During metamorphosis, larval characters are lost. Charges that occur during metamorphosis are given below: Endostyle undergoes a change into thyroid gland. Mouth becomes rounded and teeth, tongue and complex musculatures develop. Paired eyes appear, the olfactory organ becomes internally folded, and olfactory nerve and tracts become enlarged. Nasohypophysial sac grows backwards to gills. In pharynx, the gills develop into sacs opening to branchial chamber. Continuous median dorsal fin becomes divided into two parts by a median notch. Gall bladder disappears. Pronephros degenerates and the mesonephros are formed. Spinal cord becomes flattened. Yellow brown colour of larva gives place to black with silver underside of the adult.

5.5 Myxinoidea

Hag-fishes, *Myxine* and *Bdellostoma* are modified for sucking, live buried in mud or sand; eat polychaetes, other invertebrates, and scavenging dead fish. Eyes are functionless rudiments, sensitive to illumination through skin receptors. Sensory tentacles are found around the mouth, and in hag-fishes, the teeth and sucking apparatus are well developed. They burrow into bodies of fishes and find dying or dead fish. Gills are modified into pouches, opening by tubes into pharynx, and to exterior. In Myxine, all tubes are joined and open by a single posterior aperture on each side. Water enters at nostril, pumped backwards by the muscular velum through gill chambers and out behind. A single posterior oesophagocutaneous duct is found on left side, which is closed during normal respiration but is opened to allow expulsion of large particles.

Thyroid gland consists of a long series of sacs formed by evagination from floor of pharynx. There are the pairs of protective slime glands. Difference from nervous system of lampreys is that the dorsal and ventral roots join. Brain shows reduction and simplification without pineal eyes. Only one semicircular canal is found in ear. Pronephros persists in adult and is hardly marked off from mesonephros. There is a regular series of mesonephric glomeruli, a pair in each segment. Development is known only in *Bdellostoma*, where egg is yolky and cleavage partial, leading to formation of an embryo perched on a yolk mass. *Myxine* is a protandric hermaphrodite, as individuals are found in which the front end of gonad contains eggs, whereas hind part is testis-like. No ripe sperms have been found in this region, and individuals with fully testicular gonads do occur. Perhaps the *Myxine* is not a functional hermaphrodite but that double-sexed gonad shows a late persistence of indeterminate stage.

Hag-fishes live in sea and their blood differs from other chordates as it is isosmotic with sea water. The individual ions are regulated; sodium and phosphate exceed their values in sea water, and other ions are present in lower concentration. The ostracoderms are fossil forms from freshwater Silurian and Devonian deposits. They are the oldest fossils vertebrates which show affinity with cyclostomes. These are fossils that are rarely found complete, particularly the pteraspids. In the cephalaspids the head was flattened and composed of a shield. Rest of the body was fish-like with an upturned tail. A pair of flaps behind gills may have functioned like pectoral fins.

On dorsal surface of shield are two median holes, which served as naso-hypophysial opening and a pineal eye. Brain is similar to lamprey. There are paired eyes and only two semicircular canals. Long tubes lead through the shield contained the cranial nerves. On the under side of shield is a series of ridges, which outline a set of ten pairs of branchial pouches. Probably the gill pouches, canals of aorta, epibranchial arteries, and some features of veins and heart have been preserved. Mouth was a slit at the extreme front end. On dorsal surface there are sunken areas, covered by small scales, known as median and lateral fields, and supposed to contain electric organs.

Summary

1. Living agnathous animals are Cyclostomata, lampreys and hag-fishes which are characterized by developed skull, small pharynx, and relatively few gill slits. Except the most ancient form the more anterior arches between the gills form jaws.

2. Life history of Lampreys comprises ammocoete larva and adult. Larva is fresh water form, microphagous, and remains buried in mud. Of 20 lamprey species, most inhabit temperate regions. Lake Lamprey injures fishes while Brook lamprey do not. Two species spawn in identical places.

3. Life cycle involves metamorphosis. Adult passes about 3 years, exclusively by sucking fish blood. They fasten to fish and are carried along up the stream. Male builds nest by removing and placing stones with suctorial mouth. Female joins the male, and together they work until making a ditch.

4. In nests the lamprey eggs, after being fertilized, sink to bottom and adhere firmly to sand and stones. The eggs are hatched into young lampreys like small angle-worms which burrow into sand and make their way to sand along the banks of stream. Here, they remain for 2 years or longer, with their rudimentary eyes and valvular mouths. They feed on minute organisms. Adults probably die soon after spawning.

5. Larval life is spent buried in mud. There is no sucker, mouth being surrounded by an oral hood. Muscles and skin cover paired eyes. Head is slightly sensitive to light. Photoreceptors in tail are connected with lateral line nerves. In larva, these are main photoreceptors. When disturbed, they swim with the head downwards which leads them to burrow rapidly. Nasal and hypophysial sacs are poorly developed, and sense of smell hardly serves this purpose.

6. *Petromyzon*, an ectoparasite of fishes, almost worldwide distribution is found in the seas of North America, Europe, Japan, Australia, Tasmania and West Africa. Common marine species are *Petromyzon marinus* and *P. planeri*. Freshwater form, the *P. fluviatilis* is found in European rivers. It attaches itself to body of hosts with help of buccal funnel and scrapes flesh with help of its rasping tongue which bears teeth.

7. Body of *Petromyzon* is divisible into head, trunk and tail and without scales. Paired fins are absent. Median dorsal fin is divided into 2 unequal parts by a notch. A small urinogenital aperture, borne on a small papilla, is situated just behind anal opening. Eversible penis is found in male. At the anterior end, on ventral aspect of head there is a depression, the buccal funnel.

8. Hag-fishes, *Myxine* and *Bdellostoma* are modified for sucking, live buried in mud or sand; eat polychaetes, other invertebrates, and scavenging dead fish. Eyes are functionless rudiments. They sense the illumination through skin receptors. Sensory

tentacles are found around the mouth. In hag-fishes, the teeth and sucking apparatus are well developed.

9. In *Myxine*, all tubes are joined and open by single posterior aperture on each side. Water enters at nostril, pumped backwards by muscular velum through gill chambers and out behind. A single posterior oesophagocutaneous duct is found on left side, which is closed during normal respiration but is opened to allow expulsion of large particles.

Review Questions

Short Answer Questions

1. What is the meaning of the term "Agnatha"?
2. Name two classes belonging to Agnatha.
3. Name the order to which *Birkenia* belongs.
4. Name the larval form of lampreys.
5. What are chromatophores?
6. What is brachial basket?
7. What is Pineal eye?
8. Give two characters of the order Cyclostomata.
9. Give two examples of Cyclostomata.
10. Name the order to which *Cephalaspis* belongs.

Long Answer Questions

1. Describe the anatomical features of *Petromyzon* with diagram.
2. Write a note on ammocoetes larva of *Petromyzon*.
3. Classify Agnatha up to order with examples.

CHAPTER

6

The Pisces

Fish, the cold-blooded animal with backbone live in water, breathe with gills and use fins to balance and move, and most have scales. Fish are classified into 2 major groups. Skeleton of bony fish (carp, mullets) consists of hard bones. Skeleton of cartilaginous fish (sharks, rays) consists of soft and flexible cartilages. In cartilaginous species like sharks, rays, skates and chimaeras the body consists of cartilage. Cartilage supports the growth. Ray includes over 23,000 species like salmon, trout, and cave fish. Skeleton of ray-finned fish is composed of bone.Lobe-finned species like lungfish and coelacanths (a group of bony fish) with paired fins were perhaps the ancestor of the first land vertebrate. The earliest fish was the ancestor with notochord, but without jaw or teeth, now an extinct group. About 24,000 species of ray-finned fish referred as diadromous are divided into 431 families.

Some exotic like (*Ctenopharyngodon idella*) was introduced from Japan in 1959 for weed control; silver carp (*Hypophthalmichthys molitrix*) from Hong Kong in 1959 and Tilapia (*Tilapia mossambica*) from Bangkok in 1952 are now used in intensive freshwater acquaculture. Of 20,000 recorded fish species, above 2000 are found in the India's river and network of irrigation canals, reservoirs, ponds, estuaries, lakes, lagoons, backwaters, impoundments, mangroves, sea and swamps. Estuaries of west-flowing rivers like Zuari and Mandovi, Netravathi Gurupur and Kalinadi, Aghanashini and Sharavati are excellent fish habitats. Major fishes of these rivers are carps, catfishes, mullets, perches, mahaseers, and pearl spots. Indian shad, *Hilsa ilisha*, migrate into rivers from the sea for spawning and are called anadromous fish. Local fish, migrate between ecosystems in search of breeding grounds, are the mahseers, the Indian major carps and large or medium-sized catfish.

In capture fisheries, people reap without sow are practiced in the sea, rivers, estuaries, large reservoirs, and lakes. Fish stocks are replenished naturally. Culture fisheries, also known as pisciculture, are done in small water bodies which people manipulate. Here fish fry are sown, tended, nursed, reared, and harvested. Fishing gear used are shore seives, gill nets, and mini-otter trawls.

Fish food is nutritive containing protein (about 20%) and is a good source of trace elements like Cu, P, Fe and I, and vitamins A & D. Fish oil has high level of unsaturated fatty acids which are good for the heart, in preventing cellular aging and to prepare gelatin. Sustainable fishing has importance in economic prosperity. Non-consumable fish is used as organic fertilizer as it is rich in nitrogen and phosphates. Fish meal is used to feed ruminants and poultry. *Gambusia* controls mosquito larvae. Grass carp controls aquatic weeds. Fish bones are used in making combs. Shark skin is used for making leather purses, belts and shoes. Fishes are also important in research. Some fishes are provided with accessory respiratory organs to live out of water for long time. They include *Mystus seenghala* (freshwater catfish), *Clarius batrachus* (catfish), *Boleophthalmus* sp. (mud-skippers), *Periophthalmus* sp., *Anabas testudineus* (climbing perch), *Channa punctatus* (snakehead), *Heteropneustes fossilis* (stinging catfish), *Rasbora daniconius* (danio), *Glossogobius giuris* (goby).

Since the body fluid in freshwater fish body is saltier than surrounding water, there is problem of soaking up water and swelling. Freshwater fish doesn't drink. Water that comes in through skin, and gills is brought to kidneys to carry away waste materials as urine. Saltwater fish face the opposite problem as its body fluid is less salty than surrounding water, and it faces the problem of dehydration. The fish drink large amount of water to compensate the loses through its gills and skin. Some salt pass to digestive tract and is excreted. Saltwater fish seldom urinates.

Before 1938, coelacanths were thought as extinct .The first living coelacanth *Latimeria chalumnae* was discovered in 1938. Indonesian coelacanth occurs in temperate waters. The coelacanth, moves slowly near substrate, where it feeds cephalopods and fishes. It moves quickly while prey or avoiding danger. They are ovoviviparous. A rostral organ is part of the electro sensory system. Intracranial joint in skull allows anterior cranium to swing upwards to widen the gap of mouth. Notochord extends the body length.Coelacanths, marked by their distinctive shape and lobed fins consists of 90 species, were distributed in marine and freshwater. Evolutionary relationships of coelacanth are controversial. The primitive lungfishes were close living relatives of Coelacanths. Coelacanths are deep-bodied, with a 3-lobed diphycercal tail. The type first appeared in late Devonian .The osteolepids were rare in Carboniferous and disappeared after early Permian, but a line descended from them still survives today. *Latimeria* (Figure 6.1) with well-developed spiral intestine lives on other fish. Air-bladder arises by a ventral opening from esophagus and proceeds backwards and dorsally for whole length of abdominal cavity. The small lumen reduces the specific gravity. Gills perform respiration. The heart consists of sinus venosus, auricle and ventricle. There are 4 rows of valves in conus. The red cells are large. The brain lies far back in cranium occupying less than one-hundredth parts of cranium. Brain is provided with thin fore-brain roof, large striatum, and large pituitary cleft. There is no valvula to the cerebellum. Both anterior and posterior nares open on surface of head. The rostral organ, a large median sac opening to surface by 3 pairs of canals is innervated by superficial ophthalmic nerve. The eye, inner ear, and lateral line system are well developed. Its habitat is isolated with small population. It shows developments parallel to those of the Teleostei rather than to Dipnoi. Several characteristics are paedomorphic.

Figure 6.1 *Latimeria* : living bossil.

Jawed vertebrates called gnathostomes include sharks, rays, fish, amphibians, reptiles and mammals. Gnathostomes grip and tear food by jaws and teeth. Jaw evolved by modification of supporting rods of anterior pharyngeal slits. Jawed vertebrates have duplicated Hox genes, enlarged forebrain, and cartilaginous skeleton. Basic functioning of muscles, liver, hormonal control and nervous system is similar in fish and other vertebrates. Fishes obtain oxygen from water by gills. Heart and circulatory system are adapted to respiration by gills. Low oxygen tension and toxic substances in water can lead to emergency responses.

Most fish are covered with scales to protect the body. Bony plates in catfish serve the same purpose. Some have small, or no scale. Wild fish are colorful. Some rely on stripes, or brown color to be camouflaged. Attractive colour attracts mates. Some use eye-spots. Mouth brooding cichlids rely on colored egg spots for fertilization. Pigment and light reflection determine colour. Solid, dark coloration are due to pigmented skin. Silvery iridescence relies on light reflection. Some species alter colouration, showing different color at night. Healthy fish are more colorful than unhealthy one. Color-enhancing foods bring out certain colors. During territorial displays and spawning, color is enhanced. Lateral line organ located just under scales picks vibrations in water. Fish detect predators and food, and navigate efficiently without vision in darkness. The Blind Cave Fish relies entirely on its lateral line system. Swim bladder keeps fish in neutral buoyancy. Fish can sleep in mid water. Some swallow air, which is passed to swim bladder.

Characteristics

1. Marine, or fresh water forms.
2. Laterally compressed streamlined body.
3. Cold blooded.
4. Body is covered with dermal scales.
5. Paired and unpaired fins are supported by fin rays.
6. Paired fins are pectoral– and pelvic fins.
7. Unpaired fins are dorsal–, caudal– and anal fins
8. Respiration occurs by gills.
9. True jaws are present.
10. Erythrocytes are nucleated.
11. Venous heart is present.

12. Visceral arches are well developed.
13. Internal ear with three semi-circular canals is present.
14. Some possess poison glands.
15. Few are provided with electric organs.
16. Ten pairs of cranial nerves are found.
17. Kidneys are mesonephric, pronephric tubules are found in some.
18. Sexes are generally separate.
19. No allantoic bladder is formed.

6.1 Classification

Berg (1940) based his classification on the works of C.Tate Regan and E.A Stenrio. He has divided fishes into seven classes as given below:

Series: Pisces

Classes

    Acanthodii
    Pterichthyes   Placoderms
    Coccostei
    Elasmobranchii
    Holocephali
    Dipnoi
    Telcostomi
    Subclass Crossopterygii
    Actinopterygii

Class: Aacanthodii

1. Endoskeleton consists of true bone.
2. A gill slit was present.
3. Four or 5 branchial arches were present.
4. Notochord was persistent,
5. Ganoids scales were Present.
6. Caudal fin was heterocercal.
7. one to 5 pairs of spines was present.
8. An operculum was present.

Example: *0imatlus, Acanthodes, Diplacanthus*

Class: Coccostei

1. A carapace was present.
2. External gill aperture was present.
3. Persistent notochord.

4. Neural and haemal arches were ossified.
5. Pectoral fins were replaced by immovable spines.
6. Dorsal fin was present.
7. Eyes were laterally placed with a pineal aperture in between them.

Example: *Coccosteus. Arctolepis.*

Class: Pterichthys

1. A carapace was present.
2. An operculum was present.
3. Pectoral fins were paddle like.
4. Tail was heterocercal.
5. One or 2 dorsal fins were present.
6. Nostrils were two.

Example: *Pterichthys, Bothriolepts*

Class: Elasmobranchii

1. Endoskeleton is cartilaginous.
2. Placoid scales are present, or skin is naked.
3. No operculum; gill aperture numbers 5 to 7 on each side.
4. Hyostylic, or amphistylic skull.
5. No air bladder.
6. A cloaca is present.

Example: *Scoliodon, Pristis, Dasyatis.*

Class: Holocephali

1. Endoskeleton is cartilaginous.
2. Skull is holostylic; palatoquadrate fuses with cranium.
3. An operculum is present.
4. Teeth in the form of grinding plate.
5. No cloaca.

Example: *Chimaera, Callorhynchus.*

Class: Dipnoi

1. Air bladder is present.
2. Operculum, an external branchial aperture, is present.
3. Cycloid scales are present.
4. Paired fins are lobate with jointed median axis.

5. Skull is autostylic
6. Notochord is persistent.
7. Internal nostris are present.
8. Cloaca is present.

Example: *Protopterus, Lepidosiren.*

Class: Teteostomi

1. Skull is usually hyostylic. Palatoquadrate is not fused with cranium.
2. An operculum and an external branchial aperture are present.
3. Air bladder is present.
4. No cloaca.

Teleostomi are divided into 2 subclasses: (a) Crossopterygii, (b) Actinopterygii

Subclass: Crossopterygii

1. Paired fins are provided with lobe.
2. Endoskeleton consists of a jointed median axis with radials.
3. Nostrils are present.
4. Scales are cosmoid, or cycloid.

Example: *Latimeria, Undina* (Figure 6.2a-g).

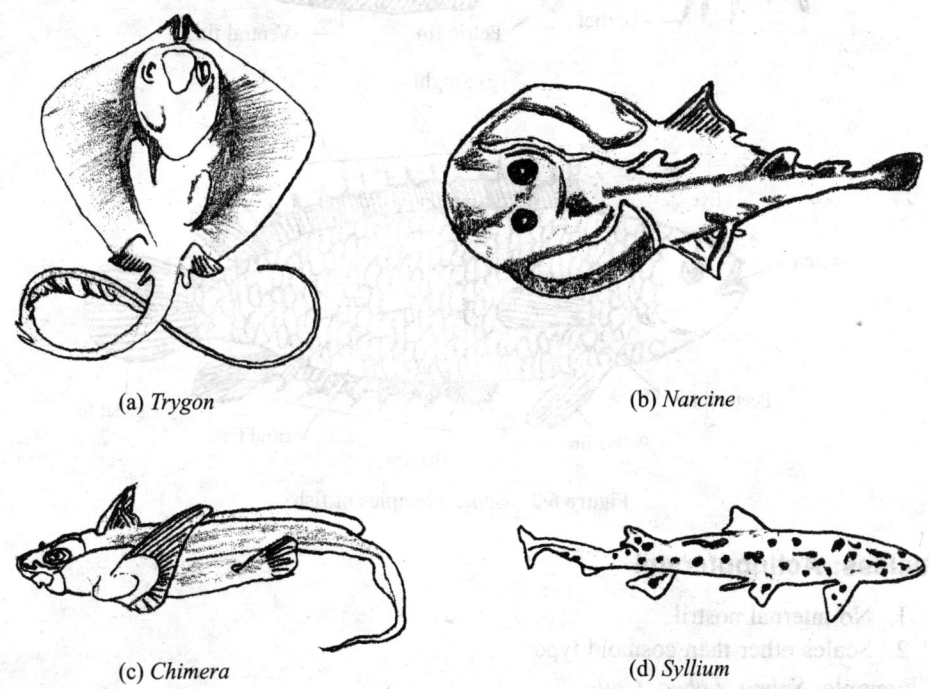

(a) *Trygon*

(b) *Narcine*

(c) *Chimera*

(d) *Syllium*

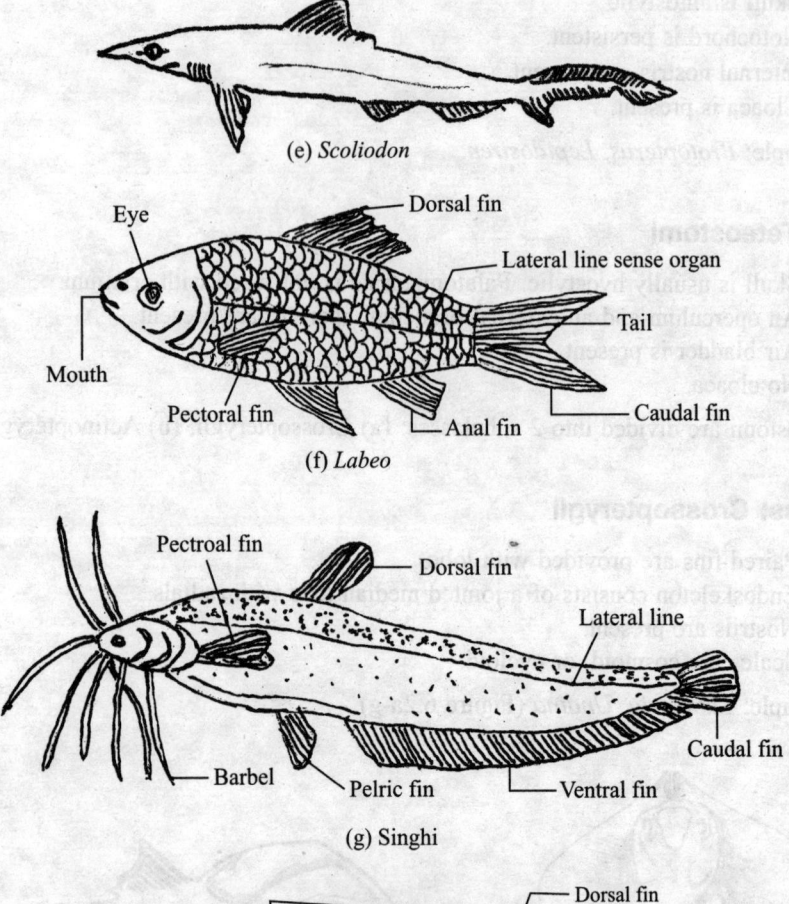

(e) *Scoliodon*

(f) *Labeo*

(g) Singhi

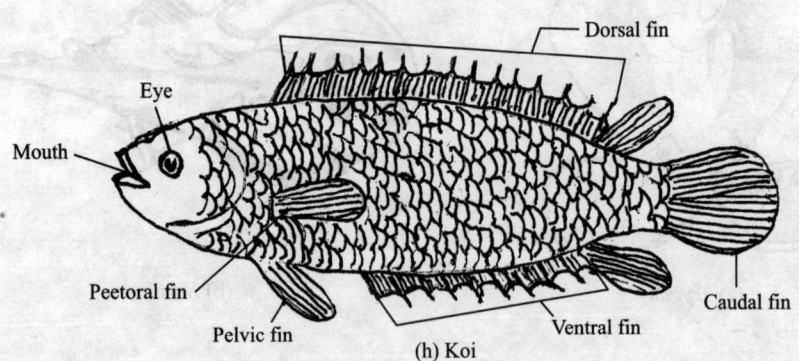

(h) Koi

Figure 6.2   Some examples of fish.

Subclass: Actinopterygii

1. No internal nostril.
2. Scales other than cosmoid type.

Example: *Salmo, Labeo, Catla*

6.2 *Labeo*

Labeo rohita, commonly called Rohu, a common bony fish inhabiting fresh water is found throughout India and Burma except in Tamilnadu and Western Coast. The fish attains a length of about 100 cm. Its body is dark or bluish black along back, silvery on sides and beneath. Scales are sometimes orange to reddish in centre.

Systematic Position

Phylum: Chordata, Subphylum: Vertebrata, Division: Gnathostomata, Superclass: Pisces, Class: Osteichthyes, Subclass: Actinopterygii, Order: Cypriniformes, Genus: *Labeo*, Species: *rohita*

External Features

Bilaterally symmetrical, moderately elongate body is covered with cycloid scales. Dorsal profile is more arched than ventral. Scales are absent in head. Snout is obtuse, fairly depressed, truncate, small and inferior without lateral lobe and projects beyond mouth. Eyes are dorsolateral. Lips are thick and fringed with distinct inner fold to each lip, lobate or entire. Maxillary barbels remains concealed in lateral groove. Teeth are absent on jaws. Pharyngeal teeth are found in 3 rows. Upper jaw does not extend to front edge of eye. Simple dorsal fin rays are 3, or 4 in number, while branched dorsal fin rays are 14. Dorsal fin is inserted midway between snout tip and base of caudal fin. Pectoral and pelvic fins are laterally inserted. Pectoral fin is devoid of osseous spine. Caudal fin is deeply forked. Lower lip joins to isthmus by narrow, or broad bridge. Pre-dorsal scales are 12-16. Lateral line is distinct, complete and runs along midline of caudal peduncle. Colour is bluish on back, silvery on belly. Soft epidermis covers body. Epithelial cells with unicellular mucus glands make epidermis. Musculature is arranged in a zigzag manner. Spindle shaped elongated body is divisible into head, trunk and tail with rounded abdomen. Head extends from snout to operculum with convex dorsal profile than abdomen and about 10 to 15 cm long. Mouth, somewhat crescentic in shape is bounded by thick lips. A pair of nostrils is present on dorsal side of snout and in front of eyes. Eyes are large, lateral and lack eyelids. Cornea is covered by a transparent layer of skin. Maxilla bears a pair of short maxillary barbels. Operculum is large, flat and bony structure. Large crescentic gill is present.

Historical Background

Rohu (*Labeo rohita*) is the important among the 3 Indian major carp species. This species is natural inhabitant of the riverine system of India, Pakistan, Bangladesh and Myanmar. It is found in almost all Indian riverine systems including freshwaters of Andaman. The species is found in Sri Lanka, the former USSR, Japan, China, Malaysia, Nepal and the Philippines. Its traditional culture goes back hundreds of years in small ponds from the early part of 20th century. Its compatibility with catla and mrigal has made it ideal for polyculture. Riverine collection of seed fulfilled the requirement for culture of species until the first half of the 20th century. Success in induced breeding in the year 1957 ensured seed supply for its culture in freshwater ponds and tanks. Because of its high growth potential, and high consumer preference, this is important freshwater species.

Habitat and Biology

In its early life stages it prefers zooplanktonlike rotifers and cladocerans, with phytoplankton as the emergency food. In fingerling stage, they consume small phytoplanktons like desmids, phytoflagellates and algal spores. Adults prefer the phytoplankton. In juvenile and adult stages, rohu is an herbivorous column feeder, and consume algae and submerged vegetation. It is a bottom feeder. Nibbling type of mouth is provided with fringed lips, sharp cutting edges. Bucco-pharyngeal region devoid of teeth aids to feed on vegetation, which need not require seizure and crushing. Modified thin and gill rakers also help to feed on plankton through sieving water. In ponds, both the fry and fingerlings show schooling mainly for feeding. This habit is not found in adults. Rohu is eurythermal and does not thrive at below 14°C. It is a fast growing and attains 45 cm total length and 800 g in one year. In polyculture, its growth rate is higher than that of mrigal but lower than catla. Minimum age at first maturity for both sexes is 2 years. Full maturity is reached after 4 years in males and 5 years in females. In nature, spawning occurs in shallow and marginal areas of flooded rivers. Spawning season coincides with the south-west monsoon. In captivity with proper feeding, it attains maturity at the end of 2nd year. Breeding does not occur in lentic pond. Fecundity varies from 226 000 to 2 794 000, depending on fish size and ovary weight. Rohu is polygamous and seems to be promiscuous. Optimum temperature for spawning is 31°C.

Trunk: Body part behind operculum which extends up to anus is called trunk. It is compressed laterally, elliptical in shape in cross section, covered with thin overlapping cycloid scales which are arranged length wise and in diagonal rows. Lateral line runs along lateral sides of body. Six to six and half rows of scales between base of ventral and lateral line are found. Each scale lies in a dermal pocket. Anus is situated ventrally just in front of anal fin. Urinogenital aperture lies behind anus.

Fins: Dorsal fin arises from mid dorsal line of trunk half way between snout and base of caudal fin. Single caudal fin is homocercal with 2 symmetrical lobes. Anal fin lies posterior to anus on ventral side. Paired pectoral fins behind opercula are borne by pectoral girdle. Each pectoral fin is supported by 19 fin rays. Pelvic fins are paired, situated on ventral side behind pectoral fins. Each pelvic fin contains 9 fin rays.

Tail: It is narrow part behind anus bearing tail fin posteriorly. Part of lateral line extends over this region.

Skeletal Elements: These are two types namely exoskeleton and endoskeleton. Exoskeletal elements are scales and finrays. Endoskeleton is divided into axial skeleton and appendicular skeleton. Axial skeleton consists of skull, visceral skeleton, vertebral column, fin exoskeleton and ribs. Appendicular skeleton consists of pectoral girdle, pectoral fin, peivic girdle and pelvic fin.

Skull: Skull is composed of the cranium, sense capsule and the visceral arches.Skull is elongated and divided into a dorsal roof, posterior occipital region, the otic region, orbitotemporal region, nasal region.The bones found in the skull are premaxilla, maxilla, lachrymal, mesethmoid, nasal opercular, parietal supraoccipital, basioccipital, exoccipital. It also contains bones like frontal, supraorbital, infraorbital.

Viceral Skeleton: The first visceral arch is the mandibular arch which is divided into paltopterygoquadrate bar and Mackel's cartilage. The second arch is called the hyoid which is divided into the upper hyomandibular and lower hyoid cornu. Of the rest five are branchial arches, the four support the gills and the fifth one forms the inferior pharyngeal bone.

Vertebral Column: This is a completely ossified structure consisting of 38 vertebrae which are amphicoelous. This is divided into anterior trunk region consisting of 21 vertebrae and posterior caudal vertebrae. The last three caudal vertebrae are modified for supporting the caudal fin. The posterior most trunk vertebra is called the urostyle.

Fin Skeleton: Median fin consists of somactidia, and dermatotrichia which are branched, jointed and are called lepidotrichia. Actinotrichia are found at the free edges of fin. The dorsal fin is supported by lepidotrichia.Pectoral fin is supported by 19 lepidotrichia.Each pelvic fin is supported by 9 fin-rays and 3 radials.

Ribs: Seventeen pairs of ribs are found between the muscles and peritoneum.

Appendicular Skeleton: It consists of pectoral girdle and pelvic girdle. Pectoral girdle consists of supracleithrum, cleithrum, mesocoracoid, coracoid, radials, scapula, and finrays. Pelvic girdle consists of pelvic bones, radials, cartilage in addition to fin rays

Digestive System: This system comprises long alimentary canal comprising mouth, buccal cavity, pharynx, esophagus, stomach, or intestinal bulb, intestine, rectum, anus and its associate glands. The alimentary canal begins at mouth and terminates at anus. The Rohu is herbivorous and its intestine is very long. Intestine is short in carnivorous fishes.Mouth is bounded by soft lips. Free margins of lips are broad and beset with 4 or 5 rows of darkly pigmented conical papillae. Inner fold of each lip is muscular, narrow and lacks papillae. Buccal cavity is a dorsoventrally compressed cavity with an arched roof and flat floor. Thick mucous membrane lines the buccal cavity. Distinct tongue is absent, although floor is lined with thick muscles. Pharynx is a dorsoventrally flattened cavity bounded by gill arches laterally, divisible into anterior and posterior pharynx. Anterior part is narrow in front and wide behind, perforated on sides by gill slits. Thick lining of mucus membrane bears papillae. Gill arches are supported by gill rakers which form a broad sieve-like structure for filtering water to retain food in pharyngeal cavity. Posterior pharynx is masticatory and its lateroventral walls are beset with pharyngeal teeth. Ventral wall is highly folded transversely. Homodont pharyngeal teeth (Figure 6.3) are arranged in 3 rows, one alternating other. Each tooth consists of basal root and crown projecting above. Crown is laterally compressed. Teeth crush solid food. Mucus-secreting cells are present in pharynx wall. Mucus lubricates food. Esophagus is a narrow thick and short tube. Mucus membrane lining oesophageal cavity is thrown into 7 prominent longitudinal and several small folds. From dorsal side of esophagus arises ductus pneumaticus leading into swim bladder. Opening of esophagus into stomach is guarded by a valve like structure. Esophageal mucosa is composed of stratified - and columbar epithelium. Mucus secreting cells are also found.

True stomach is absent. Anterior intestine is swollen to form a sac like structure behind esophagus called intestinal bulb for food storage. Mucus lining of anterior cardiac region is honeycomb in appearance, while posterior pyloric region is thrown into several thick longitudinal folds. Mucus lining of intestinal bulb contains absorptive cells and mucus

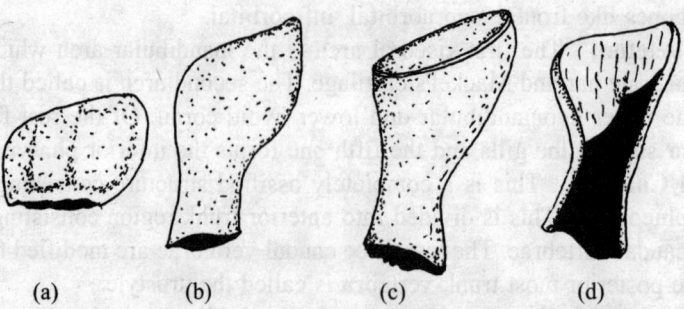

(a)   (b)   (c)   (d)

Figure 6.3 Pharyngeal teeth of *Labeo* (a) Growing teeth, (b) Fully formed teeth,
(c) Side view of B and (d) Sagittal section of the same

secreting cells. Absorptive cells possess a free striated border. Mucus secreting cells are abundant in posterior intestine.Absence of stomach may bring about by neoteny. Intestine is thin walled, elongated coiled tube, uniform in diameter. Intestine is long in herbivorous fishes. Its anterior portion shows oblique transverse folds, while posterior part exhibits distinct longitudinal folds. Pyloric caeca, characteristic of teleostean fishes are altogether absent. Rectum is thin walled sac, wider than intestine. Mucus lining of rectum presents indistinct oblique transverse folds. Mucosal folds are short and broad than those of intestine and possess many mucus secreting cells. Rectum opens to outside at anus situated in front of urinogenital opening (Figure 6.4).

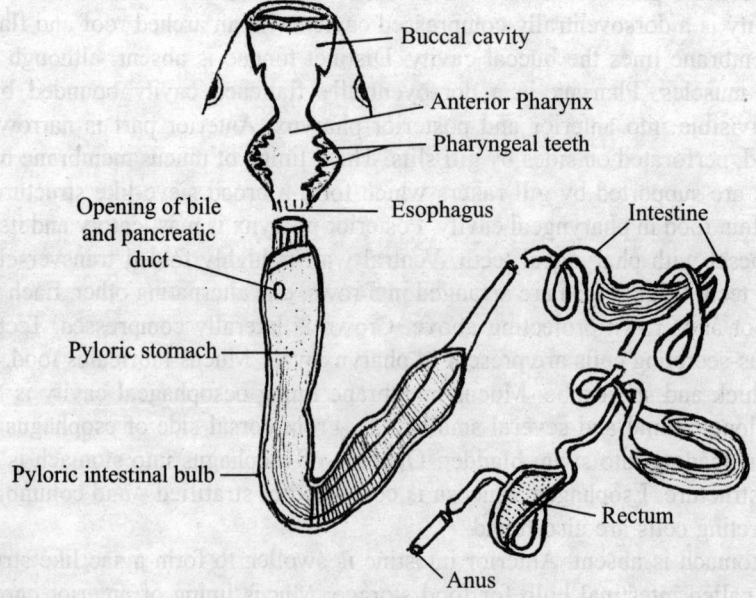

Figure 6.4 Alimentary canal of *Labeo*

Digestive Glands and Digestion

Liver is a solid bilobed gland of chocolate brown colour. Right lobe is narrow and elongated. Left lobe is broader. Two lobes are connected to each other at 3 places by transverse connections namely the anterior median lobe, middle connecting lobe and a posterior median mass. Gall bladder, an elongated sac having thin wall is situated between right lobe and anterior portion of intestinal bulb.

A cystic duct arises from anterior - ventral end of gall bladder and receives 3 hepatic ducts from liver lobes to continue as bile duct. Pancreas is a diffuse mass in body cavity among intestinal coils. It extends into liver and remains embedded in substance of spleen.Exocrine cells are large, cuboidal, or columnar with large nuclei. Basal part of each contains homogeneous cytoplasm while apical part contains zymogen granules. Pancreatic duct is formed by union of ductules and after emerging from anterior region of left hepatic lobe runs side by side with bile duct for a short distance to open into anterior intestinal bulb.Process of digestion is not clearly known. Pancreatic juice contains amylase, erepsin and trypsin. Intestinal secretion contains enterokinase, lipase and maltase. In absence of true stomach, pepsin and HCl are not produced. Most enzyme work on carbohydrates as fish is herbivorous (Figure 6.5).

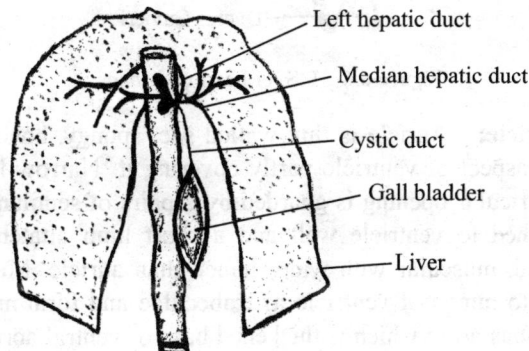

Left hepatic duct

Median hepatic duct

Cystic duct

Gall bladder

Liver

Figure 6.5   Part of the Liver, gall bladder and ducts in *Labeo*.

Circulatory System

This is closed type and consists of blood, heart, arteries and veins. There is a clear cut single circulation from body to gills and back. Blood is red in colour, consists of plasma and blood corpuscles. RBC is nucleated oval cells. WBCs may be granular, or agranular. Granulocytes are acidophils, basophils and neutrophils. Lymphocytes, thrombocytes and monocytes are agranular. Plasma is colourless. Haemoglobin is dissolved in plasma.

Heart: Heart is situated just posterior to gills. In other teleosts, it is placed relatively forward in body than in elasmobranchs. It remains in a pericardial sac adhering closely to heart and consists of sinus venosus, auricle, and ventricle. Conus arteriosus is absent and a non contractile bulbous arteriosus is present. Sinus venous is thin walled spacious sac which collects venous blood. Paired ducti cuvieri; paired hepatic veins, a posterior cardinal vein and an inferior jugular vein collect blood. Valves are absent at openings of these veins. Sinus venosus opens by a median orifice into auricle. Orifice called the sinu-auricular aperture is guarded by a pair of membranous semilunar valves.Sinus venosus has a pair of lateral appendages. It is spongy and fibrous in nature (Figure 6.6).

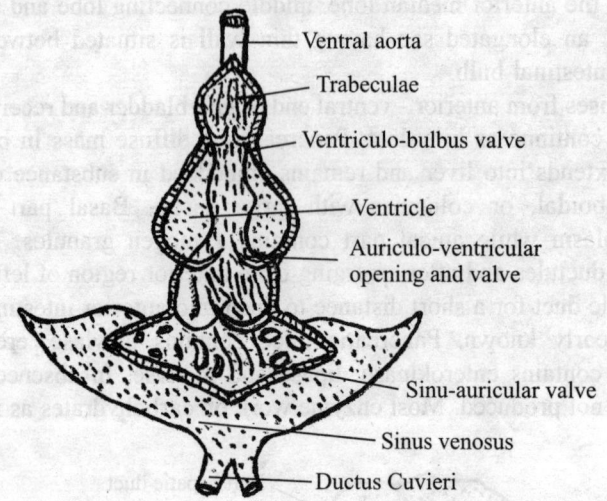

Figure 6.6   L.S. of heart of *Labeo*.

Auricle and Ventricle: Auricle is thin walled sac, spongy, honeycomb in appearance and situated on dorsal aspect of ventricle partly covering it. Narrow lumen continues up to ventricle. Auriculo ventricular opening is guarded by 2 pairs of semilunar valves. Each valve has a long limb attached to ventricle wall and a short limb attached to auricular wall. Ventricle is thick walled, muscular with wider lumen than auricle. Muscle bundles of inner spongy layer project into lumen of ventricle as trabeculae and form numerous subdivisions. Ventricle opens into bulbus aorta which is thickened base of ventral aorta. Conus arteriosus is absent. Bulbus has narrow lumen and its walls are thick. Bulbus aorta continues forward as ventral aorta. Thin trabeculae present in lumen are disposed parallel to each other.

Arteries

Afferent Branchial Arteries: Afferent arteries are 8 in number. Ventral aorta continues forward along ventral surface of pharynx up to posterior end of hyoid and bifurcate into 2 afferent branchial arteries, one on either side, called first afferent branchial. Each runs along first gill of its side supplying blood to capillaries in gill filaments. Second afferent branchial originates a little behind first and supplies blood to second gill.Third and 4th afferent branchials arise in succession behind second from ventral aorta supplying 3rd and 4th gills respectively (Figure 6.7).

Efferent Branchial Arteries: Generally only one efferent branchial is present in each gill arch in teleosts. But in *L. rohita* 2 efferent branchials are present in each gill arch which collect blood from gills and then unite to form epibranchials arterie. First 2 epibranchials open into lateral aorta while 3rd and 4th epibranchials open into dorsal aorta (Figure 6.8).

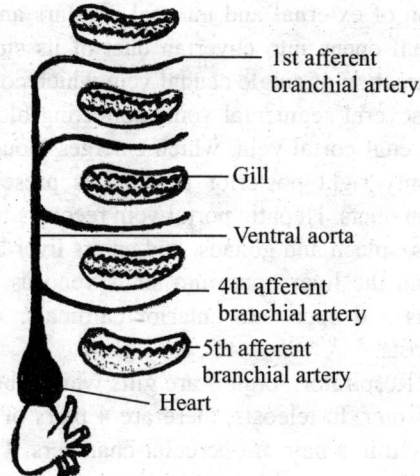

Figure 6.7 Afferent branchial vessels of *Labeo*.

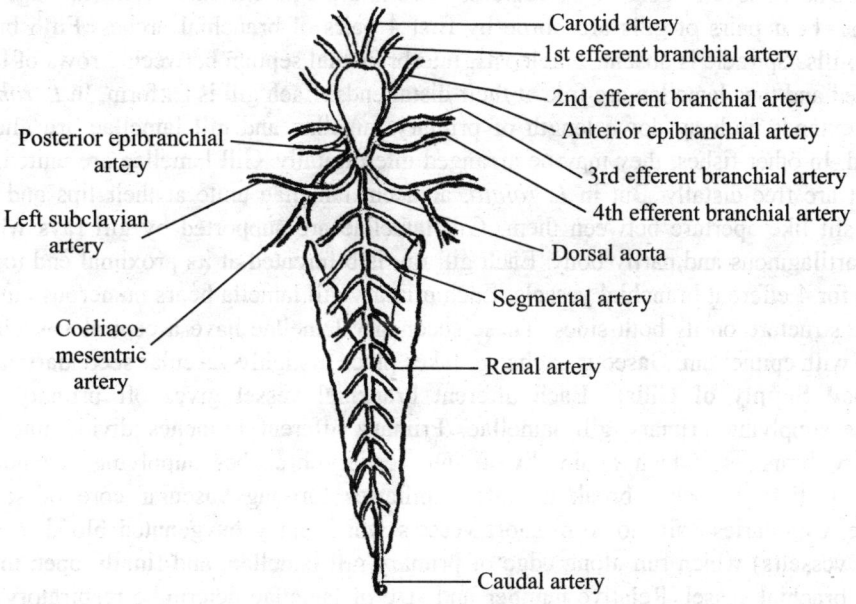

Figure 6.8 Efferent branchial system.

Principal Arteries: From dorsal aorta arise several arteries for supply blood to different body parts Among principal arteries, subclavian supply blood to pectoral fins, coeliaco-mesenteric supply parts of alimentary canal, liver, air bladder and gonads, renals supply kidneys, iliacs supply pelvic fins and caudal artery sending branches to muscles of caudal region. Cephalic region receives blood supply from external and internal carotids.

Principal Veins: Principal veins are paired anterior cardinals, paired posterior cardinals, unpaired hepatic portal, paired subclavian veins and a caudal vein. Anterior cardinal vein of each side is formed by union of external and internal jugulars and collects blood from head region. Each anterior cardinal opens into cuverian duct of its side. Right and left posterior cardinals are formed by bifurcation of single caudal vein which collect blood form tail region. Posterior cardinals receive several segmental veins collecting blood from trunk region. Left branch of caudal forms the renal portal vein, which emerges though kidneys as left posterior cardinal. In some fishes, only right posterior cardinal is present. Posterior cardinals run forward to open into cuverian ducts. Hepatic portal vein receives blood from different parts of alimentary canal, air bladder, spleen and gonads, and enters liver breaking up into capillaries. Two hepatic veins arise from the liver open into sinus venosus. Paired subclavians collect blood from the pectoral fins and open into anterior cardinals. Ventral abdominal veins in *Scoliodon* are absent in teleosts.

Respiratory System: Respiratory organs are gills which obtain oxygen from water and pass out carbon dioxide to water. In teleosts, there are 4 pairs of gills which are situated on either side of pharynx enclosed in a pair of opercular chambers. They are covered with a pair of operculum each of which is supported by opercular bones. Posterior edge of operculum is thin, membranous and is called branchiostegal membrane. It is applied to ventro-lateral body surface. Due to development of operculum, there is a single external branchials aperture on each side. Four pairs of gills are borne by first 4 pairs of branchial arches. Fifth branch is without gills. Spiracle is absent. In teleosts, interbranchial septum between 2 rows of lamellae is reduced and thus lamellae are free at their distal ends. Each gill is filiform. In *L. rohita*, gill septum extends midway down length of primary lamellae and gill lamellae are alternately arranged. In other fishes, they may be arranged interdigitally. Gill lamellae are united at their base but are free distally. But in *L. rohita*, adjacent lamellae unite at their tips and leave a narrow slit like aperture between them. Gill lamellae are supported by gill rays which are partly cartilaginous and partly bony. Each gill ray is bifurcated at its proximal end to make a passage for 4 efferent branchial vessels. Each primary gill lamella bears numerous minute flat leaf like structure on its both sides. These secondary lamellae have a central vascular layer covered with epithelium. Gaseous exchange takes place in highly vascular secondary lamellae.

Blood Supply of Gills: Each afferent branchial vessel gives off primary afferent branches supplying primary gill lamellae. Primary afferent branches divide into several secondary branches, which again divide into tertiary branches supplying secondary gill lamellae.Tertiary branches break up into capillaries forming vascular core of secondary lamellae. Capillaries unite to form short vessels which carry oxygenated blood to primary efferent vessel(s) which run along edge of primary gill lamellae, and finally open into main efferent brachial vessel. Relative number and size of lamellae determine respiratory area of gills, and this depends on habits of fish.

Mechanism of Respiration: A current of water enters buccopharyngeal cavity through mouth and pass over gills, and exit through external branchial aperture. Inhalation and exhalation of water is effected by buccopharyngeal cavity acting both as suction and pressure pump. At the beginning of inspiration, pharynx acts as suction pump drawing in water through mouth. For this, mouth is opened. Brachiostegal rays spread and lowered and buccal cavity is enlarged by lateral movement of opercula. Increased volume capacity of bucco pharyngeal cavity creates a negative pressure and water is sucked in through mouth. Then space between gills and operculum is enlarged and water flows over gills. At this time, mouth remains closed and opercular flaps are tightly pressed against body by outside water pressure. Oral valves prevent flow of water through mouth. For exhalation of water, buccal and opercular cavities begin to contract exerting pressure on contained water. Opercula are now quickly brought towards body and opercular flaps open expelling water. During its course, water current bathes gills for gaseous exchange. Direction of blood and water circulation helps in increased oxygen uptake. Afferent and efferent branchial vessels are so arranged that blood and water current flow in opposite direction constituting a counter current system.

Nervous System: Brain is more specialized than that of dogfish. Cerebral hemispheres are small and undivided. Pallium or roof is thin and non nervous. Floor forms a prominent thickening called the corpora striata, a centre of correlation. Olfactory bulbs are followed posteriorly by olfactory lobes. Diencephalons are reduced and bear dorsally the pineal body. Optic lobes are large and lobi inferiors are found on their ventral surface. Optic nerves simply cross one another without forming a chiasma. Pituitary body lies behind crossing of optic nerves. Behind pituitary lies saccus vasculosus. Cerebellum is very large and bent on itself. Anterior part of cerebellum forms valvula cerebeli (Figure 6.9).

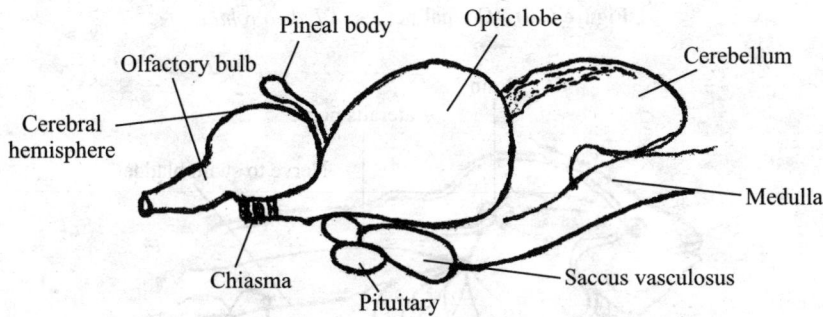

Figure 6.9   Optic lateral view of the brain of *Labeo*.

Cranial Nerves: There are 10 pairs of cranial nerves. Olfactory nerve (I), optic nerve (II), occulomotor (III), trochlear (IV), abducens (VI), and auditory nerves (VII) have same origin and distribution as in dogfish. Fifth nerve or the trigeminals divides into two branches (Figure 6.10, 6.11).

Opthalmic: This branch of fifth nerve is in intimate association with ophthalmic branch of seventh nerve and as such ophthalmic is represented by two branches (I) opthalmicus superficialis running dorsally, and (II) Opthalmicus profundus running below:

Maxillary branch runs along posterior margin of orbit and after a short distance form its origin bifurcates into maxillary proper supplying upper jaw and mandibular which supplies lower jaw. Seventh or facial nerve originates just behind fifth and divides into 3 main branches (I) opthalmicus superficialis, (II) buccalis, and (III) hyomandibular. Buccal runs obliquely downwards and forwards across orbit and supplies associated region. Hyomandibular runs backwards and divides into a mandibularis externus, mandibularis internus and hyoidean. Another branch of 7th nerve is palatine that runs across orbit to palate. Ninth or glossopharyngeal originates from ventrolateral surface of medulla and running backwards, supplies first gill where it divides into anterior pre-trematic and posterior post-trematic branches. Tenth or vagus nerve divides into 3 branches - branchialis supplying nerves to

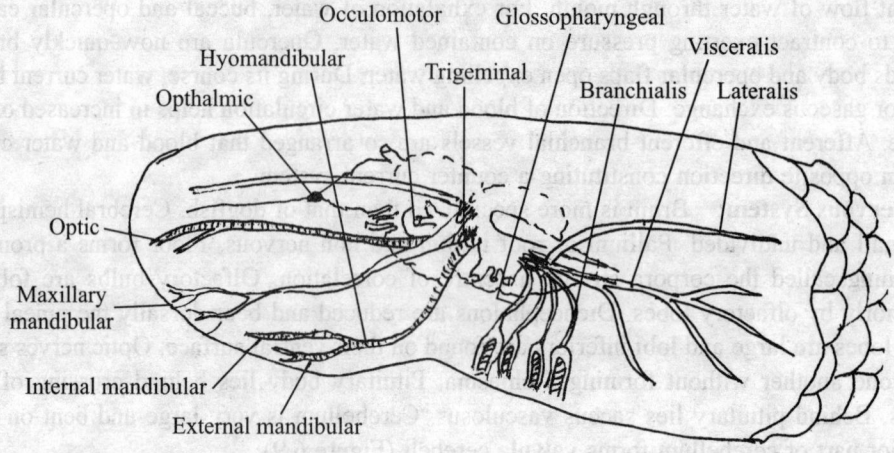

Figure 6.10 Cranial nerves of *Labeo rohita*.

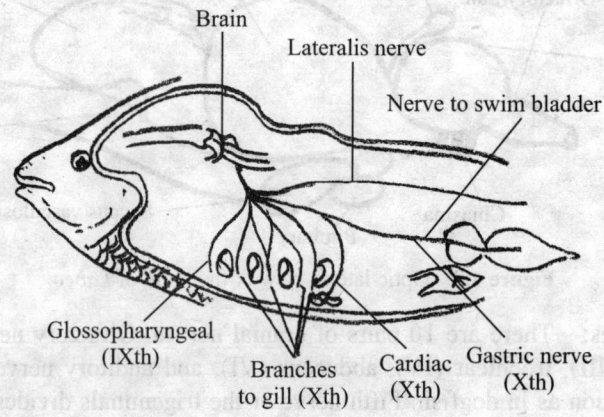

Figure 6.11 IXth & Xth Cranial Nerves

remaining gills, visceralis which supplies viscera, and lateralis which runs parallel to lateral line lying deep in tissue and supplying nerves to lateral line sense organs.

Excretory System: Kidneys are elongated, reddish brown structure occupying the whole length of abdominal cavity and are situated dorsal to air bladder. They are distinct anteriorly, but are partly fused in middle line. Kidneys are mesonephric/ opisthonephric. Since posterior region performs renal function, 2 ureters one on each side emerges from posterior kidneys and they empty into a thin walled urinary bladder, which lies ventral to cloaca and opens into urinogenital sinus. Numerous nephrons with glomeruli help in eliminating excess water .Certain tubules develop for salt reabsorption to maintain balance. Kidneys eliminate nitrogenous waste products and maintain osmoregulation. Water proof slimy skin prevents entry of water through body surface. Lymphoidal tissue fills the space between uriniferous tubules. Anterior part of kidney consists of lymphoid haemopoietic, inter renal and chromaffin tissue. Malpighian bodies and tubules are absent. In freshwater fishes, kidney excretes uric acid and creatine. Ammonia and urea are removed by gills (Figure 6.12 and 6.13).

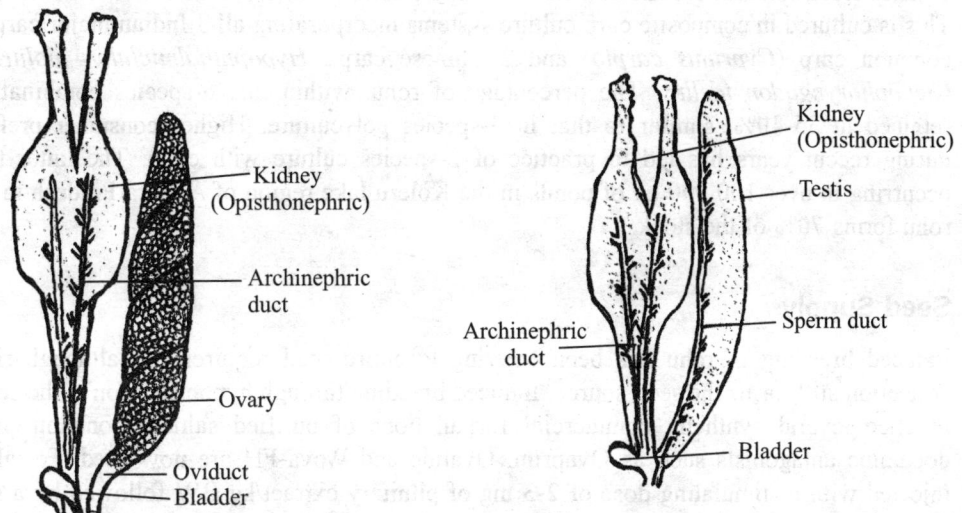

Figure 6.12   Female urinogenital organs of *Labeo*.   **Figure 6.13**   Male urinogenital system of *Labeo*.

Reproductive System: Gonads are enlarged in adults during breeding season. Testes become pinkish, elongated and flattened, extend to posterior region of kidneys. From posterior edge of each testis originates vas deferens, which runs posteriorly to open into urinogenital sinus. In some teleosts, paired seminal vesicles outgrow from posterior ends of vasa deferentia. Each testis is composed of several seminiferous tubules. Interlobular spaces of testis are filled with connective tissue, interstitial cells and blood capillaries. Testes exhibit different seasonal phases like resting phase, mature phase, and spent phase. Ovaries are paired sacs like structures, larger than testes and are situated in abdominal cavity lying ventral to kidney. Oviducts are absent though in other teleosts, they remain as short ducts. Mature ova are discharged into body cavity. A pair of genital pores is formed in anterior wall of urinogenital sinus through which ova emerge. Eggs are demersal that is they sink to bottom. In many other fishes, eggs float on water surface. Like testis, ovary also exhibit seasonal cycle of different phase.

Breeding and Development *Labeo* breeds in rivers during monsoon. Eggs are laid in large numbers, fertilized by spermatic fluid and all do not survive. Cleavage, formation of blastula, gastrula is not known. Eggs hatch within a period, 15 hours. Hatchling has prominent yolk sac suspended underneath its body and yolk takes 7 days to be absorbed. Young ones called 'fry' measures about 1.5 cm to 2.5 cm in length. They have fringed lips, dorsal fin rays and a vertical dark spot at base of tail which disappears in course of time. When fries attain a size of about 5 cm, they are called 'fingerlings' which grow up to 15 cm in length. They become sexually mature by the end of second year.

Production Systems

Rohu is the principal species reared in carp polyculture systems. Its wider feeding niche extends from column to bottom. This is usually stocked at higher levels than other 2 species. This is cultured in composite carp culture systems incorporating all 3 Indian major carps, and common carp (*Cyprinus carpio*) and 2 Chinese carps, *Hypophthalmichthys molitrix* and *Ctenopharyngodon idellus*. The percentage of rohu, within this 6-species combination, is retained at 35-40%, similar to that in 3-species polyculture. Higher consumer preference during recent years has led to practice of 2-species culture with catla. The latter type is occurring in over 100 000 ha of ponds in the Koleru lake region of Andhra Pradesh in which rohu forms 70% of the stock.

Seed Supply

Induced breeding of rohu has been catering to entire seed requirement, although riverine collection still forms the seed source. Induced breeding through hypophysation is the common practice.Several synthetic commercial formulations of purified salmon gonadotropin and dopamine antagonists such as Ovaprim, Ovatide and Wova-FH are now used. Females are injected with a stimulating dose of 2-3 mg of pituitary extract/kg BW followed by a second dose of 5 to 8 mg/kg after a gap of 6 hours. Males are given single dose of 2-3 mg/kg during second injection of female. In using synthetic formulations, a single dose of 0.4-0.5 ml/kg body weight (females) or 0.2-0.3 ml/kg (males) is injected.

The Chinese circular hatchery is the common system for seed production. Such hatchery possesses 3 components, viz., spawning/breeding tank, incubation/hatching tank, and water storage and supply system. Depth of water in breeding tank is maintained at up to 1.5 m, based on brood stock density; 3-5 kg brood stock/m³ is usually recommended. Female: male ratio is maintained at 1:1 by weight (1:2 by number). Size and number of hatching tanks vary. Optimum egg density for incubation is 0.8 million/m³. Generally, 0.15-0.2 million eggs/kg of female are obtained. Seed rearing involves 2-tier system, i.e. a 15-20 days nursery phase for raising fry, followed by 3 months phase for fingerling production.

Rearing Fingerling

Nursery Phase

Three day old hatchlings are reared up to fry of 20-25 mm in earthen nursery ponds of 0.02-0.1 ha. Pre-stocking nursery pond preparation includes removal of aquatic weeds and predatory fish, followed by liming and manuring with organic and inorganic manures. Aquatic insects are removed by soap-oil emulsion. In earthen ponds, hatchlings are stocked up to10 million/ha, but higher level of 20 million/ha is used in cement nurseries. Hatchlings receive supplementary feed of 1:1 w/w mixture of rice bran and groundnut/mustard oil cake. Survival ranges from 30 to 50%.Non-availability of commercial feed, forcing the farmers to conventional bran-oil cake mixture.

Fingerling Production

Nursery-raised fry of 20-25 mm are reared for 3 months to 100 mm fingerlings in earthen ponds. Rohu are grown together with other carp species at combined densities of 0.2-0.3 million fry/ha, with the rohu comprising 40 % of the total. Pond manuring with organic and inorganic fertilizers and supplementary feeding with conventional mixture of rice bran and oil cake are the usual practices. Dosage and form of application vary with farming intensity and inherent pond productivity. Overall survival in these fingerling rearing systems ranges from 60 to 70%.

On Growing Techniques

The grow out production confined mainly to earthen ponds, is followed in combination with other 2 Indian major carps within 3-species polyculture, and sometimes within 6-species composite carp culture system involving 3 Indian major carps, common -, grass - and silver carp in varied proportions. Large share of production still comes from extensive farming involving stocking and fertilization as the inputs and achieving more modest production levels of 1-2 tonnes/ha/yr. Practical technology includes predatory and weed fish control; stocking of fingerlings at combined density of 10 000/ha; pond fertilization with organic manures like cattle dung or poultry droppings and inorganic fertilizers; the provision of mixture of rice bran and groundnut cake as supplementary feed, and water management. Grow-out period is 1 year, during which rohu grows to about 800 g. In certain cases the farmers resort to partial harvesting of marketable size groups at intermittent intervals.

Fingerlings are used for stocking grow-out ponds. However, their inadequate availability compels farmers to stock their ponds with fry, leading to poor production. Supplementary feed forms the major input. The high price of commercial feeds force farmers to conventional bran-oil cake mixture leading to wastage and deterioration of water quality. Judicious feed management is required to enhance the profit margin. In grow out, especially at higher stocking densities an ectoparasite, Argulus is a major problem for rohu. Rohu is the important components in sewage-fed carp culture system in an area over 4 000 ha in West Bengal, India.

Harvesting Technique

As carp are cultured in small ponds and tanks, manually operated drag nets are convenient gear for harvesting. Length of these nets depends on width of pond. In most cases, fish are harvested at the end of culture period through repeated netting. Sometimes, this is followed by total draining of ponds. Cast nets are used for partial harvesting in small and backyard ponds. In water bodies with multiple stocking, the harvesting of larger sizes is initiated after 7 months of culture, and the smaller ones are returned to pond for further growth. Multiple stocking and harvesting is the common practice in sewage-fed carp culture system.

Handling and Processing

Marketing of rohu mostly relies on local markets. In commercial farms, fish after washing in water, are packed with ice at 1:1 ratio in plastic crates. Long-distance transport of these ice-packed fish in insulated vans is a common practice in countries like India. Post-harvest processing and value-addition of this species is almost non-existent at present in any of the producing countries.

Production Costs

Carp are low-valued species fetching market prices of less than USD 1/kg at the producers' level; the use of major inputs like seed, fertilizers and supplementary feed, in addition to labour costs, is kept to a lowest. Supplementary feed constitutes 50% of total input cost in carp polyculture. Judicious feed management has prime importance for increasing profits. In extensive systems, with targeted production level , production cost is about US$ 0.30/kg, while the costs increase to US$ 0.6/kg in semi-intensive culture, where targeted production is 8 tonnes/ha. India is by far the largest producer of rohu. Low level production is also reported by Lao People's Democratic Republic and Thailand.

Market and Trade

Almost all the rohu produced from aquaculture is consumed in nearby local markets. Post-harvest processing is non-existent. Rohu fetches comparatively high market prices. They are either marketed fresh in local market, or carried to urban markets with ice. Long-distance transport in insulated vans with ice, covering distances of 3 000 km is common in India. Locally-produced fresh fish fetches about one and half times higher market price than iced fish. When sold live, the market value increases over 2-fold compared to iced fish. Carp culture has not been perceived as a threat to environment. Increased emphasis on intensification for enhancing production in recent years has increased use of chemical fertilizers, feeds, therapeutics, drugs, and chemicals with some concern. Practicing countries should formulate guidelines and impose regulatory measures for judicious use of such critical inputs.

6.3   Dipnoi

Lung-fishes and their allies are considered as ancestor of amphibian. They stand close to the ancestral stock of gnathostomes. Only 4 genera of this group found now are *Neoceratodus*, *Lepidosiren*, *Protopterus*, (Figure 6.14) the lung-fishes of Australia, South America, and Africa respectively, and *Latimeria*. These are relics of a group that can be traced back with relatively little change to Devonian. There is little doubt that the first amphibia arose from some similar line at that period. Modern crossopterygians approach to the condition of the ancestors of all tetrapods. They dig into the mud, leaving a small opening for breathing, and remain in this state for at least six months.

Aestivation has been adopted by the group at least since Permian times. The 3 survivors show similar deviations from the conditions found in *Dipterus*, but *Neoceratodus* has diverged less than other 2. The tail fin is symmetrical in all 3, with no trace of separate dorsal fins.The scapula is covered by clavicles, cleithra, and post-temporals, the latter articulating with skull. The scales are reduced to bony plates. The vertebrae are cartilaginous arches, the notochord

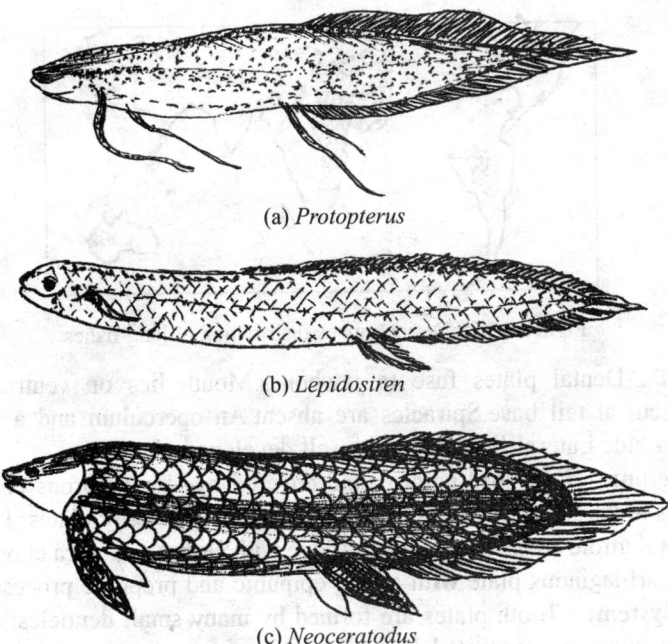

(a) *Protopterus*

(b) *Lepidosiren*

(c) *Neoceratodus*

Figure 6.14   Lung-fishes.

remaining as an unconstricted rod. In skull, ossification reduces. The dorsal bones consist of few bony plates, forming a pattern not comparable with that of other forms. Food consists of small invertebrates and decaying vegetable matter, which is eaten in large amounts. The early Dipnoi of Devonian possessed cosmoid scales. In later osteolepids and Dipnoi, thinning of the scales occurred, as among the Actinopterygii, so that the later Dipnoi are covered with thin, over- lapping, cycloid.

Bony fishes belonging to class Dipnoi are generally called' lung-fish' .Dipnoi owes its name from the presence of 2 internal nostrils. They evolved during middle Devonian and flourished well in Penman and Triassic periods, became rare after Triassic and are represented now by 3 genera. *Neoceratodus* is found in Burnett and Mary rivers of Queensland in Australia.Only surviving species is *N.forsteri*. *Protopterus* is distributed in Central Africa, found in river Senegal, White Nile, Zambesl River, Lake Tanganyika and in central Congo basin. Three species are *P.aethiopicus*, *P.annecteus*; *P.dolloi*. *Lepidosiren* is restricted in tropical South America, Amazon River and its branches, marshy and swampy areas of Chaco plains. Only species of *Lepidisiren* is *L. paradoxa*. Dipnoans had almost worldwide distribution. Their persistence in southern parts of globe and complete absence from Asia is explained assuming that southern parts became separated from Asia before evolutions of Dipnoans (Figure 6.15).

External Characters: Body is large, elongated, fish like and covered with thin cycloid scales. In *Dipterus* scales were thick. "Archipterygial" type of fins having an axis with 2 rows of radials.Tail diphycercal, dorsal, and anal and ail fins are continuous. In *Dipterus* tail was heterocercal. Skull is autostylic .Nostrils is found on ventral surface of snout. Internal nostrils

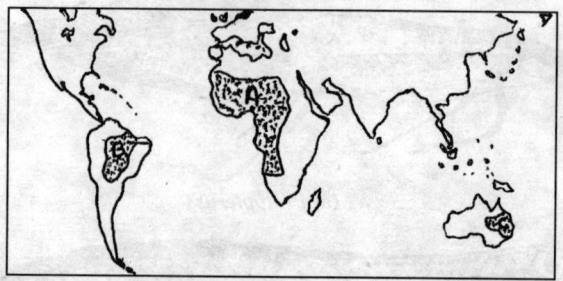

Figure 6.15 Discontinuous distribution of lung fishes.

open into mouth. Dental plates fuse to jawbones.Mouth lies on ventral surface.Cloacal aperture is present at tail base.Spiracles are absent.An operculum and a branchial openng present on either side.Lateral line system is well developed.

Axial Skeleton: Notochord persists and covered by tough fibrous covering. Vertebrae are cartilaginous arches. Dorsal bones of skull consist of few bony plates. Lower jaw acts as dentaries. Pectoral girdle consists of stout cartilage with clithra and intra clavicle. Pelvic girdle composed of a cartilaginous plate with a long epipubic and prepubic processes.

Digestive System: Tooth plates are formed by many small denticles. Cavities between stomach and intestine are separated by a flap like pyloric valve.Intestine is ciliated with a spiral valve, which ends in rectum.Rectum opens into a small cloaca. Between intestine and cloaca lies a rectal gland. Liver is divided into 2 unequal lobes. Gall bladder is large, situated on left margin of liver. Pancreas remains within gut walls. Spleen is attached to right dorso-lateral stomach wall. Gut lacks stomach and intestine is ciliated. Hepatic caeca is absent, but a spiral valve is well-developed.

Air-bladder is developed into lung-like structure, divided into many chambers. *Neoceratodus* ome to surface to breathe air and survive in foul water that kills other fishes. *Lepidosiren* and *Protopterus* obtain 98 per cent of their oxygen from the air. The wall of the air-bladder contains muscle-fibres and the cavity is divided into number of pouches or alveoli. In *Protopterus* and *Lepidosiren* the edges of slit-like glottis are controlled by muscles and there is an epiglottis. Lung is supplied with blood from the last branchial arch in *Neoceratodus*, but in other Dipnoi, there is a more elaborate arrangement (Figures 6.16, 6.17).

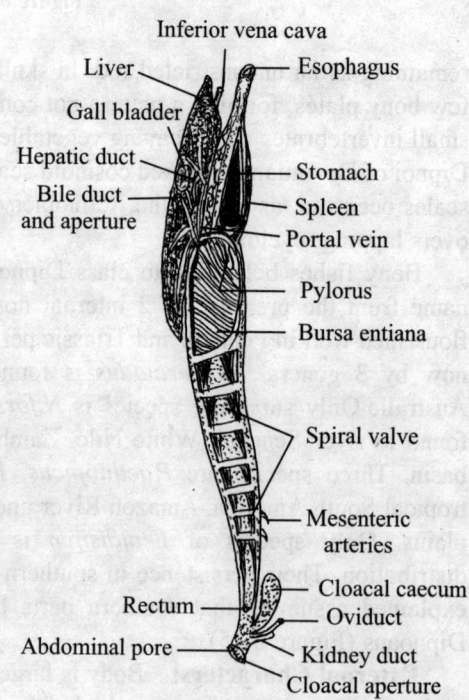

Figure 6.16 Alimentary canal and associated structures of *Protopterus*.

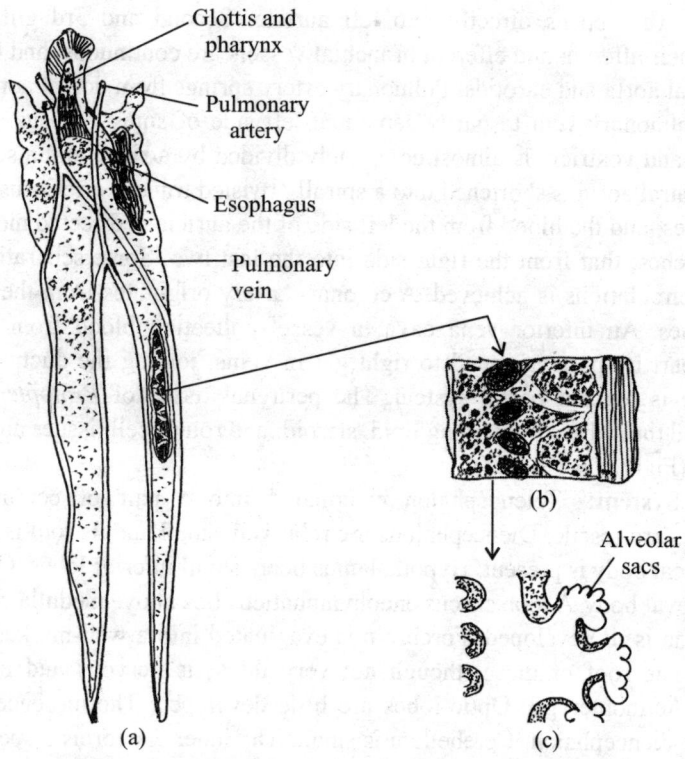

Figure 6.17 (a) Swim bladder of *Protopterus*, (b) Portion of the internal cavity of lung of *Protopterus*, and (c) A single alveolus of *Protopterus*.

Respiratory System: Respiration is both aquatic and aerial. In *Protopterus* gill slits are 5 pairs while *Neoceratodus* has only 4 pairs. *Protopterus* mainly depends on aerial respiration. Air bladder acts as lungs. *Protopteus* and *Lepidosiren* have a pair of lungs but in *Neoceratodus* it is single (Figure 6.18).

Circulatory System: Heart enclosed in pericardium consists of sinus venonus, auricle, ventricle and conus arteriosus. Sinus venosus opens into auricle by a sinu-auricular opening. Auricles are communicated with ventricle by auriculoventricular aperture, which is plugged by aurianoventricular cushion.Ventricular cavity is simple. Conus arteriosus is spirally twisted. Four afferent branchial arteries arise from conus. Two efferent branchial arteries from each gill-bearing arch join to form 4 epibranchial arteries, which join to form a single dorsal aorta. Pulmonary arteries carry blood to lungs. *Neoceratodus* lacks second efferent branchial vessels. Pulmonary veins bring oxygenated blood from so-called lungs.

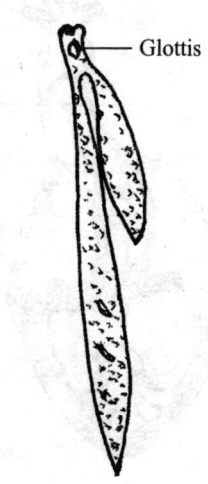

Figure 6.18 Lungs of *Protopterus* and *Polypterus*.

Pulmonary vein opens directly into left auricle. Second and 3rd gill arches bear no lamellae and their afferent and efferent branchial vessels are continuous, and blood flows from ventral to dorsal aorta and carotids. Pulmonary artery springs from dorsal aorta. Blood returns in a special pulmonary vein to partly separated left side of sinus venosus. Auricle is partly divided into 2 and ventricle is almost completely divided by a ridge and a series of muscular trabeculae. Ventral aorta is shortened into a spirally twisted truncus arteriosus, provided with a system of valves, and the blood from the left side of the auricle is directed mostly into the first 2 branchial arches, that from the right side into the last two. Some separation of pulmonary and systemic circulations is achieved. A coronary artery originates from the anterior efferent branchial arches. An inferior vena cava, a vessel collecting blood from the kidneys and reaching to heart by passing round to right gut in veins, joining the ductus Cuvieri, remain present. There is a renal portal system. The perirenal tissue of *Protopterus* is a mass of material around the kidney, containing lipid, steroid, and round-cell tissues and endothelial and pigment cells (Figures 6.19-6.22).

Nervous System: Telencephalon evaginated into a pair of cerebral hemispheres. Olfactory lobes are sessile. Diencephalons are relatively small and its roof is formed of saccus dorsalis. A pineal body is present. Hypothalamus bears small inferior lobes. Optic lobes fuse to form a single oval body. A lobe saccus endohymphaticus lies above medulla oblongata. Lateral line sense organ is ill developed. Forebrain is evaginated into a well-marked pair of cerebral hemispheres. The roof of these, though not very thick, is nervous and inverted type, not everted as in Actinopterygii. Optic lobes are little developed. The mesencephalon is hardly wider than the diencephalon. Cerebellum is small. The inner ear forms a special lobed saccus endolymphaticus, lying above the medulla oblongata (Figure 6.23).

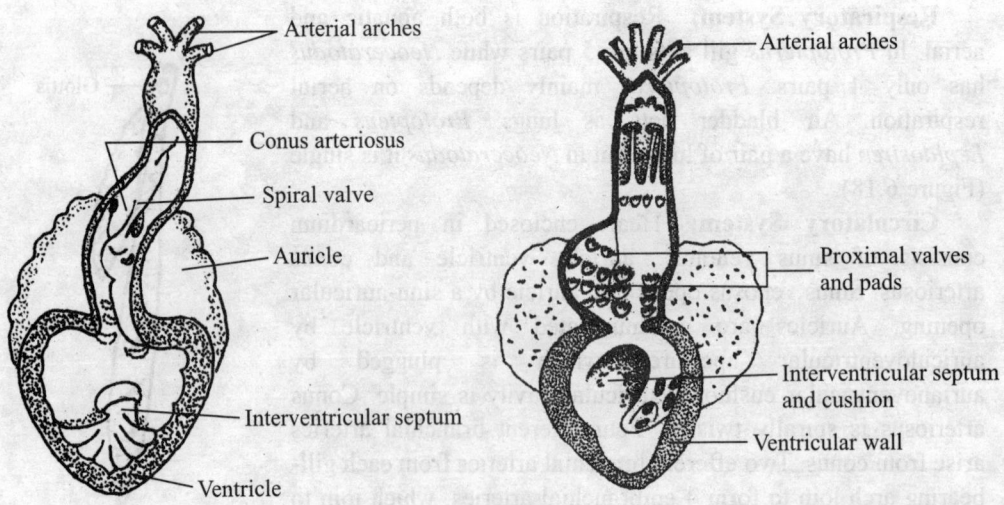

Figure 6.19 Diagrammatic view of the section of heart of *Protepterus*

Figure 6.20 Heart of *Neoceratodus* in L.S.

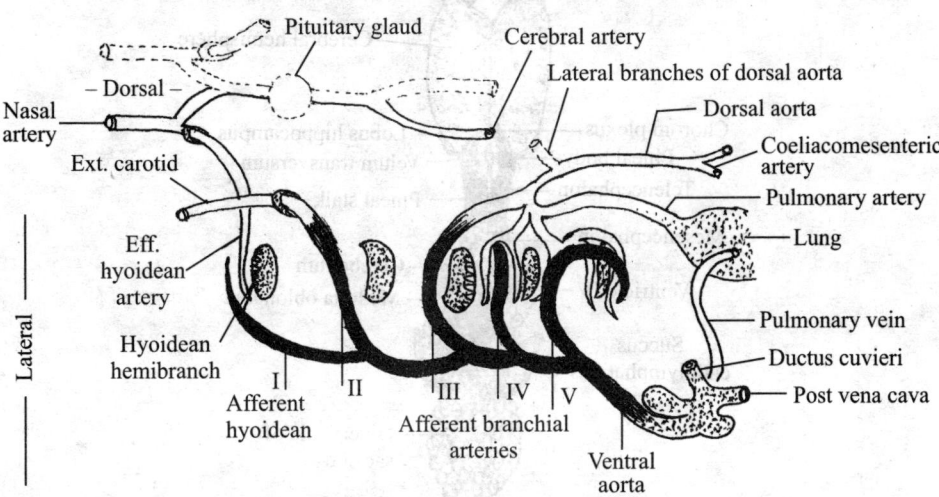

Figure 6.21 Blood vascular and respiratory systems.

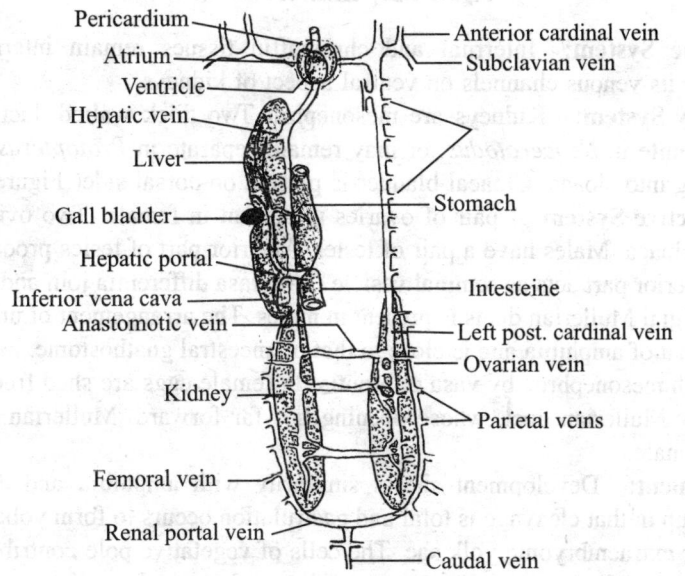

Figure 6.22 Visceral relationships of *Protopterus*.

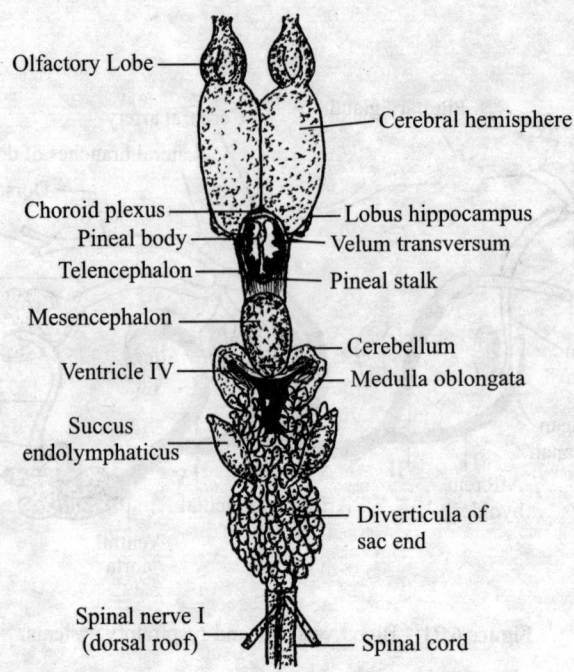

Figure 6.23　Brain of *Protopterus*.

Endocrine System: Interrnal and chromaffin tissues remain intermingled and are situated along its venous channels on ventral aspect of kidneys.

Excretory System: Kidneys are mesonephric.Two thick walled ducts, one from each kidney, may unite in *Neoceratodus*, or may remain separate in *Protopterus* and *Lepidosiren* before openmg into cloaca. Cloacal bladder is present on dorsal side(Figure 6.24a, 6.24b).

Reproductive System. A pair of ovaries is present in female .Two oviducts join before opening into cloaca. Males have a pair of testes. Anterior part of testes produces spermatozoa while the posterior part acts as seminal vesicle. Two vasa differentia join and open into cloaca. A pair of vestigial Mullerian ducts is present in males. The arrangement of urinogenital system is similar to that of amphibia and is close to that of ancestral gnathostome. In male, sperms are passed through mesonephros by vasa efferentia. In female eggs are shed free into coelom and carried out by Mullerian duct, whose opening lies far forward. Mullerian duct is very well developed in male.

Development: Development shows similarity with amphibia and dissimilarity from other fish group in that cleavage is total and gastrulation occurs to form yolk plug. There is no blastoderm or extraembryonic yolk sac. The cells of vegetative pole contribute little to shape of embryo and may form a partially separate yolk sac. Larvae show distinct similarity to those of amphibia.

Evolutionary Significance: Fossil remains of lungfishes appeared in early Devonian when air breathing habits were evolved by development of lung. Both primitive lungfishes and crossopterygians possessed functional lungs. In higher fishes, lungs migrated dorsally and

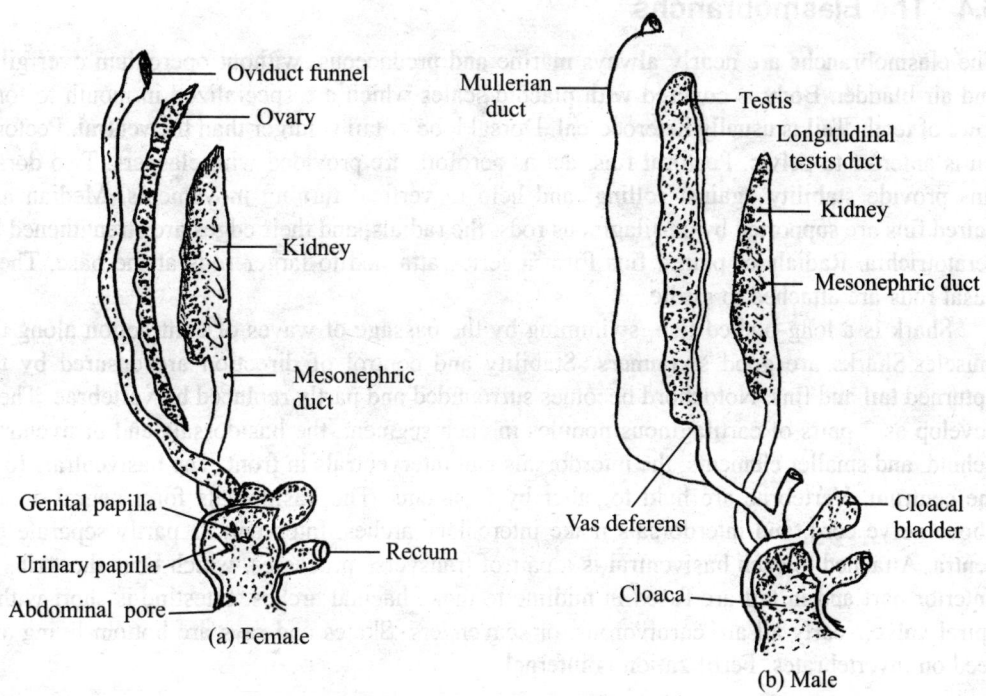

Figure 6.24 *Protopterus* : Urinogenital System.

converted into an organ with several functions (hydrostatic, gas storage, sound production) and its use for respiration was abandoned by most fishes. Lungfishes presumably arose from a common ancestral stock of Crossopterygii that led to coelacanths.

Jarvik argues that dipnoans agree in their structure with Palcoderms, Elasmobranchs and living amphibians as with crossopterygians, but a close agreement has not been established. Romer stated that lungfishes and basal stock (rhipidistians) of crossopterygii were clearly derived from a common ancestor. That dipnoans hold the direct ancestry of amphibian is difficult to interprete.Important obstacle is deviation of terrestrial limb from specialized archipterygial paired appendages of dipnoans.

In dipnoans, cloacal bladder develops from dorsal wall of cloaca, but in amphibians it develops from ventral wall. Earliest fossil amphibians show close similarities with osteolepis of Devonian than any extinct and surviving dipnoans. Amphibians have originated either directly from some rhipidistians, or from an unknown group closely related to rhipidistians, but never from dipnoan source. It is safe to state that dipnoans have diverged very early from remote basic stock from which amphibians actually originated. Rhipidistians and dipnoans are divergent offshoots of a common piscine ancestral stock. If the contention of emergence of amphibians from rhipidistian forms were taken as a fact, dipnonans can be regarded as "grand uncle" of land dwellers, but not the "grandfather". However, modern dipnoans give us a glimpse of condition that caused vertebrates to conquer.

6.4   The Elasmobranchs

The elasmobranchs are nearly always marine and predaceous, without operculum over gills, and air-bladder. Body is covered with placoid scales which are specialized in mouth to form rows of teeth. Tail is usually heterocercal. Dorsal lobe of tail is larger than the ventral. Pectoral fin is anterior to pelvic. Pectoral fins, act as aerofoil, are provided with claspers. Two dorsal fins provide stability against rolling, and help in vertical turning movements. Median and paired fins are supported by cartilaginous rods, the radials, and their edges are strengthened by ceratotrichia. Radials of paired fins form a series attached to larger rods at the base. These basal rods are attached to girdle.

Shark is a long-bodied fish, swimming by the passage of waves of contraction along the muscles.Sharks are good swimmers. Stability and control of direction are ensured by the upturned tail and fins. Notochord becomes surrounded and partly replaced by vertebrae. These develop as 2 pairs of cartilaginous nodules in each segment, the basidorsals and basiventrals behind, and smaller elements, the interdorsals and interventrals in front. The basiventrals form the centrum. Vertebrae are held together by ligaments. The basidorsals form neural arches above nerve cord, and interdorsals make intercalary arches. Interventrals partly separate the centra. Attached to each basiventral is a pair of transverse processes, which bear short ribs in anterior part and in tail are fused in midline to make haemal arches. Intestine is short with a spiral valve.Nearly all are carnivorous, or scavengers. Skates and rays are bottom-living and feed on invertebrates. Fertilization is internal.

Swimming

The propulsive forces produced by longitudinal myotomes move a fish. In some, movement of fins give the stability, help to maintain a constant course and to change its course forms the propulsion. Myotomes placed on either side of notochord, or vertebral column. Contraction of muscle-fibres bends the body. In forward swimming, the contraction of each myotome occurs after that in front of it.Waves of curvature passés down the body, alternately on each side. Number of waves per minute in steady swimming varies. A definite angle exists between surface of fish and its path of motion. Each side of fish moves like blade of an oar .The effect of movement is increased by the amplitude of oscillations passing backwards. The whole fish thus operates as single self-propelling system.

Magnitude of forward thrust depends on angle between the surface of fish and its path of motion, angle between surface of fish and axis of forward movement, and velocity of transverse movement of body. Body form of the fast moving bony fishes has advantages for swimming over that of elongated types. Bony fishes have large caudal fins, smaller body length relative to its depth, and less flexibility. Large caudal fin resist transverse movements; it keep the leading surface of body directed obliquely backwards during both phases of transverse movements and exert a steady pressure on water. Tail exerts large propulsive effect. The effect of caudal fin, combined with shortness of body and reduced flexibility, help the front part of a bony fish to do small transverse movements.

Equilibrium: A fish remains in stable motion if it varies slightly from its line of progress. Forces acting on fish are measured along 3 primary axes, longitudinal, horizontal, and vertical. Deviation from line of motion about longitudinal axis is called rolling, about the transverse axis as pitching, and about the vertical axis as yawing. Forces along these 3 axes are called drag, lateral force, and lift. Shape of dorsal fins maintains stable swimming, and prevents yawing. Turning of a fish is due to propagation of a wave through body or by asymmetrical braking with pectoral fins. In elasmobranchs and teleosts, the dorsal fins are well developed in active swimmers. In most elasmobranchs, they are fixed, but in many teleosts, the dorsal fin folded up and down, and the fin is raised during turning. This result from increasing yawing produces asymmetrical action of body muscles, or by unilateral braking with pectoral fins.

Skin: Sharks lack heavy external ormament. Skin is tough, covered by epidermis. Beneath this is a thick dermis. Denticles or placoid scales are scattered on skin. Denticle consists of a pulp cavity, around the edge of which a layer of odontoblasts exists. Odontoblast secretes the calcareous matter of scale, called dentine. Odontoblasts send fine processes throughout denticle. The outside, the dentine is covered by enamel, secreted by ectoderm. Neither denticles pierce through ectoderm, after which enamel is not added to their surface. Scales are similar to teeth, which are specialized denticles. Skin colour gives protection. Chromatophores produce colour. Many sharks have spotted, or wavy pattern. They change their colour, become darker on a dark background.

Skull and Branchial Arches: Modern elasmobranchs show the skull and jaws in modified and reduced condition. Skull of dogfish consists of cartilaginous boxes surrounding the brain and receptor organs. The nasal capsules, orbital ridges, and auditory capsules are largely fused with the main cranium, producing, the chondrocranium. Head has probably arisen by modification of a segmental arrangement. The first rudiment of skull in embryo consists of 2 pairs of cartilaginous rods, the parachordals and trabeculae. The former lies on either side of notochord, or the trabeculae in front of the notochord. These first rods fuse up to make a continuous plate; from this grow sides and roof, completing the cartilaginous neurocranium around the brain.

Cartilaginous capsules form around nose, eyes, and ears, and joined to neurocranium. Posteriorly, behind the auditory capsules, the cranium is completed by addition of several segmented elements, evidently modified vertebrae. Visceral or branchial arches are pairs of rods of cartilage developed in walls of mouth and pharynx. In dogfish, each branchial arch consists of 4 pieces, the pharyngo-, epi-, cerato-, and hypo-branchials. Ventrally some arches join a median basibranchial plate. These rods lie in pharynx wall and carry a series of projecting rods, the branchial rays and extrabranchial cartilages. Five such branchial arches differ only slightly from each other. In front of these lie the hyoid and mandibular. Hyoid resembles a typical branchial arch. Hyomandibular is a thick rod attached dorsally to skull by ligaments and at its lower end forming the support for hind end of jaw. The more ventral elements, cerato- and basihyal, are similar to corresponding members of more posterior arches.

The Jaws: The front end of palatopterygoquadrate bar is attached to cranium in dogfish. In most elasmobranchs, the hind end of upper jaw is not fixed to cranium. Hyostylic support was supposed to have been the original one. Early elasmobranchs lack a hyostylic jaw support, but upper jaw is attached to cranium and supported by hyomandibula. This amphistylic condition persists today in primitive shark *Hexanchus*. As sharks eat larger fish, those in which the hind end of upper jaw was less firmly fixed to skull were more successful and the hyostylic condition was thus evolved. Many sharks have 2 pairs of labial cartilages, which represent arches. This is the trabeculae cranii, the rods lying on each side in front of parachordals. These rods are not part of axial skeleton. The main axis of body presumably ends at front end of notochord. Branchial arches, hyoid, jaws, and trabeculae all constitute a single series suitably modified at each level.

Prootic Somites and Eye Muscles: Three myotomes, the pro-otic somites, can be recognized during development. The auditory sac becomes surrounded by cartilage. In the spinal region the dorsal and ventral roots join .In the head region the dorsal and ventral roots remain separate. The head preserves the relic of ancient condition. Branchial nerves, such as glossopharyngeal, show signs of this condition. Each has a small pretrematic branch in front of slit, a larger post-trematic branch behind it, and a pharyngeal branch to wall of pharynx. Pre-trematic branch contains sensory fibres from skin. Pharyngeal branch visceral sensory fibres include those from taste buds.

Post- trematic branch contains motor and sensory fibres. The branchial nerves provide dorsal rami to skin of back.Three prootic somites form 6 extrinsic eye muscles. Four recti roll the eye straight upwards, downwards, forwards, or backwards, and 2 obliques, lying farther forward. The superior, anterior, and inferior rectus and inferior oblique muscles are all derived from first myotome and are innervated by oculomotor nerve. Superior oblique, innervated by trochlear nerve is derivative of second and posterior rectus.

Gut: Digestive system of shark consists of true stomach. Teeth consist of rows of backwardly directed denticles. They are carried on special folds of skin lining the jaws and are replaced. The 'gill rakers' are rods attached to the branchial cartilages, and prevent the escape of prey. The basihyal supports a short non-protrusible tongue. The wall of pharynx is lined by a stratified epithelium on which numerous mucus glands open. The mucus helps the passage of food. Pharynx narrows to an esophagus, leading without sharp transition to stomach.

The stomach is formed as special portion of esophagus. It serves as a receptacle for large pieces of food. The mucus glands modify to produce acid. Pepsin digests proteins. In the gastric glands only one type of cell is found. The stomach is divided into a descending cardiac and ascending pyloric limb. The pyloric sphincter guards the region where the stomach joins the intestine powerful. Liver, a large 2 lobed organ stores glycogen and fat and plays role in destruction of red blood corpuscles. Bile is carried away to a gall bladder. The pancreas, an elongated body between stomach and intestine contains exocrine and endocrine cells. Its duct enters the intestine shortly below the pylorus. The small intestine is short but with its surface greatly increased by a spiral ridge. The intestinal contents are alkaline and contain trypsin, amylase, and lipase. Absorption occurs wholly in this organ. A short rectum is attached to the rectal gland.

Circulatory System: Heart develops as a specialization of subintestinal vessel and consists of sinus venosus, atrium, and ventricle. The conus arteriosus is provided with valves. Five afferent branchial arteries carry blood to gill lamellae, from where it is collected by 4 efferents and connecting vessels into a median dorsal aorta. Oxygenated blood is supplied to the head from 3 sources. A carotid artery leaves the efferent branchial runs towards the midline and divides into an external carotid to upper jaw and internal carotid to brain. Dorsal aorta divides at its front end into branches, which join the carotids. A hyoidean artery carries oxygenated blood to spiracle.

Heart is supplied by a cardiac artery. Blood pressure in ventral aorta is 30-40 mm Hg and there is a drop of 10-20 mm Hg across the gills. Pericardium is enclosed in a cartilaginous framework by basibranchial plate above and pectoral girdle below it. The pericardio-peritoneal canal runs from pericardium to abdominal coelom. A caudal sinus from the tail opens into a renal portal system. There are large sinuses like the anterior cardinal sinus, running above the gills, and the jugular, lateral cardinal, subclavian, and other sinuses. Heart muscles can contract in one direction only and each chamber therefore needs to be actively dilated. A cardiac branch from the vagus ends in a plexus in the sinus venosus. Stimulation of this nerve slows the heart. Abundant sympathetic fibres run to arteries. There are receptors in efferent branchial vessels and in the post- branchial plexus above the cardinal veins.

Urinogenital System: The blood has very high content of urea and is isotonic with surrounding sea water. However there is less salt in blood than in sea. Gill surfaces are not permeable to urea. In the ordinary marine elasmobranchs, the high urea concentration is due to the presence of a special urea-absorbing section of kidney tubules. Urinary apparatus is a mesonephros. The hinder part of kidney consists of a mass of tubules ending in very large glomeruli. Each tubule absorbs the urea. The tubules join to form 5 urinary ducts and these enter a urinary sinus. The sinus is similar to a bladder.

The small urinary sinus excretes small volume of liquid. The genital system is specialized for internal fertilization and produces few yolkv, well-protected eggs. There is a single large ovary. Testes are paired and sperms are collected at their front ends by vasa efferentia which consists of coiled, thick-walled, vas deferens, whose glands secrete material for aggregating the sperms into spermatophores. Transmission of sperms is done by claspers. These are inserted into female cloaca. Development is by partial cleavage, producing a blastoderm, perched on the top of a large mass of yolk. The egg is protected by an elaborate egg-case, the 'mermaid's purse', within which development proceeds until the yolk has been used up. In some, development is viviparous, the oviduct forming a 'uterus'. In Mustelus, there is a yolk-sac placenta. In Trygon 'uterine milk' is secreted into embryo by villi.

Nervous System: The brain is large and well developed. The forebrain is large and has cerebral hemispheres thickened both in floor and roof. The hemispheres are wide relative to their length. Attached to the ends of the cerebral hemispheres are large olfactory bulbs and large nasal sacs. The olfactory sense is well developed. Cortical arrangement of tissue is not found in hemispheres. The cells form thick masses around ventricle. The roof is thick and contains decussating fibres in midline. The sides and floor make up the main bulk of organ. The lateral wall is called the striatum, its upper part the epistriatum. The medial wall is known as septum and its upper portion is called the primordium hippocampi. The main efferent pathways are tracts leading to the hypothalamus and to the optic lobes.

The diencephalon is a narrow band of tissue. The lower part of between-brain, the hypothalamus, is well developed. Its hind part receives olfactory impulses via forebrain and gustatory pathways from the medulla. Its efferent fibres run to reticular centres. Anterior part of hypothalamus lies above the pituitary and contains the supraoptic nucleus. The anterior hypothalamus is a higher centre for visceral control, regulating circulation, respiration, and many metabolic activities. Attached to the hind end of the hypothalamus is the saccus vasculosus, with folded, pigmented walls. This perhaps acts as a pressure receptor.

The midbrain is very large and is perhaps the dominant centre of the brain. The optic tracts end in its roof after complete decussation below the brain. The cells of the tectum are arranged in a complicated pattern of layers. Other sensory centres that send tracts to the optic lobes are the olfactory, acustico-lateral, cerebellar, gustatory, and probably also the general cutaneous centres of the spinal cord. The cerebellum is a large organ. Its main source of sensory fibres is from the ear and from the organs of the lateral line system. The internal structure of cerebellum is uniform. From the medulla oblongata most cranial nerves spring and especially those that regulates the respiration and visceral functions.

Receptor Organs: The paired nasal sacs have much folded walls. Taste-buds remain scattered over the pharynx wall are receptors for food. Nose acts as a distance receptor. Smell and taste are different senses for a dogfish. Eyes are well developed in sharks. The retina contains only rods, and visual discrimination is probably poor. There is often a reflecting layer, the tapetum lucidum with pigment cells, which expand in light but contract in darkness.

The lens is spherical and very hard and provided with a protractor-lentis muscle. The iris is peculiar in some which hunt by day. The muscles of the iris are better developed. The sphincter iridis muscle narrows the pupil. Radial dilatator fibres receive motor-fibres from oculomotor nerve. Closure of iris when illuminated is slow. Ear contains receptors for maintenance of muscle tone, with angular accelerations, with gravity, perhaps with hearing. There are 3 pairs of semi- circular canals, each with an ampulla containing receptor cells, whose hairs are embedded in a gelatinous cupula.

6.5   Freshwater Elasmobranchs

They are known for centuries in four habitats. The earliest xenacanthiformes arose in the upper Devonian. *Orthacanthus* attained a length of about 3 m. *Xenacanths* were widely distributed in Europe, North America, and East Asia for about 200 million years in freshwater until the end of Triassic. In Permian and Triassic, freshwater and brackish Hybodontiformes replaced the Xenacanths. Deriving from protoselachians, *Hybodonts* are distributed in freshwater and marine habitat of Europe, Greenland, North and South America, and southern Asia. Freshwater hybodonts were small. *Lissodus* was only 15 cm in length. Hybodonts persisted in freshwater until the late Cretaceous. Pristoids may have arisen as early as the late Cretaceous, and the first pristid was probably euryhaline. Myliobatoids arose in sea in the late Cretaceous. Some invaded fresh waters by the early Eocene.

About 45 species of elasmobranch, in 4 families and 10 genera dominated by potamotrygonid and dasyatid stingrays comprising almost half of freshwater elasmobranchs. At least 48 additional elasmobranch species penetrate fresh water in estuaries or river mouths. Extant obligate euryhaline and freshwater elasmobranchs comprising 3 ecomorphotypes are restricted to tropical rivers and lakes. Freshwater may be a marginal habitat for elasmobranchs. Some elasmobranchs occur in the Mississippi River in USA, but most occur in tropics of both hemispheres. The rich diversity and endemism occurs in the Atlantic drainages of South America with radiation of Potamotrygonidae. Scattered endemism and diversity occur in West

Africa and in Asia. They also occur in Tigris River system. Most occur in tropical rivers and lakes of developing countries.

Increasing levels of exploitation threaten many elasmobranch stocks and obligate freshwater species. Euryhaline elasmobranchs may be less vulnerable than obligate freshwater species. Certain euryhaline elasmobranchs reproduce in fresh water and are affected by anthropogenic problems. Ray-finned fishes inhabit extreme environments including high altitude lakes and streams, desert springs, subterranean caves, ephemeral pools, and polar seas. Some fish spend considerable time outside water. Mudskippers prey on invertebrates of mudflat habitats.

Air breathing catfishes and gouramies live in stagnant, low oxygen ponds. Dragon fishes, deep-sea codfishes and spiny eels produce lights to find prey, communicate or for defense .In general, 5 developmental periods in fish are embryonic, larval, juvenile, adult, and senescent. Species in several teleostean families bear live young. Poeciliidae, Scorpaenidae, and the young in some families emerge as juveniles after hatching from egg. Two developmental characteristics that separate fish from most vertebrates are indeterminate growth and larval stage. Most fish grow by change in anatomy, ecological requirements, and reproduction. Larval stage is usually associated with a period of dispersal from the parental habitat.

Nearly all bony coral reef fishes produce pelagic young, and the length of stage is highly variable. Initially, researchers made simplistic assumptions about pelagic phase, "portray larvae as little more than passive tracers of water movement that 'go with flow,' doing nothing much until they bump into a reef by chance and settle at once" . Larvae of most coral reef fishes have good swimming power, well developed sensory systems, and fine tuned behavior. Mortality rates are high at this stage, but many larvae detect predators at a distance, and they are often transparent or cryptically colored. Young of reef fishes develop differently from most temperate fishes. While eggs of most temperate fishes hatch within 20 days after laying, the eggs of most coral reef species hatch within a day.

Lifespan of ray-finned fishes varies widely. Smaller fish have shorter lives and vice versa. Pimephales live for a year or less. European perch can live 25 years. Sturgeons live between 80 to 150 years. Rockfish species live up to 140 years. Aggressive behaviors like dominance hierarchies are found. Hierarchy is determined by size, sex, age, previous residency, and experience. Males dominate females, and dominant individuals have favorable habitats, higher feeding rates and remain dominant. Territoriality found in numerous ray-finned fishes primarily occurs along territorial boundaries.

Communication and Perception

They perceive the environment through vision, magnetic reception, mechanoreception in addition to chemoreception, and electroreception. Perception is used to communicate with individuals of the same or other species. Eyes recognize broad range wavelengths. Ability to sense various wavelengths corresponds to depth at which it lives.

Several deep sea fishes have elongate, upward-pointing, tubular eyes that enhance light gathering and binocular vision, providing better depth perception. Fishes also communicate visually through dynamic display, which involves color change and rapid, highly stereotyped movements of body, fins, opercula, and mouth. Visual communication also occurs through light production. Lantern fishes have rows of lights along the underside of body, probably for mating and foraging. Even shallow-water species of the Red Sea use internal light sources to form nighttime feeding shoals.

Mechanoreception includes equilibrium and balance, hearing, tactile sensation, and a 'distance-touch-sense'. Ray-finned fishes have otoliths in the inner ear. Many ray-finned fish have modified gas bladders and swim bladders adjacent to inner ear. Most have keen hearing ability and sound production is common which is used in mating and communication between shoal mates. Swim bladder is source of complex forms of sound production. The lateral line detects turbulence, vibrations and pressure in water, acting as a close-quarters radar. This sensation is important in formation of schools, because consistent positioning is essential for turbulence reduction and smooth hydrodynamic functioning.

Chemoreception involves both smell and taste. Many fishes use chemical cues to find food. Taste buds are scattered around the lips, mouth, pharynx, and even the gill arches. Barbels are used for taste reception. Nares detect pheromones. Pheromones are detected by conspecifics, and sometimes closely related species. Pheromones allow recognizing specific habitats, members of the same species or of the opposite sex, young, predators. Some groups in dominance hierarchies associate the scents of individuals with their particular ranking. Cyprinids identify 'fear scents,' which are pheromones.

Some fishes have specialized organs for electro reception. Somel groups can identify weak electrical currents emitted by organs, such as the heart, and locate prey buried in sediment. Elephant fishes create weak electrical field that functions like radar, help to navigate. Few highly migratory ray-finned fish can detect earth magnetic fields directly. Magnetic perception helps fish locate long distance migration routes for both feeding and reproduction. Ray-finned fishes display complexity in ability to perceive their environment and communicate with other individuals.

6.6 The Bony Fish

Osteichthyes are commonly called bony fish, the largest group of vertebrate .Most bony fish are ray-finned fish, living lobe-finned fish species are 8 in number. Dermal bones cover the pectoral girdle. A sclerotic structure of 4 small bones supports the eyeball. The inner ear contains labyrinth. The cranium is often divided into anterior and posterior sections. Bony fish have swim bladder but do not have fin spine, instead the lepidotrichia. Bony fish lack placoid scale. Ganoid, cycloid, or cytenoid scales are found which are smooth and overlapping. Mucus glands coat the body. They are omnivorous carnivorous, herbivorous, or detritivorous. Some are bisexual. Fertilization is external, or internal. Actinopterygians are the largest group of fishes comprise about 96% of all living fish species and make up half of all living vertebrates. They appeared during Devonian and in Carboniferous, they become dominant in freshwater and started to invade the seas. At present, about 42 orders are recognized. Habitat destruction, pollution and international trade, and other human activity have endangerd actinopterygians.

Calcified bony skeleton provides better muscle leverage. Lobe-finned fish appears in continental fresh water and then moved in the oceans. They developed the water-proofing and shallow-water adaptations. This group becomes slow-moving but heavily-armored. Bony fish bears the marks of mastery of water. They avoid enemies by subtle changes of colour. Elaborate eyes, ears, and chemical receptors meet emergencies. Reproductive mechanisms involve nest-building and parental care. At least 3,000 million herrings are caught in the Atlantic Ocean each year. Thousand million blue-fish collect every summer off the Atlantic coast of the United States.

Salmo trutta, the brown trout is abundant in rivers and streams throughout Europe and is about 20 cm long at maturity. It is grey above and yellowish below, with several dark spots. Body form is short, narrow in lateral plane but deep dorsoventrally. Movements do not involve bending of body into an S. Swimming is carried out by propagation of waves along body by contraction of myotomes. Tail is outwardly symmetrical. Besides caudal 'fish-tail', there are 2 dorsal fins and a ventral fin. The hinder dorsal fin has no rays called an adipose fin. Paired fins are small .There is no lobe projecting from the body. Basal apparatus of fin is found in body wall. Fin rays project outwards. Pelvic fin often lies far forward; in trout it is unusually far back.

Skin

It consists of thin epidermis and thick dermis. The former has stratified squamous layers but without keratin. It contains mucus glands. Mucus of some eels precipitates mud of turbid water. Mesodermal dermis contains smooth muscle, nerves, chromatophores, and scales. Scales are thin overlapping bony plates. Exposed part of each scale bears pigment cells. Bone of scales is absorbed at intervals by scleroblasts, making a series of rings. In adult salmon, no growth occurs in fresh water. Head shows specialized teleostean features. There are 2 nostrils on each side, but no external ears. Mouth is very large and movable bones support its edges. Maxillary and mandibular valves are folds of buccal mucosa which prevent the exit of water during respiration. Tongue has no muscles, but may have teeth and taste-buds. Behind the edge of the jaw is the operculum.

Skull

Skull includes chondrocranium and set of branchial arches. In early stages, there are cartilaginous boxes around the nasal and auditory capsules, brain and eyes, and cartilaginous rods in gill arches. Bones are then added either as cartilage bones by replacement of parts of chondrocranium, or as membrane or dermal bones. Skull bones show regular pattern in arrangement. Four classes of dermal bones called canal bones, tooth bones, 'ordinary' bones, extra bones, fill special areas (Wormian bones). At hind end of skull, the floor ossifies as basioccipital, the sides as exoccipitals, and roof over spinal cord as supra-occipital bone. These posterior bones are not well marked off from each other in adult.

Of 5 separate otic bones in auditory capsule, epiotic and pterotic can be seen externally. The floor in front of basioccipital is occupied by a basisphenoid bone and walls above this by alisphenoids.Eyes meet in midline and orbits are here separated by thin orbitosphenoid. The more anterior part of chondrocranium forms ectethmoid. Dermal bones that cover this partly ossified neurocranium are frontal, supraethmoid, parietal and small post-parietal. A ring of circumorbitals are seen around eyes, and on floor of skull parasphenoid and vomer are found. Jaw bones include both endochondral and dermal elementsr. In embryo, palatopterygoquadrate bars and Meckel's cartilages are seen. The upper jaw bears inward projections. However, the effective support in adult is provided by ossified hyomandibular cartilage. The palato-pterygo-quadrate bar ossifies in several parts. Palatine, pterygoid, mesopterygoid, metapterygoid, and quadrate bones appear, some of them partly formed in membrane. The only part of Meckel's cartilage to ossify is the articular bone at hind end.

The actual edges of jaws are supported by premaxilla, maxilla, and jugal, covering the upper jaw. Dentary covers most of the lower jaw, except for an angular. Hyomandibular runs from an articulation with otic capsule to upper end of quadrate. Symplectic is a small ossification at lower end of hyomandibula. The rest of hyoid arch is present as epi-, cerato, and hypohyals. Branchial arches are formed of several pieces. Teeth are found on vomers, palatines, premaxillae, maxillae, dentary, and on tongue. Covering the typical dentine is a layer of harder vitrodentine.

Respiration: Respiration is carried out by a current passing in at mouth and out over gill lamellae. Pumping action, produced by buccal pressure pump and opercular suction pumps results from sideways movements of operculum. Branchiostegal folds prevent inflow of water from behind. When operculum moves inward, dorsal and ventral flaps in throat prevent the exit of water forwards. Gill lamellae have reduced septum between the respiratory surfaces. This has effect of leaving lamellae as free flaps, increasing the surface for respiration. Area of gills varies greatly; being relatively larger in active species. Respiration rate is controlled by medullary centre. Respiratory surface is an important limiting factor in movement and growth of fishes.

Vertebral Column and Fins: Vertebral column prevents shortening of body, when the longitudinal muscles contract. It is complicated, extensively ossified with ribs and neural and haemal arches and forms a body form for fast swimming. There is one vertebra in each segment. Each vertebra consists of centrum, neural arch and neural spine, and in tail region, haemal arch and haemal spine. These parts are formed partly by ossification of basidorsal and basiventral, interdorsal and interventral, and partly by extra ossification in sclerogenous tissue around notochord and nerve cord and between muscles. Vertebrae are intersegmental.

Centra are concave both in front and behind, and in the hollows between them are pads. Extra processes on front and back of vertebrae ensure articulation. Ribs are pleural ribs between the muscles and lining of abdominal cavity, and more dorsal intramuscular ribs. Both sorts are attached to centrum. Bony rods attached above the neural and below haemal arches are called neural and haemal spines. They form supporting rods or radials of median fins and are usually divided into 2 or 3 separate bones in each segment. Fins are also supported by bony rods. The dermal fin rays are forked at their tips. In the tail region, internal skeleton is not symmetrical. Notochord turns up sharply at the tip. Neural spines are very shorter than haemal spines, known as hippural bones. The last portion of notochord is surrounded by urostyle. Tail with internal asymmetry but external symmetry is called homocercal. Myotomes are arranged in a complicated pattern. In fast swimmers like the tunny, each myotome may overlap as many as 19 vertebrae. Pectoral girdle consists of cartilaginous endoskeletal portion in which ossify the scapula, coracoid, and sometimes mesocoracoid, while dermal bones, large cleithrum, and one or more small clavicles, become attached superficially. The supra-clavicle and post-temporal, attach the pectoral girdle to otic region of skull. Pelvic girdle is simple, consisting only of single bone, the basipterygium.

Alimentary Canal: Esophageal sphincter guards the entrance to stomach which is divided into cardiac and pyloric portions. Wide-mouthed pyloric caeca beset the duodenum. Intestine and caeca are lined throughout by columnar epithelium. Specialized multicellular glands are absent. Exocrine pancreas consists of diffuse glands in mesentery. Endocrine portion forms a compact mass very rich in insulin. Intestine is longer and often coiled. Its internal surface may be increased by folds, but without true spiral valve. There is no gland attached to rectum.

Air-bladder: Dorsal to gut a large sac with whitish walls, the air-bladder, filled with oxygen is found. Pneumatic duct connects this with pharynx in primitive forms. It serves as hydrostatic organ, enabling them to remain suspended in water at any depth.

Circulatory System: There is a single circuit .Blood passes through at least 2 sets of capillaries. Heart contains sinus, auricle, and ventricle. Muscular conus arteriosus is absent. Thin-walled bulbus arteriosus is present. Walls of bulbus are elastic but not muscular. It is dilated by ventricular beat and then contracts. Ventral aorta is short. Blood pressure in ventral aorta is less than 40 mm Hg at rest, and in dorsal aorta about half of this.Venous pressures are around zero. Connection between pericardial and peritoneal chamber is absent. There is a vagal cardiac depressor nerve, but no sympathetic nerve to heart. Well-developed lymphatic system is found beneath the skin in muscles and viscera. Lymphoid tissue is abundant but no lymph nodes along vessels. Large spleen precedes the kidneys. Red cells are small. White cells are present; acidic and basic granules may occur in same cell.

Urinogenital System and Osmoregulation: Kidneys are mesonephric in adult. Ducts of 2 kidneys join posteriorly and form a bladder. Urinary duct opens separately behind anus. Common cloaca is absent. Nitrogenous elimination is function primarily of gills. Kidneys excrete creatine, uric acid, and trimethylamine oxide. In fresh water fish, the blood is more dilute, about 6% NaCl, but is concentrated than surrounding medium. Freshwater fish take up salts from water. Skin is little vascularized and makes an almost waterproof layer.

Mucus helps in water proofing. Abundant mucus is secreted when an eel is transferred from salt to fresh water. In marine teleosts, the problem is opposite one of keeping water in, or keeping out salt. Genital system is about completely separated from excretory in both sexes. Testes are large pair of sacs opening into base of urinary ducts. Ovaries are elongated in trout. Eggs are shed free into coelom and passed to exterior by abdominal pores. Fertilization is external. Eggs of trout are shed in small pits or depressions in sand; being sticky, they become attached to small stones. Eggs are yolky and cleavage is only partial, forming the blastoderm, which differentiates into embryo. After hatching, the young fish may carry the yolk sac and obtain food from it for sometime

Breeding Habits: In trout and salmon, the adult spent the most of life in sea but returned to rivers to breed. Trout and salmon abundant on West Atlantic coasts ascend to rivers to breed, the process called the 'run'. During breeding-season changes take place. In salmon, jaws become long, thin, and hooked, especially in male. Fishes make pairs. Males fight with others that approach female. As gonads ripen, the other parts of fish become more watery.

Spawning takes place, the female laying eggs in a shallow trough. Male sheds sperms over them. She then covers the eggs with gravel. Young male salmon may become sexually mature, accompany fully grown fish, hanging around cloacal region and shedding their sperms at same time as large male. Very young trout or salmon are called alevins or fry. When they emerge they are called parr. After 2 to 4 years, salmon acquire a silver colour and pass to sea as smolts. Young salmon returning for the first time to breed are called maidens. If they have spent only one and a half year in sea, they are called grilse, and may then return to sea as kelts.

Endocrine Glands: Neural and glandular regions are present. Adenohypophysis has 3 parts, the 2 more posterior are intermediate and anterior lobes. The most anterior glandular region is similar to pars tuberalis. Posterior lobe secretes a melanophore-dispersing hormone. There may be melanophore-concentrating one in anterior lobe. Oxytocin and vasopressin are present. Thyroid secretes mono- and di-iodotyrosine and thyroxin. Thyroid follicles are found in kidneys, heart, and eye. Rennin is found in freshwater teleosts, but not in marine ones. Gonads produce steroid hormones. Neurosecretory cells of spinal cord called urohypophysis function in salt regulation.

Brain: Brain is built on the same general plan like elasmobranchs. Forebrain is often large and has characterized chiasma opticum. This condition is called 'eversion'. Forebrain mostly a smell brain is reached by olfactory fibres. Diencephalon is not large and may contain light sensitive receptors. Light-sensitive cells are found in walls of diencephalon. Hypothalamus is well developed and receives large tracts from forebrain. Midbrain is the largest part of brain. Cells spread out over its roof are not all collected round the ventricle. Into this midbrain cortex the great optic tracts, the ascending tracts from the sensory regions of spinal cord come Large motor tracts pass back towards spinal cord. Midbrain mediates elaborate acts of learning. The base of midbrain contains motor centres. Cerebellum is very large. Medulla oblongata is well developed, with special lobes connected with entry of lateral line nerves and gustatory fibres of cranial nerves. In gurnard, Trigla, chemical receptors in elongated fins are found.

Receptors: The lateral line system provides a system of 'distant touch'. Localizing the distant objects by distant touch is assisted by echolocation. Smell is analyzed by a distinct system in the forebrain.

Eyes: The greatest sensitivity of eye is in the yellow-green. Teleosts have developed retinas with distinct rods, cones, and twin cones, and in some fovea are composed of numerous thin cones. Pupil varies little in diameter, and adjustment of sensitivity is caused by migration of pigment scleral cartilage. In bright light, the pigment expands, cones contract forward, towards the light and the rods contract back. A large, dense, spherical lens is attached to a retractor muscle, which occupies the persistent choroidal fissure in retina. Eye is myopic at rest .Trout has a round pupil. Some fishes have mobile iris. Sympathetic system sends branches into the head and its fibres contracts the sphincter of the iris. In the eel, the pupil can changes widely in diameter. Eyes may small or absent in fishes living in caves, muddy waters, or deep sea.

Ear: Ear provides receptors for maintenance of a correct position in relation to gravity and to angular accelerations. Mostly it serves for hearing. The inner ear is enclosed in otic bones. A perilymphatic space is found in species that hear well. Each ear sac is divided into 3 semicircular canals and 3 chambers, the utriculus, sacculus, and lagena. Each chamber carries an ear stone (otolith) and these are called the lapillus, sagitta, and asteriscus. Macula of the utricle lies horizontally, with the lapillus resting on it. Maculae of the saccule and lagena are vertical. These receptors with otoliths have double, or triple functions. At rest, they are static receptors, signalling the position of fish in relation to gravity and setting the fins and eyes in suitable positions. Some otolith organs respond to sonic vibrations.

Sound Production: Some fish use the sounds for echolocation. Drum-fish (*Pogonias*) of the Eastern Atlantic produce loudest sound. The 'whistling' and other noises of the 'maigre' (*Sciaend*) are also mentionable. In these fishes, the sounds are made in the breeding season. The mechanism for sound production is varied, involving stridulation by vertebrae, operculum, pectoral girdle, teeth, or phonation by air bladder. Sound production is common in some families.

Lateral Line Organs: Lateral line organs occur partly as distinct pits, and partly in canals. Besides the main canal running down the body and served by lateral line branch of tenth cranial nerve, there are lines on head, namely, supra- and sub-orbital lines, a line on lower jaw, and a temporal line across back of skull. The canals on head are innervated mainly from seventh, partly from ninth cranial nerve. Nerve fibres enter acoustico-lateral centres of medulla and valvula cerebelli. Lateral line system has great importance in aquatic life. It is found in all types of fishes.

Protective Mechanisms

Teleosts depend on swift swimming, powerful jaws, good receptors, and brain for protection. The protective devices may be the protective armour of body surface, sharp spines and poison glands, electric organs, luminous organs, coloration and scales, and other surface armour. The typical cycloid form a covering of thin overlapping bony plates, providing some measure of protection. The hinder edges of scales are sometimes provided with rows of spines. In some, scales bear upstanding spines and denticles. In the tropical globe-fishes and porcupine- fishes, spines are very long and sharp. Puffers can cause the spines to project outwards. In coffer fishes, the scales are enlarged and thickened into a rigid box.

Spines and Poison Glands: European weever has poison spines on the operculum and the dorsal fins. Dark colour of the fins serves as a warning. Some catfishes, scorpion fishes, and toad-fishes also have poison spines. Uranoscopus has powerful spines on operculum, which inflict unpleasant wound. Lophius is also armed with dangerous spines. Several catfish have large spines, sometimes serrated. In the trigger-fishes one or more of the fins make a spine that can be raised and locked in that position.

Electric Organs: Electric organs arise bilaterally from modified muscle fibres, the electroplaques. Each plate is innervated on only one surface by motor neurons. In some, there is a controlling nucleus in medulla. Electric organs can develop appreciable voltages in surrounding fluid, up to 550 volts in *Electrophorus*. These voltages are achieved by series summation of electromotive forces generated by individual cells. The columnar array of several hundreds or thousands of electroplaques in series in the strongly electric fish, *Electrophorus*, *Malapterurus* and *Torpedo* are paralleled so that the electric organs of these fish can generate considerable current at high voltage. Electric organs (EOs) are located on tail or along the side of body. Some possess accessory EOs on head. In most electric fish, they consist of modified, non-contractile muscle fibres. Gymnotiforms and mormyriforms that are preyed on by electroreceptive catfish or electric eels have adjusted the frequency of their EODs to minimize the risk of detection by these predators.

Luminous Organs: About 95% fishes live below 100 fathoms are luminescent. The luminescent organs show as shining beads of various colours on sides and ventral surface of fish. In many, light is produced by luminous bacteria. Some have self-luminous photophores and these are also found in *Spinax* and in Squalidae. Formed from modified mucus glands, they may be provided with reflectors and lenses. They can be flashed on and off by sympathetic stimulation. Luminous organs serve for recognition of sexes and show distinctive patterns.

Colour Change in Teleosts: Melanophores cause contraction of the pigment and a paling of skin colour. Colour change is produced by nerve-fibres tending to make the animals pale and secretion of the posterior pituitary to make them dark. Adrenaline induces contraction of chromatophores, and is perhaps similar to sympathetic transmitter. Fihes mostly become pale in colour on light background and vice versa, and the effect is produced pre dominantly through eyes. Change in colour begins rapidly, but its completion may take many days.

Aerial Respiration and Air-bladder: Many fishes live outside the water. Excursions to land vary from the wriggling of eel. In eel, there is no special apparatus for breathing air. Climbing perch is provided with special air chambers above the gills. Many other fishes gulp air, especially those living in shallow tropical waters, which readily become deoxygenated.

Neural Control Mechanisms: In mormyriforms and gymnotiforms, discharge patterns are generated by a hierarchy of control nuclei in the medulla, the midbrain and the thalamus. In both groups, a nucleus is found in the ventral medulla.This structure is called pacemaker nucleus in gymnotiforms, while in mormyriforms it is called command nucleus. Both these nuclei project via a medullary relay nucleus to the electromotor nucleus in the spinal cord. Inputs from other sensory modalities are integrated. This integration of signals has parallels in other groups.

Electroreceptors: All electric fish with exception of stargazers can generate and perceive electric fields with electroreceptors. These receptors are widely distributed over body surface, but densely clustered around head. Three different types of electroreceptor have been detected in fish. Ampullary receptors, found in strongly and weakly electric fish are used for passive electrolocation of prey and predators. Tuberous receptors are surrouned by an epithelial cell plug. Mormyriforms possess 2 types of tuberous receptor. Knollenorgans are important in communication and mormyromasts in electrolocation. In gymnotiforms, one type of tuberous receptor is specialized for detecting the amplitude of stimulus and another for timing, but both are used for electrolocation. Electroreceptors are innervated by lateral line nerve.

Electric Discharges: The function of powerful discharges consists in stunning of prey or defence against predators. Electric fish can use the perception of electric currents to detect prey and navigate, but as their electric sense are active. Electrogenesis in combination with electroreception allows for the active electrolocation of objects in surroundings and sophisticated electrocommunication.

General Notes

Fisher exhibit diversity in their size, color, anatomy, physioloy, and behavior. It is difficult to discuss all these diversities. Hence, the important features are discussed below as a account.

6.7 Food Habits

They are omnivores, zooplanktivores, herbivores, carnivores, and detrivores. Many are opportunistic feeders. Primary feeding habits are associated with body form, mouth and digestive apparatus, and teeth. Gars have elongate bodies, long snouts, and sharp teeth with the fins placed toward the back of body. This is the design of fast-start predator, which lurks motionless in water column, slightly camouflaged. Tunas with their rounded and highly tapered bodies are streamlined pelagic chasers. These 2 fishes are ram feeders. Other predators reduce extra energy expenditure on chasing prey, wait passively, depending largely on good vision, explosive thrust and large mouths, and effectively inhaling prey. Such sit-in-wait predators are hidden with elaborate camouflage.

Herbivorous fish posses specialized organs like extended guts. Some successful families and abundant coral reef families include herbivorous fishes. Several herbivorous coral reef species defend territories. Some parrotfishes use shoals to overwhelm the defenses of territorial species. Silversides congregate in feeding shoals numbering in the millions. Smaller shoals of zooplanktivores, like rabbit fishes are also found hovering above and around coral reefs. Zooplanktivorous fishes are small, streamlined with compressed bodies, forked tails, few teeth, and protrusible mouth. Turbid habitats are home to many fishes that use electroreception to find prey. Some predators use electrical shocks of as much as 350 volts to stun prey .Archerfishes exploit a food source that is unavailable to most other fish.

Predation

They avoid predators through behavioral adaptation and physical structures, such as spines, camouflage and scents. Primary goal of most fish is to escape detection. If detected, a fish might try to hide quickly, blend in with surroundings. Many fishes avoid the chance of attack through particular cycles of activity, shading and camouflage, mimicking, and warning coloration. Fishes avoid dusk because predators often take advantage of quickly changing light conditions that make it difficult for prey to see predators. Most feed during the daylight hours.

Zooplanktivores are abundant and conspicuous by day. Wrasses secrete a foul-smelling mucus tent or bury themselves in sediment for protection. Shoaling provides many benefits as a daytime defense. Some predator mistake shoals for large fish. When shoals detect predators, they form a school that makes synchronous motions. Zooplankton and other invertebrates come out at night. Many groups and some cardinal fishes, have luminescent organs. Some use luminescence for defense. Rows of lights along the bottom of body make them indistinguishable to benthic predators as they match the intensity of moonlight or dim sunlight shining down.

Counter shaded fishes are graded in color from dark on top to light on bottom. A common and elaborate method in tropical seas is mimicking the background of habitat, which involves variable color patterns and peculiar growths of skin that may resemble pieces of dead vegetation, corals. Bold or bright coloration serves in structural or chemical defense. Sturgeon has bold coloration to match scalpel-like and poisonous spines. Aposematic fishes advertise their inedibility by moving slowly, instead of darting away, when predators are present. Displays of aggression back up this behavior.

6.8 Ecosystem Roles

Ray-finned fishes are essential components of most ecosystems. While many ray-finned fishes prey one another, they have significant impacts on nearly all other animals. Zooplanktivorous fishes, select for specific types and sizes of zooplankton, when they feed. When non-native species invade new habitats, the fragility of this balance is dramatically changed. When alewives invaded Lake Michigan, they decimated 2 larger species of zooplankton and dramatically reduced 2 midsize species, resulting in increase of 10 smaller species and higher algal content. Later, Pacific salmon were introduced into lake and dramatically reduced alewife populations and the larger zooplankton species recovered. As the larger species grazed on algae, phytoplankton density decreases and the lake become cleared, an example of a trophic cascade.

The introduction of Nile perch, into Lake Victoria caused a precipitous decline of many planktivorous cichlids. These cichlid species exerted predation pressure on zooplankton. This introduction resulted in mass extinctions of endemic species. Many local people consumed the smaller cichlid species. When Nile perch affect local cichlid fisheries, locals consume Nile perch but this fish required firewood for drying and preservation. Because ray-finned fish are important food source to terrestrial organisms, including humans, changes in ray-finned fish communities would have ecological implications. Piscivorous ray-finned fish compete with many organisms above and sometimes are involved in symbiotic relationships with them. A com mensal relationship is found between bluefish and common terns. These 2 species interact at a critical period of the terns' feeding cycle, after mating when there are chicks to consume. This time, bluefish feed on anchovies, driving them up in water column, where terns can catch sight of anchovies. Bluefish reduces anchovies populations.

Gobies share burrows with several shrimp-like crustaceans. Importance of fish lies in linking terrestrial and aquatic ecosystems. This is true of anadromous species. During rainy periods in tropical watersheds, ray-finned fish consume seeds and disperse them throughout floodplain. Several invertebrates regularly consume various ray-finned fish. There are some unlikely predators like dinoflagellates that kill large fish. Some dinoflagellates consume scales of dead fish. Ray-finned fishes have impacts on plants. The trophic cascade example illustrated an indirect connection between microscopic plants and fish. Fish also excrete soluble nutrients like phosphorous into water. Phosphorus is essential for phytoplankton growth. More direct connection is the consumption of numerous plant species. Fish may significantly alter dynamics of their habitats.

6.9 Economic Importance and Conservation Status

Fishes are economically important. We consume fish and fish are source of protein for millions of people. The farmed salmon industry is valued at over 2 billion dollars a year, but aquaculture operations can have serious ecological consequences. Ray-finned fishes are popular in aquarium trade. Many fish are poisonous and venomous, and when disturbed, they can inflict serious wounds and death sometimes. Sometimes, people die from consuming poisonous fish. Often fishes have positive or negligible impacts on humans.

Threat to aquatic habitats has grown steadily over the course of 20th century and continues today mostly due to human intervention. There are cases of collapses in many of the world's fisheries and declines in many large species. Some species live long and have low reproduction and growth rates and removal of larger individuals can have significant impacts on populations. Another threat is excessive removal of exotic reef species using harsh chemicals like cyanide for the aquarium trade. The significant threats are to families with restricted distribution because localized threats easily eliminate all individuals of such species. Mosquito fish combined with pollution and habitat alteration have proven disastrous for groups of endemic ray-finned fishes. A total of 90 species of ray-finned fishes are extinct, 279 are critically endangered or endangered, and other 506 are listed as vulnerable or near threatened.

6.10 Swim Bladder

Swim bladder is more or less a sac-like structure lying between alimentary canal and kidney and contains a mixture of carbon dioxide, oxygen and nitrogen. It is present in all bony fishes except elasmobranchs, arises form dorsal wall of gut as a median diverticulum. In dipnoans and polypteridae, it originates as ventral outgrowths form the floor of pharynx and later rotates to occupy dorsal positions. Air bladder differs greatly in shape, size and structure. Primarily it is a tough sac like structure with capillary network. Beneath capillary network lies tunica externa made of connective tissue and tunica interna composed of smooth muscle fibers and epithelial gas gland.

Air bladder may be composed of 1 or 2 or more chambers. Generally air bladder opens into esophagus by duct know as ductus pneumatics. Pneumatic duct may remain open and air bladder is then called physostomons type. When the connection is lost and pneumatic duct is closed, air bladder is called the physoclistous type. As regards shape, air bladder may be tubular, oval, fusiform etc. In physostomons type, a vessel arising from coeliaco-mesenteric aster supplies swim bladder. Blood form it is carried to heart thought a vein joining the hepatic portal vein. This condition is found in dipnoans, bony ganoids fishes, and soft rayed teleosts.Physoclistous type of air bladder is found in spring rayed fishes. In this type of air bladder, there lies an antero-ventral secretary gas and a postero-dorsal gas absorbing region called oval. Air bladder is supplied by coeliaco-mesentenc artery and also by arteries form dorsal aorta. Blood form different parts of bladder is returned by two different routes. Blood from gas gland is returned by hepatic postal vein and from rest of bladder by posterior cardinal veins.

Gas bladder shows a number of structures. Modifications in various groups of bony fishes are summarized below:

Chondrostei: Among chondrosteans, the most primitive air bladder is found in *Polypterus* where it is a bilobed sac with a short left and long right lobe. It opens on floor of pharynx just behind gill-slits. Air bladder lies ventrally to esophagus for most part and only elongated right lobe takes up positions dorsal to gut. It leads to glottis through a muscular vestibule. Air bladder, in *Acipenser* is oval in shape having smooth walls.

Holostei: Among Holsteins such as *Lepidasteus*, gas bladder is unpaired elongated sac opening into esphagus by a glottis. Wall of bladder is provided with alveoli formed by fibrous bands. Alveoli are arranged in two rows and each is further divided into sacculi. *Amia* has a very large air bladder and its wall shown smaller but more numerous alveoli than in *Lepidosteus*.

Dipnoi: In Dipnoi, gas bladder resembles closely with lung of amphlbians in structure. In *Neoceratodus*, gas bladder is a large unpaired sac and has two fibrous saes on its inner surface. There project into cavity as two ridges, one dorsal and other ventral in position. Transverse septa divide space between ridges into numerous alveoli, each of which is subdivided into several sacculi. In *Protopterus* and *Lepidosiren*, bladder is more complicated in structure and long. In all Dipnoi, bladder is highly vascularlbur "red bodies" or "red glands" are not present.

Teleostei: In teleosts, air bladder presents a variety of structure and shape. Usually it lies below kidneys, between gonads and above gut. In Cyprinidae, it is divided into two intercommunicating chambers. In some species belonging to Notoptendae, Sparidae, Carangidae and Scombndae, gas bladder may extend into tail in form of a pair of cacca. Gas bladder is considerably reduced in fishes living in torrential waters of hills, as in *Psillorhynchus* and *Nemacheilus*, in which posterior chamber of bladder has disappear and anterior chambers very small enclosed in a bony capsule.

In same air breathing fishes also, as in *Clarias batrachus* and *Heteropneustes fossilis*, air bladder is reduced and enclose in bone. A pair of coecal outgrowths develops form air bladder of *Gadus* and extends forwards into head regions. In *otolithus*, two short tubular outgrowths develop from antero-lateral wall of bladder and each subdivides into two sacs, one directed forward and other backward. In Silusidae, cavity of bladder is drvided by a T-shaped septum. All teleosts are physostomus is in the beginning but in adult condition, same of them become physoclistic (Figure 6.25a, 6.25b).

Function of Swim Bladder

Swim bladder in fishes performs a variety of functions.

Hydrostatic Organs. It is primarily a hydrostatic organ and helps to keep the weight of body equal to volume of water the fish displaced. It also serves to equilibrate the body in relation to surrounding medium by increasing, or decreasing volume of gas content. In physostomous fishes, expulsion of gas from swim bladder is caused by way of ductus pneumations, if absent superfluous gas is removed by diffusion.

Adjustable Float: With help of air bladder fishes can swim at any depth. When fish rises up, air bladder is distended decreasing its specific gravity, and when fish sinks to lower depths specific gravity is increased by expelling air form air bladder.

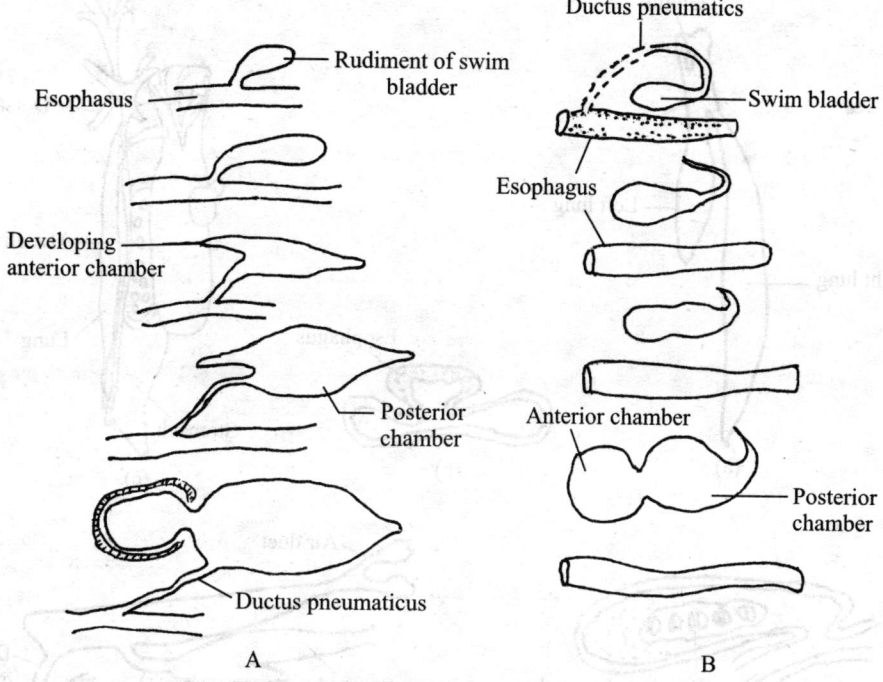

Derivation of the swim-bladder from the gut.
A. Formation of physostomous type of swim bladder in *Catostomus*.
B. Formation of physoclistus type of swim bladder

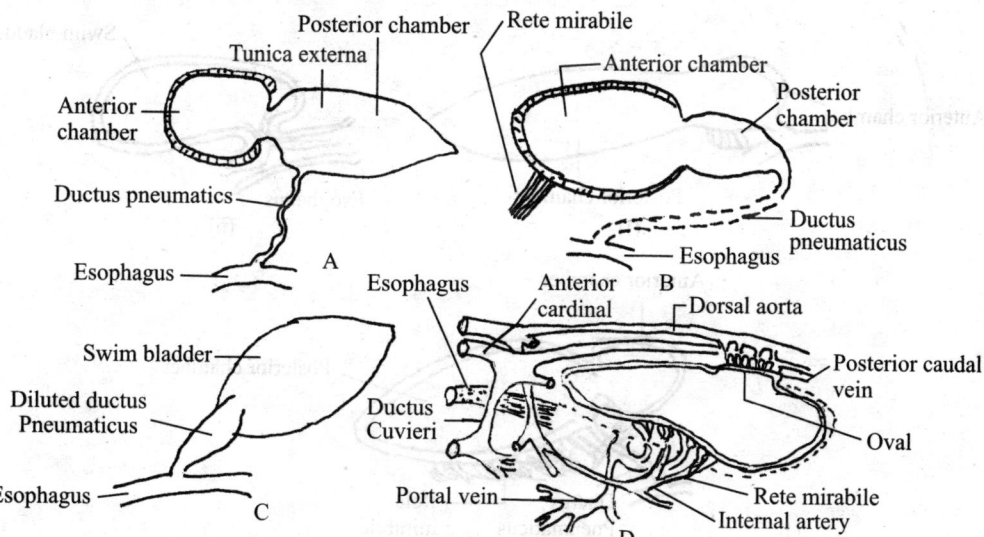

Figure 6.25a Variation in structure of swim bladder. A. *Catostomus*, B. Typical physoclistic type of swim bladder. C. Transitional swim bladder in eel, and D. Physoclistous swim-bladder with oval.

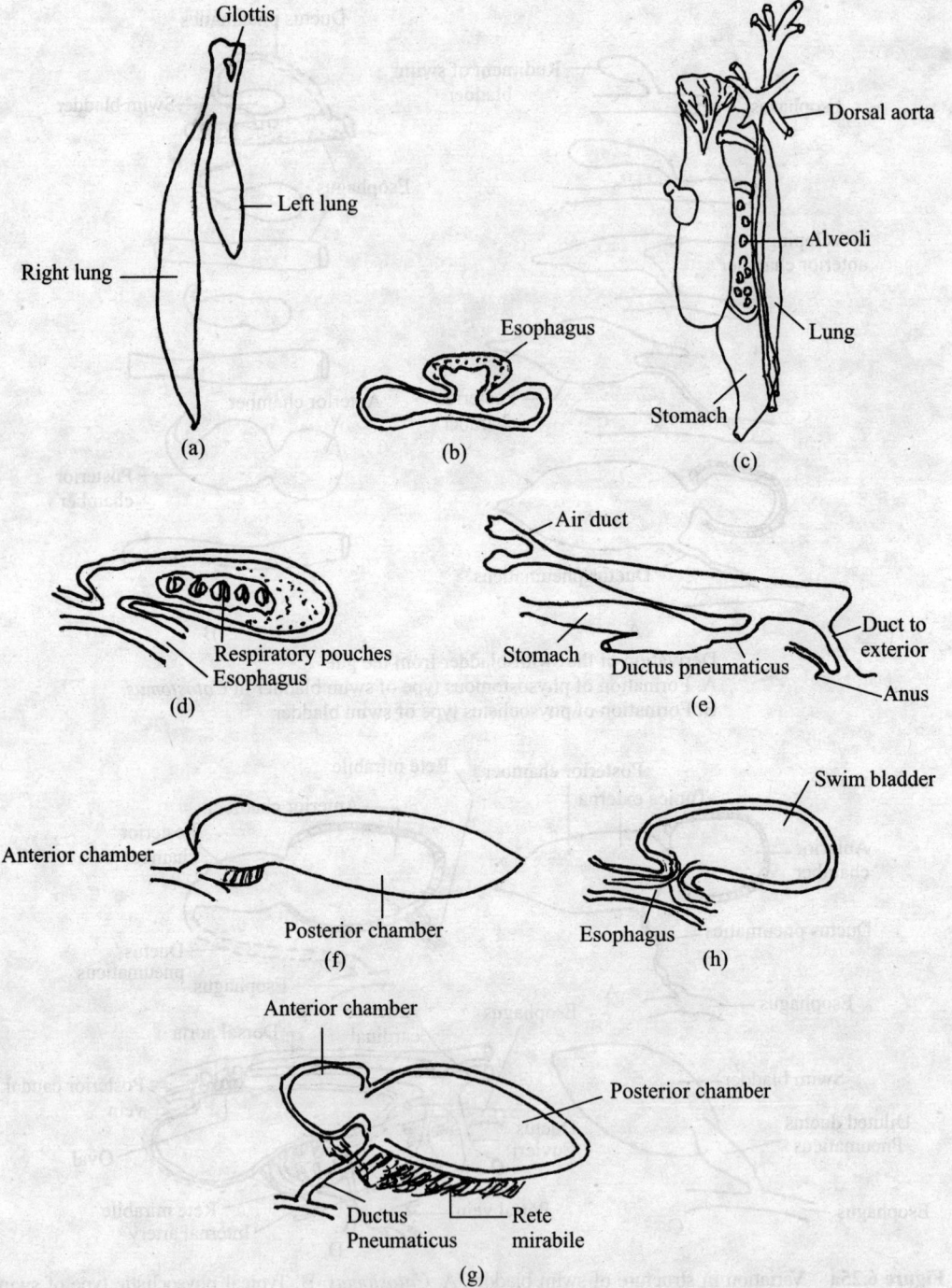

Figure 6.25(b) (a) Variations of swim bladder (a) in *Polypterus*, (b) Sectional view of swim bladder and esophagus, (c) in *Gymnarchus*, (d) in *Amia*, (e) in *Clupea* and (f) in *Esoa*. (g) in Cyprinoid fish (h) in *Acipenser*.

Maintain Proper Centre of Gravity: Swim bladder helps to maintain proper centre of gravity by shifting the contained gas form one part of it to other and this facilitates in exhibiting a variety of movement.

Respiration: It many fishes like *Polypterus*, *Amia* and *Lepidosteus*, air bladder mainly functions as a respiratory organ. In Dipnol, air bladder is modified into lung which is able to take up atmospheric air. In fishes which live in swamps or pond where oxygen content is low, air bladder functions as lung. Air bladder can also store oxygen for use when required. Oxygen produced in bladder, definitely helps in fish in respiration.

Resonator: In some fishes, air bladder acts as a resonator intensifying the sound vibrations,

Sound Production: Swim bladder plays an important part in sound production. Many fishes like *Doras*, *Platystoma*, and *Trigla* can produce grunting, or hissing, or drumming sound. Circulation of contained air inside swim bladder causes vibration of incomplete septa. Sound is produced the consequence of vibration of incompletes septa present on inner wall of swim bladder. Vibrations are caused by movement of contained air of swim bladder. Sound may also be produced by compression of extrinsic and intrinsic musculature of swim bladder. Polypterin *Protopterus* and *Lepidosiren* can produce sound by compression and forceful expulsion of contained gas in swim bladder.

Sensory Function: Air bladders functions as a pressure receptor and enable fish to maintain a steady depth.

6.11 Comparison of Elasmobranchs and Teleosts

Fishes are usually described under 2 broad categories depending on nature of endoskeleton. Fishes having cartilaginous endoskeleton are usually regarded as cartilaginous fishes or chondrichthyes while bony fishes are called osteiehthyes. Cartilaginous and bony fishes have a common piscian body construction, but they show many differences which are summarized in the following tables:

Table 6.1   Comparisons between elasmobranchs and teleosts

Elasmobranchs (Chondrchthyes)	Teleost (Osteiehthyes)
Habitat	
Mostly marine Example : *Scoliodon*, sting Electric rays.	Both freshwater and marine. Example: *Labeo Cyprinus, Catla*.
External Features	
1. Body elongated, laterally compressed and is covered by placoid scales. 2. Head is dorsoventrally compressed. 3. Mouth is situated on ventral surface of head. Position is actually subterminal. 4. Tail is heterocercal.	1. Body is laterally compressed and covered with cosmoid, ganoid, cycloid, or ctenoid scales. 2. Head is dorsoventrally or laterally flattened 3. Mouth lies terminally. 4. Tall is homocercal, diphycercal or sometimes heterocercal.

5. In most fishes, except certain primitive sharks, there are usually five pairs, of gill openings In holoce- phalans, operculum is developed

5. Gill openings are covered by operculum so that a single exit is visible on each side

6. A pair of spiracles or gill like cleft is present
6. Spiracles are absent.

7. Presence of single nostril aperture
7. Presence of double nostrils

8. Claspers are present in male
8. No claspers in male.

9. Fins with cartilaginous fin rays
9. Fin is supported by bony fin rays

Internal Features
Endoskeleton

1. Endoskeleton is entirely cartilaginous.
1. Mostly bony.

2. Membrane bones are present
2. Absent.

3. Notochord persistent
3. Usually non- persistent

4. Jaw suspension is of hyostylic type
4. Mostly amphistylic type.

Digestive System

1. Intestine is short and its lumen contains complicated spiral valve
1. Long and devoid of spiral valve.

2. Cloaca present
2. Absent.

3. Pancreas is well developed and distinct
3. Pancreas is inconspicuous.

4. Presence of elongated rectal glands
4. Rectal gland absent.

5. Presence of Bursa entiana at junction of pyloric stomach and intestine
5. Absent.

6. Pyloric cacea is absent
6. Present

D. Respiratory System

1. Gills are present on gill arches
1. Gill are present is gill pouches.

2. Usually 4 pairs of gills.
2. Usually 4 pairs of gills.

3. Air bladder is absent.
3. Present.

4. Inter branchial septum is distinctly present.
4. Inter branchial septum is much reduced.

5. Gills are lamelliform.
5. Gill is filiform or pectinate.

Circulatory System

1. A muscular conus arteriosus is present
1. Conus is replaced by a non- contractile bulbous aorta.

2. Five pairs of afferent and efferent branchial vessels
2. Only four pairs of afferent and efferent branchial vessels are present.

3. Veins are very thin-walled and occasionally form sinus
3. Such types of veins absent

4. Presence of lateral abdominal vein
4. Absent

5. Presence of 2 renal portal veins
5. Only left renal portal vein are present

E. Excretory System

1. Paired kidneys are elongated and narrow, do not fuse posteriorly
1. Paired kidneys with broad anterior portion, fuse posteriorly.

2. Anterior portion of kidney is non renal in function
2. No part of kidneys is non- renal

3. Urinary bladder is absent.
3. Present

| 4. Testes remain attached to cephalic end of the kidneys | 4. Usually free form kidneys |

F. Reproductive system

1. Tested are connected by efferent duct with epididymes at anterior end of kidneys.	1. No connection to testes with kidneys
2. In females, no connection exists between reproductive and renal organs	2. In females, sometimes eggs are discharged m coelom or rarely directed oviduct for transmission to extensor.
3. Fertilization is internal	3. Mostly external.
4. Ova large and a few in number, enclosed within a horny case after fertilization.	4. Small and numerous
5. Viviparous sometimes oviparous	5. Generally oviparous and rarely viviparous

G. Nervous System

1. Optic chaisma is present with decussated optic nerves.	1. Optic nerves do not form chiasma but are simply crossed with each other.
2. Cerebellum overlaps optic lobes and medulla oblongata.	2. Cerebellum is large and bent upon Itself.
3. Presence of olfactory bulbs with peduncle.	3. Absent
4. Optic lobe is comparatively small	4. Larger.
5. Weberian ossicle is absent	5. It is present in some cases
6. Corpora restiform bodies are present	6. Absent

H. Sense Organs

1. Presence of two olfactory sacs on snout	1. Presence of dorsal olfactory sac on snout
2. Comparatively large paired eyes are present	2. Paired eyes are present
3. Ampulla of Lorenzini is present	3. Absent
4. Retina lacks cones	4. Cones are present

6.12 Osmoregulation

Osmoregulation indicates the collective activity of variety of mechanisms used by organisms to control water movement and water volumes. Most fishes are confined to either salt or fresh water and have a limited salinity tolerance. They are known as stenohaline species. In sea, concentration of salts is more than in body fluids of fish, so that water tends to diffuse out. Marine fishes tend to drink water. Fresh water fishes lose salts, and water enters into body through semi-permeable membranes of buccal cavity and gills. This is due to .higher concentration of salts within body of fresh water fishes than surroundings. Kidneys and gills play an important part in excretion of nitrogenous wastes and in maintaining water-salt balance.

In Fresh Water Fishes

Osmotic pressure of body fluid is higher than that of surrounding water. So fishes always face hydration. Skin and its mucus reduce water permeability, although small amount of water can enter through skin. To harmonize inflow of water, fishes produces a large amount of dilute urine which is hypotonic in regard to fish. Main function of kidneys is water excretion. Kidneys are large in relation to body weight. Fresh water teleost produces urine between 5-12% of body weight .Per day rate of urine flow is much larger in fresh water teleost (Figure 6.26).

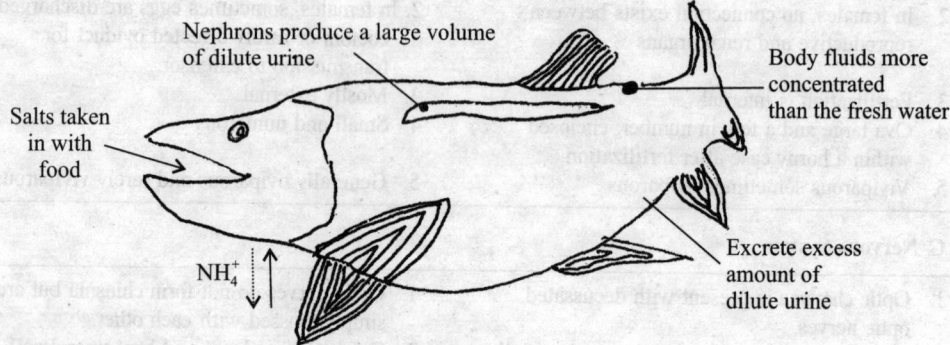

Figure 6.26   Osmoregulation in fresh water fish.

In Marine Fishes

Fishes lose water continuously by diffusion through entire tissue surface, semi permeable oral membrane and gills and little is excreted through urine. Dehydration is counteracted by retention of water, drinking sea water alongwith food, reabsorption by proximal tubules of kidney, excreting highly concentrated urine, and reabsorbion of entire water present in urine. Many marine fishes have aglomerular kidney, or only partially glomerular kidney. Neck segment, connecting tube between glomerulus and collecting tubules are absent. Distal convoluted tubes are absent. Slight water intke occurs through gills and oral membranes. Rest of the body surface is impermeable to water. Salts are excreted through urine and faeces but not through gills. Some salt are reabsorbed in kidneys tubules. No special chloride cells in gills (Figure 6.27).

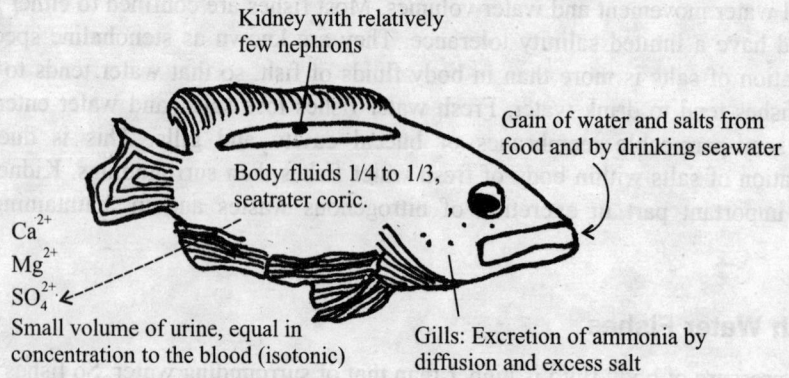

Figure 6.27   Osmeregulation in marine fish.

6.13 Parental Care in Fish

In fish, most species abandon their offspring to the vagaries of environment. Many species care only for eggs, build nests with empty shells, vegetal matter, pebbles, and sand. Eggs are defended from predators. Parents fan the eggs, spray water over brood with fin to provide oxygen. They clean eggs by brushing with fins, remove dead eggs by mouth. Some coat eggs with mucus having anti-microbial properties. In intertidal zone, parents cover egg batch with algae before low tide. Many species carry eggs either outside, or inside the body. Outside egg carriers are medakas, bagrid, and banjo catfish. Inside egg carriers are mouth-brooding cichlids, sea catfishes, lump fish, and gouramis. Only 20% of parental fish care for eggs and fry including most cichlids that lay eggs on substrate. Cichlids breed in predators rich areas. Perhaps one guardian could not protect the brood as predation pressure is high. Biparental care affords better protection to young rainbows than uniparental care.

Care is Costly

Parental males do not feed and lose about 10% of their body weight. Other small males are cuckolders. Parents estimate the risk of cuckoldry from simply seeing small males. Brian Neff has shown that parents can adjust their level of care. In *Telmatherina sarasinorum* increased risk of cuckoldry leads to increase rates of brood cannibalism by father. Eggs are eaten soon after mating. Parent fishes look after their offspring during their critical stage of life when they are defenseless. Almost every fish practices passive, or active parental care. Passive parental care is the "hereditary foresight" of females to provide more yolk for embryo. Some fish have poisonous substance in their eggs to keep predators away. In active parental care, either one or both parents take active part in caring their eggs, and sometimes the fry. This includes the preparation of suitable place for depositing eggs, selection of substrate to which the eggs can adhere, collection of nest materials, and building of the nest. "Nest cleaners" have the most primitive type of "nest", consisting of bushy roots. There is no collection of nest building materials nor nest building. "Nest builders" collect pebbles, leaves, roots, to build their nest. Some gouramis build nest out of objects which are cemented together with sticky substance. "Foam nest builders" build nest out of foam, which being unpalatable, defends the eggs hidden in foamy mass. The mouth breeding cichlids take the eggs into their mouth, and keep them in mouth till hatching.

Helpers at the Nest

In *Lamprologus brichardi*, offspring from former broods stay with parents to raise new broods. Seven to eight helpers within same natal territory share all parental duties, for their young brothers and sisters. As becoming old, helpers start their own breeding venture. Helpers resist eviction by submissive postures to parent .Most parents are not dissuaded to such pleas. Parents value the defensive function of largest helpers. Helpers grow slowly than carefree non-helpers; enjoy lower mortality as they reside inside territory.

Male Versus Female Care

Not all cichlids are biparental. In most mouth brooders, one parent provides care as brood is the mobile. When danger lurks, parent flee taking the brood. In most mouth brooding cichlids, care givers are the mothers. In bony fish, female-only care is the rarest form. The most common form is male-only care. In sticklebacks, and sunfishes males stake out territories, prepare spawning site, and court passing females. Females that spawn with male do not linger afterwards. Males release milt over eggs and defend territory along with fertilized eggs. Males usually tend to egg only. Rarely some attention is given to newly-hatched fry for a day.

Costs and Benefits of Parental Care

Most parents incur tangible costs for parental care. Weight loss in parents is due to predator-chasing activity, and lack of time to forage. Reproductive cycle reduces energy reserves. They do not breed again for a while. Parents can be stressed. In sticklebacks, roving gangs of females attack nests of parental males, for eating the eggs inside. When the male see and smell conspecific female three-spined stickleback loss much weight. De Fraipont and others called this stress imposed by egg predator, female conspecific here, the "female fatal effect": Hardships of parental car are paid through production of viable offspring. Parental species often breed in habitats that require parental attention. Harsh environmental conditions call for parental attention. Spread of disease, accumulation of debris, and lack of proper oxygenation causes eggs death.

Moving the Eggs

Substrate-brooding *Aequidens* lay eggs on submerged leaf. In disturbed condition, they move leaf by their mouth and swimming backwards with it. Parents choose leaves that can be moved easily. When the predator was visible, parents move the leaf 3 times more than usual. Usually, the leaf is pulled to deep area with cover. Scientists supported the view that leaf-brooding and leaf-moving are adaptation to reduce predator attacks on eggs.

"Broken Wing" Display

Nest-raiding is habit of female three-spined sticklebacks. Shoals of females roam and sometimes fall in nest of parental male, eating all the eggs inside. When parent male sees shoal of hungry-looking females, he often swims short distance away from nest and starts poking snout into ground. A female would perform this action while raiding nest which fools the females into believing that nest has been discovered. They rush to site and start digging there too. Male leaves this writhing mass of females and returns to his territory, hoping that cloud of sediments lifted by "feeding" frenzy will conceal his own real nest. This behaviour is similar to "broken wing" display of ground-nesting birds. In *Amia calva*, the fry follow male parent for a while after they hatch. When fry predator appears on scene, male moves away and thrashes about in water as if injured, drawing the mind of predator onto himself and away from fry (Figure 6.28a-6.28e).

Figure 6.28(a) Principal types of parental care in fishes.

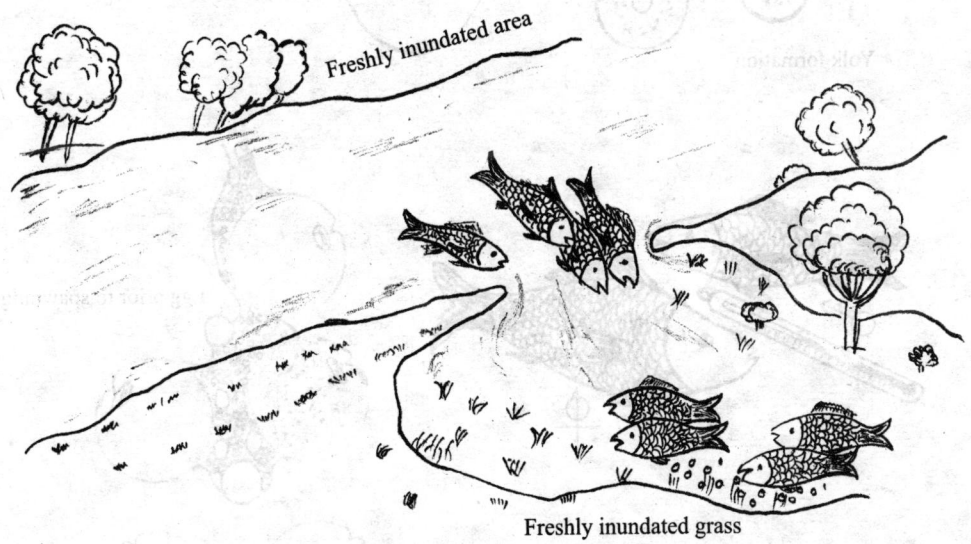

Freshly inundated area

Freshly inundated grass

Figure 6.28(b) Reproductive behaviour in fish.

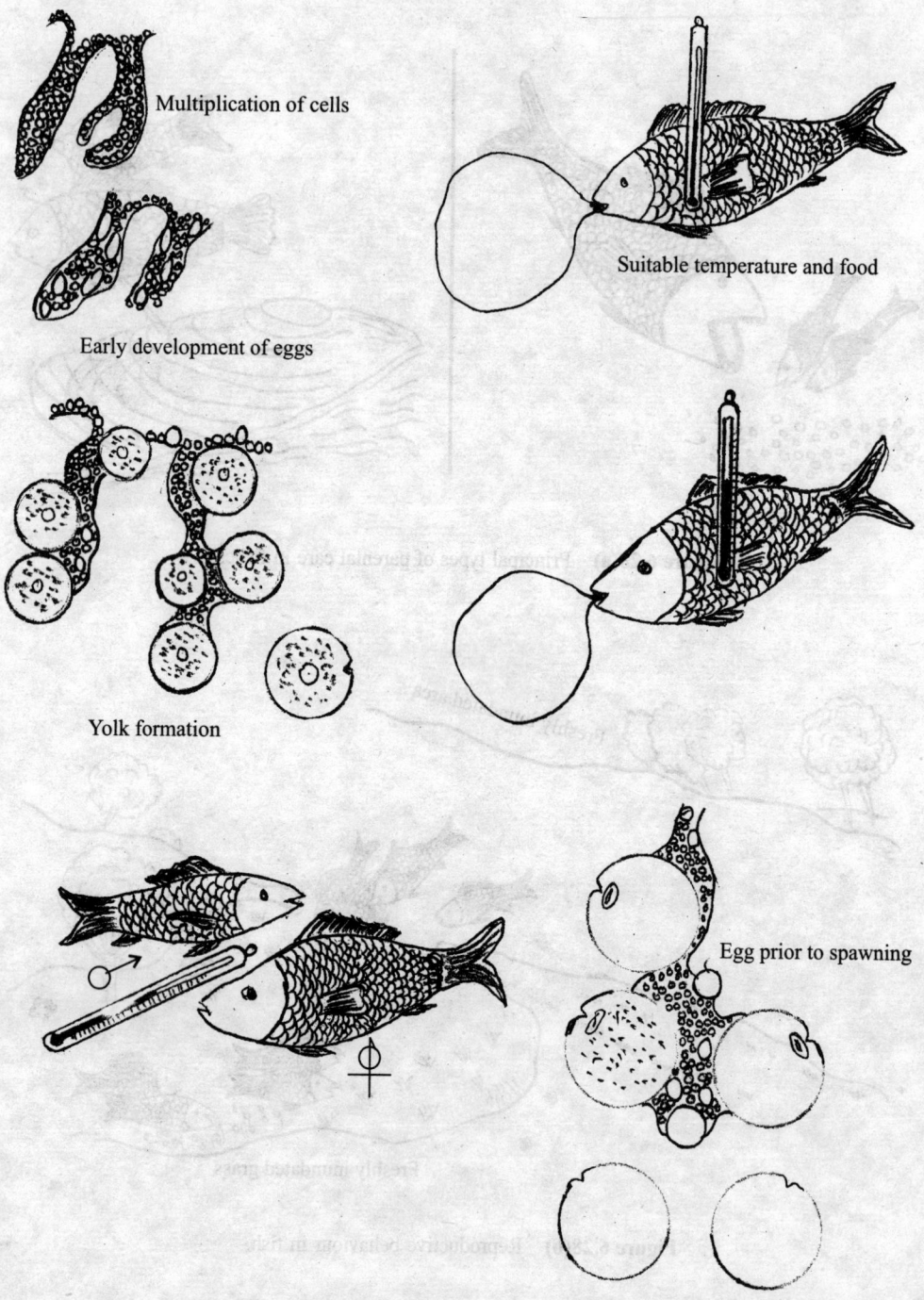

Multiplication of cells

Early development of eggs

Yolk formation

Suitable temperature and food

Egg prior to spawning

Eggs ready for fertilization

Figure 6.28(c)   Environment and maturation of fish egg.

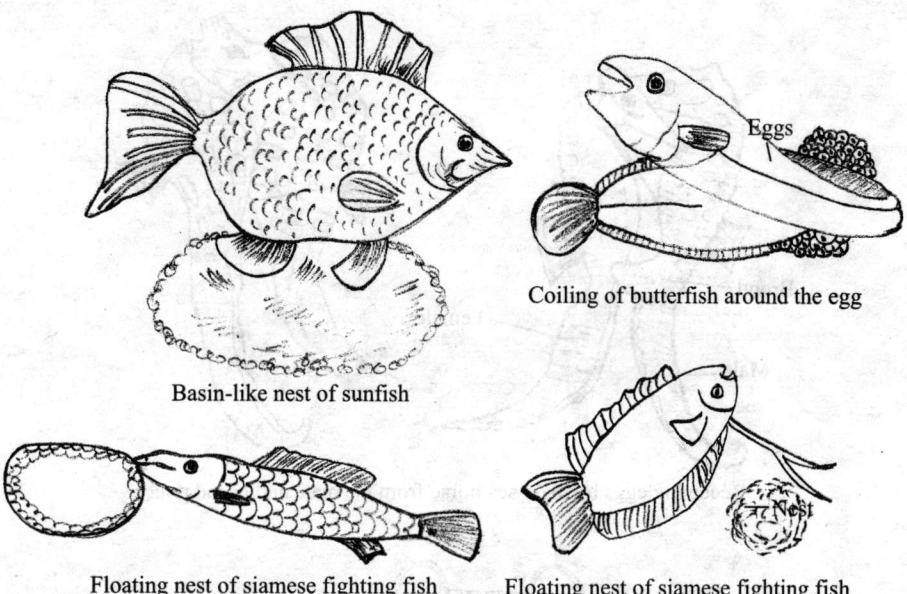

Basin-like nest of sunfish

Coiling of butterfish around the egg

Floating nest of siamese fighting fish

Floating nest of siamese fighting fish

Nest of ten-spined stickleback

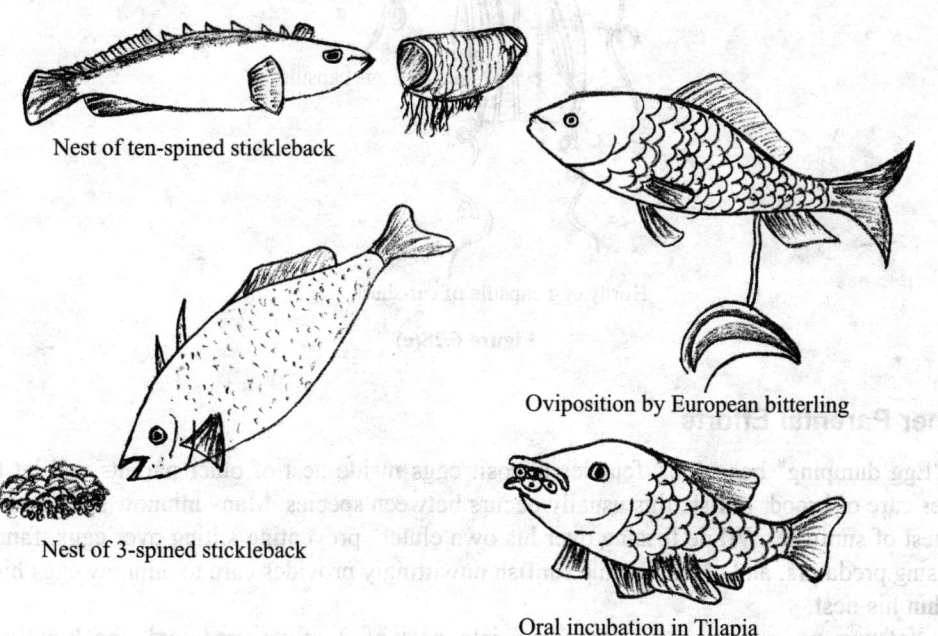

Oviposition by European bitterling

Nest of 3-spined stickleback

Oral incubation in Tilapia

Figure 6.28(d)   Nest of three spined stickleback

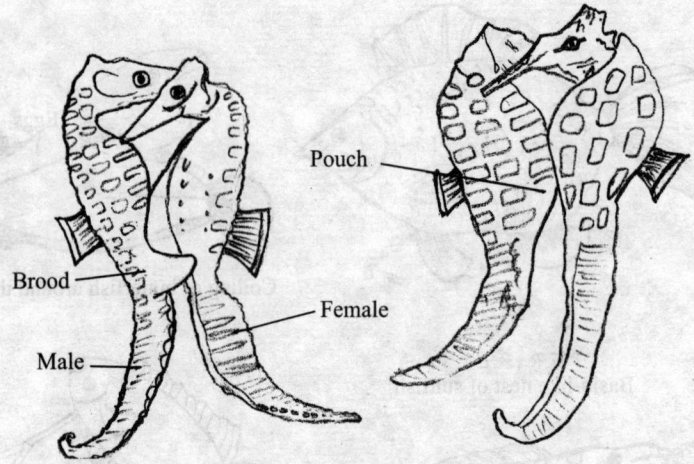

Receiving eggs by male sea horse from his mate and brood pouch

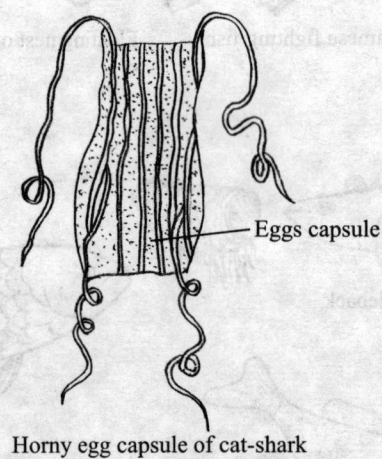

Horny egg capsule of cat-shark

Figure 6.28(e)

Other Parental Efforts

In "Egg dumping" behaviour, females deposit eggs inside nest of other parents and let them takes care of brood. In fish, this usually occurs between species. Many minnow species spawn in nest of sunfishes. While fussing over his own clutch, preventing silting over eggs, fanning, chasing predators, and parental male sunfish unwittingly provides care to minnow eggs hidden within his nest.

Notemigonus crysoleucas, drop eggs into nest of 2 of its predators, the bowfin and largemouth bass stay near tail of bowfins, avoiding dangerous mouth area. This system resembles the nest parasitism practiced by cuckoos. Sunfishes seldom try to stop minnows from spawning in their nests. Often, sunfish does not suffer from presence of minnow eggs in nest, they may even benefit. When predators break the parent's defense, they get chance to pick up the parent's eggs. *Synodontis multipunctatus* attends spawning ritual of various mouth brooding cichlids and lays eggs at same time, in same spot.

Female cichlid picks up all eggs and begins incubating them inside her mouth; some of those are catfishes eggs. These alien eggs hatch earlier than cichlid eggs. While in mouth of cichlid, catfish fry finish absorbing their yolk sac. Cichlid eggs hatch, but just in time to be devoured by baby catfish. In *Symphodus tinca*, some large males circumvent the costs of egg care by temporarily usurping the successful nests of smaller males of same species, spawning in these nests with various females for a day or so, and then abandoning the site. The original owners, whose eggs are still present among those of usurper, may not want to let their part of nest contents be wasted and so they often come back and resume guarding the nest, protecting the foreign eggs and their own. This tactic on part of big males is called piracy.

Voluntarily Caring for Other's Eggs

In some species, females spawn in nests that already contain eggs. In fathead minnows, big newly reproductive males sometimes evict the owner of an established nest rather than occupy a similar but empty site. Such usurpers do not destroy the previous owner's eggs but instead care for them. Female fatheads prefer to mate with males who are already caring for eggs.

In three-spined sticklebacks, territorial males dash over to nest of neighbour and stealing eggs from him. They surreptitiously enter the nest, take mouthful of eggs, swim back to their domain and deposit the kidnapped eggs into their own nest. This behavior is explained in females' preference for nests that already contain eggs. Thieving males make their nest more attractive. Female *Harpagifer bispinis* prepares nest site by cleaning patch of ground. After eggs are laid and fertilized, she remains on nest, cleaning the eggs and chasing predators until hatching occurs 4 to 5 months later.

Fry Care

Parental care sometimes extends beyond egg stage, into fry stage. Beside cichlids, fry care can be seen in marine catfishes, freshwater catfishes, in bowfin, and in *Acanthochromis polyacanthus*. Care consists mainly of guarding the fry against predators. Fish generally do not feed their offspring. In *Bagrus meridionalis*, females do not lay a full complement of eggs. Some eggs are unfertilized and held back within ovaries. When the fry are old, mothers force out these extra eggs and the young consume them. Every day, a mother hovers 1 m above the bottom and spreads her fins slightly downwards, at which point her young raise from nest, line up at her vent, and grab small eggs she exudes.

Several cichlids from the genus *Cichlasoma* are known to "fin-dig" and to turn leaves over when they are accompanied by fry. Evidence suggests that this is a way for parent to turn up food for their young, at least in case of fin-digging. Parents dig more than non-parents. Parents with larger broods fin dig more. Fin-digging rate increases as brood gets older and hungrier. Young gather up near the fin-digging parent and feed actively on stirred-up material. The possibility exists that fin-digging is not a means to provide food for young, but a way to obtain food for adult itself.

In many cichlids such as angelfish, orange chromide, and others, fry can feed off the skin mucus produced by their parents. Number of mucus-producing gland in parent's skin increase during the fry stage and young nips at body surface of both parents. The fry do more nibbling when they are deprived of other sources of food, which shows that mucus is indeed a dietary supplement

Signaling Danger

In most parental fishes, fry care includes the signalling of danger to the young. In Siamese fighting fish for example, the parental male can communicate danger to his young through surface waves. In their first few days of independence from air-bubble nest, young fighting fish stay in contact with surface like most anabantids, they need to breathe some air. Parental male stays nearby and if he senses danger he shakes his pectoral fins close to surface.

Surface wavelets thus generated and are perceived by the young at a distance of up to 40 cm. Young then swims in direction of source, and this action brings them close to male who can then suck them up into his buccal cavity and carry them back to nest. In cichlids, various visual displays seem to warn young in a similar fashion. Mad dashing-about by parents induces young to settle quickly to bottom and remain still. Brief jerks of head or twitches of whole body induce fry to swarm near parent. In mouth brooding species, an alarmed mother can pitch slightly head-down and swim slowly backwards, on which fry quickly dash into her open mouth for safety. Some cichlids, when alarmed, signal their young by flickering their pelvic fins up and down.

Retrieving the Young

Many cichlids retrieve their fry when danger lurks. Parents suck 2-3 young at a time into their mouth and bring them back to the old nesting site or to a safe place where they spit them out. The fry then dart or sink towards the bottom, where they stay relatively immobile. Cichlids do this regularly at the end of the day. This ensures that by time darkness comes, youngs are gathered in one place over which parents guard all night long.

Brood Mixing

Cichlid parents sometimes guard swarms of fry that are made up of several sub groups of different body size. Such parents have accepted within their brood the young from other parents. It is relatively easy to integrate experimentally foreign fry into broods of cichlids kept in aquaria. Parents accept these new fry readily, provided that newcomer are same size or only slightly smaller than their own young.

Fry Care in Presence of Predators

Duration of parental care at the fry stage is often dictated by the age of young, which after a while become too mobile for parent(s) to watch over. In mouth brooding species, parents have more direct say on end their duties. They can expel young from their mouth and refuse to take them back in. One such species has given evidence that it can extend the duration of the incubation period if it perceives that risk of predation on fry would be high. Females of mouth brooding cichlid *Ctenochromis horei* kept young in their mouth about 4 days longer when they swam in presence of another predatory cichlid. These species are both found in Lake Tanganyika. This extra effort took a toll on females. They could not feed during those 4 days, and it took them longer to breed again, as compared to females that were not exposed to the predators and that ended incubation sooner.

Filial Cannibalism

Parents who face the spectre of weight loss through the reproductive phase can counteract this effect with filial cannibalism. In many families, parents sometimes eat a small part of their brood. Many of the consumed eggs appear perfectly viable. In painted greenlings, river bullheads, bluegill sunfish, and the cardinal fish *Apogon doederleini*, the more emaciated a parental male is, the more of the eggs under his care he consumes. However, this does not explain why even males who are in very good condition still eat up few eggs. There may be an adaptive side to the behaviour of filial cannibalism, as suggested by some of the results above, but we must also recognize the possibility that was favoured by earlier fish observers even though it was non-adaptive, that cannibalizing one's own eggs is a pathological breakdown in the normal egg-eating inhibition of good parents.

Parental Care Adjustment as a Function of Brood Value

In the late 1980s and early 1990s, the scientific literature saw a burst of publications on the topic of brood value. Many of the published articles supported the notion that parents could adjust their level of care as a function of the value of their brood. Parents sometimes eat part of their brood as an insurance against starvation. Sometimes it is the whole brood that is devoured. Ethologists now consider that total brood cannibalism is a manifestation of parental investment theory. Parents eliminate a poor brood to start a new and improved breeding attempt when possible. Observations in many species have revealed that only small broods are the victim of total cannibalism. Large broods are left intact or only partially cannibalized.

6.14 Migration in Fish

Migration is the movement of animals from one place to another for feeding, reproduction, or to avoid extreme conditions and return back to native place. Migration of many fishes together is called shoaling. Migratory fish exhibit coordination in movements to bear synchronized manoeuvre for producing various shapes called schooling as found in tuna and sardine. When population density of fish is high, food resources exhaust quickly and fish migrates for food resources which are called feeding or alimental migration. Salmons, cods and sword fish migrate for food in the sea. Spawning migration occur in breeding season when spawning grounds are far away from feeding places.

Migratory fishes like eels, salmons and several riverine fishes spawn in tributaries of river and migrate for laying eggs in O_2 rich waters. Juvenile migration involves larval stages, which hatch in spawning grounds and migrate long distances to reach the feeding habitats of their parents. In recruitment migration, larvae moves from nursery habitat to habitat of adults. Adult eels live in rivers in Europe and America. Their larval stages migrate from the sea to reach rivers which take 1 to 2 years. Seasonal migration occurs in fishes inhabiting arctic areas where summer is conducive and food is abundant. When winter approaches temperatures fall below zero, food becomes scarce and fishes migrate to subtropical and tropical areas.

Types of Migration

Fishes live in freshwater and marine habitats, which pose osmotic problems for migration from one type of habitat to another. Fish migrate as one habitat rarely provides optimal conditions for all life-history stages. Areas of high food availability are rarely found in same place. In addition, some habitats are good for spawning. Fish migration from one freshwater body to another for food or spawning, is called potamodromous migration. About 8,000 species migrate within lakes and rivers, for food on daily basis as food availability varies from place to place and from season to season.

Fishes migrate to lay their eggs in places where O_2 concentration in water is suitable and food is sufficient for freshly hatched juveniles. Migration from sea water to sea water is called oceanodromous migration. About 12,000 marine species regularly migrate within sea. Herrings, sardines, mackerels, and cods migrate for food by shoaling or schooling. Migration from fresh water to sea or from sea to fresh water is called diadromous migration. About 120 fish species overcome osmotic barriers and migrate between 2 types of habitats. This migration may be divided into 3 types.

Most migrations occur from areas of large areas like oceans, seas and lakes, into and up rivers are called anadromous migration. Species like Salmon, Sturgeon and various cyprinids exhibit anadromous migrations. Catadromous migration takes place between the river or lake and ocean. Some of the best-known anadromous fishes are the Pacific salmon species, such as eel, Chinook, coho, chum, shad, pink and sockeye salmon. Salmon hatch in small freshwater streams and migrate to the sea to mature, living there for 2 to 6 years. When mature, salmon return to same streams where they hatch to spawn. Salmon undertake journey covering hundreds of kilometers upriver.

Freshwater fish migrations occurs in short range, from lake to stream or vice versa, for spawning purposes.Potamodromous migrations of endangered Colorado pike minnow of Colorado River system can be broad. Migrations to native spawning grounds are 100 km, with maximum distances of 300 km reported from radio tagging studies. Colorado pike minnow migrations display high degree of homing and fish make upstream or downstream migrations to reach very specific spawning locations in whitewater canyons.

Migration of Salmon

Sockeye Salmon migrating up the Yukon travel 40 km per day. Chum Salmon in Amur travel about 50 km per day. Atlantic Salmon that pass through the White Sea have 2 races. Autumn race enters the rivers in late autumn with poorly developed gonads. They travel up the river during which gonads develop. Spring race enters the same rivers in spring with their fully developed gonads. They feed little and travel quickly up the rivers, catching up with autumn race near spawning sites. Sockeye Salmon make inspiring migrations of 3,600 km up the river Yukon. Crucian Carp migrate less than 1 km up the Kerkinitis River to spawn. In the USA, Sockeye Salmon make journeys of more than 3,000 km up the rivers, get exhausted to return to the sea after their spawning, simply spawn and die. In Europe, Atlantic salmon migrate up shorter rivers, making journeys of less than 400 km. They retain energy to return to sea after they spawn. Atlantic salmon may spawn 3 or even 4 times before they die.

Anguilla anguilla or *Anguilla vulgaris* and *Anguilla rostrata* migrate from the continental rivers to Sargasso Sea off Bermuda in south Atlantic for spawning, crossing Atlantic Ocean and cover a distance of about 5,600 km. Pre- migratory changes in them are as follows:

1. They deposit huge fat as reserve food.
2. Colour changes from yellow to metallic silvery grey.
3. Digestive tract shrinks and feeding stops.
4. Eyes are enlarged and vision sharpens.
5. Other sensory organs become sensitive.
5. Skin modified to be respiratory.
6. Gonads become matured and enlarged.
7. They develop restlessness and urge to migrate in groups.

They migrate through the rivers and reach coastal areas. Both males and females swim together in large numbers, reaching sea in about 2 months. Each female lays about 20 million eggs which are fertilized by males quickly. .In late autumn, female salmon buries fertilized eggs in stream bottom gravel nests called redds. Eggs hatch into alevin or sac fry in late spring, and yolk sac is gradually absorbed. Three to 6 weeks after hatching, alevins emerge from gravel to seek food and are called fry. Fry develop into parr with camouflaging vertical stripes. Parr are 2 inches long, feed and grow from one to 3 years in their native stream before becoming smolts. Smolts are silver colored and about 6 inches long. In spring, smolt body chemistry changes. They now weigh about 2 ounces and are ready to enter salt water.

They migrate to ocean where they will develop in about 2 to 3 years into mature salmon weighing about 8 to 15 pounds. Adult salmon begin returning in spring to their native stream to repeat spawning cycle. Spawned-out salmon, called kelts or black salmon, return to ocean or over winter in river. Atlantic salmon spawn in November, burying their eggs in cobble areas in streams called redds. Most female lay 8,000 eggs in 2 or more redd. Supply of clean, well oxygenated water is critical for these eggs. Eggs remain in gravel throughout the winter before hatching in spring. Newly hatched salmon, the sac fry, get food from their attached yolk sac.

Salmon emerge from the redd, primarily from April to June, when yolk sac has been totally absorbed. Feeding start at this time. Salmon fry, measures one and one quarter inches long at emergence, set up feeding territories and defend it from other fish. Growing salmon like stream habitat lined with stone and clean, cool sediment free water. Fish are found in riffles and along the interface of fast moving water, under overhanging cover and toward bottom of water column. Fry spent their first summer in stream where they hatched are 3 to 4 inches long by fall now called parr. After spending one year in freshwater, the parr grow to length of 4 to 6 inches.

Parr remain in freshwater for a period of 1 to 3 years. Freshwater residence period is largely dependent on growth rate. The fastest growing parr are found in productive tributaries. Slower growing parr from colder water of fertile tributaries spend 3, or 4 years in freshwater. Most parr spend 2 years in freshwater. During first fall, parr may disperse widely from their first summer location to seek new habitat. Parr destined to leave the freshwater environment the following spring to begin a process called smoltification . Marked physical changes occur during spring after salmon reach a size suitable for migration to sea, 6 to 8 inches or more. These changes allow juvenile salmon to adapt to life in marine waters. Throughout smoltification process, series of physiological and morphological changes occur that transform young salmon from territorial, bottom-dwelling, freshwater fish to schooling, marine fish. Juvenile salmon migrating for the ocean are called smolts. Smolts lose dark vertical stripes, parr marks, and become bright silver. They migrate to Long Island from April to June. Some smolts exhibit pre-smolt movement in fall to start their migration. Few salmon called grilse return after spending only one winter at sea and others wait until after their third sea winter to return (3SW).

Average 2SW salmon grows from 6 inches long and weighing about 2 ounces as a smolt entering Long Island Sound to about 30 inches and 10 pounds as a returning mature salmon. Grilse (1SW) average about 4 pounds and 3SW salmon often weigh more than 15 pounds. Adult salmon return to the Connecticut River in June. In the freshwater environment, color of adult salmon changes from the silver to dark. Salmon reach their native streams, where they spend summer holding in deep, cold pools before spawning in fall. From the time they enter freshwater until spawning, often 6 months later, salmon do not feed; feeding begins after they return to saltwater in fall or spring. Atlantic salmon do not die after spawning, though many die due to rigours of upriver migration. Adults that survive are called kelts which return to ocean in late fall at which time they regain their silver color. Small percentage of salmon survives several spawning runs, alternating between freshwater and marine environments. Repeat spawners and grilse are valuable to salmon for maintaining genetic variability and providing buffer for all sources of mortality affecting the predominant class.

Salmon perform migrations from freshwater to marine habitats and back again. This behaviour is called 'anadromy'.Pacific salmon live in coastal and river waters from Alaska and Russia in north, to Japan and Mexico in south, while Atlantic salmon inhabit areas in the North Atlantic Ocean and Baltic Sea, including associated river ways in USA, Canada, Norway. Salmon eggs are laid in small pits that are excavated in gravel-based freshwater streams because of their specific temperature, currents and O_2 levels. Salmon eggs hatch after about 3 months, although juveniles remain dependent on yolk-sac for several weeks after hatching. Juveniles begin their downstream migration during which they develop tolerance to saline waters.

Young salmon may remain in fresh water for up to 4 years, before entering ocean. During this period, juveniles are vulnerable to predation. Entry into ocean coincides with planktonic blooms, on which juveniles feed. Depending on the species, salmon may spend between 1 and 7 years at sea. Once sexually mature, the salmon migrate back to their original hatching grounds to reproduce using combination of chemical, magnetic and celestial cues for navigation. For most species, this landward migration occurs throughout the summer and autumn months, with few species, such as Chinook, and chum salmon, continuing to migrate through winter months.

Salmons' spawning migration is both risky and energetically costly. Salmon must travel continually against the current and overcome numerous threats and barriers including predators, disease and waterfalls. As a result, many salmon die during migration, and those that survive are often bruised and battered. On arrival at nesting site, salmon spawn several times before dying, although some species can survive and may even repeat spawn. As water temperatures increase, several negative effects on salmon may arise. Direct biological impacts on salmon include physiological stress, increased depletion of energy reserves; increased susceptibility and exposure to disease and disruptions to breeding efforts. As the developmental rate of salmon is directly related to water temperature, increasing temperatures could cause the more rapidly developing juveniles to enter ocean before their planktonic food source has reached sufficiently high levels. Additional indirect effects to salmon, associated with increasing air and water temperatures, relate to negative changes to their habitat. Areas of particularly warm freshwater can present a thermal barrier to migrating salmon that requires additional energy to navigate around. Such barriers can also delay, or even prevent spawning.

As the air temperatures warm, much of the snow that feeds the river systems is expected to melt earlier. This will lead to reduction in summer flows of many rivers, coupled with an increase in freshwater inputs during the winter. A reduction in summer flow levels will serve to increase water temperatures further and is likely to reduce the overall habitat available to salmon. Increased winter flows are likely to scour the river beds, disturbing nests and causing physical damage to both salmon eggs and juveniles. Coupled with an increase in freshwater inputs, is an increase in sedimentation of river and stream beds. Such sedimentation is likely to reduce the amount of gravel substrate available for spawning, and to smother both eggs and juveniles.

Atlantic salmon (*Salmo salar*) migrates to the North American rivers for spawning while 6 Pacific salmon species (*Onchorhynchus*) migrate to rivers of Asia. During fall, Salmons enter rivers and swim against water currents, and reach tributaries in hilly areas, make pit in which female lays eggs and male releases smelt over them. Eggs take 2-3 months to hatch in following spring, when the juvenile stage called alvin emerges out but remains within nest, obtaining its nourishment from yolk sac attached to its belly. Alvin then transforms into fry which feed on planktons. Fries are denatant and feed and grow into fingerlings taking the shape of adult fish. They change into smolt which congregate at river mouth in large numbers and then enter sea water in to metamorphose into adult salmons.

Problem of Navigation

Fishes orient by positions of stars and moon in night sky and sun in daytime to find direction of swimming. Hasler proved that salmons are guided by odour of their parent stream during return journey. Eels migrate using similar odour maps. Fish achieve better growth rates and greater reproductive success by moving between areas on grand scale of hundreds or thousands of miles or simply a movement into and out of coastal waters. Most migrations are highly seasonal and result in temporal and spatial aggregations of fish.

Forage Fish

Forage fish often make migrations between their spawning, feeding and nursery grounds. Schools of a stock usually travel in triangle between these grounds. One stock of herrings have their spawning ground in southern Norway, feeding ground in Iceland, and nursery ground in northern Norway. Forage fish, when feeding, cannot distinguish their own offspring. Capelin is a forage fish of the smelt family found in the Atlantic and Arctic oceans. In summer, they graze on dense swarms of plankton at edge of ice shelf. Larger capelin eats krill and other crustaceans, move inshore in large schools to spawn, and migrate in spring and summer to feed in plankton rich areas between Iceland, Greenland, and Jan Mayen. Migration is affected by ocean currents. Around Iceland matured capelin make large northward feeding migrations in spring and summer. Return migration occurs in September to November. Spawning migration starts north of Iceland in January.

Highly Migratory Species

The term highly migratory species (HMS) has its origins in Article 64 of the United Nations Convention on the Law of the Sea. The Convention lists the species considered highly migratory by parties to the Convention. The list includes tuna, pomfret, marlin, sailfish, swordfish, saury sharks, dolphins and other cetaceans. These oceanodromous species undertake migrations of significant but variable distances across oceans for feeding, often on forage fish, or reproduction, and also have wide geographic distributions. These species are found both inside 200 mile exclusive economic zones and in high seas outside these zones. They mostly live in open ocean and do not live near sea floor, although they may spend part of their life cycle in waters. Straddling stock ranges both within an EEZ as well as in high seas. Transboundary stock range in EEZs of at least 2 countries. A stock can be both transboundary and straddling(Figure 6.29a-6.29e).

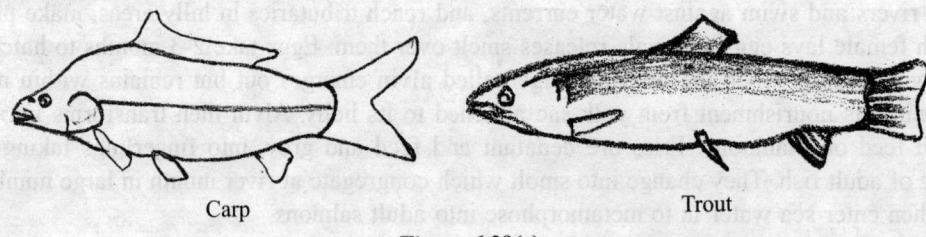

Carp   Trout

Figure 6.29(a)

Figure 6.29(b)   Tuna.

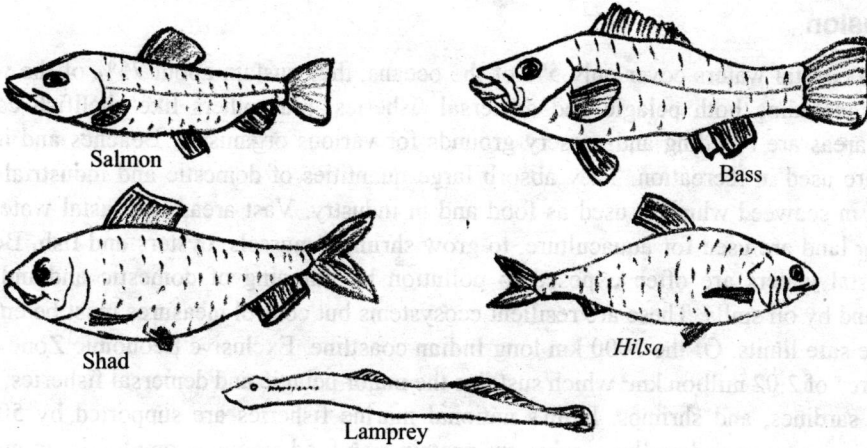

Salmon

Bass

Shad

Hilsa

Lamprey

Figure 6.29(c)

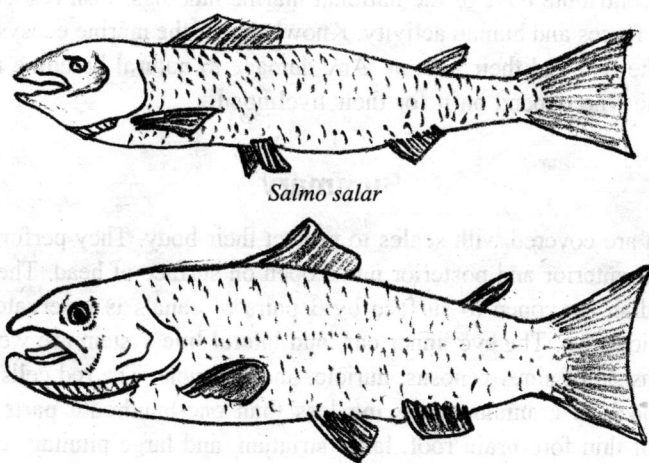

Salmo salar

Oncorhynchus

Figure 6.29(d)

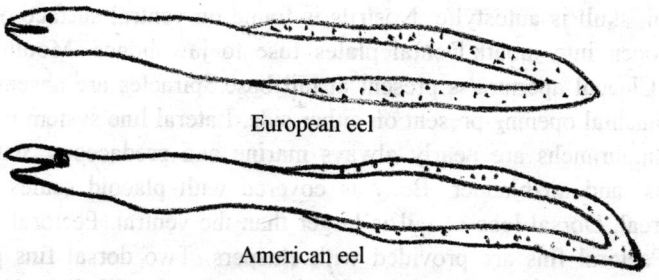

European eel

American eel

Figure 6.29(e)

Conclusion

Although coastal waters cover only 5% of the oceans, they sustain about 75% of the world's fisheries including both pelagic and demersal fisheries, plus others like shellfish catches. Coastal areas are breeding and nursery grounds for various organisms. Beaches and inshore waters are used in recreation. They absorb large quantities of domestic and industrial waste and rich in seaweed which is used as food and in industry. Vast areas of coastal waters and adjoining land are used for aquaculture, to grow shrimps, mussels, oysters and fish. Beaches and coastal waters are often exposed to pollution by dumping of domestic and industrial wastes and by oil spills. These are resilient ecosystems but control measures must be enforced to ensure safe limits. Of the 7000 km long Indian coastline, Exclusive Economic Zone (EEZ) has an area of 2.02 million km² which sustains the major pelagic and demersal fisheries, which produce sardines, and shrimps. India's national marine fisheries are supported by 500 fish species, crustaceans and molluscs whic are protein-rich food resource, employ large numbers of people and provide substantial foreign exchange through exports of marine products. States of the west coast contribute 65% of the national marine landings. Fish resources are affected by several natural forces and human activity. Knowledge of the marine ecosystems can help in conserving both the fish and their habitat. Any damage to natural breeding habitat threatens fish and the people who depend on it for their livelihood.

Summary

1. Most fish are covered with scales to protect their body. They perform respiration by gills. Both anterior and posterior nares open on surface of head. The rostral organ, a large median sac, open to surface by 3 pairs of canals is innervated by superficial ophthalmic nerve. The eye, inner ear, and lateral line system are well developed.

2. Heart consists of sinus venosus, auricle, and ventricle. The red cells are large. Brain lies far back in cranium occupying less than one-hundredth parts of cranium and consists of thin fore-brain roof, large striatum, and large pituitary cleft. There is no valvula to the cerebellum.

3. *Labeo rohita*, commonly called Rohu, a common bony fish inhabiting fresh water is found throughout India and Burma except in Tamilnadu and Western Coast. The fish attains a length of about 100 cm.

4. In Dipnoi, skull is autostylic .Nostrils is found on ventral surface of snout. Internal nostrils open into mouth.Dental plates fuse to jaw bones. Mouth lies on ventral surface. Cloacal aperture is present at tail base.Spiracles are absent. An operculum and a branchial opemng present on either side. Lateral line system is well developed.

5. The elasmobranchs are nearly always marine and predaceous, without operculum over gills, and air-bladder. Body is covered with placoid scales. Tail is usually heterocercal. Dorsal lobe of tail is larger than the ventral. Pectoral fin is anterior to pelvic. Pectoral fins are provided with claspers. Two dorsal fins provide stability against rolling, and help in vertical turning movements. Median and paired fins are supported by cartilaginous rods, the radials, and their edges are strengthened by

ceratotrichia. Shark is a long-bodied fish, swim by the passage of waves of contraction along the muscles.Sharks are good swimmers.

6. Osteichthyes commonly called bony fish lack placoid scale. Smooth and overlapping ganoid, cycloid, or cytenoid scales are found. Mucus glands coat the body .They are omnivorous carnivorous, herbivorous, or detritivorous.

7. Their dermal bones cover the pectoral girdle. A sclerotic structure of 4 small bones supports the eyeball. Inner ear contains labyrinth. Cranium is often divided into anterior and posterior sections. Bony fish have swim bladder but do not have fin spine, but instead the lepidotrichia.

8. Actinopterygians are the largest group of fishes comprise about 96% of all living fish species and make up half of all living vertebrates. They appeared during Devonian and in Carboniferous, they become dominant in freshwater and started to invade the seas. Habitat destruction, pollution and international trade, and other human activity have endangerd actinopterygians.

9. Swim bladder is a sac-like structure lying between alimentary canal and kidney and contains a mixture of carbon dioxide, oxygen and nitrogen. It is present in all bony fishes, arises form dorsal wall of gut. In dipnoans and polypteridae, it originates as ventral outgrowths form the floor of pharynx and later rotates to occupy dorsal positions. Air bladder primarily is a tough sac like structure with capillary network. Beneath capillary network lies tunica externa made of connective tissue, and tunica interna composed of smooth muscle fibers and epithelial gas gland. Generally air bladder opens into esophagus by duct know as ductus pneumatics.

10. Most fishes are confined to either salt, or fresh water and have a limited salinity tolerance. They are known as stenohaline species. In sea, concentration of salts is more than in body fluids of fish, so that water tends to diffuse out. Marine fishes tend to drink water. Fresh water fishes lose salts and water enters into body through semi-permeable membranes of buccal cavity and gills. This is due to .higher concentration of salts within body of fresh water fishes than surroundings.

11. In fish, most species abandon their offspring to the vagaries of environment. Many species care only for eggs, build nests with empty shells, vegetal matter, pebbles, and sand. Eggs are defended from predators. Parents fan the eggs, spray water over brood with fin to provide oxygen. They clean eggs by brushing with fins, remove dead eggs by mouth. Some coat eggs with mucus having anti-microbial properties. Many species carry eggs either outside or inside the body.

Review Questions

Short Answer Questions

1. Define fish.
2. Give two examples of elasmobranchs.
3. Give two examples of dipnoi.
4. What do you understand by lung fish?
5. Name the scales found in the living fish.
6. Give two examples of dipnoi.

7. What is swim bladder?
8. What type of scales is found in elasmobranch?
9. What is osmoregulation?
10. Differentiate between oviparous and viviparous?
11. What is food web?
12. Define induced breeding
17. What is a hatchery?
18. Name a fish possessing an electric organ
19. Name an organ, which regulates buoyancy
20. What is composite fish culture?
21. Define migration.
22. Define oceanodromous migration.
23. What is diadromous migration?
24. Name an amphidromous fish species
25. Mention the function of gill.
26. Name the two highly migratory fish.
27. Define parental care.
28. What do you understand by physostomous swim bladder?
29. What do ypou understand by physoclistous swim bladder?
30. What is ductus pneumaicus?

Long Answer Questions

1. Describe the process of digestion in *Labeo* with suitable diagrams
2. Describe and distinguish between the classes Chondrichthyes and Osteichthyes, noting the main traits of each group.
3. Identify and describe the main subgroups of the class Osteichthyes.
4. Draw and describe the circulatory system of *Labeo*.
5. Describe the anatomical features of dipnoi.
6. Describe various modifications of swim bladder in fish.
7. Describe the various types of parental care in fish.
8. Decribe various types of fish migration with examples.

7

The Amphibians

Amphibians link terrestrial and aquatic vertebrates. Their larval stage is found in water and larvae breathe by gills. Mature animals use lungs, or skin to breathe. The smooth skin contains several sweat glands. The amphibian skeleton is mostly made of bones. The backbone is made of vertebrae with their prominence. There is no chest and thorax breathing muscles and they inhale by swallowing. Amphibians' body temperature depends on the temperature of the environment. The amphibians also have middle ear and inner ear.

Amphibians lack scales, or hair. They were first vertebrate to inhabit land and water with two major subgroups: the frogs and toads, and the salamanders and newts. Frogs and toads are hoppers with large back legs and short front legs. Frogs, less resistant to drying than toads, prefer wet environments. Various species differ in size. Salamanders and newts are not scaly but lizard-shaped crawlers. Adult salamanders live on land. Newts have flat tails to swim and spend sometime on land before living in water as adults.

There are about 5,420 species of tailless amphibians. In many species, specialized glands secrete toxins to deter predators. They do not survive long except in proximity of water. Desert toad, *Chiroleptes* makes burrow and holds large amount of water. *Rana*, abundant throughout world, except south of South America, oceanic islands, and New Zealand, are specialized. Anurans inhabit marshes or ditches, live for most of their life in grass or undergrowth and feed on insects. Vocalization common in males attract females, or defend territory. Migration occurs to specific breeding habitats. Complex social behaviours especially during breeding are noticeable. External fertilization occurs mostly. Eggs are typically laid in water. Some are ovoviviparous, or viviparous and carry the eggs till hatching.

Extinct Amphibians

Eyes, of the extinct amphibians called labyrinthodonts, were situated at the top of skull and had special sense organs in the skin, forming a lateral line organ. They had massive stapes, likely anchoring the brain case to skull roof. They possessed well developed internal gills and primitive lungs to breathe air. Air was inflated into lungs by contractions of throat sac. There was no diaphragm. Many aquatic forms retained larval gills in adulthood. The loss of rhomboid scales allowed additional oxygen uptake through skin. They laid eggs in water. They would remain in water throughout the larval life till metamorphosis. Metamorphosed individuals would venture onto land on occasion. There is a gap (the "Romer's gap") in the fossil record in the early Carboniferous. However, they appear to be composed of several nested clades. Ichthyostegalia and the reptile-like amphibians are best known. The early labyrinthodonts known from the Devonian possibly extended into the Romer's Gap are often grouped together as the order Ichthyostegalia. Ichthyostegalians were predominately aquatic and had functional internal gills throughout life. Their polydactylous feet had more than the 5 digits and were paddle-like. Tail had true fin rays. Vertebrae were complex.

An early branch was the reptile-like amphibians, called Anthracosauria or Reptiliomorpha. *Tulerpeton* was perhaps the earliest member of the line, indicating the split before the Devonian-Carboniferous transition. Their skulls were relatively deep and narrow. Front and hind feet bore 5 digits. Except the diadectomorphs, the terrestrial forms were moderate sized creatures in the early Carboniferous. The vertebrae had small pleurocentra, which grew and fused to become the true centrum in later vertebrates. The best known genus is *Seymouria*.

The diverse group of labyrinthodonts, the Temnospondyli appeared in the late Devonian. Their fore limb had only 4 toes, and the hind limb had 5. Vertebral column had the small pleurocentra. The intercentra was large and forming a complete ring. All were flat-headed with vertebrae and limbs. The Lepospondyli were characterized by spool-shaped vertebrae formed from a single element. Most were aquatic with external gills. The best known genus is *Diplocaulus*.

7.1 Skin

Scales of ancient amphibia are now absent except some Apoda. Dermal plates on back fuse to neural spines in some frogs. Skin contains cornified outer layers. Epidermis of several layers is renewed by moulting in adult frog. Local thickenings of epidermis in warty skin of toads form horny teeth to aid in feeding. In larva, skin is ciliated. Skin glands either mucus or poison glands consisting of small sacs are highly developed. Mucus keeps skin moist and regulates temperature. Frogs in dry air are colder than their environment. Poison glands are less developed in *Rana* than in *Bufo*. Poison causes eye and nose irritation; rarely affect skin of hands on man. Poison of *Dendrobates* acts on nervous system and is used on arrows. Some produce smells to attract sexes to each other. In some male newts, special gland cells occur below the chin. Glandular secretions keep eyes and nostrils free from obstruction.

7.2 Colour

Posterior lobe of pituitary produces intermedin to expand melanophores. Temperature, incident illumination and humidity influence colour. Colour change is slow and is caused by 3 layers of pigment cells. Melanophores are found in deepest. Guanophores full of granules produce a blue-green colour by diffraction. Yellow lipophores overlying these filter out the blue. Animals are often greenish. Expansion of pigment in melanophores and movements of other chromatophores change colour. Yellow is produced by disarrangement of guanophores, while blue by absence of lipophores.

Changes in melanophores may be primary or direct, and secondary or visual. Primary response depends on direct effect of light on skin. Secondary effect contracts pigment on illumination on a light-scattering surface, but expand pigment on illumination from above on a light-absorbing background. In frogs, contact with water accentuates the black background response and in darkness produces expansion. Drying contracts the melanophores even on the black background. Brilliant green colour is uniform in tree-frogs. Conspicuous colour is associated with development of poisonous parotoid glands and is warning coloration. Many frogs make brightly coloured patches suddenly on thighs during jumping. In some anurans, colours are irregular dark marks.

7.3 Skull of Amphibians

Labyrinthodontia (extinct amphibians), includes all extinct amphibians and is ancestral to all extant terrestrial vertebrates. The name signifies the infolding of dentin and enamel of teeth. They had heavily armoured skull roof (hence the name "Stegocephalia"), and complex vertebrae. Skull was flat and skull roof had the openings for nostrils, eyes and parietal eye. Otic notch at the back edge of skull was prominent. Complex vertebrae made of 4 pieces, an intercentrum, 2 pleurocentra, and a vertebral arch/spine consisting of numerous, often poorly ossified elements. Relative sizes and ossification of the elements was variable.

Jaws were lined with small, sharp, conical teeth. They had a second row of teeth on roof of mouth. All teeth were labyrinthodont. Bones of the limbs were short and broad and the ankle had limited mobility. The toes lacked claws. The Mesozoic labyrinthodonts were mostly aquatic with more cartilaginous skeleton. Stegocephalians, first known from the Devonian, are distinguished by reduction in number of skull bones, neck, and replacement of fin rays by segmented bony digits. The pelvis was in contact with at least one vertebra. Skull of Devonian and Carboniferous amphibian was like that of osteolepid fishes in arrangement of bones. Preoptic region was large. Nasals and frontals were long in stegocephalians (Figures 7.1, 7.2, 7.3) while the parietals were shorter and post-parietals were absent in the later forms. Bones identified as frontal in fish were parietals, while the parietals were the post parietals, which have lost completely from most amphibians. Opercular apparatus covering the gills was lost early. The reduction of whole posterior part of head was perhaps effected by a single morphogenetic change.

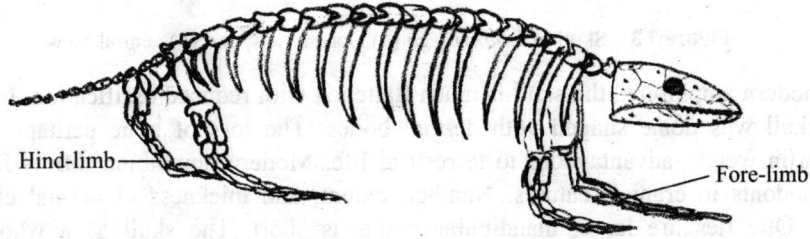

Hind-limb

Fore-limb

Figure 7.1 Skeleton of *Tuditanus*.

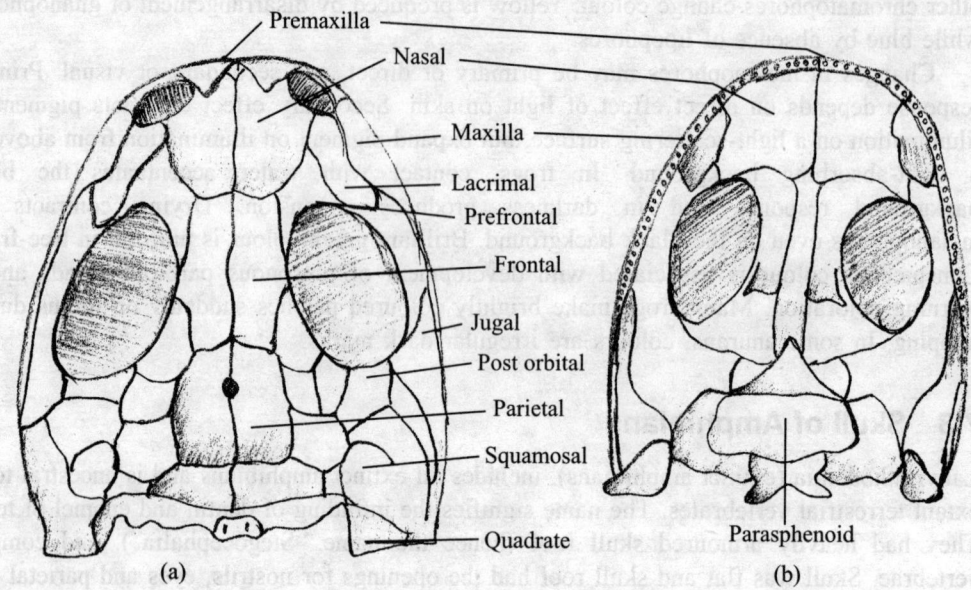

Figure 7.2 Skull of *Doleserpeton* (a) Dorsal view, and (b) Ventral view.

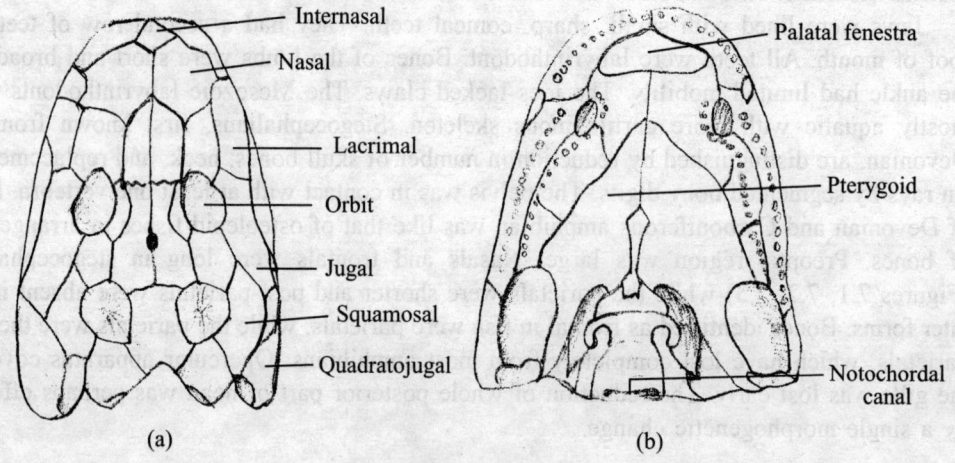

Figure 7.3 Skull of *Ichthyostega* (a) Dorsal view, and (b) ventral view.

In modern amphibian, the skull is much flattened with reduced ossification. In the earlier forms, skull was dome shaped with dermal bones. The loss of bone perhaps produced a reduction in weight advantageous to terrestrial life. Modern amphibian differs from typical labyrinthodonts in cranial features. Number, extent, and thickness of dermal elements are reduced. Otic ties are large, mandibular ramus is short. The skull as a whole is much flattened. Occiput is shortened. Hypoglossal nerve emerges behind skull. Parietal foramen has been lost. Skull of frog reduces and specializes from the early amphibian. It consists of cartilaginous boxes or capsules, the central neurocranium around brain, and olfactory and auditory capsules. Ossifications occur especially around foramen magnum, where the auditory capsule joins cranium and at the base of nasal capsules.

Paired occipital condyles found only in modern amphibia are formed by failure of basioccipital to become ossified. Dermal bones cover skull roof are nasals and frontoparietals, while on floor there are large dagger bone, the parasphenoid, and small tooth-bearing vomer. Remains of cartilaginous palato-pterygoquadrate bar is found as rod, covered in front by premaxillae and maxillae, and divides behind into an otic process fixing it to skull and a cartilaginous quadrate region articulates with the lower jaw. This region is covered by pterygoid ventrally, quadratojugal laterally, and squamosal dorsally. Palatines, the membrane bones form the anterior wall of orbit. Upper jaw is supported by struts formed from nasals and palatines in front and squamosal and pterygoid behind, that forms a large mouth for respiration and eating insects (Figure 7.4).

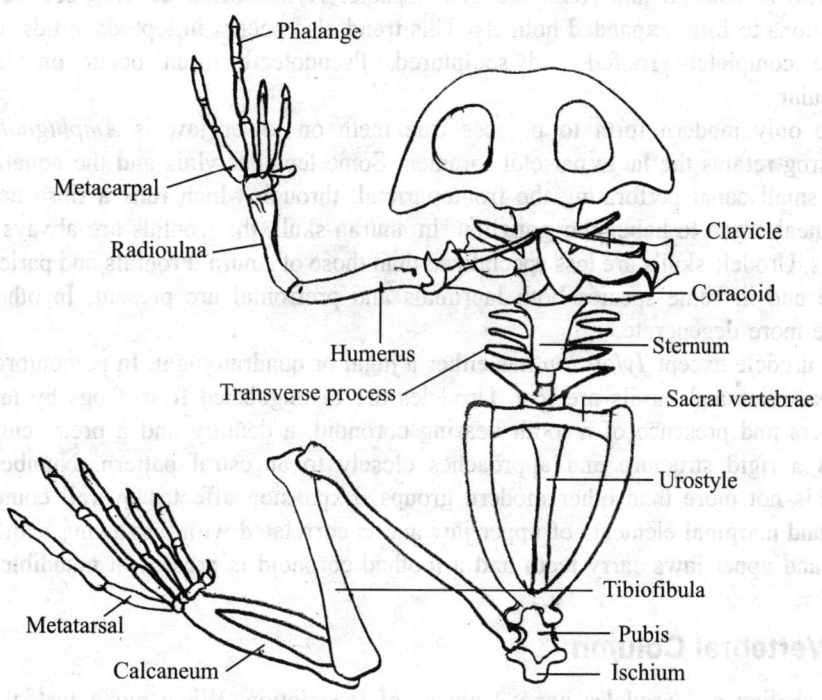

Figure 7.4 Skeleton of *Hyla*.

Lower jaw consists of Meckel's cartilage, covered on its outer surface by a dentary and on its inner by an angulo-splenial bone. Anterior tips of cartilages ossify as mento-Meckelian bones. Visceral arches are well formed in tadpole, but are modified in adult frog. In tadpole, skeleton of hyoid arch consists of a large pair of ceratohyals attached to basal hypohyal. In subsequent metamorphosis, the ceratohyals form the long anterior cornu of hyoid, attached to prootic bone.

Body of hyoid is a plate lying in floor of mouth and formed from hypohyal and from hypobranchial plate at the base of remaining arches. Posterior cornua support the floor of mouth and the whole apparatus help in respiration. Sixth and 7th of the series of branchial arches give rise arytenoid and cricoid cartilages of larynx respectively. Lateral plate muscles of branchial arches are well developed. Certain muscles of scapula are innervated from vagus and recall the sternomastoid and other muscles innervated by spinal accessory nerve in mammals.

Muscles of hyoid arch innervated by facial nerves, as depressor mandibulae run from back to angle of jaw and lower the floor of mouth. Jaw closing muscles are innervated by trigeminus. They run from hind end of jaw to surface of skull and squamosal. Skull and jaws of frog protect brain and special sense organs. Front part of skull concerned with nose, eyes and brain, is increased in size and hind part, originally concerned with gills and pharynx, is reduced.

Bufonid skulls are completely devoid of teeth but possess a supratemporal bone, which fuses with squamosal and roofs the otic capsule. Hylids often develop secondary dermal ossifications to form expanded helmets. This trend also occurs in leptodactylids, where skull may be completely roofed and sculptured. Pseudoteeth often occur on dentary and prearticular.

The only modern form to possess true teeth on lower jaw is *Amphignathodon*. No recent frog retains the large parietal foramen. Some leptodactylids and the aquatic xenopids have a small canal perforating the frontoparietal, through which runs a fibro nervous tract from pineal organ to habenular ganglion. In anuran skull, the frontals are always fused with parietals. Urodele skulls are less specialized than those of Anura. Frontals and parietals remain discrete and in some species both lacrimals and prefrontal are present. In other respects, they are more degenerate.

No urodele except *Tylotriton* has either a jugal or quadratojugal. In perennibranchs, even the maxillaries and nasals are lost. Urodeles are distinguished from frogs by large size of prevomers and presence of a tooth bearing coronoid, a dentary and a prearticular. Apodan skull is a rigid structure and approaches closely to ancestral pattern. Number of bones present is not more than other modern groups. Expansion affectes overall compactness of nasals and marginal elements of upper jaw and is correlated with burrowing habits of group. Lower and upper jaws carry teeth and a toothed coronoid is present in mandible.

7.4 Vertebral Column

Stegocephalian and urodeles have 2 means of locomotion. When move fast, they wriggle along with belly on ground, by serial contraction of myotomal musculature. During deliberate movement, newt raises its body on legs, and then propels it along as movable levers, by humerus or femur. Distal muscles of limbs keep the digits pressed against ground. Carrying the weight on 4 legs places new set of stresses on vertebral column which as a girder carry the body weight and transmits it to legs. These results in column with largely bony parts articulated together.

Many urodeles with vertebrae often lacking ossification spend much time in water. Parts of notochord persist to provide the compression for swimming. In anurans, the skeletal and muscular system is specialized for swimming and jumping, by extensor thrusts of both hind limbs. Frogs and toads walk on land by myotactic reflexes depending on muscle contraction against an external resistance. Jumping and walking are performed by changing arrangement of skeleton and muscles. Myotomal muscles do not produce metachronal waves of contraction.

Vertebral column attaches to pelvic girdle to act as a support by which movement of hind limbs is transmitted to rest of body. The urostyle is found. Shortening of body is a change from aquatic to terrestrial life. Second to 8th vertebrae of *Rana* are concave in front, convex behind with large transverse processes. In others, they may be amphicoelous or opisthocoelous. They fit together by zygapophyses. First vertebra has 2 concave facets for articulation with 2 condyles of skull; its centrum and transverse processes are reduced. Large transverse processes of 9th vertebra articulate with ilia of pelvic girdle. There are free ribs in primitive frogs *Ascaphus* and *Leiopelma*.

7.5 Pectoral Girdle

In the earliest labyrinthodonts (Figure 7.5), shoulder girdle was like that of their osteolepid ancestors except that interclavicles were added to the ventral surface. Perhaps a cartilaginous structure was present between the hindermost margins of epicoracoid cartilages. The girdle consisted of a primary or endochondral component evolved from basal fin elements of ancestral fish to provide an articulatory surface for limb and points of attachment for limb musculature, and a dermal ring of bony elements which sunked inwards and applied to ventro-anterior surfaces of endochondral girdle.

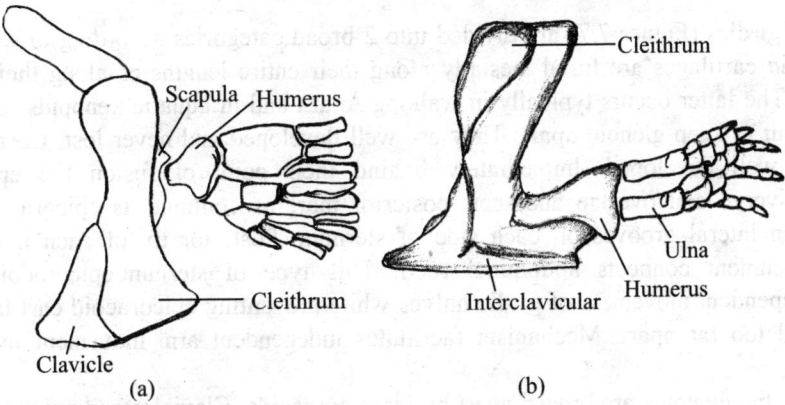

Figure 7.5 Shoulder girdle and Pectoral fin of (a) an osteolepiform, (b) a fossil amphibian to illustrate resemblance in limb pattern.

Endochondral girdle consisted of 2 overlapping half rings in ventral midline. Each half was a unit but by topographical comparison with girdles of later tetrapods. It is often divided into 2 regions, dorsal scapula and ventral coracoid. Between these 2 regions, a screw shaped glenoid received humerus. Endochondral ossification is usually homologized with scapula of amniotes. Later forms possessed a second bony element called the precoracoid.

Endochondral girdle was small in the earliest amphibia. In later genera, its size increased to withstand the greater thrust transmitted by larger limbs and to provide attachment for increased mass of brachial musculature.

Dermal girdle consisted of paired cleithra, clavicles, and interclavicle. The latter lay often between beneath clavicles and together with sternum formed a locking mechanism preventing complete separation of epicoracoid cartilages. In the earliest rhachitomes, the dermal girdle was attached to post-temporal region of skull. This connection was lost in later forms to provide mobility to head. This foreshadowed the reduction and loss of dermal and shoulder girdle elements (Figure 7.6). Of the modern amphibia, Salientia nearly approach the condition of fossil forms. They alone of recent tetrapods have retained a cleithrum. Each half of endochondral girdle consists of dorsal, bony scapula with a cartilaginous suprascapula, and ventral coracoid bone connected to an anterior precoracoid cartilage by a mesial epicoracoid cartilage. Precoracoids are invested by clavicles and, as in all modern amphibians the interclavicle is absent.

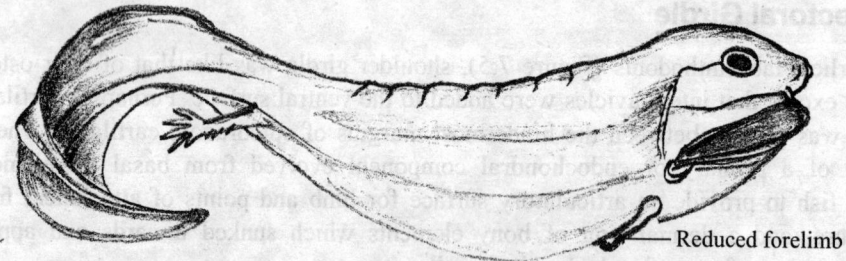

Figure 7.6 Reduced fore-limb in *Crassgyrinus*.

Anuran girdles (Figure 7.7) are divided into 2 broad categories according to whether the 2 epicoracoid cartilages are fused mesially along their entire lengths or along their anterior edges only. The latter occurs typically in walking Anura and in aquatic xenopids. Clavicle is the main strut to keep glenoid apart. They are well developed and never lost. Coracoids are moderately well developed. Immediately behind their point of fusion the epicoracoid cartilages diverge and overlap and their posterior margins continue as epicoracoid horns, which run in lateral grooves on each side of sternum. Posterior tip of each horn with a muscle attachment connects abdominal recti. This type of sternum/epicoracoid system permits independent movement of girdle halves while preventing epicoracoid cartilages from being forced too far apart. Mechanism facilitates independent arm movement in arciferal frogs.

In frogs, the glenoids are braced apart by large coracoids. Clavicles and precoracoids are deprived of their strutting function and become reduced or even completely lost. Epicoracoid horns are absent and sternum, no longer involved in locking girdle halves, serves for attachment of pectoral muscles. This function is also performed by a prezonal element, an extension of precoracoid cartilages. Shoulder girdles of modern urodeles are largely simplified, the only ossification being a scapulo-coracoid encircling glenoid. Two epicoracoids overlap broadly and anteriorly are free of each other. Posteriorly they are weakly locked by a cartilaginous sternum. Apoda retain no vestiges of either limbs or limb girdles.

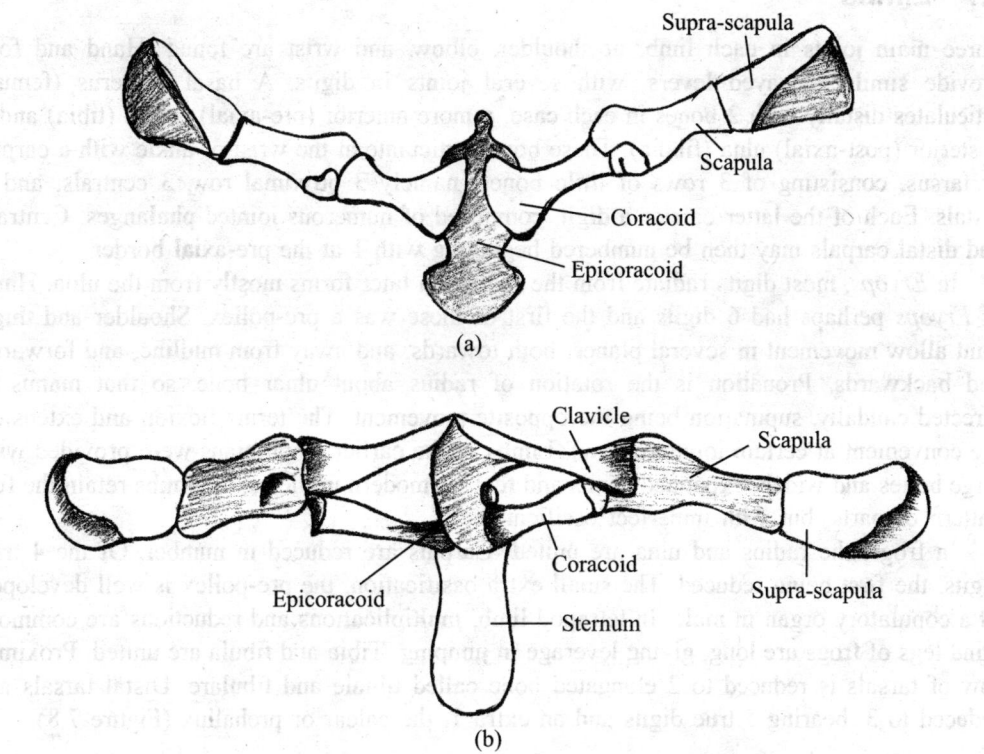

Figure 7.7 Pectoral girdle in anuran (a) in *Kaloula,* (b) in *Bufo.*

7.6   Pelvic Girdle

This girdle is formed of 3 main bones. Dorsal ilium attaches to modify transverse processes of one, or more sacral vertebrae. Ossification along a line of compression stress occurs in ilium due to weight bearing. Ventral portion of girdle consists of an anterior pubis and posterior ischium, the 3 bones meeting at the acetabulum, where the femur articulates. Girdle provides a plate to which muscles that brace the limb can be attached to balance the body on leg. In urodeles, the girdle is reduced and mainly cartilaginous, while in anurans it is highly specialized. Ilia are long, directed forward to articulate with transverse processes of sacral vertebrae. Base of ilium make dorsal portion of a disk, of which pubis is anterior, ischium the posterior, with acetabulum at the centre.

Girdle is developed into a long lever to transfer force from limb to vertebral column during jumping. Considerable movement is possible at the ilio-sacral joints in Salientia. In *Rana*, ilia may rotate through an angle of over 90 on sacral ribs in vertical plane which is used during a strong leap. In *Discoglossus*, sacrum turns laterally on pelvis through 20°. Movement is used both in turning to take food and in locomotion. In *Xenopus*, sacrum slide backwards and forwards on pelvis, producing a shortening and lengthening of whole animal. This movement is used in driving into mud.

7.7 Limbs

Three main joints in each limb, at shoulder, elbow, and wrist are found. Hand and foot provide similar 5 rayed levers, with several joints in digits. A basal humerus (femur) articulates distally with 2 bones in each case, a more anterior (pre-axial) radius (tibia) and a posterior (post-axial) ulna (fibula). These bones articulate at the wrist or ankle with a carpus or tarsus, consisting of 3 rows of little bones, namely 3 proximal row, 3 centrals, and 5 distals. Each of the latter carries a digit, composed of numerous jointed phalanges. Centrals and distal carpals may then be numbered beginning with 1 at the pre-axial border.

In *Eryops*, most digits radiate from the radius, in later forms mostly from the ulna. Hand of *Eryops* perhaps had 6 digits and the first of these was a pre-pollex. Shoulder and thigh joint allow movement in several planes, both towards, and away from midline, and forwards and backwards. Pronation is the rotation of radius about ulnar bone, so that manus is directed caudally, supination being the opposite movement. The terms flexion and extension are convenient at certain joints (elbow). Limbs of the earlier amphibians were provided with large bones and widely expanded hands and feet. In modern urodeles, the limbs retain the full pattern of parts, but with imperfect ossification.

In frogs, the radius and ulna are united. Carpals are reduced in number. Of the 4 true digits, the first being reduced. The small extra ossification, the pre-pollex is well developed as a copulatory organ in male. In tetrapod limb, multiplications and reductions are common. Hind legs of frogs are long, giving leverage in jumping. Tibia and fibula are united. Proximal row of tarsals is reduced to 2 elongated bone called tibiale and fibulare. Distal tarsals are reduced to 3, bearing 5 true digits and an extra 1, the calcar or prehallux (Figure 7.8).

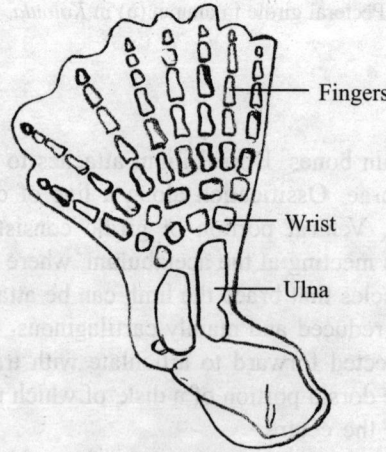

Fingers

Wrist

Ulna

Figure 7.8 Development of forelimb from the common tetrapod ancestor.

7.8 Back and Belly Muscles

In urodeles, dorsal musculature is well developed. In anurans, dorsal portions of myotomes, the epaxial musculature, do not produce locomotory effect by lateral flexion. In frogs, the musculature bends body dorsally to brace vertebral column on sacrum. Short muscles run between vertebrae and dorsal to these longissimus dorsi muscle run from head to sacral vertebra and urostyle. This muscle is crossed by tendinous intersection. At the hind end, coccygeosacralis and coccygeoiliacus muscles brace the urostyle.

Pectoral girdle is attached to axial skeleton by muscles. Rhomboid and levator scapulae muscles run from suprascapula to vertebrae and skull.

Cucullaris muscle run from skull to suprascapula. It is derived from lateral plate musculature and innervated by vagus. Hypaxial musculature is developed than in fish and differentiated into several parts for slinging the viscera. The plan found in all tetrapods, include four sets of fibres. The external oblique runs caudally and ventrally. Inside this layer the internal oblique run in opposite direction and within this transversus abdominis run nearly dorsoventrally. Rectus abdominis consists of fibres in midline, run antero-posteriorly.

In adult frog, 3 sets of fibres are found. In mid-ventral region, the longitudinally arranged fibres of rectus abdominis make a sling between the sternum and pubis. Fibres are interrupted at intervals by transverse fibrous tendinous inscriptions. In the mid-ventral line, there is tendinous linea Alba. Sling formed by rectus abdominis is supported laterally by thin sheets of muscle-fibres which run up to vertebral column, the obliquus externus and transversus abdominis. In the anterior region, hypaxial muscles is restricted to throat, form hyoid musculature, which raise and lower the floor of mouth. Submaxillary muscle runs transversely between rami of jaw. Deep to this lie other muscles including sternohyoid, close to midline, which is a continuation of rectus abdominis.

7.9 Limb Muscles

Muscles of limbs were perhaps derived from radial muscles. In modern amphibian, the limb musculature is partly formed from myotomes. In modern urodeles, the limb is brought forward and its joints flexed as epaxial muscles at the level of its front end contact, and then pass back and extend as the wave of contraction moves past. This makes the limb lever. Limb muscles are derived from those that raise and lower the fins of fishes, modified to brace the limbs and move them to allow standing and walking.

Muscles that run from girdles to humerus and femur draw leg forward and backward. Actions of various bundles are not confined to a single plane. All the muscles running from back to humerus can raise the upper limb. But the more anterior members also protract, while the more posterior retract it. Anterior members of a ventral series work with anterior dorsal muscles as protractors, although they antagonize the action of raising whole limb. Many muscles have a rotating action on humerus and femur. Muscles of arm and leg fall in 2 major groups. More anterior and ventral set draw the limb mainly forward and towards midline and to flex its more distal joints. More posterior, dorsal mass which draw the limb backwards and away from body and extend its joints. In fore limb, the proximal members of ventral group make a sheet of fibres running transversely to main body axis and attached to sternum and hypaxial muscles at one end and to humerus at other. Within this sheet, there are deltoideus, pectoralis, coracoradialis, and coracobrachialis muscles.

In limb, this group is continued, there being a set of muscles in each segment to flex it on next. Brachioradialis flexes the elbow joint and in forearm the flexor carpi radialis and flexor carpiulnaris flex wrist. Flexor digitorum longus muscle arises from medial epicondyle of humerus and is inserted by tendons to carpus and terminal phalanges. Flexor digitorum brevis muscles arise from this tendon for insertion on digits. Of the dorsal muscle mass, the latissimus dorsi and dorsalis scapulae are the most proximal, running from middle of back to humerus and abduct and draw back the whole limb. Triceps extend the elbow. In forearm, there is extensor carpi ulnaris and radialis and extensors for fingers. Protractor muscles lie mainly anterior to retractors. Spinal roots innervate flexor muscles roots anterior to those for extensors.

Movements of limbs show the passage of an excitation wave backwards along spinal cord. Many muscles produce rotation and movement in the body. In hind limb, muscles of the same 2 general types are anterior muscles which draw the limb forward and flex and adduct its joints, and posterior ones which draw it back and extend and abduct. In thigh, the muscles of anterior group are pectineus and adductors, running from pelvic girdle to femur and move the whole limb inwards.

The sartorius, biceps, semimembranosus, and semitendinosus are 2-joint muscles mainly producing flexion at knee and at hip. The more posterior and dorsal group of muscles includes the gluteus and tensor fascia lata from girdle to femur, extending the thigh joint, and the very large cruralis running from girdle and femur to tibia. This is the main extensor of knee, being helped by gracilis and semimembranosus, and plays important part in jumping of frog. In the shank, the arrangement of flexors and extensors into anterior and posterior groups is modified. The tibialis anterior and peroneus run from femur to tarsus to flex the ankle joint. Long and short flexors move the toes as in forelimb. At the back of tibiofibula, the gastrocnemius runs from femur to be attached by the tendo achillis to tarsus. It extends the ankle in jumping and swimming. Tibialis posterior runs from the tibia to tarsus.

7.10 Respiration in Frog

Lungs are paired sacs, open to a short laryngeal chamber, which communicates with pharynx by a median aperture; the glottis.Glottis and laryngeal chamber are supported by arytenoid and cricoid cartilages. Arytenoids guard the opening of glottis and are moved by special muscles. During breathing, the mouth is kept tightly closed. The lips make an air-tight junction. Air is sucked in through nostrils by lowering the floor of mouth by hypoglossal musculature, and can then either be breathed out again or forced into lungs by raising the floor.

External nares are closed by a special pad on anterior angle of lower jaw, supported by mento-Meckelian bones. This pad is thrust upwards and pushes premaxillaries apart, to alter the position of nasal cartilage for closing the nostrils. This mechanism is found only in anurans. In urodeles, valves are provided with smooth muscles close the nostrils. Such valves in frog are functionless. After a period of slight movements, the nostrils are kept closed while throat is lowered. Air is drawn from lungs and then returned to them once or twice before the nostrils are reopened. The whole procedure ensures the maximum gaseous exchange.

Basic rhythmic mechanism, centred on nerve cells of medulla oblongata, is no doubt the same throughout, but anurans have improved on it. Skin is very vascular and plays part in respiration, by removing more CO_2 than the lungs. There is regulation of exchange in lungs. Rate of breathing depends on effect of CO_2 tension of blood on respiratory centre in medulla. There is also a vasomotor control of blood supply to lungs and, through vagus nerve, of the state of contraction of latter. By such means, respiratory exchange is greatly increased during breeding season.

Respiratory Adaptation

Skin and lungs show variations according to habitat, special devices being adopted to live in particular environments. Lungs vary from the well vascularized sacs with a highly folded surface as found in frogs, to small simple sacs in some stream-living amphibia. Lung lifts the animal in water. It is reduced in *Ascaphus*. In newts, hydrostatic function of lungs is predominant. Lung is entirely lost in stream-living salamanders. Coldness of the water lowers the need for respiratory exchange and it can be fully met by skin.

Skin shows increased circulation in forms with reduced lungs, capillaries reaching nearly to the outermost layers of epidermis. In *Astylosternus*, the lungs are vestigial and the male develops vascular papillae on waist and thighs during breeding season. Gills are extensions of branchial arches, and carry branched villi, richly supplied with blood and present in larvae and in certain adult urodeles. Where the main trunk is long, the gill is the external. In other cases, the filaments are directly attached to arch and are called internal. There are no profound differences between the 2 types.

Male frog produces sound to attract female. Both sexes have vocal organs, but those of female are small. Vibration of the laryngeal chamber, the vocal cords, produces noise. Air is passed backwards and forwards between lungs and large pair of sacs, the vocal pouches, formed below mouth. These serve as resonators, and are developed only in male.

7.11 Digestive System

Nearly all adult feed on invertebrates, mainly insects, partly worms, slugs and snails, spiders, and millipedes. Larval stages are usually omnivorous, may be cannibalistic, feeding on tadpoles of the same, or other species. Tongue catches food and is reduced in aquatic amphibia. In *Rana*, it is attached to floor of mouth anteriorly and flicked outwards by its muscles. To keep the skin moist and sticky, intermaxillary gland is found. From the shape of premaxillae it is assumed that this gland was present in labyrinthodonts. Saliva contains weak amylase and some protease. Special tracts of cilia carry the secretion from intermaxillary glands to vomcronasal organ and palatal taste buds. Presence of cilia keeps the fluids moving over the oral surfaces. Cilia are absent in aquatic amphibia.

Teeth (Figure 7.9) on premaxillae, maxillae, and vomers prevent the escape of prey. Few amphibia bites. Biting teeth are present in adult *Ceratophrys ornata*. *Amphignaihodon* has teeth in lower and upper jaw and presumably has redeveloped them. Teeth are present on lower jaw of most urodeles. Esophagus is not sharply marked off from either mouth or stomach and the latter is a simple tube. Its lining epithelium of mucus secreting cells is folded and simple tubular glands open at the base of folds. These glands are composed of a single type of cell, which secretes both acid and pepsin.

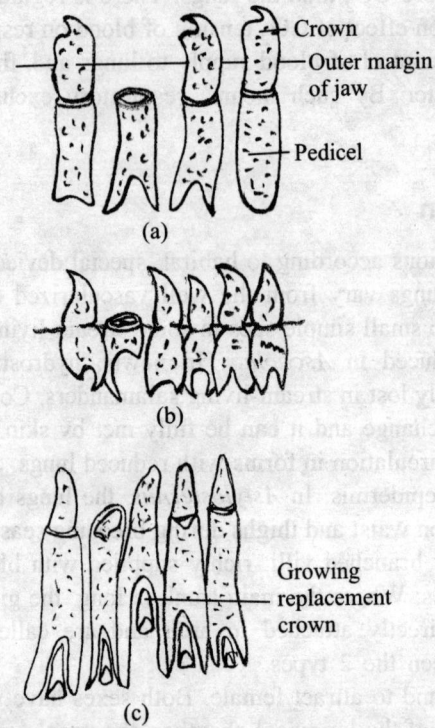

Figure 7.9   Teeth of (a) Salamander, (b) Caecilian, and (c) Anuran.

Intestine is marked off from stomach by pyloric sphincter. It is relatively short and dilates into a large intestine, there being a valve interposed in frogs, though not in all amphibia. Liver and pancreas produce juice. Intestine of omnivorous tadpole is coiled than that of adult frog. Most amphibians are not particular feeders. They learn to avoid distasteful insects. Frogs and toads devour large numbers of insects. If the common insects available are pests, the amphibians play a role in controlling their number works to the advantage of man.

7.12 Circulatory System

Venous and arterial systems are less fully separated. Auricles are completely divided by an inter-auricular septum, venous blood returning to right, arterial to left auricle. A single ventricle with spongy projections in its wall prevents the mixing of blood. Ventral aorta is very short, springs from right side of ventricle and receives first the venous blood. Conus arteriosus has transverse and longitudinal valves. Arches are much modified in adult. Of the original 6 arches the first 2 disappear, the 3rd on each side gives rise carotid artery, the 4th remains complete and forms systemic arch. The 5th remains in some urodeles, but disappears in anurans. The 6th arch becomes the pulmonary artery and loses its connection with dorsal aorta.

Special cutaneous arteries carry deoxygenated blood to skin. Pulmonary arches probably offer the lesser resistance than the systemic and carotid ones. The pressure in the latter increased by carotid gland, which is a receptor, connected with regulation of blood pressure. The first blood leaving the ventricle contracts flows to lungs. In anurans, this separation may be helped by an arrangement such that the pulmonary arteries join at their base and open to dorsal part of truncus arteriosus, which is partly separated from more ventral cavum aorticum, leading to carotid and aortic arches.

The classical view suggests that as the pressure raises the truncus contracts and spiral valve moves to force all the blood that leaves the ventricle during the later part of its contraction into ventral portion and to systemic and carotid arches. In this way, separation of blood from the right and left auricles would be achieved. Since the blood from skin returns to right auricle, it is not clear that a separation of streams would be advantageous.

Spongy walls of ventricle allow metabolic exchange as the heart is provided only with small coronary arteries. The posterior cardinal veins are replaced early in life by vena cava inferior. Most blood from hind limbs passes through renal portal system. An alternative path through pelvic veins and a median anterior abdominal vein, which breaks into capillaries in liver, is also suggested.

Extrinsic nerves regulate blood pressure of heart, fibres from vagus tend to slow and from the sympathetic nervus accelerans tend to speed the beat. The latter nerve is new development, as there is no sympathetic innervation of heart in fish. Diameter of arteries throughout body is under control from sympathetic vasoconstrictor and perhaps also vasodilatator nerves. Arterioles in web of foot constrict, when medulla oblongata is stimulated. Substances extracted from posterior lobe of pituitary and from adrenal medulla cause constriction of arteries and perhaps also of capillaries. There is a complex mechanism for ensuring that pressure of blood is maintained and flow directed into part of body that requires it for the time being.

Lymphatic System

Transduction affects the transfer of substances between cells and blood stream through capillary walls into tissue fluids. Under the pressure of heart beat, water and solutes leave capillaries; pass through their walls, while proteins remain behind. Blood pass into venous ends of capillaries has a high colloid osmotic pressure and this suck back fluid from tissues. Acirculation from capillaries into spaces around cells is produced.

Lymphatic system consists of a set of spaces which communicate with tissue spaces around capillaries. Lymph spaces in tissues join to form larger channels and great sinuses, such as that below the loose skin of back of frog. Lymph is kept circulating by action of lymph hearts. In frog, anterior and posterior of lymph hearts open into veins. The more posterior pair lies on either side of coccyx. Lymphatic vessels assist in repairing. If, after injury, red cells come to lie in tissues, the lymphatics send out sprouts to pick them up and return them to blood stream.

Blood

Red corpuscles are larger, formed mainly in kidney, and are destroyed after about 100 days by spleen and liver. Bone marrow is a source of red cell formation in *Rana temporaria* but not except during the breeding season in *R. pipiens*. Haemoglobin of frog has lower affinity for oxygen. The affinity of blood to combine with CO_2 is great, but there is a less delicate regulation of reaction of blood. White cells are of 3 types lymphocytes with a lytic macrophages, and polymorphonuclear granulocytes. These last may be neutro-, eosino, or basophil and are migratory and phagocytic. Blood of frogs contains numerous small platelets which probably break down, when in contact with foreign surfaces to produce thrombin that combines with fibrinogen of blood plasma to produce clotting.

7.13 Urinogenital System

Excretory organs of adult are always the tubules of mesonephros. In *Rana*, these extend over only a small number of segments. The kidneys are compact. In urodeles, the kidneys are elongated. Mesonephros consists of tubules leading from nephric funnels to Wolffian duct. In frog, the funnels do not open into tubules, but into veins. They form independently of the rest of tubule. In adult, there are some 2,000 glomeruli, from each of which a short ciliated tube leads to proximal convoluted tubule. Their follows a 2nd short ciliated region leading to a distal convoluted tubule, which joins Wolffian duct.

Blood supply of kidney differs from that of mammals as blood arrives from 2 distinct sources; branches of renal artery run mainly to glomeruli, those of renal portal vein to tubules. This corresponds to function now well established for those 2 parts, namely that the glomerulus filters off water and crystalloids, some of which are then reabsorbed by tubule. Frog having a moist skin remains in constant danger of osmotic flooding with water when it is submerged, and of desiccation when on land. Flooding is prevented by efficient functioning of glomeruli; they allow frog to excrete as much as one-third of its weight of water per day. Mechanisms for resistance to desiccation are less perfect. There is no long water reabsorbing segment, the part of tubule corresponding to Henle's loop being short.

Water can be reabsorbed from a large cloacal (allantoic) bladder. *Chiroleptes* conserve water by losing glomeruli altogether. *Rana cancrivora* is euryhaline and may have 2-9% of urea in blood. Mullerian duct develops separately from Wolffian system in frog, but arises from the latter during development in urodels. In this, the frog shows a greater degree of specialization of its developmental processes. Ovaries are the folds of peritoneum having no solid stroma. Follicle cells around each egg presumably produce ovarian hormones. Ripening of eggs proceeds under influence of a hormone of anterior pituitary which is controlled by external environmental factors to ensure breeding in spring. Walls of oviduct are glandular and secrete albumen. They are dilated at the lower end to form uterine sacs, in which eggs are stored until laid. Testes discharge directly through mesonephros by special ducts, the vasa efferentia, formed by outgrowths from mesonephros into gonad. Sperms pass through kidney. In *Alytes*, the sperms do not pass through kidney.

In some frogs, a special diverticulum, the vesicula seminalis lead by several small channels to lower end of Wolffian duct. It contains spermatozoa during breeding season.

Most amphibia have failed to complete transfer to land life and return each year to water to breed. Special modifications of reproductive system for land life are not found. Secondary sexual differences are marked in many species. In frogs, the males precede the females to water and then attract the latter by vocal apparatus. Male clings to back of female by nuptial pad, developed as an extra digit, prepollex, on the hand. In newts, an elaborate courtship ensures fertilization.

Sperms are made into spermatophores by special pelvic and cloacal glands. Abdominal glands produce a secretion attractive to female. After a courtship, the spermatophores are picked up by cloaca of female and stored in a spermathecal chamber. In caecelians, males are provided with intromittant organ for internal fertilization. Some species are oviparous, aquatic breeders. Some species are viviparous: young feed on "uterine milk". In caudata, internal fertilization occurs via spermatophore. Gelatinous cap containing sperm is picked up by the cloaca of the female. Salamanders are oviparous. In anura, external fertilization is most. Elaborate advertisement vocalization in males (species specific calls). Male grasps female from behind, stimulates oviposition.

7.14 Nervous System

Plan of spinal cord is like that of fishes, but dorsal and ventral horns are well-marked. The large motor cells of cord have dendrites that spread widely in white matter, where their synaptic connections are made in a complicated 'neuropil'. Development of limbs modifies arrangement of spinal nerves. Ten spinal nerves are found but since an embryonic first one is missing they are sometimes numbered 2-1 1. Two spinal segments contribute to brachial and 4 to sciatic plexuses in frog. From these plexuses, fibres are distributed to muscle and skin of limbs. Brain resembles that of Dipnoi very strikingly.

Prosencephalon is based on an inverted plan. The large evaginated cerebral hemispheres have a thick nervous roof and floor. In the frog, there is only a short unpaired region of forebrain (diencephalon) but this is longer in urodeles. Walls of each hemisphere may be divided into dorsal pallium, medial ventral septum, and latero-ventral striatum. Cell bodies lie around ventricle in all parts of hemisphere. Cells are pyramidal in shape and the connections are made in outer 'white' matter. Nearly all parts of hemisphere are reached by olfactory tract fibres, the axons of mitral cells of olfactory bulb. In frog, there are regions at the hind end of hemispheres that receive forwardly directed fibres. Some probably connected with tactile and others certainly with optic impulses. Hemispheres act as correlating centres. Their backward projections are made by 2 large tracts, the lateral and medial forebrain bundles, but these reach only to thalamus, hypothalamus, and midbrain, not back to cord.

Diencephalon is interesting for considerable number of glomerulus in which fibres of olfactory nerve contact dendrites of mitral cells. In anurans, but not in urodeles, there is a partial division into separate thalamic sensory nuclei, for touch, sight, and other receptor modalities. Pineal organ shows evidence of retina in few amphibia. Pituitary body is well developed and the usual parts, anterior, intermedia, nervosa, and tuberalis can be recognized. Midbrain is very well developed. Cells it contains do not lie round the ventricle. Many have moved out to make an elaborate system of cortical layers. Most fibres of optic tract end here and there are other pathways from olfactory, auditory, medullary, and spinal regions. Efferent fibres leaving the tectum pass to midbrain base, medulla, and perhaps back into cord. This region has wider connections than other part of nervous system and nearly reaches status of a dominant integrating organ. Cerebellum is very small.

Skin Receptors

Lateral line organs are present in skin of all aquatic amphibian larvae and in some aquatic adults. They are of simple form, consisting of groups of cells in an open pit. In newts, they are present in larvae, which are aquatic, but are covered by epidermal layers during first post-larval stage during which newt lives on land. Skin, also contains tactile organs, and is often sensitive to chemical stimuli. This chemical sense is mediated by fibres running in spinal nerves, not by special elements. Skin is sensitive to heat and cold. Senses are served by fibres different from those that mediate touch, pain, or chemical senses. All the nerve endings are free nerve-endings, except for a few touch corpuscles on special regions like feet. Taste buds on tongue and palate respond to presence of 2 types of substance.

Chemoreceptors in tongue of frog respond to salt and sour substances, but that no reaction is given to substances that in mammals are classed as sweet or bitter. Olfactory organ functions both on land and water, special mucus glands keep it moist when in air. A continual circulation of water, or air is maintained over olfactory epithelium by cilia, or movements of respiration. Internal nostril makes a circulation around olfactory receptors possible. Jacobson's organ is a special diverticulum of olfactory chamber to test the 'smell' of food in mouth. Apoda, being blind, have the sense of smell.

Eyes

Upper lid is fixed; the lower is very mobile and folded to make a transparent structure. The nicitating membrane moves rapidly across the eye surface. Eyeball is almost spherical with a rounded cornea. Lens is farther from the cornea and is flattened, more so in anurans than in urodeles. These modifications focus more distant image. There is an iris with a rapidly moving aperture, operated by powerful circular and radial muscles. These muscles are partly actuated by a nervous mechanism and are directly sensitive to light, and pupil of isolated eye of frog shows wide excursions with change of illumination.

Accommodation is effected by protractor lentis muscles, attached to fibres by which the lens is supported. These muscles move the lens forward. Other fibres, the musculus tensor chorioideae, run radially and around lens. They may help protractors, and are probably the ancestors of ciliary muscles of higher forms. In amphibia living in water, the eye is based more on fish plan and the lens is round. There are no lids or lachrymal glands and eye make an image by a thickening of inside cornea. Rods and cones are present in retina, the former containing visual purple, which may be red or greenish.

Two sets of receptor are found throughout the retina in urodeles, but in *Rana macular* region contains cones in excess. Six types of detector operate on information provided by rods and cones. Contrast detectors give a sustained response when a sharp edge moves into visual field. Convexity detectors respond to object that are curved. These 2 types together may be called 'on' fibres. Moving-edge detectors respond with a frequency proportional to velocity of movement. Dimming detectors respond on reduction of illumination. Darkness detectors fire with frequency inversely proportional to illumination. These types of fibre project to different depths in tectum as sheets of endings, and arrangement of retina is accurately reproduced there, although the fibres are interwoven in nerve. The sixth type of fibre is sensitive to blue light and is connected with thalamus. Skin is sensitive to light. Frogs react to light even after removal of eyes and cerebral hemispheres. This skin sense is especially developed in certain cave-living urodeles, *Proteus*, in which eyes are not functional. The similar degeneration also occurs in Apoda.

Ear

Inner ear is divided into a utricle, from which semicircular canals arise, and a saccule from which arises a diverticulum, the lagena, part of whose surface is covered with a tectorial membrane. There is no coiled cochlea. Middle ear of frog consists of a funnel- shaped tympanic cavity communicating with pharynx and closed externally by a tympanum. Sound waves are transmitted across cavity by a rod, the columella, fitting by an expanded foot, the otostapes, into the hole fenestra ovalis. This hole is also partly occupied by a second plate, the operculum, which is joined to scapula. Operculum and otostapes develop within wall of auditory capsule and middle part of columella forms as an outgrowth from otostapes.

The outer part of columella and tympanic ring develop close to quadrate and probably from its cartilage. Columella shows no developmental relationsips to hyoid arch. Tympanic cavity is developed from spiracular cleft. Original cleft degenerates 6 days after hatching but about 6 of its lining cells persist and at the end of tadpole stage form a tympanic vesicle, which becomes connected with pharynx by a rod of cells. This rod then degenerates again and an open air passage to vesicle of drum is not established until some 30 days after emergence from the water. Hyomandibular nerve, which divides above middle ear of amniotes lies behind tympanic cavity of frog and branches below it. Arrangement for conveying vibrations to ear varies considerably among amphibians.

In urodeles, there is no tympanum. In some of them, columella is attached to squamosal, perhaps in connection with a semi-aquatic, or burrowing habit. A similar arrangement may have been present in the earliest amphibians, which have a columella but no oval window. In other urodels, the columella is attached to quadrate and there may be a second ossicle, operculum, working in parallel, with its inner end in oval window caudal to columella and its outer end attached by a muscle to scapula. In terrestrial forms, columella becomes fused with window at metamorphosis and its function is taken over by operculum, probably receiving vibrations from fore-legs.

Cryptobranchus retain the larval condition and never develop an operculum. Tympanum and columella are also reduced in some terrestrial anurans (*Bombinator*) but in aquatic *Xenopus* and *Pipa* the operculum and its muscle are lost, perhaps a paedomorphic feature. Sense of hearing is well developed especially in Anura, which respond to vibrations from 50 to 10,000 a second. Hearing is used especially in breeding-season. Prey may also be located by sound. Urodeles give no response to ringing of a bell suspended from ceiling, which produces reactions in *Rana* and *Bufo*. A peculiar feature of many Anura is an immense backward development of perilymphatic space of inner ear, forming a sac extending above brain and on either side of spinal cord as far back as sacrum. Portions of this sac emerge between vertebrae. Calcium salts in sacs diminish during metamorphosis and they then refill. The system may serve as a calcium reserve for adult.

7.15 Origin of Amphibia

Geobatrachus from the Permian was proposed as the ancestor of lissamphibians. Modern amphibians, the Lissamphibia perhaps originated from the Labyrinthodont stock. The Lepospondyli has been favored as lissamphibian ancestors. They were small with simple vertebrae, resembling lissamphibians in external anatomy. It was believed that labyrinthodonts gave rise to leopspondyls, and lepospondyls to lissamphibians. Some studies favour the lepospondyl link considering Lepospondyli as close relatives of reptile-like amphibians. Some scientists argued that Amphibia as a whole is biphyletic, based on their nasal capsule and cranial nerves. In their view, lepospondyls are ancestors of frogs. Salamanders and caecilians have evolved independently from porolepiform fish. Carroll suggested that the tailed amphibians are originated from the microsaurs and frogs from temnospondyls.

The gaps in Palaentological record of Amphibia are enormous. The combined researches of Gregory, Watson, Williston, have thrown light on beginnings of land life among vertebrates. The outstanding Palaentological contribution proves how slight a structural alteration changed the primitive fish ancestor into the first land vertebrate. Three groups of tetrapods viz. labyrinthodonts, lepospondyls, and phyllospondyls had developed in Carboniferous times. The first two were present in Lower Carboniferous. Footprints are known from Devonian of Pennsylvania. Hence, the tetrapods must have arisen in at least Devonian and possibly Silurian times.

Piscine Ancestors

Air breathing fish, the dipnoans and crossopterygians exhibit a close relationship with amphibians. So, chance for dipnoans and crossopterygians to hold the significant position in amphibian ancestry needs consideration.

Dipnoi-foreshadowed amphibian organization

Pectoral and pelvic girdles of dipnoans foreshadowed some amphibian features in several ways. These are:

(i) Internal skeleton of paired appendages is well developed in dipnoans.

(ii) Paired appendages articulate with respective girdles by a single proximal bony piece, which can be compared with humerus, or femur.

(iii) Outward extension of myomeric muscle into paired fins is suggestive of arrangement of musculature of paired appendages of Amphibia.

(iv) Ability to breathe air by lungs,

(v) Pectoral girdle of *Necturus* resembles closely that of dipnoans.

Remarks: Dipnoans exhibit too many specialized features and such a specialized group can not possibly hold the ancestry of another group of animals. Striking similarities in them are possibly due to physiological convergence. Dipnoans, today, give an idea of form that probably linked the fishes with amphibians. Dipnoans are regarded as collateral uncle of amphibia.

Crossopterygian Ancestry: There are such close resemblances between skulls of earliest amphibians and those of Devonian crossopterygian fishes that there can be no doubt of relationship. At present, there is no detailed fossil evidence of the stages of transition from one type to other. Early crossopterygians, osteolepis and *Eusthenopteron* furnish the strongest support in this connection. They posses many features, which are amphibian, or lead towards amphibia. Features are bony pattern of jaws and skulls are comparable to that observed in early amphibians, 2 large bones on top of skull can be homologized as amphibian parietal bones. Pectoral girdle presents certain features which are prerequisites for amphibian forelimbs. Skull of *Eusthenopteron* (Figure 7.10) contains almost all elements observed in early amphibians. Pectoral fin of *Eusthenopteron* can be compared to forelimbs of amphibia. The single proximal piece of bone can be homologized with humerus and next 2 pieces can be compared to radius and Ulna.

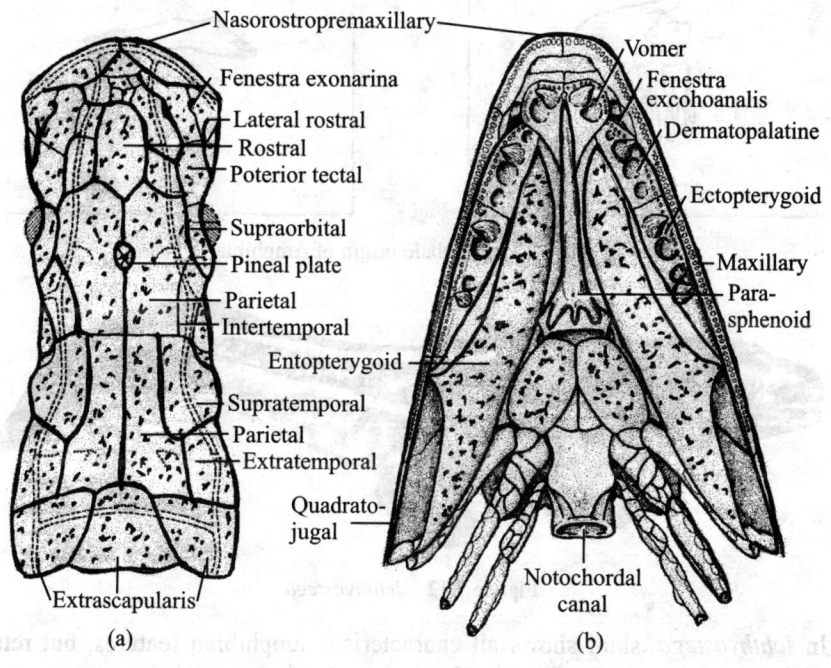

Figure 7.10 Skull of *Eusthenopteron*, an Osteolepiform (a) Dorsal view, (b) Ventral View.

(vi) Various wrist and ankle bones, bony elements of hand and foot have evolved from distal bony complex of crossopterygian fins.

Remarks: Thus, in many respects, crossopterygians show close similarities with the amphibian and it is expected that these fish are the direct progenitors of the early amphibians.

Ichthyostega -a Transitional Form

Ichthyostega and similar forms are found in fresh water beds of the Greenland. These are dated as very late Devonian, or early carboniferous (Figures 7.11a, b; 7.12 a, b). They are the oldest members yet found of group of Labyrinthodontia. Their enamel and dentine surrounding pulp cavity at the base of tooth was folded into a labyrinthine pattern. All were fish eating. This condition was also present in their crossopterygian ancestors. They flourished and developed many different lines, one giving rise reptiles. They were mainly aquatic or semiaquatic forms and only a few seem to have been completely terrestrial.

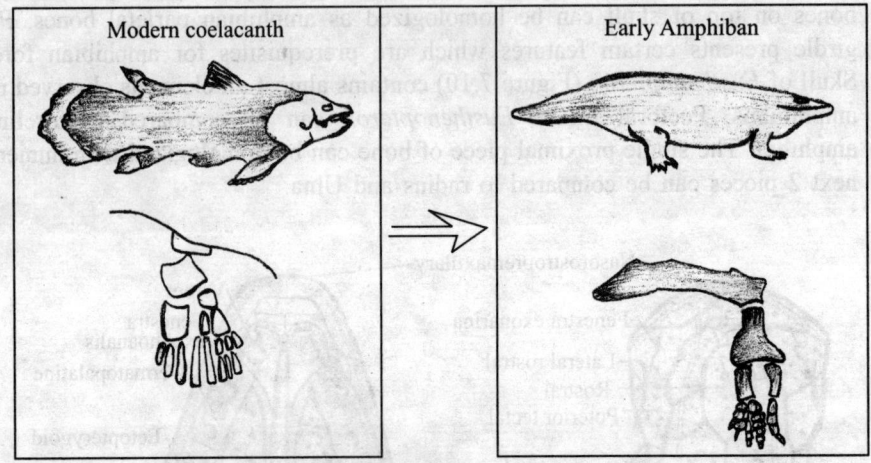

Modern coelacanth   Early Amphiban

Figure 7.11   Probale origin of Amphibia.

Figure 7.12   *Ichthyostega.*

(i) In *Ichthyostega*, skull shows all characteristic amphibian features, but retain traces of fish ancestry in its shape with short wide snout and long posterior region and presence of preopercular bone.

(ii) There was some sign of separation of 2 parts of brain case.

(iii) Nostril lies on very edge of upper lip apparently partly divided by a flange of maxilla into internal and external opening.

(iv) There being no gills, the operculum had been lost and second gill arch lay in an otic notch between the tabular and squamosal bones and presumably carried a tympanum.

(v) These creatures were about 100 cm. long and possessed a long tail with a dorsal fin, but had strong legs, a strong, well ossified "rhachitomous" vertebrae. They must have been not very for from ancestors of all tetrapods.

Origin of Modern Amphibia

Modern forms are generally considered to be a single monophyletic group, but this is not certain and their relationship to the Paleozoic Amphibia is unclear. They show some characters typical of later temnospondyls, like flattening of skull and loss of ossification in roof and palate. The vertebrae are simple rings like those of lepospondyls.

Earliest known anuran, *Protobatrachus* comes from lower Triassic and has a frog like skull but many vertebrae with ribs and a segmented tail. Ilia and tarsal bones are long. Urodeles are superficially more like early amphibians. Fossil salamanders are known back to Jurassic, but resemble modern forms. An independent origin has been suggested from fish, *Porolepis*, supposed to have an internal nostril. No fossil Apoda are known but their retention of scales suggests an early separation from the rest, and they may be derived from lepospondyls rather than from the labyrinthodonts. Evolution of various groups from early tetrapods is given in Figures 7.13, 7.14, 7.15.

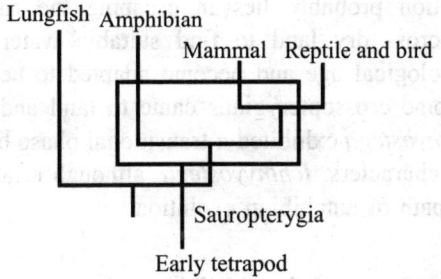

Figure 7.13 Evolutionary pathway of vertebrates.

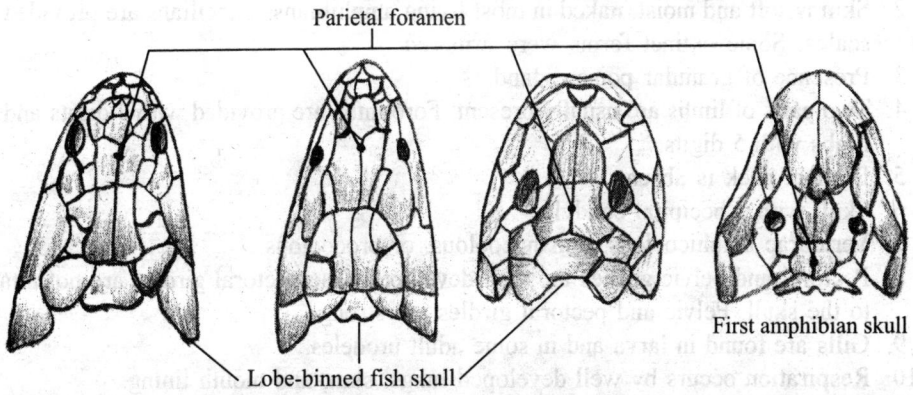

Figure 7.14 Similarities of dermal bones of lobe-finned fish and first amphibian.

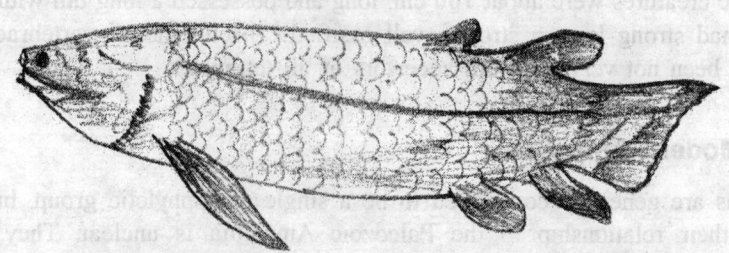

Figure 7.15   Rhipidistian : the probable ancestor of the Amphibians

Causes for Transition from Water to Land

According to a view, Devonian was a period of seasonal droughts. Certain rhipidistian crossopterygians had sufficiently developed lobe fins to move over land to ponds. Periodic escapes resulted in conversion of their lobe fins into pentadactyl tetrapod limbs. Another view contrasts between aquatic and terrestrial habitats during Devonian compared to water, land had more O_2 in air, food, shelter and breeding places. In addition, land had less impurity and fewer predators.

Real cause of transition probably lies in escaping the risk of survival for which crossopterygians had to cross dry land to find suitable water. From such a start, the amphibians evolved in geological age and become adapted to new terrestrial environment. During Devonian time, some crossopterygians came to land and became transformed into primitive amphibians. *Ichthyostega* exhibited a transitional phase by possessing an admixture of piscine and amphibian characters. *Ichthyyostega*, although retained many distinct piscine characters has paved the path of amphibian evolution.

7.16   General Characters of Amphibians

1. Amphibious, live on land as well as in water.
2. Skin is soft and moist, naked in most living amphibians. Caecilians are provided with scales. Some extinct forms were armored.
3. Presence of granular poison gland.
4. Two pairs of limbs are usually present. Forelimbs are provided with 4 digits and hind limbs with 5 digits.
5. Distinct neck is absent.
6. Skull with 2 occipital condyle.
7. Vertebrae amphicoelous, opisthocoelous, or procoelous.
8. Pectoral and pelvic girdles are well developed. The pectoral girdles are not attached to the skull. Pelvic and pectoral girdles are strong.
9. Gills are found in larva and in some adult urodeles.
10. Respiration occurs by well developed lungs, skin, and mouth lining.
11. R.B.C is oval and nucleated.
12. Poikilothermous.

13. Heart is composed of sinus venosus, 2 auricles, a ventricle, and a conus arteriosus.
14. Renal and hepatic portal systems are well developed.
15. The "papilla amphibiorum," a membrane allows amphibians to hear acoustic signals of less than 1,000 Hz.
16. Chromosome number is highly variable.
17. Kidneys are mesonephric.
18. Urinary bladder is large. Urinary ducts open into cloaca.
19. Brain with 10 pairs of cranial nerves.
20. Cerebral hemisphere is elongated and large; cerebellum is well developed.
21. Sexes are separate, fertilization is usually external.
22. Neoteny is exhibited by some urodels
23. Amnion and allantois are absent.
24. Parental care is exhibited by some.
25. Development occurs through metamorphosis.

7.17 Classification

G. Kinsley Noble (1931) classified the class into six orders. Classification Followed here is of Noble (1931).

Extinctct Orders

Order: Labyrinthodontia
    Order: Phyllospondyli.
    Order: Lepospondyli
Living orders:
    Order: Gymnophiona (or Apoda)
    Order: Caudata or Urodela
    Order: Salientia (or Anura)

Order: Labyrinthodontia

Characters

1. Skull was completely roofed over bone.
2. Teeth were enlarged with greatly folded dentine.
3. Many bony elements were present in skulls than modern amphibians.
Example: *Eryops, Cacops, Ichthyostega, Acanthostega, Platyhystrix, Eryops*

Order: Phyllospondyli

Characters

1. Vertebrae tubular with spinal cord and notochord lying in one cavity.
2. Well-marked transverse processes with stout ribs.

3. Pubis is cartilaginous.
4. Four fingers and 5 toes were present.
5. Coracoid remained cartilaginous
6. Skull roof had a separate quadrato-jugal and lacrimal.

Example: *Branchiosaurus*.

Order: Lepospondyli

Characters

1. Vertebrae composed of single piece. Neural arch was continuous with centrum.
2. Ribs articulated with column intervertebrally.
3. Small, mostly aquatic.

Example: *Dipocaulus, Siren, Amphiuma*

Order: Gymnophiona or Apoda

Characters

1. Limbless, burrowing, tropical amphibians
2. Circumtropical (except Madagascar).
3. Adults are almost entirely subterranean
4. Within transverse grooves are found a series of small scale
5. Modified for burrowing life.
6. Protrusible tentacle is found.
7. A protrusible copulatory organ is found in male.
8. Short tail, vent being terminal,
9. Scales may be present, or absent.
10. Larval life may be aquatic, or not.
11. Aquatic, semi-aquatic, terrestrial, in some respects most primitive.
12. Internal fertilization, plus some very interesting parental care and reproductive modes occur.

Example: *Ichthyophis, Hypogeophis*, Slamanders, Newts,

Order: Caudata

1. Possesses tails.
2. Aquatic larvae resemble their parents closely.
3. With 2 pairs of sub equal limbs.
4. Tongue is fixed and immovable.
5. Intestine forms several loops.
6. Respiration by gills in larval stages, but adult respires mostly by lungs and skin.
7. Erythrocytes are the oval, nucleated.
8. Eyes are small and functionless especially in cave-dwelling forms.
9. Columella in middle ear is absent.

10. Mullerian duct is present in male.
11. Fertilization is generally internal.

Example: *Cryptobranchus, Ambystoma, Triton.*

Order: Salientia

1. The tailless amphibians; postsacral vertebrae are fused into tailbone.
2. Various adaptations for jumping, involving elongated, fused hindlimbs.
3. Most frogs have no ribs.
4. More prevalent in tropics than elsewhere.
5. Most are nocturnal.
6. Hind limbs are elongated for hopping.
7. Vertebral column is short and inflexible.
8. Pelvic girdle is enlarged, strengthened and anchored to vertebral column.
9. Legs are long.
6. Posterior limbs have 4 segments.
7. Eyelids are well formed.
8. Tympanum is distinct.
9. Frontal and parietal bones form frontoparietals.
10. Mandible is devoid of teeth.
11. Almost all have external fertilization, and are oviparous.
12. Most mate in a posture called amplexus, in which the male grasps female in armpits, or at the waist.
13. Length of breeding season is limited by climate, or seasonal availability of breeding sites.
14. Temporal patterns of reproduction can be divided into explosive breeding and prolonged breeding.
15. Neoteny is not observed in larval forms.

Example: *Alytes, Pipa, Xenopus,* and *Pelobates* (Figures 7.16a-l).

7.18 Metamorphosis

Metamorphosis is a rapid and complete transformation from larval to adult form. Amphibians furnish the best example of metamorphosis in vertebrates. In them, metamorphosis incorporates ecological, morphological, physiological and biochemical changes. From morphological point of view, anurans undergo much extensive metamorphic changes. The degree of difference between their adults and larvae remain much profound. Metamorphosis in anuran includes regressive metamorphic changes, (II) progressive metamorphic changes.

Regressive Metamorphic Changes of Anura

1. Ventral suckers, external gills and tail along with fine folds of tadpole larvae are reabsorbed.

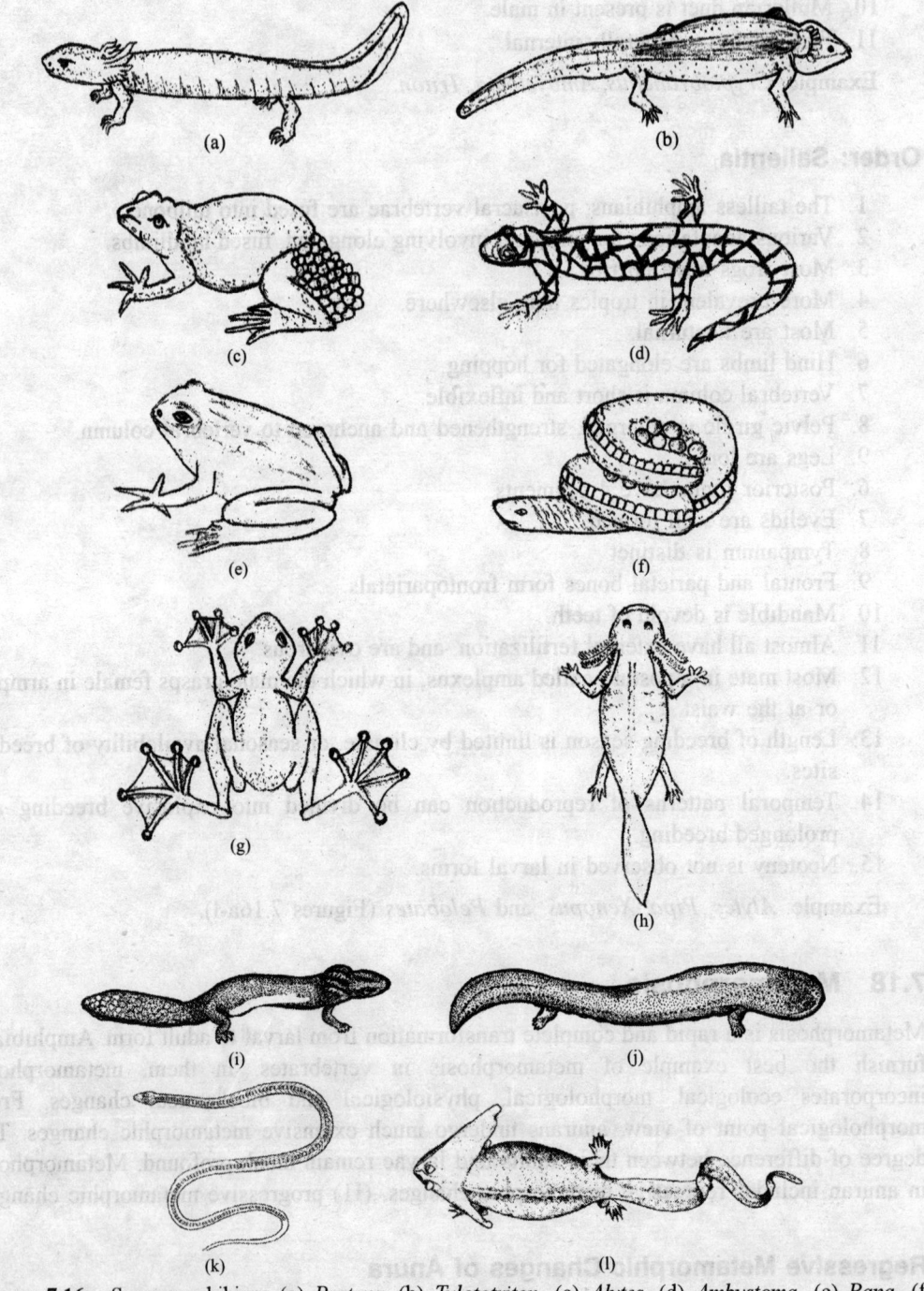

Figure 7.16 Some amphibians (a) *Proteus*, (b) *Tylototriton*, (c) *Alytes*, (d) *Ambystoma*, (e) *Rana*, (f) *lcthyophis*, (g) *Rhacophorus*, (h) Axoltl larva, (i) *Necturus*, (j) *Megabatrachus*, (k) *Aornerpeton* and (l) *Suaropleura*.

2. Gill clefts are closed.
3. Peribronchial cavities disappear.
4. Horny teeth of perioral disc shed.
5. Horny linings of jaws shed.
6. Shape of mouth changes.
7. Cloacal tube becomes shortened.
8. Lateral line organs of skin of tadpole disappear.

Progressive or Constructive Metamorphic Changes of Anura

1. Progressive development of limbs, in size and differentiation.
2. Fore limbs, in frogs develop under the cover of opercular membrane and break through to exterior.
3. Gill arches modifies into hyoid apparatus.
4. Middle ear develops in connection with first pharyngeal pouch.
5. Tympanic membrane develops and is supported by circular tympanic cartilage.
6. Eyes protrude on dorsal surface of 'head and develop eyelids.
7. Tongue is developed from floor of mouth.

Urodeles undergo less ecological and morphological metamorphic changes.

Regressive Metamorphic Changes in Urodeles

1. Tail is retained, only fin folds disappear.
2. Branchial apparatus is reduced.
3. External gills become resorbed.
4. Gill clefts are closed.
5. Visceral skeleton becomes much reduced.
6. Head becomes oval.

Progressive Metamorphic Changes in Urodeles

1. Skin becomes cornified.
2. Multicellular skin glands are differentiated.
3. Skin pigmentation changes.
4. Eyes bulge and develop lids.
5. Legs and intestine suffer no change.

In urodeles, metamorphosis is more gradual and takes up to several weeks.

Morphological and physiological changes during metamorphosis

1. Body forms and structure.
2. Formation of dermal glands.
3. Restructuring of mouth and head.
4. Intestinal regression and reorganization.
5. Calcification of skeleton.

Appendages

1. Degeneration of skin and muscle of tall.
2. Growth of skin and muscle of limbs.
3. Nervous system and sense organs change
4. Increase in rhodopsin in retina.
5. Growth of extrinsic eye muscle.
6. Formation of nictating membrane.
7. Growth of cerebellum.
8. Growth of preoptic nucleus of hypothalamus.

Respiratory System

1. Degeneration of gill arches and gills.
2. Degeneration of operculum that cover the gills.
3. Development of lungs.
4. Shift from larval to adult hemoglobin.

Organs

1. Pronephric resorption in kidney.
2. Induction of urea-cycle enzymes in liver.
3. Reduction and restructuring of pancreas.

The said account of metamorphosis is based on White and Nicoll (1985)

Comment: Lungs do not change drastically during metamorphosis in both anuran and urodeles. They develop gradually and become fully functional in larval state. Long before metamorphosis, larvae of frogs and salamanders start coming up to surface and gulping air into their lungs and thus supplementing their aquatic respiration. This may be of considerable importance where larva develops in stagnant and polluted waters, as is often in the ease (Figure 7.17)

Control of Metamorphosis: Metamorphosis is achieved rapidly by hormonal control. Main effector hormone is thyroxine T4. Removal of thyroid prevents metamorphosis. Addition of iodine or T4 to water accelerates it. Change is a result of increasing sensitivity to T4 and increased amounts of it secreted. Change involves (i) Suuppression of purely larval organs (tail), (ii) conversion of some larval feateres to adult state (jwas), and development of adult characters (legs).

Tissues that first become sensitive to T4 are tail fins. Jaws become sensitive next and hind limb bud latest. Increase in thyroid secretion is triggered by thyrotrophic hormone (TSH) from pituitary, whose removal prevents metamorphosis. Prolactin plays a part in control, by promotion of larval growth and by inhibition of changes. Whole sequence of metamorphosis is probably regulated by hypothalamus. Median eminence becomes developed only at premetamorphosis. A graft of pituitary away from hypothalamus produces some TSH, but not enough to allow complete metamorphosis.

Processes of change involve almost every part of body. Genes that have been operating in larva are switched off. For instance, hemoglobin is coded by different loci in larval and

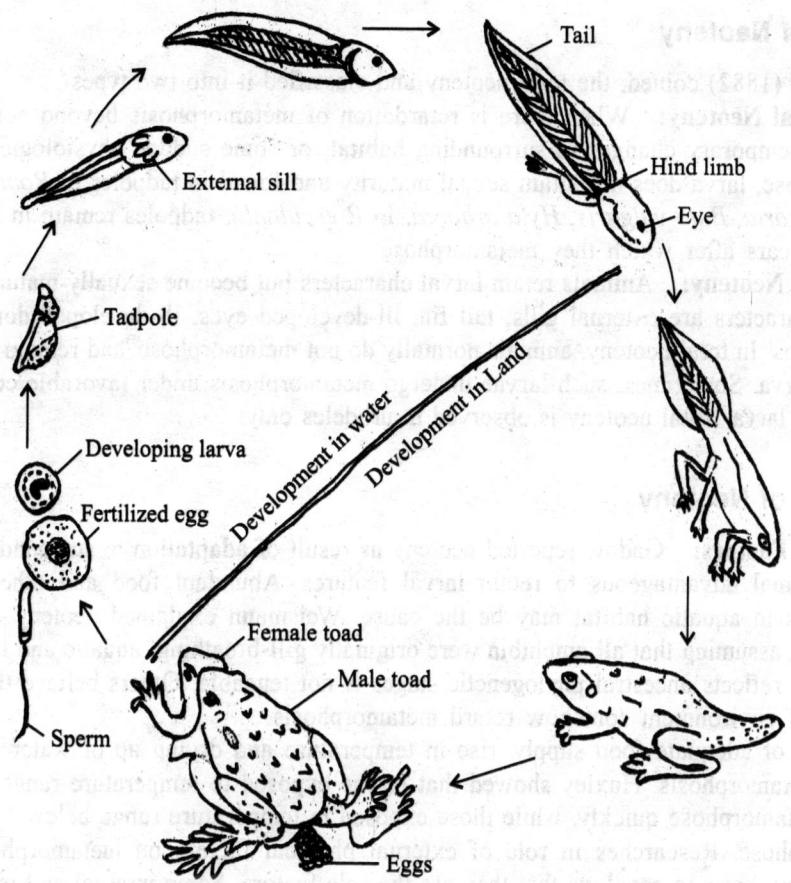

Figure 7.17 Metamorphosis of *Bufo*.

adult *Rana*. Larval gills, tail and beak are resorbed, while limbs, eyelids and lungs are developed. Intestine becomes shorter and there are many changes in liver and kidneys and in metabolism. Mauthner cells disappear from brain and lateral line organs from skin.

7.19 Neoteny

Neoteny or Paedogenesis is a phenomenon of retention of larval characters in sexually mature state. Metamorphosis of larva is retarded and larval characters are retained beyond normal period. In urodeles, this phenomenon is common. It also occurs in anurans. *Ambystoma tigrinum*, in Mexican plateau lives in cool mountain ponds, retains external gills and many other larval features, reproduces as a larva, and never completes its metamorphosis. This permanent larva is called as axoltl. At lower elevations in the United States, *Ambystoma* normally metamorphoses. *Ambystoma* is an example of facultative neotenic species. *Necturus* is an obligatory neotenic species and does not metamorphose under any circumstances.

Types of Neoteny

Kollmann (1882) coined, the term neoteny and classified it into two types

Partial Neoteny: When there is retardation of metamorphosis beyond normal period due to totemporary changes in surrounding habitat, or some sudden physiological disorder. In such case, larva does not attain sexual maturity and found in tadpoles of *Rana esculanta*, *R. temporaria*, *Bufo vulgaris*, *Hyla arborea*. In *R esculanta*, tadpoles remain in larval stage for two years after which they metamorphose

Total Neoteny: Animals retain larval characters but become sexually mature. Retained larval characters are external gills, tail fin, ill-developed eyes, ill-developed dorsal fin and weak limbs. In total neoteny, animals normally do not metamorphose 'and remain as sexually mature larva. Sometimes, such larvae undergo metamorphosis under favorable conditions as the axoltl larva. Total neoteny is observed in urodeles only.

Causes of Neoteny

External Factors: Gadow reported neoteny as result of adaptation to surroundings, which make animal advantageous to retain larval features. Abundant food and other favorable conditions in aquatic habitat may be the cause. Weismann explained neoteny as cause of reversion, assuming that all amphibia were originally gill-breathing, aquatic and limbless and that larva reflects ancestral phylogenetic stages is not teneable. Others believe that physical factors of environment somehow retard metamorphosis.

Lack of adequate food supply, rise in temperature and drying up of water bodies may retard metamorphosis. Huxley showed that larvae exposed to temperature range above 5°C could metamorphose quickly, while those exposed to temperature range below 5°C failed to metamorphose. Researches in role of external physical factors on metamorphosis do not provide any basis to conclude that they are the sole factors. Some internal and physiological factors may play role in controlling metamorphosis.

Internal Factors: Metamorphosis is mainly influenced by level of thyroxine and the degree of responsiveness of larval tissue to the hormone. Prolactin plays an effective role in metamorphosis, as its level is high in early stages before metamorphosis. Genetic explanation advanced by Etkin and his coworkers is that formation of prolactin is increased by concerned genes. Genes responsible for synthesis of thyroxine are switched off by concerned operator genes in early phase of larval life. As a result, hypothalamus becomes sensitive to low concentration of thyroxine and secretes throtropin-releasing factor (TRF), which stimulates anterior lobe of pituitary to produce thyroid stimulating hormone (TSH), which in turn increases rate of thyroid-secretion. Increase in level of thyroxine brings about initiation of metamorphosis.

When level of TSH increases, the level of prolactin drops. Thyroxin secreting alveoli of thyroid gland in neotenous larva remain in underdeveloped condition. If rudiments of hypophysis in late embryos is removed or destroyed, they fail to metamorphose, but metamorphosis can be again initiated if pieces of hypophysis from adult frogs are implanted. Even fully developed thyroid glands sometimes fall to secrete adequate thyroxine and under such circumstances transplantation of a few more thyroid glands induces metamorphosis.

Significance of Neoteny: Weissmann regarded neoteny as a case of atavism, a phenomenon of reversal to ancestral characters. This implies that all amphibians were originally gill-breathing, aquatic creatures which are not true. External gills of urodeles are now regarded as secondary specialization serving as additional respiratory organs. Other larval features of neotenous larvae do not represent atavism, but are characters secondarily acquired for aquatic life. Noble (1954) pointed out that retention of larval characters is not connected with phylogeny of amphibians. Great heterogeneity of perennibranchiate forms, which are all neotenous, proves this point. So, it is likely that larval features are retained due to some intrinsic factors combined with environmental factors, and are advantageous for neotenous individuals.

7.20 Toad

This is a typical example of the class Amphibia. Various features of toad are described below:

Habit and Habitat

Toad is a cold blooded or poikilothermic vertebrate. During colder months, toads undergo through a phase called hibernation or winter sleep. An adult toad generally lives on land, but if necessary, it can live in water for sometime. It is truly amphibian both in biological and literal sense. Toad is quite at home in water and on land. On land, the toad remains in dark and moist places obligatorily, preferably near water. Skin is an accessory respiratory organ and kept moist.

It tries to remain away from bright light. This is carnivorous and catches insects by it sticky tongue. They are nocturnal, usually come out of their hides at dusk. Males produce croaking sound by the help of vocal sac, especially during breeding season. They live on land but breed in water. All their organ systems are adjusted to land life except the reproductive system. For reproduction it had to go back to primal aquatic abode.

External Structures: A short bilaterally symmetrical body lacks exoskeleton over the skin which is rough in texture. Dorsal side of body is blackish -gray, while ventral side is yellowish grey. Toad can change body colour to match with the hues of its surrounding. Ordinarily its body colour is same as that of earth. Body is divisible into head and trunk without distinct neck. A postanal tail is absent in adult. Tail is well developed in larval condition.

Head is semicircular in outline, broad, depressed with a blunt snout. Mouth is a wide opening at the terminal end of head. At the anterior dorsal side of head, there is a pair of rounded openings called nostrils or external nares. Eyes are very large, prominent, protruding and are situated one on either side of head. Each eye is provided with a thick upper eyelid and an ill developed lower eyelid. A transparent nictitating membrane is stretched to cover the eye ball. Behind each eye, a circular area called eardrum is present. Just behind each tympanum, there are elongated elevation called parotid glands which secrete a whitish, pungent and sticky juice. It is poisonous and causes nausea and affects the heart in man, if swallowed. When the secretion falls in eyes and nose, it causes irritation, but rarely affects skin. These glands are organs of offense and defense.

Skin on the floor of buccal cavity becomes inflated to form vocal sac in males. Trunk is broad, short, and flattened. Numerous small elevations called warts are present on dorsal side of body. Cloacal aperture or vent is located on dorsal side of posterior end of body between two hind limbs. Two pairs of limbs are unequal in size. Forelimbs are small than hind limbs. Each forelimb consists of brachium and ante-brachium, a wrist and the manus. Four digits follow the hand. In male individual, a cushion like thumb pad develops during breeding season at the basal part of inner finger. These pads facilitate the males grip during amplexus.

Hind limbs are strongly built and are longer than forelimbs. These two limbs are modified for jumping and swimming. Each hind limb is composed of a proximal sector called femur (thigh) which is followed by crus. Distal to the shank lies the pes. Foot consists of a long tarsal region and five elongated slender digits. Webs unite the digits, which help in swimming (Figures 7.18, 7.19).

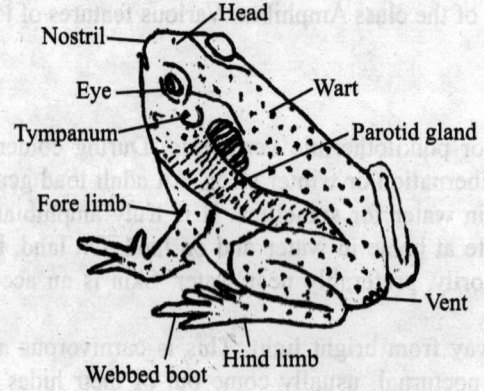

Figure 7.18 External features of toad.

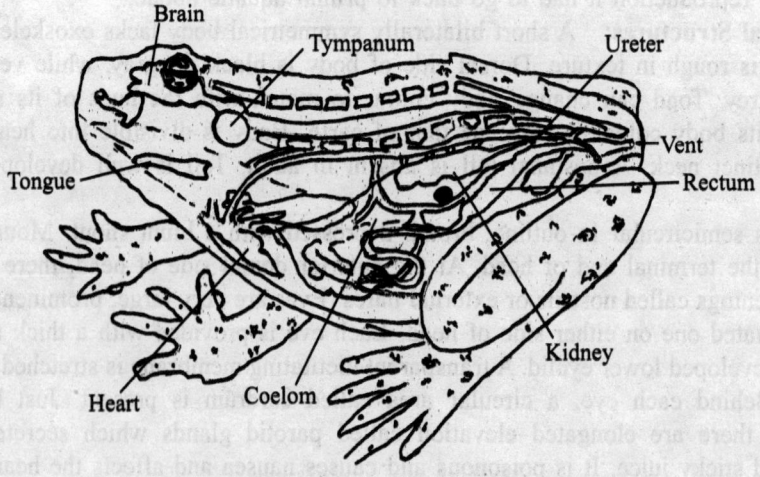

Figure 7.19 Organ system of male toad.

Skin

Skin is kept moist and frequently slimy. Besides its protective function, the skin serves as an additional respiratory organ. During hibernation, toad respires entirely by the skin. Skin is composed of an outer epidermis and an inner dermis. These two layers are separated by a basement membrane and a fibrous layer containing the chromatophores. The epidermis is a compound structure made up of several layers. The outer most layer called stratum corneum is a thin, scaly and dead cornfield layer and shed periodically. Periodic shedding of stratum corneum is called ecdysis.

In other amphibians, the stratum corneum is thin and delicate, while in toad it is thicker and heavily cornfield. The inner most layer of epidermis composed of columnar cells with prominent nuclei is called stratum germinativum, which sits on the basement membrane. Cells lying between stratum germinativum and stratum corneum constitute the transitional layer. These cells from several layers and decrease in size from blow up wards.

Dermis is thicker than epidermis and is divisible into two layers. Outer layer accommodating most glands is called stratum spongiosum, which is composed of loose network of connective tissue matrix, blood vessels and lymphatic spaces. The superficial part of this layer contains the pigment cells. The inner most layer is called stratum compactum, which is composed of dense connective tissue, smooth muscle fibers, nerves and blood vessels. Beneath the stratum compactum lies a loose subcutaneous connective tissue layer containing fatty tissue.

Skin marked by presence of numerous bumps all over the dorsal body surface. Warts may be either due to underlying poison glands, or sensory papillae. Two types of skin glands are mucous glands secreting mucus and poison glands producing poison. Mucous glands are smaller than poison glands. Poison glands are composed of granular secretary cells. Skin glands usually lie in stratum spongiosum of dermis, but poison glands may more deeply locate. These glands are unsheathed by connective and muscular tissue which help in squeezing the secretory products. Products of the glands come out through duct to outside.

Colouration

Colour depends on presence of pigment cells or chromatophores in dermis. Some chromatophores also invade epidermis. Depending on types of contained pigment granules, the chromatophores may be melanphores, guanophores, and lipophores contain yellow pigment and situated in the deepest layer. The guanophores are intermediate in position while lipophores are present in upper layer.

Melanophores produce blackish colour and lipophores cause yellowish effect. Guanine crystals in the guanophores produce diffraction effect. Colour change is a slow process and is controlled by Melanophore Stimulating Hormone (MSH) of the pituitary gland and is not under nervous control. In fish, instantaneous colour change is caused through the nervous system, while the slower change is under hormonal control.

Muscular System: This system is composed of muscles which are used primarily for movement of body. There are three types of muscles: Striated -, Unstriated - and (3) Cardiac muscles. Striated muscles can be contracted at will and are attached to skeleton forming the main mass of external musculature. There are many skeletal muscles in the body. Muscles possess the power of contraction and relaxation. All the muscles are mostly arranged in opposing groups in such a fashion that when one set contracts, the opposing set remains in a relaxed state. This coordination is controlled by nervous system.Skeletal muscles in *Bufo* are grouped into the following general types depending on mode of action:

Flexor Muscle: This type of muscle bends one part on another. Biceps flexes forearm towards upper arm.

Extensor Muscle: This muscle extends or straightens a part. Triceps extends forearm on upper arm.

Abductor Muscle: Such a type of muscle draws a part away from the axis of body or of a limb.Deltoid muscle draws arm forward.

Adductor Muscle: This type of muscle draws a part toward a limb. Latissimus dorsi muscle draws arm up and back.

Depressor Muscle: This muscle lowers a part. Depressor mandibular muscle moves lower jaw down to open mouth.

Levator Muscle: This muscle raises or elevates a part. Masseter muscle raises lower jaw to close mouth.

Rotator Muscle: This muscle rotates a part. Pyriformis muscle raises and rotates the femur (Figure 7.20).

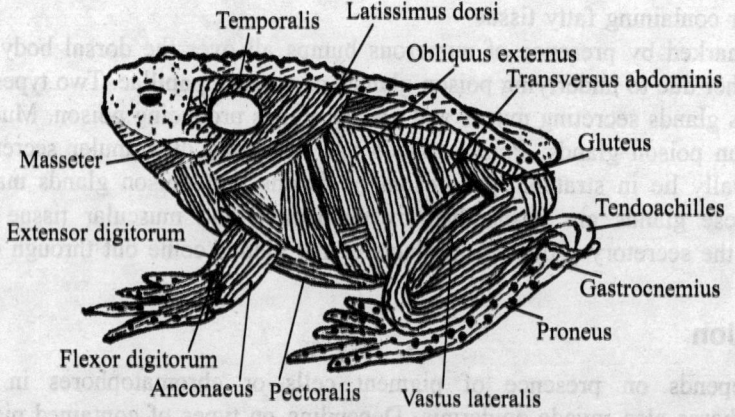

Figure 7.20   Skeletal museles of toad in side view.

Locomotion: The entire muscular and skeletal systems have become specialized for jumping and swimming. Movements are caused by the thrust of both the hind limbs. Besides jumping and swimming, toad is able to walk on land, which is caused by a set of proprioceptor reflexes. In resting state, the anterior part of body is supported by hind limbs which are folded in manner of "Z" from, a sitting or squatting posture. Toad jumps by a sudden extension of hind limbs. Forelimbs manipulate and adjust direction before each jump. Swimming is done by the activity of limbs which act as propellers. Hind limbs are long and digits are webbed. These limbs act like oars and enable the animals to swim.

Skeletal Structures: Exoskeleton is absent. Skeleton that supports the soft parts lies internally and is called endoskeleton. It is chiefly made up of bones and cartilages which are associated with one another to form the internal framework. Endoskeleton is described under two broad heads: (1) axial skeleton and (2) appendicular skeleton.

Axial Skeleton: This skeleton comprises of skull and vertebral column.

Skull: Skull is the flat, broad, contains a tubular cranium and is pierced posteriorly by a large aperture called foramen magnum. Through this aperture, the spinal cord passes. On each side of the foramen, there is an exoccipital bone which bears convex occipital condyle. Two occipital condyles fit into two concavities of the first vertebra. Occipital condyles developed from the exoccipitals. The roof of cranium is made up of two flat bones called frontoparietals. Each frontoparietal is formed by fusion of two bones, the frontal and parietal. The floor of skull is formed of a dagger like parasphenoid. A ring like sphenethmoid bone is present at the anterior end of cranium. This bone is completely covered by frontoparietals on dorsal side.

Nose is covered dorsally by a triangular nasal bone and floor is provided with vomer. Cartilaginous otic capsules are loosely attached with cranium. Auditory capsules are situated in front of exoccipitals. Floor of auditory capsule is supported by lateral extension of parasphenoid and dorsal side is covered by prootic bone. A small hammer like squamosal connects posterior part of upper jaw with otic capsule. Each half of upper jaw is made up of small premaxilla, long slender maxilla, and quadratojugal. Behind quadratojugal, a very small Y- shaped supports the quadrate. Other two supporting bones are pterygoid and palatine. Palatine is rod like and connects maxilla with sphenethmoid bone. Lower jaw is composed of two halves which are united anteriorly by ligament. Each half is developed from a Meckel's cartilage and consists of three bones, the dentary, angulosplenial, and mento-meckelian. Posterior part of angulosplenial articulates with upper jaw. Both the upper and lower jaws are toothless.

Hyoid Apparatus: This is a cartilaginous structure which supports the floor of buccal cavity and forms the supporting frame work of attachment of tongue. Body of hyoid constitutes the main bulk of apparatus. Two pairs of prolongations from body of hyoid apparatus are found. The anterior pair is long and extends up to auditory capsules. These are called anterior cornua. Similar pair on posterior side is called posterior cornua which enclose laryngotracheal chamber (Figure 7.21).

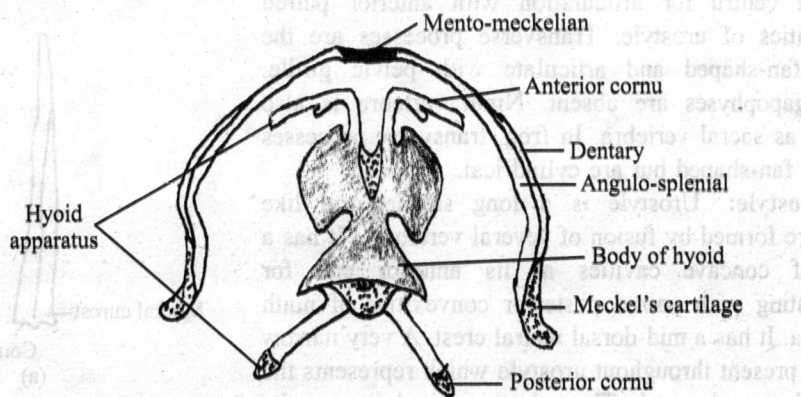

Figure 7.21 Lower jaw and hyoid apparatus of toad.

Vertebral Column: The vertebral column is composed of nine vertebrae and a terminal rod-like structure called urostyle

Typical Vertebra: A typical vertebra has a solid cylindrical part called centrum, which is procoelous i.e. concave anteriorly and convex posteriorly ondorsal side. The centrum bears a ring like neural arch which encloses neural canal. The roof of neural arch possesses a median elevation called neural spine. The lateral side of neural arch carries transverse processes. Articulating processes are present on neural arch. The anterior pair of processes is called prezygapophyses and the posterior pair is called postzygapophyses (singular-postzygapophysis). All the vertebrae excepting the first and the last one have typical structural construction (Figure 7.22).

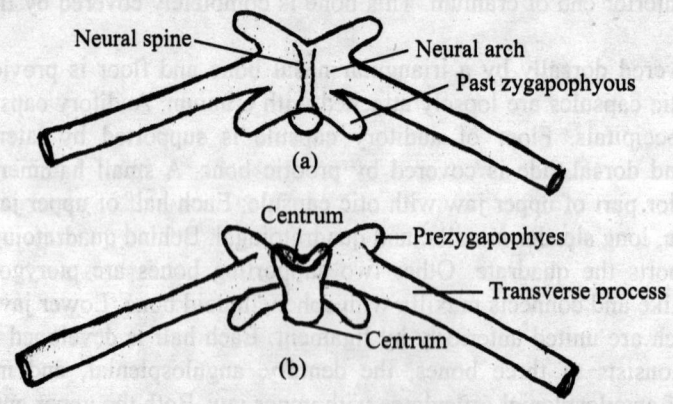

Figure 7.22 Typical vertebra of toad (a) Dorsal view, and (b) Ventral view.

First Vertebra or Atlas: It articulates with occipital condyles of skull. It is ring like in appearance. Transverse processes and prezygapophyses are absent. Anteriorly, it possesses two concave facets which fit with paired occipital condyles of skull. The centrum is greatly reduced.

Ninth Vertebra: The vertebra is peculiar and has two rounded condyles on posterior side of centru for articulation with anterior paired concavities of urostyle. Transverse processes are the stout, fan-shaped and articulate with pelvic girdle. Postzygapophyses are absent. Ninth vertebra is also known as sacral vertebra. In frog, transverse processes are not fan-shaped but are cylindrical.

Urostyle: Urostyle is a long slender rod like structure formed by fusion of several vertebrae. It has a pair of concave cavities at its anterior end for articulating with paired posterior convexities of ninth vertebra. It has a mid-dorsal neural crest. A very narrow hole is present throughout urostyle which represents the reduced neural canal. Through this canal passes the filum terminale of spinal cord (Figure 7.23).

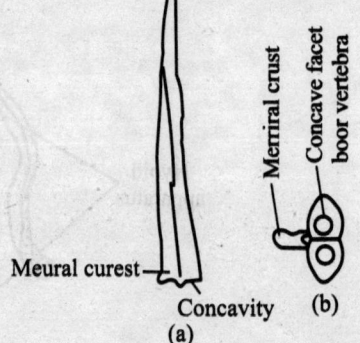

Figure 7.23 Urostyle of toad (a) Side view, (b) Anterior end.

Appendicular Skeleton: Skeletal frame of paired limbs and girdles constitute the appendicular skeleton.

Pectoral Girdle

It consists of two symmetrical halves which are united at the midventral line, but are free dorsally forming a bony framework for encircling the anterior part of trunk. Each half is made up of a broad dorsally placed and partly cartilaginous plate like structure known as suprascapula attached to it. With a strong bone called scapula, two rod like bones is connected. The anterior one is called clavicle, while posterior one is called coracoid. Clavicle encloses the precoracoid cartilage which is hardly visible. The clavicle and coracoid is joined with partly overlapping pieces, the epicoracoids. Space present between these three bones is called coracoid fontanella. Just at the junction of clavicle, coracoid and scapula, there is a cup shaped depression called glenoid cavity into which the head of humerus fits. Projecting posteriorly from the united posterior end of the epicoracoid there lies a sternum which has a terminal flattened cartilaginous xiphisternum (Figure 7.24).

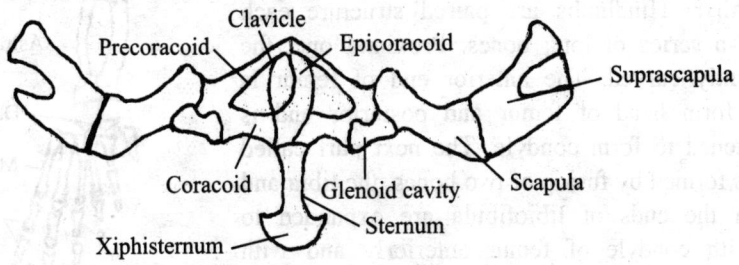

Figure 7.24   Pectoral girdle of toad (ventral view).

Pelvic Girdle: The girdle is a V-shaped bony structure having a disc like posterior end, which is formed by union of three bony units on each side, the ilium, ischium, and pubis. A cavity called acetabulum is present on each side of the disc into which the head of femur fits. Ilia are elongated, curved rod like structure attached with the transverse processes of ninth vertebra. They form the anterior and dorsal sectors of the disc. Two pubes are represented by the triangular cartilage on ventral side of disc. Posterior sectors of the disc are formed by the ischium (Figure 7.25).

Forelimbs: There are two forelimbs. Each forelimb is composed of several long

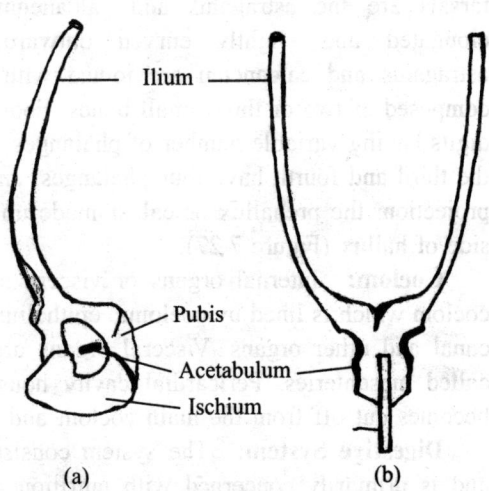

Figure 7.25   Pelvic girdle of toad (a) Side view, (b) Dorsal view.

bones arranged end to end. The first one is called humerus whose middle region is slightly curved. The proximal end is called head of humerus, which fits with glenoid cavity of pectoral girdle. Distal end carries a pulley like trochlea. A prominent crest called deltoid ridge extends from the head to middle region of humerus. The second is the radio-ulna formed by the fusion of two separate bones, radius and ulna. The anterior end of radio-ulna is concave and fits with trochlea of humerus. The proximal end is drawn out into an olecranon process. Distal end is flattened to give attachment of six carpal bones which are arranged in two rows. Four slender rods like bones called metacarpals are connected with digits. Of the four digits; third and fourth digits have three phalanges and the first and second have two phalanges (Figure 7.26).

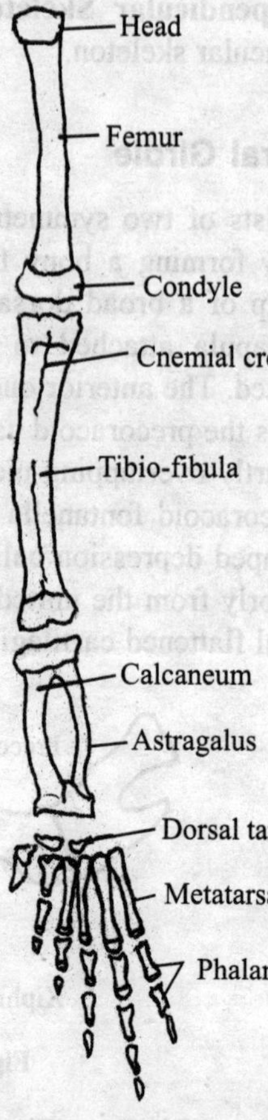

Hindlimbs: Hindlimbs are paired structure each made up of a series of long bones. Proximal one, the femur is gently curved. The anterior end of femur is rounded to form head of femur and posterior end is slightly flattened to form condyle. The next part called tibiofibula is formed by fusion of two bones, the tibia and fibula. Both the ends of tibiofibula are expanded to articulate with condyle of femur anteriorly and with tarsal bones distally.

Tarsal bones are arranged in two rows. Proximal tarsals are the astragalus and calcaneum which are elongated and slightly curved outward. Both the

Figure 7.26 Left hind limb of toad.

astragalus and calcaneum are joined with one another at both ends. Distal tarsals are composed of two or three small bones. Foot is constituted of five metatarsals.There are five digits having variable number of phalanges. The first and second digits have two phalanges, the third and fourth have four phalanges, and in the fifth there are three phalanges. A bony projection, the prehallux or calcar made up of two small bony nodules is present on outer side of hallux (Figure 7.27).

Coelom: Internal organs or viscera are lodged in a large and undivided cavity the coelom which is lined by coelomic epithelium called peritoneum, which encircles alimentary canal and other organs. Visceral organs are suspended to body wall by fan-shaped folds called mesenteries. Pericardial cavity housing the heart is a coelomic derivative which becomes cut off from the main coelom and exists as a separate cavity to house the heart.

Digestive System: The system consists of an alimentary canal and digestive glands and is primarily concerned with nutrition. Alimentary canal, a long tube starts from the mouth and ends in cloaca. Canal with a basic histological picture throughout its length regionally modified to perform specific functions. Canal is made up of four layers. The thin

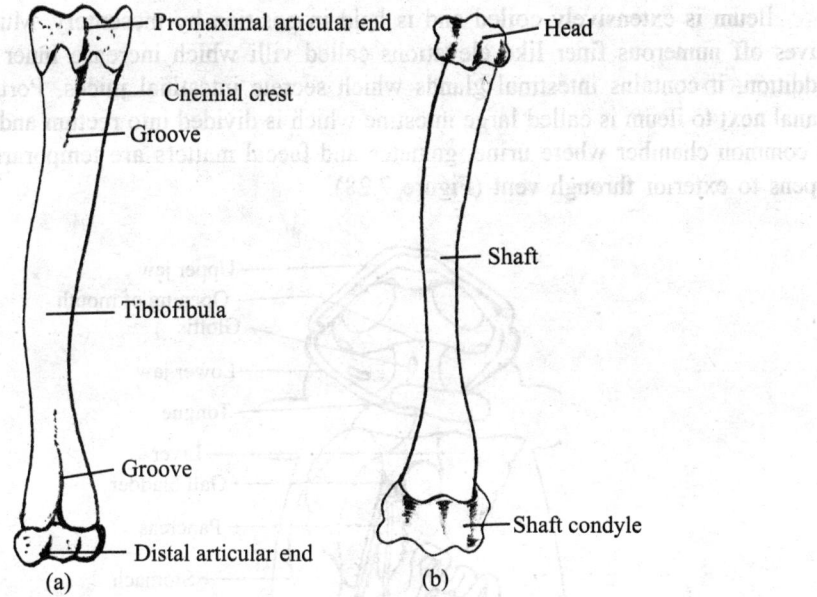

Figure 7.27 (a) Tibiofibula of *Bufo*, and (b) Femur of *Bufo*.

outermost layer is called serous coat. Next to serous coat lies muscular layer, which is composed of outer longitudinal muscles and inner circular muscles. The innermost layer lining the lumen is mucous coat, which is composed of epithelial cells and gland at certain, regions. Region between the muscular coat and mucous coat is filled up with connective tissue matrix with blood and lymphatic vessels. This layer is called sub-mucus coat.

Mouth is a wide aperture and is bounded by upper and lower jaws. Both the jaws are toothless. Mouth leads into a spacious buccal cavity. Roof of buccal cavity contains a pair of internal nostrils, impressions of eyeballs, and a pair of Eustachian apertures. Floor of buccal cavity bears a large fleshy tongue which is attached to hyoid arch. Tongue is very peculiar, as it is fixed anteriorly but fee behind. Tongue is sticky by secretion of intermaxillary glands. It can quickly be protruded to capture insects. Behind the tongue, a longitudinal slit called glottis leads into laryngotracheal chamber. Mucus glands in the buccal cavity do not produce any enzyme. A system of cilia in buccal cavity keeps oral fluid in circulation. Buccal cavity narrows towards the pharynx, which leads into a wide tube, the esophagus or gullet opens into stomach which is a thick walled slightly curved spacious sac.

Broad anterior part of stomach is called cardiac end and other end is called pyloric end. Pyloric part opens into intestine. Opening is guarded by circular sphincter muscle called pyloric valve, which regulates the exit of food from stomach. Mucus coat of stomach which secretes digestive juices and unicellular parietal glands (oxyntic cells) secrete hydrochloric acid opens into the intestine. Besides, the longitudinal and circular muscle layers in the muscular coat, oblique muscles are present in stomach. Part of intestine next to stomach, the small intestine is again divided into an anterior short duodenum and posteriorly into an ileum. Duodenum receives duct from liver and pancreas.

Ileum is extensively coiled and is held in position by mesentery. Mucus coat of ileum gives off numerous finer like elevations called villi which increase inner surface areas. In addition, it contains intestinal glands which secrete intestinal juices. Portion of alimentary canal next to ileum is called large intestine which is divided into rectum and cloaca. Cloaca is a common chamber where urine, gametes and faecal matters are temporarily stored. Cloaca opens to exterior through vent (Figure 7.28).

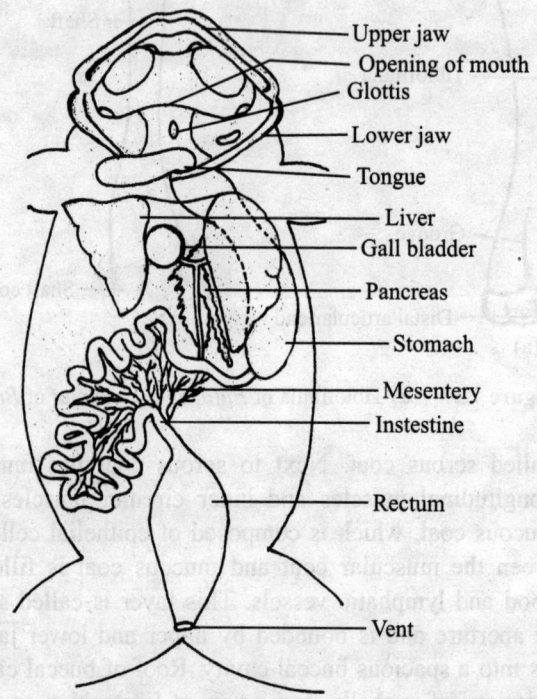

- Upper jaw
- Opening of mouth
- Glottis
- Lower jaw
- Tongue
- Liver
- Gall bladder
- Pancreas
- Stomach
- Mesentery
- Instestine
- Rectum
- Vent

Figure 7.28 Digestive system of *Bufo*.

Besides gastric and intestinal glands, the liver and pancreas are important digestive glands. Liver is the largest gland and consists of two main lobes, left and right. Lobes are connected with one another by a bridge. Left lobe of liver is the larger one and is subdivided into two lobes. Secretory and excretory products of liver are called bile. Bile comes out by hepatic ducts and is stored in gall bladder. Cystic duct from the gall bladder and hepatic duct from liver unite to form the common bile duct, which passes through pancreas and receives numerous minute pancreatic ducts. The common bile duct is called hepatopancratic duct which ultimately opens into duodenum to pour hepatopancreatic juices into duodenal cavity and shows the relationship of liver and pancreas with duodenal part of intestine. Pancreas is an irregular yellowish gland. It is both exocrine and endocrine gland. Exocrine part secretes the pancreatic juice. Endocrine part islets of langerhans, produces the hormone insulin.

Physiology of Digestion: Toad is a carnivorous animal which feeds on small animals preferably the insects. Food consists primarily of carbohydrates, proteins, fats, vitamins, mineral salt, and water. Vitamins, mineral salts, and water are absorbed as such but carbohydrates, proteins and fats are complex organic compounds which cannot be absorbed unless broken down into simpler forms. Digestion is effected by the digestive juices, which contain specific enzymes to act on a specific food by causing hydrolysis. Three groups of enzymes are amylolytic or diastatic enzymes such as amylase of pancreatic juice, lipolytic enzyme, the lipase of pancreatic juice and proteolytic enzymes, like pepsin of gastric juice, trypsin of pancreatic juice, erepsin of intestinal juice.

Thoroughly mixed with mucus in buccal cavity, the food comes into stomach. Parietal glands or oxyntic cells in mucus coat of stomach secrete hydrochloric acid. In acidic medium, pepsin acts on proteins and converts them into peptones. After remaining for sometime in stomach, the half digested acid chime passes into duodenum, where it comes in contact with bile. Bile being alkaline neutralizes the acid. Trypsin then converts the peptones into soluble amino acids, which are readily absorbable. Erepsin of intestinal juices quickens the process of conversion. Lipase converts fats into the soluble fatty acids and glycerol. Amylase of pancreatic juice and maltase of intestinal juice convert fats into soluble fatty acids and glycerol and convert starch into glucose.

Soluble food is absorbed and assimilated into body and digestible products are egested as faeces through vent. Surplus glucose is converted into glycogen and stored in liver and skeletal muscles. Fatty acid transforms into fat and remins in body as fat bodies. Amino acids are not stored, the excess being transformed into urea and excreted as urine.

Spleen: Spleen is a ductless glandular body of dark red colour, spherical and remains morphologically connected with mesentery near junction of ileum and rectum. It acts as a storehouse of blood and destroys old and worn out erythrocytes. Leukocytes are believed to be manufactured in spleen.

Respiratory System: Respiration is a physicochemical process in which oxygen is taken in and carbon dioxide and water and is given out. During this process, stored food within cells becomes slowly oxidized; the oxidation energy is liberated as heat which is necessary for vital activities. Carbon dioxide and water vapour are given out. In toad, two modes of respiration are observed. Terrestrial respiration involves the utilization of oxygen present in atmospheric air and in aquatic respiration the oxygen dissolved in water is used (Figure 7.29).

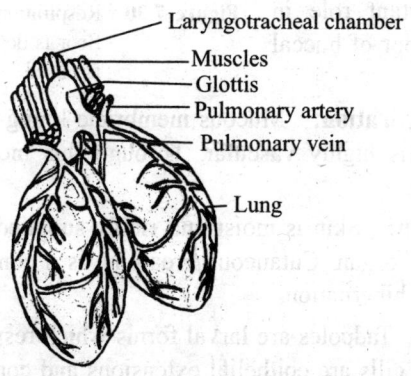

Figure 7.29 Structure of the lungs of *Bufo*.

Terrestrial Mode of Respiration: Toad is capable to breathe air by lungs and skin. Respiration with help of lung is called pulmonary respiration, and that with the skin is called integumentary or cutaneoous respiration.The organ involved in pulmonary respiration are external nares or nostrils, internal nares or nostrils, buccal cavity, glottis, laryngotracheal chamber, bronchi and lungs. Buccal cavity communicates to laryngotracheal chamber, which is a stout box like structure with the two elastic vocal cords stretching across the cavity.

Wall of laryngotracheal chamber is supported by arytenoids and cricoid cartilages. Sound is produced by vibration of vocal cords, which also control intensity and pitch of sound. Vocal sac in male helps to intensify the sound by acting as resonator. Vibration produced sound. Of two extremely short bronchi, each opens into a thin walled spongy lung. Internally, the lungs have innumerable simple sacs called alveoli. Each alveolus is a highly vascular structure and alveolar epithelia are actually the centres of exchanges of gases.The physical mechanism of pulmonary respiration involves three successive stages: aspiration, expiration, and inspiration. During aspiration, the toad closes its mouth but external nostrils are kept open. Floor of buccal cavity is then lowered. As a result of lowering of floor, partial vacuum is created and thus fresh atmospheric air rushes into buccal cavity through nostrils. Glottis remains closed, so the air cannot enter the lungs.

Aspiration is quickly followed by expiration. Trunk muscles contract on the lungs and causes expulsion of air form lungs to buccal cavity through glottis. This expelled air is rich in carbon dioxide. As a result of inhaled and expelled air, the buccal cavity becomes highly distended. Mixture of these two types of air now goes out through open nostrils.

Expiration is immediately followed by inspiration. External nostrils are then tightly closed and floor of buccal cavity is raised forcibly as a result of which mixed air form the buccal cavity is pushed into lungs through glottis. Hyoid apparatus plays an important role in lowering and raising the floor of buccal cavity (Figure 7.30).

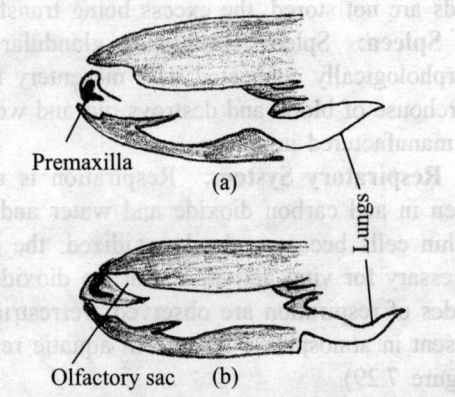

Premaxilla (a)

Lungs

Olfactory sac (b)

Figure 7.30 Respiration in frog (a) When the mouth floor is depressed, and (b) When the same is raised.

Buccopharyngeal Respiration: Mucous membrane lining the buccopharyngeal cavity is always kept moist and is highly vascular. Through this membrane exchange of gases occurs normally.

Cutaneous Respiration: Skin is moist and richly supplied with blood vessels. It acts as an additional respiratory organ. Cutaneous breathing is a continuous process and is very important especially during hibernation.

Aquatic Respiration: Tadpoles are larval forms which respire by external and internal gills besides skin. External gills are epithelial extensions and contain blood capillaries. Gills can absorb oxygen dissolved in water and give out carbon dioxide.

Exchange of Gases: Oxygen is conveyed to different tissues by blood. Oxygen combines with haemoglobin to form an unstable compound oxyhaemoglobin. In this state, oxygen reaches cells and oxyhaemoglobin on reaching the places release oxygen and regains its former state. This circle is repeated. Blood plasma transports carbon dioxide in form of bicarbonates and finally expelled through respiratory surface.

Circulatory System: This system includes two systems: blood vascular system and lymphatic system. Circulatory system is the internal transport mechanism by which nutritive materials, hormones, waste product, carbon dioxide and oxygen are conveyed to the different parts of the body. Cardiovascular system is well developed. This system is composed of three main components: blood, heart, and blood vessels.

Blood: Blood is the main circulatory fluid. It consists of a straw coloured fluid called blood plasma and different blood corpuscles suspended in plasma. Plasma is a watery liquid and contains mineral salts, food, wastes and hormones. Corpuscles are of three types– red blood cells, white blood cells and thrombocytes. Erythrocytes are the oval, biconvex and nucleated, ranging from 15to 20 micra in size. They contain the red coloured iron containing protein, the haemoglobin. Affinity of haemoglobin for oxygen is lower than that in mammals. Leucocytes are colourless, nucleated and amoeboid cells. They are phagocytes, destroy bacteria and protect the animal from invading microbes. Thrombocytes are small spindle-shaped, nucleated and play role in coagulation of blood. When there is shedding of blood, platelets release enzymes which help in coagulation.

Heart: Heart is the central pumping organ in cardiovascular system. It is a pear shaped, muscular structure situated in the anterior part of the body cavity. It remains enclosed by a transparent protective membrane called pericardium. The space between heart and the pericardium is called pericardial cavity. Heart is composed of 1) receiving parts. Two auricles and a sinus venosus are receiving parts of hearts, 2) forwarding parts. Ventricle and conus arteriosus are the forwarding parts of heart, 3) intercommunicating apertures and valves between the different parts of heart. Auricles are two in number and are of unequal size. Both the auricles are sharply marked off form ventricle by a narrow constriction called coronary sulcus (Figures 7.31a, b, c).

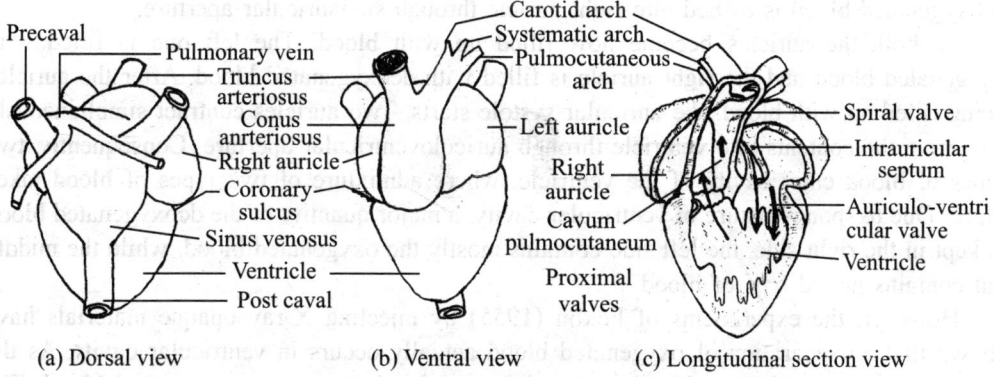

(a) Dorsal view   (b) Ventral view   (c) Longitudinal section view

Figure 7.31 Structure of heart of *Bufo*.

Left auricle is smaller than right auricle. Two auricles are completely separated by interauricular septum. The sinus venosus is a triangular, thin walled chamber formed by the union of three major veins, two precaval and one postcaval. It is situated on the dorsal side of right auricle. It communicates with right auricle through a sinuauricular aperture which is guarded by sinuauricular valves. Deoxygenated blood, collected in the sinus venosus from precaval and postcaval veins, enters the right auricle but the back flow is prevented by the sinuauricular valves.

The left auricle receives oxygenated blood through a small opening of the common pulmonary veins. This aperture is also guarded by valves. The right and left auricles communicate to the ventricle by a common auriculo-ventricular aperture. This aperture is guarded by membraneous valves known as auriculo-ventricular valves. The free ends of the valves are attached with the wall of the ventricle by fine thread like chordae tendineae. The auriculo-ventricular valves give one way traffic of blood from the auricles to ventricle.

The ventricle is a highly muscular chamber. It is conical in shape. Its cavity is greatly reduced by many interlacing muscles called columnae carnae. Conus arteriosus is a stout tubular body situated on the ventral side of the heart. It is continued anteriorly into the truncus arteriosus which is the basal stem of three main anteries. It remains uncovered by the pericardium and lacks cardiac muscles. A set of pocket like semilunar valves demarcate the truncus from the conus. A similar set of three semilunar valves guard the opening between the conus and ventricle, and prevent back flow of blood from conus to ventricle. A twisted longitudinal spiral valve divides the cavity of the conus into two channels. The left channel is designated the cavum pulmocutaneum and the right one is cavum aorticum. Deoxygenated blood passes through cavum pulmocutaneum and oxygenated blood travels through cavum aorticum by the manipulation of the spiral valve.

Mechanism of Circulation Through the Heart: Periodic contraction (systole) and relaxation (diastole) is an innate property of heart. During diastole, the sinus venosus receives deoxygenated blood from 2 precavals and one post caval veins. Left auricle becomes also simultaneously filled up with oxygenated blood from lungs carried through common pulmonary vein. Just with the on set of systole, the sinus venosus contracts and deoxygenated blood is rushed into right auricle through sinuauricular aperture.

So both the auricles become now filled up with blood. The left one is filled with oxygenated blood and the right auricle is filled with deoxygenated blood. After the auricles being filled up with blood, the auricular systole starts. Two auricles contract simultaneously and drive the contents into ventricle through auriculoventricular aperture. Consequently, two types of blood enter cavity of the ventricle, where admixture of two types of blood takes place. Due to spongy nature of ventricular cavity, a major quantity of the deoxygenated blood is kept in the right side, the left side contains mostly the oxygenated blood, while the middle pat contains mixed type of blood.

However, the experiments of Foxon (1955) by injecting X-ray opaque materials have shown that no separation of oxygenated blood actually occurs in ventricular cavity. As the skin sub serves respiratory function, the right auricle also receives oxygenated blood. The spongy ventricular walls presumably help in metabolic exchanges and have nothing to do in the separation of two types of blood. Next the ventricle starts contraction. Back flows of blood into auricles are prevented by auriculo-ventricular valves. Blood from ventricular cavity finds its way through conus.

As the conus arise from right side a large quantity of deoxygenated blood from the ventricle enters first. This blood is now conveyed through cavum pulmocutaneum by spiral valves. From the cavum pulmocutneum, blood goes to pulmocutaneous arteries for oxygenation. The deoxygenated blood is aerated in lungs and brought back to left auricle through common pulmonary vein and thus completes the pulmonary circuit. With enhancement of contraction force of ventricle, the mixed blood from middle region of ventricle is pushed into systemic arches through cavum aorticum. Lastly, the pressure exerted by carotid labyrinth is overcome and mostly the oxygenated blood from left side of ventricle is force fully pumped into carotid arches. Spiral valves direct the entry of blood into different arches.

Blood Vessels: Arteries and veins are the blood vessels through which blood circulates. These vessels are different in structures and functions. Arteries take blood away from the heart to different parts of body and veins return the same towards heart. Arteries supply blood to different parts, and break up into finer branches, the arterioles, which finally end into a network of capillaries, which again reunite to form small venules. Venules unite to form veins. A typical blood vessel is composed of three distinct layers. The outermost covering is made up of fibrous connective tissue called tunica adventitia. The middle layer is composed largely of involuntary muscles called tunica media. The innermost layer made up of endothelium and elastic fibres is called tunica interna.

Arterial System: The arteries and their branches constitute the arterial system. Truncus arteriosus divides into two main branches each of which again splits into three arches: (1) anterior carotid arch, (2) middle systemic arch, and (3) posterior most pulmocutaneous arch. In the embryonic state, six pairs of arterial arches are present. Of these paired arches, the first and second pairs disappear in an adult. The third pair is converted into the carotid arches. The fourth pair is transformed into systemic arches. The fifth pair disappears and the sixth pair persists as the pulmocutaneous arches in adult.

Carotid Arches: There are two carotid arches, each of which proceeds forward and outward. It soon divides into an outer branch called internal carotid artery supplying blood to brain, meninges and into an inner branch called external carotid artery which supplies blood to the outer side of head, thoracic musculature, tongue and buccal cavity.

Just at the point of bifurcation and towards the base of internal carotid artery, there is a small swelling called carotid labyrinth which is derived from the remnants of gill, and connects blood vessel between the first afferent and efferent branchial arteries. The inner cavity of carotid labyrinth contains a network of small vessels and forms a spongy structure, though this structure is a receptor, it physiologically connects with the regulation of blood pressure. It controls blood pressure in internal carotid artery. The internal carotid artery, after its origin, takes a backward course and comes very close to systemic arch and is being tied with it by fibrous carotid ligament.

Systemic Arches: Of two systemic arches, each takes the median position and sweeps outward to surround esophagus. It then comes to dorsal side and joins with its fellow of opposite end to form dorsal aorta. Following branches arise from each systemic arch: (1) a laryngeal artery which is very short and supplies the larynngotracheal chamber,

(2) an occipitovertebral artery, which gives branches to the pharynx, back of the head vertebral column and spinal cord, (3) a stout sub clavian artery supplies blood to the shoulder and forelimb. All these branches exhibit bilateral symmetry, but left systemic arch gives off an additional branch called esophageal artery which supplies blood to esophagus. This branch is not present on right side. The dorsal aorta occupies middorsal position and is situated ventral to vertebral column and ends posteriorly into 2 iliac arteries.

Dorsal aorta gives of the following branches antero posteriorly: (1) a stout coeliacomesenteric artery which emerges out just from origin of dorsal aorta. Artery immediately breaks up into a celiac branch to supply blood to stomach, liver, gallbladder and pancreas, and a mesenteric artery supplying blood to mesenteries, intestine, cloaca and spleen. (2) Four, or five pairs of renal arteries supply the kidneys. From the anterior renal arteries, additional branches arise for gonads. These small branches are called genital arteries. Renal arteries and genital arteries are collectively known as urinogenital arteries. (3) Iliac arteries are the last branches of dorsal aorta. From each iliac artery, an epigastricovesicalis artery is given off to supply the urinary bladder and ventral body wall. Iliac artery then enters hind limb and divides into femoral and sciatic arteries (Figure 7.32).

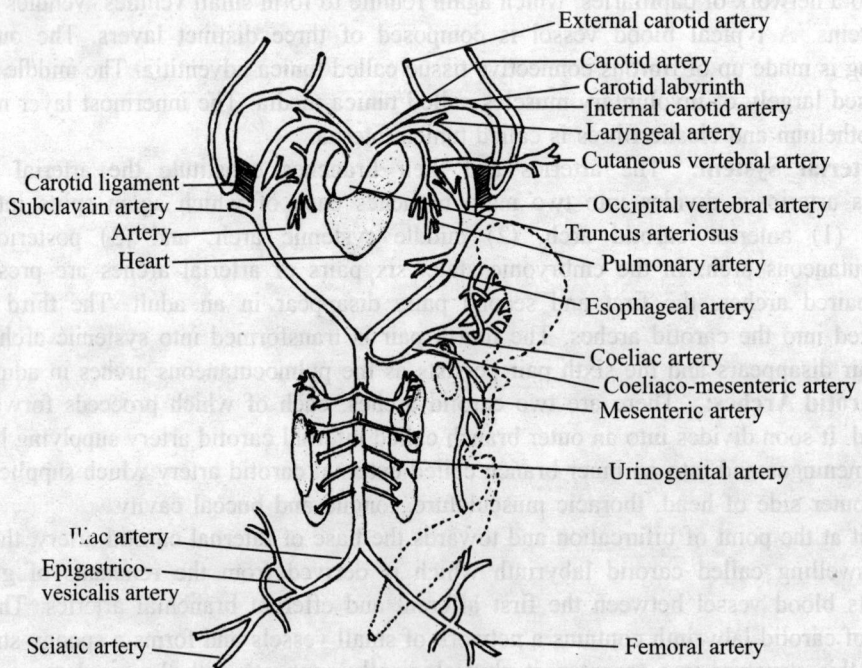

Figure 7.32 Arterial system of *Bufo*.

Pulmocutaneous Arches: These are the shortest and the hindermost arches which carry deoxygenated blood to lungs and skin. Each main arch enters lung as pulmonary artery and a very slender branch from its goes to skin as cutaneous artery.

Venous System: Veins and their branches constitute venous system which can be described under three headings: systemic, portal and pulmonary.

Systemic Veins: Three large veins or venae cavae open into sinus venosus represent systemic veins. Systemic veins carry deoxygenated blood from almost all parts of body excepting lungs. The anterior two venae cavae are known as left and right precavals, and single posterior one is called postcaval. Each precaval vein is formed by union of three branches. These are external jugulars vein; innominate vein and subclavian vein. The external jugular vein is formed by two veins, a lingual carrying blood from tongue, and a faciomandibular from snout and jaws.

The innominate vein is formed by union of 2 veins, an internal jugular bringing blood from head and a subscapular from back of shoulder. Subclavian vein is formed by 2 veins, a brachial vein bringing blood from forelimb, and a musculocutaneous vein from muscles and skin. As the skin acts as an accessory respiratory organ, the musculocutaneous vein brings oxygenated blood. Post caval vein is formed by 4 or 5 pairs of renal veins which receive blood from kidneys. Blood from reproductive organs is also poured into renal veins by the genital veins. Postcaval vein then ascends to enter the sinus venosus.

Portal Veins: A portal vein has its origin in capillaries and it ends in capillaries. Blood from the portal vein returns to heart through an intermediate organ. Hepatic and renal portal systems are two portal systems. Capillaries from gut unite to form a hepatic portal vein. This again breaks up into capillaries in liver. Capillaries from posterior part of body unite to form two renal portal veins which in turn break up into capillaries in kidneys on their way to heart.

Hepatic Portal System: This system comprises of hepatic portal vein and anterior abdominal vein or epigastric vein. Hepatic portal vein carries blood from stomach, intestine, pancreas and spleen. The main vessels receive the anterior abdominal vein under liver and enter substance of liver. Anterior abdominal vein is formed by union of 2 pelvic veins in mid-ventral line. Pelvic vein arises as an off shoot of the femoral vein. On its anterior, the abdominal vein receives small branches from urinary bladder and ventral body wall.

Renal Portal System: Blood form the hind limbs is carried by femoral and sciatic veins. Each femoral vein on entering the body cavity gives off a pelvic vein. The main trunk of femoral vein receives the sciatic vein above the level of pelvic vein to form renal portal vein. Renal portal veins proceeds by side of corresponding kidney and enters it to break up into capillaries. Each renal portal vein receives 2 or 3 dorsolumbar veins carrying blood from body wall (Figures 7.33, 7.34).

Lymphatic System: While circulating through body, blood does not come directly into cells. During the transit of blood through capillaries, plasma exudes from capillary wall into intercellular spaces as lymph. It is a colourless fluid with few leucocytes. Lymph from intercellular spaces is collected by lymphatic vessels which in turn lead into lymph sacs.

Some contractile lymph vessels become lymph hearts which pump back the lymph into veins. A pair of lymph hearts, opening into femoral vein is present near urostyle and another pair is situated below scapulae which open into subscapular veins. Some openings of small dimension in peritoneum communicate with lymphatic vessels or lymphatic. Transport of materials inside body is performed by circulatory system, which includes heart, arteries, capillaries, veins and lymphatics together with fluid components- the blood and lymph.

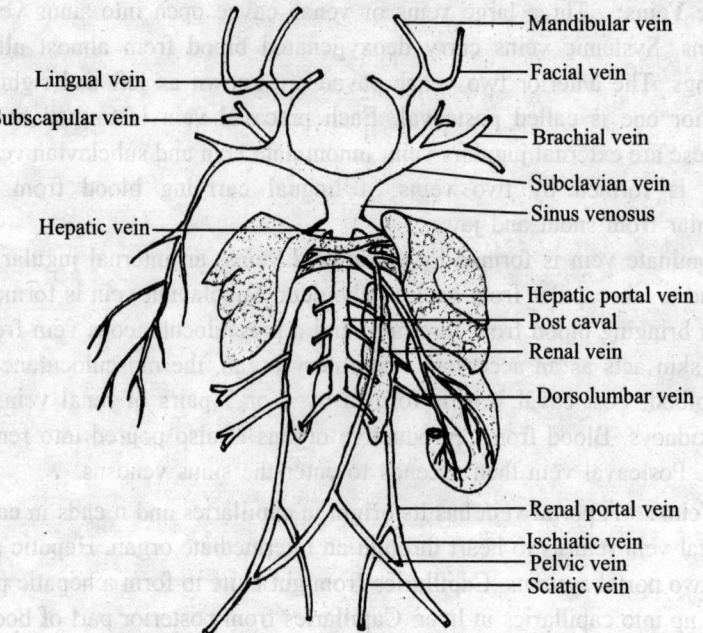

Figure 7.33   Venous system of *Bufo*.

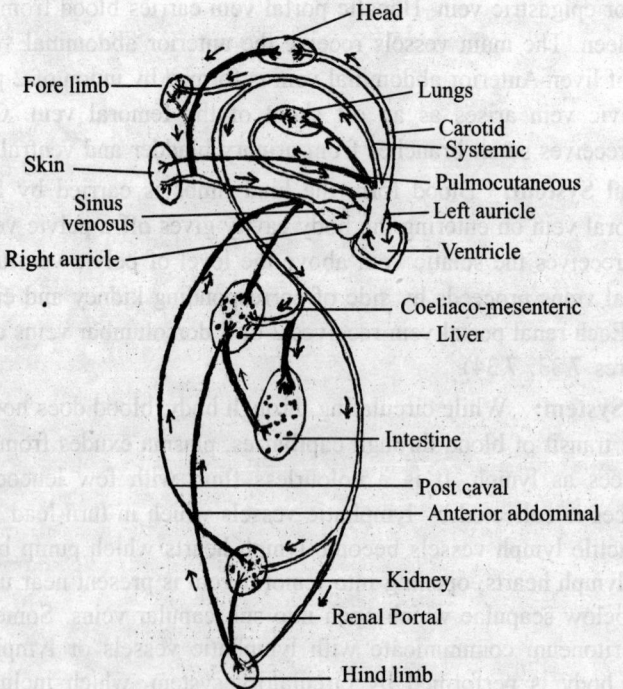

Figure 7.34   Course of blood circulation in toad.

Nervous System: This includes central nervous system, peripheral nervous system and autonomic nervous system. Nervous system controls and co-ordinates various activities of body.

Central Nervous System: Central nervous system includes brain (Figure 7.35) and spinal cord. It is a hollow tube whose anterior portion becomes brain, and posterior part narrows down to form spinal cord and is filled up with cerebrosphinal fluid. Brain and spinal cord are made up of nerve cells and nerve fibres. A collection of the extended processes of neurons enclosed by a connective tissue sheath constitute nerves.

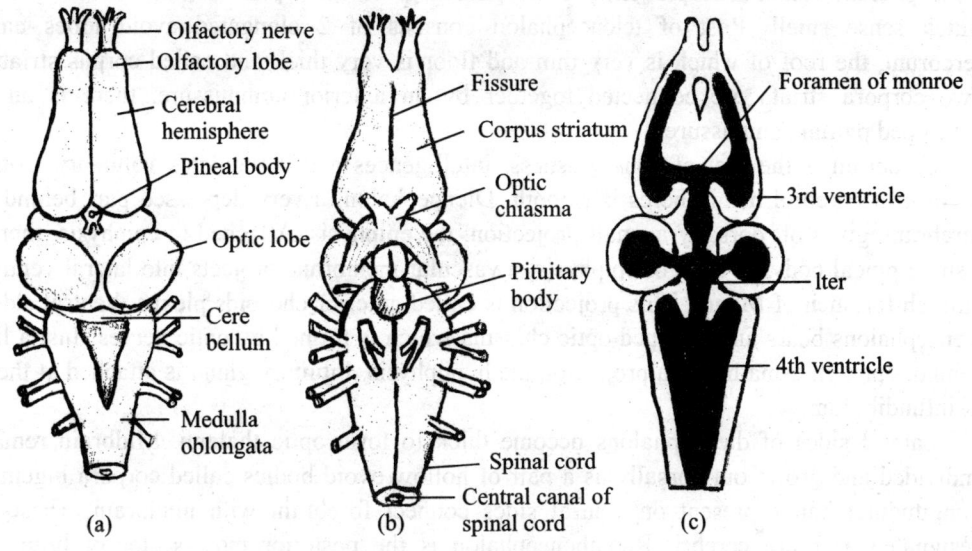

Olfactory nerve
Olfactory lobe
Cerebral hemisphere
Pineal body
Optic lobe
Cere bellum
Medulla oblongata
(a)

Fissure
Corpus striatum
Optic chiasma
Pituitary body
Spinal cord
Central canal of spinal cord
(b)

Foramen of monroe
3rd ventricle
Iter
4th ventricle
(c)

Figure 7.35 Brain of toad (a) Dorsal view, (b) Ventral view, and (c) Ventricles in the brain.

Structure of Neuron: A neuron has a very large cell body with several processes. Nucleus of cell body is conspicuous and cytoplasm contains numerous deeply stainable granules called Nissl granules. The processes called dendrites are usually branched, short. One of the cell processes is long and unbranched, the axon which from the axis cylinder of nerve fibre. The solid part of central nervous system is composed of grey- and white matter. Aggregation of nerve cells with their nuclei gives a grayish appearance and is recognized as grey matter, while collection of nerve fibres giving a white colour is white matter.

The relative arrangement of grey matter and white mater differs in different regions. In brain, the grey matter is situated on outer side. White matter lies towards the lumen. Arrangement is just reverse in spinal cord i.e. grey matter is situated on inner side, and white matter on outer side. The whole central nervous system is enclosed by 2 protective coverings or meninges. The outer one is dura mater which is thicker and fibrous in nature. The inner one is thin highly vascular and called pia matter.

Development of Central Nervous System: Nervous system develops from the embryonic ectoderm. Along the middorsal line of the whole length of body above notochord, the ectoderm forms a thickened neural plate. The sides of neural plate rise up as lateral folds to form a neural groove. As development goes on, the edge of groove fuse along middorsal line and is converted into a neural tube. Anterior part is transformed into brain and posterior part forms spinal cord. Brain was primarily a tube like structure, which due to unequal growth, torsions and bending transforms into a most complicated adult structure.

Brain: Brain is located inside cranium primarily differentiated into three parts by development of two primary constrictions: forebrain (prosencephalon), midbrain (mesencephalon) and hindbrain (rhombencephalon). Prosencephalon constitutes the anterior most part of brain and becomes subdivided into an anterior telencephalon and posterior diencephalons (thalamencephalon). Telencephalon gives off a pair of small olfactory lobes which sense smell. Rest of telencephalon consists of 2 elongated avoid lobes called cerebrum, the roof of which is very thin and floor is very thick and called corpus striatum. Two corpora striata are connected together by an anterior commissure, there is an ill-developed pallial commissure.

Cerebrum is the seat of consciousness, intelligences and it regulates voluntary motion. Surface of cerebral hemispheres is smooth. Diencephalon, a very depressed part behind the cerebrum, gives off dorsally a small projection, the epiphysis. Attached to epiphysis, there is a small pineal body. In front of epiphysis, a vascular membrane projects into lateral ventricle through foramen of Monro. This projection is called anterior choroids plexus. Ventral side of diencephalons bears an X-shaped optic chiasma which is formed by optic nerves. Just a little behind, optic chiasma hangs a projection the hypophysis. Pituitary gland is attached at the tip of infundibulum.

Lateral sides of diencephalons become thick to form optic thalami. Midbrain remains undivided and grows out dorsally as a pair of hollow ovoid bodies called corpora bigemina. Longitudinal bands present on ventral sides connect forebrain with hindbrain. These are designated as crura cerebri. Rhombencephalon is the posterior most sector of brain and becomes divided into metencephalon and myelencephalon. Metencephalon is represented by a thin transverse band like cerebellum. Floor and sides of the myelencephalon become very thick to form medulla oblongata. Cerebellum coordinates body movement. Medulla oblongata controls and regulates some of vital processes viz., regulation heart beat, metabolism and respiration. Roof of myelencephalon is formed by a thin vascular membrane called posterior choroids plexus (Figure 7.36).

Cavities in Brain: Brain is not solid but contains well formed internal cavities called ventricle. Cavities in cerebral hemispheres constitute the lateral ventricles (first and second ventricles). Cavity in diencephalons constitutes the third ventricle, and that of medulla oblongata is the fourth ventricle. Lateral ventricles communicate with third ventricle by a small opening called foramen of monro. Third ventricle communicates with the four ventricles by a narrow passage called aqueduct of Sylvius (or iter). Fourth ventricle leads into cavity of spinal cord. Cavity in optic lobes is called optocoel and that in olfactory lobes as rhinocoel (Figure 7.37).

Spinal Cord: Spinal cord is a hollow tube having a mid dorsal groove called dorsal fissure and a similar ventral fissure on midventral line. The cavity of spinal cord called neurocoel is also continuous with ventricles of the brain. It extends posteriorly from medulla

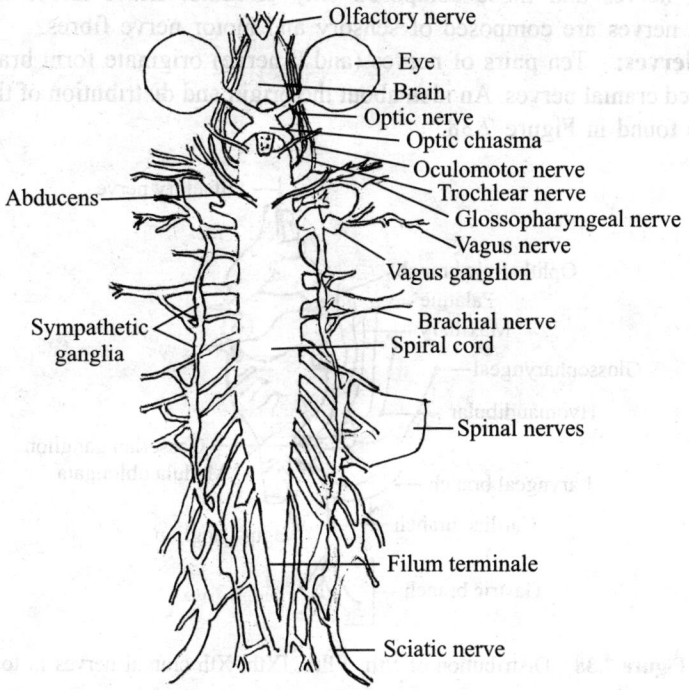

Olfactory nerve
Eye
Brain
Optic nerve
Optic chiasma
Oculomotor nerve
Trochlear nerve
Glossopharyngeal nerve
Vagus nerve
Vagus ganglion
Brachial nerve
Spiral cord
Abducens
Sympathetic ganglia
Spinal nerves
Filum terminale
Sciatic nerve

Figure 7.36   Ventral view of nervous system of toad.

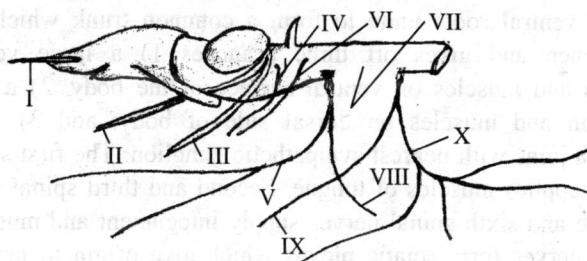

Figure 7.37   Origin of cranial nerves from brain in toad (Diagrammatic) (I) Olfactory nerve, (II) Optic nerve, (III) Occulomotor nerve, (IV) Trochlear nerve, (V) Trigeminal nerve, (VII) Facial nerve and (VIII) Auditory nerve, (IX) Glossopharyrgeal nerve, and (X) Vagus nerve.

oblongata and remains encased within neural canal of vertebral column. It extends even within urostyle as a slender filament, the filum terminale.

Peripheral Nervous System: The system includes cranial and spinal verves arising from cerebrospinal axis. A nerve is a bundle of nerve fibres, but in cases, an aggregation of nerve cells causes swelling on nerve fibres. This swelling is called ganglion. Nerve fires are of two types. Afferent or sensory fibres convey information from receptor organs to central. Nervous system, and efferent or motor fibres carry impulses from central nervous system to effector organs. When the nerves are exclusively made up of sensory nerve fibres, these are called sensory nerves and those composed only of motor nerve fibres are called motor nerves. Mixed nerves are composed of sensory and motor nerve fibres.

Cranial Nerves: Ten pairs of nerves (and 0 nerve) originate form brain. These paired nerves are called cranial nerves. An idea about the origin and distribution of the cranial nerves in toad can be found in Figure 7.38.

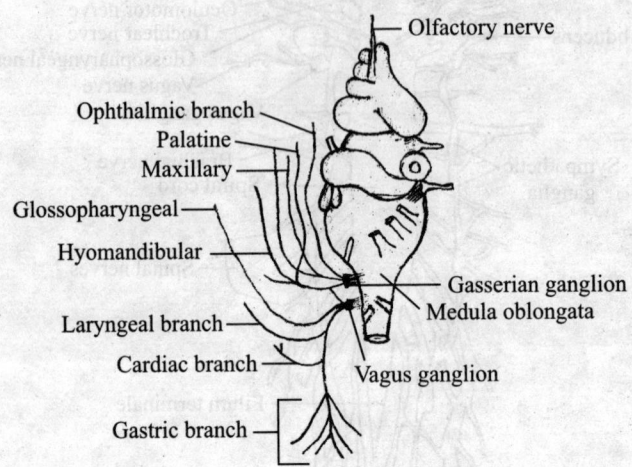

Figure 7.38   Distribution of Vth, VIIth, IXth, Xth cranial nerves in toad.

Spinal Nerves: There are ten pairs of spinal nerves of mixed nature. Each spinal nerve has a dorsal sensory root and a ventral motor root. Dorsal root possesses a ganglion near its origin. Dorsal and ventral roots unite to form a common trunk which comes out through intervertebral foramen and gives off three branches 1) a large ventral branch which innervates the skin and muscles on ventral surface of the body, 2) a small dorsal branch which supplies skin and muscles on dorsal side of body, and 3) a very small ramus communicans which joins with nearest sympathetic ganglion. The first spinal verve is known as hypoglossal. It supplies muscles of tongue. Second and third spinal nerves form brachial plexus. Fourth, fifth and sixth spinal nerves supply integument and muscles of trunk region. The rest of spinal nerves form sciatic plexus which give origin to large sciatic nerve and supply hind limb. The last spinal nerve is insignificant.

Autonomic Nervous System: The systems consists of two sympathetic trunks one on either sides of dorsal aorta. Sympathetic trunks start near the point of formation of iliac arteries and each trunk contains ten small ganglia connecting to the corresponding spinal nerve by ramus communicans. Anteriorly, the sympathetic trunk divides to encircle the subclavian artery. It then enters cranium connects by a twig with vague ganglion and terminates after connecting with Gasserian gangalion.

Sympathetic trunks give out branches to innervate cardiac muscles blood vessels and alimentary canal. These branches sometimes unite to form ganglionated plexuses such as cardiac plexus and solar plexus. Autonomic nervous system regulates involuntary activities like heart beats and peristalsis of alimentary canal. It is an autonomic nervous system in the sense that its activities are somewhat independent of central nervous system.

Sense Organs: Sense organs are receptors for external stimuli and are avenues through which central nervous system is kept informed of the outside world. Each receptor organ responds to a particular type of stimulus and produces its own specific sensation. Receptors for cold, heat, pain and touch are present beneath the epidermis of skin. These are microscopic in structure and are usually regarded as cutaneous receptors.

Receptors: The receptor are scattered in nasal passage. Mucous membrane lining nasal passage contains peculiar olfactory cells and slender supporting cells. Olfactory cells are connected with nerve fibres form olfactory nerve and are receptors for smell. Taste buds are present in tongue and mouth cavity. Each taste bud is made up of taste cells and supporting cells. Taste buds are located in tongue in association with minute elevations called papillae.

Receptors for Vision: eyes are two compact photosensitive organs lodged in orbits, each with a spherical body called eyeball which can rotate inside orbit within its limit by six extrinsic muscles namely superior rectus, inferior rectus, external rectus, internal rectus, superior oblique and inferior oblique. Eye can be protruded by levator bulbi and can be withdrawn by retractor bulbi to a certain limit. Eyeball is composed of three layer arranged in a regular sequence. Sclerotic layer the outer most layer is very tough and thick, consists of 2 parts.

An anterior transparent circular portion called cornea and an opaque posterior portion called white. Cornea permits entry of light rays inside eye. The outer surface of cornea is covered by a thin transparent membrane called conjunctiva. Sclera is composed of cartilage and fibrous tissue and protects the delicate parts of the eye. Uvea or Middle Layer: This layer is divided layer into three parts Choroid, iris and suspensory ligament. Choroids are a highly vascular pigmented layer. At the anterior end just behind the cornea, choroids are modified to form a circular pigmented disc called iris. Iris contains and aperture at the centre which is called pupil acts as a diaphragm and contains circular and radial muscle fibres which help pupil to contract or dilate. Excepting pupil the eyeball is totally light proof. Iris regulates entry of light into eyeball. Just behind the iris, sensory ligaments keep the lens in position. (c) Retinal layer: retinal layer forms the innermost light sensitive screen. it contains two types of photosensitive cells called rods and cones.

Cones are primarily concerned with colour vision in bright and light and rods are chiefly useful in colour less vision at low light intensities. Photosensitive cells are connected with nerve fibres of optic nerve. A crystalline lens is situated just behind pupil and is kept in position by suspensory ligaments. Lens divides the cavity of eye into two chambers. Anterior chamber between the lens and cornea is filled up with a watery fluid. Aqueous humor and the posterior one behind lens is filled up with a transparent jelly-like substance the vitreous humor. Just at the point of entry of optic nerve into retinal layer there is a depression called blind spot where no image is formed. Two muscles, one dorsal and other ventral are connected with suspensor ligament of lens and with cornea. These muscles are called protractor lentis. Contraction of protractor lentis draws the lens closer to cornea and when relaxed the lens is pushed away from cornea.

Mechanism of Vision: Light rays, on way through cornea, pupil and lens, are converged to retinal layer. In retinal layer an inverted and reduced image is formed which is transmitted to brain via optic nerves. This inverted image is translated into a corrected one by brain. Adjustment of image on retinal lever is done by forward and backward movements of lens by protractor lentis. Photochemical basis of image formation relates that vision depends on photosensitive pigments of photosensitive cells. Visual purple (rhodopsin) is abundant in rods and visual violet (iodopsin) is present in cones. Synthesis of both these pigments depends on presence of vitamin A. Pigments are decomposed by light and various products are produced which can create impulses in photoreceptor cells of retina. Two types of vision are encountered- one is binocular vision which means that two eyes can be focused on same object and other is monocular vision when each eye has a different visual field. In toad vision is monocular, because eyes laterally placed and cover different visual areas.

Receptors for Hearing and Balancing: Ears sub serve dual functions, hearing and balancing and consists of three parts: external-, middle - and internal ear. External ear is a tightly stretched membrane called tympanum. Middle ear is a tube like cavity which communicates with buccal cavity by Eustachian tube which equalizes atmospheric pressure on two sides of tympanum. A bony rod, the columella connects tympanum with membranous partition separating the middle and internal ear. Internal ear represented by membranous partition separates the middle and internal ear. Internal ear, the membranous labyrinth floats in perilymph and cavity of labyrinth is filled called endolymph which is enclosed by auditory capsule.

Auditory capsule is sealed from all sides by a membranous partition. Membranous labyrinth is made up of two chambers. The upper chamber is called utriculus and lower one is sacculus. Utriculus gives out narrow tubular semicircular canals. Of 3 semicircular canals, one is horizontal in position and other 2 are vertical. Three semicircular canals are arranged at right angles with one another. Both ends of semicircular canals open into utriculus. Each canal bears an ampulla at its one end only. Sacculus produces short projection called lagena.

Patches of sensory receptors are presents in inner wall of membranous labyrinth. Each patch is composed of sensory- and supporting cells. Receptor cells are connected with nerve fibres from auditory nerve. Sound waves directly impinge on tympanum and vibrations are conveyed by columella to perilymph. From perilymph, vibrations are conveyed to endolymph to stimulate sensory cells of sacculus and lagena. Impulses are transmitted to brain through auditory nerve and are perceived as sound in brain. Semicircular canals maintain body equilibrium. Calcareous particles (otoliths) inside semicircular canals strike on bristles of receptor as animals lose balance. Semicircular canals are arranged in such a fashion, these can easily detect changes of centre of gravity during movement.

Endocrine System: The endocrine glands of toad are:

1. Pituitary gland has two lobes an anterior lobe and a posterior lobe. Anterior lobe of gland secretes a growth stimulating hormone, a thyroid-stimulating hormone, a gonad stimulating hormone and many hormones for regulating the activities of other endocrine glands,

2. Thyroid glands are thy pair of small oval thyroid glands situated on floor of buccal cavity one on each side of hyoid apparatus, produce thyroxin.

3. Parathyroid glands are oval yellowish glands situated near thyroid glands.

4. Thymus gland is small yellowish tympanum

5. Adrenal glands are narrow orange coloured bands situated on ventral side of kidneys. Cortex produces a cortin and medulla secretes adrenalin.

6. Pancreas contains two types of glandular structure. Exocrine part secretes pancreatic enzymes while endocrine part secretes hormones.
7. Besides production of germ cells, the gonads produce sex hormones.
8. Bidder's organ: Biological nature of Bidder's organ is uncertain. Many workers claim the endocrine nature of Bidder's organ.

Urinogenital System: Excretory and reproductive system is functionally different. These two systems are described under the urinogenital system because of close association of the systems. During metabolism various nitrogenous waste products are formed which are eliminated as useless and detrimental. Removal of nitrogenous wastes from the body is called excretion. Nitrogenous waste products such as urea, uric acid and excess of water with some salts are excreted as urine. A pair of kidneys, a pair of ureters, a urinary bladder, cloaca and vent performs this function. Kidneys are retroperitoneal in position i.e. not enclosed within peritoneum and mesonephric. Each with compact elongated structure is attached to dorsal side of body wall.

Edges of kidneys are marked by notched. Each kidney is composed innumerable microscopic uriniferous tubules, each with a proximal two walled cup like structure called Bowman's capsule. A small afferent vessel from renal artery enters capsule and breaks up into a network of capillaries to form glomerulus. From the glomerulus and efferent vessel emerges out and pours the blood into renal vein. The Bowmen's capsule together with glomerulus constitutes Malpighian body. The other end of Bowmen's capsule communicates with tubule supplied by capillaries from renal portal vein. Tubules open into collecting tubules which open into ureter. Numerous minute openings called nephrostomes are found on ventral side of kidneys. Ureters unite posteriorly to form a common duct which opens into cloaca. There is a bilobed urinary bladder which is formed by fusion of two urinary bladders.

Others excretory products are carbon dioxide and water which is eliminated by lungs and skin. Blood containing excretory products is brought to kidneys by renal artery and renal portal vein. On its way through glomerulus, water containing nitrogenous waste products and other substances is filled out into capsular space. The filtrate then passes through convoluted tubule where useful substances are reabsorbed and go back into circulating blood. Fluid from the tubules is collected by collecting tubules and is drained into ureters as urine. Urine then drips into cloaca and is stored temporarily in urinary bladder. When fully filled with urine, urinary bladder contracts to eject urine through vent. With all probabilities, nephrostomes drain away excess of water and wastes from body cavity (Figure 7.39).

Reproductive System: Sexes are separate. Sexual dimorphism is present. During breeding seasons a cushion like thumb pad appears at the base of each innermost finger of hand of male. Vocal sac present in male opens into buccal cavity. Impression of vocal sac can be seen from outside.

Female Reproductive System: It consists of two ovaries and two oviducts. Each ovary is an irregular much folded sac and is attached with ventral side of kidney by a thin fold of membrane called mesovarium. During breeding season, ovaries become enlarged and abdomen of females becomes distended to accommodate the increasing number outgrowths called fat bodies which are found anterior to kidneys. Fat bodies, the storehouse of fatty substances are utilized during hibernation and during germ cell formation.

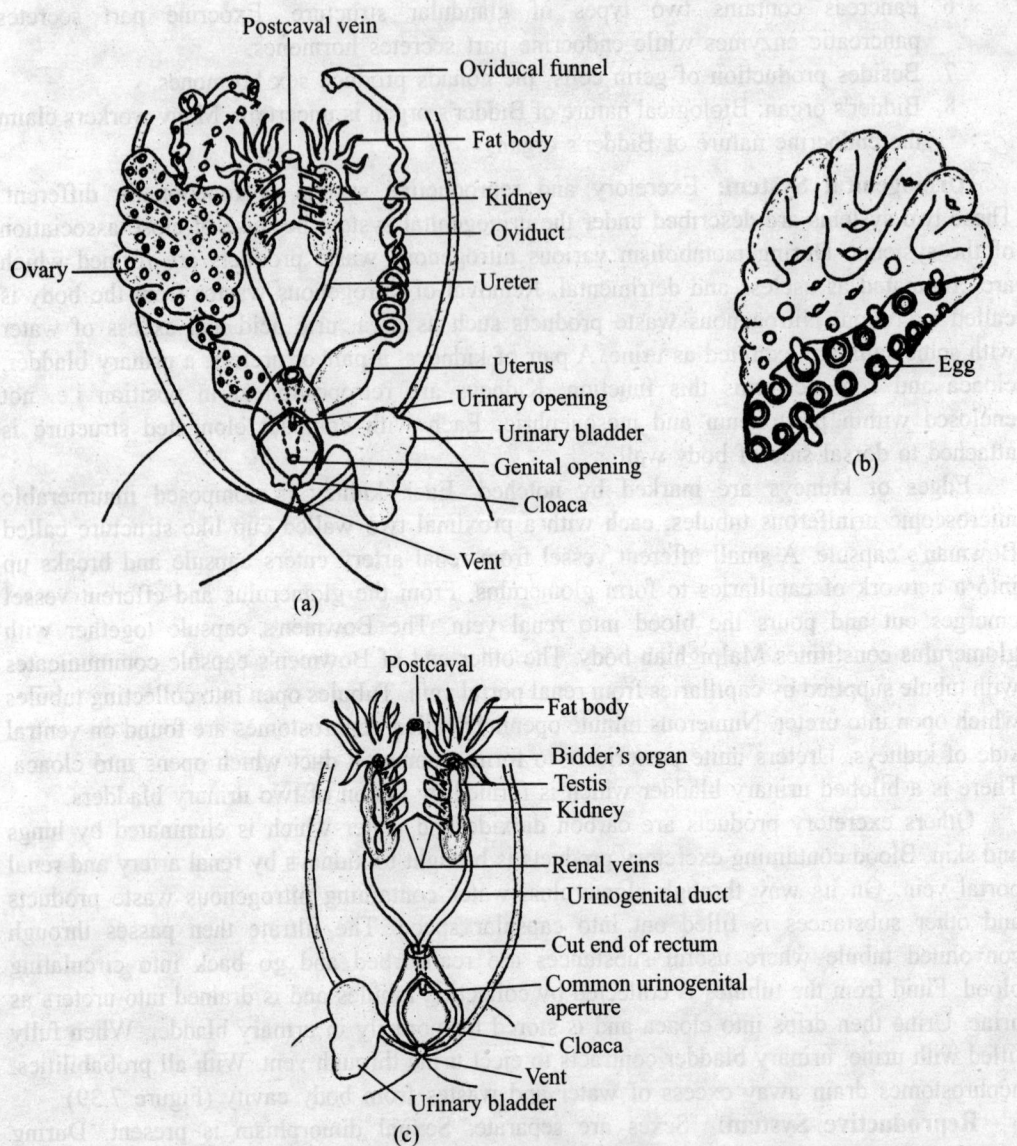

Figure 7.39   (a) Excretory & genital systems of a female *Bufo* sp., (b) An enlarged sectional view of ovary showing the discharge of eggs and (c) Urinogenital system of a male *Bufo*.

Female gonoduct is a very long coiled tube placed on lateral side of body. Each oviduct is differentiated into three parts. Anteriorly, it opens directly into coelom by a funnel like opening called oviducal funnel without direct connection with ovary. Median portion is extensively coiled and glandular. The posterior part dilates to a thin walled uterus. Two uteri unite to form a common median tube which opens to dorsal side of cloacal chamber. Eggs after maturation are discharged into oviduct through opening of oviducal funnel.

Male Reproductive System: It comprises of two elongated white bodies called testes. Each testis is attached with ventral side of kidney by a thin fold of membrane called mesorchium. Each testis is made up of a collection of fine seminiferous tubules, which are connected with collecting tubules of kidney by fine ducts called vasa efferentia which open into ureter, and in male is called urinogenital duct. Through this duct, male gametes and urine are poured into cloaca in front of testis and a small rounded body called Bidder's organ remains attached to anterior end of each kidney. It is an undeveloped ovary and may be functional if testes are removed. This peculiar structure regarded by some as an ovotestis which is also present in female in ill developed condition.

Vocalization: Voice of toads and frogs attract mates.In American toad the voice of male has a strong influence in attracting females. Vocalization of male *Bufo melanostictus* can be classified into 2 types.

1. Advertisement Call.

(i) Premating call: Before amplexus, male begins to croak with low pitch which is quickly followed by other males in breeding site. This call continues until the females approach closely. The call is made up of series of identical notes.

(ii) Mating call: During amplexus, males often produce a guttural croak with high pitch whose duration is very short.

(iii) Territorial call functions as spacing mechanisms among males and is common in species with prolonged breeding seasons. In *B. melanostictus*, one male begin a single note call, is followed quickly by other males in same breeding ground area to form a chorus. If one of croakers stops, other gradually stops crocking. This call is used as advertisement to attract females for mating.

2. **Distress Call:** *B. melanostictus* when escape from their enemies, rarely croak or chirp. But when their predators like snakes try to swallow by seizing their hind legs or middle part of body, they produce a screaming sound.

7.21 Amphibian Declines

Amphibian declines primarily in higher altitude zones: Western US, Costa Rica, Panama, Venezuela, and Australia. In general, species with entirely aquatic life histories are most affected. Typical amphibious life histories are the intermediate; and those without aquatic phases the least affected. The chytridiomycete fungus is the proximate cause of recent declines. The fungus has been found in sick and dead adult anurans collected during mass mortality events in Central America, Australia, and Europe; and most recently, in the Sierra Nevada in California. Climate change may raise the transmission rate of many diseases-effects amphibians, too. Seventeen years ago, in the mountains of Costa Rica, the Monteverdeharlequin frog (*Atelopus* sp.) vanished along with golden toad (*Bufo periglenes*).

An estimated 67% of 110 or so species of *Atelopus*, which are endemic to American tropics, have met the same fate, and chytrid fungus (*Batrachochytrium dendrobatidis*) is now implicated. Large-scale warming is a key factor in the disappearances of amphibians. Temperatures at many highland localities are shifting towards the growth optimum of

Batrachochytrium, thus encouraging outbreaks. Climate change is thus promoting infectious disease and eroding biodiversity. Both *Rana lessonae* and *R. esculenta* exihibit some phenotypic plasticity in response to environmental changes. Triphenyltin (TPT) compounds are used worldwide as agricultural fungicides. TPT is known to kill fish and arthropods, including mosquito larvae, odonate larvae, and snails. Accumulation of TPT is pH dependent. TPT slows growth rate, lengthens time to metamorphosis in both parental species and hybrids. As lower pH is better for parental species; hybrids do better at high pH.Survival of parental species much lower in polluted habitats; hybrids are better generalists, and are not as affected.

7.22   Parental Care in Amphibians

Parental care includes activities that are directed by parents for protection, maintenance and increasing the chances of survival of offspring. Such care reduces the parent's ability to invest for additional offspring. Parental care means care of eggs or juveniles which has evolved to reduce the energy expenditure on reproduction. Lower animals produce large number of eggs and do not exhibit parental care. Higher animals show varied degree of parental care.

Amphibians were the first vertebrates to have evolved different kinds of parental care. Evolution of parental care are of central importance in understanding the sexual selection. Parental care includes all the ways in which parents contributed to the fitness of their offspring. The care of fertilized eggs by parents is found in many groups of animals. Care of eggs laid directly on substrate or attached to it or in nests or burrows as in many amphibian is known. Care of eggs attached externally to, or carried by parents, in some amphibian is also recorded. The retention of fertilized eggs with the female's reproductive tract, often associated with ovoviviparity or viviparity is remarkable. Eggs are usually retained in ovary, oviduct or specialized brood pouch. Care of eggs in the parent's body cavity as in the stomach (Australian frog) draws special attention. Amphibians exhibit an array of parental care through caring and transporting eggs, or larvae, and feeding the larvae.

Care of Eggs

Parental care is associated with species that laid their eggs in single clusters, never in species that scatter their eggs in water. Some lay their eggs in moist land. *Rhacophorus schlegli* lays eggs in hole on muddy area with foamy mucus to prevent the eggs from drying. *Gyrinophilus* lays eggs under stones in stream. *Hylodes* lay eggs on under surface of leaves hanging above water. Eggs are fixed with aquatic weeds by glues in *Triton*. *Rhacophorus maculates* make the surrounding water frothy by limb movements, after the eggs are laid, which prevent the eggs from desiccation and escaping from the predator's eyes.

Male *Rana clamitans* defend their eggs by defending intruders in their territories. Male *Mantophryne robusta* holds cluster of eggs in gelatinous envelop in forelimb. *Ichthyophis* females guard the eggs after laying by coiling body till the eggs hatch. Male performs the same function in *Megalobatrachus*. *Pelobates* hold the newly hatched tadpoles in their mouth. Female *Desmognathus* fuscus carry cluster of eggs glued to their body. In

Desmognathus, females carry the eggs and live in underground hole. In *Rhacophorus reticulates*, the eggs are glued to abdomen of females. In *Alytes obstericans*, the male entangles the eggs around his hind legs.

In *Hyla goeldii*, females carry eggs on their back. In *Pipa pipa*, eggs are carried by females on the back. In *Cryptobatrachus evansi*, dorsal skin contains many small pockets for lodging of eggs. In Pipa dorsigera, eggs are developed in pits on back of females. During breeding season, the dorsal skin becomes soft, spongy. Embryonic development occurs within pits. In *Arthroleptis*, the larvae are attached to males and are carried. *Rhinoderma darwinii* keeps fertilized eggs in vocal sacs. In *Hylambates breviceps*, the female carries eggs in her buccal cavity.

Formation of Nests

Amphibians build nests for egg deposition. *Hyla faber* digs small holes in mud for deposition of eggs. *Phyllomedusa hypochondrales* folds the margin of leaves and glued together which acts as nest for eggs. *Triton* makes shoot nest by fixing the shoots with gelatinous secretion.

Viviparity

In *Salamandra atra* and *S. maculosa*, the eggs are placed inside uterine cavity where entire development takes place.Uterine wall functions physiologically as primitive placenta.

In Apoda

The female coils around the egg clutch and periodically rotates it, till the eggs hatch. Mother does not take any food during the parental care. *Ichthyophis beddomei* lays 25 to 38 eggs in an egg clutch. Egg size ranges from 6 mm at the time of laying. Female *Boulengerula taitanus* provide their own cast skin as food to their offspring. In *Boulengerula taitanus*, the skin of female is transformed to provide supply of nutrients for developing larvae, which are equipped with a specialized dentition. Embryos of *Dermophis mexicanus* feed on the egg yolk supply for about 3 months of gestation. Viviparity is reported in *Gegeneophis seshachari* in which the oviducts are highly vascularized.

In Urodela

Hynobius retardatus female attaches an egg sac on branches that run horizontally touching the water surface. egg sacs are mainly set off about 3 cm under the water surface. Each egg sac contains up to102 eggs. In *Paradactylodon mustersi*, fertilization of eggs in paired egg sacs is external. These sacs are attached to underside of rocks and are guarded by males. During mating, *Paradactylodon gorganensis* female produces paired gelatinous egg sacs, each containing 35-70 eggs. Male grasps these and fertilizes them externally. Males *Cryptobranchids* prepare nests below large, submerged stones or logs. Females lay paired strings of several hundred eggs which are fertilized externally by male. Males guard the eggs until they hatch in 2-3 months.

In Anura

Some frogs carry the eggs and tadpoles on their hind legs or back (*Alytes*). Some frogs protect their offspring inside their bodies. Male *Assa darlingtoni* has pouches along the side of body in which tadpoles reside. Female *Rheobatrachus* swallow its tadpoles, which then develop in stomach. Darwin's Frog (*Rhinoderma darwinii*) puts the tadpoles in vocal sac. Strawberry Poison Dart Frogs (*Oophaga pumilio*) avoid laying eggs in ponds and streams. The mother carries the tadpoles on back. *O. pumilio* tadpoles are egg feeders. After mating, the female lay an average of 3 to 5 eggs on a leaf. Male ensure that the eggs are kept hydrated by transporting water in his cloaca. After about 10 days, eggs hatch and the female transports the tadpoles on her back to some water body. Care about the young reaches the highest degree in case of *Rheobatrachus silus* and *Rheobatrachus vitellinus*. These species carry about 20 young in the stomach, during which they do not feed. Female swallows eggs after the male fertilizes them. After 8 weeks, fully formed frog lets come out of stomach to mouth of mother, sit on its tongues, and jump out from it to water. Tadpoles secrete special chemical substance - prostaglandin E2, which suppresses secretion of acid by mother's stomach.

All the Eleutherodactyls have direct development from heavily yolked eggs to small froglets. *Eleutherodactylus cundalli* guards the egg clutch until the young hatch. Jamaica's hylid frogs breed in the water-filled leaf-axils and produce rapidly developing eggs and by laying further eggs which are eaten by the first-born larvae. The eggs laid during the first few days are fertilized and later the unfertilized eggs are laid, which the larvae consume rapidly. *Leptodactylus fallax* lays eggs in foam nests, and the tadpoles develop without water. Female *Eleutherodactylus cundalli* carry frog lets from cave to rain forest.

Dendrobates auratus female lays up to 6 eggs which are encased in gelatinous substance for protection. During the 2 week development period, the male returns to eggs periodically. Once the tadpoles hatch, they climb onto the males back and to be transported for further development. Tadpoles are attached to the males back by a mucus secretion. In *Rhinoderma darwinii*, the female lays her eggs in moist leaf litter. She then hops off, leaving the male to attend to them. Male guards the eggs. *A. cisternasii* and *A. obstetricans* attach the fertilized eggs to their hind legs. This starts by the male wrapping them first around his ankles. Sometimes they carry up to 200 eggs on their bodies. Father frogs then keep eggs moist by settling into shallow pools, allowing the eggs to double in size. Male Leptodactylid, *Thoropa petropolitana* cares for eggs in same way. The microhylid males *Breviceps adspersus* and *Synapturanus salseri* also do the same (Figure 7.40).

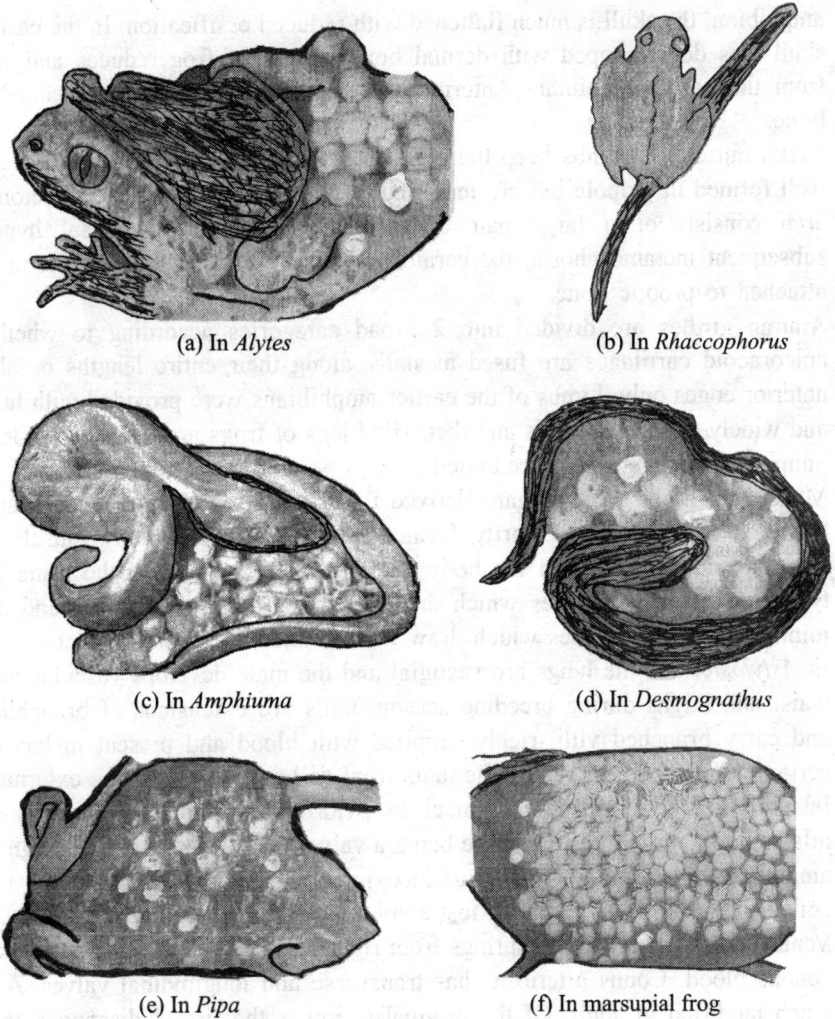

(a) In *Alytes*

(b) In *Rhaccophorus*

(c) In *Amphiuma*

(d) In *Desmognathus*

(e) In *Pipa*

(f) In marsupial frog

Figure 7.40 Parental care in amphibia.

Summary

1. Amphibians live on land as well as in water. Their skin is soft, moist and naked. Caecilians are provided with scales. They have granular poison gland. Two pairs of limbs are usually present. Amphibians link terrestrial and aquatic vertebrates. Their larval stage is found in water and larvae breathe by gills. Mature animals use lungs or skin to breathe.

2. Amphibian skeleton is mostly made of bones. They also have middle ear and inner ear. Eyes of the extinct amphibians called labyrinthodonts were situated at the top of skull and had special sense organs in skin, forming lateral line organ. Skull of Devonian and Carboniferous amphibian was like that of osteolepid fishes. In modern

amphibian, the skull is much flattened with reduced ossification. In the earlier forms, skull was dome shaped with dermal bones. Skull of frog reduces and specializes from the early amphibian. Anterior tips of cartilages ossify as mento-Meckelian bones.

3. Distal muscles of limbs keep the digits pressed against ground. Visceral arches are well formed in tadpole but are modified in adult frog. In tadpole, skeleton of hyoid arch consists of a large pair of ceratohyals attached to basal hypohyal. In subsequent metamorphosis, the ceratohyals form the long anterior cornu of hyoid, attached to prootic bone.

4. Anuran girdles are divided into 2 broad categories according to whether the 2 epicoracoid cartilages are fused mesially along their entire lengths or along their anterior edges only. Limbs of the earlier amphibians were provided with large bones and widely expanded hands and feet. Hind legs of frogs are long, giving leverage in jumping. Tibia and fibula are united.

5. Muscles of limbs were perhaps derived from radial muscles. In modern amphibian, the limb musculature is partly formed from myotomes. Many muscles produce rotation and movement in the body. In hind limb, muscles of the same 2 general types are anterior muscles which draw the limb forward and flex and adduct its joints, and posterior ones which draw it back and extend and abduct.

6. In *Astylosternus*, the lungs are vestigial and the male develops vascular papillae on waist and thighs during breeding season. Gills are extensions of branchial arches, and carry branched villi, richly supplied with blood and present in larvae and in certain adult urodeles. Where the main trunk is long, the gill is the external.

7. Intestine is marked off from stomach by pyloric sphincter. It is relatively short and dilates into a large intestine, there being a valve interposed in frogs, though not in all amphibia. Liver and pancreas produce juice. Intestine of omnivorous tadpole is coiled than that of adult frog. Most amphibians are not particular feeders.

8. Ventral aorta is very short, springs from right side of ventricle and receives first the venous blood. Conus arteriosus has transverse and longitudinal valves. Arches are much modified in adult. Of the original 6 arches the first 2 disappear, the 3rd on each side gives rise carotid artery, the 4th remains complete and forms systemic arch.

9. *Chiroleptes* conserve water by losing glomeruli altogether. *Rana cancrivora* is euryhaline and may have 2-9% of urea in blood. Mullerian duct develops separately from Wolffian system in frog, but arises from the latter during development in urodeles. In this, the frog shows a greater degree of specialization of its developmental processes.

10. External fertilization occurs mostly. Eggs are typically laid in water. Some are ovoviviparous or viviparous and carry the eggs till hatching. Ovaries are the folds of peritoneum having no solid stroma. Follicle cells around each egg presumably produce ovarian hormones.

11. Ten spinal nerves are found but since an embryonic first one is missing they are sometimes numbered 2-1 1. Two spinal segments contribute to brachial and four to

sciatic plexuses in frog. From these plexuses, fibres are distributed to muscle and skin of limbs. Brain resembles that of Dipnoi very strikingly.

12. Metamorphosis in anuran includes regressive metamorphic changes and progressive metamorphic changes. Neoteny or Paedogenesis is a phenomenon of retention of larval characters in sexually mature state. Metamorphosis of larva is retarded and larval characters are retained beyond normal period. In urodeles, this phenomenon is common. It also occurs in anurans. *Ambystoma tigrinum*, in Mexican plateau lives in cool mountain ponds, retains external gills and many other larval features, reproduces as a larva, and never completes its metamorphosis. This permanent larva is called as axoltl.

13. Parental care includes activities that are directed by parents for protection, maintenance and increasing the chances of survival of offspring. Such care reduces the parent's ability to invest for additional offspring. Parental care means care of eggs or juveniles which has evolved to reduce the energy expenditure on reproduction. Lower animals produce large number of eggs and do not exhibit parental care. Higher animals show varied degree of parental care. Amphibians were the first vertebrates to have evolved different kinds of parental care.

Review Questions

Short Answer Questions

1. Define amphibian.
2. Give the name of two living orders of Amphibia.
3. Give the name of two extinct orders of Amphibia.
4. Give two characters of Urodela.
5. Give two characters of Apoda.
6. What is the significance of the term "Apoda?"
7. Mention the significance of the term "Amphibia".
8. Give the significance of the term "Urodela".
9. Comment on the mucus gland.
10. Comment on the poison gland.
11. What is the function of hyoid apparatus?
12. What do you understand by endoskeleton?
13. What is urostyle?
14. What do you understand by procoelous vertebrae?
15. What is deltoid ridge?
16. What is uvea?
17. What is blind spot?
19. What are nephrostomes?
20. Comment on the distress call.

Long Answer Questions

1. Classify Amphibia up to living orders with examples
2. Write a note on the respiration in frog.

3. Give an illustrated account on the origin of Amphibia.
4. Give an account of metamorphosis in Amphibia with suitable illustrations.
5. Describe the neoteny in amphibian along with a note on its significance.
6. Draw and describe the digestive system of toad.
7. Draw and describe the mechanism of circulation through heart in toad.
8. Draw and describe the female reproductive system of toad.
9. Discuss the phenomenon of parental care in Amphibia.

8

The Reptiles

Reptiles are ectothermic tetrapods with clawed toes. Skin is dry with epidermal scales, or plates, but without mucous glands. They respire by lungs and regulate body temperature .Turtles living in water have flippers. Land-living tortoises are plant-eaters .Lizards have limbs, external ears, and developed vision. They detach their tails to escape from predators. Vipers sense temperature changes to locate animals. Snakes are related to lizards. Their jaws disconnect to swallow large prey.

The crocodilians moved back from land into water. Shells of turtles have highly, modified scales supported by modified rib cage. Distinctive box like shell of upper and lower shields connects to vertebrae, collarbones and ribs. They inhabit water, deserts and lands. Reduced shells and pronounced flippers are used for swimming in sea turtles. Fertilization is internal. Eggs with leathery shell are laid on land. Some snake and lizard are viviparous. Embryo develops within egg. No larval stage. Modern forms are placed in Order Squamata. *Sphenodon*, of New Zealand is surviving with little change from Triassic. Crocodile, an older offshoot of stock from which modern birds has evolved. Tortoises and turtles preserve some organization of earlier times, through the special protection of their shells.

8.1 Skin

Skin is dry with few, or no glands. Malphigian layer of epidermis produces horny scales, which are periodically shed in flakes, or cast as single slough. Many reptiles have bony plates in dermis beneath the horny scales which may be restricted to head, or may cover most of body. Tortoise's shell has both horny and bony components. Horny scales are often modified to form crests, spines, and other appendages. Bold and elaborate colour patterns help for concealment. In lizards, there are marked colour differences between the sexes. Colour change is pronounced in certain lizards than other known reptiles

8.2 Posture, Locomotion, and Skeleton

Head is carried off the ground on a well developed neck. The first and second cervical vertebrae are atlas and axis. Atlas is a ring of bone without centrum, but with facet in front for occipital condyle and one behind for odontoid process. Vertebrae articulate with each other by interlocking processes. Each centrum is concave in front, covers the convex hind end of vertebra next to it, a condition called procoelous. The centra articulate by flat surfaces in amphicoelous vertebrae. Vertebrae are united by zygapophyses, the facets on neural arches. Ribs are well developed in middle or trunk region; each articulates with body of vertebra by single capitular facet.

Cervical vertebrae are variable in number and have short ribs. Sternal ribs occur in thoracolumbar segments. Two sacral vertebrae are short and broad and articulate with ilia. Caudal vertebrae are reduced in all parts, especially towards tip of tail. Chevron bones attached to caudal centra called ossicles are the reduced intercentra. Girdles and limbs have the general features like amphibia. Limbs form the main locomotor system. Humerus and femur are held in position with their outer ends lie higher than inner in position of abduction. Radius, ulna and tibia, fibula proceeds downwards towards ground.

Hand and foot are turned outwards at right angles to rest on ground. The main muscles draw the humerus and femur backwards and forwards and downwards. The ventral regions of girdles are large and flattened to receive muscles. Pectoral girdle has a dorsal scapula and a large ventral coracoid, which may be fenestrated. Distinct pro- and post-coracoid elements are found in extinct reptiles. Dermal components are paired clavicles and median interclavicle. A cleithrum is found in few very primitive forms.

In pelvic girdle, the usual dorsal ilium, anterior pubis, and posterior ischium are found, the last 2 meet their fellows in midline symphysis. The general plan is like that of primitive amphibians. Except the most primitive reptiles, hole develops in temporal region to provide space for bulging temporal muscles. Skull roof made up of dermal bones, including nasals, prefrontals, frontals, supraorbitals, and parietals. Side of skull is less complete, composed of tooth-bearing premaxilla, maxilla, lacrymal, jugal, postorbital, squamosal, supratemporal and quadrate. Margins of palate are formed by flanges of premaxillae, maxillae and small ectopterygoids.

Internal nostrils lie forwards between maxillae, vomers, and palatines. More posteriorly, the floor of skull is made up by pterygoid and parasphenoid, which is partly fused with lower surface of basisphenoid. Occipital bones surround the foramen magnum and make up the single occipital condyle, which in some produces 3 partly distinct lobes. In many reptiles, an epipterygoid on either side of brain case behind orbits is an ossification in ascending process of palatoquadrate.

Lower jaw consists of 6 bones, the articular forms joint with quadrate, and dentary carries teeth. Anterior part of chondrocranium, surrounding the front of brain and nasal capsule may be membranous. There may be small ossified orbitosphenoids and farther back pleuro- or laterosphenoids, which develop in pila prootica uniting orbital cartilage with otic capsules. Between the eyes, thin sheet of cartilage called interorbital septum is partly ossified by small presphenoid elements. Posterior part of chondrocranium ossifies to form occipital complex, basisphenoid, and ossifications in otic capsule (Figures 8.1, 8.2, 8.3).

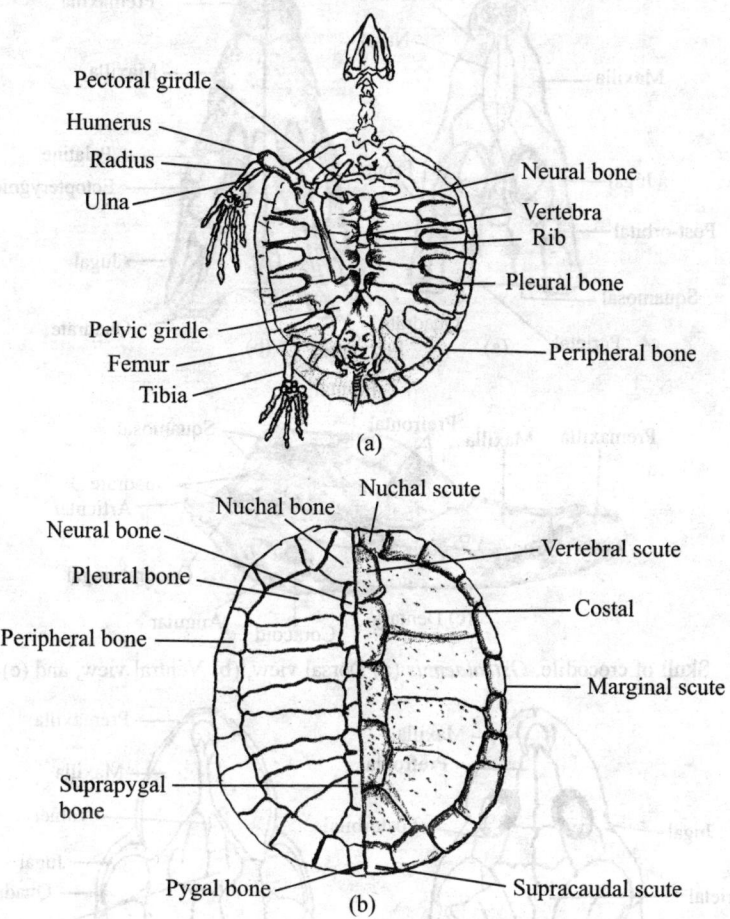

Figure 8.1   Turtle skeleton (a) Ventral view, and (b) Dorsal view showing the bony components on the left & the overlying epidermal scutes on the right.

In many reptiles, upper jaw and front part of skull move in relation to occipital region and cranial base.Such movement called kinesis is often associated with mobility of quadrate. Kinesis helps to widen gap and provide shock absorbing effect, when jaws are snapped together. Postmandibular visceral arches play no part in jaw support, but are incorporated into ear and hyoid apparatus.

A rod-like columella auris occurs with a small cartilaginous element at its outer end. Columellar system behind quadrate conducts vibrations from tympanum to fenestra ovalis and inner ear. In some, the tympanum is absent and outer end of columella is applied to quadrate. Such forms may be deaf to air-borne sounds, but sensitive to ground vibrations, transmitted through bones of jaw. Hyoid apparatus consists of basal plate, which projects into tongue, and 3 pairs of ascending horns which are the remains of hyoid and branchial arches.

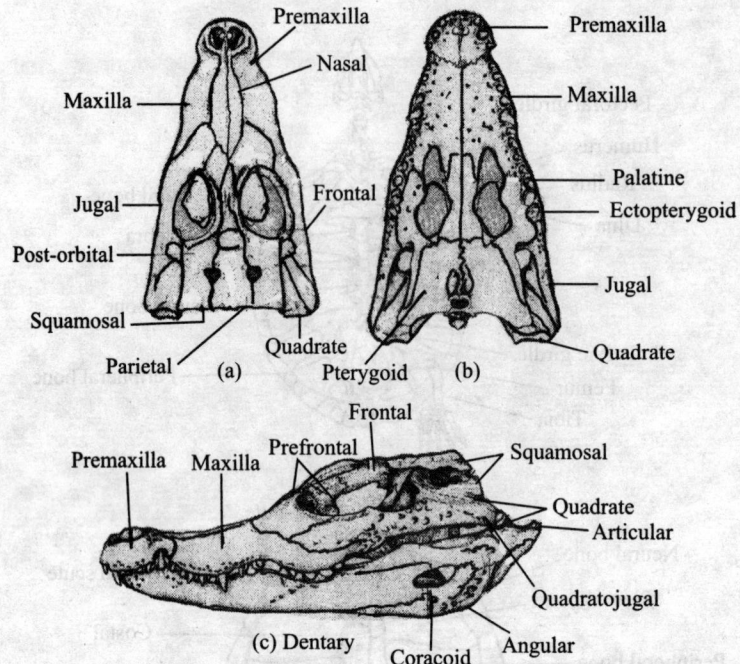

Figure 8.2   Skull of crocodile, *Osteolaemus* (a) Dorsal view, (b) Ventral view, and (c) Lateral view.

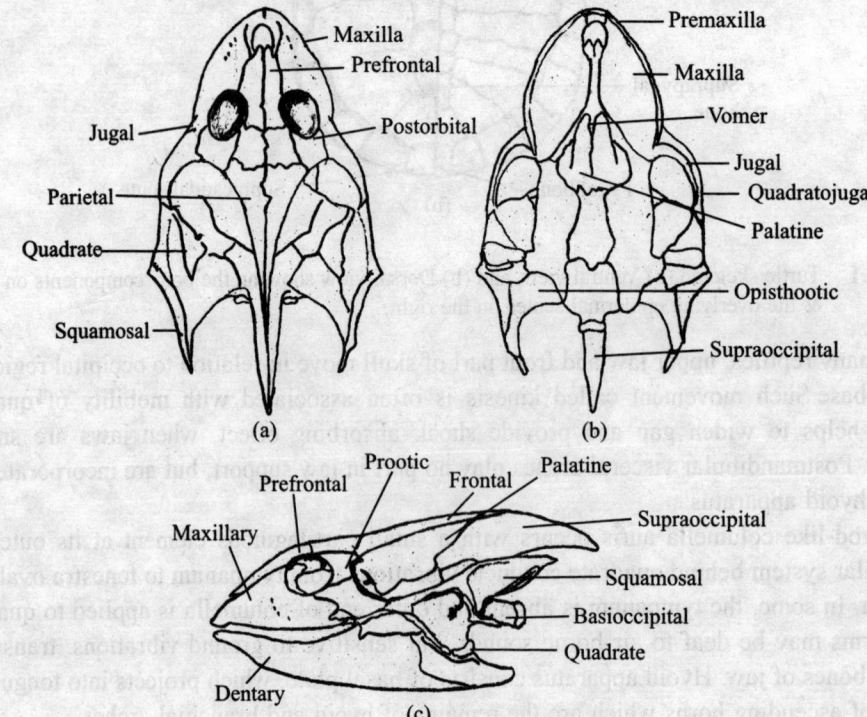

Figure 8.3   Skull of *Trionyx* (a) Dorsal view, (b) ventral view, and (c) lateral view.

8.3 Feeding and Digestion

Food is seized either by teeth, or with elongated tongue. Teeth are situated along edges of jaws and often on bones of palate. They are conical in shape, may be slightly serrated, or modified to form crushing plates, poison fangs, and other devices. Tooth succession is continuous throughout life except the lizards. Salivary glands are well developed in some. In snakes and one genus of lizards, some of them are modified to form poison glands. Tongue is variable, being hardly movable in some reptiles but long, forked, and highly mobile in others.

Digestion begins in stomach. Alimentary canal is built on typical vertebrate plan, with a tubular stomach, short small intestine, and wider large intestine, leading to short caecum. A well-marked cloacal chamber is subdivided into coprodaeum for faeces, and urodaeum for products of kidneys and genital organs. These 2 chambers open into a final common proctodaeum, closed by a cloacal sphincter. This division of cloaca helps in retention of water. The cloacal chambers reabsorb water from both faeces and urinary excreta.

8.4 Respiration, Circulation, and Excretion

Typical respiration is a backward movement of ribs, produced by muscles attached to them. No complete separation of thorax from abdomen, but a partial diaphragm may be present. Glottis, a slit at the back of mouth, leads into larynx with supporting cricoid and arytenoid cartilages. Many reptiles produce small sounds with less developed voice box. Lungs are sacs whose wall is folded into ridges, separating several chambers or bronchioles. Hinder part of lung is smooth. In some lizards, it develops into air sacs. More anterior portion of lung tends to become effective vascular and respiratory region, allowing air stream to be drawn across it at both inspiration and expiration.

Circulation of lizards shows a partial separation of venous and arterial blood. Two auricles and one ventricle are partly divided by a septum into right and left sides. Three arterial trunks right and left aortae, and pulmonary trunk arise directly from ventricle. Opening of the latter lies opposite right side of ventricle and receives predominantly venous blood. Left systemic arch opens opposite to incomplete ventricular septum and receives mixed blood. Right systemic arch opens from left side of ventricle and carries almost pure arterial blood. Carotid arteries of both sides arise from right systemic arch.

Venous system is based on plan as that of frog. Pelvic veins receives blood from tail and hind legs and returns it to heart through either an anterior abdominal vein or renal portals and inferior vena cava. The metanephros leaves mesonephric (Wolffian) duct to serve as vas deferens in male. There is sometimes an endodermal bladder. Waste nitrogen is largely excreted as uric acid with the reabsorption of much of water in urodaeum, with precipitation of organic matter. *Chelonia* produce ammonia and urea, but relatively little uric acid, whereas the last is the main excretory product of fully terrestrial types, like Grecian tortoise which live in almost desert conditions.

8.5   Reproduction

Fertilization is internal. In modern reptiles except Sphenodon , special organs of copulation derived from cloacal wall are developed in male. In crocodiles and tortoises, there is a single median penis, but in lizards and snakes there is a pair of these structures. Erection involves both muscular action and vascular engorgement. Sperms pass from the vasa deferentia into urodaeum, and are carried into a groove along each penis. In snakes, the sperms survive within female for long. Some individual lay fertile eggs after months.

Eggs of oviparous reptiles are laid on land. Egg shell is secreted by walls of oviduct and often hardened by lime impregnation. The amnion and allantois form and large quantity of yolk remains enclosed in a bag, the yolk sac. Embryonic cleavage is affected by amount of yolk, and is only partial. Albumen is present in eggs of crocodiles and tortoises and serves as reservoir of water. In eggs of lizards and snakes, the albuminous layer is poorly developed or absent. Formation of amnion and allantois is remarkable in development.

8.6   Nervous System and Receptors

Brain shows interesting developments. Cerebral hemispheres are relatively larger than in amphibians. The increased bulk lies mainly in basal parts of hemisphere. The roof is little developed and lacks elaborate cortical differentiation. Thalamus is well developed and receives connections from optic tracts, which no longer run to midbrain. Many fibres run from thalamus to cerebral hemispheres. Brains of modern reptiles are more like those of birds. Eyes are the main exteroceptive sense organs with movable eyelids, including a third eyelid or nictitating membrane. Lacrymal and Harderian glands provide secretions that keep surface of cornea moist. Eye is supported by a scleral cartilage and a ring of bony scleral plates. Accommodation is produced by striated ciliary muscles which cause ciliary process to squeeze lens, making its anterior surface more rounded. In many, retina possesses both rods and cones; the latter predominates in diurnal types.

Pineal complex is often well developed. In Sphenodon and many lizards, pineal or parietal eye is present. Pineal foramen in parietal bone lies near frontoparietal suture. In lizards pineal registers solar radiations, and perhaps by secretion of hormones influences the thermoregulatory behaviour. Pineal complex plays role in reproduction. In majority of reptiles, olfactory region of nose is well developed. Except in crocodiles, there is a single nasal concha. Organ of Jacobson, a specialized and sometimes separate region of nose, innervated by a separate branch of olfactory nerve, is present in turtles, Sphenodon, and Squamata. In the latter, it is very highly developed. Tympanum, when present, lies at the back of jaws, sunk a little below the surface. Response to sound waves is unknown in lizards, but ears of certain tortoises are very sensitive to sound over a narrow range of about no cycles per second; thus apparently there is some resonating mechanism, perhaps columella auris. Hearing is best developed in Crocodilia and certain lizards.

8.7   Classification

It is convenient to classify them by the skull that shows the main lines of evolution within the class. In the cotylosaurs, the dermal bones of temporal region of skull presented an unbroken surface without temporal fossae. Arches or 'apses' of bone in temporal region were absent. Such forms are placed in subclass Anapsida. Jaw muscles took origin from the deep surface of temporal side wall, between it and braincase and they passed down through holes in palate to insert in lower jaw. This is the most primitive condition, and resembles that of early amphibians. It is still seen often in a modified form in *Chelonia*, which are placed in subclass Anapsida.

In advanced groups, fossae bounded by bony arches appear in temple region, enabling the jaw muscles to extend through them onto the outer surface of skull that increases their mechanical advantage. In many reptiles, 2 such fossae appeared, the condition being termed diapsid as in subclasses Lepidosauria and Archosauria, perhaps the most successful groups of reptiles. In lepidosaurs of the Order Squamata, lower temporal arch is always incomplete, without quadratojugal and jugal separated from squamosal. In some lizards and in snakes, the upper arch is also lost.

In other groups, a single fossa and arch is present. When this is situated high on skull, the condition is called parapsid as occur in Subclasses Ichthyopterygia and Synaptosauria. Formerly, these 2 subclasses were placed together in Parapsida .In the Subclass, the Synapsida, there is also a single fossa, but in the earlier forms at least it is placed low down, and is bounded below by the jugal and squamosal. The term synapsid, meaning 'fused arch', is actually a misnomer. Early workers believed wrongly that the single arch was formed from fusion of 2 as seen in diapsids. Synapsids comprise the mammal like reptiles, but in later members of group like *Cynognathas*, and in their descendants the mammals, the temporal fossa has greatly enlarged, and has lost its primitive relationships (Figure 8.4).

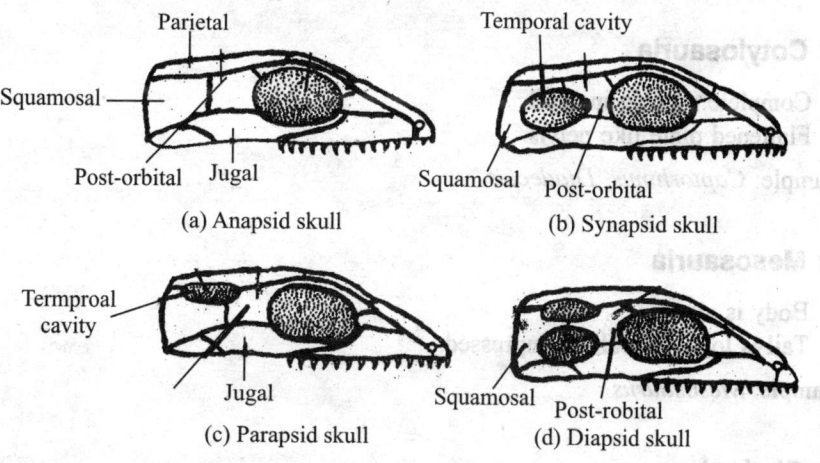

Figure 8.4 Skull types in reptiles.

General Characters

1. Ectothermic.
2. Skin is dry and waterproofed.

3. Skin contains few, or no glands.
4. Presence of crests, spines, and other appendages.
5. Presence of holes in temporal region of skull.
6. Single occipital condyle.
7. Vertebrae are procoelous, or acoelous.
8. Teeth are conical and modified to crushing plates, fangs, and other devices.
9. Cloacal chamber is subdivided into a coprodacum and an urodacum.
10. Heart is usually three chambered. Four chambered in crocodiles.
11. R.B.C. is nucleated.
12. Kidneys are metanephric.
13. Exertion is uricotelic.
14. Salt glands are present in some marine reptiles.
15. 12 pairs of cranial nerves are present.
16. Presence of third eyelid.
17. Organ of Jacobson is present.
18. Fertilization is internal.

Classification followed here is adapted from Young (1981).

Subclass: Anapsida

1. No arches or apses of bone in temporal region.
2. Dry skin.

Order: Cotylosauria

1. Complete roofing in skull.
2. Flattened plate like pelvis.

Example: *Captorhinus, Diadectes.*

Order: Mesosauria

1. Body is slender
2. Tail is long, laterally compressed.

Example: *Mesosaurus.*

Order: Chelonia

1. Bony plates form a box.
2. Box (shell) includes a dorsal carapace and ventral plastron.
3. Five rows of bony plates are present in carapace.
4. Pectoral girdle has three prongs.
5. Complete roofing of skull occurs.
6. Teeth are absent.

7. Kidney is metanephric.
8. Single copulatory organ is present.

Example: *Chelonia, Testudo, Emys, Trionyx, Geochelone.*

Subclass: Synaptosauria

1. Parapsid skull with a single temporal fossa.

Order: Protorosauria

1. Earlier forms were lizard-like.
2. Triassic forms had a long neck and short body.

Example: *Araeoscelis, Tanystropheus.*

Order: Sauropterygia

1. Limbs were developed into paddles.
2. Upper temporal fossa was enlarged.
3. Neck was longer, or shorter.

Example: *Pliosaurus, Placodus, Lartosaurus*

Order: Placodontia

1. Large grinding teeth.

Example: *Henodus.*

Subclass: Ichthyopterygia

1. Streamlined body
2. Vertebrae were amphicoelous discs.
3. Paired fins were small.
4. Head with very long snout.

Order: Ichthyosauridae

1. Characters are same as that of Ichthyopterygia.

Example: *Ichthyosaurus, Mixosaurus.*

Subclass: Lepidosauria

1. Diapsid skull.
2. No antorbital vacuity

Order: Eosuchia

1. Teeth on palate and jaws.
2. No opening between bones of the snout.

Example: *Younginat, Prolacerta.*

Order: Rhynchocephalia

1. Two complete temporal fossae.
2. Well-developed pineal eyes.
3. Amphicoelous vertebrae with intercentra.
4. Teeth are of accordant type.
5. Beak is present on upper jaw.
6. *Sphenodon* has no copulatory organ.

Example: *Sphenodon, Scaphonyx, Rhynchosaurus.*

Order: Squamata

1. Skull is highly kinetic.
2. Organs of Jacobson are elaborated.
3. Paired copulatory organs of unique type occur in male.
4. Lachrymal duct opens into the duct of Jacobson's organ.

Example: *Gekko, Hemidactylus, Draco, Agama, Varanus, Calotes.*

Subclass: Archosauria

1. Diapsid skull.
2. Ischium and pubis were elongated.
3. Tibia long, strong, sometimes fused with proximal tarsals.

Order: Thecodontia

1. Sharp teeth set in sockets.
2. Antorbital vacuities were present.
3. No pineal foramen.

Example: *Saltoposuchus, Mystriosuchus, Euparkeria*

Order: Crocodilia

1. Parts of skull are pneumatized.
2. Teeth are sharp.
3. Nostrils at the tip of snout.

4. Ribs are two headed.
5. Proatlas element is present between skull and atlas.
6. Esophagus can be distended.
7. Stomach consists of a gizzard a pyloric region.

Example: *Crocodylus, Alligator, Gavialis.*

Order: Saurischia

1. Triradiate pelvis,
2. Tail is long, developed.
3. Growth rings are found in dentine.

Example: *Ornitholestes, Diplodocus*

Order: Ornithischia

1. Four radiate pelvis.
2. Teeth were restricted to hind part of jaws.
3. Predentary in lower jaw.

Example: *Camptosaurus, Iguanodon, Hadrosaurus*

Order Pterosauria

1. Head and neck are long.
2. Short body and hind limbs.
3. Fur on the legs and wings.

Example: *Pteranodon, Rhamphorhynchus.*

Subclass: Synapnida

1. Skull with a single temporal vacuity.
2. Supraoccipital was broad.
3. Lower jaw was flat.
4. Heterodont type of teeth

Order: Pelycosauria

1. Skull with temporal vacuity.
2. Limbs were short.
3. Neural spines were elongated.

Example: *Dimetrodon, Varanosaurus.*

Order: Therapsida

1. Occipital condyle was double.
2. Big temporal opening in skull.
3. Quadrate and quadratojugal ere reduced

Example: *Cynognathus* (Figure 8.5).

8.8 Thermoregulatory Mechanisms of Desert Vertebrates

The terms ectotherm and endotherm are not synonymous with poikilotherm and homeotherm. Instead of referring to variability of body temperature, they refer to sources of energy used in thermoregulation. Ectotherms gain their heat largely from external sources by basking in sun, or by resting on a warm rock. Endotherms largely depend on metabolic production of heat to raise their body temperature.

Desert air temperature climb up to 40° or 50°C during summer and ground temperature may exceed 70°C. Instead of losing heat to environment, an animal absorbs heat. That heat plus metabolic heat must be dissipated to maintain body temperature in normal range. Three major responses in endotherms for desert conditions are avoiding desert conditions, behavioral means that is living in deserts but rarely exposing to full stress of desert conditions; tolerating greater range of body temperature, or body water content; and specializations like torpor in response to shortages of food, or water.

In Camel

Dromedary camel (*Camelus dromedarius*) travels more than 500 kms during which they did not get scope to drink. Watered camel shows a minimum of 36°C in early morning and a maximum of 38°C in mid afternoon. When a camel is deprived of water, temperature falls to 34.5°C at night and climbs to 40.5°C during day. Camel expends water to prevent 6°C by evaporative cooling. With a specific heat of 3.4 kilojoules/ (kilogram degree centigrade), a 6°C increase in body temperature for 500-kilogram camel, means the storage of 1 0.000 kilojoules of heat. Evaporation of a kilogram of water dissipates approximately 2427 kilojoules. Thus, a camel would have to evaporate more than 4 liters of water to maintain stable body temperature at night time level. By tolerating hyperthermia during day, it can conserve that water. When deprived of water, camel reduces evaporative water loss by 64% total daily water loss by half.

In summer, camels have hair 5 or 6 cms long on back and up to 11 cms long over hump. On ventral surface and legs, hair is only 1.5 to 2 cms long. In early morning, they lie down on surfaces that have cooled overnight. Legs are tucked beneath body. Ventral surface, with short covering of hair, is placed in contact with cool ground. In such position, a camel exposes only its well-protected back , sides to sun, places its lightly furred legs and ventral surface in contact with cool sand, which may conduct away same body heat.

Camels assemble in small groups and lie pressed closely together through day. And thus reduces its heat gain because it keeps its sides in contact with other camels (both at about

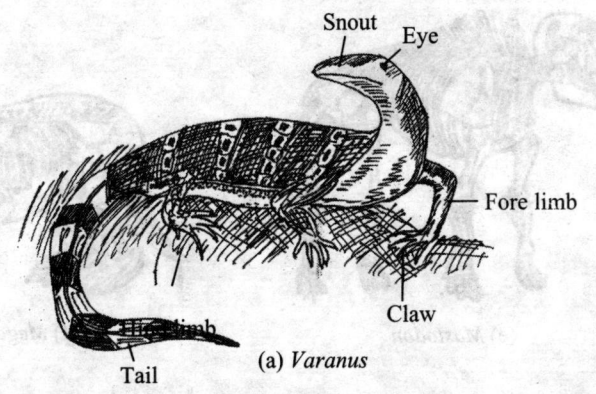

(a) *Varanus*

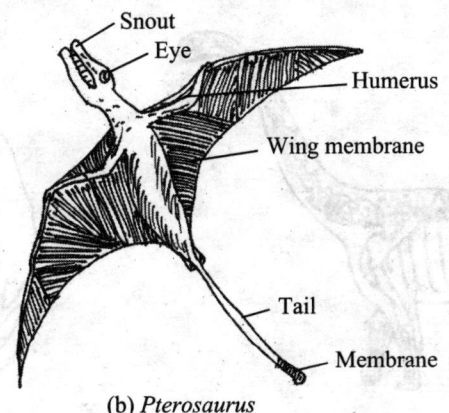

(b) *Pterosaurus*

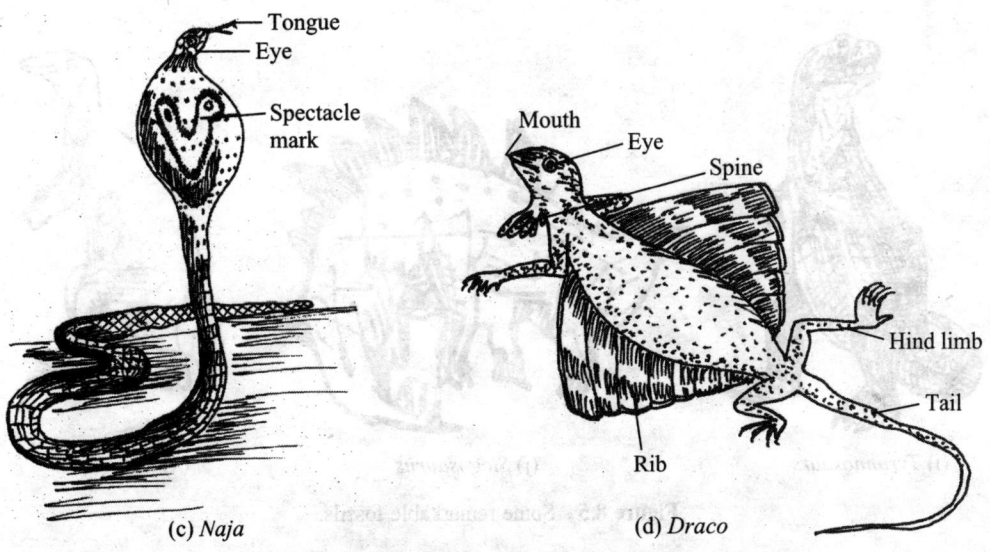

(c) *Naja*   (d) *Draco*

Figure 8.5 Some examples of reptiles.

(e) *Mastodon*

(f) *Megatherium*

(g) Dinosaur

(h) *Brontosaurus*

(i) *Tyrannosaurs*

(j) *Stegosaurus*

(k) *Iguanodon*

Figure 8.5   Some remarkable fossils.

40°C) instead of allowing solar radiation to raise fur surface temperature to 70°C or above. Reducing the time, spent during drinking is one method of reducing the risk of predation. A camel drinks as much as 30% of its body mass in 10 minutes, while a very thirsty human can drink about 3% of body mass in same time.

Other Large Mammals

Body temperatures above 43°C rapidly cause brain damage in most mammals. Grant's gazelles (*Gazella thomsoni* and *G. granti*) maintain rectal temperature of 46.5°C for 6 hours without apparent ill effects. They keep brain temperature below body temperature by counter current heat exchange to cool blood before it reaches brain. In ungulates, blood supply to brain passes via external carotid arteries. At the base of brain, these arteries break into a rete mirabile that lives in venous sinus. Blood in sinus is venous blood, returning from walls of nasal passage where evaporation cools it. This cool venous blood cools arterial blood before it reaches brain. Such mechanism is widespread among mammals like jack rabbits (*Lepus*) and kit fox (*Vulpes velox*). Large ears reduce water evaporation to regulate body temperature and function as efficient radiators to cooler desert sky. *Lepus* could radiate heat through its 2 large ears (400 cm).

In Birds

Bird's body temperature is normally variable. They tolerate moderate hyperthermia without distress which is preadaptations to desert life found virtually in all birds. Neither the body temperatures nor the lethal temperatures of desert birds are higher than those of related species from non-desert region.

In Small Mammals

Rodents live in burrows during the day, emerge at night and escape desert heat. Their greatest temperature stress may be cold. Squirrels minimize their exposure to highest temperature by running across open areas. They seek shade or their burrows to cool off. On a hot day, squirrel can maintain body temperature below 43°C by retreating every few minutes to a burrow deeper than 60 cms, where the soil temperature ranges between 30 to 32°C. Tails of many desert ground squirrels are wide and flat. Ventral surfaces of tails are usually white. Tail is held on squirrels back with its white ventral surface facing upward, acts as a parasol, shades the squirrel's body and reduces standard operative temperature.

8.9 Anatomical Peculiarities of Snakes

Snakes appeared during the Cretaceous, about 125 million years ago, became common when their main food source, the small mammals became abundant. India has 218 species of widely distributed snakes. They are not often seen because of their living habits. Snakes do not have eyelids and efficient sense of hearing. They lack external and middle ears and hear by detecting vibrations carried through ground. They use their forked tongue for smelling.

Sensory pit (thermal detector on each side of head) helps the snake in finding warm-blooded prey. Snakes are exclusively carnivorous.

The important snake in the Western Ghats is the King Cobra; *Cophiophagus hannah* which is also the longest poisonous snake of the world. It feeds on other poisonous and non-poisonous snakes and is the only snake to build a nest of decaying matter to lay eggs. Many pit viper species like the Bamboo Pit Viper, Hump nosed Viper and Malabar Rock Viper are common in the Western Ghats. Snakes are often feared. But they play important role in ecosystem by controlling rodents. A pair of rats can produce nearly 800 offsprings per year. India's estimated 8,000 million rats cause a loss of Rs 50,000 million per year by damaging crops. Rat Snake (*Ptyas mucosus*) controls rats in these godowns. A solution of cobra venom is useful in treating pain from cancer, neuritis, and other illness. Snake venom itself is used to produce anti-venom for treating snake bites. Snake skins have been used since time immemorial for various products. Snakes are source of food for people of Indochina. Modern commercial use and illegal tradeoff result in depletion of snakes beyond its natural ability to recover.

Snakes with their limbless elongated form represent one of the most specialized groups of animals. They are found in almost every kind of habitat. Due to fossorial adaptation, snakes show many peculiar features.

External Features

1. Body is extremely elongated and cylindrical.
2. Complete absence of limbs except boas and pythons.
3. Distinct neck is absent.
4. Head is more or less conical.
5. Nostrils are terminally placed.
6. Eyelids are fused over the eyes.
7. Elongated transverse ventral scales called gastrosteges are mostly found.
8. Shields are usually large and join each other by their margins.
9. Head is covered mostly by head shields and named according to their placement.
10. Shields situated in the mental groove on ventral side between rami of lower jaws are called mental shields.
11. Other shields are frontal, superciliary, rostral, upper labial, nasal, preorbital, loreal, median labial, jugular, ventral etc.
12. Cloacal aperture is a transverse slit.
13. Males are provided with paired copulatory organ (Figures 8.6, 8.7, 8.8, 8.9).

Internal Structure

1. Shell is streptostylic (quadrate is highly moveable) in nature and occipital condyle is distinctly tripartite.
2. Fossae and interparietal foramen are absent.
3. Parietals are fused together to form a single piece.
4. Frontals are provided with lateral projections.

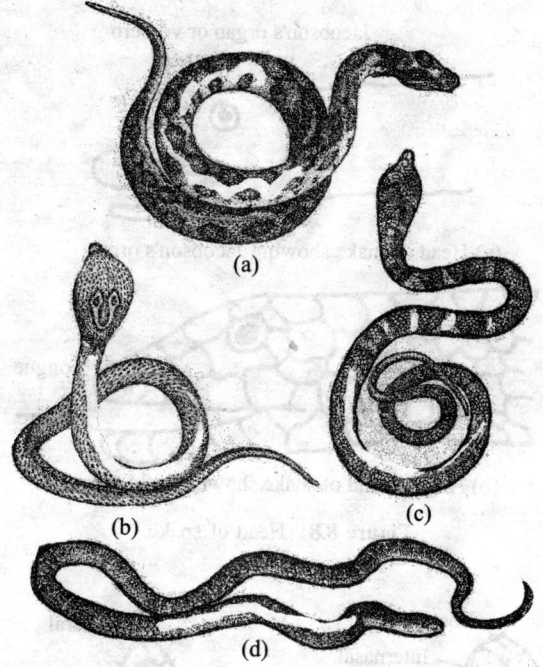

(a)

(b)   (c)

(d)

Figure 8.6 Snakes (a) Viper, (b) Cobra, (c) King cobra, (d) Krait.

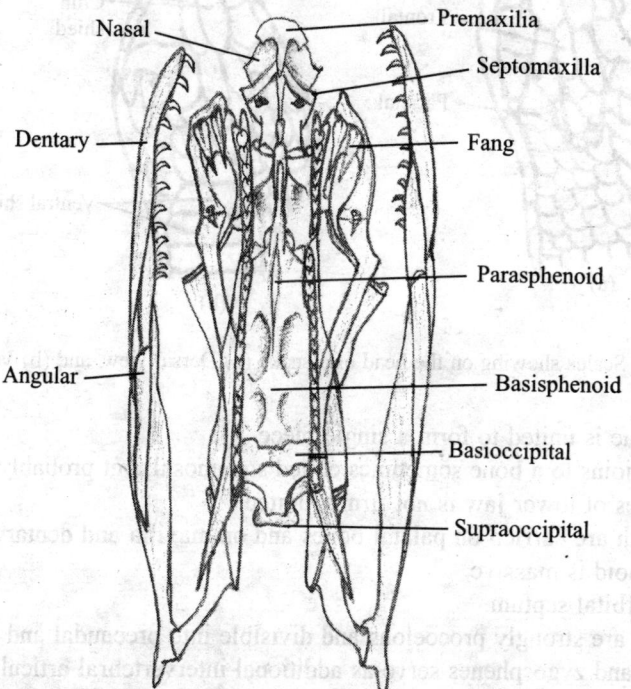

Figure 8.7 Ventral view of skull of *Naja*.

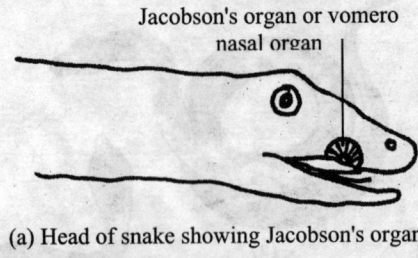

(a) Head of snake showing Jacobson's organ

(b) Anterior end of snake showing bifid tongue

Figure 8.8 Head of snake.

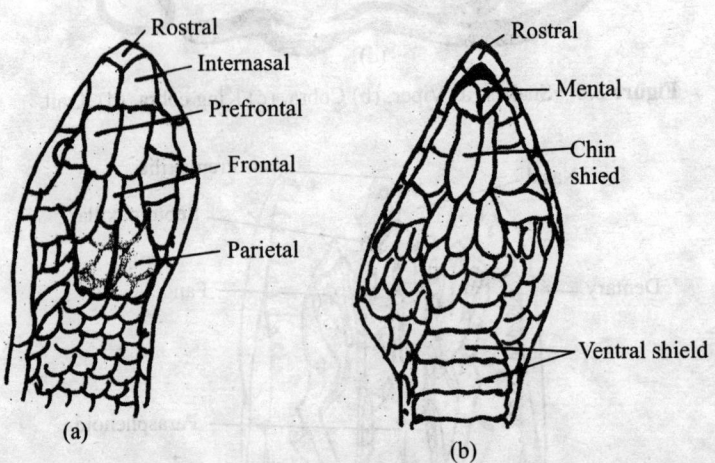

Figure 8.9 Scales showing on the head of a snake (a) Dorsal view, and (b) Ventral view.

5. Premaxiliae is united to form a single piece.
6. Quadrate joins to a bone sometimes called squamosal, but probably a supratemporal.
7. Two halves of lower jaw is not firmly united.
8. Sharp teeth are carried on palatal bones and on maxilla and dentary.
9. Parasphenoid is massive.
10. No interorbital septum.
11. Vertebrae are strongly procoelous and divisible into precaudal and caudal region.
12. Zygantra and zygosphenes serve as additional intervertebral articulations.
13. Transverses processes are rod shopped, well developed and bifid at caudal region.

14. Ribs are hollow, single headed.
15. Girdle bones are completely absent.
16. Alimentary canal is relatively small.
17. Mouth makes a wide gap by divarication of jaws.
18. An elongated and forked tongue is present, the base of which is provided with a muscular sheath.
19. Teeth are conical and backwardly slanting.
20. Superior labial glands are modified into poison glands.
21. Stomach is elongated.
22. Liver is elongated with unequal lobes.
23. Presence of a rudimentary caecum at junction of small and large intestines.
24. Lungs are asymmetrical. Right lunge is always very long and in marine species, it may reach the level of cloaca. Left lung is often rudimentary.
25. In primitive snakes, both lungs are functional and of almost equal size.
26. Some snakes posses a tracheal lung.
27. Heart is typically reptilian.
28. Arterial arches are asymmetrically disposed.
29. Ductus caroticus is absent.
30. Kidneys are asymmetrical and elongated.
31. Urinary bladder is absent.
32. Gonads are asymmetrical.
33. Apex of male copulatory organ may bear spiny projection.
34. A very distinct sexual segment is found in male kidney. A feebly developed corresponding segment occurs in female.
35. Anal glands are characteristic of snakes.
36. Brain is narrow, elongated and projects between eyes.
38. Organ of Jacobson is well developed.
39. Scleral cartilage is absent in eyes.
40. Retina and iris muscles are degenerates.
41. Visual cells include cones of a peculiar type, which are derived from rods.
42. Some diurnal snakes protect their retinas by a yellow tinted lens.
43. Ear drums, tympanic cavities and Eustachian tubes are reduced, or absent. (Figure 8.10, 8.11)

Snakes are regarded as eel type degenerates, but some workers view snakes as advanced type of specialization for swallowing whole large prey. According to them loss of limbs, especially fore limbs, greatly elongated trunk, readily dislocated jaws, hollow fangs and associated venoms are all progressive adaptative features.

8.10 Poison Apparatus and Biting Mechanism of Snakes

Many snakes, but a small percentage of the whole group is more or less venomous. India is heavily populated by snakes and many people are killed by poisoning yearly. This enormous death rate is probably matched in no other country, and is perhaps largely due to presence of a mainly bare footed population. All poisonous snakes have in their heads a poison apparatus, which is absent in non-poisonous snakes. It consists of a pair of poison glands, their ducts and a pair of fangs.

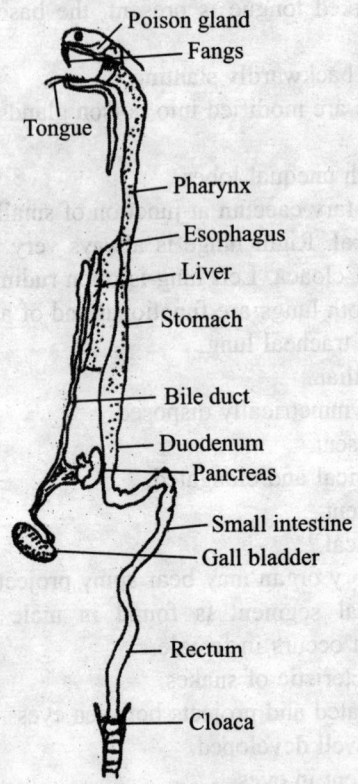

Figure 8.10   Digestives system of viper (*Echis carinatum*).

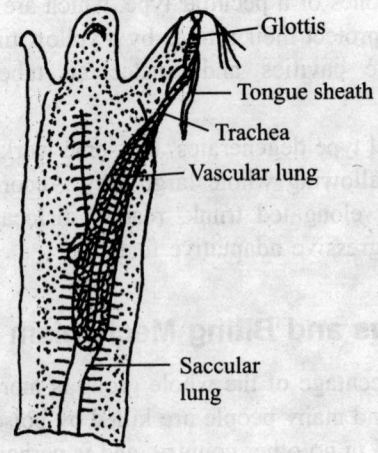

Figure 8.11   Respiratory system of viper.

Poison apparatus is associated with specialized bands of muscles. Situated one on either side of upper jaw, poison glands are possibly the modified superior labial glands or parotid glands. Each poison gland is a sac like, provided with a narrow duct at its anterior end and held in position by ligaments. An interior ligament attaches anterior end of gland to maxilla. A posterior ligament extends between gland and quadrate. Fan shaped ligaments are situated between side walls and squamosoquadrate junction. Each gland is thickly encapsulated with fibrous connective tissue and mostly covered by temporal muscle. Poison glands secrete poison and serves as a reservoir of venom (Figure 8.12).

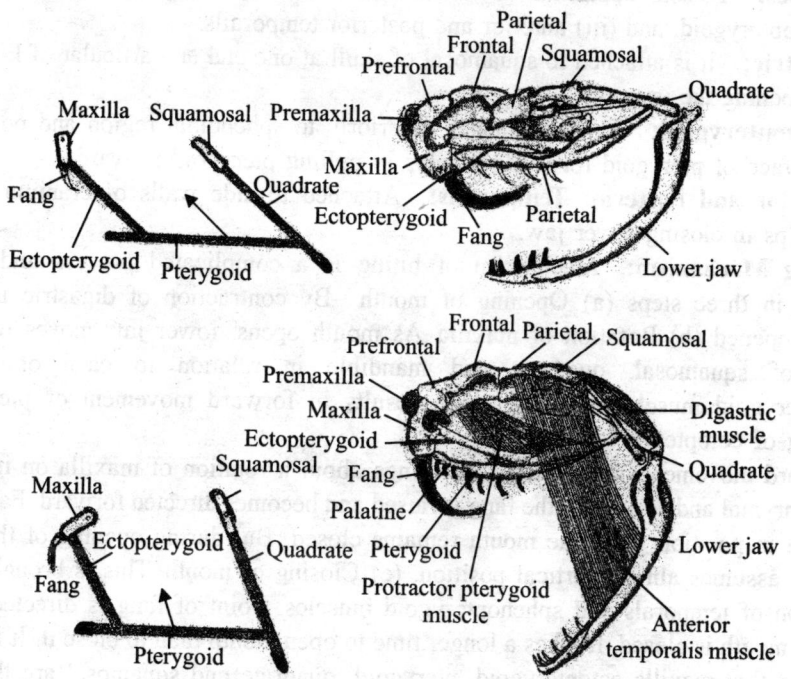

Figure 8.12 Mechanism of bite of poisonous snake.

Poison Ducts: A narrow poison duct leads anteriorly from each poison gland to base of a poison fang to enter its groove or canal. In *Naja nigricollis*, duct has an "L" shaped bend, just before exiting fang. The duct carries poison from poison gland to fang.

Fangs: Fangs are specialized teeth attached to maxillary bones. They are long, curved, sharp and pointed, enlarged maxillary teeth, which regenerate when lost. There is an intake aperture at basal part of fangs and a discharge aperture, which is subterminal. A single, large, poison fang on maxilla with small reserve fangs at its base is found in Viperidae. On the basis of structure and poison, 3 types of fangs in poisonous snakes are as follow.

(a) **Solenoglyphous:** In vipers and rattlesnakes, a large functional fang occurs on the front of each maxilla. Its base is covered on all sides by a sheath containing a few reserves and developing fangs. Fangs are movable and turned inside to lie close to roof of mouth, when it is closed. A hollow poison canal lined with enamel runs through fang opening at the tip (solen, pipe; glyph, hollowed).

(b) **Proteroglyphous:** In cobras, Kraits, coral snakes and sea snakes, fangs are small at the front of maxillae and permanently erect. Each fang is grooved all along its anterior face (Protero, first).

(c) **Opisthoglyphous:** In some poisonous snakes in family Colubridae, fangs are small, lie at back of maxillae and each one is grooved along its posterior border (opistho, behind).They serve as hypodermic needles for injecting poison into body of the victim.

Muscles: Poison apparatus is associated with following muscles: (i) Digastric (ii) Sphenopterygoid, and (iii) anterior and posterior temporails.

Digastric: It is attached to squamosal of skull at one end and articular of lower Jaw. It helps in opening the jaws

Sphenopterygoid: This is attached anteriorly to sphenoidal region and posteriorly to dorsal surface of pterygoid forward. It helps in pulling pterygoid forward.

Anterior and Posterior Temporalis: Attached to side walls of cranium and lower jaw, it helps in closing lower jaw.

Biting Mechanism: Mechanism of biting is a complicated process and it can be described in three steps (a) Opening of mouth –By contraction of digastric muscles the mouth is opened (b) Rotation of maxilla–As mouth opens, lower jaw moves forward and rotation of squamosal, quadrate and mandible in relation to each others occurs. Shepnopterygoid muscles contract which results in forward movement of pterygoid and uppushing of ectopterygoid.

Upward movement of ectopterygoid brings about a rotation of maxilla on its own axis round lachrymal and as a result the fang is raised and becomes directed forward. Fang is nearly horizontal in position when the mouth remains closed. But during opening of the mouth to bite, fang assumes almost vertical position. (c) Closing of mouth–This is brought about by contraction of temporals and sphenopterygoid muscles. Point of fang is directed backward while the mouth is closed. It takes a longer time to open mouth than to close it. It is important to mention that maxilla ectopterygoid, pterygoid, quadrate and squamosal are the principal bones involved in erection of fang.

Transference and Nature of Venom: During the contraction of digastrics muscle, posterior ligaments are relaxed and during rotation of squamosal bone, fan shaped ligaments are stretched to squeeze the walls of gland. This makes the poison to come out of gland through duct and fang. Venom is clear, transparent, pale yellow or straw-colored fluid and is a pure solution of two or more protein materials. It is acidic in reaction and soluble in water and glycerine. Poison is precipitated in reagents like silver nitrate and permanganate of potash. It contains many enzymes like proteinase, hyaluronidase L-Ariginases, hydrolases, transaminases, cholinesterase, alkaline phophatase, acid phosphates etc. Venoms are mostly tasteless but cobra venom's taste is slightly bitter. It is destroyed by alcohol, strong alkalis like NaOH, KOH. Venom contains metals like zinc, sulpher, and copper. It is thermolabile and can be dried as crystals. Dried venom is thermo stable. It is a good digestive enzyme and contains several non enzymatic proteins.

Effect of Poison: Venom is generally introduced in subcutaneous tissue and then it reaches general circulation. When introduced directly into vein, the effect is instantaneous.

Venoms act in a variety of ways. It may act on nerve cells and cause respiratory paralysis; destroy the endothelium of smaller blood vessels and allow blood to seep into the tissues; destroy erythrocytes and cause blood coagulation. However, it is customary to regard two categories of snake venoms-neurotoxic and haemotoxic. Neurotoxins are typical of elapids (cobra) and sea snakes haemotoxins are typical of vipers. The effects of venom in case of most common poisonous snakes of India are as follows.

Cobra Poison: Poison of cobra is most virulent. It is a neurotoxin attacking nerve centers and causing paralysis of muscles, especially those of respiratory muscles. Symptoms include pain and burning sensation ending in numbness of bitten part, which turns bluish or blackish. Person suffers from giddiness, weakness in legs, high pulse rate, speechlessness, dropping of saliva and eyelids, contraction of pupils, vomiting and labored breathing, Death results within a few hours due to failure of respiration (asphyxia) or of heart activity.

Krait Poison: Their bite injects a very large quantity of poison (three times more than cobra). Symptoms are very similar to those of cobra bites, except that the victim complains of unbearable abdominal pain due to internal hemorrhage. Destruction of R.B.C and paralysis of trunk and limbs occur followed by death within 6 to 24 hours.

Viper Poison: Venom of viper is mainly a heamotoxin affecting the circulatory and nervous systems more severely. Symptoms include local swelling and discoloration of bitten part with acute burning pain. Red fluid oozes out from wound, pupils dilate, pulse rate increases, profuse vomiting occurs and person loses consciousness. Death may result due to paralysis of vasomotor centers and exhaustion from profuse bleeding.

Indian Poisonous Snakes

Common Krait (*Bungarus caeruleus*):   Its bite is fatal to man and is considered to be second most poisonous in the world and first in Asia.

Banded Krait (*Bungarus fasciatus*):   Venom is less toxic than cobra poison.

Indian cobra (*Naja naja*):   12 mgm dose of venom is lethal for man. Poison gland stores 317mg of venom.

Russell's viper (*Vipera russelli*):   Fangs are largest in length among Indian poisonous snake.

King cobra (*Ophiophagus hannah*):   Poison is less virulent than cobra.

Saw scaled viper (*Echis carinatust*):   Venom is 5 times virulent than that of cobra and 16 times as toxic as Russell's viper venom.

8.11 The Dinosaurs

In the 10 million or so years at the end of the Triassic, some descendants of pseudosuchians became very successful and numerous, and many of them were very large. Large size was not a characteristic only of one line but of two distinct ones, each with several subdivisions. The term dinosaur is applied to all of them, but two main lines have little in common beyond characters common to all archosaurs.

Order Saurischia

These include forms with a triradiate pelvis, very like that of pseudosuchians. The earlier types, like their ancestors, were bipedal carnivores of no great size, such as *Compognathus* from the Jurassic of Europe and *Ornitholestes* from North America. Front legs were short, with 3 or 4 digits, provided with claws. The pectoral girdle was reduced to scapula and small coracoid, with no trace of clavicles. Some members of this line, the theropods, soon developed into large carnivores, like *Allosaurus*, over 30 ft long. These animals swallowed their food whole and to help this quadrate was movable and there was a joint between the frontals and parietals as in many lizards.

The skull was very similar to that of pseudosuchians. At the end of Cretaceous, this theropod line produced the largest carnivores that appeared on earth, such as *Tyrannosaurus rex*, nearly 50 ft long and 20 ft high, from North America. All the previously mentioned tendencies were here accentuated, producing creatures with bipedal habit, very powerful head and jaws, and much reduced forelimbs. They presumably preyed on the large herbivorous dinosaurs of Cretaceous and became extinct with their prey, either from the common inability to meet the rigors of climate or in competition with mammals and birds. Throughout most of Jurassic and Cretaceous, the theropods were the dominant carnivores of the world, taking the place occupied earlier by synapsid reptiles and later again by descendants of synapsids, the carnivorous mammals.

In Cretaceous, the organization of this saurischian line also produced some exceedingly bird like forms, *Struthiomimus* and *Ornithoithnus*, walking on three toes and having three also in hand, one opposable and used for grasping. Skull became very lightly built and the teeth disappeared, possibly in connection with an egg-eating habit.

All these carnivorous, bipedal saurischians may be grouped into a suborder Theropoda. Another line of organization, starting from bipedal, carnivorous Triassic theropods, adopted an herbivorous diet and reverted to quadrupedal habit. These animals belonging to the suborder Sauropoda culminated in immense Jurassic forms, *Apatosaurus* (*Brontosaurus*) and *Diplodocus*, the largest of all terrestrial vertebrates. Several stages of transition from bipedal to quadrupedal habit can be traced. *Yaleosaimis* from Triassic was a bipedal creature 6 ft long but with rather long front and short hind legs. *Plateosaurns*, also of Triassic, was 20 ft long, but still bipedal. Soon the front limbs became larger and more used for walking, though the disparity always remained.

Neck was immensely elongated but the head was very small, with a lightly built skull. Nostrils lay on top of head and in *Diplodocus* formed a single opening. This seems to show that the animals were aquatic or amphibious, as would in any case be suspected from very large size, making it unlikely that the legs could bear the full weight. *Diplodocus* and *Brachiosaurns* were over 80 ft long and the weight of the latter must have been nearly 50 tons. Structure of vertebral column showed that much weight was carried on legs.

The vertebrae were strong though hollowed in places. Footprints of animals have been found. One or more of digits bore claws. Skull (Figure 8.13) became relatively short and broad, and among many features was the weakness of jaws and small size of teeth, mostly crowded towards front of mouth. These teeth would have served well enough for cropping, but there are no teeth on hind part of jaws and no provision for grinding food. Animals of large size could only have been supported by this feeble apparatus if some very nutritious food was readily available. This perhaps agrees with the small size of brain, which was several times smaller than lumbar enlargement of the cord.

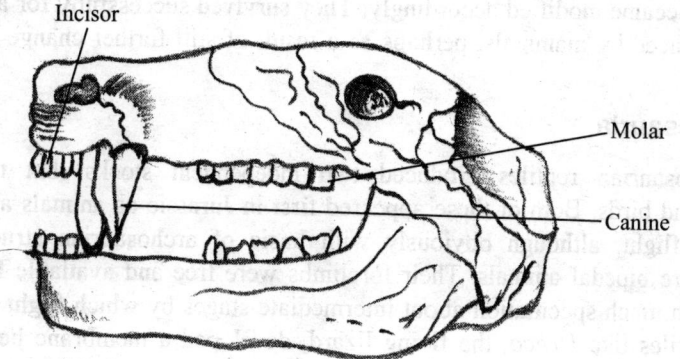

Figure 8.13 Skull of Saurischian.

Order Ornithischia

The second main group of dinosaurs appeared later than the sauropods and possessed a 4-radiate pelvis, with pubis directed backwards and an extra prepubic bone pointing forwards. Teeth were restricted to hind part of jaws, the front bearing a beak. At the front end of lower jaw there was an extra bone. These herbivorous forms appeared in Jurassic and achieved their maximum in Cretaceous, by which time the sauropods had become less common. The earliest of the ornithischians were bipedal animals, included in a suborder Ornithopoda from Jurassic and Cretaceous. These animals, like *Iguanodon* were built on same general lines as pseudosuchians, from which they were presumably derived.

Skull was heavily built and adapted for herbivorous diet, with powerful muscles attached to a coronoid process of lower jaw. Bipedalism was less marked than in saurischians and the fore-limbs were less reduced. Several separate lines then reverted to a quadrupedal habit. Trachodonts were very successful group of amphibious forms in Cretaceous, with webbed feet. Teeth were suited for grinding, parallel rows being present, making as many as 2,000 teeth in one animal. In several types of hadrosaur, the top of head was prolonged in various ways, giving a structure that perhaps allowed the nostrils to remain above water while animal was feeding below. These animals reached 30 ft in length and may have supplanted the sauropods as marsh-living forms, possibly when soft foods gave place to harder plants.

Other lines of ornithischians became fully terrestrial and quadrupedal and were mostly heavily armoured. Thus stegosaurs of Jurassic carried immense spines on back and tail had sharp spikes. Hind legs were much longer than front, a relic of bipedal ancestry. Feet carried hoof-like structures. Skull was very small and brain much smaller than lumbar swelling of cord. Teeth were in a single row and small. *Ankylosaurs* of Cretaceous were covered all over bony plates.

Ceratopsians such as *Triceratops* of late Cretaceous developed enormous heads, with huge horns and a large bony frill, formed by extension of parietals and squamosals to cover neck. These later Cretaceous animals appeared to live on dry land and to walk on all four legs, although bipedal ancestry is shown in shortness of front legs. There are several indications that climate at the close of Cretaceous was becoming drier and organization of giant reptiles became modified accordingly. They survived successfully for a while, but were ultimately replaced by mammals, perhaps as a result of still further change in climate.

Order Pterosauria

Triassic archosaurian reptiles produced two independent stocks that took to air, the pterodactyls and birds. Both of these appeared first in Jurassic as animals already were well equipped for flight, although obviously with basic of archosaurian structure. The early archosaurs were bipedal animals. Their forelimbs were free and available for use as wings. There has been much speculation about intermediate stages by which flight was produced.

Other reptiles like *Draco*, the flying lizard, developed a membrane between limbs and body for soaring jumps. Flight of pterodactyls and birds may have originated thus or, as suggested by Nopcsa, by flapping of fore limbs during rapid running on ground, the animals then becoming airborne for longer periods. Stages of evolution of flight may have been different in two cases. The birds are obviously bipedal animals and their similarity to reptiles as *Strnthiomimus* and *Ornithomimus* is obvious. The pterodactyls probably could not walk on their hind legs and may have used the wing for soaring than for flapping flight.

In spite of great differences, there are interesting parallelisms in structure of fully evolved fliers of two groups, for instance limb bones became light, skull bones fused, and jaws toothless and beaked. This parallelism in lines are distinct, although of remote common origin, is similar to that which we have noticed before in aquatic animals, and it can be interpreted as showing that populations with similar genotypes respond to similar environmental stimuli in same way.

Pterodactyls are most commonly found in Jurassic strata, less often in Cretaceous. Many specimens were found in marine deposits and seem to be fish-eaters. Characteristic features that have produced pterodactyl structure from a thecodont ancestry may be a lengthening of head and neck, shortening of body and ultimately of tail, lengthening of arms and especially of fourth digit, shortening of legs, and development of ventral parts of limb girdles. These are changes in bony parts available for study; no doubt there were many others in soft parts also paralleling evolution of birds, for instance animals may have been warm-blooded. There is no evidence that they possessed feathers; wing was a membrane (patagium).

Rhamphorhynchns of Jurassic is still recognizable by archosaurian structure, especially in skull, which has two fossae and large forward-sloping teeth. Forelimb was elongated, but carpus still short, with an extra pterygoid bone in front, presumably to support wing. The first three digits were short and hooked, the fourth long, supporting the wing, and the fifth absent. Hind limb was slender with five hooked digits. A long tail end was present in an expanded fin. Both girdles had well developed ventral regions and a large sternum, keeled in front. Scapula articulated directly with vertebral column.

Pteranodon of Cretaceous showed further modifications. Trunk became shortened to ten or fewer segments and fore-limb further lengthened, carpus being long and fourth digit much longer than other three. Hind-limb remained small and tail became very short. Very large and elongated head gradually lost its teeth, presumably acquiring a horny beak. In the latest forms, skull was drawn out backwards into an extraordinary process. Some earlier related forms were only a few inches long, but *Pteranodon* itself, of the late Cretaceous, had a wing span of 25 ft. Membrane, which stretched between both legs and body, and perhaps also included head, must have been easily torn.

Feathers of birds can be ruffled without breaking and loss of a few does no great harm: the bat's wing can be torn, but at least many digits support it, whereas that of pterodactyl was a huge continuous membrane supported by a single finger. It is possible that they always came to rest hanging from cliffs, which they could leave by soaring. Even the flight itself presents many difficulties. Although there is a sternum and a strong humerus, neither suggests the presence of muscles sufficiently strong to carry a creature as large as *Pteranodon*. The biggest pterodactyls were mostly, if not all, marine. The largest flying birds alive today are the albatrosses, which use their great weight to gain height with increasing velocity of wind a few feet above the sea. It is possible that pterodactyls used a similar method of soaring. They were presumably unable to compete with birds, and died out at the end of Cretaceous, along with so many other reptiles.

Sir Richard Owen used the word Dinosaur to designate certain fantastic Mesozoic archosaurian reptiles. They mastered the land and ruled the earth majestically for more than 100 million years during Mesozoic. Dinosaurs were readily divisible by clearly marked characters into two orders viz. Saurischia and Ornithischia. Of the present day living forms, crocodile on one hand and birds on other hand stand nearest the dinosaurs. According to Friedrich von Huene, dinosaurs have arisen form cotylosaurian stock which arose in Carboniferous and continued until Triassic. The smallest dinosaur, *Compsognathus* was about two and a half feet with bulk of a domestic cat. The other extreme was exhibited by *Gigantosaurs* of East Africa which was about 120 feet or more in length and weighed about 40 tones. Initial evolution of Dinosaurs possibly took place under the stress of semi aridity of climate. With changed climate, dinosaurs adapted to humid conditions and some become amphibious in habit. Dental analysis shows that some were carnivorous where majority become herbivorous.

Duration: Known record of dinosaurs extends form middle Triassic to very close of cretaceous. This is particularly true of saurschian dinosaurs. Plant feeding ornithischian do not appear in fossil record until late Triassic but are doubtless older, their subsequent duration being coextensive with that of others. Dinosaurs exhibited two distinct groups on basis of their feeding habits. First group includes carnivorous forms which could run swiftly on their hind legs and use their forelegs for grasping and tearing the prey and second group of herbivorous forms.

Distribution: Though dinosaurs were first found in Germany but this was not their original radiation centre. North Atlantic basin connecting Europe and North America is the probable place of origin. Dinosaurs began their worldwide march of conquest and extend the world over except New Zealand. In India, herbivorous dinosaurs had been first located near river the Godavari, bordering Andhra Pradesh and Maharastra. It is stated to be about 170 million years old. Fossils of this species have also been found in Mangolia, Canada and United States. On December 20, 1988, scientists of Indian statistical institute and Geologists of Jabalpur University have found fossils of world's first carnivorous dinosaur near Gun carriage factory hill at Jabalpur. These fossils date back to about 60 million years. A part of mouth of dinosaur had earlier been found by Brown in 1992 form Gun carnage factory hill.

Anatomical Pecularities of Various Dinosaurian Groups

1. Mesozoic reptiles becoming extinct by the end of Cretaceous.
2. Bi- or quadripedal.
3. Herbivorous, or carnivorous.
4. Terrestrial, or aquatic.
5. Size variable extending between 40-50 tons.
6. Scales absent, ill developed, or well developed.
7. Diapsid pterygoid was large usually with a foramen.
8. Palate was well developed, with two pairs of palatal vacuities.
9. Mandible composed of 6 hones.
10. Vertebrae were amphicoelous, usually differentiated into cervical, dorsal, sacral and caudal
11. Ribs were double headed.
12. Pectoral girdle with well developed scapula sternum was probably cartilaginous.
13. Pelvic girdle was well developed, usually both pubis and ischiatic symphysis were present.
14. Limbs were well developed. Hind limbs were more developed than fore limbs

Order: Saurischia

1. Tail was long, well developed.
2. Skull was similar to that of thecodonts.
3. Vertebrae were amphicoelous, or acoelous in earlier form, while opisthocoelous in later forms.
4. Pectoral girdle was reduced with round coracoid,
5. Pelvic girdle was triradiate type with open acetabulum.
6. Reduction of digits in manus and pes.

Suborder: Theropoda

1. Bipedal, carnivorous.
2. Skull was elongated with narrow pterygoid.
3. Vertebrae were amphicoelous except a few anterior ones.
4. Ribs were double headed in anterior region of a vertebral column.
5. In pectoral girdle scapula was small, coracoid was large, clavicle and interclavicles were present.
6. Pelvic girdle was well developed. Femur was very long.

It has two infraorders whose characters are tabulated below (Figure 8.14)

(a) Ornithischian   (b) Saurischian

Figure 8.14   Pelvic girdle of dinosaur.

Coelurosuraurus	*Carnosaurus*
(1) More primitive.	(1) Less primitive.
(2) Very lightly built.	(2) Strongly built.
(3) Neck long and slender.	(3) Neck short and heavy.
(4) Three functional and two vestigial digits in hand. Example : *Coelophysis Compsognathus*	(4) All 5 digits are functional. Example : *Megalosaurus Tyrannosaurus.*

Suborder: Sauropodomorpha

Jurassic to Cretaceous, large quadripedal herbivorous. Skull was small, in proportion to the size of animal. Teeth were small. Quadrate was short and palate was slightly developed. Anterior vertebrae were small. Procoelous, posterior vertebrae were large. Pubis was ill developed, while symphysis was well developed. This has two infra-orders- Prosauropods of Triassic, and Sauropods of Jurassic and Cretaceous. The former is believed to give rise the later.e.g. Prosaurapodia, Tenanosaurus.

Order Ornithischia

Ornithichians existed during Jurassic and Cretaceous. Mostly bi-pedal and herbivorous reptiles were provided with slender elongated premaxilla, small well developed maxilla with thecodont dentition. Pelvis was tetraradiate type. Pubis was provided with a long pubic process and ischium was well developed with broad symphysis.Horn like projections in head region were present.

This order is divisible into 4 suborders whose characters are given below

(a) **Ornithopoda:** Late Jurassic and early cretaceous of Europe and North America. Skull was long, heavy, well developed coracoid process; front part of mouth was toothless and covered by horny beak. Occipital condyle was projected downward. Cervical vertebrae were opisthocoelous. Example: *Camptosaurus, Anatosaurus, Heterodontosaurus.*

(b) **Stegosaurus:** Mostly Jurassic quadripedal with small forelims.Series of roughly triangular plates and spines were arranged in a double alternating row along neck, trunk and tail .Tip of the tail was provided with two pairs of long spikes .Vertebrae were amphicoelous. Example: *Stegosaurus*

(c) **Ankylosaurus:** Cretaceous, entire back covered by bony plates. Tail armed with bony spines and encased in rings of bones (defensive). Skull was large broad with dentition or on teeth. Example: *Ankylosaurus, Edmontonia*

(d) **Ceratopsia:** Upper Cretaceous of mostly North America and Asia. Head was very large, about 113 of total body length. However, half of it was not actual skull but a large portion of bone formed by extension of parietal and squamosal. A pair of large horns on post-orbital bones and a median horn over nasal region was found. Vertebrae were acoelous. Example: *Triceratops, Ortoceratops.*

Probale Causes of Extinction: One of the most inexplicable events is the dramatic extinction of this mightly race that practically ruled the world. Various authorities have put forwarded various reasons behind their extinction. The important reason seems to be the followings.

Internecine Warfare Amongst Dinosaurs

Destructive slaughter of the young, possibly while yet in egg by blood thrusty mammals is one cause. Draining of great inland Cretaceous seas along low lying shores where dinosaurs had their home is another cause. According to Cowles, large bodies cool off much more slowly than do small ones and he suggested that in an increasingly hot climate, reptiles like dinosaurs might only rarely cool off sufficiently to permit spermatogenesis. Thus great size of ruling reptiles in combination with an increasingly hot climate could lead to their extinction by sterilizations of males.

Remarks: Large dinosaur possessed special arrangement for cooling brain and spinal cord. Cooled blood from nasal turbinals may have passed to a rete mirabile in cavernous sinuses. Nasal crests of hardosaurs probably provided large nasal surfaces either for special olfactory epithelia or for cooling surfaces or both. There were sinuses in bony plates along back for instance in *Stegosaurus*. Freed to cool central nervous system could arise in either an ectotherm endodermous animals. Thus theory of male sterilization cannot be accepted totally because dinosaurs may have possessed special mechanism for continuing spermatogenesis.

At the close of Mesozoic, rapid environmental changes in topography due to post Cretaceous mountain building activity and climates due to incoming ice age and consequently reduction in vegetation began to take place. Further mountain building activity resulted in outpouring of lava and dust storms. Dinosaurs failed to adjust and become extinct. Few interesting theories have been forwarded to explain dinosaurian extinction.

According to the theory formulated by McGhee, Luis, Alvarez and Walter Alvarez, dust storms might have been caused by a large asteroid which hit earth and caused total darkness among dinosaurs and in turn prevented photosynthesis in plants upsetting the ecosystem and subsequently lead to death of dinosaurs in absence of their food. According to Hsu, a huge comet collided earth that lead to extensive rise of temperature and release of poisonous gases. Hazardous affects persisted for about 50 years on earth. As a result of this, dinosaurs became extinct.

8.12 *Calotes versicolor*

This is a common Indian garden lizard, often called a blood sucker but in reality, it does not suck blood. Its body colour is changeable and in males the head is often blushed red. *Calotes* passes much of its time lying on boughs and twigs and often seen running swiftly on the ground. It can swim, if necessary. It lays eggs in holes in the ground. *Calotes* is widely distributed in India and Southern China and like many lizards such as chameleons, possesses a peculiar power of changing the skin colour. Colour change is due to contraction or expansion of chromatophores possibly under nervous control. Colour as a protective device serves in protecting the animals form enemies.

External Structures

Body is divisible into three regions, head, trunk and tail. A small narrow neck joins the head with trunk. Head is more or less triangular. Snout is short, pointed and bears a pair of apertures called external nares. Eyes are provided with movable nictating membrane. Mouth is a transverse aperture and is terminal in position. Trunk bears tow pairs of appendages, the forelimbs and hind limbs.

Each forelimb is divisible into brachium, antibrachium, and manus. Hind limb is divisible into femur, crus, and pes. Limbs end in digits and are provided with sharply pointed claws. Cloacal aperture or vent is a transverse opening situated in the postero-ventral part of trunk and at base of tail. Tail is long, slender, tapering and exhibits alternate dark and light annuli. Body colour is brown or grayish on dorsal surface and dirty whitish ventrally. Young females often show two light yellow dorsolateral stripes. Fully grown males have greenish tinge (Figure 8.15).

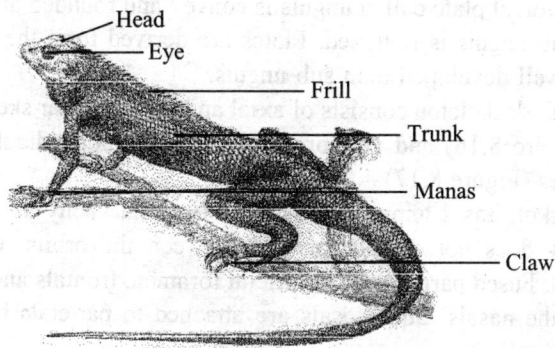

Figure 8.15 *Calotes* : External structure.

Skin

Skin is highly modified for living on land. To safeguard against frictional contact with dry ground and to prevent desiccation in dry air, the epidermis becomes greatly elaborated than dermis. Skin glands are almost absent. Only skin glands in Calotes are the femoral glands in males found along the posterventral margin of thigh and become functional during reproductive period.

Skin develops first as a simple cuboidal ectoderm. Ectoderm then differentiates into an outer periderm and an inner basal germinative layer. Epidermis becomes stratified due to addition of outer layers and scales appear subsequently. Epidermis is a compound layer composed of an inner layer of columnar cells called stratum germinativum which is arranged as a basement membrane. Outermost layer of epidermis is the dead stratum corneum. Between the stratum corneum and stratum germinativum lies the transitional layer. Stratum corneum becomes thickened over the scales, while between scales this layer becomes thinner.

Exoskeleton

Scales and claws are described below.

Scales: Two types of scales, larger scales and smaller scales are found. Smaller scales are present between the larger scales .Sensory structures called prototrichs as sensory bristle are present on some large as well as small scales. Prototrichs have been claimed to be precursors of mammalian hair. Scales on ventral surface of body are smaller in size than those on dorsal side.

Scales develop from stratum germinativum (or Malpighian layer) of epidermis. Scales on head region are large and do not overlap but touch each other by their edges. These larger scales are called head shields. Few scales situated on mid-dorsal line and on border line of neck have modified into large and pointed structures. These can move and are usually called frills. Similar structures are present on postero-lateral border of head is periodically shed by a process called ecdysis.

Claws: Digital tips are provided with sharp claws. Each claw is made up of dorsal and ventral scales-like horny plates which are so placed that they converge at the end of digit to make the tip pointed. Dorsal plate called unguis is convex and rounded at tip and lateral sides. Ventral plate called sub unguis is flattened. Plates are derived form the Malpighian layer of epidermis. Unguis is well developed than sub unguis.

Endoskeleton: Endoskeleton consists of axial and appendicular skeleton. Axial skeleton includes the skull (Figure 8.16) and vertebral column while appendicular skeleton includes girdles and limbs bones (Figure 8.17).

Skull: Diapsid skull has 2 temporal fossae. Ant-orbital vacity and secondary plate are absent. Cranial cavity does not extend forward between the orbits. Cranium covered by investing bone is small. Fused parietals bear parietal foramen, frontals and nasals. Premaxillae are situated between the nasals. Squamosals are attached to parietals by sutures. There are prefrontal and post frontal.

The basl bone extends as a slender bar called parasphenoid. Basipterygoid processes articulate with pterygoid. A stout os transversum arises from the junction of pterygoid and palatine. Vomer is small. Exoccipitals are continuous with the prootic processes. There is a single occipital condyle. Quadratojugal is not ossified. The lower jaw consists of two rami, each of which is made by 6 pieces of investing bones namely articular, angular, supraangular, coronoid, splenial and dentary.

Vertebral Column: This column is divided into cervical region consisting of 10 vertebrae, thoracolumbar region consisting of 22 vertebrae, sacral region containing 2 vertebrae. The first cervical vertetebra is called atlas, while the second one is called axis. The midventral line of axis bears hypapophysis. The rest of the cervical vertebrae are typical. The hypapophysis is absent in thoracic vertebrae. Sacral vertebrae bear expanded ribs. Caudal vertebrae bear chevron bones.

Sternum: This is a rhomboidal plate like structure and supported by T- shaped interclavicle. It bears sternal aperture at its middle and bears ribs.

Hyoid Apparatus: This structure supports the floor of the buccal cavity and consists of basihyal, anterior cornua, middle cornua, and posterior cornua. The anterior cornua represent the hyoidean arch, while the middle cornua represent the vestiges. The posterior cornua originate from the postero-lateral edge of the basihyal.

Digestive System: This is composed of the alimentary canal and digestive glands.

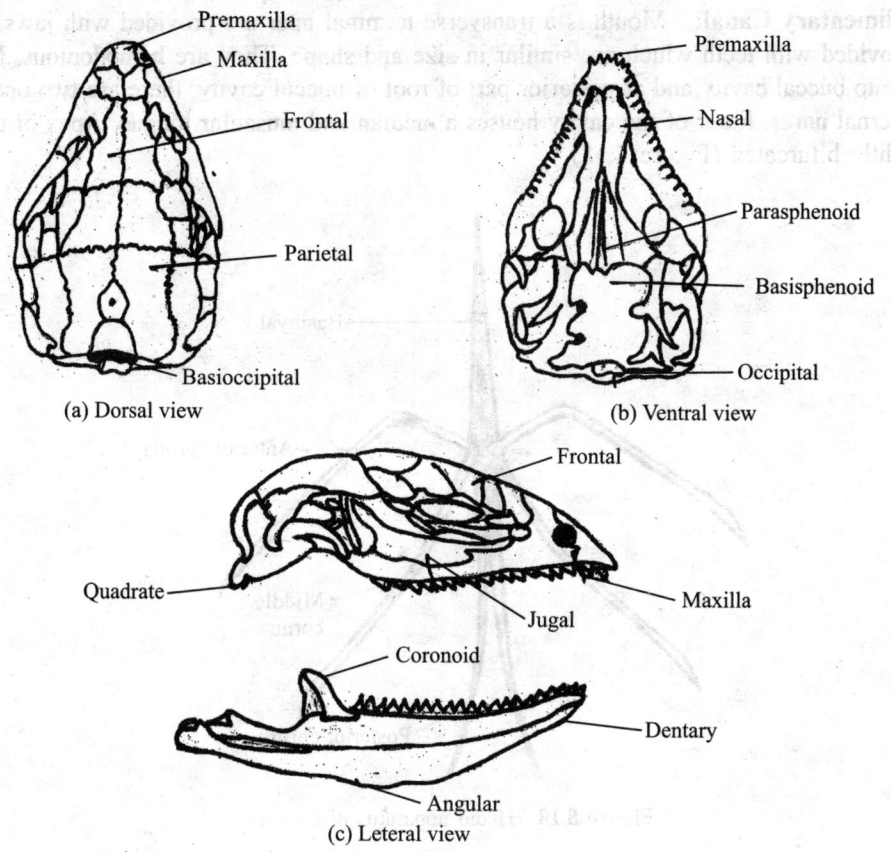

(a) Dorsal view

(b) Ventral view

(c) Leteral view

Figure 8.16 Skull of *Calotes*.

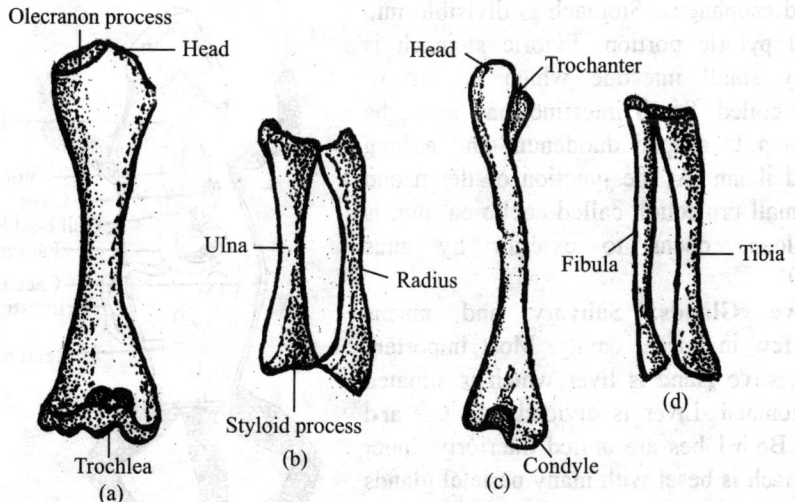

Figure 8.17 Bones of fore limbs and hind limbs of *Calotes* (a) Humerus; (b) Radius-ulna; (c) Femur, and (d) Tibio-fibula.

Alimentary Canal: Mouth is a transverse terminal aperture provided with jaws. Jaws are provided with teeth which are similar in size and shape. They are homodontous. Mouth leads into buccal cavity and at posterior part of roof of buccal cavity, there are two openings for internal nares. Floor of the cavity houses a median and muscular tongue. Apex of tongue is slightly bifurcated (Figure 8.18).

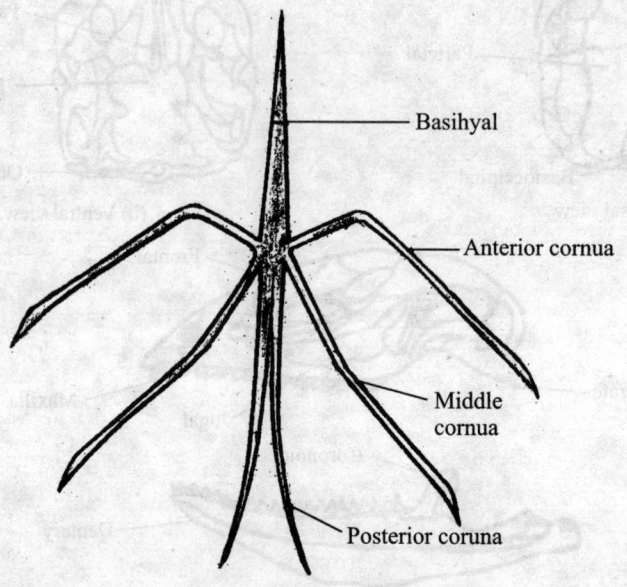

Figure 8.18   Hyoid apparatus of *Calotes*.

Buccal cavity passes to stomach through pharynx and esophagus. Stomach is divisible into cardiac and pyloric portion. Pyloric stomach is followed by small intestine which is narrow, tubular and coiled. Small intestine may again be divided into a U shaped duodenum and a long much coiled ileum. At the junction of ileum and rectum, a small projection called coelic caecum is present. Cloaca opens to exterior by anus (Figure 8.19).

Digestive Glands: Salivary and mucus glands are few in buccal cavity. Most important massive digestive gland is liver which is situated dorsal to stomach. Liver is divided into left and right lobes. Both lobes are united interiorly. Inner wall of stomach is beset with many parietal glands and gastric glands. Other important digestive gland is pancreas (Figure 8.20).

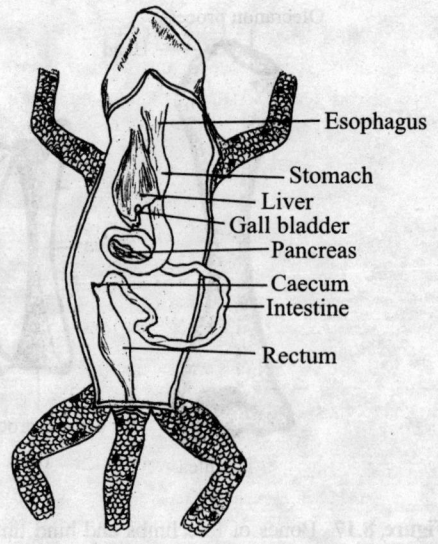

Figure 8.19   Digestive system of *Calotes*.

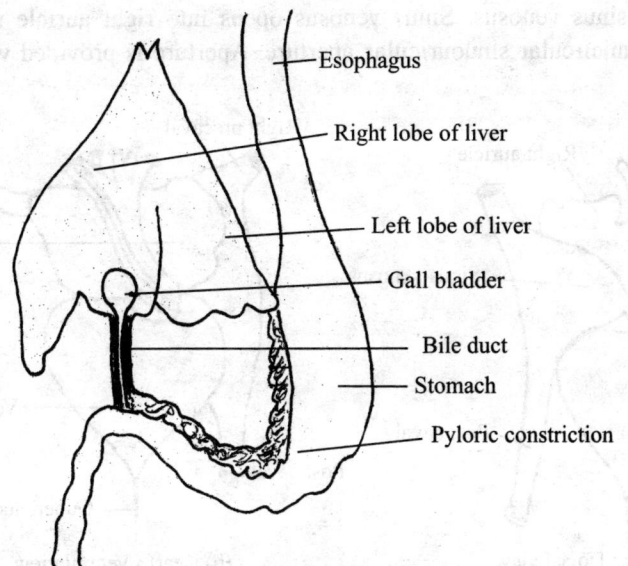

Figure 8.20 Structural relationship of liver and pancreas with duodenum.

Respiratory System: Respiratory structures include a pair of nostrils situated a little ahead of the eyes. Nostrils lead to nasal passages, which open into the roof of buccal cavity. Glottis opens into larynx, which opens into trachea. Trachea is bifurcated into two and forms a pair of narrow passages called bronchi. Each bronchus enters a lung. Lungs are elongated sac-like structures. Right lung is slightly larger than left one. Internally, each lung is incompletely divided into small chambers by development of many incomplete septa. These chambers are called alveoli. Non- partitioned posterior part of lung is the reservoir for residual air. It constitutes about one-third part of whole lung (Figure 8.21).

Circulatory   System: This   system consists of cardiovascular system and lymphatic system. Cardiovascular system includes a heart which is an efficient machine to propel, the fluid vehicle, the blood into the pipelines of arteries and veins. Heart is triangular (Figure 8.22) and is covered by a thin and transparent pericardial membrane. Space between heart and pericardium is filled with pericardial fluid. Auricular region is wider than ventricular region. Heart is composed of a sinus venous, two auricles, and an incompletely divided ventricle. Sinus venosus is reduced and its right half is larger than its left counterpart. A constriction marks off the right

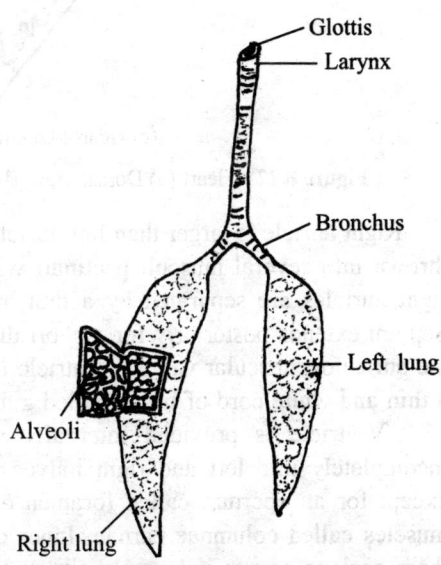

Figure 8.21 Respiratory system.

and left parts of sinus venosus. Sinus venosus opens into right auricle near the region of constriction by semicircular sinuouricular aperture. Aperture is provided with sinuoauricular valves.

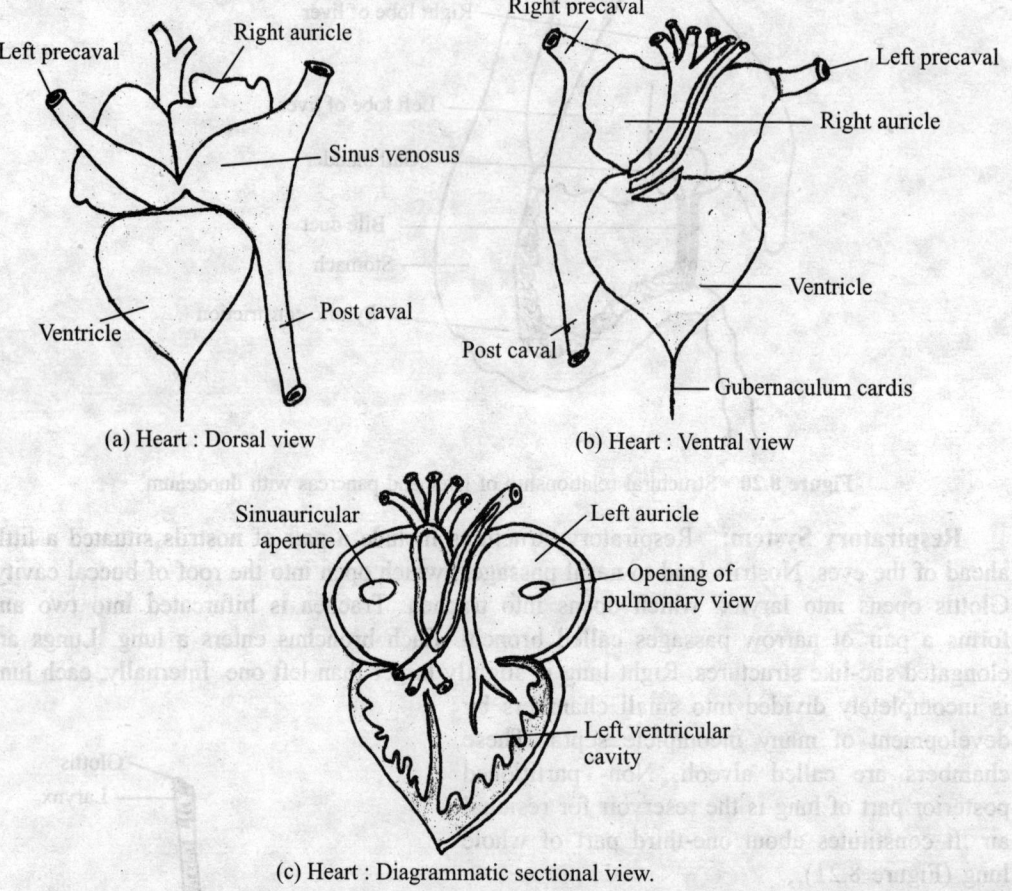

(a) Heart : Dorsal view

(b) Heart : Ventral view

(c) Heart : Diagrammatic sectional view.

Figure 8.22   Heart (a) Dorsal view, (b) Ventral view, and (c) Diagrammatic secional view.

Right auricle is larger than left auricle. Wall of right auricle is thick and its inner lining is thrown into several musculi pectinati which are projected within lumen. Internally left and right auricles are separated by a thin, muscular, and non-perforated interauricular septum. Septum extends posteriorly for a short distance within ventricle and bears at its posterior tip the auriculoventricular valves. Ventricle is muscular, spongy and triangular and its apex bears a thin and white cord of tissue called gubernaculum cordis.

Ventricle is provided internally with an interventricular septum which divides it incompletely into left and right halves. This partition has become complete in crocodiles except for an aperture called foramen of panizza. Inner wall of ventricle is provided with muscles called columnae carnae. Inner cavity of ventricle has been arbitrarily divided into three regions cavum pulmonale situated in right side, cavum venosum (or middle portion) and cavum arteriosum (or left hand portion).

Mechanism of Circulation Through Heart: Circulatory circuit is double with pulmonary or lesser circulation and systemic or greater circulation. Pulmonary circulation is conducted by pulmonary arteries which carry deoxygenated blood to lungs. In lungs, blood becomes oxygenated and returns to left auricle by pulmonary vein. Left auricle pours its content into ventricle through auriculo-ventricular aperture. In greater circulation, deoxygenated blood returns to sinus venosus by two precaval and one postcaval veins. Sinus venosus opens into right auricle.

Right auricle empties its content into ventricle. Ventricle sends blood for circulation into different parts of body through systemic and pulmonary arches. Entry and exit of blood in ventricle are so beautifully arranged that a major quantity of oxygenated blood is always forwarded to brain region. As ventricle is incompletely divided, admixture of oxygenated and deoxygenated blood occurs thrice, once in cavum venosum, once in dorsal aorta, and another in left ductus caroticus. Thus, though ventricle is morphologically incompletely divided, there is a tendency for physiological separation of two types of blood at least in two auricles complety and in ventricle partially. From this point, heart is biologically more advanced than that of Bufo.

Blood: Blood is red in colour and made up of plasma and blood cells. Red blood corpuscles are biconvex, elliptical in an outline and each bears an elliptical nucleus. White blood corpuscles are irregular in outline, non-pigmented and each bears a spherical nucleus.

Arterial System: Of the six pairs of arterial arches joining the dorsal aorta to the ventral aorta during embryonic development of arteries, the third, fourth and six pairs persists in adult. As the ventricle tends to divide into left and right ventricles by the development of incomplete interventricular septum, the base of ventral aorta splits into three parts, two of which remain in right part of ventricle and third goes to leave part of ventricle. Thus from ventricle arise arterial system including three aortic arches. Thus arches are: one pulmonary aorta and two systems aortae, right and left.

Pulmonary Aorta: It arises independently from right portion of ventricle and soon splits into two branches, each entering a lung (Figure 8.23).

Left Systemic Aorta: This aorta originates independently form right (left to pulmonary aorta) portion of ventricle and moves forward for some distance. Then it curves round heart and goes downwards to meet right systemic aorta, a little posterior to apex of ventricle. It carries mostly deoxygenated blood. From left systemic arch, four oesophageal arteries arise. The first of these arteries arises near the point of insertion of left ductus caroticus, while the origin of fourth one is very close to point of union of two systemics. Parietal arteries do not originate from left systemic arch.

Right Systemic Aorta: It curves to right side of heart, meets left systemic aorta posteriorly to form dorsal aorta. It carries oxygenated blood. From the apex of curvature of right systemic aorta arises a single and common carotid artery which advances anterioly and then splits into four arteries. The inner pairs of these 4 branches form external carotid, while outer pair forms internal carotid arteries. On both sides before its division into left internal and external carotids and right internal and external carotids one thyroid artery is given of form the common carotid. The main branches from right and left external carotids are laryngotracheal artery (one from each) and three buccal arteries (form each).

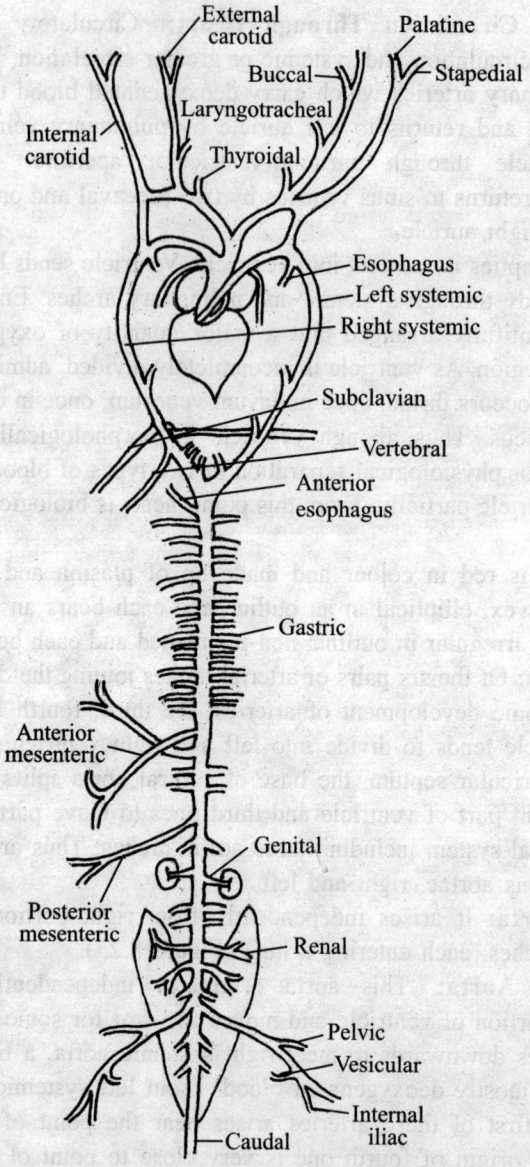

Figure 8.23 Arterial system of *Calotes*.

Internal carotids both left and right bifurcate at their tips into inner palatine artery and outer stapedial artery. Two internal carotids are connected to systemic aorta of corresponding sides by ductus caroticus. From right systemic aorta, three oesophageal arteries are given out. Right systemic aorta before its meeting with left systemic aorta gives rise to a subclavian artery which bifurcates into two and supplies forelimbs. From right systemic aorta, a vertebral artery also originates to send blood to vertebral column. Right and left systemic aortae unite a little behind the heart and give rise to dorsal aorta, which runs posteriorly and gives branches to visceral organs and posterior parts of body. The following arteries originate form dorsal aorta chronologically along the antero-posterior axis. These are anterior oesophageal artery, first pair of parietal arteries, first and second pairs of gastric arteries, second pair of parietal arteries, third, fourth and fifth pairs of gastric arteries followed by third pair of parietal arteries, sixth and seventh pairs of gastric arteries followed by fourth pair of parietal arteries, eight pair of gastric arteries followed by 5th and 6th pair of parietal arteries, anterior mesenteric artery,coeliac artery and seventh and eight pair of parietal arteries, posterior mesenteric artery or hepato-intestinal artery, ninth pair of parietal arteries, right and left genital artery, tenth and eleventh pairs of parietal arteries, left and right renal arteries. These may be more than one pair. Twelth and 13th pairs of parietal arteries, one pair of iliac arteries; dorsal aorta now enters tail as caudal artery.

Venous System: Central meeting arena of all veins in the body is sinus venosus. Sinus venosus receive left and right pre-cavals and single median post-caval. Each precaval vein has been formed by union of three veins. These are external jugulars which brings back blood form floor of mouth and tongue, internal jugular which drains blood form brain, and subclavian which draws blood from forelimb. Right precaval gets an azygos vein. Postcaval is constituted by the large median vein which is formed by union of right and left efferent renal veins emerging from two kidneys. Genital veins join left and right efferent renal veins before their union. A pair of stout and short hepatic veins joins median postcaval before its entry into sinus venosus.

A median caudal vein carries blood form tail region. Caudal vein ultimately bifurcates into two veins which enter kidneys. Each vein gives rise to renal portal vein to kidneys and pelvic vein which receives femoral and sciatic veins from hind limb. Pelvic veins unite to form a median epigastric (or anterior abdominal) vein which ultimately opens into left liver. Anterior abdominal vein and postcaval are free of each other, except through renal portals in kidneys. Blood from visceral organs i.e. stomach, intestine, pancreas etc enters left lobe of liver by a hepatic portal vein. Pulmonary venous circuits comprises of pulmonary veins. From each lung, two pulmonary veins carry blood to heart. Of these veins, one comes out from anterior part, while other comes form posterior part of lung. Near left auricle, all these four branches unite and open into left auricle. Pulmonary veins bring oxygenated blood to heart from lungs (Figure 8.24).

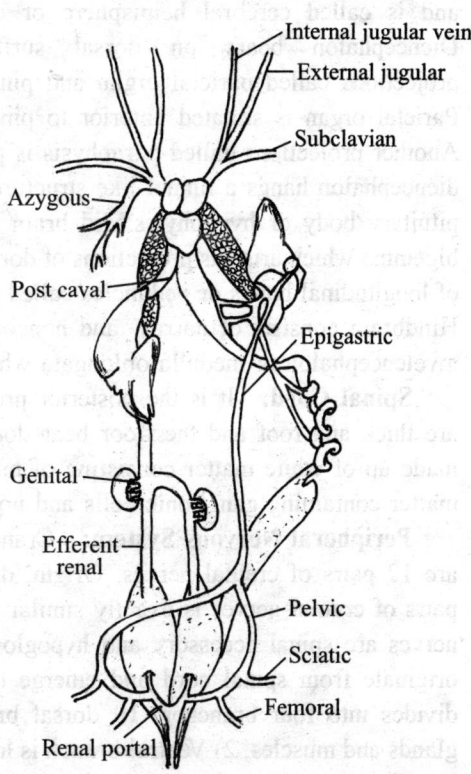

Figure 8.24 Venous system of *Calotes*.

Nervous System: It consists of central nervous systems, peripheral nervous system consisting of cranial and spinal nerves which originate from brain and spinal cord respectively, and autonomic nervous system .

Central Nervous System: Brain (Figure 8.25) and spinal cord constitute the central nervous system. Brain is encased in cranium. Nervous tissues of brain are protected by 2 meninges called piamater and duramater. Two coverings remain separated form each other and space between them is called subdural space. Brain of an adult is differentiated into forebrain, midbrain, and hind brain. Forebrain consists of telencephalon anteriorly and diencephalon posteriorly. From side wall of telencephalon emerges a pair of saclike projections called olfactory lobes.

Posterior part of telencephalon is elongated and is called cerebral hemisphere or cerebrum. Diencephalon bears on dorsal surface two projections called parietal organ and pineal body. Parietal organ is situated anterior to pineal body.

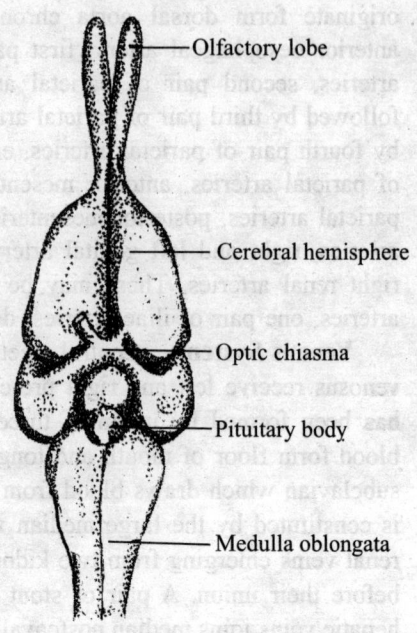

Figure 8.25 Brain of *Calotes* : Dorsal view.

Another projections called paraphysis is present in reduced conditions. From ventral side of diencephalon hangs a funnel like structure called infundibulum on apex of which is situated pituitary body or hypophysis.Mid brain consists of a pair of oval optic lobes or corpora bigemina which arise as projections of dorsolateral walls. Ventral to optic lobes, there are pair of longitudinal bands or peduncles called crura cerebri which connect hindbrain to midbrain. Hindbrain consists of narrow and nonconvoluted metencephalon or cerebellum and a long myelencephalon or medulla oblongata which continues posteriorly with spinal cord.

Spinal Cord: It is the posterior prolongation of brain through neural canal. Its walls are thick and roof and the floor bear dorsal and ventral furrows. Externally spinal cord is made up of white matter consisting of medullated nerve fibres and internally, there is grey matter containing ganglionic cells and non medullated fibres.

Peripheral Nervous System: Cranial and spinal nerves constitute this system. There are 12 pairs of cranial nerves. Origin, distributions, and biological nature of first to tenth pairs of cranial nerves is exactly similar to that Bufo. Eleventh and twelfth pairs of cranial nerves are spinal accessory and hypoglossal. There are several pairs of spinal nerves that originate from spinal cord and emerge out from it between vertebrae. Each spinal nerve divides into four branches: 10 dorsal branches, is thin, short and supplies sense organs, glands and muscles, 2) Ventral branch is long thick and supplies hypoxial organ, 3) meningeal branch is small and goes back to supply neural canal and 4) ramus communicans communicates with autonomic nervous system (Figure 8.26).

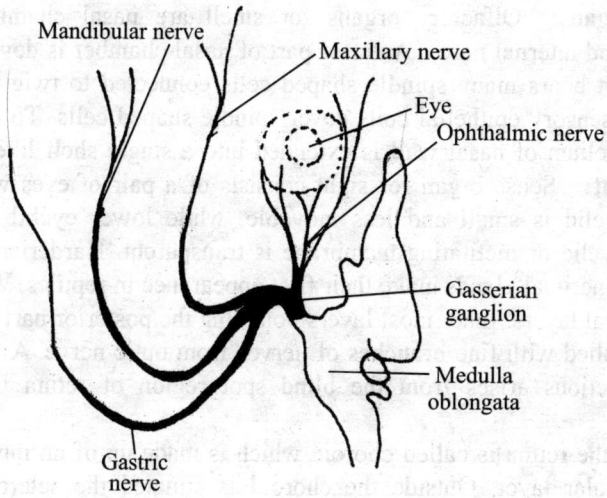

Figure 8.26(a)   Origin and distribution of fifth and seventh cranial nerves.

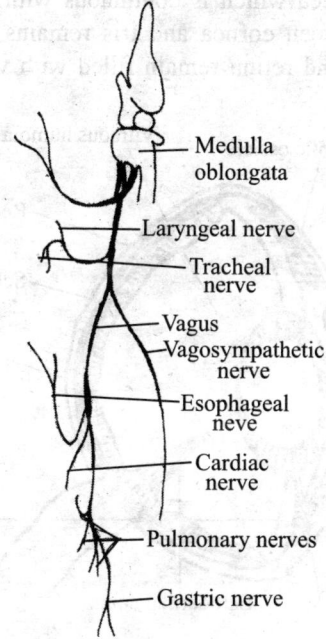

Figure 8.26(b)   Origin and distribution of ninth and tenth cranial nerves of *Calotes*.

Autonomic Nervous System: This system consists of a chain of complicated ganglia with receptor and effector nerves formed mostly by offshoots from the ventral branches of spinal nerves. Branches from these ganglia innervate muscles of heart, lungs, digestive system and glands, which work continuously and are not controlled by will.

Sense Organs: Sense organs are highly developed.

Olfactory Organs: Olfactory organs for smell are nasal chambers which extend between external and internal nares. Anterior part of nasal chamber is devoid of sensory cell, while posterior part bears many spindle shaped cells connected to twig of olfactory nerve. Highly pigmented sensory epithelial cells cover spindle shaped cells. To enlarge the surface of exposures, epithelium of nasal wall is extended into a single shelf like concha.

Organ of Sight: Sense organ for sight consists of a pair of eyes which are lateral in position. Upper eyelid is small and less movable, while lower eyelid is large and more moveable. Third eyelid or nictitating membrane is transparent. Harderian glands supply the third eyelid, while lacrimal glands make their first appearance in reptiles. Wall of globular eye is made up of several layers. Inner most layers bounding the posterior part of eye are made up of retina. It is supplied with fine branches of nerves from optic nerve. A vascular pigmented cushion like projections arises from the blind spot region of retina is called pecten or ectodermal conus.

Layer outside the retina is called choroid which is made up of an inner pigmented layer and an outer vascular layer. Outside the choroid is situated the sclerotic layer which is protective in nature. The front of eye in occupied by lens which is a clear ball derived from the skin. Iris, a prolongation of choroid is a pigmented and muscular ring in front of lens. Front part of eye is protected by cornea which is continuous with sclera. Cornea is protected by conjunctiva. Lymph space between cornea and iris remains occupied by aqueous humour, and inner space between lens and retina remain filled with vitreous humour (Figure 8.27).

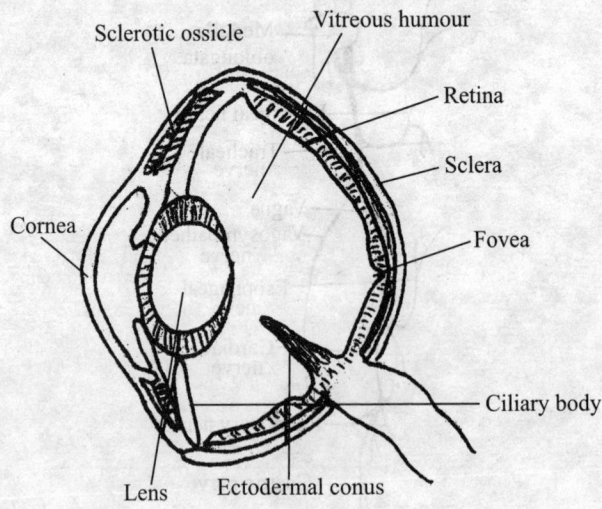

Figure 8.27 Eye.

Organ of Jacobson: An accessory structure called organ of Jacobson or vomeronasal organ is associated with olfactory organ. It is a small saclike structure with heavily pigmented walls supported by cartilages. An organ of Jacobson is present below each nasal cavity. It acts as a chemoreceptor and is highly developed. This sense organ appreciates scent particles introduced into it by it tongue tips.

Auditory Sense Organs: Ear is made up of two parts- 1) Middle ear, and 2) Internal ear. External ear is absent.

Middle Ear: It is a tubular cavity with its outer opening being covered by tympanic membrane situated behind the eye. Membrane is slightly depressed and is perforated by very minute apertures. Inner end of tube is covered by a very thin membrane. A bony rod like structure called columella auris is situated inside tube of middle ear. Role of this structure is to transmit the sound waves. Middle ear is communicated to buccal cavity by a narrow but long tube called Eustachian tube. Middle ear communicates with internal ear through a circular structure called fenestra ovalis.

Internal Ear: This is situated in a bony capsule. Bony capsule has two fenestrae - fenestree ovalis and fenestra rotunda. Columella auris is fixed internally into the membrane which covers fenestra ovalis. Space between capsule and internal ear is filled in with perilymph. Internal ear is demarked into an upper utriculus which is tubular and slightly bent and lower sacculus which is sac-like. Lagena or rudimentary cochlea is attached to ventromedian side of sacculus. Lagena is of simple type and contains papilla basilaris a special organ of hearing. From the utricles arises a tubular structure called ductus endolymphaticus which opens beneath the duramater of brain. Three semicircular canals attached to utriculus by both ends.

Each semicircular canal is swollen at one end. Swollen part is called ampullae. Ampullae of anterior semicircular canal and horizontal semicircular canal are situated side by side, while those of posterior semicircular canal and horizontal canal are situated at a distance. Minute masses of Caco3 presents inside the ampullae help in maintaining balance. These pieces of Caco3 are called satoliths or autoliths. Sacculus utriculus and three semicircular canals remain filled with a liquid called endolymph.

Excretory System: Consist of a pair of metamorphic kidneys situated in abdominal cavity. Each kidney is dark red in colour and is lobed. They are free at anterior end but united along inner margins at posterior part. Posterior ends of kidneys run over the cloacal chamber. A pair of ureter arise one from each kidneys. Ureters are short and open into cloaca separately. From lateral wall of cloaca arises a single urinary bladder. Urine is semisolid in consistency and contains uric acid (Figure 8.28).

8.13 *Sphenodon*

Squamates and *Sphenodon* have transverse cloacal slits. They show caudal autotomy that is loss of the tail-tip. *Sphenodon* was worldwide in distribution in the Mesozoic and is reported from the middle Triassic deposits of Africa, Europe, Asia and the Americas. Living *Sphenodon* is now restricted to The New Zealand mainly in the 3 North Islands. They live in association with petrels in burrows. They are nocturnal, insectivorous and carnivorous. Similarities between lizards and *Sphenodon* are superficial.

Sphenodon was classified as lizards in 1831. Albert Günther noted features similar to birds, and turtles in *Sphenodon* and placed it in the order Rhynchocephalia (meaning beak head). The word *Sphenodon* is derived from the Greek spheno meaning wedge and dont meaning tooth. *Sphenodon* has undergone changes throughout the Mesozoic. Cold weather adaptations enable them to live on the islands of New Zealand. Two species are *S. punctatus*

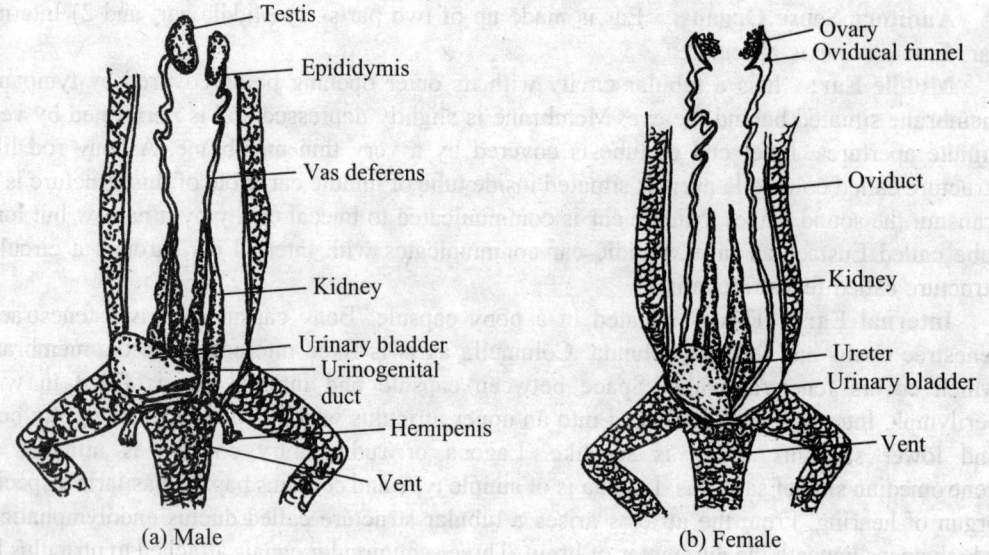

Figure 8.28   Urinogenital system of *Calotes*.

and *S. guntheri*. The name *punctatus* signifies the spotted appearance, and *guntheri* refers to the worker Günther.The species name *punctatus* is misleading as the two species are spotted. *S. guntheri* possesses olive brown skin with yellow patches. In *S. punctatus*, the colour varies from olive green through grey to pink. *S. guntheri* is small than *S. punctatus*. *S. punctatus* is divided into 2 subspecies: the Cook Strait tuatara and the northern tuatara (*S. punctatus punctatus*).

Sphenodon punctatus
Systematic position

Kingdom: Animalia; Phylum: Chordata; Class: Reptilia; Order: Rhynchocephalia; Family: Sphenodontidae; Subfamily: Sphenodontinae; Genus: *Sphenodon*. Species: *S. punctatus*, *S. guntheri*; *S. diversum*.

Sphenodon endemic to New Zealand resembles lizards. Their common ancestor is the squamates. *Sphenodon* gives evolutionary picture of lizards and snakes. Adult *S. punctatus* male and female measures 6o cm and 45 cm in length respectively. Males and females weigh up to 1 kg and 0.5 kg respectively. *S. guntheri* weighs up to 660 g. It is a separate and archaic type of reptile. The underparts are covered with large epidermal scales, while those of back are granular in appearance. Two unique rows of teeth in upper jaw overlap one row on lower jaw. Photoreceptive eye is associated with circadian and seasonal cycle. They can hear without external ear. *Sphenodon* called the living fossils is protected by law. *S. guntheri* unrecognized until 1989 are now threatened by habitat loss and predators. Their population is confined to offshore islands.

Sphenodon, the most unspecialized living amniote possess heart which is more primitive than other reptile. Their brain and locomotion resemble those of amphibians. Lungs with a single chamber lack bronchi. Both species are sexually dimorphic. They shed their skin once per year in adult state, and 3 or 4 times in juvenile state. Spiny crest on back, made of triangular, soft folds of skin, is larger in males, and becomes stiff for display. Abdomen of male is narrow than the female (Figure 8.29).

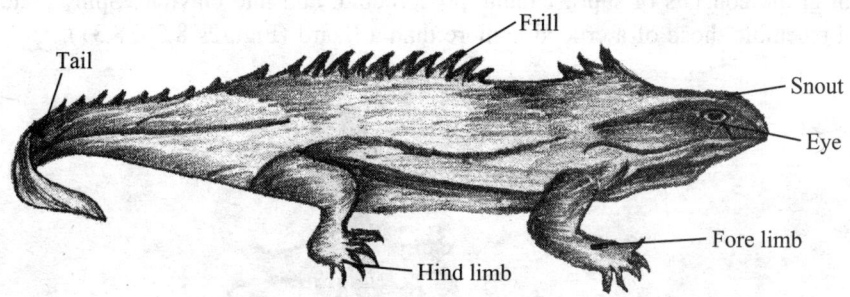

Figure 8.29   External structure of *Sphenodon*.

Skull: All the original features of skull are retained. The skull is much stronger and firmer than that of lizards and snakes; its parts are joined by bony arches. Skull has 2 openings on both sides with arches. The temporal region is bridged by 3 bony arches. The large vomer, palatine, and pterygoid form a broad bony roof to the mouth. The large quadrates are firmly fixed by the pterygoid, squamosal, lateral occipital, and by the jugal bridge.

The vertebral column is primitive. There is a series of ventral abdominal ribs (Chevron bones) that are reminiscent of the chelonian plastron. The vertebrae have an unbroken series of intercentral wedge bones. There is an elaborate system of abdominal ribs. The humerus has an entepicondylar foramen and an ectepicondylar foramen for the passage of radial nerve. The carpus has 10 bones, all of which remain separate including the intermedian. The Supratemporal Bridge is formed by the squamosal and postorbital, the latter fuse with the postfrontal. The postorbital joins the ascending branch of the jugal, both together forming the hind border of the orbit. This is bordered below chiefly by the maxillary, which is long, while the anterior process of jugal is much reduced. There is no post -orbital fossa. The nares are terminal and lateral, well separated by the premaxillaries. The posterior temporal bridge is formed by the squamosal and parietal.

The upper jaw is attached to the skull. The tip of upper jaw is beak-like and separated from the rest of the jaw by a notch. Single row of teeth in lower jaw and a double row in upper jaw are found, with the bottom row fitting between the 2 upper rows in closed mouth. Teeth are sharp projections of jaw bone and are not replaced. As teeth wear down, *Sphenodon* prey soft animals like earthworms, larvae, and slugs. Brain fills only half of its endocranium. Teeth are acrodont, ankylosed in one series with the supporting bones. The lateral edges of palatines carry teeth. The jaws, joined by ligament, chew combined with a shearing up and down action. Bite shears chitin and bone. The whole vertebral column consists of 25 presacral, 2 sacral and about 30 caudal vertebrae.

Gastralia, also called abdominal ribs are the ancestral trait. In *Sphenodon,* they are not attached to spine or thoracic ribs. True ribs are small. Ribs are single headed with uncinate processes. *Sphenodon* is the only living tetrapod that possesses gastralia and uncinate processes. Pelvis and shoulder girdles are arranged differently from those of lizards. Ilia are blade like and pubis is united by symphysis. Epipubic bone is cartilaginous. Ischio-pubic foramen is present. Hypoischium is cartilaginous and is attached to the ischium behind. Pectoral girdle consists of suprascapula, procoracoid, and interclavicle. Spiny plates on back and tail resemble those of a crocodile more than a lizard (Figures 8.30, 8.31).

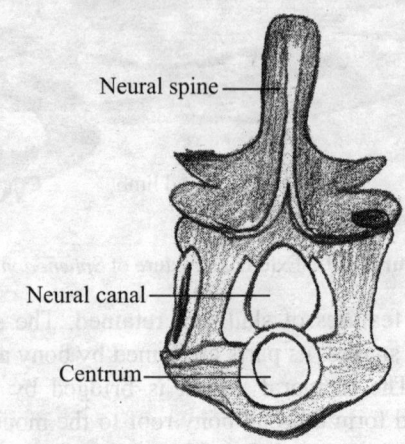

Figure 8.30   Vertebrae of *Sphenodon*.

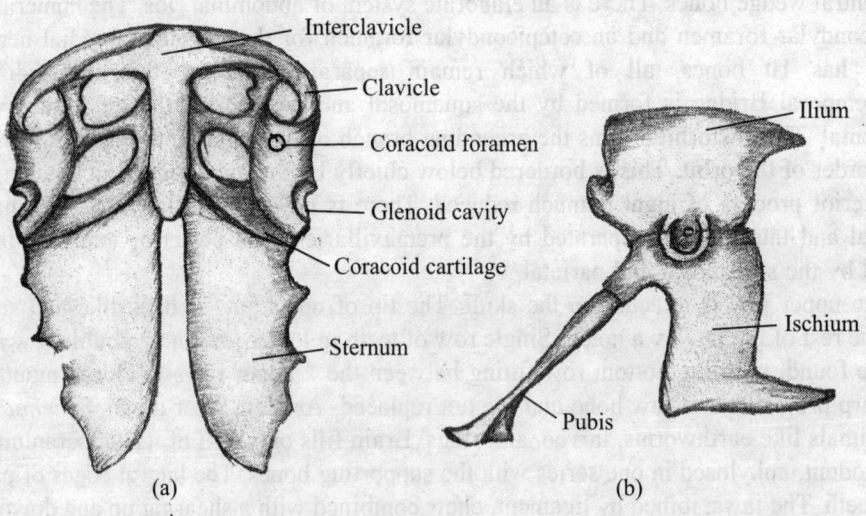

(a)   (b)

Figure 8.31   (a) Pectoral girdle of *Sphenodon,* and (b) Pelvic girdle of *Sphenodon*.

Digestive and Circulatory Systems: Stomach is simple, tubular, tapers posteriorly and ends at pyloric valve. Small intestine is found posterior to stomach. Cloacal aperture is transverse. Ductus caroticus and ductus arterisus are present. Three main aortic arches emerge from the ventricle and form a short common trunk homologous to amphibian conus arteriosus.

Nervous System and Sensory Organs: Brain is simple. Parietal organ is located in a socket of parietal bone. The organ of Jacobson is present as a vomeronasal organ in a special region. Eyes specialized with a duplex retina contain 2 types of visual cells for day and night vision and can focus independently. Tapetum lucidum reflects onto the retina to enhance vision in dark. A third eyelid on each eye is found. The third eye called the parietal eye has lens, cornea, retina, and degenerated nerve. Parietal eye is visible in hatchlings. After 4 to 6 months, it becomes covered with scales and pigment, and absorb UV rays to produce vitamin D, and to detect light/dark cycles, and help in thermoregulation. Parietal eye is part of pineal complex. The hearing organs have no eardrum and no ear hole. Middle ear cavity is filled with loose tissue. Stapes remains in contact with immovable quadrate, hyoid and squamosal. Hair cells are unspecialized (Figure 8.32).

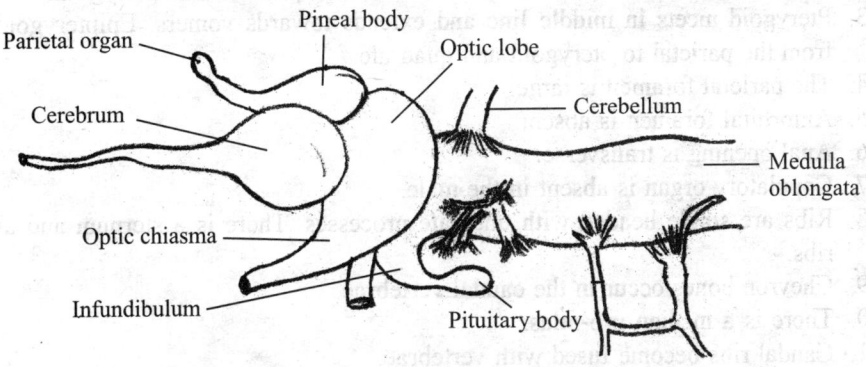

Figure 8.32 Lateral view of brain of *Sphenodon*.

Excretory and Reproductive System: Urinary bladder is single and opens into urodaeum, lying ventral to cloaca.Kidney is metanephric and excrete both urea and uric acid.Copulatory organ is absent in male. Females are provided with a single ovary and single oviduct.A rudimentary epididymis is present in female.

Anatomical features may be summarized as follows

1. Lizard like body is 2 feet long. Tail measures about 1 /3 of the body length.
2. Nocturnal.
3. Body and tail are laterally compressed.
4. Spines are present dorsally.
5. Limbs are pentadactyl. Legs are short and primitive.
6. Ten-11 carpals are present.
7. Hind limbs are plantigrade.
8. Upper body surface is covered with granular scales
9. Lower surface is covered with transverse scales.
10. Head is large with prominent ridge.
11. There is a foramen above the outer and one above the inner condyle of humerus.
12. Eleven carpal elements include 4 in proximal row, 2 centrals and 5 in distal row.
13. Tibial and fibular elements are distinct

14. Intermedium and centrale are fixed to tibiale. Three tarsal bones are found.
15. Teeth are pleurodont, pointed, triangular, laterally compressed and arranged in 2 parallel rows, one along the maxilla, and the other along palatine. Each premaxilla bears a prominent, chisel-shaped incisor, represented in young by 2 pointed teeth.
16. Pubes are united in a symphysis.
17. Oval foramen intervenes between the ischium and pubis.
18. Cartilaginous hypoischium is attached to the ischia.
19. Complete lower temporal arch is present in skull.
20. Quadrate is immovably fixed, wedged in by the quadratojugal, squamosal, and pterygoid.
21. Premaxilla is not fused together.
22. Broad palate is formed by the plate-like vomer, palatine and pterygoids.
23. Pterygoid meets in middle line and extends towards vomers. Epipterygoid extends from the parietal to pterygoid and quadrate.
24. The parietal foramen is large.
25. Antorbital foramen is absent.
26. Anal opening is transverse.
27. Copulatory organ is absent in the male.
28. Ribs are single headed with uncinate processes. There is a sternum and abdominal ribs.
29. Chevron bones occur in the caudal vertebrae.
30. There is a median pro-atlas.
31. Caudal ribs become fused with vertebrae.
32. Lateral temporal fossa is bounded by an inferior temporal arch.
33. Jacobson's organ is present in a primitive form.
34. A urinary bladder is present.
35. A T-shaped interclavicle is present.
36. Coracoid lacks fenestra.
37. It lays 10-13 shelled eggs which are buried few inches below the soil surface .
38. Youngs are hatched 13 months later.

Behaviour: Adult is terrestrial and nocturnal, though they bask in the sun. Hatchlings hide under logs and stones, diurnal, probably because adults are cannibalistic. They live in temperatures lower than tolerable to most reptiles, and hibernate in winter. They are active at temperatures up to 5°C, while temperatures above 28°C are fatal to them. Optimal body temperature ranges between 16 to 21°C. Low body temperature slows metabolic rate. Petrels share the *Sphenodon's* habitat. They use the birds' burrows for shelter.

Sphenoid prey on beetles, crickets, and spiders. Their also feed on frogs, lizards, and bird's eggs and chicks. Eggs and young of seabirds serves as seasonal food .Both sexes defend territories. Sphenodon reproduces slowly, take up to 20 years to attain sexual maturity. Mating occurs in summer. Females mate and lay eggs once every four years. In courtship, male skin becomes dark. Male raises crests and walks in circles, the female with stiffened legs. Female allows the male to mount, or retreat to burrow. Male lifts the tail of female and place his vent over hers. Sperm is then transferred. Eggs (Figure 8.33) of *S. punctatus* have a soft, parchment-like shell.

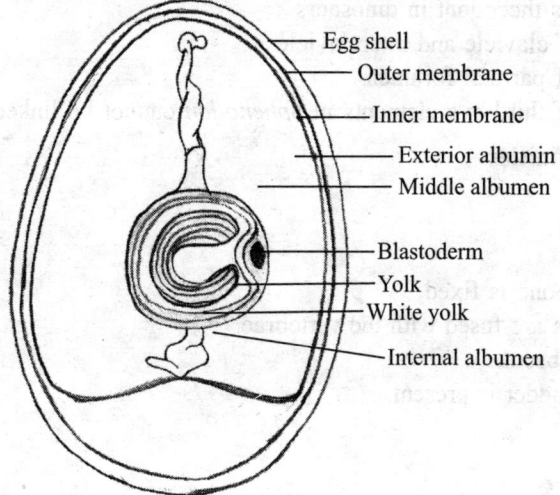

Egg shell
Outer membrane
Inner membrane
Exterior albumin
Middle albumen
Blastoderm
Yolk
White yolk
Internal albumen

Figure 8.33 Egg of *Sphenodon*.

Affinities

Sphenodon possesses several characters which are present in different groups of animals. So, position of *Sphenodon* in the animal kingdom was controversial for sometime. Its affinities with different groups have been discussed below:

Affinities with Amphibia

Circulatory system of *Sphenodon* shows resemblances with *Caudata*.

1. Aortic arches originate from a common stalk which is similar to conus arteriosus of amphibians.
2. Presence of ductus arteriosus and ductus caroticus.
3. Distribution of blood vessels is almost similar.

Remark: Due to several reptilian features, amphibian affinities cannot be accepted.
Affinities with Dinosaurs

Resemblances

1. Diapsid skull.
2. Quadrate bone is fixed
3. Abdominal ribs.
4. Uncinate process of ribs.

Dissimilarity

1. In dinosaurs, double headed ribs are present - not single headed like *Sphenodon*.

2. Dentition is thecodont in dinosaurs
3. Absence of clavicle and interclavicle
4. Absence of parietal foramen
5. Absence of third eye elements in *Sphenodon* cannot be linked with the dinosaurs.

Chelonian Affinities

Resemblances

1. Quadrate bone is fixed.
2. Caudal ribs are fused with the vertebrae.
3. Pecten is absent.
4. Urinary bladder is present.

Differences

1. *Sphenodon* is land dweller but chelonians are aquatic.
2. Only horny beaks are present in chelonians.
3. Parietal foramen in chelonians.
4. Vomer is unpaired.
5. Sternum is absent in *Chelonia*.
6. Cloacal opening is longitudinal in chelonia, but transverse in *Sphenodon*.
7. Males have penis in *Chelonia*.
8. Opening of oviduct lies on the ventral side.

Affinities with Crocodilia

Resemblances

1. Diapsid skull.
2. Quadrate is fixed.
3. Presence of Pro-atlas and abdominal ribs.
5. Caudal ribs fuse with vertebrae.
6. Uncinate process of ribs.
7. Chevron bones are present.
8. Cochlear process is tubular.

Differences

1. Thecodont dention in crocodiles.
2. Nostril is single in crocodile, but double in *Sphenodon*.
3. Vertebrae is procoelous in Crocodile and amphicoelous in *Sphenodon*.
4. Clavicles are absent in Crocodile.
5. Pecten is present in Crocodiles.

6. Penis is present in Crocodile.

Affinities with Lacertilia

Resemblances

1. Body plan is similar.
2. Pro-atlas is present.
3. Vertebrae are amphicoelous in certain geckos.
4. Ribs are single headed.
5. Chevron bone is present in both.
6. Structure of the respiratory organs.
7. Parietal organ is common.
8. Cloacal glands are present.

Differences

1. Quadrate is fixed in *Sphenodon*.
2. Procoelous vertebrae are found in most lizards.
3. Rami of the jaw are united by symphysis in lizards.
4. Erect ilium is found in *Sphenodon*.
5. Clavicles and interclavicles are found in *Sphenodon*.
6. Absence of conus arteriosus in Lacertilia.
7. Presence of pectin.
8. Uncinate process of true ribs are absent in Lacertilia.
9. Copulatory organs are found in Lacertilia.
10. Presence of lower temporal arch.

Many primitive characters of Rhynchocephalia resembles chelonians, crocodilia and dinosaurs in many aspects. *Sphenodon* is closely related to the lacertilia than any other group as discussed above. Because of the occurrence of certain peculiar features, it seems justified that the Rhynchocephalia should be kept as a separate order of Reptilia as suggested by Romer.

8.14 Crocodiles

India has three species of crocodiles: the mugger or marsh crocodile (*Crocodylus palustris*), estuarine or salt water crocodile (*C. porosus*) and the gharial (*Gavialis gangeticus*). The first 2 species are found in the Western Ghats. The gharial is confined to rivers of northern India. In Goa, marsh crocodiles inhabit the mangrove-lined Cumbarjua Canal. This is the only crocodile found in this part of the Western Ghats. Marsh and the estuarine crocodile rarely occur together in nature. It is difficult to identify the two. *Crocodylus palustris* lives in variety of habitats like hill streams, rivers, lakes, ponds, swamps and estuaries. They are olive-black coloured back with yellowish-white belly. They are distinguished from estuarine crocodile by broad snout and presence of 4 distinct raised scales behind the head.They are timid.

Estuarine crocodile, *Crocodylus porosus* is confined to estuaries and coastal sea water. They have striking resemblance with mugger. Their snout is pointed than mugger. They have distinct ridges in front of eyes. They are irritable and attacks people habitually.

Fisher folk regard crocodiles as threat to traditional fisheries. All crocodiles are thought to eat people. Religious scriptures, like the story of Gajendra Moksha of the Vishnu Purana, project the animal as being at par with evil spirits. Mangrove forest, the natural environment of crocodiles is steadily disappearing. Embankments built to reclaim land interfere with crocodile breeding. Fishing nets are death traps for juveniles. Catching crabs and mudskippers disturb crocodile habitat. Nomadic tribes hunt crocodiles and trade the skin, meat, and eggs. Tribals use various parts of animal to treat diseases. Crocodiles are wounded or killed by boat propellers. Construction noise can disturb the crocodiles. Poorly planned wildlife tourism can also disturb the reptiles. Industrial effluents and fecal contamination can degrade the habitat.

Crocodiles are at the apex of food chain in their habitat and play the ecological role of a predator as well as that of scavenger. Crocodiles hunt predatory fishes which feed on shoals of commercially important fish. By checking these predators, they may increase the fish catch. Crocodiles scavenge on dead animals and fishes that otherwise would pollute the water and depress fish numbers. Crocodiles are a valuable natural resource with immense potential for commerce, if used on sustainable basis .Crocodile farming and crocodile ranching can generate employment-especially for tribals who have field knowledge of their behaviour. Crocodile skin can be made into bags, belts, and has a export potential. Crocodile habitats have a tremendous potential for wildlife tourism.

It is possible to manage the environment to conserve crocodiles. Retaining and managing the habitat is the ideal way to manage all wildlife. For crocodile conservation, mangrove cover should be monitored. Farming could be done by the Forest Department, or tribal cooperatives. Restocking the wild population by rearing animals in captivity and releasing them in suitable environments is important conservation measures. Tribal practice of crocodile worship could be promoted for creating public awareness. Generating public awareness through the mass media and signboards near the crocodile habitat, research and monitor wild and captive crocodile populations along with captive breeding of crocodiles are mention-worthy in this regard.

Summary

1. Reptiles are ectothermic tetrapods with clawed toes. Their skin is dry with epidermal scales or plates, but without mucus glands. They respire by lungs and regulate body temperature. Head is carried off the ground on well developed neck. Caudal vertebrae are reduced especially towards tip of tail. Chevron bones attached to caudal centra are called ossicles. Limbs form the main locomotor system .Posterior part of chondrocranium ossifies to form occipital complex, basisphenoid, and ossifications in otic capsule.

2. In many reptiles, upper jaw and front part of skull move in relation to occipital region and cranial base which is called kinesis. Kinesis helps to widen gap and provide shock absorbing effect, when jaws are snapped together. Postmandibular

visceral arches play no part in jaw support, but are incorporated into ear and hyoid apparatus.

3. Teeth are situated along edges of jaws and often on bones of palate. They are conical in shape, may be slightly serrated, or modified to form crushing plates, poison fangs, and other devices. Tooth succession is continuous throughout life except the lizards.

4. Salivary glands are well developed in some. In snakes, sometimes they are modified to poison glands. Poisonous snakes have poison apparatus, which is absent in non-poisonous snakes. It consists of a pair of poison glands, their ducts and a pair of fangs. Poison apparatus is associated with specialized bands of muscles. Situated one on either side of upper jaw, poison glands are possibly the modified superior labial glands or parotid glands. Poison glands secrete poison and serves as a reservoir of venom

5. Three arterial trunks right and left aortae, and pulmonary trunk arise directly from ventricle. Opening of the latter lies opposite right side of ventricle and receives predominantly venous blood. Left systemic arch opens opposite to incomplete ventricular septum and receives mixed blood. Right systemic arch opens from left side of ventricle and carries almost pure arterial blood. Carotid arteries of both sides arise from right systemic arch.

6. Dromedary camel travels more than 500 kms during which they did not get scope to drink. Watered camel shows a minimum of 36°C in early morning and a maximum of 38°C in mid afternoon. When deprived of water, camel reduces evaporative water loss by 64% reduces total daily water loss by half.

7. At the close of Mesozoic, rapid environmental changes in topography due to post Cretaceous mountain building activity and climates due to incoming ice age and consequently reduction in vegetation began to take place. Further mountain building activity resulted in outpouring of lava and dust storms. Dinosaurs fail to adjust and become extinct. Few interesting theories have been forwarded to explain dinosaurian extinction.

8. *Calotes* is often called blood sucker but in reality, it does not suck blood. Its body colour is changeable and in males the head is often blushed red. Colour change is due to contraction or expansion of chromatophores possibly under nervous control .Calotes passes much of its time lying on boughs and twigs and often seen running swiftly on the ground.

9. Squamates and *Sphenodon* have transverse cloacal slits. They show caudal autotomy that is loss of the tail-tip. *Sphenodon* was worldwide in distribution in the Mesozoic and is reported from the middle Triassic deposits of Africa, Europe, Asia and the Americas. Living *Sphenodon* is now restricted to The New Zealand mainly in the 3 North Islands. They are nocturnal, insectivorous and carnivorous. Cold weather adaptations enable them to live on the islands of New Zealand. Two species are *S. punctatus* and *S. guntheri*.

Review Questions

Short Answer Questions

1. Give two primary characters of reptiles.
2. What do you understand by the term "ectotherms"?
3. Give two characters of Anapsida.
4. Give two characters of Chelonia.
5. What do you understand by thermoregulation?
6. Name two muscles associated with the poison apparatus of snakes.
7. What do you understand by solenoglyphous fang?
8. Comment on opisthoglyphous fang.
9. Give two anatomical features of dinosaur.
10. Give two characters of Ornithischia.

Long Answer Questions

1. Classify living reptiles up to order with examples.
2. Describe the mechanism of thermoregulation in camel.
3. Draw and describe the anatomical peculiarities of snake.
4. Describe the poison apparatus and biting mechanism of snakes.
5. Discuss the probable reasons for the extinction of dinosaur.
6. Describe the mechanism of circulation through heart in Calotes.
7. Discuss the anatomical peculiarities and systematic position of *Sphenodon*.
8. Write a note on the conservation of crocodiles.

CHAPTER

9

The Birds

About 10,000 bird species inhabit the globe, from the Arctic to Antarctic. Birds are feathered, winged, bipedal, heat-absorbing vertebrates. They lay hard-shelled eggs. They are characterized by high metabolic rate, 4-chambered heart, and lightweight strong skeleton. Wings are the evolved forelimbs. Flightless birds are ratites, penguins, and several endemic island species. Modern birds possesses beak but they lack teeth. Skin is thin, loose, dry, and lack sweat glands. The only cutaneous gland at base of tail is called the uropygial gland or preen gland. This gland is well developed in aquatic birds. Digestive and respiratory systems are well-adapted for flight. Bird lacks external pinnae. The ear is covered by feathers. In some birds like *Asio*, *Bubo* and *Otus* owls, these feathers form tufts which resemble ears. The inner ear lacks spiral cochlea.

Several birds manufacture and use tools. Many species undertake migration. Many perform short irregular movements. Birds are social, communicate using visual signals and through calls. They exhibit cooperative breeding and hunting, flocking, and mobbing of predators. Most birds are monogamous, some are polygynous, or rarely polyandrous. Eggs usually laid in nest are incubated by the parents. Most birds exhibit parental care. Many birds are used as food. Songbirds and parrots are popular pets. Guano droppings are used as fertilizer. Few species use chemical defenses against predators. Some Procellariiformes eject unpleasant oil against an aggressor. Some species of pitohuis have a neurotoxin in their skin and feathers.

Bird of Paradise has elaborate body covering to impress females. Ninety five percent of bird species are monogamous. Marriage permits for biparental care. Extra-pair mating generally happens between dominant males and females paired with subordinate males. Males of species that have interaction in extra-pair copulations can closely guard their mates to make sure the parentage of the offspring that they raise. Alternative sexual union systems also occur.

9.1 Feathers and Scales

Feathers are characteristic and facilitate flight; provide insulation for thermoregulation, used in display, camouflage, signaling, sexual display, and sensation of touch. Of 3 types of feathers, each serves its own purpose. Down feathers or plumules form the covering of nestling, may found in adult, are simpler than contour feathers or pennae and flight feathers. A third type, the filoplumes are fine, hair-like. Firstly the nestling feathers, then juvenile feathers like prefiloplumes, preplumules, and prepennae and finally the adult feathers (teleoptiles) are produced.

Feather is produced from dermal papilla or follicle, the surface of which produces keratin. In down-feathers, the surface of papilla is ridged to produce fine threads or barbs of keratin to cover the body with a coat of fluff. This coat serves in heat insulation by preventing air circulation. Feathers are moulted at some stage in life-cycle or seasonally. A new generation of feather is produced from the old papillae.

Most birds moult after breeding season. Some moult second time during the year. Down feathers of nestling are partly replaced by contour feathers when the follicle, instead of producing equal barbs, forms 2 large at one side, which together become the central axis for carrying a series of barbs that spread at right angles to it to form vane. Each feather consists of central rachis, forming the hollow calamus or quill below and carry barbs to make vane. Calamus opens at the base by inferior umbilicus, the entrance of mesodermal papilla, and at the beginning of vane a second hole, the superior umbilicus is found. At this point, a loose tuft of barbs or an extra shaft, the after shaft, perhaps represents the down-feather.

Barbs or rami make up the vane, held together by rows of barbules running nearly at right angles to barbs and carry hooks by which barbules of one radius become fixed to grooves in those of next. Such connections can be broken down to separate the barbs, but can be joined by preening the whole feather.

Muscles found at the base of feather control their position. In owls and other night-birds, special vibrissae are present. Specialized feathers are used for eyelashes, ornament, and other purposes. Patches of special feathers without rachis, break up to make a greasy powder down in some birds. Feathers are not uniformly spread over the body rather localized to certain tracts, the pterylae, separated by bare areas, the apteria. Contour feathers are remiges of wing and rectrices of tail. Remiges are divided into primaries on hand and secondaries on forearm. Each large feather is usually covered above and below by several rows of upper and under coverts. In many birds, there is a gap in secondary feathers of wing, the fifth remex feather being absent. The condition in which this feather is present is called eutaxis.

Feathers having a flexible structure can assume various shapes. Shape of quill and barbs varies. Small covert feathers at the front of wing stand up vertically with a right-angle bend to produce the wing camber. Rectrices vary, almost absent in birds like wrens that live near the ground, but are large in fast-moving birds that change direction quickly. In the latter, outer rectrices are enlarged for steering. Rectrices in woodpeckers make a rigid brace, or in peacocks, whose display feathers are tail coverts.

Feathers are epidermal growths attached to the skin and arise only in specific tracts of skin called pterylae. The distribution pattern of these feather tracts is used in taxonomy and systematic. In passerines, the primaries are replaced outward, secondaries inward, and the tail from center outward. Before nesting, the females of most bird species gain a bare brood patch by losing feathers close to belly. The skin of that place is well supplied with blood vessels and helps the bird in incubation.

Bird cleans feathers with beak, using oil from preen gland. Keratin producing powers of skin are mostly used to make feathers.

Scales on legs, feet, and elsewhere are found. Bill and claws the specialized scale like structures is sometimes moult. Nerve-endings are found throughout skin. Cere at the base of bill is perhaps an organ of touch and may have special endings, as corpuscles of Grandry of ducks.

9.2 Food and Feeding

Wherever the food is specific, substitutes are accepted in starvation. Birds quickly learn the way of process of obtaining food. Food choice depends on species-characteristic motor patterns and structures. Young chaffinches or tits peck spots of many sizes, only if they're not hungry. Sometimes they beg food from the parents.

Birds' diets embrace nectar, fruit, plants, seeds, carrion, and varied little animals. Their system is customized to take unmasticated food. Several birds glean for insects, invertebrates, fruit, or seeds. Nectar feeders like hummingbirds, sunbirds, lories, and lorikeets have brushy tongues and in several cases bills designed to suit co-adapted flowers. Flamingos and a few ducks are filter feeders. Geese and dabbling ducks are primarily grazers. Frigate birds, gulls, and skuas, exhibit kleptoparasitism in stealing food from other birds. Kleptoparasitism is believed to be a supplement to food obtained by searching, instead of a big a part of any species' diet.

Other birds are scavengers. Vultures are specialized carrion eaters. Gulls, corvids, or other birds of prey, are opportunists. Toothless birds rely largely on internal processes to break up food. Beak is modified due to food habits. Birds with a moderately long bill like song-thrush eat either flesh or fruit. Seed eating birds like finches have short, thick, strong bills. Bills are large strong in hornbills and toucans. They push through dense foliage to obtain fruit. In parrots, beak is moved on skull, pushed up by upper jaw, when the latter is pulled forward by digastric muscle.

Carnivorous birds like eagles have short and sharp beaks, whereas fish-eating results in long jaws. Flattened bill of some ducks sift out food from water, or mud. The long, thin beak of curlew selects food mostly worms from mud. Lesser flamingos feed on blue-green algae and microscopic phytoplankton, using a sucking mouth and piston-like tongue.

Some insectivorous birds have long beaks for finding their prey under bark. Woodpeckers have a strong beak for excavating in wood, and modifications for purpose of licking up insects; there is an enormously long protusible tongue and special hyoid. Woodpecker finch on Galapagos Islands probes insects from bark by a cactus spine. Most specialized feeders are humming-birds eating nectar in which the beak is long or short according to type of flower visited, and tongue is provided with a special tubular tip.

9.3 Recognition and Social Behaviour

Birds develop specific means for recognizing their fellows, enemies, and competitors; from this power an elaborate social life. Many birds live together in flocks. Species feeding on ground like rooks move about in groups during winter, and the alertness of each single bird serves to warn for many. Lack of procryptic coloration in some social birds is a measure of effectiveness of protection afforded by society. In Starlings, as many as 100,000 individuals may be found in one roost. Birds flying home from their feeding-grounds every night for distances of many miles. Different means are adopted for recognition of other members of species and of same and opposite sex. Bird-life contains elaborate social and sexual rituals. Relief of the one bird by the other at the nest is accompanied by a peculiar wing-flapping ceremony in herons. Greeting ceremonies are common.

9.4 Colour

Vivid colour patterns are used for concealment, recognition, and sexual stimulation. Pigments produce colours by reflection and diffraction effects and vary with bird's habit. Pigments are melanins, ranging from black through brown to yellow, and laid down in feathers by special cells in papilla. The processes of amoeboid chromatophores convey pigment to epidermal cells. Carotenoid pigments like yellow xanthophyll of duck's bill and feet and red astaxanthin of pheasant wattles are also found. White is usually produced by reflection. In blue colour, incident light is reflected from a turbid porous layer overlying a deposit of melanin. In iridescent feather, interference of light in thin surface films produce colours like soap bubbles.

Specialized iridescent feather produces Newton's rings, with colours of second and even third orders. Turacos or plantain-eaters of Africa contain a copper-containing red porphyrin turacin, which is soluble in mild alkaline water and dissolves out in rain, and the green, iron-containing turacoverdin. Concealing coloration is common; the brighter colour serves this purpose by breaking up the outline of bird when at rest or in motion. Most birds are dark above and white below. Feathers often show mottled or speckled patterns. Birds in sunlit upper tree branches show bright yellow, yellow-green, and blue colours, singly or in combination. Thrush living in thickets is dull brown or black. Disruptive coloration is found in white patch on throat of thrush.

Coloration is a compromise between concealment and conspicuousness. Sometimes female is cryptic, the male is conspicuous. In hole-nesting shelducks, both sexes are conspicuous. In other birds, bright colours are concealed most time. Black and white pattern of magpie, which is seldom preyed upon partly for its large size, is sematic coloration. Conspicuous black of rooks and starlings make it easy for birds to follow each other and the group is protected by combined receptors of its many members and quick response of all to escape movements by any one. Protective functions of colour often give place in one or both sexes to communicate between individuals, for pair formation, aggression between males, nest site selection, or rearing the young.

9.5 Skeleton

Skull, jaws, beak, pectoral and pelvic girdles together with limbs constitute the skeleton (Figure 9.1).

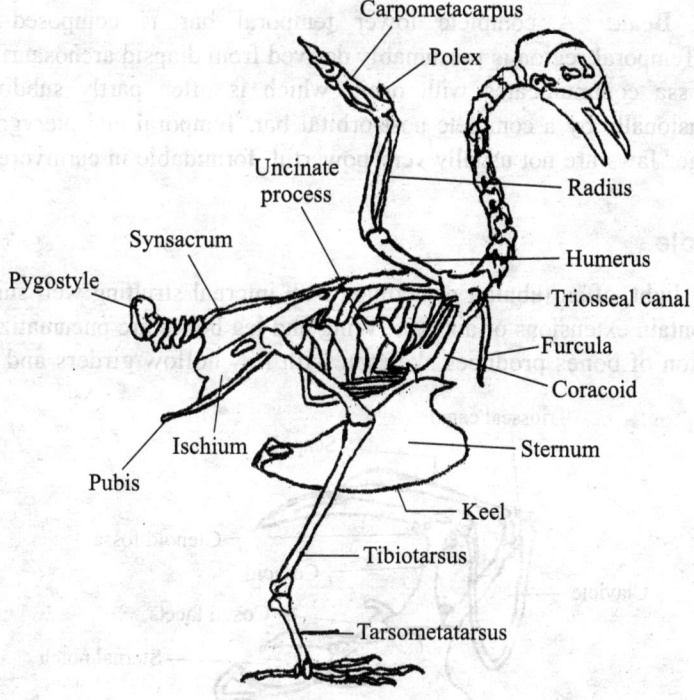

Figure 9.1   Avian skeleton

Skull: Individual bones can be marked in young. But they unite in adult to form a continuous thin-walled structure. Most birds are microsmatic. Nasal passages are simple. Turbinals are reduced. There is seldom a complete bony secondary palate. Internal nostril opens into mouth relatively far forward. The large brain and reduction of olfactory portions cause rounded form of top of head. Large orbits at sides remain separated by ossified septum. Base of skull is formed by a basioccipital behind, carrying a single occipital condyle. A large basisphenoid, covered ventrally by a pair of basitemporals, probably is the parasphenoid, the front part of which makes a basisphenoid rostrum.

Jaws are slender and elongated. Upper part of front of skull is composed of enlarged premaxillae. Nostrils lie very far back. Nasal bones are small. Palatines are long and fused far forward with maxilla; articulate movably behind with pterygoids and base of skull. Pterygoid is a slender rod, movably articulated with skull and with quadrate, which is triangular with clearly separate otic and basal articular processes. Upper jaw, a long thin bar composed of maxillae, quadratojugal, and jugal, exhibits considerable movement. It is raised when lower end of quadrate moves forwards and is well developed in parrots, where beak is freely hinged on skull. Such palatal arrangement is called neognathous. In flightless ratites, palatines are shorter, vomer large, and pterygoids are less movable, a condition called palaeognathous. Lower jaw is elongated and consists of articular bone and 4 membrane bones.

Jaws and Beak: A complete lower temporal bar is composed of jugal and quadratojugal. Temporal region is presumably derived from diapsid archosaurian condition. A single large fossa communicates with orbit, which is often partly subdivided by bony processes; occasionally by a complete post-orbital bar. Temporal and pterygoid muscles are moderately large. Jaws are not usually very powerful, formidable in carnivores.

Pectoral Girdle

Bones are very light, often tubular, sometimes with internal strutting well suited to stresses. Many bones contain extensions of air sacs. Wing and leg bones are pneumatized in fliers like albatross. Fusion of bones produces skeleton with few hollow girders and large plates of

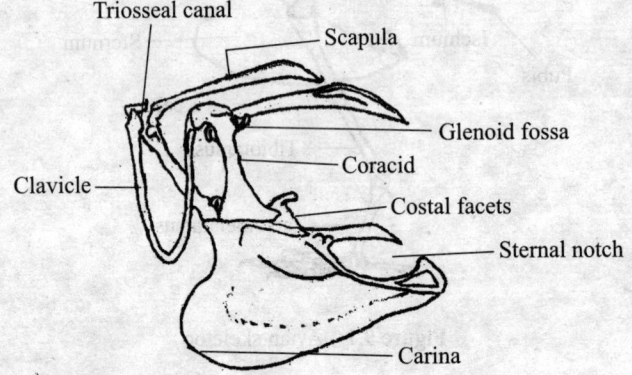

Figure 9.2   Avian Pattern of pectoral girdle.

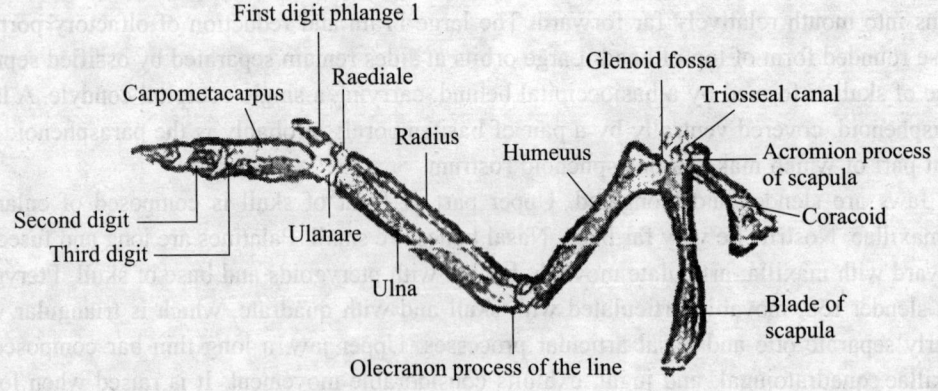

Figure 9.3   Bones of the fore-limb.

special shape. Long bones ossify to from a single diaphysis. Epiphyses are absent at ends. Backbone and limb girdles are modified to carry the body weight on wings, or legs. In Pelicans and frigate birds, small keel is concentrated near the front of breastbone. The wishbone and breastbone are fused together and the keel appears as extension of hypocleideum. The bones are inflated and play role in heat regulation. In Turkey, vultures, the large inflated wishbones have numerous openings for connection to air sacs. Hawks have well-developed wishbones. In some owls, the wishbone is little more than a loop of ossified ligament. Only 3 reduced fingers are present in birds. Only 2 distinct carpals remain in the adult wrist. The other carpals and the metacarpals of the posterior 2 fingers have fused to form carpometacarpus. The most anterior digit can move independently of the other two. Scapula is a narrow blade. Anterior coracoid is a strong strut that braces the shoulder joint. Two clavicles have fused together ventrally to form the wishbone or furcula. Major flight muscles arise from the broad, keeled sternum (Figures 9.2. 9.3).

Pelvic Girdle

Pubis has turned caudally beside the ilium. All the pelvic bones are united. The long ilium is fused with the synsacrum. The midventral pelvic symphysis has been lost. Pelvic canal between 2 halves of girdle is larger, which allows the passage of large eggs with fragile shells. Head of the femur shifts to medial side of proximal end of the bone. Tibia is the primary supporting element and accurately described as a tibiotarsus. Fibula is reduced to a thin splint. The ankle joint is a mesotarsal joint. The fifth toe has been lost. Metatarsals of remaining toes and distal tarsals are fused to form tarsometatarsus.The first toe is turned caudad. At the hip, the femur is held parallel to the ground. The second long bone in leg is the tibiotarsus. The third long bone in leg is the tarsometatarsus. It is unique to birds and consists of the fused remnants of metatarsal bones and some small anklebones. In many birds, the tarsometatarsus extends beyond the skirt of feathers. At its outer tip, the tarsometatarsus meets a fan of toes. Typically there are 4 toes, 3 pointing forward and one back, but sometimes there are 2 forward and 2 back. Some walking and running birds have lost the hind toe (hallux). Ostrich has lost a third toe (Figures 9.4, 9.5).

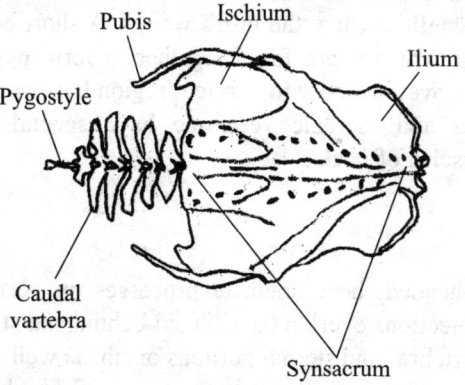

Figure 9.4   Avian pattern of pelvic girdle.

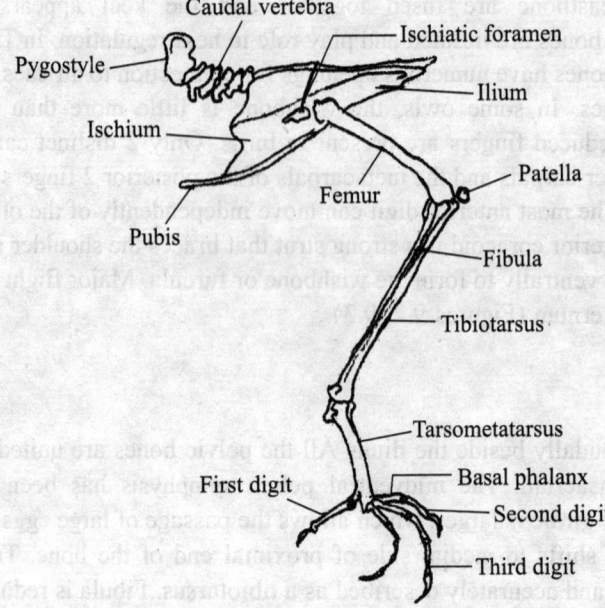

Figure 9.5 Bones of hind limb

Vertebrae

Number of cervical vertebrae is greater in birds with longer necks; 14 in pigeon including 2 that bear ribs but not articulating with sternum. Cervical centra with saddle-shaped surfaces allow great mobility in all directions and exhibits concavity running from side to side on front and up and down behind. Of 4, or 5 thoracic vertebrae, all except the last united into a single mass. The last thoracic vertebra is united with about 5 lumbars, 2 sacrals, and 5 caudals to make a synsacrum, which is fused with ilium. This produces a thin plate-like structure, whose ridged shape gives it strength to carry the bird's weight. A short bony tail consists of about 6 free caudal vertebrae, of which 4 are fused together to form pygostyle. Joints of vertebral column are reduced for movement only in cervical region between thorax and synsacrum and in tail. Hinder cervical and thoracic vertebrae have special ventral hypapophyses for attachment of flexor muscles of neck

Ribs

Ribs are large, double-headed, bear uncinate processes on vertebral portions and join to vertebrae. Hook-like projections overlap the rib next behind and strengthen the whole thoracic cage. A joint between vertebral and sternal portions of ribs is well-marked. The latter are bony and are joined with sternum, a large keeled structure in flying birds, which carry the body weight to wings by attachment of main wing muscles.

Muscles

Femoro-tibial muscles make up with longer muscles, the extensor system of knee. The lateral side of hip joint is supported by small abductor braces, ilio-trochanteric muscles, and act as medial rotators, opposed by obturator and ischio-femoral muscles, which work as lateral rotators. Pectoralis major attached to edge of sternum depresses wing in flight. The depth of keel increase length and mechanical advantage of muscle fibres and strengthen the sternum. Axial muscles are reduced .Other back muscles, except those of tail, are reduced. The whole back forms single rigid strut to carry weight of breast and viscera through ribs and abdominal muscles. Weight of bird in resting is carried by pectoralis major as a tension member,. The main locomotor muscles are posterior braces or retractors, lying behind hip joint and including posterior iliotibial, iliofibular, caudilio-flexorius, pubischio-femoral, ischiofemoral, and caudischiofemoral muscles. Some muscles act with obturator muscle as lateral rotators, and those placed more medially act as adductors or medial braces.

Knee joint is stabilized by lateral, medial, and cruciate ligaments and contains lunate cartilages or menisci. The joint allows movements of flexion and extension.. The intertarsal joint allows movements of flexion and extension. It has a very strong capsule and lateral cruciate ligaments rather like those of knee; even a meniscus on lateral side. The back of tibia is occupied by gastrocnemius and flexor muscles of toes. At the front, a tibialis anterior act across the inter-tarsal joint and as extensors of toes. Calf muscles produce flexion of toes in act of perching and form an elaborate system of tendons attached to phalanges. Such tendons act as single unit.

Flexion is passively maintained by body weight. Many muscles allow support of the joint whether in flexed or extended position. The ilioflbular muscle passes through a sling for this purpose. Flexor muscles of toes are inserted largely above the knee and tend to tighten as the bird sinks with assistance by ambiens. The muscle belly lies on medial side of thigh, and its tendon runs beneath patella to lateral surface of lower leg, where it is attached to upper end of muscles that flex the toes. This arrangement provides a single string crossing hip, knee, and ankle and allows the body weight to flex toes as the joints bend.

The second mechanism for maintaining the bird on its perch is a locking device that holds the toes flexed. The upper side of tendon sheath is ribbed and under surface is ridged at the metatarso-phalangeal joint, where body weight presses it against a branch. As the bird settles on its perch, 2 sets of ridges interlock. Feet show a variety of adaptations. Hopping produces quick movement in small birds on ground and in trees. Webbed feet are used for swimming. Digits may be enclosed in a coat of feathers in birds of cold areas. Birds of prey develop long raptorial talons. In perching birds, one digit is directed backwards, to grasp a branch. In climbing birds, 4th digit is directed backwards as well as the first, and the foot forms a sort of pincer, with long curved claws.

9.6   Wing

Wing is designed to have a minimum moment of inertia about an axis parallel to sagittal plane and passing through shoulder joint. Mucles lying outside the arm or in its proximal part with long tendons produce movements. Wing feathers are carried along post-axial border of humerus, ulna, and hand. Shape of wing depends on position in which feathers are held by their muscles, and on membranes, the pre- and post- patagia developed where the limb joins the body. Pectoral muscles produce active movements. Joints and muscles of wing spread wing, adjust its shape during each beat.

9.7 Wing Muscles

The huge pectoralis major make up as much as 1/5th of whole body weight. It runs from sternum (Figures 9.6a, b) and furcula to under side of humerus, to which it is attached, by a complicated tendon of insertion. Fibres of this muscle are very red in good fliers and contain numerous lipoid inclusions. In fowl, fibres are white and contain glycogen, but little lipoid. Depression of wing is produced mainly by pectoralis major .Elevation of wing is produced by pectoralis minor attached to sternum, lying deep to pectoralis major, but properly supracoracoideus. Its tendon passes through foramen triosseum, between furcula, scapula, and coracoid, to be inserted on upper side of humerus. It is assisted by latissimus dorsi and deltoid muscles (Figure 9.7).

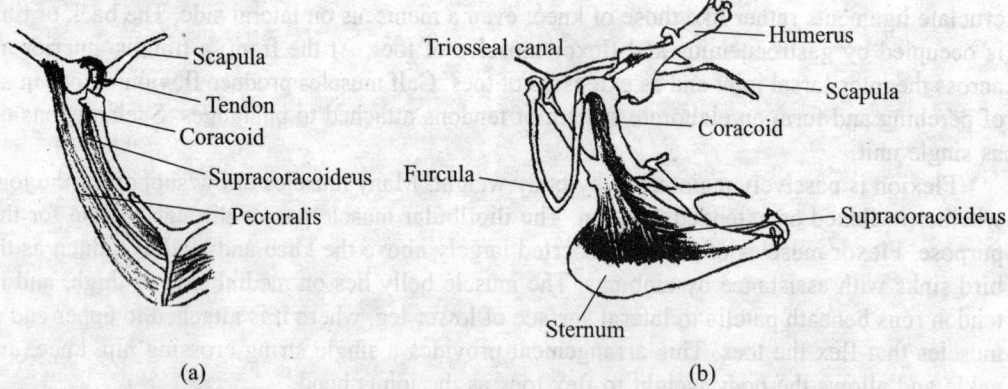

Figure 9.6 **(a)** Flight muscle in section through sternum **(b)** Lateral view.

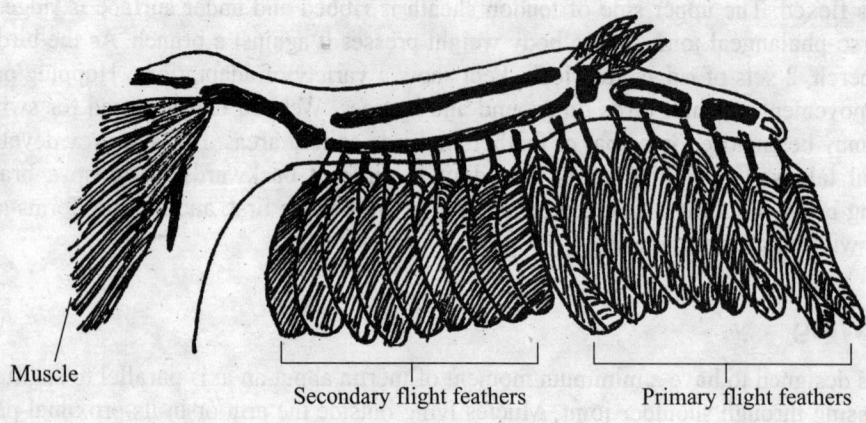

Figure 9.7 Arangment of flight feathers in pigeon.

The chief muscles of shoulder are a massive set to raise and lower the wing. Bird balances on its wings mainly by pectoralis major as the chief brace, between sternum and humerus, with coracoid as a compression member between. Stresses arise in other directions besides those tending to produce a vertical fall and these are met by muscles that produce rotation of humerus and other movements of wing, especially a pronation, depressing the leading edge. Scapula is held to vertebral column by small rhomboid muscles and there is a short series of slips attached to ribs, the serratus anterior. Other muscles running from body to humerus rotate humerus at the glenoid and adjustments of patagia, movements that are very important in flight. A scapulo-humeral muscle produces adduction and lateral rotation of humerus, raising the hinder edge of wing. Coracohumeral muscle produces the opposite effect of abduction and medial rotation, lowering the hinder aspect of wing. Deltoid muscle is divided into several parts.

Muscles known as long and short tensors reach the anterior patagium. Besides a tensor accessorius, run from the surface of biceps to skin of leading edge of wing. Muscles within arm extend or fold whole wing and alter positions of parts, by pronation and supination during flight. Large triceps and smaller biceps muscles act at the elbow. In the forearm a large extensor carpi radialis and an extensor carpi ulnaris keep wing extended at wrist. Flexor carpi ulnaris folds the wing. Two large pronators, brevis and longus, rotate the radius medially and lift the back of wing. Digital flexors and extensors, inserted into distal phalanx of the main digit, keep the wing tip spread out or fold it. Position of individual feathers is controlled by tendons and muscles along back of hand.

9.8 Flight

When the bird is in air, the sternum carries large part of weight and centre of gravity is kept well below the centre of pressure, giving stability. Most birds fly. Flight is the primary means for locomotion and is employed for breeding, feeding, predator rejection and escape. Varied modifications for flight embrace a light-weight skeleton, 2 massive flight muscles, the musculus pectoralis, and supracoracoideus. Wing serve as an aerofoil .Wing form and size governs flight. Several birds mix steam-powered, flap flight with soaring flight. About sixty extant bird species are wingless. Though flightless, penguins use musculature and movements to "fly" through water like auks, shearwaters and dippers.

Principles of Flight: Flight is central to bird evolution. Flight surfaces are composed of feathers. Wing is shaped like an airfoil, thick in front and thin and tapering behind. As the air stream flows across the wing, stream moves faster along longer upper surface than shorter lower surface. A plane surface moved through air in a direction inclined at an angle to this plane is called an aerofoil. Forces generated resolve into a lift force (Figure 9.8) acts upward and a drag force tends to stop the motion.

Both lift and drag forces are proportional to square of speed. Requirement for sustained flight in still air is that the object shall have sufficient speed to generate a lift force equal to its weight. Flow of air over the upper surface of wing reduces pressure there and provides main portion of lift. Tilting increases the wing pressure on underside, but air flow now tends not to follow the upper surface, become turbulent especially at hind edge and destroy the lift.

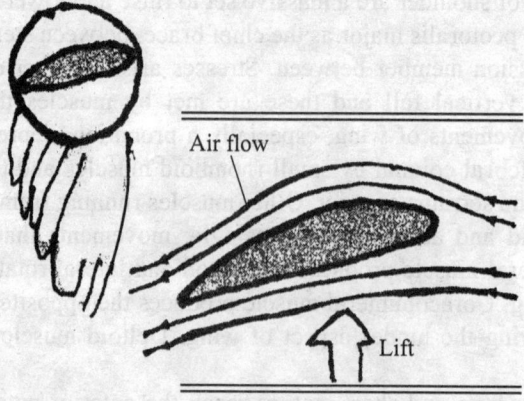

Figure 9.8   Moving air around wing produces lift.

When an aerofoil falls below this critical speed it stalls, drops suddenly, being no longer supported. Smooth flow of air over wing tends to be especially disturbed at its hinder edge and by eddies round the end. Proportion of length to breadth in wing suitable for a particular type of flight depends on need to provide a sufficiently large undisturbed area. Shape of aerofoil determines its aerodynamic capacities. Different shapes of wing enables birds to undertake various flights. First and most obvious is flapping flight which is a screw like motion of wings, providing forward and upward components .In still air, the only alternative to flapping flight is to glide downwards, which cannot continue indefinitely.

Gulls, albatrosses and buzzards, condors soar for many minutes, gain height without flapping the wings. Bird use three types of air movement: ascending currents, usually thermal; variations in wind velocity at any one level (gusts); differences in wind velocity at different levels. The first method is adopted by many soaring land birds. Gustiness of wind is probably turned to advantage by gulls, rooks, and many other birds, and the decrease in wind velocity near sea surface is used by marine soaring birds.

Bernoulli's low states that in a fluid stream pressure is least where velocity is greatest. Differential air speed decreases pressure above wing in relation to underside. This produces a lift force acting perpendicular to plane of wing motion and a drag force parallel to this plane. Both lift and drag forces are proportional to square of speed. Requirement for sustained flight in still air is that the object shall have sufficient speed to generate a lift force equal to its weight.

For flying, lift force must equal force of gravity on bird and a propulsive force must overcome drag force. Somewhat teardrop shape of wing allows a smooth flow of air across surface and minimizes lift-reducing eddies. Some of airflow however/does roll up as a vortex, which is shed from trailing margin and tips of wings as a pair of vortex lines. This often can be seen m high-flying aircrafts as a pair of vapor trails because rapid rotation of air and consequent low pressure in core of vortex lines causes condensation.

In many birds, jump provided by legs is adequate for take off. Heavier birds first acquire speed by running and heaviest birds like condors probably cannot take off at all this way. Eagles are unable to rise without a long run. Many large birds nest on a cliff or tree, which gives them, an up current for take off. Swifts usually come to rest high up and can only rise off the ground with difficulty. Albatross is unable to take off from sea surface in dead calm. Landing is achieved by lowering and fanning out tail. Tail acts as flap, provides both lift and braking. Legs are thin lowered; often one further wing stroke is given to bring the bird forward to drop onto pearch. Breaking adjustment operates through special system of coordination. Rocks make a roll and side slip to ground.

At the take-off, bird acquire sufficient forward momentum to provide lift, and leave the air sufficiently undisturbed for effective subsequent beats. In many smaller birds, jump provided by legs is adequate for take-off. Large birds run or swim rapidly to obtain sufficient speed.

Factors Governing Flight

Wing Area and Loading: Wing shape depends on wing area, aspect ratio (wing length/breadth), wing outline and taper, presence of holes or slots, camber or curvature of wing. A small wing area is necessary for fast flight, at least for high speeds. In flapping flight, wings move relatively fast and for this small size is an advantage.

Large wing area allows slow flight. Loading of wing varies considerably. Since weight increases with cube but wing area with square of linear dimensions. Large birds must have relatively larger wings than small. The larger birds usually have heavier loading of wing, for instance, 10 kg/m^2 in duck, 20 in swan, 1 in gold crest, 3 in crow. A considerable 'safety margin' remains in most birds. Pigeons can fly until as much as 45% of the wing surface was removed.

Fast flying birds like swift have relatively smaller wings than slower flying like hawks, vultures, and storks. Wing loading varies considerably. Since weight increases with the cube but wing area only with to square of linear dimensions, large birds must have relatively larger wings than small. Larger birds usually have a heavier loading of wing, for instance, 10 kg/m2 in the duck (*Anas*), 20 in swan (*Cygnus*), 1 in *Regulus*, 3 in *Corvus*.

Aspect Ratio: It means wing length: wing width. High aspect ratio is found in birds that fly fast by flapping flight (*Swifts*, swallows, albatross) or slide fast to obtain sufficient kinetic energy.

Wings Tips, Slots and Camber: Pointed wing stall first at its tip and is suitable for last flier with hand feathers. Birds built for slower flight have a short, broad wing with long arm feathers. Through slots air moves very rapidly, reduce turbulence and make very high lift. Certain feathers supported by alula can produce a slot through separation at front of wing. Slots may be formed along trailing margin of wing and at wing tip. Latter slots reducing turbulence known as tip vortex. Nearly all wings are cambered, slightly concave on under surface and convex on upper surface. High camber like low aspect ratio reduces speed of bird.

Speed of Flight: Air and ground speed varies. Some may cover 160 km per hour in wind. Racing pigeons can average 64 km per hour for considerable periods. Air speeds of 50 -80 km / hr is found in many. Swifts reach 160 km /hr in still air. Estimation of speed of flight involves distinguishing between air and ground speed. Speed relative to ground may be very high. Birds cover over 100 land miles in an hour. Racing pigeons average 40 miles an hour or more. Air speeds of 30-50 mph can be reached by many birds. Swifts reach 100 mph in still air.

Aspect Ratio: Small wing area reduces drag. Many fast-flying birds have large wing-span. Aerodynamic advantages of this allow low rate of descent when gliding, reducing the expenditure of energy necessary to sustain flight. High aspect ratio is found in fast flying birds by flapping flight and in albatross that glide fast to obtain sufficient kinetic energy to convert into altitude. Wings with very high aspect ratio stall at relatively high speeds. Birds that soar slowly on thermal up-currents over the land mostly have a low aspect ratio.

Some figures for aspect ratios are: Albatross . . .25, Gull . . . .11, Swift . . . .11, Shearwater . . .10, Vulture . . .6, Rook . . . .6, Sparrow 5,

Wing tips, slots, and camber: Pointed wing tends to stall first at its tip and is suitable for fast fliers which show development of hand feathers, producing long narrow wing. Birds built for slower flight and manoeuvre have shorter broader wing with long arm feathers. Condition of the air around wing is important for maintenance of lift. In absence of smooth stream over upper and under surfaces, air becomes turbulent, and aerofoil stalls. This happens when speed falls too low or if angle of wing relative to line of motion increases above 20. Turbulence is mitigated by openings called slots, which let through part of air and provide necessary smooth stream.

Spaces that occur between feathers, towards wing tip function as slots. The arrangement provides a series of apertures to produce an efficient high-lift device. Such slots are conspicuous in slow fliers and in birds that soar on thermal up-currents. Feathers of such birds are individually tapered. Slots are also found in fast fliers large birds, the wing being liable to stall in certain phases of down strokes. It is possible that bastard wing acts as a slotting device.

Shape of wing has very important influence on air stream. In most birds, there is a stiff leading edge and a thinner trailing edge. Nearly all wings are cambered, i.e., they taper from leading to trailing edge, especially in region of forearm. Such arrangement directs air stream over the upper surface of wing to provide an extra lift by creating a 'suction zone' of reduced pressure. High camber, like low aspect ratio, reduces speed of bird.

Kinds of Flight

Gliding or skimming: This is found in shore-birds coming in for landing; in ducks, gulls and herons over water; in swallows and swifts in air; in pigeons gliding or in a falcon swooping upon it quarry. This is the simplest kind of flight. Wings provide lift, and forward motion comes from falling through air. This flight can be exhibited for a short time.

Comment: In some cases before a high wind, bird partly flex the wings and permit itself to be carried by wind. This flight is known as flex gliding.

Soaring or Soiling: Observed in bird; like turkey, vulture or osprey. Birds engage in static soaring have relatively short, broad wings and low aspect ratio. Flight is slow and wings need a large surface area to provide a lift adequate to support the bird. This is called low wing ratio. Additional lift is generated by slotting of wing particularly near tip. Oceaning birds engage in dynamic soaring make use the increase air speed with increasing elevation above ocean surface. They possess long and narrow wings.

Flapping Flight: This flight is complex than gliding or soaring. Here wings are not stationary. Wings are extended, move downwards and forwards during down stroke. They are also inclined from horizontal plane with their leading edge lower than trailing edge. This changes direction of local lift and drag forces acting on each part of wing and gives lift force a and also reduces retarding component of drag force. In this way, wing achieves a propeller effect. Flight feathers on distal part of wings are separated and may act as individual propellers.

On up stroke, wing is flexed and moves upward and backward. This is simply a recovery strokeand generates no useful aerodynamics forces. Other lifting and propelling forces are generated by movements caused by wings. Flapping flight involves a complex, screw-like motion of wing, downwards and forwards then upwards and backwards, rapidly upwards than downwards. Action of wings varies during take-off from sustained flight.

In former, when speed is slow, forward velocity is due to backward movement of wings. During each stroke, beginning with wings raised, they are first moved downward and then forward, giving lift which is produced mainly by pectoralis major. During upstroke, wing is first adducted, folded, and flexed, and supinated at wrist by pectoralis minor and other muscles. Backward flick produced by upward and forward rotation of humerus follows, extension of wing, and pronation of manus. Effect of such movements are produced by triceps. Other extensors provide a forward component. This flight involves mainly the primary feathers. It is very tiring and continues for a few seconds.

In sustained flight, down stroke is as in slow flight but upstroke is simple, with only slight backward flick of primary feathers and inner part of wing provides lift, the tip propulsion. Upstroke in fast flight is passive, produced by pressure of air against under surface. Major effort for lift and forward propulsion is given by pectoralis major. Other muscles give extra lift when needed. Feathers are held by tendons and in some they twist when wing is being raised and barbs are arranged to open like vanes of a blind, when under pressure from above, but close when pressure comes from below.

In gulls and swans, wing is rigid on up and down strokes, twists to produce forward and upward components on up stroke vinculum is maintained straight not following a wavy path. In small birds, wing works nearly as a whole and flight differs in several respects from that of large birds. In general, wing is a very labile system and regulates itself automatically with changes in aerodynamical forces. This regulation is done by feather plasticity, joint mobility, and reflex muscular adjustments. Whirring flight of humming birds enables them to remain almost in same place in air, or even to move backwards. Wings beat backwards and forwards, often as fast as 200 times a second, and pectoralis minor is almost as large as major.

Hovering Flight: This is achieved by holding the body vertically. Humming birds remain in same place in air through such flight. Wings beat backwards and forwards, as fast as 200 times a second.

Soaring Flight: Many birds economize energy for flapping flight by use of possibilities presented by movement of air. All birds glide for short distances, some small birds with wings folded, others with wings outstretched. Sustained gliding and soaring upwards without flapping wings is found in large birds, as considerable weight is necessary for kinetic energy sufficient for continuous flight and efficient use of wind variations. There are 2 distinct types of soaring birds: land birds using thermal up-currents, marine birds using variations in wind above sea level.

Soaring on Up-currents: Up-currents of air arise in neighbourhood of large object on ground and particularly from variations in rate of warming of earth's surface in sun, over rocks, vegetation, mountain shadows. Birds use such currents and proceed upwards in a series of small circles as seen in buzzards and other hawks and vultures, which may ascend in such way above 1,000 ft. Thermal soarers, possess large wing area, low aspect ratio, wings broad at the tip and usually hare well-marked slots.

Use of Vertical Wind Variations: The decreasing effect of surface friction causes wind to blow faster at greater heights which is used by some sea birds. Albatross proceeds in a regular movements, without flapping the wings, downwind losing height and gaining speed and then upwind gaining height and passing into a faster-moving layer. Each downwind tack is longer than upwind one. During upwind tack, wings spread forwards, downwind backwards.

Albatross remains all the time within 50 ft of sea surface, as variations in wind velocity are marked only at low levels. Albatross is suitable for such flight due to large size, great wing span (11 ft), high aspect ratio (25), and pointed wing tips, without slots. Gulls take advantage of variations in horizontal wind velocity, including gusts at a level. Bird moves upwards as it meets an accelerating gust and turns when wind decelerates. Gulls use up- currents at cliff faces and air movements of all sorts are used by large birds. Wing equipment allows bird a limited range of choice and even the slightly different wing shapes of related species depend on various conditions they have to meet. A vulture could no more zoom backwards and forwards over the waves than an albatross could circle slowly on gentle thermal up-current. A pigeon cannot equal a gull at steady gliding and soaring, but can rise more steeply or descend more rapidly without stalling.

9.9 Digestive System

Food in mouth is manipulated by long, thin tongue, moistened with saliva consisting of mucus and diastatic enzyme in seed eating finches. Food swallowed down the esophagus may be stored in crop, found in grain-eating birds. True stomach is divided into two parts, a glandular proventriculus, and a muscular gizzard. Structure of anterior chambers of gut varies with diet. In grain eating birds like pigeon, the crop is large and seeds are first macerated by storage there. They are then mixed with peptic enzymes in proventriculus and ground up in muscular gizzard, which in pigeons has a horny lining and contains numerous small stones.

In insectivores and carnivores, the crop is small or absent, but is very large in some fish-eating birds. In carnivores, the gizzard is comparable to stomach. Herring gulls, normally living on fish, readily take to eating grain and that after a year or so of this diet the gizzard becomes muscular with horny walls. Peptic juice has powerful digestive powers. Many carnivorous and fish-eating birds dissolve the bones of their prey. These are regurgitated with fur or feathers, making characteristic pellets.

Crop of pigeons produces milk to nourish the young. Special glands become active in breeding-season under influence of prolactin. Milk is probably protein. Prolactin causes regression of testes and ovaries and involution of secondary sexual characters, but induces brooding behaviour in female. Duodenum and coiled intestine are relatively short, somewhat long in grain-eating birds. Bile and pancreatic ducts open into distal limb of duodenum. In pigeons, the left bile-duct enters close to pylorus. Food enters the caeca located at junction of rectum and intestine. Caeca perhaps absorb water.

Rectum opens into a coprodaeum which receives an urodaeum, a terminal portion of urinary and genital ducts. A final chamber, proctodaeum opens at anus. Urinary products are solid due to subtraction of water in urodaeum and walls of other chambers. Bursa Fabricii, a blind sac with much lymphoid tissue, opens into proctodaeum and function probably to protect locally against infection and to produce lymphocytes. Hence it is called a cloacal thymus which is prominent in young but reduced in adult. The large surface area, high temperature, and great activity of birds necessitate a high food intake, especially in smaller types. A shrike (*Lanius*) digest a mouse in 3 hours, and hens take only 12-24 hours over the most resistant grain. Amount of food taken per day may reach nearly 30 per cent of body weight (6 g) in the small gold crest (*Regains*) but is about 12 per cent in a starling weighing 75 g (Figure 9.9).

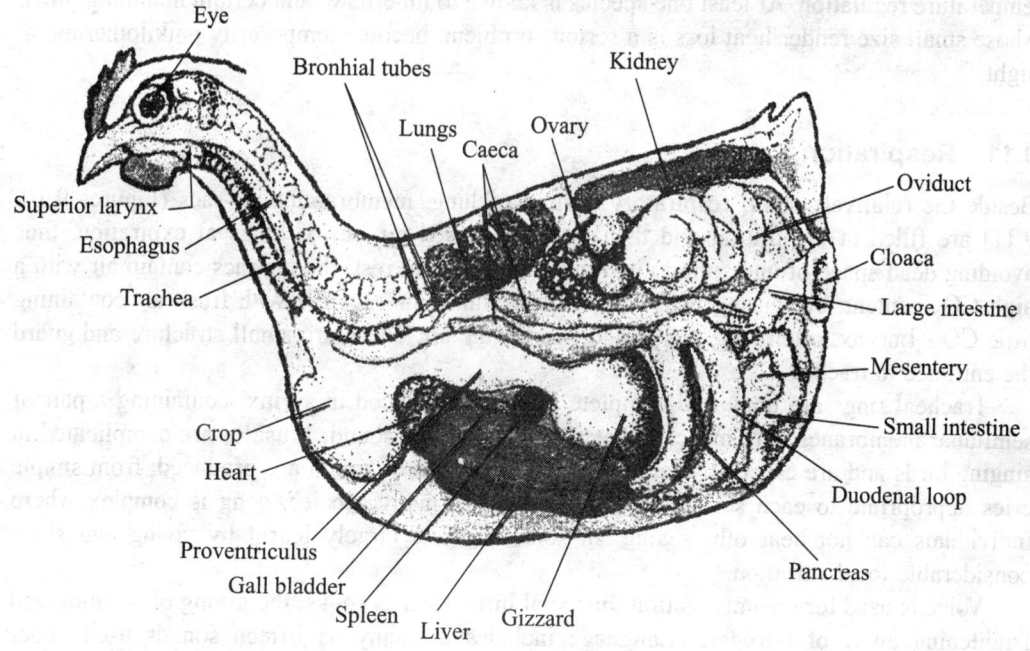

Figure 9.9   Various organ of bird

9.10 Circulatory System

Ventral aorta is split to its base into aortic and pulmonary trunks. The former arising from left ventricle curls round pulmonary trunk to form a single right aortic arch. Heart has lost sinus venosus. Ventricles are large, especially the left. Right auricle and ventricle are separated by a flap-like valve, the left side having valves with chordae tendinae. Enormous innominate arteries supply pectoral muscles. There are renal portal veins. Size of heart and rate of heart beat vary with size and activity of bird, larger birds having relatively small and less rapid hearts.

In turkey, rate of beat may be less than 100 per minute, in a hen about 300, and in a sparrow nearly 500. Red corpuscles are oval and nucleated, smaller in actively flying birds than in larger flightless ratites and carry a large amount of haemoglobin. Haemopoetic tissue is widespread in young, restricted to marrow in adult, but may found in liver and spleen. White corpuscles are numerous including neutrophils laden with crystals, and thrombocytes.

Lymphatic tissue is dispersed rather than aggregated into nodes. A pair of lymph hearts in sacral region of embryo may persist in adult. Basal metabolic rate is high and a temperature usually about 42°C reaches 45°C in some cases. Heat loss is minimized in absence of vascularized extremities, the feet being little more than keratin and collagen. Air-sacs conserve heat by providing an air cushion for viscera, with alternative possibility of losing heat in this way, by ventilation. There is a system of direct arteriovenous connections in feet, and elsewhere. Anastomotic regions have powerful muscles, whose contraction close them and force blood through capillary system. Nervous pathways control upward and downward temperature regulation. At least one species is known to hibernate, and certain humming birds, whose small size render heat loss is a serious problem, become temporarily poikilothermic at night.

9.11 Respiration

Beside the relatively small respiratory portion of lung, membranous air-sacs (Figures 9.10, 9.11) are filled at inspiration and then sweep the used air out of lungs at expiration, thus avoiding dead space of unrespired air. When the bird is at rest, the air-sacs contain air with a high CO_2 content, but during active period abdominal air sacs fill with fresh air containing little CO_2. Larynx, often long and coiled, warms the air. These are small structure and guard the entrance to trachea.

Tracheal rings are bony and complete. Voice is produced in syrinx, containing a pair of semilunar membranes with muscles that alter the pitch of sound. Muscles are complicated in singing birds and are especially large in males. Varieties of sound are produced, from simple cries appropriate to each sex to elaborate songs. In many species, song is complex where individuals can not hear others sing. In some, song is largely learnt by young and show considerable local variation.

Voice is used for communication. In social birds such as rooks, the giving of warning and frightening away of intruders. Language includes as many as fifteen sounds used under different circumstances. Elaborate song of male birds is used in courtship both as a sexual stimulant and as a threat to other birds invading the chosen territory.

Lungs are small spongy organs with little elasticity. Air passes backwards in a large bronchus running through lungs and giving off branches to lung substance, but continuing beyond inspiratory air sacs. Thin walled chambers are divided into two sets, the posterior inspiratory and anterior expiratory. Posterior inspiratory air sacs are abdominal and posterior thoracic are filled by air rushing into them through main bronchus. Anterior or expiratory air sacs include an anterior thoracic, median interclavicular, and cervical. These often communicate with spaces in bones.

At expiration, air passes from posterior sacs through lungs, by special recurrent bronchi, into anterior sacs. From this, air may be expelled to exterior. Return to lungs is prevented by closure of sphincters. In diving birds, air may be passed backwards and forwards through lungs several times until all its oxygen is used (Figure 9.12). Branches of bronchi in lungs do not end blindly in alveoli, but make lung capillaries. Air sweeps through larger channels at

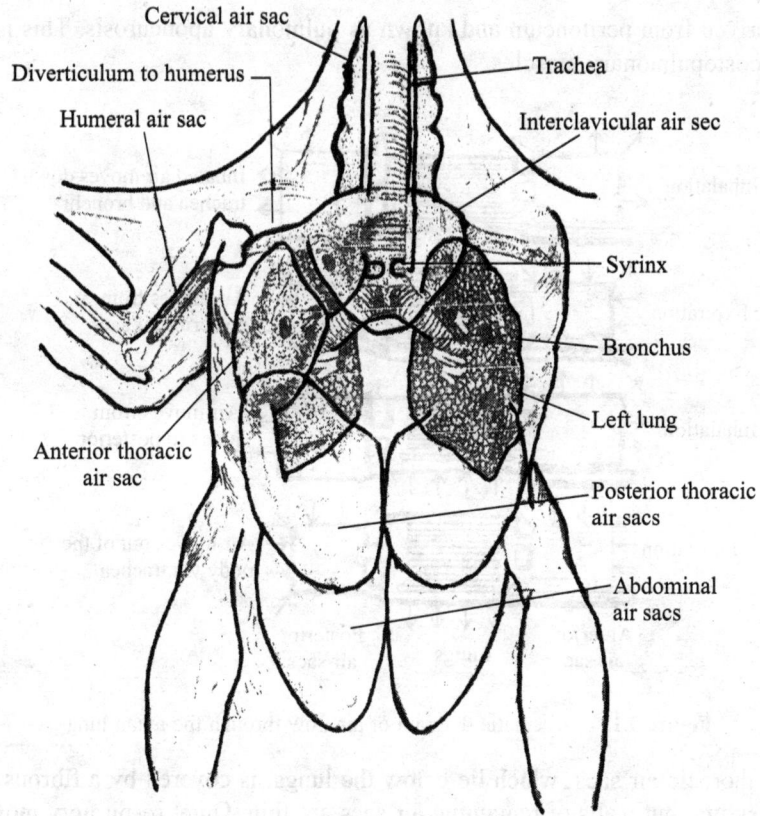

Figure 9.10 Air sacs in relation to other organs.

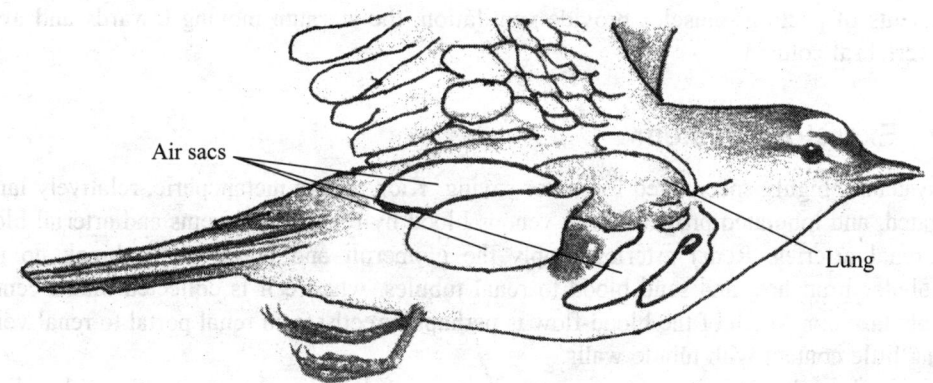

Figure 9.11 Position of lung and air sac.

inspiration and expiration, and probably reaches finer capillaries by diffusion. Mechanism for ventilation production is complicated and depends largely on movements produced during locomotion. The upper surface of lung adheres to ribs. Its lower surface is covered by a special membrane derived from peritoneum and known as pulmonary aponeurosis. This is connected with ribs by costopulmonary muscles.

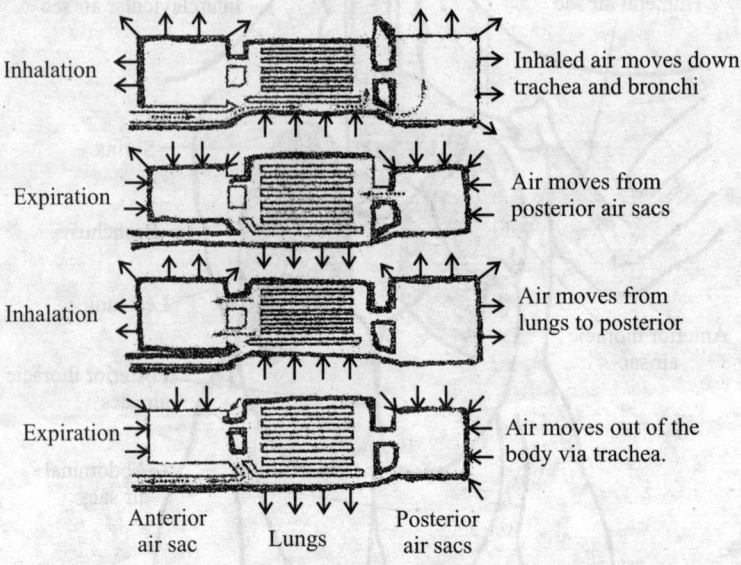

Figure 9.12   Schematic diagram of air flow through the avian lung.

Floor of thoracic air sacs, which lie below the lungs, is covered by a fibrous membrane, the oblique septum, but walls of remaining air sacs are thin. Quiet respiratory movements are produced by intercostal and abdominal muscles which act on thoracic and abdominal cavities to enlarge and contract thorax, drawing air in and out of air sacs through lungs. During flight, movements of pectoral muscles provide ventilation, the sternum moving towards and away from vertebral column.

9.12   Excretory System

The system is highly specialized for water saving. Kidneys are metanephric, relatively large, elongated, and lobulated provided with venous blood by renal portal veins and arterial blood from renal arteries. Renal arteries supply the glomeruli and portal veins, break up into interlobular branches, and send blood to renal tubules, whence it is collected into a central intralobular vein. Much of the blood-flow is perhaps directly from renal portal to renal veins, making little contact with tubule walls.

The end product of nitrogenous metabolism is insoluble uric acid, synthesized in liver; probably from ammonium lactate. Urine is concentrated in cloacal chambers. Uric acid precipitates as whitish granules. Urinary bladder is absent in adult .Glomeruli are numerous. Urinary tubules effect the concentration of urine by long loops of Henle. Viscous fluid that enters urodaeum passes into coprodaeum, where further water is absorbed. Mixed faeces and urinary products are excreted. Some desert birds survive for many weeks without water.

9.13 Reproductive System

Testis consists of coiled tubules which join to form a long epididymis. Vas deferens opens into urodaeum by an erectile papilla which is the copulatory organ. During copulation proctodaea of male and female are everted and pressed together. Sperm is ejaculated direct into female urodaeum and finds its way up the oviduct. A definite penis is found in ratites, anseriformes, and few other birds. Weight of the testes is 1,000 times greater in breeding season than in non-breeding.

Right ovary remains present as a rudiment. Complete sex reversal can occur in some races of domestic fowl, and transformed bird may acquire cock plumage and tread and fertilize hens. Sex reversal rarely takes place in opposite direction. Of a number of oocytes only few ripen to make the enormous follicles. After bursting, each follicle quickly regresses. Corpus luteum is absent.

Egg is taken up by ciliated and muscular funnel of left oviduct, and passes down a tube with circular and longitudinal muscles and a glandular, ciliated mucosa. Albumen of egg is produced by long tubular glands, opening to lumen. Oviduct has various parts, the upper part secrets mainly albumen, the lower part produces the shell, and the lowest part the mucus, to assist laying. After ovulation, oestrogen level falls, calcium is mobilized from bones, and its concentration in blood becomes very high, until used by the eggs.

9.14 Brain

Brain is larger relative to body .High temperature allow for an elaborate nervous organization and complicated behaviour. Forebrain is especially large in rooks, crows, and in parrots. They show signs of outstanding intelligence. In spinal cord, the relatively small size of dorsal funiculi and their small nuclei in medulla are remarkable. Movement of feathers provides impulses leading to reflex actions, although the loose covering does not allow elaborate organization of sense of touch. The finer senses are restricted to eyes, ears, and bill.

Large spinocerebellar tracts are perhaps proprioceptive and associated with flight adjustments. Spinal cord is controlled by large efferent tracts from brain including cerebellospinal, vestibulospinal, and tectospinal pathways. There is no direct tract from forebrain to spinal cord, but influence of large corpora striata is exercised through fibres running to red nucleus and tegmentum of midbrain, from which others pass to cord. Cerebellum is large.

Besides large spinocerebellar and vestibulocerebellar pathways, tectocerebellar and striocerebellar tracts exist, the latter perhaps conduct in both directions. Effect of cerebellum on other brain parts is exercised through cerebellar nuclei, cells of which give origin to cerebellospinal tract. Optic tracts are completely crossed and end in midbrain. A considerable portion of optic tracts passes to thalamus. Midbrain and thalamus are highly developed and have intimate and reciprocal connexions with striata of cerebral hemispheres.

Optic lobes receive ascending fibres from trigeminal nuclei and spinal cord. Their efferent pathways run to oculomotor nuclei, to underlying tegmentum, medulla and spinal cord. They play a large part in correlating visual and other afferent impulses. Thalamus is large and its dorsal part is well differentiated into nuclei. It receives optic fibres, and projections from tactile, pain, temperature, and perhaps auditory sources. Large thalamostriatal tracts conduct in both directions.

Ventral thalami receive impulses from striatum and send them to tegmentum, the main efferent pathway of forebrain. Hypothalamus is rather small, because of reduction in olfactory system. Cerebral hemispheres are larger than any other part of the brain. Ventro-lateral portions are enormously developed, whereas medial ventral walls are thin. Pallium is small, thin, and not folded. Olfactory regions of brain are small, including hippocampus.

Corpus striatum is a huge solid mass of tissue, receiving projections forward from thalamus and sending them back through latter, to midbrain roof and floor, to cerebellum, and thence to medulla and spinal cord. The striatum can be divided into various regions. The part representing the original or lower striatum is called paleostriatum; other parts lying above this are known as mesostriatum and hyperstriatum. Corpora striata do not control individual muscle movements. The striata are silent areas to stimulation.

Complete removal of both hemispheres does not reduce a pigeon to a helpless state. It may remain inert for long periods, and then become aimlessly restless for a while. Forebrain is larger in many other birds than in pigeon. The large masses of nerve-cells in striata are concerned in some way with elaboration of more complex behaviour. Birds show more stereotyped patterns of instinctive behaviour. Once they have embarked on a line of action, even a complex one like nest-building, they are supposed to pursue it in a given manner, without ability to adapts themselves to unusual happenings.

Birds frightened or disturbed may proceed to the actions of bathing, preening, feeding, or drinking, performed in a ritual and cursory manner for a long time. Such displacement activities show that the organization of bird's nervous system provides for some strange deviations, whose study may reveal much about the method of working of brain.

Much complex behaviour is responses to only limited parts of natural stimulus situation. Much of elaborate social life of birds depends on such sign stimuli displayed by one bird and serving as releasers setting off particular actions or trains of action in another bird. Many elaborate forms of display evolved by birds are releasers of this sort, and structures and actions on part of young release the appropriate behaviour of parent. The red breast of the robin is the agent that releases attacks by other birds.

9.15 Eyes

Birds depend more on their eyes than on other senses. They are fully visual. Eyes are extremely large. The shape is not spherical, lens and cornea bulge forwards in front of posterior chamber, this form being maintained by a ring of bony sclerotic plates. In most birds, the whole eye is broader than it is deep, but in those with very acute sight it is longer, and in some eagles and crows becomes almost tubular. The great distance between lens and retina broaden image, improve fine two-point discrimination needed by diurnal birds.

Shape of back of eye is such that the retina lies almost wholly in image plane, so that all distant objects within visual angle are sharply focused on photosensitive cells. Lens is soft and accommodation is effected by changing its shape and especially the curvature of its anterior surface by pressure on it of ciliary muscles behind. These, like the iris muscles, are striated, presumably allows for quick accommodation necessary in a rapidly moving bird.

Ciliary muscle is divided into 'anterior' and 'posterior' portions, muscles of Crampton and Briicke. The latter draws the lens forward into anterior chamber and the shape of eye is fixed by sclerotic plates, the lens becomes more curved and accommodated for near vision; contraction of iris sphincter assists in process. Crampton's muscle is arranged to pull on cornea, shortening its radius and assisting in accommodation. This double method of active accommodation for near vision is fully developed in diurnal predators, less so in night-birds.

In aquatic birds, Crampton's muscle is reduced, and cornea has little importance in image-formation. Special arrangements are found in diving birds. In the cormorants Briicke's muscle is large and there is a very powerful iris muscle, which assists ciliary muscles to give the change in shape of soft lens, allowing accommodation of 40-50 diopters .Kingfishers possess an amazing arrangement of double foveas, placed at different distances from lens, so that as the bird dives under water the image is transferred from one fovea to other without any change in dioptric apparatus.

Retina of day-birds consists largely but not wholly of cones. These animals are more fully diurnal. The high resolving power and hence high powers of discrimination and of movement detection depend on density of cones, as many as 1 million in each square millimetre in fovea of a hawk. Nocturnal birds have retinas composed mainly or completely of rods. One or more regions of retina usually consist of tightly packed receptors. In marine birds, the area often has the form of an elongated horizontal band whereas in tree living birds, it is circular.

Some birds have 2areas, a central one in optic axis and a second placed on temporal surface of eye, so that the image of objects in front of head falls on temporal area of both eyes. Density of cones is so high in diurnal birds, even outside the area, that they probably obtain a good detailed picture in all directions. They do not scan the world with central area of retina. The eyes move relatively little, but the bird detects very small movements anywhere in its surroundings.

Bird's-eye view usually lacks stereoscopic solidity and it is possible that in compensation for this, the animals appreciate distance by movements of the intrinsic eye-muscles. There is, often within the central area of eye, a fovea or pit. In many birds, sides of this pit are steeply curved. Since the vitreous humour and retina differ in refractive index, this curvature magnify the image and increase acuity. Foveas with steep sides are found in birds of prey, kingfishers.

Some birds have one convexiclivate and one flatter fovea, the latter being on temporal surface of retina and used in binocular vision. Birds discriminate colours apparently on trichromatic basis. Cones of birds often contain red and yellow droplets, which may heighten visual acuity by reducing effects of chromatic aberration. Droplets in central area are always yellow. Presence of droplets of various colours in adjacent cones increases powers of discrimination. Sometimes the droplets are arranged to allow accentuation of different contrasts in parts of visual field.

The lower part of pigeon's retina contains red, the upper yellow filters, increasing the contrast of blues and greens respectively, as required for vision against the sky in one case and ground in other. Although eyes of some birds are directed forwards, so that their fields overlap, they are said not to have binocular vision and decussation of optic tracts is complete.

Perception of distance must be performed in other manner. In many birds, eyes are directed sideways, and fields of view overlap behind the head. This may serve to give warning of predators. Pecten, a pleated highly vascular fold, projects from retina into vitreous. Irregular shadow cast by this organ provides, as it were, numerous small blind spots and hence by a stereoscopic action increases number of on-and-off effects produced by a small object in visual field, increasing contrast and allowing detection of its movement.

Original function of pecten was to bring nourishment to vitreous and retina. Pecten is connected with accommodation; it is not likely that it actually assists in focusing, by pressing forward lens, and no changes have been seen in it during accommodation. It might possibly assist by adjusting the intraocular pressure, which must be increased by extensive changes in lens during accommodation.

9.16 Ear

Both vestibular and auditory parts of ear are well developed. The former are not known to possess special peculiarities. There is a distinct cochlea, slightly curved and especially well developed in owls and in parrots. In this, there is a basilar membrane, with fibres increasing in length towards tip and carrying an organ of Corti with hair cells in contact with a tectorial membrane.

At the tip of cochlea, a special sensory region, the lagena perhaps responsible for reaction to lower notes, the basilar membrane respond to higher frequencies. Birds are more sensitive to distant gunfire and other low-frequency vibrations. Transmission of vibration from tympanum to inner ear is effected by columella auris, derived from cartilages of hyoid arch.

The inner portion of columella is rod-like (stapes), but outer end makes contact with tympanum by means of three processes. Hearing is acute and song-birds discriminate between simple tunes; some of them are surprisingly good mimics. Ability to localize sound is high in owls and other night birds probably find their prey largely by ear. They have developed an asymmetrical arrangement of ear cavities or asymmetrical external ears. A few birds that live in caves have power of avoiding obstacles by echolocation. They emit up to 5 to 6 clicks a second. Rate varies inversely with amount of light and increases when obstacles are met.

9.17 Other Receptors

Corpuscles of Grandry in bill of ducks are probably touch receptors. The corpuscles of Herbst, found in dermis elsewhere in body, resemble Pacinian corpuscles. They may be receptors for vibration and are numerous in certain situations, for example in feather follicles, beak, between the tibia and fibula, and in tip of tongue of a woodpecker. Chemoreceptors for taste and smell are little developed. There are few taste-buds on tongue. Nasal cavity is large but olfactory epithelium is restricted. Whether most birds use nose as a distance receptor or to test air coming from the internal nostril is not certain. In kiwis, which are nocturnal and terrestrial, the olfactory sense is well developed.

9.18 Breeding-habits

Bird of Paradise has elaborate breeding plumage to impress females. Ninety-five percent of bird species are monogamous. They pair for at least the length of the breeding season or for several years or until the death of one mate. Monogamy allows for biparental care, which is especially important for species in which females require males' assistance for successful brood-rearing.Extra-pair copulation is common .Such behaviour typically occurs between dominant males and females paired with subordinate males, but may also be the result of forced copulation in ducks and other anatids. For females, possible benefits of extra-pair copulation include getting better genes for her offspring and insuring against the possibility of infertility in her mate.

Males of species that engage in extra-pajr copulations closely guard their mates to ensure the parentage of the offspring that they raise. Other mating systems, including polygyny, polyandry, polygamy, polygynandry, and promiscuity, also occur. One hundred bird species, including honeyguides, icterids, estrildid finches and ducks, are obligate parasites, though the most famous are the cuckoos. Some brood parasites are adapted to hatch before their host's young, which allows them to destroy the host's eggs by pushing them out of the nest or to kill the host's chicks; this ensures that all food brought to the nest will be fed to the parasitic chicks.

In birds, the eggs and young cannot be left cold for long and it is desirable that the father should help. In birds, the breeding-habits involve development of elaborate systems of mutual relations to bring and keep the parents together throughout period of incubation and while feeding the young. Building of nest may be intricate in which both birds collaborate and the pair occupies a territory around the nest, which they defend against others of same species. Type of association of the sexes varies greatly.

In ruff and certain game birds, there is no pair formation: display and copulation occur at communal display grounds. Wren and a few other birds are polygamous, each male forming continuous association with several females. Most birds form pairs throughout a single season; occasionally they change mates for the second brood. The same pair may mate in successive years and a few birds stay together through the year. Breeding is nearly always seasonal, even in tropics where conditions are apparently almost uniform throughout the year.

In temperate latitudes, breeding begins in spring as gonads develop probably under the influence of increasing illumination. Changes in behaviour with ripening of gonads vary with species. Birds that have been social through the winter begin to leave their flocks, and voice of male changes from simple winter notes to more complex breeding song. Production of secondary sexual characters of plumage and other features used in display is partly by direct genetic effects and, partly through hormones.

In most birds, there is a breeding-season, initiated, by the effect of increasing length of day in spring, acting through pituitary on gonads. Degree of activity and food taken play their parts, especially near equator, where there is little seasonal variation. Song is one feature of display and courtship which has different functions from species to species. In the simplest case, display bring the sexes together, enable recognition, and at a later stage as a stimulus to copulation.

In some birds, display serves as part of stimulus to ovulation. Aggressive displays are different from courtship displays. When female does not flee or fight back, as a male would do, he gradually changes over to courtship display. Finally, some forms of courtship, especially those that are mutual, keep the partners together for incubation and feeding. In courtship, three elements, first sexual stimulation, secondly threat to other males, and thirdly mutual stimulation while rearing a family are recognized There are also begging displays, given by young to parents, various displays given to potential predators, and others.

9.19 Origin of Birds

Some scientists believe that the ancestors of all birds evolved between 65 and 53 million years past, independently of dinosaurs. However, the dinosaur-to-birds theory took surprising flip with the invention of 2 species of feathered dinosaurs in China. Birds have exploited their quality to full and acquire means of life in varied ways. A swallow might pass a part of its life in tropics, half close to the polar circle. A gull might nest on rock; eat grain during field, and fish in ocean, all at intervals of hours. Bird's show 2 options namely, freedom to maneuver to totally different conditions and living in unfortunate circumstances. Every species nests during a restricted sort of habitats. Birds may be the relatives of the deinonychosaurs that embrace dromaeosaurids, troodontids and probably archaeopterygids. *Archaeopteryx lithographica* displays each clearly reptilian characteristic: teeth, clawed fingers, and a long, lizard-like tail, and wings with flight feathers a dead ringer for those of recent birds. Well developed pectoral muscles hooked up to furculum power the flap motion of wings. Long feathers of wings act as airfoils, to assist generate elevate for flight. Birds have an outsized, four-chambered heart that proportionately weighs half-dozen times over an individual's heart. This combined with a fast heart-beat satisfies rigorous metabolic demands of flight. Lungs of birds stay inflated in any respect times, with air sacs acting as bellows, to produce lungs with a continuing supply of contemporary air. In North America, most migration routes square measure adjusted north-south for 2 reasons.Pterosaurs were among the first vertebrates in the air. *Archaeopteryx* the oldest known bird fossil was possibly the most interesting prehistoric remain ever dug up.

Based on evidence, most scientists agree that birds are a specialized subgroup of theropod dinosaurs. Birds are perhaps the closest relatives of the deinonychosaurs, which include dromaeosaurids, troodontids and possibly archaeopterygids. Together, they form the Paraves. Unlike *Archaeopteryx* and the feathered dinosaurs, which primarily ate meat, the first birds were herbivores. *Archaeopteryx* displays both clearly reptilian characteristics: teeth, clawed fingers, and a long, lizard-like tail, and wings with flight feathers identical to those of modern birds. It is not considered a direct ancestor of modern birds, though it is possibly closely related to the real ancestor. Many features of birds show close resemblance to those of reptiles and in particular to archosaurian diapsids.

In the early Triassic period, the small pseudosuchians such as *Euparkeria* showed the essential characteristics of bird, especially those associated with a bipedal habit. From such form, the birds have certainly been derived, by a series of changes parallel in many cases to those found in other descendants of the pseudosuchians like crocodiles, dinosaurs, and pterosaurs. Jurassic birds and the origin of flight in *Archaeopteryx* had achieved some powers of flight, but they were less specialized than modern birds. The whole body axis was still elongated and lizard-like. Vertebrae articulated by simple concave facets, without saddle-shaped articular facets of centrum. Dorsal vertebrae were not fixed and only about 5 made the sacrum. There was a long tail, with feathers arranged in parallel rows along its sides. Fore-limb ended in 3 clawed digits, with separate metacarpals and phalanges, the hallux being opposable. Limb was used as wing, for the fossils show feathers on back of ulna and hand, but wing area was small and the shape was rounded, like that of bird's wing that fly for short distances only. There was a furculum and a small sternum. Ribs were slender and had no uncinate processes. Pelvic girdle and hind limb resembled those of archosaurs, with elongated ilium and backwardly directed pubis. Only 6 vertebrae were fused to form sacrum. Fibula was complete and proximal tarsals were free, but the distal ones were united with metatarsals.

In skull of *Archaeopteryx*, there were teeth in both jaws. Shape was more reptilian than bird-like, with rather small eyes and brain. Premaxillae and frontals were much smaller than in modern birds. There was a large vacuity in front of eye and probably there were post-frontal and post-orbital bones. Brain-case was large and many bones were united but not pneumatized. Cerebral hemispheres were elongated and cerebellum was small. These fossils suggest that birds arose from a race of bipedal arboreal reptiles, living in forests and accustomed to running, jumping, and gliding among the branches.

Fossil Birds: These Jurassic fossils are distinct from other birds and are placed in subclass Archaeornithes. *Hesperonis* probably possessed teeth. Two aquatic birds are placed in a superorder Odontognathae. The former was a diver that had lost power of flight. Some birds had lost the teeth in Cretaceous and can be referred to orders found alive today. Birds are not commonly found as fossils.

Flightless Birds: Flightless birds or ratites like ostrich, cassowary, and kiwi, with reduced wings and no sternal keel, long legs and curly feathers, have previously placed in a distinct group. They diverged early from ancestral avian stock and never passed through a flying stage. Arrangement of palate bones manifests neoteny and do not indicate a truly primitive condition. Some neognathous birds pass through a palaeognathous stage during development.

Ratites have been descended from flying birds and represent several different evolutionary lines. Various ratite birds have been placed in eight distinct orders. Ostriches now limited to Mesopotamia are the largest living birds. Rhea is limited in South America, the emu and cassowary in Australasia. Several species of moas lived in New Zealand until recent times.

Elephant birds were similar, with several species in Madagascar in Pleistocene. Some were larger than ostriches, with eggs estimated to weigh more than 10 kg. Kiwis of New Zealand are smaller, terrestrial birds, nocturnal and insectivorous or worm-eating, with a long beak and small eyes. Sense of smell and the parts of brain related to it are better developed than in other birds. Palate shows large basipterygoid processes. There is a penis.

Superorder Impennae: Penguins have lost the power of flight and became specialized for aquatic life with a common ancestry with the petrels. They swim chiefly by means of fore-limbs, modified into flippers; the feet are webbed. Penguins are confined to southern hemisphere. They come ashore to breed; many make no nests, but sometimes carry the one or two eggs on the feet throughout the incubation period. The emperor penguin breeds in winter on Antarctic ice and is only bird that never comes on land. Egg is supported on the feet.

9.20 Classification

CLASS AVES
General Characters
1. Presence of the feather.

2. Feathers have 2 primary functions. Secondary functions include display, sensory, protection.
3. Bipedal.
4. Bones are thin, hollow.
5. Loss of teeth & heavy jaws are replaced by horny beak
6. Birds are endothermic--maintain a constant body temperature.
7. Air sacs connected to lungs cools the animals, especially during flight. These are well developed in good fliers.
8. Presence of rhamphotheca over beak.
9. Clawed digits on foot.
10. Forelimbs are modified to form wings.
11. Posterior caudal vertebrae form a pygostyle.
12. Single occipital condyle.
13. Teeth absent in Tertiary and Recent birds.
14. Esophagus is dilated into a crop.
15. Stomach is divided into proventriculus and gizzard.
16. A pair of caeca is present.
17. Lungs are spongy, non-distensible.
18. Air sacs are present.
19. Syrinx is present.
20. Fur chambered heart.
21. R B C is oval and nucleated.
22. Urinary bladder is absent.
23. All birds are oviparous - amniote egg
24. Sophisticated behavior is associated with courtship, mating, nest construction, and raising of the young
25. Excrete uric acid which requires less H_2O for storage than ammonia or urea
26. Well developed brain, relatively intelligent
27. Well developed optic lobes. Sight is the most important sense for birds.

Classification followed here is according to Young (1981).

Subclass: Archaeornithes

1. Pygostyle was absent.
2. A tail of many vertebrae was present.
3. Enamel crowned teeth was present.
4. Bones had no air spaces.
5. Hallux was small and opposable.

Example: *Archaeopteryx*.

Subclass: Neornithes

1. Tail is shortened.

2. Teeth are absent with few exceptions.
3. Metacarpals fuse with carpal.
4. Claws are absent in manus.
5. Sternum is keeled.

Super-order: Odontognathae

1. Teeth were present.
2. Possessed a brain more avian than reptilian,

Example: *Hesperornis, Enaltornis, Baptornts*

Super-order: Palaeognathae

1. Usually flightless.
2. Pygostyle is small, undeveloped.
3. Sternal keel is vestigial or absent.
4. Coracoid and scapula are completely anklyosed.
5. Wing is reduced or absent.
6. Old jawed condition (palaeognathous).
7. Cerebellum is developed.

Example: Struthio, Rhea, Casuarius.

Super-order: Impennae

1. Air sacs are absent.
2. Plumage holds little air.
3. Wing bones form paddle.
4. Hind limbs are modified.
5. Feet are strongly webbed.
6. Body is streamlined.

Example: Aptenodytes.

Superorder: Neognathae

1. New jaw arrangement.
2. Palatines extend posteriorly.
3. Vomers are brief.
4. Pterygoid is short.

Example: *Columba, Geopsittacus, Phalaropus.*

Characters of *Archaeopteryx*

1. Teeth were present.

2. Clawed wing. Only living bird with this character is the hoatzin in which the young birds use them for climbing.
3. Hands were reduced to 3 digits.
3. Caudal vertebrae were (tail) present.
4. Keel on sternum was not large.
5. Solid bones were present.
6. Furcula was present, as in modern birds. (Figures 9.13a-r)

Modern Birds

These birds have characteristic palate, sternum, and are placed in superorder Neognathae. Birds of this type probably existed in Cretaceous. Existing birds show great variety in details of structure and habits. Classification of vast number of genera involves recognition of over forty distinct orders and even then one of the orders, the Passeriformes, contains about half of all species.

Order Gaviiformes: Divers are aquatic birds retaining some primitive characteristics. They are found on open waters, feeding mainly on fishes. *Gavia* live mostly on sea, but breed by lakes throughout the holarctic region.

Order Columbiformes: Aquatic birds, almost unable to walk on land, resemble the divers in some ways. They nest on lakes, laying a small number of white eggs in a floating nest.

Order Procellariiformes: Petrels, shearwaters, and albatrosses are highly modified for oceanic pelagic life. Some are very large, lay one white egg, often in burrows. Their long narrow wings are specialized for soaring flight.

Order Pelecaniformes: Aquatic, modified for diving and fishing and including cormorants, pelicans, and gannets. They nest in colonies on rocks or trees, make spectacular dives when fishing. Gannets may plunge from more than 50 feet. Eggs are usually unspotted and covered with a rough chalky substance.

Order Ciconiiformes: Storks, herons, and flamingoes are large, long-legged, living mostly in marshes and feeding mainly on fish. They are strong flyers. Some perform extensive migrations. Nests are usually in colonies and may be used year after year. Eggs are few and unspotted.

Order Anseriformes: Ducks and swans are specialized for aquatic life. Flattened bill is used to feed on various diets. Some are vegetarians, a few filter- feeders. Some eat molluscs, others fish. Numerous eggs are white or pale. Nest is built on ground.

Order Falconiformes: The birds of prey that hunt by day have sharp, strong, curved bills and powerful feet and claws. Retina contains mainly cones. Many types are found. Most feed on birds or mammals, some on carrion, and a few on fish or reptiles. Typical examples are kestrel, eagle, buzzard, and vulture. Eggs are few in number, usually spotted and the nests are generally made on cliffs, tree-tops, or other inaccessible places; some are on ground.

Order Galliformes: Mainly terrestrial, grain-eating birds, capable only of short and rapid flights. Some of their structural characters and habits are primitive. Palate differs from both that of ratites and of most modern birds. There is often a marked difference in plumage, and sometimes in size, between the sexes. Nest usually made on ground is simple.

(a) *Ardea*

(b) *Pavo*

(c) *Anser*

(d) *Bubo*

(e) *Psittacula*

(f) *Aply*

(g) *Trogon*

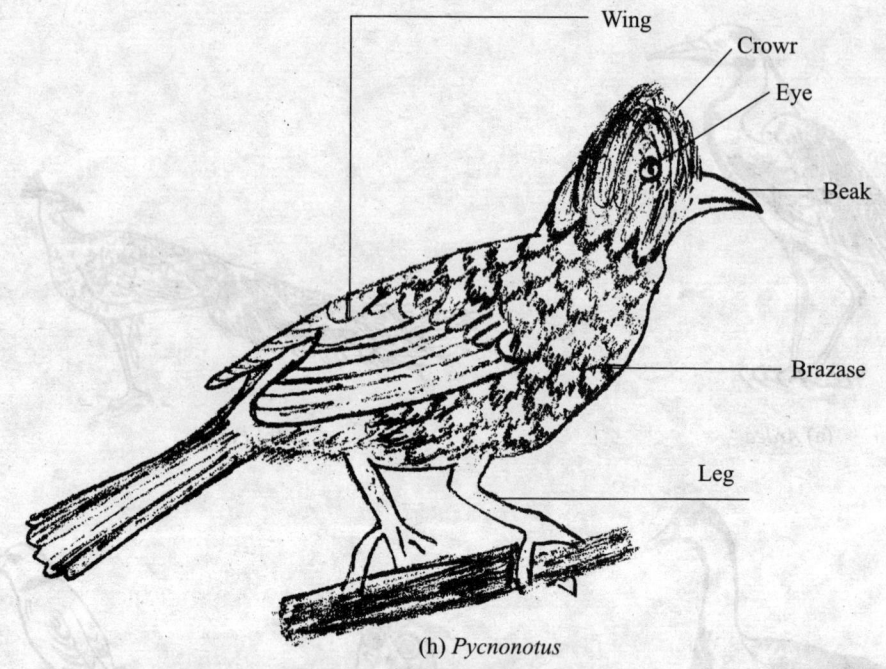

(h) *Pycnonotus*

(i) *Passer*

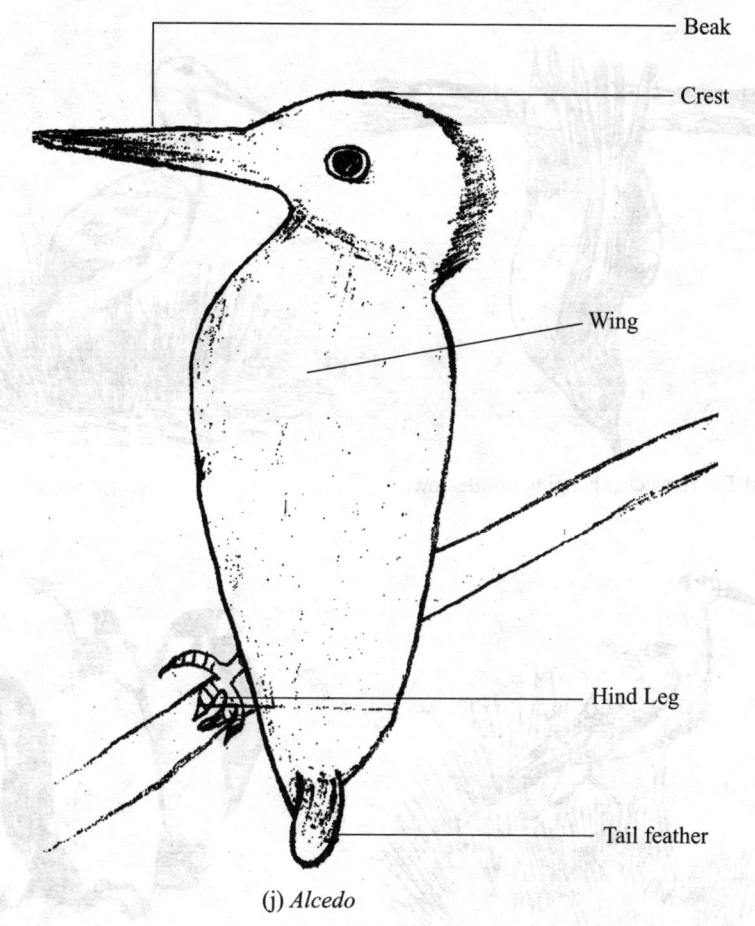

(j) *Alcedo*

(k) Red wattled laping   (l) *Archaeopteryx*

(m) The flycatcher hanging upside down

(n) Stork

(o) White stork

(p) Penguin

(q) Kiwi

(r) Ostrich

Figure 9.13 Some examples of birds.

Eggs are numerous, white, or spotted. Young develop very quickly after birth. It includes *Gallus*, the jungle-fowl of India, and all its domesticated descendants, also *Phasianus* and other pheasants, *Perdix, Lagopus, Meleagris, Numida* and *Pavo*. Megapodes or mound-builders of Australasian and East Indian regions lay their eggs in mounds of decaying leaves and earth. Claws are usually considered to be a secondary development; their resemblance to the claws of *Archaeopteryx* is remarkable.

Order Gruiformes: Rails are secretive, omnivorous terrestrial, compressed laterally and often living in marshy country. They run, swim, and dive easily, but are poor flyers. They build rather simple nests and lay numerous, often dark-spotted eggs. *Crex* and other landrails are more terrestrial in habit. Cranes are long-legged birds found in swamps and allied to rails rather than to waders. Phororhacos reaches 6 feet high. Diatryma was an even larger flightless carnivorous bird, found in Eocene of Europe and North America.

Order Charadriiformes: This order includes wading birds, gulls, terns, and auks. The typical waders live mainly on ground, often inhabiting open watery places or marshes, usually gregarious out of the breeding- season and are often numerous on sea-shores. They often have long legs and long bills and feed chiefly on small invertebrates. Curlews, snipe, and sandpipers are well-known examples. Lapwings and related plovers are found on drier land. Woodcocks inhabit swampy woods. Gulls usually have a grey or white colour, often with black head and wing-tips.

Young are usually darker than adults and mottled with brown. Guillemots and little auks are more fully marine animals, breeding in very large colonies on cliffs.

Order Columbiformes: Pigeons are tree-living, grain- or fruit-eating, mostly good flyers with worldwide distribution. There is little sexual dimorphism. Nest is usually simple and eggs are normally one or two and white. Young are born little developed and are nourished by milk secreted by crop. Dodo, a pigeon adopted a terrestrial habit in island of Mauritius and grew to a large size, but was exterminated by man in 17th century.

Order Cuculiformes: Cuckoos build nests. Many lay their eggs in those of other birds. In common cuckoo, individual female lays mostly in nests of a single foster species, in England often meadow-pipit or hedge-sparrow. She watches the building of nest and lays her egg on same day as foster parent, removing one of clutch before she does so. Often about 12 eggs are laid in this way, each in a different nest. Eggs are usually strongly mimetic with those of host, variable in colour, more so when varied host nests are available. Young hatch before host eggs, which are then ejected from nest by young cuckoo.

Order Psittaciformes: Parrots found mainly in warm climates, live in the trees. They are predominantly vegetarian. Some make use beak for breaking open hard shells. Eggs are usually laid in holes and are white and round. Period of parental care after hatching is unusually long (2-3 months).

Order Strigiformes: Owls specialized for hunting at night; resemble hawks by convergence in their beaks, claws, and in other ways. Food is swallowed whole. They detect their prey mainly by sound, and show various specializations in ears. Eyes contain mostly rods and are directed forwards. They are very large, cannot be moved in orbits, movements of neck compensating for this restriction. Feathers are so arranged as to make very little noise in flight. Eggs are white and laid in holes or in old nests of other birds, some on ground. Many genera are recognized including the barn owls and the eared owls.

Order Caprimulgiformes: Nightjars are a isolated group of crepuscular birds, feeding on insects taken on the wing. Two mottled eggs are laid on bare ground.

Order Micropodiformes: Swifts and humming-birds are fully adapted to air than any other birds. Wings are very long, composed of a short humerus and long distal segments. Swifts are insectivorous with large mouths, adapted for feeding on wing. Nests are made in holes, eggs are white, and young helpless at birth.

Order Coraciiformes: This includes bee-eaters, mainly tropical and often brightly coloured. Three anterior toes are united. Nests are usually made in holes and the eggs are white. Kingfishers are modified for diving into water to catch fish.

Order Piciformes: Woodpeckers are highly specialized climbing, insectivorous, and wood-boring birds. Bill is very hard and powerful. The tongue is long, protrusible and used for removing insects from beneath barks. Tail feathers support the bird as it climbs the tree-trunk. Nest is made in a hole in a tree. Eggs are white.

Order Passeriformes: These perching birds contain about half of all the known species, mostly live close to ground, small with varied habits. Four toes allow gripping the perch. Display and nesting is complicated, with a well-developed song in male. Many species build very complicated nests. Eggs are often brightly coloured and marked. Young are helpless at birth. Rooks and jackdaws are the largest passerines. They are mostly colonial. Starlings are also partly colonial and nest in holes. Finches are seed-eating with a short, stout, conical bill.

House-sparrows are closely related to finches. Larks make nests on ground. Pipits and wagtails are largely terrestrial birds with slender bills. Tree-creepers are tree-living, insectivorous with long bills show convergent resemblance to wood- peckers. Tits are woodland birds and chiefly insectivorous. Shrikes are mainly carnivorous; use their strong bills to eat other birds, amphibia, reptiles, and large insects. Warblers live in trees or scrub. Thrushes blackbirds, British robins, and nightingales, mainly eat small invertebrates, also fruits. They are widely distributed. Hedge-sparrows are small omnivorous. Wrens are small, mainly insectivorous. Swallows are suited for powerful flight and feed on insects.

9.21 Pigeon

Birds are easily recognized group of vertebrates. In birds, every part of body is modified to suit their aerial mode of life. Birds possess feathers, beak and feet modified in relation to their aerial life. Pigeons are flying birds (carinate). They are known both as wild and domesticated forms. Pigeons are seen both in tropical and temperate zones. About 10 species of Pigeons are found in India. Pigeons fly in flocks and roost together. Domestic pigeons have many varieties, namely panter, fantail and tumblers. They differ in size; colouration and feather arrangement.

All of them are descendants of the rock pigeon-*Columba livia*. Pigeons were domesticated more than 5,000 years ago. Rock Pigeons carried messages for the U.S. Army Signal Corps during World War I and II. They are typical blue-gray with two dark wing bars. Flocks with plain, spotted, pale, or rusty-red are not uncommon. Pigeons find their home, even when released from a distant location; navigate by sensing the earth's magnetic fields, sound and smell, and use cues based on position of sun. Domestic pigeons are utilized as homing pigeons and carrier pigeons.

The war pigeons have served roles during war. Rock Pigeons were used in maritime rescue operations in spotting shipwreck victims at sea and in experiments in biology, medicine and cognitive science. Scientists have researched exemplar and prototype memory, category-based and associative concepts. Habitats include various open and semi-open environments. Cliffs and rock edges are used for roosting and breeding in wild. Originally found wild in Europe, North Africa, and western Asia, feral Pigeon are now found in cities around the world.

Rock Dove belongs to the family Columbidae, often simply referred to as the pigeon (Figure 9.14). Visible differences between males and females are few. The species is monogamous, with two squeakers (young) per brood. Both parents care for young for a time. Habitats embrace varied open and semi-open environments. Cliffs and rock edges are used for roosting and breeding in wild. Originally found wild in Europe, north asia, and western Asia, feral pigeons found in cities a round the world. Pigeon belongs to the Columbidae, commonly referesed as the pigeon (Figure 9.14). Aristophanes and other scientists use the word (kolumbis) "diver" for its swimming motion in air. Twelve subspecies recognized by Gibbs are *C. l. livia, C. l. atlantis, C. l. canariensis,* C. *l. gaddi, C. l. gymnocyclus, C. l. intermedia C. l. livia., C. l. targia, C. l. dakhlae, C. l. schimperi, C. l. palaestinae, C. l. palaestinae, C. l. neglecta, C. l. nigricans.*

Figure 9.14 External appeerence of pigeon.

Their wide range of distribution includes western and southern Europe, North Africa, and southwest Asia. They were introduced to North America from Europe in 1606 .Its domesticated form was introduced in other areas. They are now common over most parts of the world. In cities, they breed in any covered space within buildings or bridges. They averages 30-35 cm in length with a wing span of 62-68 cm.

Head and neck of adult are darker blue-grey than back and wings with orange-colored eyes. Irises are white-grey. Eyelids are orange. A grey-white eye ring encircles the eye. Immature Rock Pigeons' plumage is overall duller. A distinct operculum is found on top of beak. Adult female is similar to male, but iridescence on neck is less intense and restricted to rear and sides, while that on breast, it is very obscure. A black band is present at the end of tail. The outer web of tail feathers are margined with white.

Young birds are duller. The feet are red to pink. When circling overhead, the white under wing becomes conspicuous. As strong flier, it often glides, holding its wings in a V shape. Pigeons feed on ground in flocks or individually. They roost together in buildings or on walls or statues. When drinking, they take small sips and tilt their heads backwards to ingest water.

They dip their bills in water and drink continuously without tilting their heads back.

When disturbed, a pigeon in a group will take off with a noisy clapping sound. Contact with pigeon droppings poses a minor risk of contracting histoplasmosis, cryptococcosis, and psittacosis. Pigeons are not the major factor in spreading West Nile virus. Pigeons are at potential risk for carrying and spreading avian influenza and are susceptible to other strains of avian influenza like H7N7.

Scientific classification: Kingdom: Animalia: Phylum: Chordata Class: Aves, Order: Columbiformes, Family: Columbidae, Genus: Columba, Species: livia Binomial name: Columba livia Gmelin, 1789

Nest: During nest building, the female sits on nest and makes a platform of straw, stems brought by male. Pigeons reuse their nests. They don't carry away the feces of their nestlings. Males typically select the nest site, sit in place and attempt to attract a mate. They may nest in stairwells, in abandoned buildings, or rain gutters. They peck food from ground and drink by their bill. When threatening a rival, they may walk in a circle. Male courts his mate by bowing, cooing, and strutting in a circle around female. The pair may preen one another and male may grasp female's bill, regurgitating food as courtship gesture.

Before mating, the female crouches and males jump on her back. Male brings twig or stem to female to build a nest. Male incubates eggs from mid-morning to late afternoon. Female incubates in late afternoon and overnight to mid-morning. Both parents feed them by regurgitating a milky liquid secreted by lining of birds' crops. The Rock Pigeon's life span varies between 3 - 5 years in wild; and up to 15 years in captivity.

Breeding / Nestling: The pigeon breeds at any time of year, but peak times are spring and summer. Nesting sites are situated along coastal cliff faces, and the artificial cliff faces. Nest is a flimsy platform of straw and sticks, put on ledge, under cover. Two white eggs are laid. Incubation period varies between 17 to 19 days. Nestling has pale yellow down feather, flesh-coloured bill along with a dark band. It fed on "crop milk". Fledging period is about 30 days. As part of courtship ritual, male puff up their nape feathers to make themselves appear bigger to impress and attract females. They pick female and approach her at rapid walk - often bowing as they get closer to her. Females walk away and males follow them persistently. Female tolerate the male at point he continues the bowing motion and makes full- or half-pirouettes in front of female.

Mating occurs shortly afterwards and only lasts for few seconds. Male mounts female and balance on top of her back, flapping his wings to maintain balance. Sometimes the pair's beaks are locked together. Female lays up to 3 eggs. Young leave the nest when they are about 25 - 32 days old. During courtship and nesting period, their well-known "cooing" calls can be heard continuously.

External Features: Body is spindle shaped. Their size varies from 20-25 cm. They are covered by coloured feathers leaving beak and a small portion of the hind limbs. The body is divisible into head, neck, trunk and a small, conical tail. The head is round and drawn out anteriorly into a strong, hard, pointed beak. Mouth is a terminal wide gape, guarded by elongated upper and lower beaks. Beaks are covered with a horny sheath or rhampotheca.

A swollen area of soft skin, the cere, surrounds the nostril. It is present on each side of upper beak. Eyes are large and guarded by upper and lower eyelids and a transparent nictitating membrane. A pair of ear openings is situated at a short distance behind the eyes.

Each opening leads into a short external auditory meatus, ending in tympanic membrane forming the ear drum. Neck is long and mobile. It helps in movement of head in various directions.

Trunk is compact, heavy and bears a pair of wings and a pair of legs. Cloacal aperture is at its hind end on the lower surface. Projecting behind the cloacal aperture is the tail. Above the tail is a knob on which opens an oil gland or preen gland or uropygeal gland. It secretes a fluid used for preening feathers.

Wings: Forelimbs as modified wings are located in anterior region of trunk. Limbs are of pentadactyl type. Wing has three typical divisions as the upper arm, forearm and hand. Hand has three imperfectly marked digits. While the pigeon is at rest the three divisions of wing are bent upon one another in form of the letter 'Z'. During flight wings are straightened and extended.

A fold of skin, the alar membrane or prepatagium, stretches between upper and forearm along the anterior border of limb. A smaller fold known as postpatagium is present between trunk and upper arm. While the pigeon is not flying, the whole weight of body has to be supported by hind limbs, In order to balance the heavy trunk, the hind limbs are attached for forwards. Each hind limb or leg (Figure 9.15) has three typical divisions, the thigh, shank and foot. Thigh with out being free is enclosed within boundaries of trunk. Each hind limb has four digits. The first toe is directed backward. The feet are naked and covered with horny epidermal scales. Each digit is provided with a horny claw. Tail is small and concealed by feathers of trunk. It bears tail feathers or rectrices.

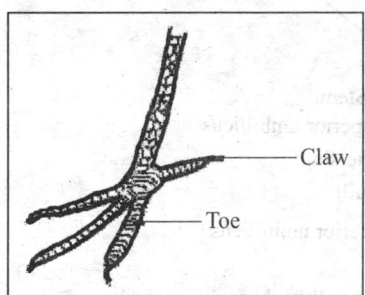

Figure 9.15 Leg of pigeon

Exoskeleton: Feathers are integument structures. They are characteristic of birds. Feathers are derived from epidermis. They are seasonally or periodically cast off and the new one being developed from old papillae. They are arranged on skin in definite tracts, called feather tracts or pterylae.The interspaces without feathers are known as apteria or featherless tracts. There are three types of feathers in pigeon. They are the large quill feathers found on wings and tail for flight, the contour feathers, forming a covering for body, and filoplumes, lying between the contour feathers.

Quill feather: Each quill feather has a central stem or scapus. It is divided into lower hollow part called quill or calamus, and a solid upper part termed rachis. Quill has at its lower ended an opening called inferior umbilicus, through which vascular processes or papilla of dermis project into growing feather. Another opening, the Superior umbilicus occurs at the junction of quill and the rachis on inner face of feather. Close to this opening, there is a small tuft of soft feathers called after shaft. Attached to the rachis are small filaments or barbs.

Rachis with barbs constitutes the vane or vexillum. Each barb is provided with barbules and hooklets. The barbs remain attached with one another to form a continuous blade for striking the air in flight. There are twenty three quill feathers or remiges in each wing. Eleven of these are known as primaries. They are attached to the hand. The remaining twelve fixed on forearm are called secondaries.

Attached to thumb is a small tuft of feathers known as ala spuria or bastard wing. Tail bears twelve tail feathers or rectrices which are arranged in the form of fan. Contour feathers are soft and barbs are plume like with no interlocking mechanisms. These help to keep the body warm and lock air pockets. Filoplumes have delictae hair like long axis and a few barbs devoid of barbules. Down feathers have small axis and a few barbs devoid of locking structures at the distal end. Nestlings are covered with down feathers (Figures 9.16-9.19).

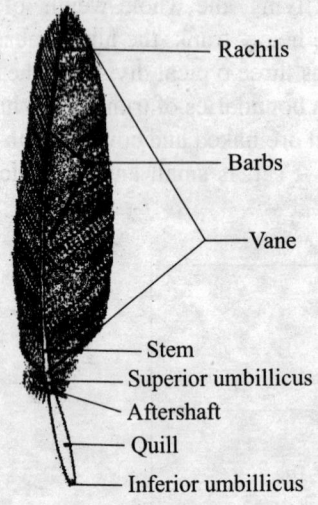

Rachils

Barbs

Vane

Stem
Superior umbillicus
Aftershaft
Quill
Inferior umbillicus

Figure 9.16   Quill feather.

Figure 9.17   Filoplume.

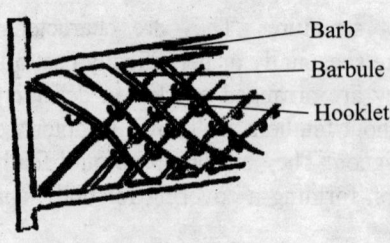

Barb
Barbule
Hooklet

Figure 9.18   Barbs and barbules.

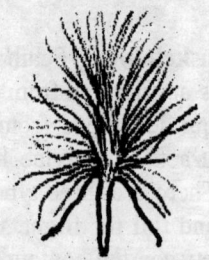

Figure 9.19   Down feather.

Endoskeleton

The endoskeleton is strong but lightly built. Texture of the bone is often spongy. Bone marrow is absent. Air spaces from lungs may continue into bones, making them light. Bones are more or less devoid of bone marrow. These are called Pneumatic bones. Most bones except those of tail, forearm, hand and hind limb contain air spaces. In general, there is a tendency for reduction and fusion of bones. It gives rigidity to skeleton. Cranial bones are marked in young skull.

The single condyle is made up by the basi- and ex-occipitals. Parietals are short but wide. Basi-temporals underlie the base of skull. Anterior part of Parasphenoid underlies the mesethmoidal septum. Palatine extends forwards to maxilla and articulate behind with rostrum. Pterygoids articulate in front with palatines and basipterygoid processes of rostrum, and they articulate with quadrate behind. The maxillae have short maxillo-palatine processes. Squamosal and jugal are connected by ligament. Vomer is absent. Internal nares open between rostrum and palatines. There is no hard palate. Skull is schizognathous owing to the palatal plates of praemaxilla and maxilla not meeting in middle line. Cervical, thoracic, lumbar, sacral, and caudal vertebrae are present in addition to sternum, and ribs.

There are 3 median bones in hyoid, one tongue-shaped, formed by union of ceratohyals, followed by basihyal and basibranchial. The first branchial arch is well developed consisting of an upper epi- and a lower cerato-branchial. The joints between several parts are synovial.

Sternum is formed from right and left plates of cartilage, constituted by fusion of ventral ends of ribs. Carina is formed from a single band of tissue continuous with clavicles. The upper epicleidium and hypocleidium is small. A coraco-clavicular ligament unites each clavicle to inner border of coracoid, and a sterno-clavicular ligament unites the hypocleidium to carina.

Shoulder girdle consists of a scapula, coracoid, and furcula. Scapula is sword-shaped and thin. There is no separate suprascapula. A small conical process internal to glenoid facet represents the meso-scapula or acromion. Coracoid is firmly united by ligament to scapula. A prominent clavicular process rises in front of its glenoid facet. There is a thin curved subclavicular process, on the internal or true anterior border in contact with acromion. A rough line runs downwards from it to broad sternal end of bone and gives attachment to coraco-clavicular membrane. Coracoid fits into a groove in sternum. Furcula is formed by fusion of ventral ends of 2 clavicles. At its upper end, each clavicle expands into a disc or epicleidium, which is tied by ligament to acromion and to sub-scapular and clavicular processes of coracoid. There is a foramen triosseum through which tendon of second pectoral muscle passes to its insertion on humerus.

Humerus lies parallel to the axis of body. Its true ventral surface turned outwards. Fore-arm is flexed on humerus and the hand is adducted. The glenoid head of humerus is transversely elongated: on its upper margin at the proximal end is a conical process to which the first pectoral or depressor of the wing is attached, and dorsally to it is the facet for insertion of second pectoral. On the ulnar margin proximally and dorsally is a deep pit, at the bottom of which is a pneumatic foramen.

Surface of articulation for radius is long, oblique. Radius is rod-like. Ulna is stout, somewhat curved, and with a short olecranon. Its outer surface is pitted by the sacs of secondary wing feathers. Two carpal bones in proximal row are scaphoid and a fused lunar and cuneiform. Distal carpal are fused to heads of metacarpal, forming a carpo-metacarpal .The first metacarpal carries a single phalanx. The second is stout, long and carries 2 phalanges. The third is slightly curved and fused distally to second, and carries one phalanx.

Pelvis has ileum, ischium, and pubis. The first extends along the whole extent of sacrum. Ischium lies parallel to backward extension of ileum; the pubes to ischium and neither of the 2 have a ventral symphysis. All 3 unite in acetabulum. The centre of this cavity is membranous. The anti-trochanter works against the base of neck which carries the head of femur. Ileum and ischium fuse distally, and thus enclose an ileo-sciatic foramen. Obturator foramen between the ischium and pubes is long and narrow, and subdivided partially by obturator process of ischium. Femur is short. Its head is prominent.

Condyles are large and separated by deep patellar groove. It has proximally a cnemial crest on anterior surface, subdivided into a pro- and ecto-cnemial process; and distally there are 2 condyles formed from a cartilage. Fibula is slender and pointed distally. Third section of the limb is the tarso-metatarsus, formed by union of a bone representing the distal tarsal to heads of the second, third, and fourth metatarsal, of which the third is the longest. The first metatarsal is small, incomplete proximally, and united to second by ligament. There are 4 digits in all - the first, or hallux is turned inwards and backwards and carries 2 phalanges; the three remaining digits carry phalnges increasing successively in number from 3 to 5. The third is the longest digit (Figures 9.20-9-26).

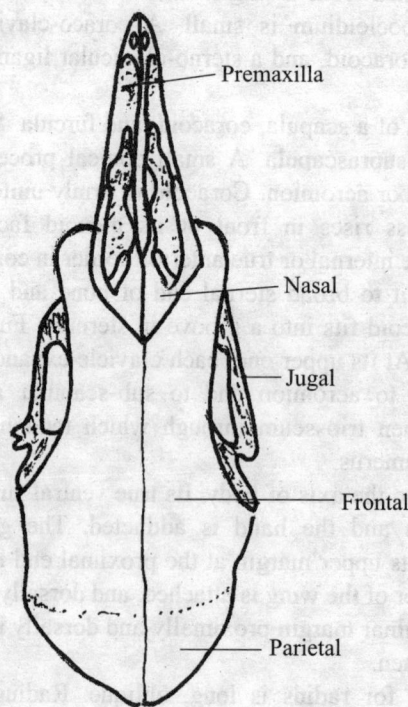

Premaxilla

Nasal

Jugal

Frontal

Parietal

Figure 9.20 Dorsal view of skull

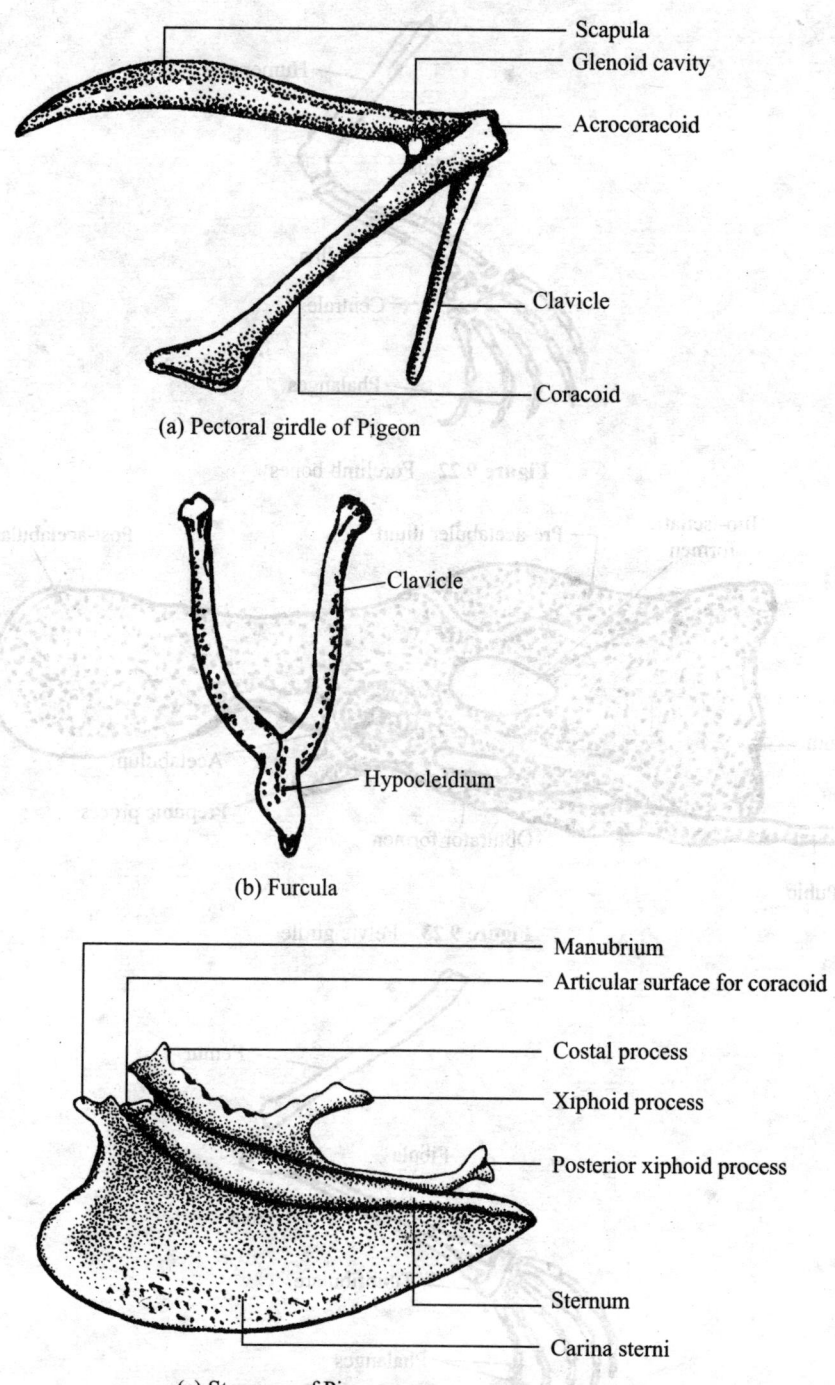

(a) Pectoral girdle of Pigeon

Scapula
Glenoid cavity
Acrocoracoid
Clavicle
Coracoid

(b) Furcula

Clavicle
Hypocleidium

(c) Stermum of Pigeon

Manubrium
Articular surface for coracoid
Costal process
Xiphoid process
Posterior xiphoid process
Sternum
Carina sterni

Figure 9.21

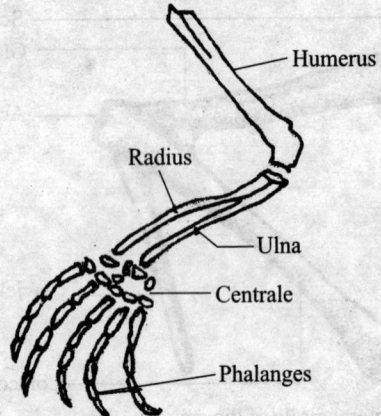

Figure 9.22   Forelimb bones

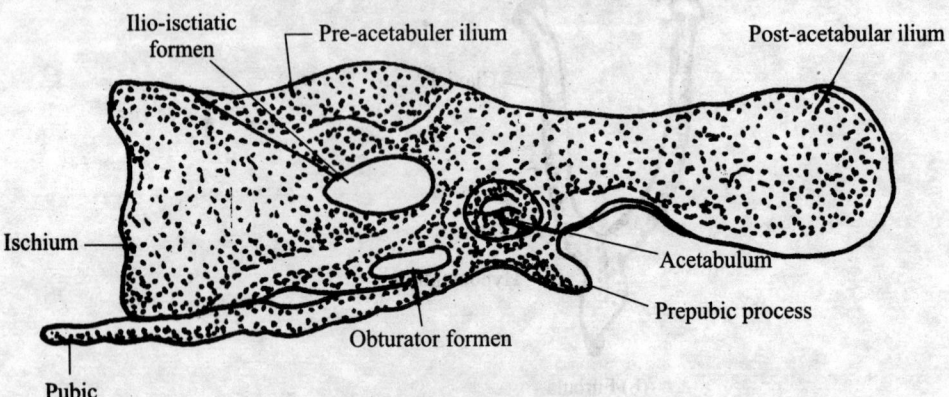

Figure 9.23   Pelvic girdle

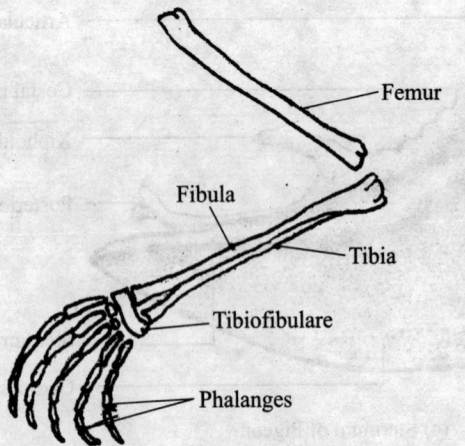

Figure 9.24   Hindlimb bones.

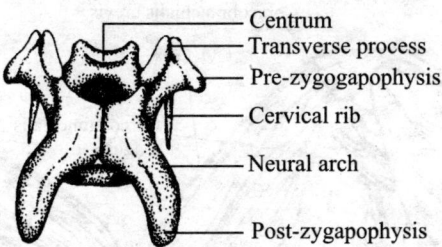

Figure 9.25 Typical cervical vertebra.

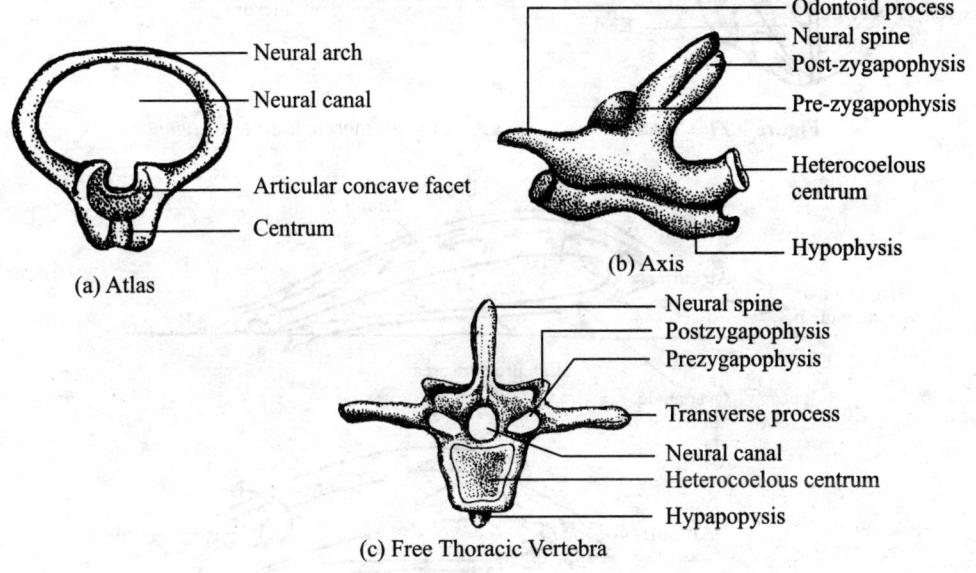

Figure 9.26

Flight Muscles: Wings are modified forelimbs. They are organs of flight. Musculature of forelimbs is greatly modified in response to the function they perform. Flight is the coordinated effort of a number of paired muscles of which the following is most important.

Pectoralis Major (Depressor muscles): These are the largest breast muscles. They are about one fifth of body weight. By the contraction of this muscle, the wings are lowered during flight.

Pectoralis Minor or Subclavius: These are smaller but longer than pectoralis major. By their contraction the wings are raised in flight.

Coracobrachialis: These small flight muscles pull the wing downwards in flight.

Flight mechanism is not easy as described earlier. Various factors associated with the flight and various types of flight are also given in detail. Understanding the flight mechanism necessitates the understanding about the aerodynamic principle, the drag and lift. (Figure 9.27, 9.28).

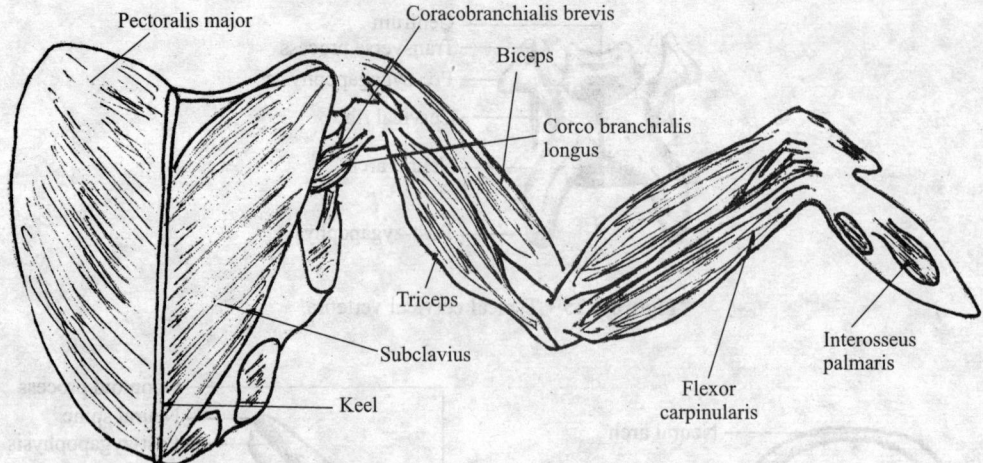

Figure 9.27   Ventral view of breast and wing musculature of *Columba*.

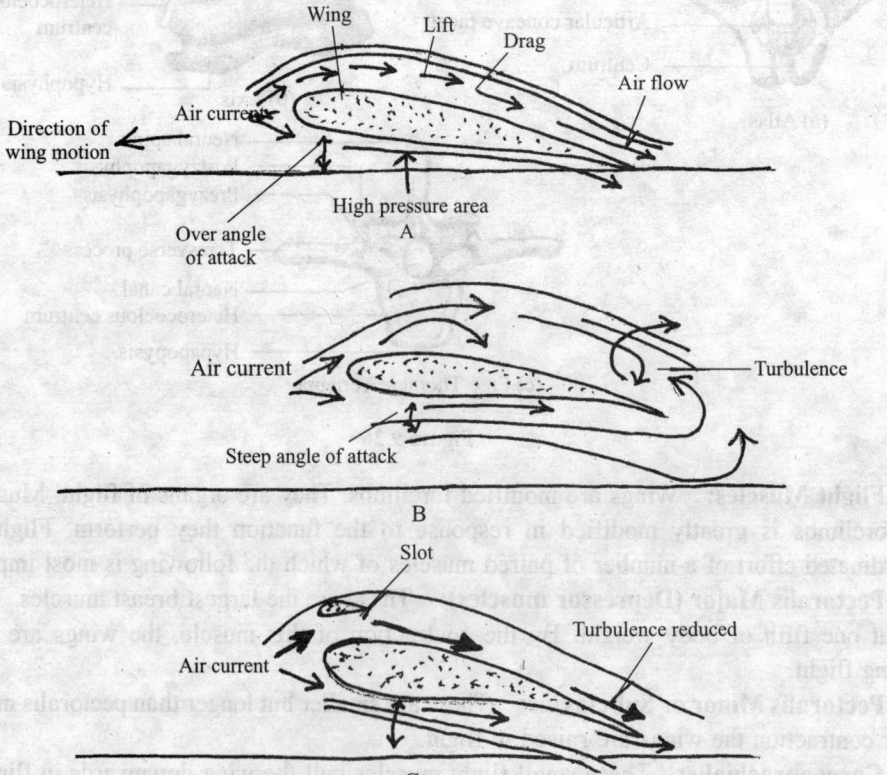

Figure 9.28   Effect of wing on air stream. (A) Air moving rapidly along upper surface of wing reduces the pressure on upper surface and generates force that is resolved into lift and drag forces. (B) Wing tilted sharply because air causes turbulence in low pressure area and reduction in lift. (C) Turbulence can be reduced and form a slot through which air moves quickly and smoothly.

Digestive System: The two jaws of mouth are modified into beak. Both the jaws are devoid of teeth. Mouth leads into buccal cavity. The floor of buccal cavity is provided with a narrow, triangular tongue. It has a horny covering and is provided with sensory papillae. Buccal cavity narrows behind into Pharynx. The salivary glands are absent in buccal cavity. Three pairs of buccal glands are present in mouth. Their secretion is mainly mucus. Alimentary canal proper starts from the Pharynx which leads into a long oesophagus that runs back through neck. At the base of neck region, it enlarges into a thin walled, distensible sac known as crop containing mucus glands. It serves as a store house for food. The crop is followed by stomach which is divisible into two parts, the anterior tubular proventriculus containing gastric glands and a posterior laterally compressed gizzard.

The gizzard has a thick muscular wall and a horny inner lining. Its cavity is small and contains small stones which are helpful to grind the food. Thus the gizzard acts as a grinding mill. This type of arrangement is necessary because of absence of teeth in buccal cavity. The intestine arises from right side of gizzard. It is divisible into an anterior U- shaped duodenum, and a posterior long coiled ileum. The ileum enlarges posteriorly into a short rectum or large intestine.

Anteriorly, the rectum bears a pair of small rectal caeca. Rectum opens to exterior by the cloaca. Internally, the cloaca is divided into three chambers, the anterior coprodaeum, the middle urodaeum, and the posterior proctodaeum. The rectum opens into the coprodaem. Urinogenital ducts open into the urodaem. The proctodaem opens to the exterior by a transverse slit like aperture called cloaca. At the proctodacum, there is a dorsal glandular sac known as Bursa of Fabricii. Its function is unknown. The digestive glands associated with alimentary canal are the liver and pancreas. Liver is bilobed with a large right and a small left lobe. It is devoid of gall bladder. There are two bile ducts, one from each lobe. They open into duodenum independently. Pancreas lies between the two limbs of duodenum. It has three ducts, all opening into distal limb of duodenum (Figure 9.29).

Respiratory System: Flight activity requires a continuous and abundant supply of oxygen. Hence; the respiratory system is highly developed and well differentiated. Respiratory system consists of external nostrils, glottis, larynx, trachea, bronchus, and lungs. External nostrils are a pair of slit like apertures occurring at the base of upper beak. They communicate to pharynx by internal nostrils. A glottis lies behind tongue. It opens into larynx which opens into a trachea.

Trachea is a long, cylindrical and flexible tube running backward through neck. On entering the thoracic cavity, trachea expands into a syrinx or voice box. Later it divides into two bronchi, one for each lung. Walls of tracheal and bronchial tubes are supported by a series of closely set cartilaginous rings. Each bronchus enters a bright red lung. Bronchus divides and subdivides into smaller branches, ultimately ending in fine air capillaries.

Lungs are solid spongy organs. They do not hang freely in the thoracic cavity, but are lodged firmly in ribs. Some branchial tubes pass through lungs and communicate with air cavities in bone. There are nine air sacs. They are a median interclavicular, a pair of cervical, two pairs of thoracic, and a pair of abdominal air sacs. Air sacs help to maintain high body temperatures. They make the body lighter and help in flight.

Mechanism of Respiration: In birds the expiration is an active process. The process of inspiration is passive. In a resting bird, the sternum is moved up and down with help of intercostal and abdominal muscles. During flight, the sternum is rendered immovable due to support of wings, but body cavity is raised and lowered by action of wings and by lowering of vertebral column (Figures 9.30, 9.31, 9.32, 9.33).

Circulatory system: Heart is four chambered, with two auricles and two ventricles. There is complete separation of oxygenated and non-oxygenated blood. Birds have two distinct circulations as arterial and venous systems.

Arterial system: Arteries are blood vessels that carry blood away from heart. Since the heart pumps blood with force into arteries; the vessels have slightly muscular and strong walls. Two major arteries that leave heart are pulmonary artery and systemic artery.

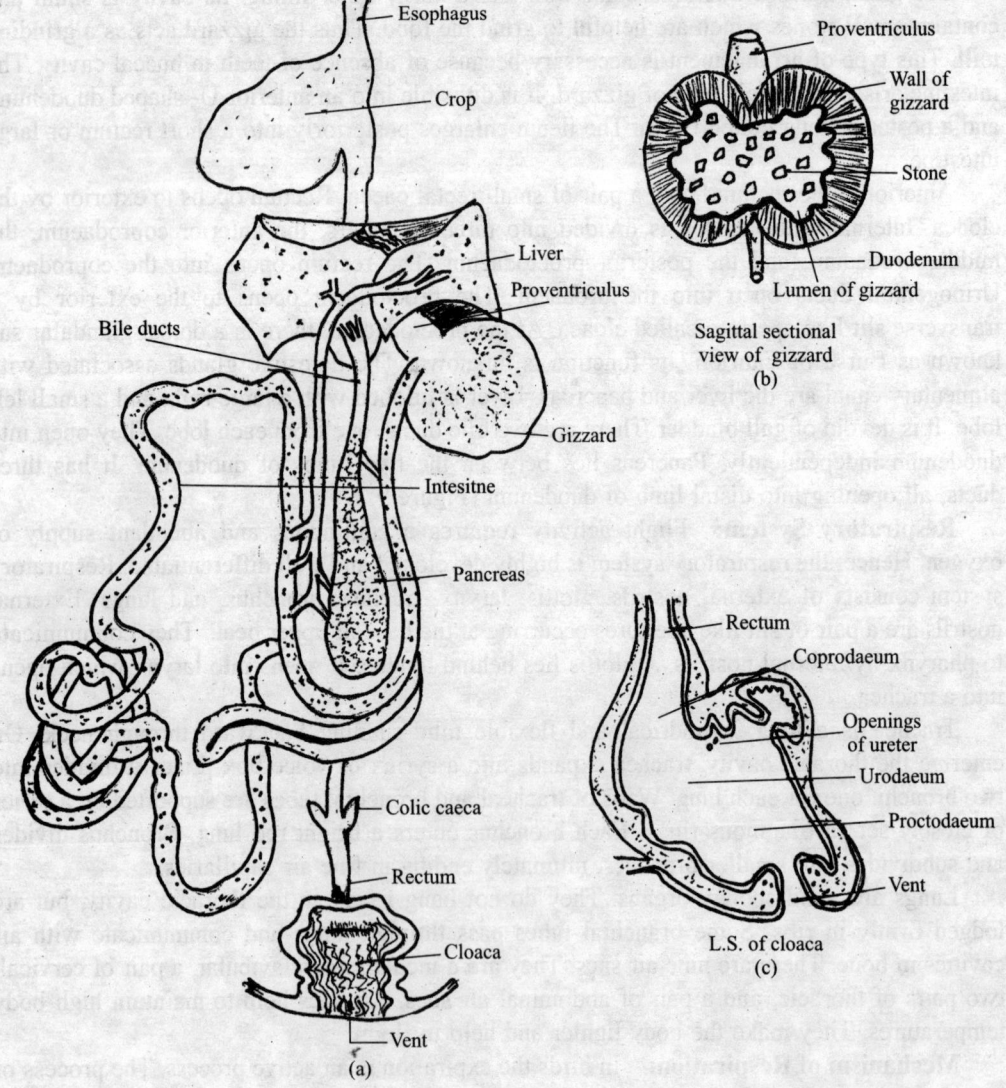

Figure 9.29   Digestine system of *Columba*.

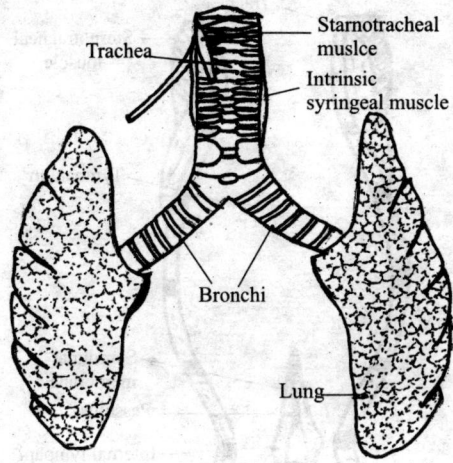

Figure 9.30 Trachea, bronchi and lungs.

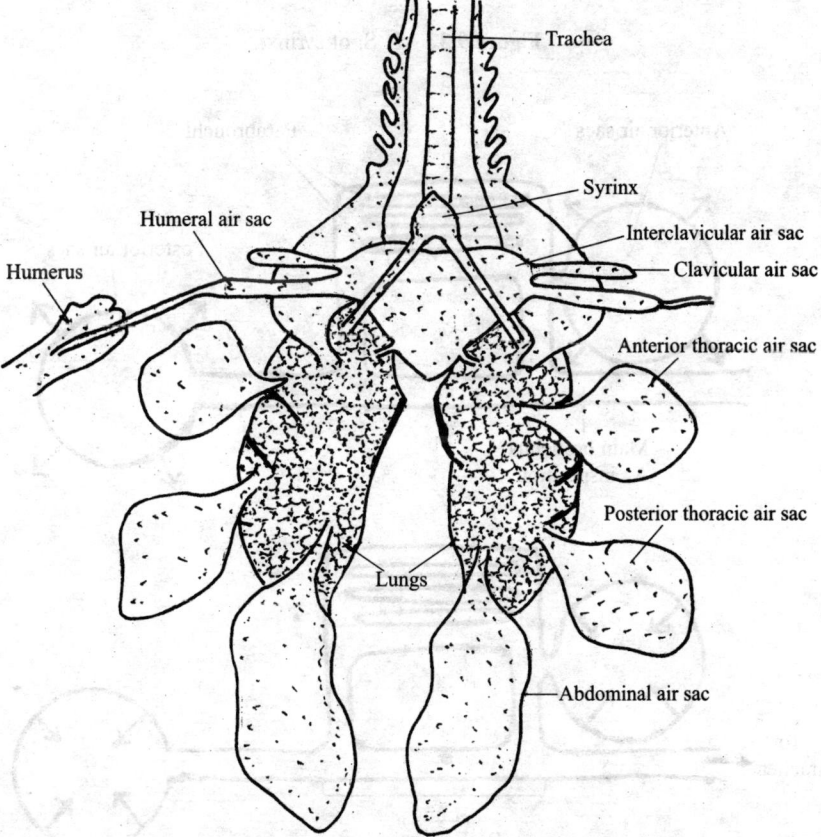

Figure 9.31 Air sacs.

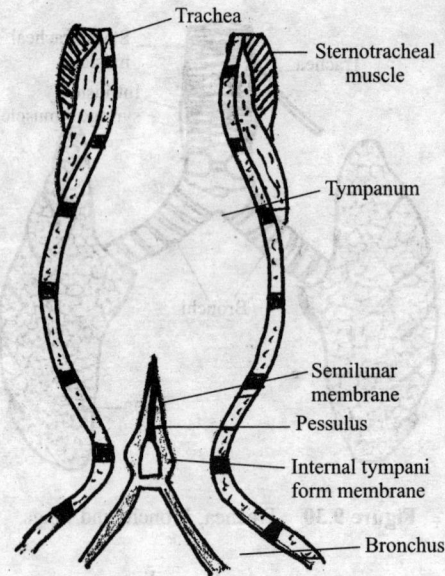

Figure 9.32 L. S. of syrinx.

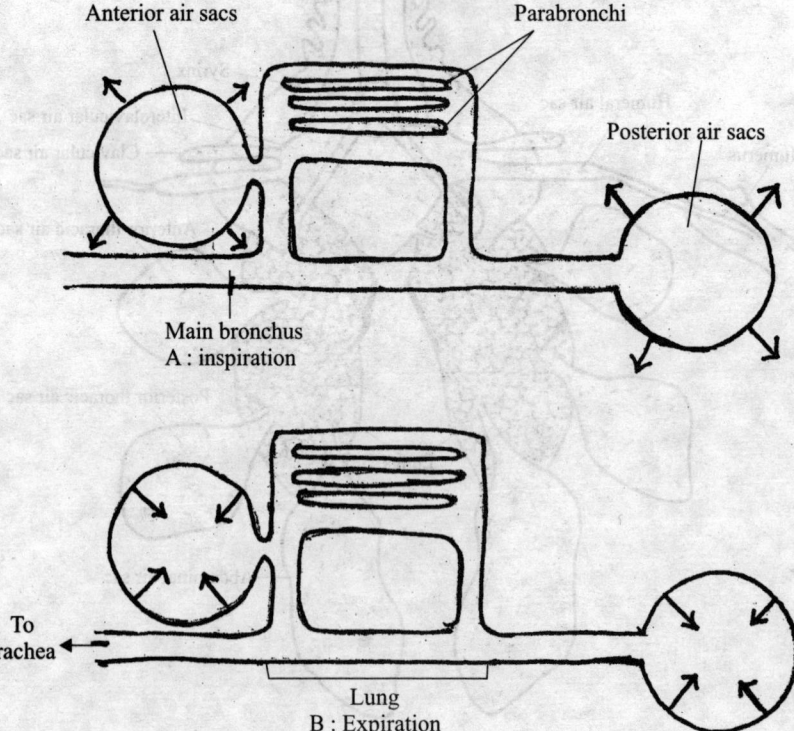

Cycle 2

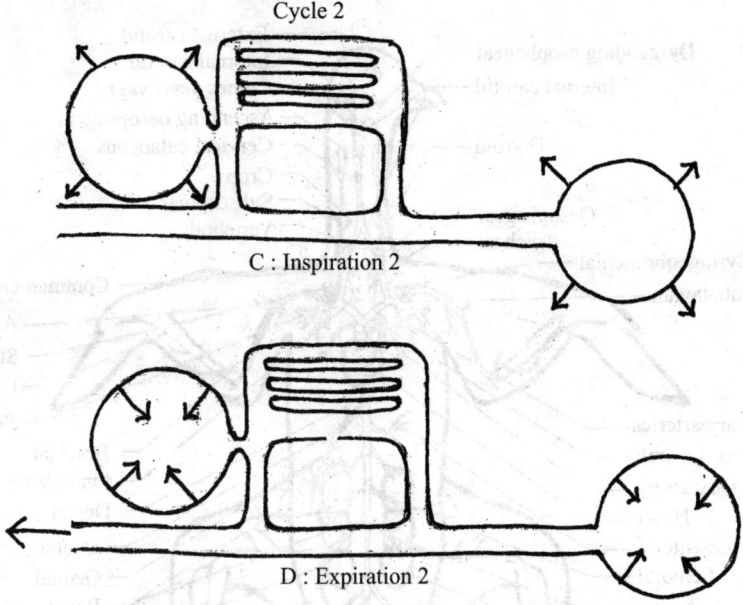

C : Inspiration 2

D : Expiration 2

Figure 9.33   Movement of a volume of gas through respiratory system.

Pulmonary artery originates from right ventricle. It carries deoxygenated blood towards the lungs for oxygenation. As this artery leaves heart, it divides into right and left pulmonary arteries carrying blood to right and left lungs. The systemic artery is a part of right aortic arch. In birds, the left aortic arch is absent. The right aortic arch begins from left ventricle.

The left ventricle pumps oxygenated blood. The right systemic artery is formed as a constituent of right aortic arch. The systemic artery forms right and left innominate arteries. Each innominate artery forms a carotid artery and a subclavian artery. The subclavian arteries in turn divide into brachial and pectoral arteries. They carry blood to muscles around thoracic region and flight muscles. The carotid arteries carry blood to head region. The right systemic artery further runs posteriorly and forms the dorsal aorta which is a major artery supplying blood to abdominal organs and posterior regions of body. The most prominent blood vessels are

1. The coeliaco-mesenteric artery and anterior mesenteric artery: These arteries carry blood to various regions of the alimentary canal and certain visceral organs.
2. Renal arteries carry blood to the kidneys.
3. Femoral arteries (2) supply pure blood to pelvic muscles and external thigh region.
4. Sciatic arteries (2) supply pure blood to internal thigh region.
5. Iliac arteries (2) supply blood to hip region.
6. Posterior mesenteric artery supplies blood to the most posterior visceral organs.
7. Caudal artery carries blood to the tail region (Figure 9.34).

Venous System: Deoxygenated blood from various regions of body is collected by several veins. Finally these veins take the blood to right auricle through two precaval and a

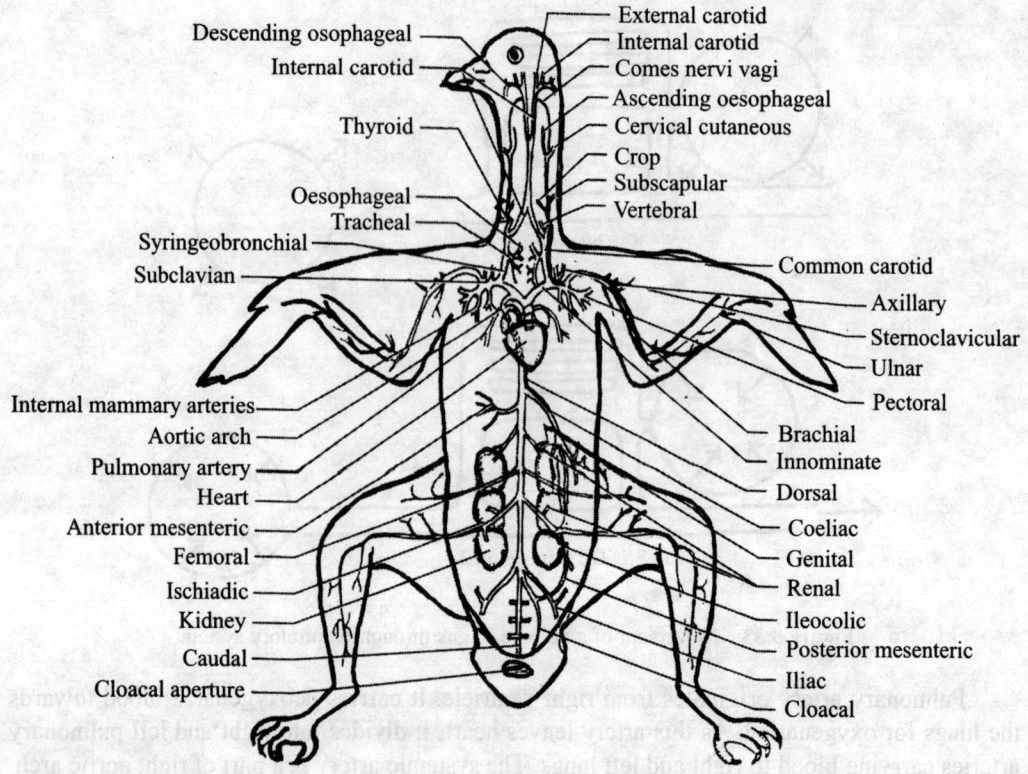

Figure 9.34 Arterial system of pigeon.

single postcaval veins. Precaval and postcaval veins receive blood from several veins originating from various organs of body. From the tail region, the blood is drained by caudal vein. It runs up to renal portal veins. Each portal vein runs through kidney and joins femoral vein. Renal portal vein provides renal veins. It also receives a pair of sciatic veins draining blood from legs. Right and left femoral veins are united to form the posterior part of post caval vein. Precaval veins are formed by jugular and brachial veins from neck shoulder and head regions (Figure 9.35)

Nervous System: Brain is divisible into the fore-, mid- and hind brains. Cerebral hemispheres are distinct. They are round and large in size. Olfactory lobes are very small and they do not contain cavities. Diencephalon is hidden from the view by forward prolongation of cerebellum. Diencephalon has pineal body dorsally and infundibulum and pituitary body ventrally. Optic lobes are lateral inposition owing to large size of cerebral hemispheres and cerebellum.

Medulla oblongata instead of being continued backwards as in other tetrapods, descends almost vertically from cerebellum.

Sense Organs: In pigeon the olfactory sense is poor. There is no external ear. Tympanum is slightly sunken from surface of skin. Eyes are large. During flights the eyes and

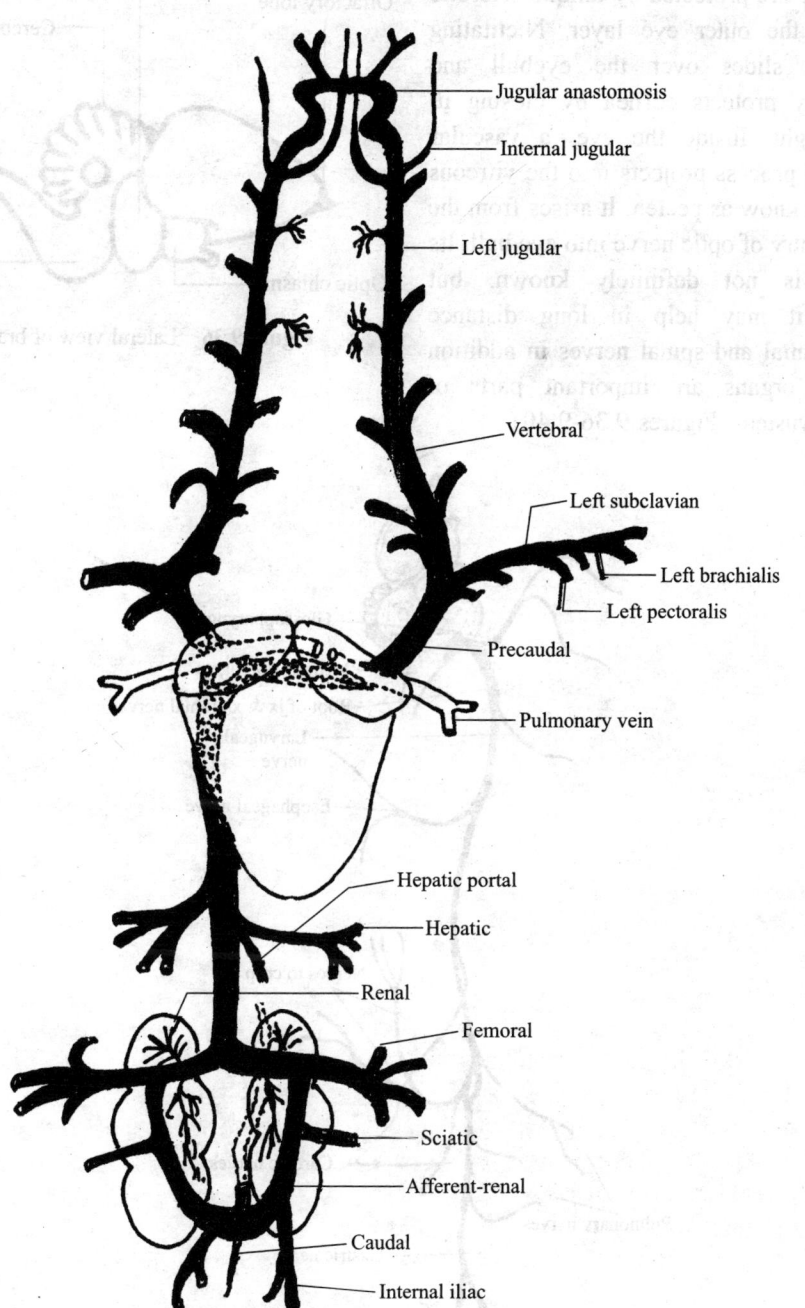

Figure 9.35 Venous system.

their shape are protected by unique sclerotic plates of the outer eye layer. Nictitating membrane slides over the eyeball and presumably protects cornea by closing it, during flight. Inside the eye, a vascular pigmented process projects into the vitreous body. It is know as pecten. It arises from the point of entry of optic nerve into eye ball. Its function is not definitely known, but possibly it may help in long distance vision.Cranial and spinal nerves in addition to sense organs are important parts of nervous syustem (Figures 9.36-9.40)

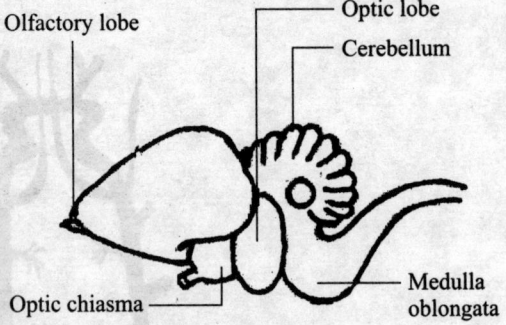

Figure 9.36   Lateral view of brain.

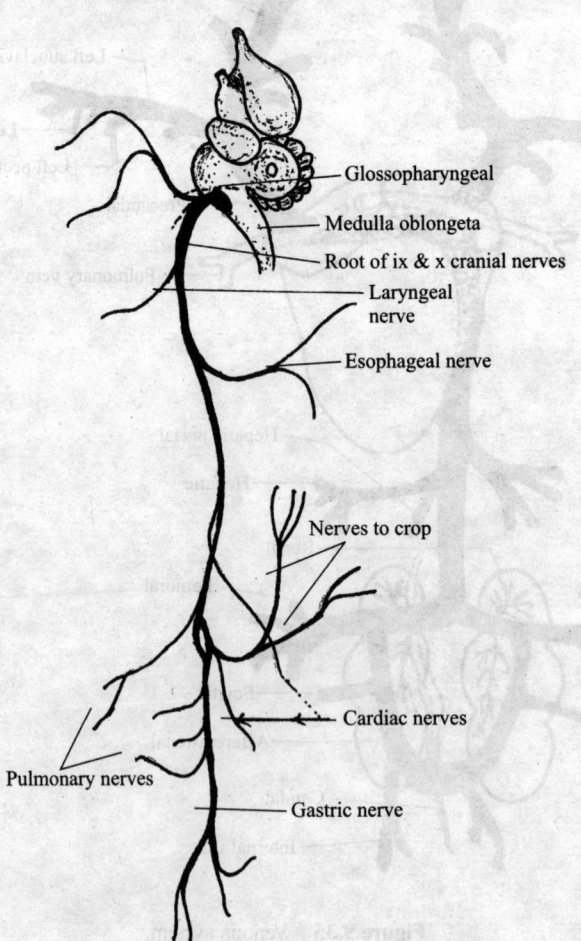

Figure 9.37   Ninth and tenth cranial nerves.

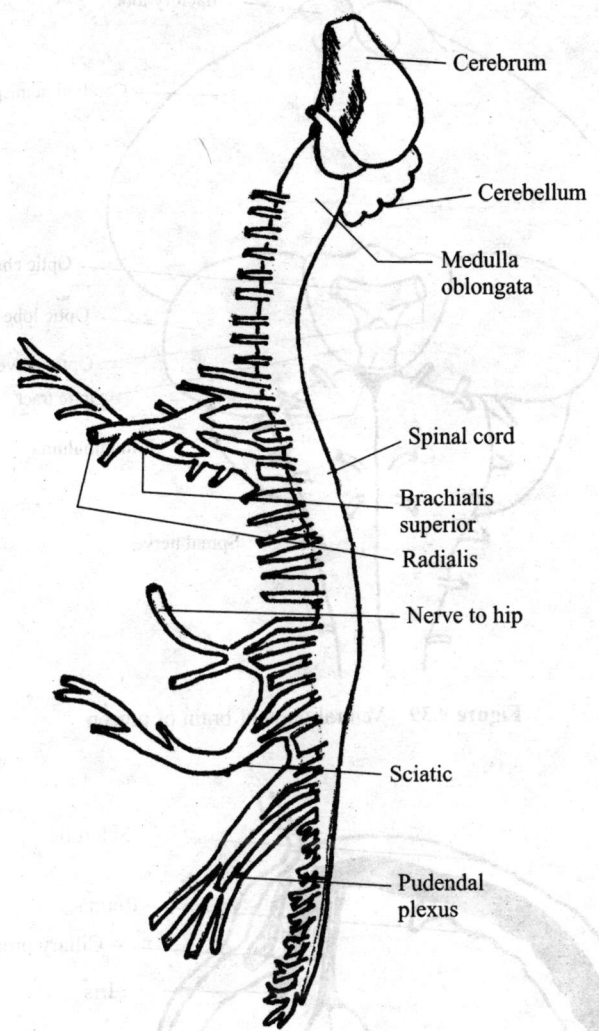

Figure 9.38 Spinal nerves.

Urinogenital System: Excretory organs are a pair of kidneys. They are dark red, three lobed structures. They open separately into urodaeum of cloaca through two different ureters. There is no urinary bladder. Urine is excreted in form of uric acid, a semi solid white mass discharged along with faeces through cloacal aperture.

Reproductive System: Male has a pair of oval testes. From each testis, a duct, the vas deferens, passes back and opens into cloaca. Vas deferens is dilated at its posterior end into a seminal vesicle. There is no copulatory organ. Only the left ovary persists in adult. Right ovary disappears during development. Ovary and oviduct of only one side are functional during breeding season (Figures 9.41, 9.42).

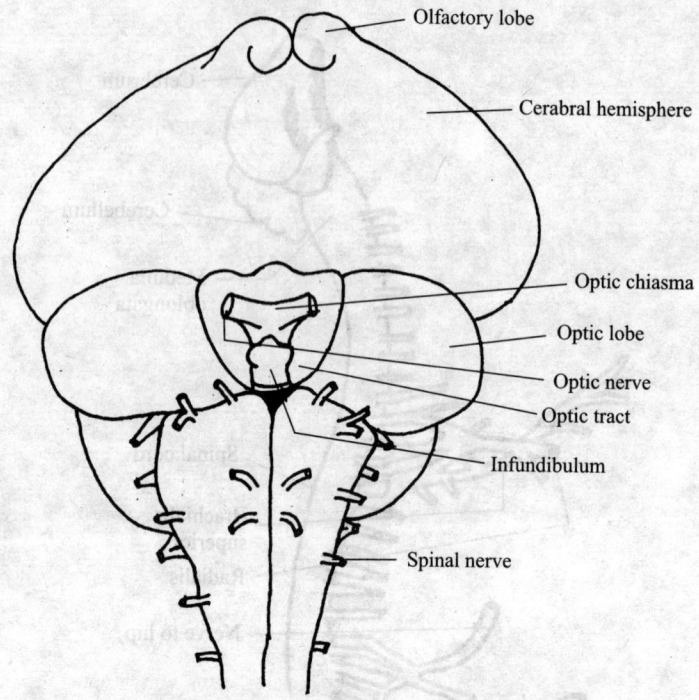

Figure 9.39   Ventral view of brain of pigeon.

Olfactory lobe

Cerebral hemisphere

Optic chiasma

Optic lobe

Optic nerve

Optic tract

Infundibulum

Spinal nerve

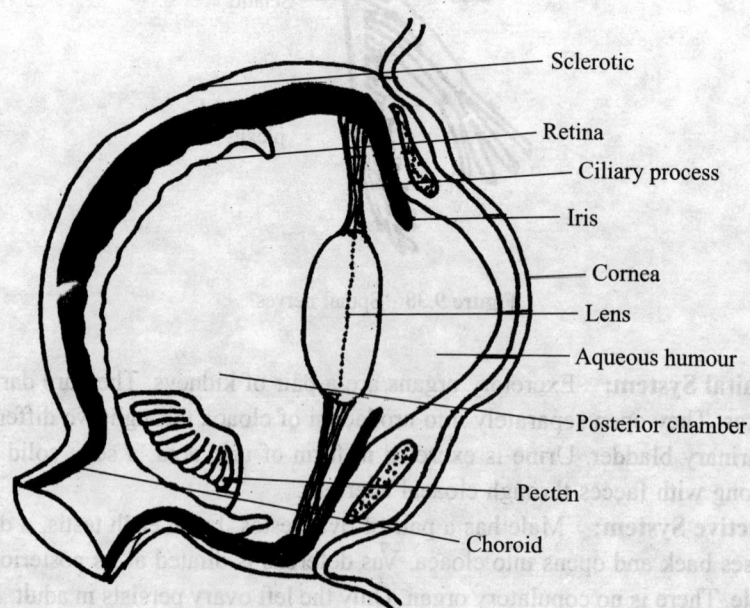

Sclerotic

Retina

Ciliary process

Iris

Cornea

Lens

Aqueous humour

Posterior chamber

Pecten

Choroid

Figure 9.40   Sagittal section of eye.

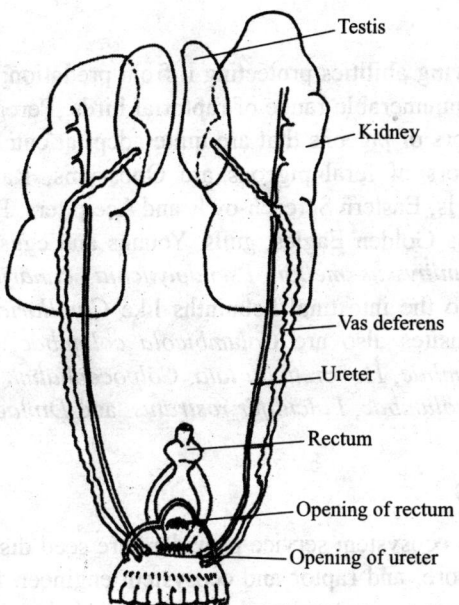

Figure 9.41 Male urinogerital system.

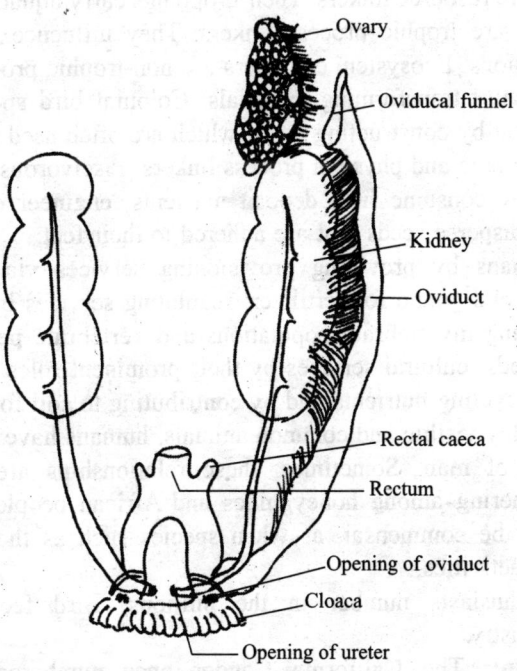

Figure 9.42 Female urinogenital system.

Predators

Predators with only its flying abilities protecting it from predation; they are a favorite almost around the world for an innumerable range of raptorial birds. Peregrine Falcons and Sparrow hawks are natural predators of pigeons that are quite adept at catching and feeding upon this species. Common predators of feral pigeons are Opossums, Raccoons, and Red - tailed Hawks, Great Horned Owls, Eastern Screech-owls and Accipiters. Predators in North America include American Kestrels Golden Eagles, gulls. Youngs and eggs remain at risk from feral and domestic cats. *Tinaminyssus melloi*, *Pseudolynchia. canariensis* also attack pigeon. Pigeons provide shelter to the intestinal helminths like *Capillaria columbae* and *Ascaridia columbae*. Their ectoparasites also are *Columbicola columbae*, *Campanulotes bidentatus compar*, *Bonomiella columbae*, *Hohorstiella lata*, *Colpocephalum turbinatum*, *Dermanyssus gallinae*, *Dermoglyphus columbae*, *Falculifer rostratus*, and *Diplaegidia columbae*.

9.22   Role of Birds

Eight main types of avian ecosystem service providers are seed disperser, pollinator, nutrient depositor, grazer, insectivore, and raptor and ecosystem engineer. Birds are mobile links for maintaining ecosystem function, memory and resilience. Avian ecological functions are three linkages namely genetic, resource, and process. Seed-dispersing frugivores and pollinating nectarivores are genetic linkers, as they carry genetic material from one plant to another.

Piscivorous birds are resource linkers. Their droppings carry aquatic nutrients to terrestrial environments. Grazers are trophic process linkers. They influence plant, invertebrate and vertebrate prey populations. Ecosystem engineers are non-trophic process. They modify their environment by physically transforming materials. Colonial bird species and woodpeckers modify their environment by constructing nests, which are often used by other species. Many bird species are both trophic and physical process linkers. Piscivorous bird colonies carry out all these linkages. They consume fish, deposit nutrients, engineer ecosystems via burrow construction and even disperse seeds that are adhered to their feet.

Birds benefit humans by providing provisioning services via game meat for food, materials for garments and guano for fertilizer; regulating services by scavenging carcasses and waste, by controlling invertebrate populations and vertebrate pests, by pollinating and dispersing the plant seeds; cultural services by their prominent roles in art and religion and supporting services by cycling nutrients and by contributing to soil formation.

Since birds are highly visible and common animals, humans have had a relationship with them since the dawn of man. Sometimes, these relationships are mutualistic, like the cooperative honey-gathering among honeyguides and African peoples such as the Borana. Other times, they may be commensal, as when species such as the House Sparrow have benefited from human activities.

Amateur bird enthusiasts number in the millions. Bird feeding has grown into multimillion dollar industry.

Bird Conservation: The California Condor once numbered only 22 birds, but conservation measures have raised that to over 300 today. Governments and conservation groups work to protect birds, either by passing laws that preserve and restore bird habitat or by establishing captive populations for reintroductions. Such projects have produced some successes; one study estimated that conservation efforts saved 16 species of bird that would otherwise have gone extinct between 1994 and 2004, including the California condor and Norfolk Parakeet. A variety of bird types have appeared and disappeared, ultimately resulting in the existing 9000 different species.

Migration in Birds

Migration found in many animals is an interesting phenomenon, and is most remarkable in birds. Many birds span the earth over mountain, desert and sea. They undertake the troublesome journeys, and get rewards. About 5 billion terrestrial birds of about 200 species leave North America for central and South America every fall. Many birds migrate from Europe and Asia to reach Africa.

Migration is predictable seasonal movements of individuals in response to seasonal variation in climate and resources from native home to other places and backward journey to home place again. Migration is a form of dispersal which involves movement away from and subsequent return to the same location, typically on an annual basis.

In flight, birds maintain their compass orientation by polarized skylight pattern through "compass sense". Migration is found in species that live in open edge habitat rather than forest living species that live in forest. Birds use earth's magnetic field during migration. Individuals of the same species sometimes choose to migrate one year and not in the next. Fox Sparrows reduce competition by migrating to different spots.

Migration of geese is an annual, large-scale movement between breeding homes and non-breeding grounds. Of more than 650 bird species of North America, some are permanent residents, the majority is migratory. Birds migrate from low resources areas to areas of high resources. Two primary resources are food and nesting sites. Birds that nest in northern hemisphere migrate northward in spring to get insect populations as food, budding plants as nesting sites. As winter approaches, the insect population drops and birds migrate south. Avoiding the cold for many, including hummingbirds is important factor.

Short-distance migrants move short distance, from higher to lower elevation on mountainside for food. Medium-distance migrants cover one to several states. Long-distance migrants cover extending from the United States and Canada to Mexico. This is more complex and includes the involvement of genetic make-up. Day length, changes in food supply and genetic predisposition trigger migration. Various birds and even part of population may follow various migratory patterns.

Cost-benefit Analysis: Evolution of migration necessitates examining its costs and benefits. More than half of all Northern Hemisphere migratory terrestrial birds are reported to not return. Of 100 million waterfowl that reach southern wintering grounds, only 40 million are estimated to return .Migration is risky as it makes migrants susceptible to face extreme weather events like hurricanes and sandstorms. Survival costs in migration are high. Migrants spare more times in journey and therefore get less time to reproduce. Smaller clutch and few breeding attempts are reported in migrant birds than residents breeding at similar latitudes. Birds need to double in weight before migration. Extra weight slows down take-offs and makes them vulnerable to predators.

Benefits

Migrants exploit favourable seasonal opportunities and resources. Birds migrate south to avoid winter, and to avail the comfort of benign climates besides higher food availability. Then they fly north to take advantage of the spring bloom of abundant resources. Many land bird migrant species are believed to evolve from tropical species that move small distances following changing resource levels. Warblers, tanagers and vireos fly to North to take chance of less competition, less predation, but abundant food that feature the temperate spring and summer.

Age and Sex of Migrants

Cost-benefits of migrants, temperate residents and tropical residents can also be applied to different migrations strategies within a species. Often, males and females winter at different latitudes. Usually, males winter farther north than females, but in raptors, the reverse is true. Usually, immature birds winter farther North in some species, but farther south in other .Adult females winter farthest south; immature females and adult males in middle, and immature males farther north. Males usually depart north for breeding grounds before females, as they need to get back soon to establish territories. Young in songbirds depart before adults on fall migration. Young in other species go with their parents or after. The greater migrational mortality favours young to fly shorter distances and male's reproductive success is dependent on time of return to breeding grounds, so they winter farther North to get back earlier. Survival is easier in farther South, so adult females that are unaffected by the previous two factors winter the farthest South.

Fuel and Stimulus for Migration

Hyperphagia: Fat stores more energy than either carbohydrate or protein. Birds can double their weight in few days. They accumulate more fat before migration, and replenish it en route. How quickly birds replenish their fat reserves, which foods they prefer, and risk of predation varies.

In spring, the premigratory state is characterized by change in neural centers in lower brain controlling hunger and satiety. Bird gains weight by overeating. This increased energy income is stored as large fat deposits under the skin, in flight musculature, and in abdominal cavity. Sparrows and warblers gain about 1 to 1.5 g per day. They retain the ability to gain weight during stopover periods of migration. In nonmigratory periods, fat comprises about 3-5% body weight. Short distance migrants enhance their fat load to 15% of their weight, and in long-distance migrants, fat is 30-50% of their weight. Fat fuels the aerobic contraction of flight muscles, allowing flights with minimal fatigue.

Day length as stimulus results in premigratory weight gain. Light directly affects the hypothalamic feeding centers, stimulates adjacent centers in brain to change the bird's endocrine secretions, increasing prolactin, corticosterone, and sex steroids. Hormonal changes facilitate the fat deposits resulting from the greater food intake caused by increased appetite. Premigratory state is characterized by increased activity before migration. They become restless. Intensity and duration of migratory restlessness in captives are correlated with distance and period of migration in wild birds. Migratory restlessness is influenced by long days through effect of light on hypothalamus.

Light controls the secretion of melatonin which is necessary for expression of this behavior. Light stimulus is function of length of light period rather than of change in day lengths. Both external and internal aspects of light stimulation reflect their geographic distributions. Temperature is also associated. Thus, when spring comes late, birds do not reach early. When spring is advanced, the birds arrive early. Development of vegetation can influence light-caused reproductive development.

Flight Ranges: Scientists have attempted to derive formulas for "gas mileage" based on the depletion of weight, timing of arrival, and distance between points in migration. They vary considerably. However, small migrants can cover about 2500 km in 100 hours, if 40% of their weight is in fat. That will get them across any of world's major geographical barriers to migration.

Important stopover locations: Shorebirds use key stopover locations during migration. Study over 600 sites and thousands of census reports, states that many species of shorebirds rely on few, important staging areas. Loss of such areas could be detrimental to shorebird. Prime example is the current situation at Delaware Bay in New Jersey. In mid-May, thousands of shorebirds including Red Knots, Ruddy Turnstones, and Sander lings stop on shores of Delaware Bay to refuel before continuing their journey to their breeding grounds on Arctic Tundra.

Internal Rhythm's Influence

Photoperiod and circannual cycles are clearly important, because even in captivity under constant temperature and food. Migratory birds exhibit zegunruhe - migratory restlessness. They get fidgety during night and bump into their cages to south or north depending on season. External factors also play role, especially at fine-tuning migration. Many migrants wait at staging areas for favorable weather conditions to initiate their flights.

Time of day: Soaring birds fly during the day to take advantage of rising columns of hot air - thermal soaring. Aerial insectivores migrate by day to feed along the way. Most other passerines fly at night, when temperature drops, and air is more humid.

Speeds: Most birds migrate at speeds very close to the "speed of maximum range" that is 6 hours at 15 mph covers 90 miles, but 5 hours at 20 mph covers 100. Speed of maximum range is always slightly higher than speed of minimum power. Speeds vary from 20-40 mph; higher for species with high wing loadings, lower for species with lower wing loading.

Altitudes: Raptors and passerines fly in relatively low altitude (800-3000 m).Shorebirds and waterfowl fly much higher, where air is thin but very cold and low on O_2. Their large bodies and adapted morphologies handle the cold better. The record is for swan ranges at altitude of 29,000 feet. Generally, birds climb as they get lighter.

Migration Hazards

Physical stress of trip, lack of adequate food supplies, bad weather, and increased exposure to predators are the hazards. Communication towers and tall buildings are also becoming hazardous. Many birds are attracted to lights and are killed in collisions with the man made structures. Migrating birds are found in certain areas in larger than normal numbers during bad local weather conditions. Small songbirds migrate north in spring, flying directly over the Gulf of Mexico to coastlines of Texas. Under favorable conditions, they continue inland for many miles before stopping to rest. Storms and headwinds cause the birds exhausted. In such cases, they search for nearest location that provides food and cover. Many motts along the gulf coast become temporary home called migration traps to many birds in short time.

Migratory Patterns

Migratory patterns vary by species, or within same species. Permanent residents like Northern Cardinal and Northwestern Crow do not migrate and find adequate food in winter in their home. Short distance migrants cover short distances. They are permanent residents in most of their range, but with migratory tendencies on edges of their range. Long distance migrants undertake journeys for weeks to cover thousands of miles. About 350 Neotropic migrant species including raptors, vultures, waterfowl, shorebirds, and passerine species breed in United States and Canada, winter in Caribbean, Mexico, Central America and South America. The occasional great invasions beyond the limits of normal range of certain birds, especially species breeding in the far North, are different from general migration pattern as found in periodic flights of crossbills is called vagrant migration. Sometimes they extend well south into southern States. Snowy Owls are noted for periodic invasions in correlation with declines in lemmings. Band recoveries, netting records, and personal observation can determine migration routes and provide insight into the origin and evolution of these pathways.

Routes and patterns of migration vary among species. In North America, most migratory routes are North-South as climatic conditions vary consistently in North-South than East - West. Birds fly to South for warmer winter weather and North for abundant food resources. Migration routes may reflect recent range expansions. Some species migrate within few miles, up and down slope with seasons. Altitudinal migrants is well developed in tropical species exploiting seasonal resources like fruit and nectar that vary with respect to wet and dry seasons. They travel slowly up and down mountain slopes. Many species along Central America coast are short-distance migrants breed in Pacific NW and wintering in South California and N. Mexico. Other species are true long-distance migrants. Their breeding and wintering ranges are widely separated. Arctic terns are the champions, flying 12,000 km one way from Arctic breeding colonies to Antarctic wintering waters.

Buff breasted Sandpipers breed on North Slope of Alaska and winter on pampas grasslands of Argentina. Swainson's thrushes winter in Central– and S. America. Certain deviations from the expected north and south movements are recorded. Some routes are not pole ward rather proceed in many directions.

Loops

Many species do not return north in spring over same route, they used in fall. They fly a loop or ellipse. Cooke believed that individual returning following the same route and not finding sufficient food either did not return or did not breed. Individuals that took different route with adequate resources survived and left progeny. Others considered the prevailing winds as the major selective factor which give an advantage to individuals who returned north on different route, if the prevailing winds were in suitable directions than along the path used during southward flight.

Loop migrations, as found in adult American Golden Plover, has evolved separately to satisfy its particular needs, and this occurs throughout the world among unrelated species. Golden-plovers leave their winter quarters, cross northwestern South America and the Gulf of Mexico to reach North American mainland .Then they proceed slowly up the Mississippi Valley and, by the early June reach their breeding grounds, with journey as an big ellipse with the minor axis about 2,000 miles and the major axis 8,000 miles stretching from the Arctic to South Temperate Zone.

Dog-legs

In Dog-leg migration, a prominent bend characterize migration patterns in the route. Some indirect pathways connect wintering and breeding areas. When species extend their range, many continue to follow the old route from original range. The new extended routes are simply added to old routes. The Northern Wheatear has extended its range into Greenland and Labrador where local population has transformed into separate race. When the Labrador individuals depart from breeding grounds, they migrate north to Greenland, their ancestral home, and then east to Europe and south to Africa.

Pelagic Wandering

Many pelagic birds of coasts appear to be nomadic, when they are not breeding. Their movement is not random, as there is seasonal shift in population, often for long distances and in specific directions, away from the breeding area after completion of nesting cycle. They have regular migration routes. Movements of tubenoses are correlated with ocean currents, prevailing winds, and temperature. Commercial fishermen are aware that ocean currents are important in supplying nutrients, plankton, and forage fish for larger fish. These foodstuffs attract pelagic birds.

Leap-frogging

When two or more races of same species occupy different breeding ranges during migration, the races breed in the farthest north winter the farthest south. Thus, a northern race "leap-frogs" over the breeding and wintering range of southern populations. This has been recorded in the Fox Sparrow.

Vertical Migration

Many North American birds fly hundreds of miles. Others reach target areas by moving down the sides of a mountain. Here few hundred feet of altitude correlates to hundred miles of latitude. Such altitudinal migrations occur where there are large mountain ranges. Birds in winter come to plains to be warmed. During summer they migrate for coolness to hills. In vertical migrations, some mountain-dwelling bird's annual journey is made on foot from breeding to wintering ground.

Premigratory Movements

Many migrants, especially young, tend to disperse after fledging. Such premigratory movements called post-fledging dispersal, reverse migration, and post breeding northward migration which relates to locality-faithfulness, range extension, and gene flow. These movements are not true migrations even though they are repeated annually by species of the same age class but not by the same individuals. The young of some species wander late in summer and fall for many miles north of the area in which they were hatched. Molt migration is exhibited by many waterfowl species which travel considerable distances away from their nesting area to traditional molting sites, where they spend a flightless period in eclipse plumage. At this time, they may move well into breeding ranges of other geographic races of their species. Such movements may be governed by food availability.

   Migration by Night: These journeys are wonderful in many ways. Here again, food probably play an important part. Most of the daytime has to be devoted by birds in search for food. Few birds accomplish the journey. By crossing at night, many difficulties are obviated. Birds migrate by night to escape the attacks of gulls. Powerful birds like Cranes and Storks migrate during the hours of darkness

Migration Route

Each migratory species has often broad route between its nesting and winter ranges. Waterfowl follow restricted path based on stopover habitat. It was thought that birds followed specific flyways, like the Mississippi Flyway. Migrating songbirds fly across broad areas and are not strictly grouped into specific flyways. Some general patterns are noted. In North America, many songbirds and shorebird follow an elliptical route. Several shorebirds winter in South America, take northern route through Central America to reach their summer homes in Northern Canada. In fall, the birds fly towards southeast, to wintering grounds. Different bird groups fly at different altitudes. Soaring migrants like hawks and vultures take advantage of thermals and migrate at 3,000 feet. Migrating waterfowl uses wide range of altitudes, from 300 feet to 10,500 feet. Most passerine species migrate at night. Over land, they fly within 2,400 feet. Over water, migration occurs within 12, 000 feet. Weather influences the migratory altitude. Some birds are found at extreme high altitudes. Bar-headed Geese migrate over the Himalayas and fly at 27,880 feet.

Orientation and Navigation

Bird's homing instinct allows them to return to same area year after year. Young birds have the innate knowledge of direction and distance they travel. After its arrival on wintering grounds, the young bird select winter range to which it imprints during that winter. After the first year, bird return to same area. Adults possess more homing skills. Migrants cover many miles in annual travels, often the same route year after year with small deviation. First year birds migrate to winter home they have never seen without escort and return the following spring to home. Their amazing navigational skills are largely hidden. Birds navigate by using position of stars, by sensing changes in earth's magnetic field, and even smell. Some follow selected pathways in annual migrations in relation to stopover sites that provide food supplies. In spring, Sand hill Cranes use the Central Platte River Valley at Nebraska as staging habitat during migration.

Factors in environment influence the expression of migratory behavior that leads to evolution of migratory pattern. Such factors provide direct, proximal stimulation for physiological preparation for migration and provide information to navigate during migratory passage. Navigation requires knowing about present location, destination, and direction to travel to get from the present location to destination. Birds have navigated for eons using information from surroundings. Of the 3 kinds of information necessary for navigation, environmental cues are used to orient migratory flight in proper direction. Some evidence suggests that birds use neither the positions of the sun or the stars. They learn both the location of the wintering area and location of breeding area to navigate.

Orientation Cues

Understanding the migratory flight has been directed toward environmental cues that bird maintain a particular flight direction. These cues are landmarks on earth's surface, magnetic lines of flux, both the sun and the stars in celestial sphere, and perhaps prevailing wind direction and odors. Landmarks are useful as primary navigation reference only if the bird has been there before. Birds used the river as reference to shift their orientation.

Gustav Kramer showed that migratory European Starlings oriented to azimuth of the sun when he used mirrors to shift the sun's image by 900 in laboratory and obtained a corresponding shift in birds' orientation. Since birds maintain constant direction even though the sun traversed from east to west during day, the compensation for this movement demonstrated that birds were keeping time. Birds can detect polarized light from sunlight's penetration through atmosphere. Radar studies have shown that birds do migrate above cloud decks where landmarks are not visible, under overcast skies where celestial cues are not visible, and even within cloud layers where neither set of cues is available(Figure 9.44,9.45).

Iron-containing magnetite crystals are associated with the nervous system in pigeons, but their association, with sensory receptor for geomagnetic cue, is unknown. An alternate hypothesis for sensory receptor suggests that response of visual pigments in eye to electromagnetic energy is the basis for geomagnetic orientation. Previous exposure to celestial orientation cues enhances ability of bird to respond suitably, when only geomagnetic cues are available. Since there are characteristic patterns of wind circulation around high and low pressure centers at altitude most birds migrate, it is hypothesized that bird could use these prevailing wind directions as an orientation cue. Sense of smell has long been considered to be poorly developed, although some species can discriminate odors well.

Homing Pigeon Studies

Homing pigeons are used to develop an understanding of migration and homing abilities. Hans Wallraff transported homing pigeons to very distant location under strict conditions, in closed, airtight cylinders. Light was turned on and off and loud white noise was played. Cylinders were enclosed in magnetic coils for changing magnetic field and were mounted on tilting turntable connected to computer that varied both rotation and tilt of cylinders. After release at distant and unknown area, the birds came home to their roost, without trouble. This shows that birds have both an internal compass and an internal map. Some possible explanations are as follows:

Internal Maps

The nose knows theory suggests that homing pigeons use an olfactory map. In theory, a gradient map of odors could be produced for giving some information for detecting direction, even if the pigeon were dropped in new site. Olfactory navigation may extend to distance of 310 miles.

Magnetic Map Theory

Birds use the earth's magnetic field to obtain partial map of its position and might be able to estimate its latitude based on strength of magnetic field. While the change in strength is very small from one site to next, homing pigeons identify small changes in magnetic field.

The Compass

That navigation requires the compass is advocated. Internal map provides with general location of where it is relative to its homing and its internal compass guides its flight and keeps it on course. Migrating birds probably use several different compasses.

The Sun Compass

Gustav Kramer discovered the sun compass by placing European Starlings in orientation cages and then using mirrors to shift the apparent location of the sun. Birds shifted their migratory restlessness to match the compass direction showed by new position of the sun. Bird's sun compass is tied to its circadian rhythm and has time compensation ability for changes in the sun's position over course of day. This theory is supported by another experiment in which pigeons were placed in closed room with an altered cycle of light and dark.

The Star Compass

The sun compass plays role in homing and used during day time migration. Many songbirds migrate at night. Birds use stars for navigation as evidenced from experiments in which birds were placed inside an enclosed planetary dome. Birds perhaps have genetically coded map of stars.

The Magnetic Compass: Robins in migratory mood were placed in covered cages to eliminate sun, star and other light clues and were observed hopping in correct migratory direction. Helmholtz coil was placed around the covered cages. The coil allowed shifting the direction of the earth's magnetic field. When direction of magnetic field was changed, the robins changed their hopping direction. Birds sense the north and south ends of compass, but they cannot differentiate the two. Birds apparently can sense that magnetic lines of force align toward earth poles. They can also identify the lines of force as they approach earth and, through some method, detect and make navigational decisions based on dip angle.

The sunset cue: Patterns of polarized light also appear to play a key role in navigation. Many nocturnal migrants initiate flights at sunset. Birds seem to use the polarized light to provide information on initial migratory flight directions.

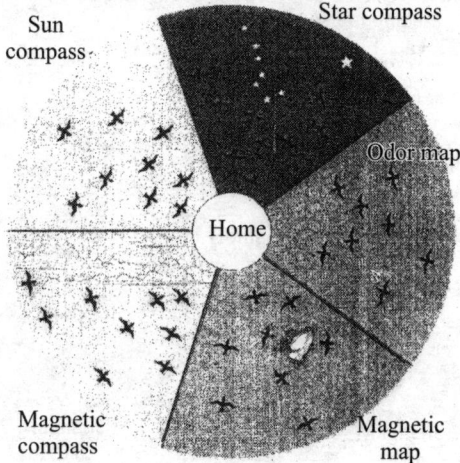

Figure 9.43 Compasses used by birds.

Vector Navigation

Perdeck transported starlings from The Hague to Switzerland, releasing banded starlings in a geographic location for which the population had no experience. The recapture of banded birds showed that the adults, who had previously made the migratory flight, knew they had been displaced and returned to their normal wintering range by flying a direction about 900 to their usual southwesterly course. The juveniles continued to fly southwest and were recaptured on the Iberian Peninsula. Coupled with another displacement of starlings to the Barcelona coast in Spain, Perdeck ended that the proper direction of the migratory flight was innate, that is, inherited in DNA.

Landmarks: Birds that migrate during the day follow, and recognize natural landforms like mountain ranges, rivers, and lakes. Bird uses multiple compass methods and calibrates them against one another. Some use one type of compass. Others rely on different primary system. Complexity of migration and the way it is accomplished is a marvel. Individual birds show amazing consistency to their migratory pathways and their nesting locations from year to year. Bird may track their migration path almost exactly from year after year and often return to same field or nesting location. This site fidelity extends to stopover points along migration route and to wintering locations. Individual Wood Thrushes winter in same area each year in Veracruz, Mexico, and demonstrate consistently site fidelity in their U.S. breeding grounds. Fidelity to stopover points is noted in some larger bird species. Sand hill gather into large flocks at stopover points each year. Spectacular event occurs each spring when 80-90 % of mid-continent population of Sand hill Cranes stops in North Platte and Platte River Valleys of Nebraska.

BOX 9.1

Wonders of bird migration

Dr. Fatik Baran Mandal

Many species of fish, mammals, birds and insects undertake amazing migratory journeys. Birds are most mobile animals on earth. Recorded history of bird migration goes back to about 3000 years to the times of Homer, Herodotus and Aristotle. Regular, recurrent, seasonal movement from one geographic location to another and back again of a population is called migration. Inhabiting two different regions when each region provides favourable conditions is the benefit derived by migratory species in lieu of rigors of migratory journeys. Arctic Terns move yearly as far as from the Arctic to the Antarctic with subsequent return. Upland Sandpipers never experience winter. They breed in North American grassland and spend winter on the Pampas of Argentina. By departing wintering ranges to arrive breeding areas during springs, migrants probably reduce competition for adequate space and food for themselves and their offspring. Various Wood Warblers and Flycatchers are totally migratory, while most Woodpecker

species are permanent resident. Some species have both migratory and non-migratory individuals. Such partial migrant species like Blue jays create problem in suggesting simple, singular explanation for the origin of migration.

Adequate food for the young appears to be the primary factor in determining where and when a bird species will breed. Coldness and stress are believed to prompt their migratory departure. Movements to higher latitudes, where day lengths are larger, provide ample time for feeding young and shorter exposure in the nest where there is risk of predation. At lower latitudes, breeding seasons are longer, allowing multiple attempts to produce young ones. This longer breeding season is related to a higher probability that nests will suffer losses to predators. Fall departure from higher latitudes removes individuals from climatic conditions that will eventually exceed their physiological tolerance limits. The arrival of migrants on the winter range, however, increases the chances for greater inter-specific competition with resident species, when resource avilability might be reduced. This cost plus the hazards associated with the migratory journeys decreases adult surivorship. Evolution of migratory behaviour must on an average offer a favourable balance between various costs and benefits.

Migratory population becomes more abundant, when the resident non-migratory population becomes smaller. If changing environmental conditions become increasingly disadvantageous for the resident population of inter-specific competition becomes more servere, the resident population could eventually disappear, leaving the migrant population.

There occurs change in the neural centre in hypothalamus controlling hunger and satiety in the pre-migratory state of birds in spring. Birds gain weight by

overeating. Small perching birds like warblers gain weight about two weeks prior to migration. Fat comprises about 5% of a bird's body weight during non-migratory period. Short and middle distance migrants increase their fat load to about 15% of their body weight while in case of long distance migrants it is 30-50%. In migratory state, most birds become restless during night, perhaps in anticipation of the migratory flight. Like premigratory weight gain, migratory restlessness is stimulated by long days through the effect to light on the hypothalamus, corticosterone, and sex steroids. Additionally, light stimulates the release of melatonin, which is necessary for the expression of migration behaviour.

Birds wintering in the tropics have evolved a response to photoperiod which results in premigratory changes similar to that of birds wintering in the North

Temperate zone under increasing day length. Even birds wintering in South America initiate pre-migratory preparation in March and April under the decreasing day lengths. Birds generally travel in waves, the magnitude of which varies with populations, species, year, and time of the year. Characteristically, a few early individuals come into an area followed by a much larger volume of migrants. Smaller birds such as Rails, Shorebirds, Flycatchers, Orioles and Thrushes are tropical nocturnal migrants. Small birds migrate by night to avoid their enemies. Day migrants include Ducks and Geese, Loons, Cranes, and Swifts. Some birds migrate only during the day, as they are depedent on updrafts created either by thermal convection on the deflection of wind by topographic features like hills and mountain ridges.

Flight velocity of birds ranges from 20 to 50 miles per hour. For sustained flight larger birds typically fly faster than smaller birds. The American Robin is a slow migrant, taking an average of 78 days to make the 3000 mile trip from Iowa to Alaska. Gray-cheeked Thrush cover 4000 miles trip at an average rate of 130 miles per day. Ninety five percent of migratory movements occur at less than 10,000 feet although the bulk of movements occure below 3000 feet. Bar-headed Geese have been observed flying over the highest peaks of 29,000 feet. For most small birds the favoured altitude appears to be between 500 and 1000 feet.

Most of the effort applied to understanding how birds make a migratory flight has been directed toward environmental cues that birds use to maintain a particular flight direction. These cues are landmarks on the earth's surface, the magnetic lines of flux that longitudinally encircle the earth, both sun and the stars in the celestial sphere arching over the earth and perhaps prevailing wind direction and odours. Proper direction of the migratory flight is probably innate, inherited in their DNA, and birds were also genetically programmed to fly a set distance.

Source: Science India, May 2010

Summary

1. Birds are feathered, winged, bipedal, heat-absorbing vertebrates. They lack external pinnae. Several birds manufacture and use tools. Many species undertake migration. Many perform short irregular movements. They are social, communicate using visual signals and through calls. Pigments produce colours by reflection and diffraction effects and vary with bird's habit. Most birds are dark above and white below. Feathers often show mottled or speckled patterns. Birds in sunlit upper tree branches show bright yellow, yellow-green, and blue colours, singly or in combination.

2. Digits may be enclosed in a coat of feathers in birds of cold areas. Birds of prey develop long raptorial talons. In perching birds, one digit is directed backwards, to grasp a branch. In climbing birds, 4th digit is directed backwards as well as the first, and the foot forms a sort of pincer, with long curved claws.

3. Birds quickly learn the way of process of obtaining food. Food choice depends on species-characteristic motor patterns, and structures. Several birds glean for insects, invertebrates, fruit, or seeds. Birds with a moderately long bill like song-thrush eat either flesh or fruit. Seed eating birds like finches have short, thick, strong bills.

4. Backbone and limb girdles are modified to carry the body weight on wings or legs. In Pelicans and frigate birds small keel is concentrated near the front of breastbone. Number of cervical vertebrae is greater in birds with longer necks; 14 in pigeon including 2 that bear ribs not articulating with sternum. Ribs are large, double-headed, bear uncinate processes on vertebral portions, and join to vertebrae.

5. Hook-like projections overlap the rib next behind and strengthen the whole thoracic cage. Femoro-tibial muscles make up with longer muscles, the extensor system of knee. The lateral side of hip joint is supported by small abductor braces, ilio-trochanteric muscles, and act as medial rotators, opposed by obturator and ischio-femoral muscles, which work as lateral rotators. Pectoralis major attached to edge of sternum depresses wing in flight.

6. Varied modifications for flight embrace a light-weight skeleton, 2 massive flight muscles, the musculus pectoralis and supracoracoideus. Wing serve as an aerofoil. Wing form and size governs flight. Several birds mix steam-powered, flap flight with soaring flight. About sixty extant bird species are wingless.

7. Flight is central to bird evolution. Flight surfaces are composed of feathers. Wing is shaped like an airfoil, thick in front and thin and tapering behind. As the air stream flows across the wing, stream moves faster along longer upper surface than shorter lower surface.

8. Food swallowed down the esophagus may be stored in crop, found in grain-eating birds. True stomach is divided into two parts, a glandular proventriculus, and a muscular gizzard. Structure of anterior chambers of gut varies with diet. In grain eating birds like pigeon, the crop is large and seeds are first macerated by storage there. They are then mixed with peptic enzymes in proventriculus and ground up in muscular gizzard, which in pigeons has a horny lining and contains numerous small stones.

9. Size of heart and rate of heart beat vary with size and activity of bird, larger birds having relatively small and less rapid hearts. Red corpuscles are oval and nucleated, smaller in actively flying birds than in larger flightless ratites and carry a large amount of haemoglobin. Elaborate song of male birds is used in courtship both as a sexual stimulant and as a threat to other birds invading the chosen territory.

10. Lungs are small spongy organs with little elasticity. Air passes backwards in a large bronchus running through lungs and giving off branches to lung substance, but continuing beyond inspiratory air sacs. The end product of nitrogenous metabolism is insoluble uric acid, synthesized in liver; probably from ammonium lactate. Urine is concentrated in cloacal chambers. Uric acid precipitates as whitish granules. Urinary bladder is absent in adult.

11. Albumen of egg is produced by long tubular glands, opening to lumen. Oviduct has various parts, the upper part secrets mainly albumen, the lower part produces the shell, and the lowest part the mucus, to assist laying. Much complex behaviour is responses to only limited parts of natural stimulus situation. Much of elaborate social life of birds depends on such sign stimuli displayed by one bird and serving as releasers setting off particular actions or trains of action in another bird.

12. Many elaborate forms of display evolved by birds are releasers of this sort, and structures and actions on part of young release the appropriate behaviour of parent. The red breast of the robin is the agent that releases attacks by other birds. Birds depend more on their eyes than on other senses. They are fully visual. Eyes are extremely large.

13. Some scientists believe that the ancestors of all birds evolved between 65 and 53 million years past, independently of dinosaurs. However, the dinosaur-to-birds theory took surprising flip with the invention of 2 species of feathered dinosaurs in China.

14. Monogamy allows for biparental care, which is especially important for species in which females require males' assistance for successful brood-rearing. Extra-pair copulation is common .Such behaviour typically occurs between dominant males and females paired with subordinate males, but may also be the result of forced copulation in ducks and other anatids.

15. Ratites have been descended from flying birds and represent several different evolutionary lines. Various ratite birds have been placed in eight distinct orders. Ostriches now limited to Mesopotamia are the largest living birds.

16. Rock Dove belongs to the family Columbidae, often simply referred to as the pigeon. Visible differences between males and females are few .The species is monogamous, with two squeakers (young) per brood. Both parents care for young for a time. Habitats embrace varied open and semi-open environments.

17. About 5 billion terrestrial birds of about 200 species leave North America for central and South America every fall. Many birds migrate from Europe and Asia to reach Africa. Birds shifted their migratory restlessness to match the compass direction showed by new position of the sun. Bird's sun compass is tied to its circadian rhythm and has time compensation ability for changes in the sun's position over course of day.

Review Questions

Short Answer Questions

1. Mention two characteristic features of birds.
2. What is rhamphotheca?
3. What is auricular feather?
4. What is uropygium?
5. Mention the function of preen gland.
6. What are remiges?
7. What are rectrices?
8. What are barbules?
9. Mention the function of feathers.
10. What do you understand by schizognathous skull?
11. Define heterocoelous vertebra.
12. What are uncinate processes?
13. What is pygostyle?
14. What is synsacrum?
15. Which bone is called wish bone?
16. Comment on Pigeon's milk.
17. What is syrinx?
18. What are parabronchi?
19. What do you understand by double respiration?
20. Comment on altitudinal migration.

Long Answer Questions

1. Discuss the factors governing bird flight.
2. Describe the evolution of birds.
3. Classify the Class Aves up to living order with examples.
4. Draw and describe the feathers of pigeon.
5. Draw and describe the digestive system of pigeon.
6. Describe the respiratory system of pigeon with suitable diagrams.
7. Describe the mechanism of circulation through heart in pigeon.
8. Describe the evolution of migration in birds.
9. Describe various types of avian migration.

10

The Mammals

About 5000 mammal species have 3 chief features in general viz., presence of 3 middle ear bones, hair, and mammary gland. Malleus, incus, and stapes transmit vibrations from the tympanic membrane to inner ear. Hair serves for insulation, color patterning, and in sense of touch. Female produces milk to nourish offspring and invest much energy for caring their offspring. The small mammals are the shrews and bats. The largest mammal is the blue whale .Mammal occupies diverse ecological niches and exhibit different lifestyles. Mammals fly, glide, swim, run, burrow, or jump.

They are found in all continents, oceans, many oceanic islands, deserts, tropical rainforests, and polar icecaps. Some are arboreal. Many are partially aquatic. Whales and dolphins are found in polar, temperate, and tropical waters. Hair slows the heat exchange with the surrounding. Whiskers are sensory. Vibrissae are richly innervated. Hair affects color and provides protection; by providing an extra protective layer or by deterring predator. It camouflages predators or prey, to warn predators or communicate social information. Differentiated teeth are replaced just once throughout a human life. Lower single jaw bone is called the dentary. Four-chambered heart, secondary palate, well-developed brain, and homeothermy are their important characters.

Monotremes are the most primitive mammal which lay eggs. Marsupials give birth to altricial young after 43 days gestation period. Young born at an early stage of morphological development attach to mother's tit. Gestation lasts longer in placental. Eutherian young remains connected with mother through placenta. Newborn depends on mother's milk. Mammals are either polygynous, or promiscuous. Females incur high costs in gestation and lactation. Male attempt to produce many offspring in a mating season. In polygynous mating system, few males fertilize many females and many males fertilize none. Many mammals are marked by sexual dimorphism. About 3% mammals are monogamous where males also provide care to offspring. Mating systems vary within species depending on environment. When resources are low, males mate with single female and care for offspring. When resources are plenty, the mother care for young on her own and males attempt to sire offspring with many females. Some display cooperative breeding. Naked mole rats are eusocioal. Many are seasonal breeders. Females, of some, store sperm until favorable conditions. In some, eggs are fertilized shortly after copulation. Development of embryo may be arrested for sometime in some species. Small mammals live short lives. Large mammals live longer lives. Bat, the small mammal live for more than a decade. Thermoregulation play important role in behavior. Mammals exhibit fossorial, aquatic, terrestrial, and arboreal lifestyle. They live in groups of tens, hundreds, thousands, or more individuals. Some are solitary, may be nocturnal, diurnal, or crepuscular.

Olfaction, hearing, tactile perception, and vision are sensory functions. Olfaction plays role in foraging, mating and communication. They use pheromones and olfactory cues to inform about their territory, individual, or group identity and even reproductive status. Hearing is well-developed. Echolocation has evolved in several groups. They perceive through hair and skin. Touch serves in communication. Vision is well-developed in many. Nocturnal mammals have well-developed eyes. They are carnivores, herbivores or omnivores. Predation is a major cause of mortality for many. They are also preyed on by many animals. Some domesticated mammal provides meat, milk, or fiber. They are important for the eco tourism and help control crop pest population and serve in research.

10.1 Milk

Female produces milk for young. The name "mammal" refers to mammary gland and mammae. Nipples, the small, outer protrusion of mammary gland, are generally found. Number of nipples shows the size of litter. Teats are elongated, extend from the mammary gland and are characteristic of members that bear precocial young including the family Bovidae and Cervidae. Precocial young not born in nests are covered with hair. Their eyes open just after birth. Sometimes eyes open before birth. These species follow mother soon after birth. Young obtains milk which is composed of water, fats, carbohydrates, protein, minerals, and vitamins. The composition varies among species. In some species, the young rapidly increase body mass by consuming the milk which contains large amount of fat. In species with rapid skeletal growth; milk contains high level of protein and some minerals. The first milk produced by female and ingested by young, the colostrum is high in fat, nutrients, and vitamins, and immunoglobulin. Within hours, or days of birth, the immunoglobulin does not pass through gut of young. Immunoglobulin obtained earlier from mother remain in newborn's blood until the young develops its own immune system.

10.2 Hair and Colouration

Hair is an elongated rod of keratinized cells. Keratin is the major component of epidermal structures. Hair cells area lives only in active growth site in root located at the base of hair follicle. In root, the insoluble keratin is secreted to form the shaft. As the new growth pushes hair cells of shaft outward, the hair dies. Pigment is added in this formative period. Visible hairs contain no living tissue and cannot change color. Hair color can fade. The term pelage refers all hairs. The new pelage often has the same color as the old. In some, juvenile hairs may be characterized by one color, whereas adults exhibit different color. Hair color may vary seasonally. In snowshoe hare, the pelage is brilliant white in winter, but brown in warm months. Outer cell layer on hair strand constitutes the cuticle, one of 3 hair layers. The cuticular pattern often is unique in some species. Hair identification indicates the diet of predator species.

The "fur" is the soft, dense hair. Hair of sheep is very fine, often crimped. Cuticle possesses small barbules that hold hairs together. Many dead air spaces in hairs makes wool a fine insulator. Sebaceous glands associated with hair follicles, also present in hairless areas, produce sebum. Sebum maintains the hair and skin by keeping them wet. Most mammals possess 2 kinds of hair. Dense hair provides insulation. The outer guard hairs are longer, thicker, and contain the pigment. Longer hairs lie over dense hair under fur and protect it from abrasion. Hairs are generally directed posteriorly. Muscles cause the hair to stand up or fluff by pulling on follicles. Whiskers on face called vibrissae have sensitive receptors at base of their follicles. Nocturnal species have long vibrissae to move among obstacles in night. Porcupine quills and the hard, sharp spines of some are uniquely modified guard hairs. Quills are effective antipredator structures because small barbules allow them to penetrate only in one direction-deeper into face and mouth of predator.

Coloration is relatively drab, indistinguishable at night. Nocturnal species have brown to gray colors for blending with their night-time haunts. Hairs on back and sides are rarely solid brown, or gray, or reddish. Alternating pattern of light and dark pigments called agouti works in species that are active in night, or day, like voles and cotton rats. The stripes of chipmunk do not enhance its visibility but provide camouflage. Colouration of most carnivores provides camouflage. Dorsal and ventral surfaces of mammals, such as on the head, body, and tail, often display different colors.

10.3 Skull

The cranium contains and protects the brain and provides openings for ears. Receptors for critical senses of vision and olfaction, and openings associated with gas exchange and nutrition, occur within rostral area of skull. Beginning in center of jaw, the front-most teeth are incisors for gnawing or nipping; next are long and sharp canines for piercing; and finally the premolars and molars, which vary in size, number, and appearance. Rodent lacks canines and a wide gap called diastema is found between the incisors and cheek teeth. In rodents, only 2 upper and 2 lower incisors exist instead of 4 or 6. Incisors in rodents grow continuously throughout life and their roots originate in skull. Skulls of raccoon and cat possess the full set of teeth and both have evident canines, 6 incisors above and 6 below. Dentition adapted for 3 general diet types: rodent = herbivore; raccoon =omnivore; and cat = carnivore. Molariform teeth differ among rodents based on diet. Cheek teeth of omnivorous raccoon are unspecialized. Humans are omnivores with teeth like those of raccoon. Cheek teeth of carnivores are specialized for processing flesh. Carnivores have fewer cheek teeth than others, but with piercing and slicing edges. Cheek teeth of canids are intermediate between those of raccoon and cat; piercing and slicing cheek teeth are present, but they have teeth for grinding and crushing. Tooth wears, or lack there of, determine the relative age. An older adult show more teeth wear, whereas a young show no tooth wear, or the presence of temporary teeth. In young zigzag sutures are evident. As animals age, these sutures fill in and harden. The occipital condyles located on both side of foramen magnum articulate with atlas (Figure 10.1).

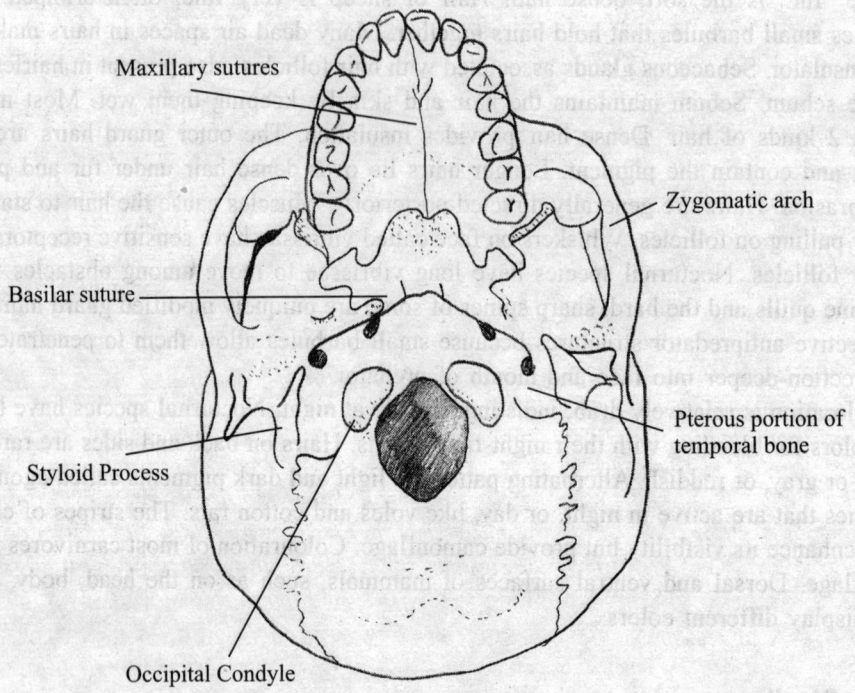

Figure 10.1 Skull base in human.

Nails, Claws, and Hoofs

Nails, claws, and hoofs are epidermal hard structures .Keratin is their primary component. Nails are found on tops of digits. Claws are found at the ends of digits, curved and sharp at tips, and serve many functions. Claws are used to dig, or to hold prey. Claws fit into irregularities of tree bark and may dig in for better grip. Hooves are weight-bearing structures at ends of digits. Bones of foot of hoofed mammals are modified and in some elongate to enhance running.

Antlers and Horns

Antlers and horns are partially epidermal and include keratin. Antlers often called horns during deer-hunting season. However, antlers and horns are different structures. Antlers, the feature of male of family Cervidae, but in caribou and domesticated reindeer, antlers are found in females. Antlers made of solid bone are shed annually, and often become larger and elaborate with age. During spring and summer, new antlers grow, become enveloped in skin layer covered by fine hair called velvet. In late summer, antlers attain full growth; the blood supply to skin is cut off, skin dries. The animal works hard to shed or rub off this dead skin, exposing the solid bone of antler. Antlers are attached to skull by bony pedicels on frontal bones, which are not shed. Antler size depends on deer age, genetic factors, and quantity and quality of diet.

Horns are present in cattle, bison, true antelopes, goats, sheep, and in family Bovidae. Except for special breeds of hornless cattle, horns are present in both sexes. In cattle, differentiation between male and female based on presence, or absence of horns is impossible. Horns originate on frontal bone with bony base that is hollow and covered by highly keratinized sheath that antlers lack. This sheath is used to create the powder horns to store black powder for muzzle-loading weapons. Horns are never branched and shed. An antelope-like mammal possesses neither true horns nor true antlers, but shows a hybrid structure, the pronghorn which is present in both males and females. Pronghorns have horn sheath that is shed annually, but the small, bone core is not shed (Figure 10.2).

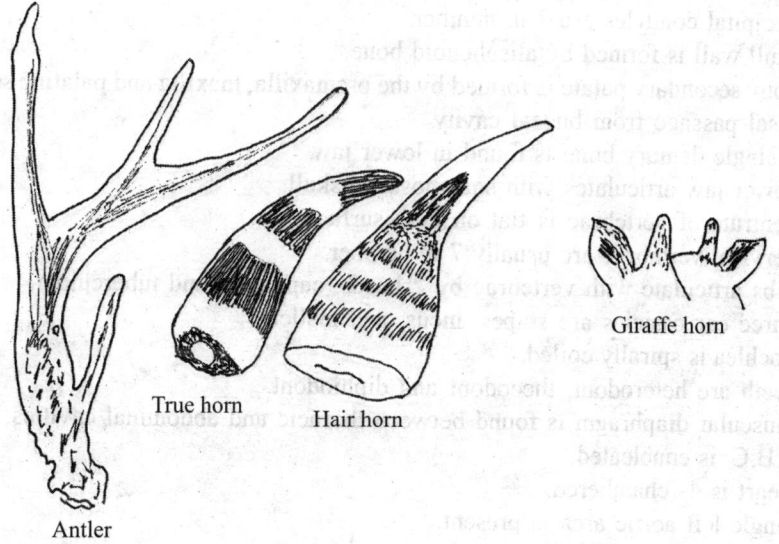

Figure 10.2 Antler and horn.

10.4 External Ear and Tail

Burrowing moles have reduced pinnae. Fossorial mammals lack pinnae completely. In semi-aquatic fur bearers, the pinnae are reduced. Large pinnae would be disadvantageous to semi-aquatic animals, as large ears hinder movements in underwater. Tails of fossorial species like meadow voles are typically short. Deer mice and rats have long tail for balance. The elongated tail of opossum is prehensile. Young opossums ride on back of their mother, hanging onto mother's tail with their own tails. Opossums use tail to carry leaves for nesting. Tail of beaver is wide and flat and acts like third leg. Tail of white-tailed deer is very obvious as it runs away, but when the tail is not erect, the deer blends in with forest background.

Cattle have long tail with brush on end. Virginian bats have tail membrane or uropatagium that serves in flight. In most bats, the uropatagium is naked.

10.5   Classification

Scheme of classification followed here is based according to scheme of classification by Young (1981).

Characters

 1. Skin is hairy and water proofed.
 2. Presence of sweat and sebaceous glands.
 3. Mammary glands are functional in adult females.
 4. Eyes are provided with lower and upper eyelids.
 5. External ears are present.
 6. Occipital condyles are 2 in number.
 7. Skull wall is formed by alisphenoid bone.
 8. Bony secondary palate is formed by the premaxilla, maxilla and palatine separates the nasal passage from buccal cavity.
 9. A single dentary bone is found in lower jaw.
10. Lower jaw articulates with squamosal to skull.
11. Centrum of vertebrae is flat on both surfaces.
12. Cervical vertebrae are usually 7 in number.
13. Ribs articulate with vertebrae by 2 heads, capitulum and tuberculum.
14. Three ear ossicles are stapes, incus, and malleus.
15. Cochlea is spirally coiled.
16. Teeth are heterodont, thecodont and diphiodont.
17. Muscular diaphragm is found between thoracic and abdominal cavities.
18. R.B.C. is enucleated.
19. Heart is 4- chambered.
20. Single left aortic arch is present.
21. Warm blooded.
22. Brain is large with expansion of cerebral hemisphere.
23. Corpus callosum connects 2 cerebral hemispheres.
24. Cranial nerves are 12 pairs.
25. Kidney is metanephric.
26. Ureters empty into urinary bladder.
27. Penis is always present.
28. Fertilization is always internal.
29. Viviparous, young one born alive and are nourished by milk.

Subclass: Prototheria

 1. Absence of pinna.

2. Teeth are present only in young. Adults are provided with horny beak.
3. Cloaca is present.
4. Pectoral girdle possesses separate clavicles and an interclavicle.
5. Ribs articulate only with bodies of vertebrae, not with transverse processes.
6. Mammary glands lack nipples.
7. Cochlea is partly coiled.
8. Testes are abdominal.
9. Oviparous.

Order: Monotremata

General characters are same as that of Prototheria. Example: *Echidna, Ornithorhynchus.*

Subclass: Theira

1. External pinna is present.
2. Teeth are present both in young and adults.
3. Cloaca is usually absent.
4. Mammary glands are provided with nipples.
5. Testes are usually found in scrotal sac.
6. Vasa differentia and bladder open through a common urethra in penis.
7. Oviducts open into vagina.
8. Viviparous.

Infraclass: Metatheria

1. Marsupium or brood pouch is present in females.
2. Mammary glands are sebaceous and bear nipples.
3. Epipubic bones (marsupial) are attached with pubis.
4. Separate coracoid and interclavicle are absent.
5. Vertebrae are provided with epiphyses.
6. Corpus callosum is feebly developed or absent.
7. Vagina and uterus are double (didelphic condition).
8. Viviparous.

Order: Marsupialia

General characters are same as that of Metatheria.
   Example: Macropus, Paramele, Didelphis.

Infraclass: Eutheria

1. Marsupium is absent.
2. Epipubic bones are absent.

3. Mammary glands are well developed with nipples.
4. Ribs bear 2 heads, tuberculum and capitulum.
5. Cloaca is absent.
6. Testes are contained in scrotal sac.
7. Vagina is single.
8. Viviparous.
9. Young always nourished by allantoic placenta.

Cohort: Unguiculata

It includes placental with nails or claws and derived directly from primitive insectivores.

Order: Insectivora

1. Snout is long, tapering.
2. Feet 5- toed with claws, inner toe is not opposable.
3. Teeth are sharp and pointed; dental formula is i3/3, c1/1, pm4/4, and m3/3.
4. Placenta is discoidal.
5. Nocturnal and terrestrial.

Example: *Paraechinus* (Hedgehog), *Talpa* (Mole), *Echinosorex* (Shrdus).

Order: Demoptera

1. Commonly known as "flying lemurs".
2. Membranes stretch between fore-and hind limbs, and between tail and hind limbs.
3. Incisor teeth 2/3
4. Nocturnal.

Example: *Galeopithecus*

Order: Chiroptera

1. Fore limbs are modified for flight.
2. Second and fifth digits are greatly elongated.
3. Ears are provided with large pinnae.
4. Eyes are small.
5. Teeth are sharp. Dental formula is i2/3, c1/1, p3/3, m 3/3.
6. Sternum has a keel.
7. Mostly nocturnal.

Example: *Pteropus, Rhinolophus, Desmodus*.

Order: Edentada

1. Incisors and canines are absent.
2. Feet with well developed claws.
3. Testes are abdominal.

4. Clavicle is present.
5. Coracoid fuses with acromion.
6. Ischium unites with sacrum.
7. Lumbar vertebrae have extra articulating surfaces.

Example: *Dasypus, Myrmecopha, Bradypus.*

Order: Pholiodota

1. Large horny scales are present.
2. Hairs are found between scales.
3. Snout is elongated.
4. Teeth are absent.
5. Tongue is long, sticky, and protrusible.
6. Ears are reduced.
7. Limbs are short, each bears 5 digits.
8. Fore limbs have well developed claws.

Example: *Manis* (Pangolin).

Order: Primates

1. Completely hairy.
2. Hands and feet are prehensile.
3. Five digits are present on each hand and foot is provided with flat nails.
4. Mode of walking is plantigrade.
5. Orbits are directed forward and surrounded by bony ring.
6. Clavicles are always present.
7. Testes are enclosed in scrotum.
8. Two pectoral mammae are provided with teats.
9. Cerebrum is usually large and well convoluted.
10. Placenta is discoidal or metadiscoidal and haemochorial.

Example: Lemur, Loris, *Macca, Homo.*

Cohort: Glires

The cohort is now divided into order Rodentia and Lagomorpha.

Order: Rodentia

1. Usually small mammals.
2. Limbs are provided with 5 toes and claws.
3. Single pair incisors are long and chisel like.
4. Canines are absent.
5. Diastema is present.
6. Testes are usually abdominal.

7. Placenta is discoidal and haemochorial.

Example: *Ratus*, *Funambulus* (squirrel)

Order: Lagomorpha

1. Small to moderate sized mammals.
2. Canines are absent.
3. Two pairs of incisors are present in upper jaw.
4. Diastema is present.
5. Tail is short.
6. Soles of feet are hairy.
7. Toes bear claws.

Example: *Lepus* (hare), *Oryctolagus* (rabbit)

Cohort: Mutica

It includes highly specialized mammals those show complete divergence from their primitive eutherian ancestors.

Order: Cetacea

1. Medium sized to very large aquatic animals.
2. Body is usually spindle-shaped and fish like.
3. Hairy covering of skin is reduced to few bristles on muzzle.
4. Head is long, often pointed without neck.
5. Eyes are small.
6. Forelimbs are modified as paddles.
7. Digits are enclosed in the integument devoid of claws.
8. Hind limbs are absent.
9. Long tail ends in 2 broad transverse fleshy flukes notched in mid-line.
10. Thick layer of fat (blubber) is found beneath the skin.
11. Bones of skull are spongy and contain fat.
12. Stomach is complex and consists of 3 or 4 chambers.
13. Testes are abdominal.
14. Placenta is non-deciduate and diffuse.

Example: *Platanista* (Ganges dolphin), *Balaenoptera* (Blue Whale), *Sousa* (white dolphin).

Cohort: Ferungulata

It includes herbivorous and carnivorous forms, which appear to arise from a common mammalian stalk.

Super order: Ferae

Order: **Carnivora**

1. Carnivorous, terrestrial, arboreal and aquatic.
2. Canines are well developed; incisors are small and always 3 on each side in each jaw.
3. Carnassials teeth are well developed.
4. Clavicles are incomplete or absent.
5. Stomach is simple; caecum, if present is small.
6. Mammae are abdominal.
7. Placenta is deciduate and zonary.

Example: *Panthera, Herpestes* (mongoose), Hyaena.
Super order: Protoungulata

Order: **Tubulidentata**

1. Skin is very thick and covered with sparse hairs.
2. Snout is long and tubular with round nostrils at tip.
3. Pinnae are long, erect and pointed.
4. Tongue is slender and protrusible.
5. Milk teeth are numerous, and permanent teeth are fewer.
6. Digits are 4 or 5 with heavy claws.
7. Placenta is zonary.

Example: *Orycteropus* (Hard-vask)
Super order: Paenungulata

Order: **Proboscidia**

1. Largest, highly specialized and terrestrial.
2. Skin is thick and sparsely covered with hairs.
3. Trunk bears nostrils at tip.
4. Eyes are small; the ears are large.
5. Legs are large and pillar like .Five functional digits are found on fore and hind limbs.
6. Canines are altogether absent. Molars are lophodont. Dental formula is i1/0, c0/0, p3/3, m3/3.
7. Clavicle is absent.
8. Testes are abdominal.
9. Placenta is non-deciduate and zonary.

Example: *Elephas* (Elephant)

Order: **Hyracoidea**

1. Small rabbit - like animals with a split snout.

2. Tail is much reduced.
3. Dental formula: i1/2, c1/0, pm4/4, m3/3
4. Four digits in fore limbs and 3 in hind limbs are found.
5. Clavicles are absent.
6. Testes are abdominal.
7. Mammae are 6 pairs, 4 pairs are inguinal and 2 pairs are axillary.
8. Placenta is zonary.

Example: *Hyrax, Dendrohyrax.*

Order: Sirenia

1. Aquatic mammals are commonly known as sea-cows.
2. Body is streamlined or spindle-shaped.
3. Hind limbs are absent.
4. Fore limbs are modified as paddles.
5. Tail is flattened with lateral flukes.
6. Two pectoral mammae are present.
7. Placenta is non-deciduate and zonary.

Example: *Dugong* (Sea cow) *Manatus* (manatee),
Super order: Mesaxonia

Order: Perissodactyla

1. Large sized hoofed mammals.
2. Teeth are lophodont.
3. Dorso-lumbar vertebrae are 23.
4. Stomach is simple; caecum is large and sacculated.
5. Mammae are inguinal.
6. Brain is well convoluted.
7. Placenta diffuse and epitheliochorial.

Example: *Eauas* (horse), Zebra, *Tapirus* (tapir).
Super order: Paraxonia

Order: Artiodactyla

1. Fore and hind limbs bear 2, rarely 4 digits.
2. Teeth are solenodont, or bunodont.
3. Dorso-lumbar vertebrae are 19.
4. Stomach is complicated and caecum is small.
5. Mammae are few and inguinal, or many and abdominal.
6. Placenta diffuse or cotyledonary.

Example: Hippopotamus, *Sus* (Pig) (Figure 10.3a-g).

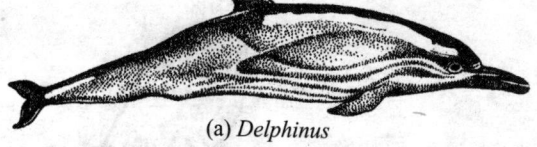

(a) *Delphinus*

(b) *Balaenoptera*

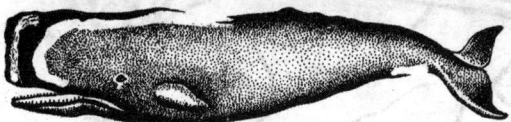

(c) *Physeter*

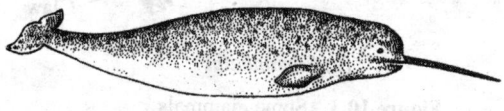

(d) *Monodon*

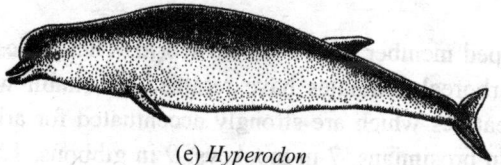

(e) *Hyperodon*

Ciwes digire

Hind limb

Patagium

Mouth

Nostril — Snout

Eye

Pinna — Head

Fur

(f) *Pteropus*

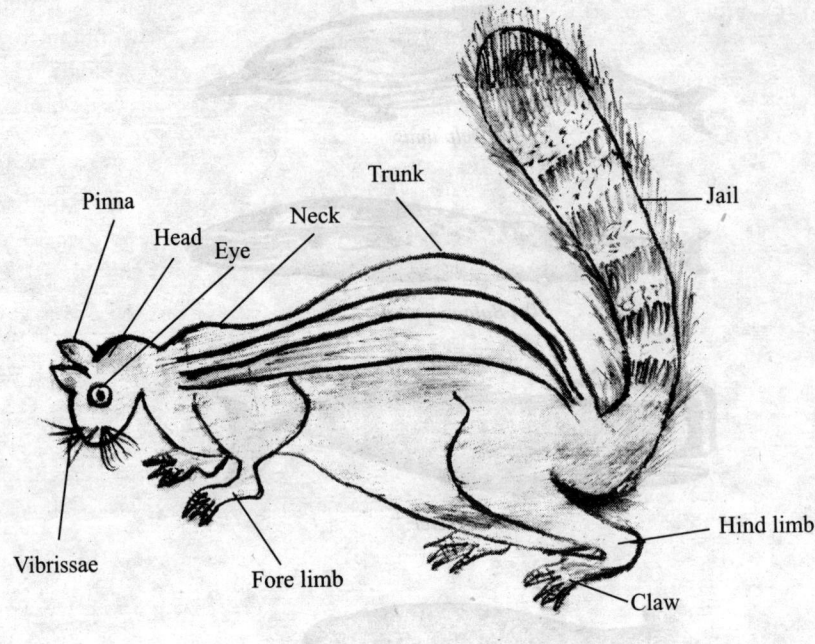

Pinna
Head
Eye
Neck
Trunk
Jail
Vibrissae
Fore limb
Hind limb
Claw

(g) *Funambulus*

Figure 10.3   Some mammals.

10.6   Primates

Primates are the first developed member of the animal world. The earliest eutherians of the Cretaceous were probably arboreal. Primates have retained this habit with features like 5 fingers, toes, clavicle, and features which are strongly accentuated for arboreal life. Growth continues for about 3 years in prosimians, 7 in monkeys, 9 in gibbons, 12 in other apes, and 20 in man. Eyes and ears are developed, at the expense of nose. Primates are microsmatic, with reduction of number and length of turbinal bones and hence of the long snout. Eyes come to face forwards and their fields overlap. Vision is binocular and central areas appear in retinas. Monkeys are more dependent on vision.

Brain is very large in later primates, with cerebral hemispheres reaching far backwards. Olfactory bulbs and rhinopallium are small. Neopallium is very large, differentiated into areas and with large corpus callosum. Occipital pole and frontal areas are well developed in man. Stereoscopic eyes with numerous cones discriminate shapes, and retain the impression of past situations. At early development, all primates have the same relative brain weight, but in adults, the brain is absolutely larger in man. Facial portion is short with large and round brain-case. Foramen magnum comes to face downwards. As the eyes are directed forwards, the orbits become closed off from temporal fossae behind. Head is clearly marked off from body. Neck is very mobile. Skeletal and muscular systems allow jumping, swinging, and grasping. The pentadactyl plan is retained, without loss of digits or fusion of bones. Hand and foot are suitable for grasping. Digits have sensitive pads. The original claw is replaced by flat nail. Clavicle is large to allow mobility of forelimb. Muscles allow rotation of scapula. Teeth are not as specialized like ungulates. Hands are often used to obtain food. They are omnivorous or frugivorous. Molars are quadritubercular, the upper adding hypocone and lower losing paraconid of original pattern, leaving metaconid and protoconid, while hypoconid and entoconid raise to make a posterior pair, sometimes with extra hypoconulid. Cusps are not sharp, but with low (bunodont) cones and extra ones, or cusps join to make ridges. Uterus retains signs of its double nature in the earlier types, but later becomes single chamber. Number of young produced is small with large brains. There is often only single pair of pectoral teats. Primates extend parental care for long time (Figure 10.4).

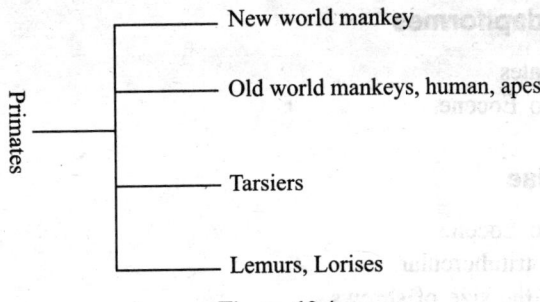

Figure 10.4

10.7 Classification of Primates

Classification followed here is based according to scheme of classification of Young (1981).

Characters

1. Completely hairy and generally arboreal mammals.
2. Head is clearly marked off from body than is usual in mammals.
3. Neck is very mobile.
4. Pendactyl limb.
5. Digits are mostly elongated and with sensitive pads and flat nails.
6. Hallux and pollex are often opposable to allow grasping.
7. Short clavicles are present.
8. Scapula could be rotated.
9. Orbits are directed forward and surrounded by bony ring.
10. Radius and ulna are separate and jointed.
11. Tibia and fibula are separate except in tarsioids.
12. Molars become quadritubercular.
13. Often with single pair of pectoral teats.
14. Tooth row is short usually with 2 or, 3bicuspid molars.
15. Braincase is relatively larger and rounded.
16. Omnivorus, or frugivorus.

Order : Primates

1. Head is provided with snout.
2. Some toes are provided with claws, others with nails.
3. Tail is long (rarely absent), never prehensile.
4. Fur is usually thick, wooly.
5. Solitary.
6. Mostly nocturnal.
7. Paleocene to Recent.

Infra-order: Plesidapiformes

1. Earliest primates.
2. Cretaceous to Eocene.

Family: Paromyidae

1. Cretaceous to Eocene
2. Molars were tritubercular.
3. Animals had the size of shrews.
4. Probably herbivores.

Example: *Puragalorius*.

Family: Plesiadapidae

1. Abundant during Paleocene to early Eocene.
2. Upper and large procumbent lower incisors were chisel-like.
3. Heavily built skeleton was probably used for vertical climbing and leaping.
4. Large compressed claw was present.

Example: *Plesiadapis*

Family: Carpolestidae

1. Paleocene animals.

Example: *Carplestes*.

Family: Picrodontidae

1. Paleocene animals.

Example: *Picrodus*.

Infraorder: Lemuriformes

1. It includes lemurs of Madagascar and their fossil allies.
2. Paleocene to recent.

Family: Adapidae

1. Eocene animals without procumbent incisors.
2. Brain case was small but carried temporal crests.
3. There was typically a full dentition (2.1.4.3.)
4. Tympanic ring was included in bulla.
5. Central hemispheres were small.

Example: *Adapis*, *Noyhrectus*.

Family: Lemuridae

1. Mostly nocturnal.
2. Nose is provided with numerous well developed turbinal bones.
3. Pollex and hallux are used for grasping; most digits have nails, but 2nd digit of foot has a toilet claw.
4. There is a post-orbital bar, but temporal fossa opens widely to orbit.
5. Tympanic bone forms a ring, lying within petrosal bulla, but not fused with it.
6. Upper incisors are very small.
7. First lower premolar is caniniform. Molars are triangular, or square shaped, typical tuberculosectorial type, with heel.
8. Snout is long.
9. Cerebral hemispheres are small.
10. Cerebral sulci tend to run longitudinally.
11. Females are polyoestrous.
12. Uterus is bicornuate.
13. Placenta is epitheliochorial and diffused.
14. Feed on fruits, flowers and leaves but not insects.

Example: *Megaladapis*, Lemur.

Family: Indridae

1. Plesitocene to Recent.
2. Small to large lemurs with bare face and densely wooly pelage.
3. Tail is short to long.
4. Hind limbs are large than forelimbs.
5. Thumb is slightly opposable.
6. Salivary glands are greatly enlarged.
7. Dental formula is 2.1.2.3.

Example: *Indri*, *Propilhecus*.

Family: Daubentoniidae

1. Small primate with long bushy hair.
2. Face is short.

3. Ears are large and membranous.
4. Hind toe is opposable.
5. Incisors are rodent-like and permanently growing with enamel on anterior surface.
6. Dental formula is 1 0.1.1.3.

Example: Daubentonia

Infraorder: Lorisiformes

Family: Lorisidae

1. Small primates with dense wooly fur.
2. Face is shorter.
3. Tail is short or absent in lorises, long in galages.
4. Eyes are very large.
5. Ears are small to large and membranous.
6. Opposable thumb and toe.
7. Dental formula is 2.1.3.3.

Example: Loris, *Galago*, *Perodicticus*.

Infraorder: Tarsiiformes

1. Paleocene to Recent.
2. Represented by one living genus *Tarsius*.

Family: Anaptomorphidae

1. Eocene to Miocene.
2. Eyes were large and face short.
3. Temporal and occipital lobes were enlarged.
4. Olfactory regions are better developed.

Example: *Tetanias*, *Omomys*.

Family: Tarsiidae

1. Very small, may be arboreal, nocturnal, or insectivorous.
2. Eyes are enormous, relatively larger than in any other primate.
3. External ears are large.
4. Tail is long and naked,
5. Hind limbs are elongated.
6. Digits terminate in disk like pads.
7. Hind toe is opposable.
8. Dental formula is 2.1.3.3.
9. Molar retains simple tritubercular pattern.
10. Olfactory regions are small and cerebral hemispheres are large.

11. Corpus callosum is small and anterior commissure is large.
12. Cerebellum is small and simple.

Example: *Tarsitus*.

Suborder: Anthropoidea

1. Diurnal and microsmatic.
2. Large forwardly directed eyes are present.
3. Well marked central area in retina is found.
4. External ears are small and edge is usually rolled over.
5. Rounded head is carried on mobile neck.
6. Snout is short.
7. Premolars bicuspid, 2 to 3 in number.
8. Two mandibles are united by fusion of symphysis.
9. Incisors are spatulate.
10. Brain is large.
11. Central hemisphere is well developed.
12. Extended neopallium is present.
13. Surface of neopallium is highly fissured and well marked central sulcus separates motor and sensory areas.

Super family: Ceboidea

1. Arboreal.
2. Limbs are long
3. Thumb is not fully opposable.
4. Flat nosed.
5. Presence of tail for balancing.
6. Facial vibrissae are present.
7. Caecum is large.
8. Tympanic bone is a ring fused with petrosal.

Family: Callitrichidae

1. Squirrel like in appearance.
2. Presence of thick fur and prehensile tail.
3. Digits are provided with claws except of foot.
4. Pollex is not opposable.
5. Incisors and canines are partly procumbent.
6. Presence of 3 premolars and 2molars.
7. Cusp pattern is tritubercular.
8. Live in families of 3-8, but may also form large groups.

Example: *Callithrix*.

Family: Cebidae

1. Small to medium sized.
2. Generally possess a long tail, which in some species is prehensile.
3. Lack ischial callosities and cheek pouches.
4. Thumb nail is opposable.
5. Big toe is opposable.
6. Dental formula is 2.1.3.3.
7. Feed on fruit and nuts or insects.

Example: *Aotus, Ateles, Cebus*.

Super family: Cercopithecoidea

1. Internasal septum is narrow.
2. Tail, if present is never prehensile.
3. Often with internal cheek pouches.
4. Premolars are 2.

Family: Cercopithecidae

1. Nostrils point downwards and catarrhina or drop nose.
2. Absence of prehensile tail.
3. Presence of cheek pouches for storing food.
4. Laryngeal sacs are complicated.
5. Presence of bony tympanic tube.
6. Dental formula is 2.1.2.3.
7. Molars are quadrangular.
8. Colon usually has sigmoid flexure and small caecum and appendix.
9. Placenta is haemochorial.
10. Cerebral cortex is large and fissured and its frontal regions well developed.

Example: *Macaca, Papio, Mandrillus, Colobus*.

Family: Parapithecidae

1. Face is short.
2. Dental formula is 2.1.3.3.
3. Oligocene animals.

Example: *Apidium, Parapithecus*.

Super family: Hominoidea

1. Anthropoid apes and man.
2. No tail or cheek pouch.

Family: Pongidae

1. Head is large.
2. Neck is longe.
3. Limbs are large.
4. Hands with a short thumb and long metacarpals and digits with curved phalanges.
5. Both thumb and big toe are opposable.
6. Arms are larger than legs and are provided with powerful muscle.
7. Chest is wider.
8. Absence of tail.
9. Cervical and sacral region are larger while lumber region is shorter.
10. Canines are large especially in males and lower front premolar forms sectorial blade.
11. Molars carry grinding tubercles.

Example: Pongo, Gorilla, *Pan, Hylobates*.

Family: Hominidae

1. Long with upright posture.
2. Hair is long and or continuous growth on head, but sparse and short on body.
3. Legs are 30 percent larger than arms and straight.
4. Thumb is opposable, but toe is not big.
5. Tail is absent.
6. Prolonged infancy and skeletal maturation.
7. Face is flatter and vertical; brow ridges reduced lower jaw less protruding.
8. Skeleton and soft parts exhibit different configuration and proportions; body is provided with subcutaneous fat.
9. Tooth row is bow shaped, canines are small, and premolars are bicuspid.
10. Large brain with greater functional ability.

Example: *Ramapithecus, Australopithecus, Homo*.

10.8 *Cavia* sp

Guineapigs lack tails, but have 4 toes on each front paw and 3 toes on each hind paw. They measure 20 to 40 cm from head to rump. Adult weigh 500 g to 1500 g. Females weigh slightly less than males. Their incisors and molars grow continuously and wear down. Exoskeleton structures are well-developed .Hair with central core and an outer cortex develops in hair follicle. Claws present at digital tips are typically reptilian. Endoskeleton is completely ossified except some parts in nose and ear. Cartilage is present at tip of some bones. Endoskeleton comprises of appendicular, axial and heterotopic skeleton (Figure 10.5).

Axial Skeleton: This comprises the following

Skull: This is made up of cranium, sense capsules, and visceral skeleton. Skeletal pieces are distinct, united with each other with district sutures and are as follows: Occipital bone is unpaired, encircles the foramen magnum and made up of supraoccipital, basioccipital and 2 exoccipital bones. Each exoccipital is drawn back as par occipital process. Occipital

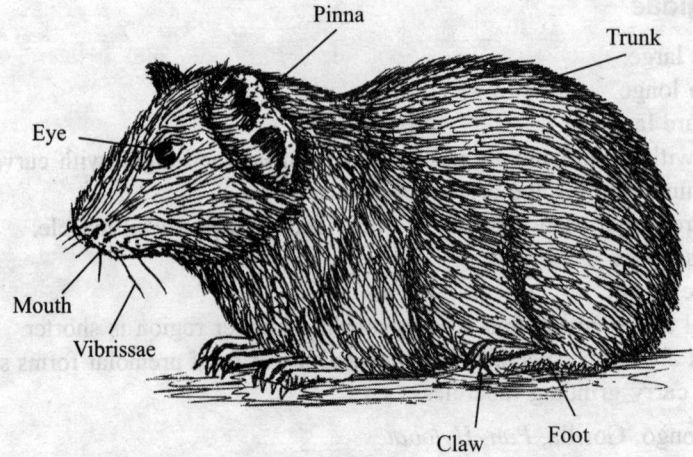

Figure 10.5   The Guinea-pig : External appearence.

condyles are 2 in number, located one on either side of foramen magnum, articulates skull with atlas. Parietal bones are paired, rectangular, and form cranium roof. Interparietal present in young stage fuses with parietal in adult. Frontal bones form anterior most portion of cranium. Concave outer border of each frontal forms the dorsal wall of orbit. Sphenoid bone is unpaired having anterior end called presphenoid and posterior end called basisphenoid and forms cranium floor. Alisphenoid and orbit sphenoid form the lateral wall of cranium. Squamosals are 2 in number, each present one on either side of lateral wall of cranium and takes part in formation of zygomatic arch. Cribiform plate is perforated form the inner anterior wall of cranium and separates cranium from olfactory capsule (Figure 10.6).

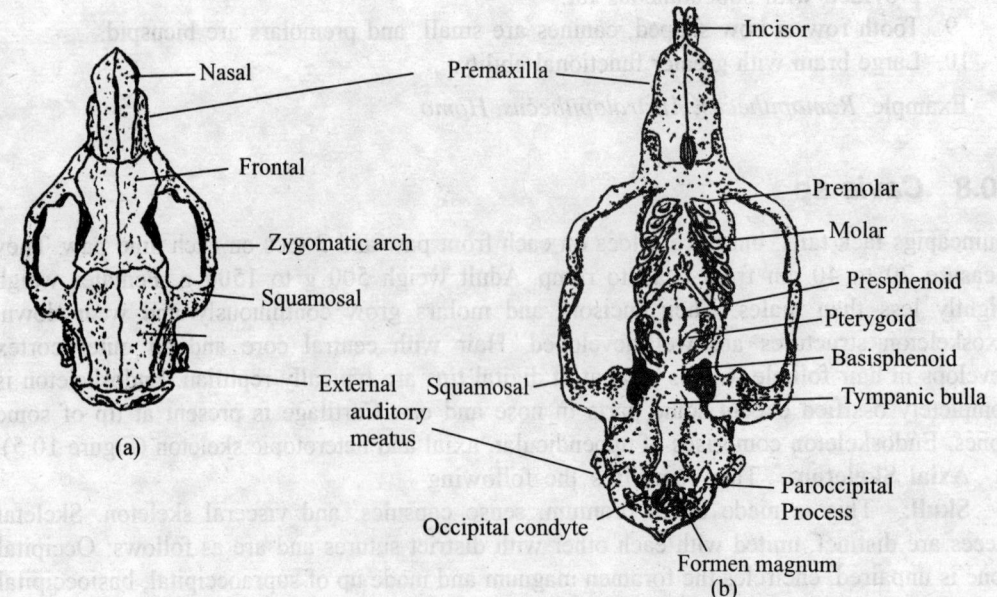

Figure 10.6   Skull of Guineapig (a) Dorsal view, (b) Vertral view.

Sense Capsules: Three types of paired sense capsules are as follows:

(a) **Olfactory Capsule:** Roof, posterior lateral wall and anterior lateral wall is formed by a pair of nasals, maxilla, and premaxilla. Vomer is situated along midventral line and articulates posteriorly with presphenoid. Two nasal cavities are separated by vertically place cartilaginous plate. Mesethmoid is small. Spongy turbinals are found inside nasal cavity.

(b) **Optic Capsule:** Anterior wall and ventral wall are formed by small lachrymal and zygomatic arch. Maxilla, jugal, and squamosal takes part in formation of zygomatic arch. An interorbital septum is present between 2 orbits. Upper and lower part of septum is formed by orbitosphenoid and presphenoid.

(c) **Otic Capsule:** Periotic bone is formed by union of 3 small bones viz. malleus, incus, and stapes, together known as ear ossicles and covers the membranous labyrinth. Tympanic bone remains firmly attached with outer surface of periotic. Tympanic bulla is flask- shaped. Base of flask is called tympanic bulla and neck is called external auditory meatus.

Visceral Skeleton: First and 2nd pair of visceral arches form upper and lower jaw and help in articulation of jaw with skull. Each half of upper jaw is composed of premaxilla, maxilla, and jugal. Premaxilla and maxilla bears teeth. Root of mouth cavity called the palate is formed by inward projection of premaxilla, maxilla, palatines, and pterygoids.

Lower Jaw: Lower jaw or mandible, the largest bone of skull is made up of 2 incompletely united halves, called dentary which is made up of body and ramus. Broad part of ramus is provided with 3 projection viz. coronoid, condylar, and angular process. Mandible is articulated with squamosal at posterior region of zygomatic arch (Figure 10.7).

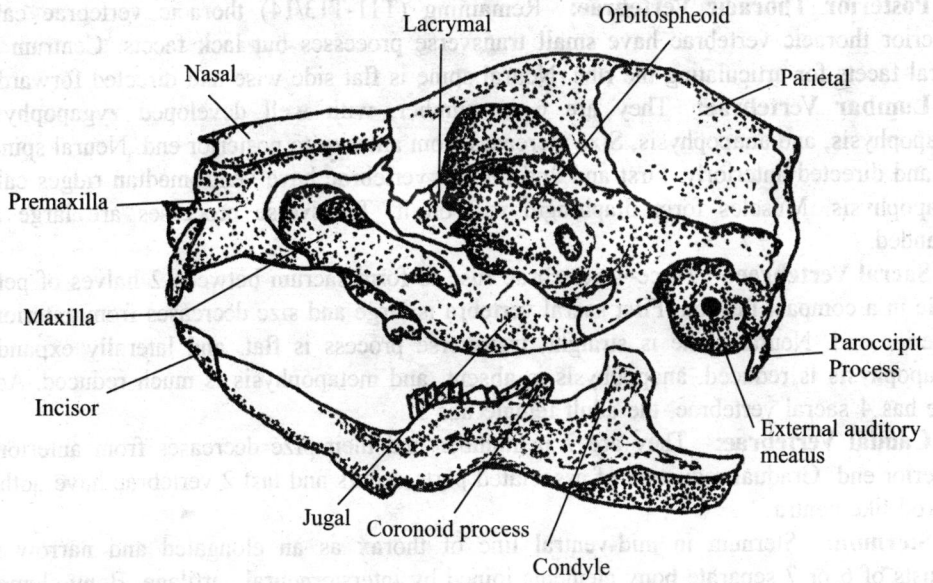

Figure 10.7 Skull and Lower jaw of Guineapig.

Hyoid Apparatus: This is located below the tongue, consists of a median body and 2 pairs of projections called cornua. Greater cornua extends from dorsolateral process of basihyoid articulates with tympanic bulla. It consists of cranio-caudally- epihyoid, stylohyoid and tymparohyoid. Lesser cornua extends as cartilaginous thynohyoid dorsomedially form basihyoid and articulates with thyroid cartilage of larynx.

Vertebral Column: Vertebrae forming vertebral column are cervical, thoracic, lumbar, sacral and caudal. Intervertebral disc remains present between the successive vertebrae.

Cervical Vertebrae: They are 7 in number with small centrum and short neural spine. Craniocaudal base of transverse process is perforated by arterial foramen through which carotid artery passes. It is pierced dorsally by alar foramen. Ring like first cervical vertebra with insignificant centrum is called atlas. Neural spine and transverse process of atlas is small. Zygapophyses are absent. A pair facet at the anterior surface receives occipital condyle. Neural canal is divided horizontally by fibrous ligament. Second cervical vertebra called axis is provided with broad centrum which is drawn anteriorly into peg like odontoid process. Odontoid process rests in dens fosse of atlas. The remaining cervical vertebrae vary slightly. Transverse processes are stubby, those of C4, C5 and C7 are bifid while those of C6 bearing 3 terminal prominences.

Thoracic Vertebrae: They are 13-14 in number and are grouped into anterior thoracic vertebrae and posterior thoracic vertebrae.

Anterior Thoracic Vertebrae: The first 10 thoracic vertebrae (T1-T10) called anterior thoracic vertebrae are provided with small and compact centrum and concave facets on sides for articulation of head of capitulum of rib. Neural spine of vertebra is elongated, narrow and directed backward. Transverse process is strongly built and bears a facet on ventral surface for articulation with head, or tuberculum of rib.

Posterior Thoracic Vertebrae: Remaining (T11-T13/14) thoracic vertebrae called posterior thoracic vertebrae have small transverse processes but lack facets. Centrum has lateral facets for articulating the ribs. Neural spine is flat side wise and directed forward.

Lumbar Vertebrae: They are 6 in number, with well developed zygapophyses, metapophysis, and anapophysis. Size increases from anterior to posterior end. Neural spine is flat and directed anteriorly. First and 2nd lumbar vertebrae have ventromedian ridges called hypapophysis. Muscles form diaphragm rest on it. Transverse processes are large and expanded.

Sacral Vertebrae: Three-4 vertebrae fuse to form sacrum between 2 halves of pelvic girdle in a compact fashion. First sacral vertebra is large and size decreases from anterior to posterior end. Neural spine is straight; transverse process is flat, and laterally expanded. Zygapophysis is reduced, anapophysis is absent, and metapophysis is much reduced. Adult male has 4 sacral vertebrae, the adult female has 3.

Caudal Vertebrae: They are 7 in number and their size decreases from anterior to posterior end. Gradual reduction of associated parts occurs and last 2 vertebrae have nothing but rod-like centra.

Sternum: Sternum in mid-ventral line of thorax as an elongated and narrow rod consists of 6 or 7 separate bony elements joined by intersternebral cartilage. Bony elements are presternum, manubrium, and sternebre (3-4). Xiphisternum shows craniocaudal arrangement. First presternum is largest and articulates with ill-developed clavicles. All but the last sternebrae has the articulating surfaces for ribs. Tip of xiphisternum has a cartilaginous plate called xiphoid cartilage (Figure 10.8a, b).

Ribs: Of 13 or 14 pairs of ribs, the first six pairs extend between vertebral column and sternum are called true ribs. Ribs (7-9) are false ribs which articulate with 6th rib cartilage. Remaining ribs are either with indirect or without any connection with sternum are floating ribs. A typical rib is made up of a bony

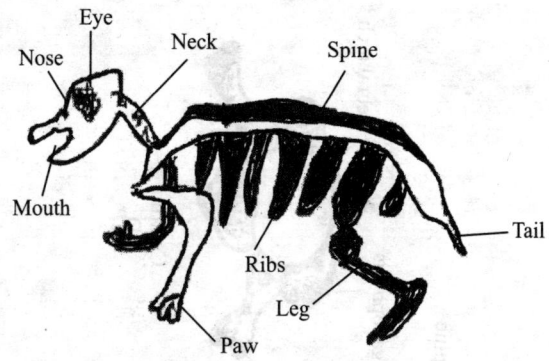

Figure 10.8 Skeleton showing ribs.

vertebral part and a short and cartilaginous sternal or costal part. The first nine ribs articulate with vertebral column by 2 heads. Capitulum articulates with centrum, and tuberculum articulates with transverse process of vertebrae. Tuberculum is absent in last 3 ribs and they cannot articulate with transverse process.

Appendicular Skeleton: Pectoral and pelvic girdles together with forelimb and hind limb make up appendicular skeleton.

Pectoral Girdle: This is small without hard structure between 2 halves of girdle. Ossification is total. Each half of girdle is made up scapula and clavicle. Scapula is large and triangular with glenoid cavity at the apex which articulates with head of humerus. Small coracoid process is present in front of glenoid cavity .Suprascapula is present as a strip of cartilage .At outer region of scapula, there is a bony projection or spine which extends as acromion that gives out a downward process called metacromion. Clavicle is slender, rod like, reduced and remains embedded within muscles (Figure 10.9).

Forelimbs: From proximal to distal end, the forelimbs are made up of humerus, radius and ulna, carpals, metacarpals and phalanges. Humerus is strongly built. Proximal end of humerus is round to form head. Greater and lesser tuberosities and bicipital groove are located at proximal end. A deltoid ridge runs along side of humerus. Distal end of humerus is provided with a pulley like trochlea which articulates with radius and ulna. Depression located in front is called coronoid fossa and depression at the back is called olecranon fossa. A supratrochlear foramen runs between depressions. Radius and ulna is completely separated rod like bones but 2 bones lie close together. Radius is shorter than ulna and is located on inner side. Proximal end of ulna is provided with a sigmoid notch which fits with trochlea of humerus. Tip of ulna extends backward as olecranon process and tips of both radius and ulna are provided with articulating surfaces. Carpal bones are 9 in number and are arranged in 2 rows. Carpal bones are irregular nodular with flat surfaces. Proximal row, when seen from dorsal side, consists of radiale at the centre, ulnare at the left and centrale at the right of it. Distal row contains 4 small bones. Carpale is located at the right margin on dorsal side. There are 2 sesamoid bones- falciforme and pisiforme. Four metacarpal bones and 3 sesamoid bones form palm region of forelimb. There are 4 digits and each digit is made up of 3 phalanges. Terminal phalanx in each digit ends in a claw.

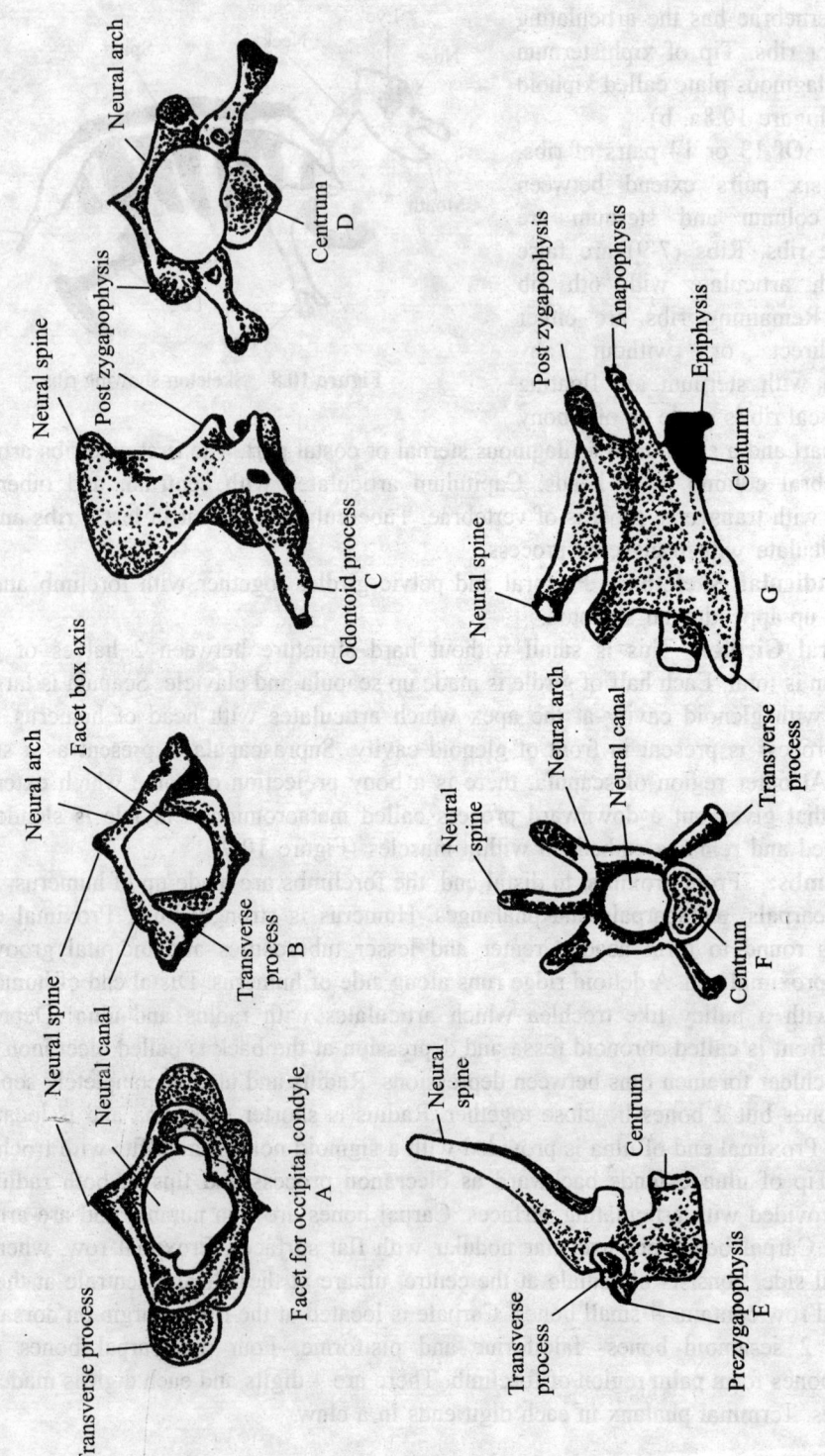

Figure 10.9(a)   Vertebrae of Guineapig. A : Atlas (dorsal view), B : Atlas (ventral view), C : Axis, D : Typical cervical vertebra, E : Posterior thoracic vertebra, F : Typical Lumbar vertebra (anterior view) and G : Typical Lumbar vertebra (lateral view).

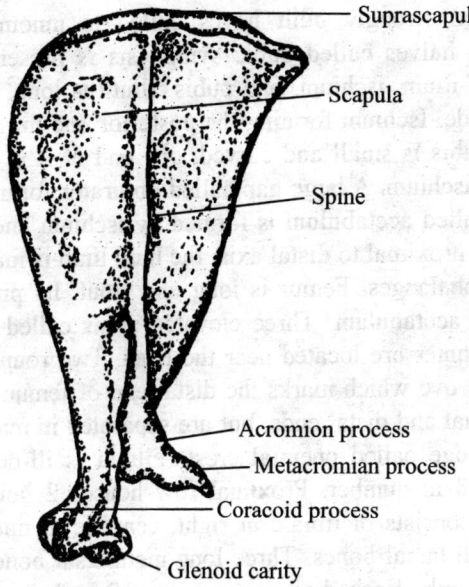

Figure 10.9(b) Pectoral girdle.

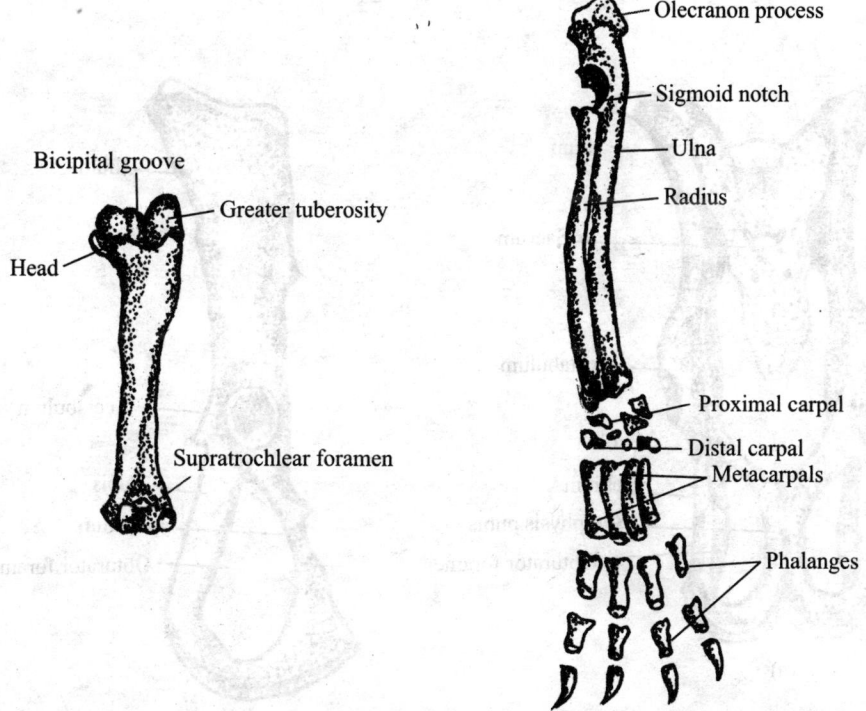

Figure 10.9(c) Humerus   **Figure 10.9(D)** Fore limb bones.

[Figure 10.9(a), 10.9(b), 10.9(c) and 10.9(d)]

Pelvic Girdle: Two strongly built haves called is innominatum form this girdle. Symphysis between two halves called pubic symphysis is present at the region of pubis. Each half is made up of ilium, ischium, and pubis. Ilium is long, blade like and articulates with sacrum along the side. Ischium forming the posterior one third of girdle is the posterior continuation of ilium. Pubis is small and curved; one end of it is united with ilium and the other end is united with ischium. A large gap called obturator foramen occurs between pubis and ischium. A socket called acetabulum is formed by ischium and pubis.

Hind Limb: From proximal to distal axis, the hind limb is made up femur, tibia, fibula, tarsals metatarsals and phalanges. Femur is long and stout. Its proximal end is rounded to form head which fits in acetabulum. Three elevated areas called greater trochanter, larger trochanter, and 3rd trochanter are located near the head. Two round condyles separate from each other by patellar groove which marks the distal end of femur. Rod like tibia and fibula are united at their proximal and distal ends, but are separated in middle. Tibia is strong, well developed and with a ridge called cnemial crest. Fibula is ill-developed and thin. Tarsal arranged in 3 rows are 8 in number. Proximal row houses 2 bones called astragalus and calcaneum. Middle row consists of tibiale at right, centrale at middle, and cuboids at left. Distal row houses 3 small tarsal bones. Three long metatarsal bones form foot. Three digits are present in the hind limb. Each digit is made up of 3 phalanges. Terminal phalanx is provided with claw. Two tarsals are sesamoid. Six metatarsals are sesamoid and 6 interphalangeal sesamoid bones are also present (Figure 10.10).

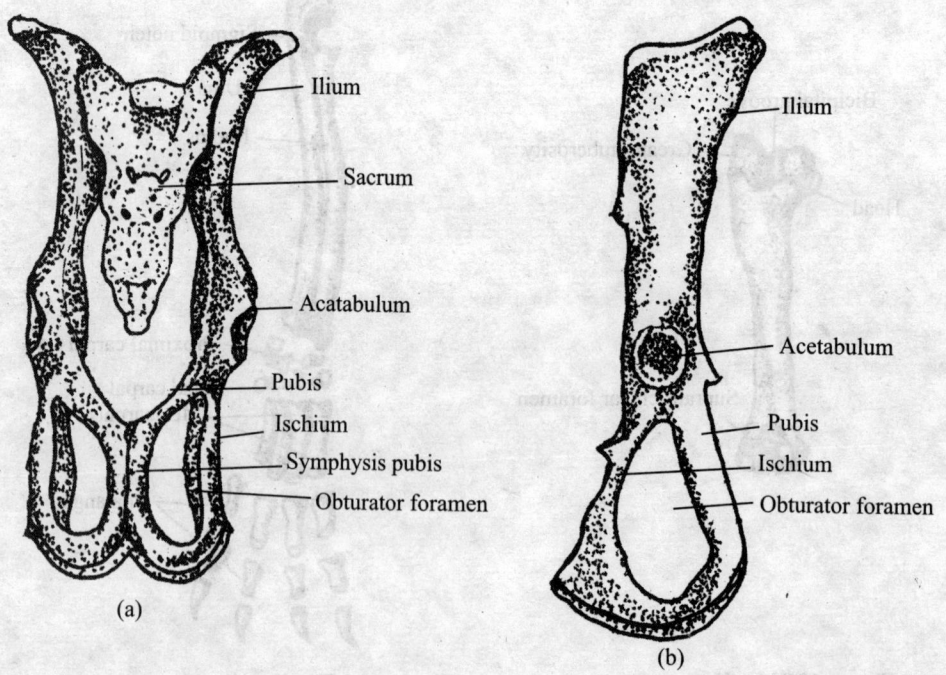

(a)

(b)

Figure 10.10   Pelvic girdle of Guinea-pig (a) Vetral view witrh sacrum (b) Side view of a single innominate.

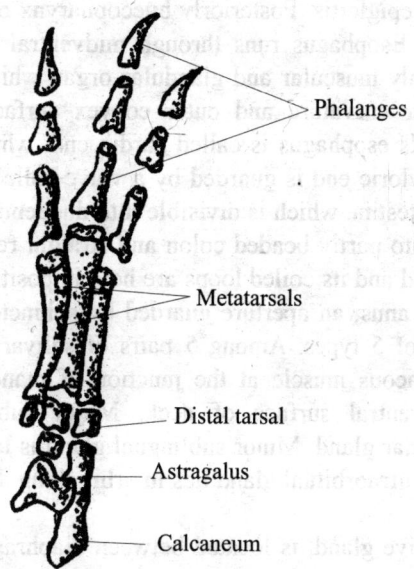

Figure 10.10(c) Right hindlimb of Guineapig.

Labels on figure: Phalanges, Metatarsals, Distal tarsal, Astragalus, Calcaneum

Heterotopic Skeleton: Bones of unusual location constitute this skeleton and involve various sesmoid bones like OS penis within glans. OS penis is a small, thin and dorso ventrally flattened bone.

Locomotion: They run on soles of their feet. Young plays by jumping, bucking, and throwing up their heads. In danger or fright, they freeze or run away. Groups may respond by suddenly scattering in all directions.

Digestive System: It consists of alimentary canal and digestive glands. Alimentary canal starts from mouth which is transverse aperture bounded by 2 lips. Upper lip is provided with a cleft in middle. Mouth leads to buccal cavity, the roof of which is formed by a palate. The supporting bones present at anterior portion of palate makes it hard which is called hard palate. Supporting bones absent in posterior region of palate is called soft palate. Palate separates mouth cavity from nasal passage. Muscular and moveable tongue is located at floor of buccal cavity. Anterior end of tongue is free while posterior end is attached with floor. Upper surface of tongue is rough and contains taste buds. Both jaws are provided with teeth. Teeth are thecodont, heterodont, and diphyodont. Dental formula is 1.0.1.3. Long chisel shaped incisors can be seen from outside .Canines are absent resulting in a gap called diastema, which occurs between incisory and premolar. Tooth is made up of dentine. Dentine of root region remains covered by cement, while the crown region of tooth remains covered by enamel.

Buccal cavity leads to pharynx, dorsal part of which is nasopharynx and ventral part is buccopharynx. Paired internal nostrils and Eustachian tubes enter nasopharynx region. Posterior margin of soft palate extends into nasopharynx as velum. Two sides of velum are provided with lymphoid tissue called tonsil. A shit called glottis is present on buccopharynx floor, just posterior to tongue. Glottis communicates with respiratory tube and is guarded by a cartilaginous flap called epiglottis. Posteriorly buccopharynx opens into esophagus through an aperture called gullet. Esophagus runs through midventral line of neck and ultimately opens into stomach, a highly muscular and glandular organ which is provided with an inner concave side called lesser curvature and outer convex surface called greater curvature. Portion of stomach towards esophagus is called cardia end, while its opposite end is called pyloric end. Opening at pyloric end is guarded by a valve called pyloric sphincter.

Stomach opens into intestine which is divisible into duodenum, ileum, and large intestine which is again divisible into partly beaded colon and straight rectum. Duodenum forms "U" shaped loop. Ileum is coiled and its coiled loops are held in position by mesenteries. Last part of alimentary canal is the anus, an aperture guarded by sphincter muscle (Figure 10.11).

Digestive glands are of 5 types. Among 5 pairs of salivary glands, parotid glands are situated beneath the cutaneous muscle at the junction of mandible and neck. Mandibular glands are present on ventral surface of neck. Major sublingual glands are located ventromedially to mandibular gland. Minor sublingual gland is located between last 2 molars and tongue. Zygomatic or infraorbitual gland lies in orbit along dorsomedial rim of zygomatic arch.

Liver, a 4-lobed massive gland, is located between diaphragm and stomach. The largest and square lobes, quadrate lobe are divided into 2 equal sized lift and right sub lobes. A rectangular left lobe lies dorsal to quadrate lobe. An oval right lobe lies to right of midline. The smallest caudate lobe lies dorsomedially in association with lesser curvature and in angular notch of stomach. Bile is the secretary product of liver. Whitish, elongated and irregular shaped pancreas is located between lines of duodenum. Secretion of pancreas is called pancreatic juice. Gastric juice is produced by numerous gastric glands present along inner lining of stomach. Glands located in the inner lining of duodenum and intestine (intestinal glands) help in digestion.

Spleen: This is situated on dorsal wall of stomach by a fold of mesentery. Spleen is devoid of duct, elongated, brown coloured, and believed to destroy old rbc.

Respiratory System: Respiration is aerial. Lungs are site where gaseous exchanges occur. Air form environment passes through long tract before entering lungs. Following are different regions of respiratory tract:

External Nares: These are paired openings situated at tip of snout. **Nasal cavities** are paired cavities separated from each other by a nasal septum. **Internal nares** are paired opening of nasal cavities. Nasopharynx is the part of pharynx where internal nares open. **Glottis** is an aperture on floor of buccal cavity. A cartilaginous flap called epiglottis forms a cover glottis. **Larynx** is a chamber formed by 4 cartilages. A thyroid cartilage forms ventral and lateral sides. A pair of small arytenoids cartilages from dorsal wall and a circular cricoids cartilage forms posterior part. Cavity of larynx bears a pair of elastic bands called vocal cords which are kept separated by a narrow rim, a glottis. Sound production is done with help of vocal cords. Trachea is a relatively non collapsible long tube. It emerges from larynx and runs midventrally through neck. It is encircled by series of 35-40 transverse cartilaginous rings. Rings are incomplete on dorsal side and their free ends are joined by fibrous and muscular tissue.

Trachea, after entering thorax, bifurcates into two branches forming right and left principal bronchi. Each bronchus enters lung of corresponding side and then breaks up into

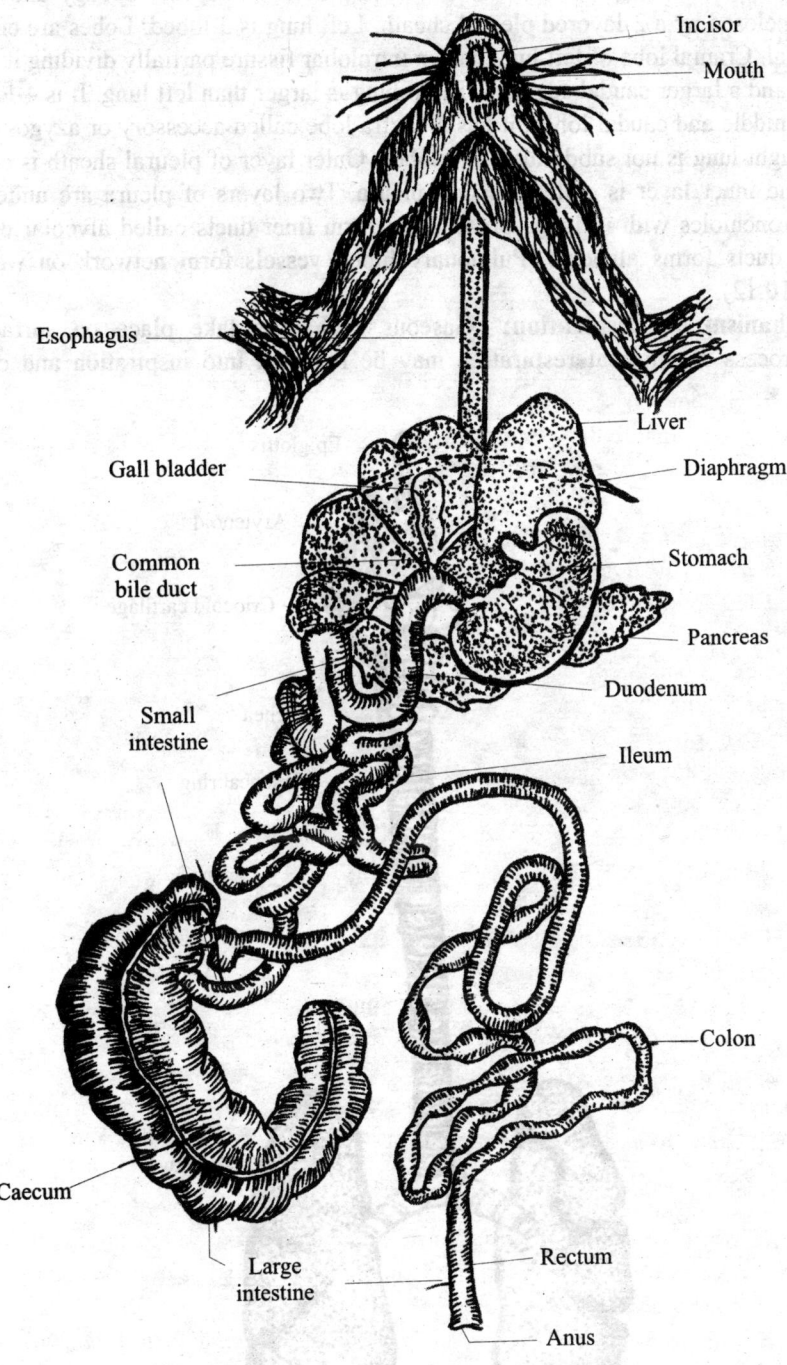

Figure 10.11 Disestire system of Guineapig.

finer bronchioles. **Lungs** are housed in thoracic cavity. They are spongy and elastic. Each lung is enclosed by a 2-layered pleural sheath. Left lung is 3 lobed. Lobes are cranial, middle and caudal. Cranial lobe of left lung has an intralobar fissure partially dividing it into a cranial segment and a larger caudal segment. Right lung is larger than left lung. It is 4-lobed. Besides cranial, middle and caudal lobes, it has an extra lobe called accessory or azygos lobe. Cranial lobe of right lung is not subdivided into lobes. Outer layer of pleural sheath is called parietal pleura and inner layer is called visceral pleura. Two layers of pleura are united at apex of lungs. Bronchioles with in lungs break up to form finer ducts called alveolar ducts. Wall of alveolar ducts forms alveolus. Pulmonary blood vessels form network on wall of alveoli (Figure 10.12).

Mechanism of Respiration: Gaseous exchanges take place on surface of lungs. Entire process of physical respiration may be resolved into inspiration and expiration. In

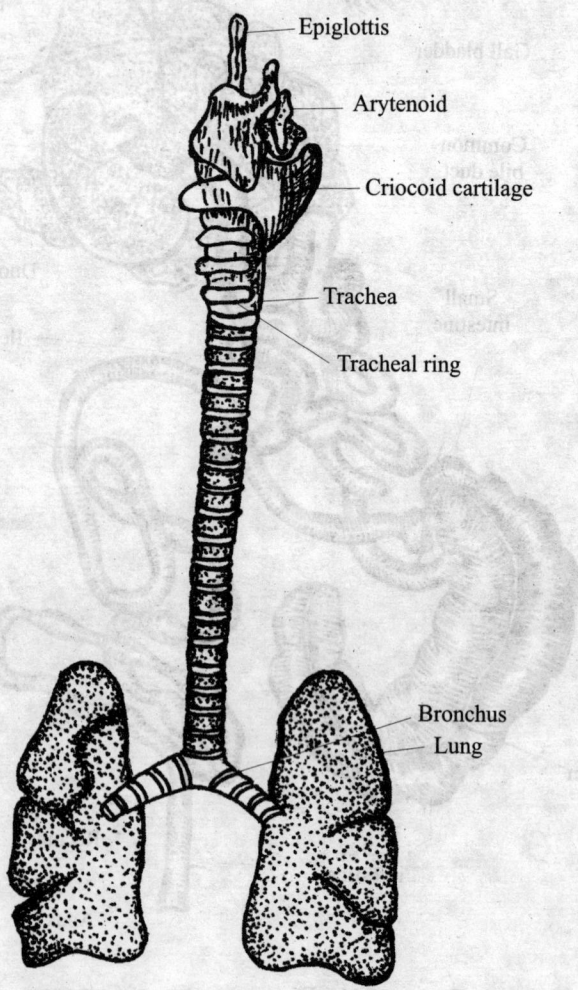

Epiglottis

Arytenoid

Criocoid cartilage

Trachea

Tracheal ring

Bronchus
Lung

Figure 10.12   Respiratory organs of Guineapig.

inspiration, intercostals muscles contract to raise ribs which increases thoracic cavity. Diaphragm at the same time flattens to increase the chest cavity antero-posteriorly. Increase of thoracic cavity permits lungs to expand. By expanding lungs air form outside is drawn in. Air from outside rushes through respiratory tract and ultimately reaches alveoli for aerating blood. In expiration, retraction of intercostals muscles and diaphragm brings thoracic cage back to its normal state and thus exerts pressure on lungs. This pressure drives air form lungs to outside. Some air called residual air always remains captive inside lungs.

Circulatory System: Two distinct circulatory fluids namely blood and lymph circulate through circulatory system. Blood flows through blood vessels and is pumped by heart. Lymph flows through lymph vessels and intercellular spaces. Several contractile lymph hearts force flow of lymph. Blood vascular system is formed by blood, heart, blood vessels, while lymph, lymph vessels and lymph heart constitute lymphatic system.

Blood: Blood consists of liquid plasma and corpuscles floating in plasma. Plasma is pale yellow fluid and contains various inorganic salts, vitamins and hormones. Different types of blood cells are RBC, WBC and thrombocytes. RBCs are round and biconcave. Mature erythrocytes are non- nucleated and contain red pigment, haemoglobin. Haemoglobin renders red colour to blood. It has a strong but loose affinity for oxygen. RBC carries O_2 to different parts of body and carries back CO_2 from different parts of body to lungs through heart. WBC is larger than RBC and their number is much less than RBC. Leucocytes are of different types depending on structure of nucleus and satability of cytoplasm granules. WBC move in amoeboid fashion and act as scavengers of body. Thrombocytes are small and non nucleated, occur in groups in case of injury .These cells break down and produce an enzyme which coagulates blood and thereby prevents loss of blood.

Heart: Heart is located in a space called the mediastenum within thorax between 2 pleural bags and is covered over by thin peritoneal membrane called pericardium. Pericardium is 2 layered. Outer layer is fibrous pericardium and inner layer is serous pericardium. Serous pericardium is again made up of 2 layers. Inner most layer adherent to heart is visceral layer of epicardium and the other one is parietal layer. Space between fibrous layer and serous layer is called pericardial cavity. Heart is a muscular, internally hollow, 4-chambered, cone-shaped that occupies most of thoracic cavity. It lies roughly in midline of cavity with its base located cranially and apex caudally. On the surface of heart, there are 3 grooves or sulci. Coronary sulcus encircles heart transversely and externally represents line of separation between auricles and ventricles. Sulcus is fitted up with coronary vessels and little amount of fat. Dorsal and ventral inter-ventricular sulci mark externally the line of separation between 2 ventricles. There are shallow and small vessels that travel within them. (Figure10.13) Heart consists of 2 auricles and 2 ventricles. Sinus venosus is absent. Right auricle lies cranial to right ventricle and receives venous systemic blood and most of coronary venous return. An internally located ridge within right auricle divides it. Sinus venarum cavarnum is smooth walled but auricular region is lined by 5 muscular ridges called pectinate muscles. There are 4 main openings in right auricle. Blood enters right auricle through 3 of these openings while through other blood passes onto right ventricle. Three opening through which blood comes to right auricle are.

(a) **Coronary Sinus:** It drains blood of heart and is located between opening of caudal vena cava and atrioventricular opening.

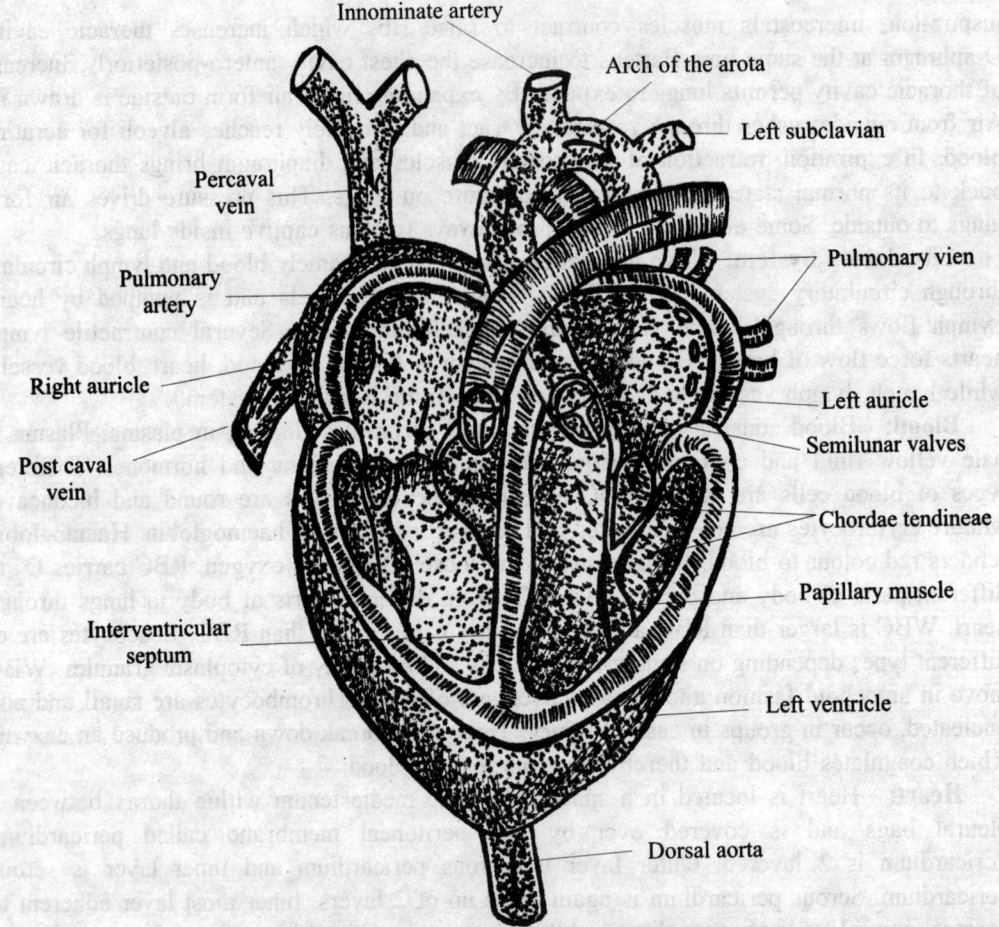

Figure 10.13　Heart of Guineapig.

(b) **Larger Ostium of Caudal Vena Cava:**　it is located in caudal aspect of auricle near interarterial septum.

(c) **Smaller Ostium of Cranial Vena Cava:**　It is located on dorsocranial aspect of auricle. Blood from right auricle flows to right ventricle through right artioventricular ostium. Dorsolateral wall of right auricle is demarked by interauricular septum which at its right auricular face bears a small crescent shaped depression called foossa ovalis. During embryonic stage, an aperture called foramen ovale remains present at this spot. This aperture becomes closed before birth of animal. Left auricle is smaller in size than right auricle and is separated internally from it by interauricular septum inside left auricle, there is left auricular (auricular sinister) which is very similar to that of right auricula. It receives blood through large pulmonary opening (ostium venerum pulmonalium) located on its dorsal wall. It opens into left ventricle through left atrioventricular ostium.

Right ventricle lies caudal to right auricle. It is thick walled and receives blood from right atrioventricular ostium. Ostium is an oval ring surrounding right atriovntricular valve which is called tricuspid valve as it is formed by 3 delicate and transparent cusps. Depending on location, cusps are called ventral angular cusp, dorsal parietal cusp, and medial septal cusp. Bases of these cusps remain attached to border of atrioventricular ostium and their apexes project into ostium. Each cusp is held in position by chordae tendineae and papillary muscles to prevent backflow of blood to right auricle. Chordae tendineae are strong. White and delicate fibres attached to cusps and anchored to muscular wall.

Conus arteriosus is fused with right ventricle and is present as funnel-shaped cranial portion of right ventricle leading to ostium of pulmonary trunk internally, and bordered externally on right by right auricular. Lumen of ventricle is provided with several muscular ridges called trabeculae carnae. Pulmonary trunk arises from right ventricle. Round ostium of pulmonary trunk lies close to inter-ventricular septum. Ostium is provided with pulmonary valves. Three semi lunar transparent cusps constitute valves. Cusps are known as right, left, and intermediate cusps. Valves prevent backflow of blood to right ventricle. Left ventricle is larger than right ventricle and is provided with thick walls. Left atrioventricular ostium is provided with bicuspid or mitral valve, which is composed of 2 unequal sized cusps- a larger septal cusp and a smaller parietal cusp. Cusps are held in position by chordae that remain anchored to ventricular wall. Valves prevent the backflow of blood. Lumen of left ventricle is provided with trabeculae carneae resembling that of right ventricle. Left aortic arch arises from left ventricle. Aortic ostium is opening of aorta and it lies near centre of base of heart. Ostium is provided with aortic valves. Three semilunar cusps similar to those of pulmonary cusps are present. Valves prevent backflow of blood.

Mechanism of Circulation Through Heart: Heart works day and night by alternate contraction (systole) and relaxation (diastole). Two auricles begin their systole at the same time and blood from both auricles is arced into both ventricles. Deoxygenated blood goes to right ventricle and oxygenated blood comes to left ventricle. Back flow of blood is prevented by auriculoventricular valves. Ventricles thus fled up with blood start systole. In this phase the auricle valves become closed and blood from right ventricle is forced through pulmonary arch, while that from left ventricle is forced thorough left aortic arch. Ventricular systole is followed by phase of diastole of whole heart. At this stage semilunar valves remain closed, deoxygenated blood from caval veins enters right auricle and oxygenated blood from pulmonary veins enters left auricle.

Blood Vessels: Blood is conveyed from heart to different parts of body through well developed blood vessels. These vessels form a well knit circuit inside body. This type of circulation is known as closed circulation. Oxygenated blood is carried away from heart by arteries (excepting pulmonary arteries) and deoxygenated blood is carried to heart by veins. (excepting pulmonary veins) Arteries break up into arterioles which in turn break up into arterial capillaries. Arterial capillaries unite with venous capillaries and make up a capillary network. Venuies form veins and thus a close circuit is built.

Arterial System: Pulmonary and aortic arches come out from the heart. Pulmonary arch emerges from conus arteriosus of right ventricle and left aortic arch emerges from left ventricle. Semilunar valves are present at the base of each of openings. Pulmonary arch carries deoxygenated blood. Soon after its emergence from right ventricle, it bifurcates into right and left branches. Right pulmonary artery passes dorsal to ascending aorta and then divides into 4 main branches. Each lobe of right lung receives one branch. Left pulmonary artery passes ventral to descending aorta and divides into 3 branches to enter 3 lobes.

Aortic Arch: Only left aortic arch is present. Originating from the left ventricle, it is the largest artery in body and main trunk of systemic arterial system. It is divided into 3 portions the ascending aorta, arch of aorta, and descending aorta. Aorta begins as ascending aorta and then curves dorsally and to left to form arch or aorta. From level of 2nd and 3rd thoracic vertebrae, it is known as descending aorta. It runs caudally parallel to vertebral column and bifurcates at lumbar region of abdomen forming iliac vessels.

Branches from different portions of aortic arch are

A. Ascending Aorta: This aorta gives rise to 2 or 3 coronary arteries. Artery that supplies the right and dorsal surface of heart is called right coronary artery and which supplies left side is called left coronary artery.

B. Arch of Aorta: From cranioventral aspect of arch of aorta arise 2 brachiocephalic trunk and left subclavian artery. These arteries supply blood to neck, head, and left forelimb. Arch of aorta is bound to pulmonary trunk by a fibrous band called ligamentum arteriosum immediately after origin of left subclavian artery. Brachiocephalic artery is large, stout and courses cranially. At the level of first rib, it gives rise to left common carotid. Brachiocephalic artery continues cranially and then bifurcates into right common carotid and right subclavian artery. Two common carotid arteries lie lateral and parallel to trachea. From both arise similar types of arteries. Both carotids at the level of cricoids cartilage of larynx bifurcate to form external and internal carotids. Internal carotids are smaller and they supply blood to deep structures of head and brain, while external carotids supply respective sides of head. Of 2 subclavian arteries left one is larger then right and both courses craniolaterally towards left and right. Subclavian arteries give following branches.

(a) Costocervical trunk arises from craniodorsal aspect of branch and courses dorsally;

(b) Vertebral artery arises from craniodorsal border or from costocervical trunk. It courses cranially for some distance and then curves to take a dorsal position;

(c) Internal thoracic artery originates from caudoventral border of subclavian, lies opposite to cost cervical trunk and courses caudally. Right internal thoracic artery gives rise to percardionephric, bronchial, mediastinal and phrenic arteries of both sides. While from left internal thoracic artery arises only one branch on left side to mediastinium.

(d) Superficial cervical trunk is distal branch of subclavians. It is a long artery and lies between shoulder and neck.

C. Descending Aorta: Part of left aortic arch from level of 2nd and 3rd thoracic vertebrae and rest is called descending aorta. It runs caudally parallel to vertebral column and bifurcates in lumbar region to form left and right iliac arteries. It is discussed under 2 heads - thoracic subdivision and abdominal subdivision. Thoracic subdivision is the part of descending aorta lying between 2nd or 3rd thoracic vertebrae and 2nd to 4th lumbar vertebrae is called thoracic subdivision or thoracic aorta. It runs dorsocaudally along left of vertebral column. Eight pairs of dorsal intercostals arteries arise form dorsal border of aorta and these extend laterally. Each dorsal intercostals artery sends a dorsal branch and a spinal branch. Dorsal branch supplies dorsal and lateral vertebral muscles and overlying skin, while spinal branch supplies spinal canal. Several short twigs called cranial phrenic arteries arise as separate vessels from aorta near its point of entry into diaphragm. They supply dorso-lateral peripheral border of diaphragm. Abdominal subdivision, the last part of descending aorta from level of diaphragm is known as abdominal subdivision. Branches from abdominal subdivision or aorta are divided into paired and unpaired visceral arteries and paired parietal or lumbar arteries (Figure 10.14, 10.15).

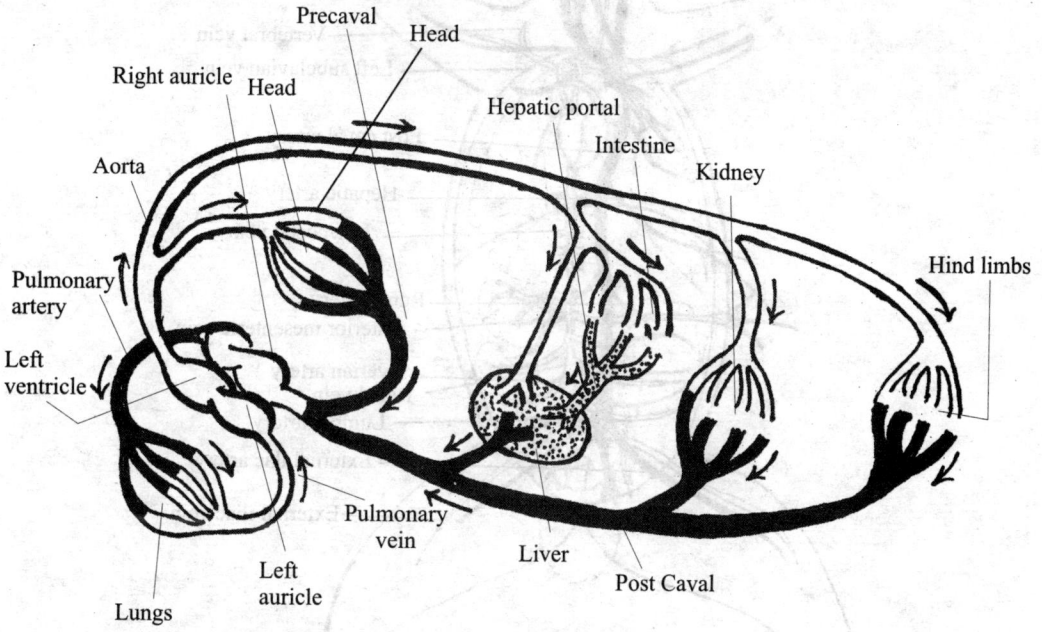

Figure 10.14 Course of circulation of blood in the Gunea-pig.

1. Unpaired Visceral Arteries

I. Coeliac trunk: This is the largest stout artery arising from abdominal aorta. It gives following branches. i) Gastropancreaticosplenic artery is the first branch of coeliac artery. Near its origin, it gives the left gastric artery which courses cranially and supplies blood to ventral and dorsal surfaces of lesser curvature of stomach. The artery continues laterally on craniodorsal side and from it arises the splenic artery which sends many branches (pancreatic branch to the pancreas). ii) Accessory middle celiac artery arises from a point just distal to gastropan creaticosplenic artery. It is short and supplies blood to transverse and proximal distal colon. iii) Hepatic artery is located caudad to accessory middle celiac artery. From it arises cystic artery that supplies the gall bladder. iv) Cranial mesenteric artery is the last and largest branch from the celiac artery. It supplies most of the intestine and mesenteries.

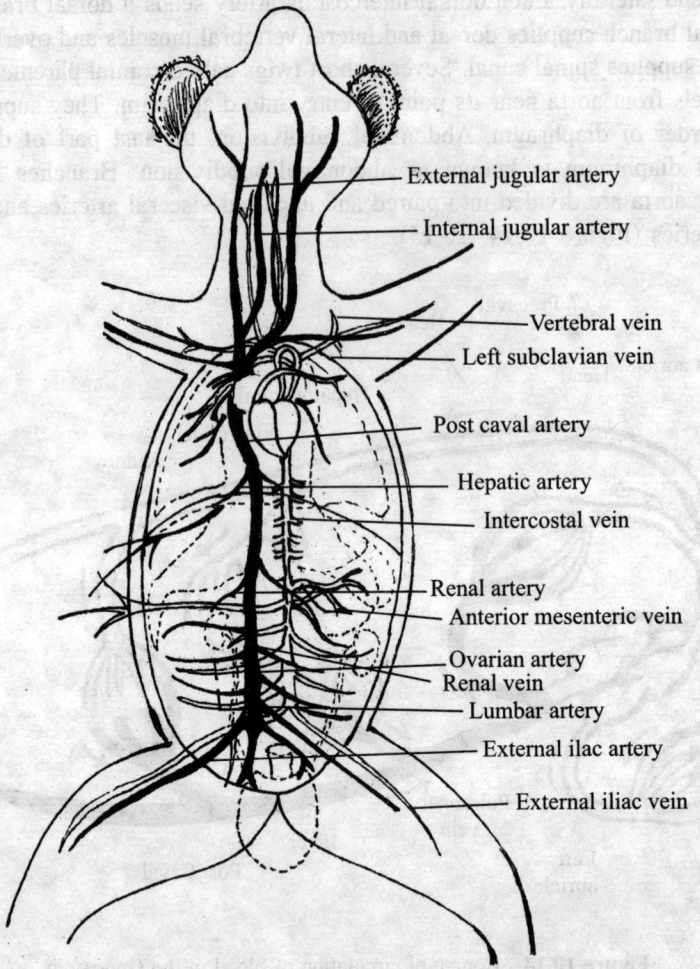

External jugular artery

Internal jugular artery

Vertebral vein

Left subclavian vein

Post caval artery

Hepatic artery

Intercostal vein

Renal artery

Anterior mesenteric vein

Ovarian artery

Renal vein

Lumbar artery

External iliac artery

External iliac vein

Figure 10.15(a)   Circulatory system of Guineapig.

II. **Caudal Mesenteric Artery:** It is a small branch and arises form ventral surface of the aorta from a point just caudad to kidneys. It courses ventrocaudally and supplies blood to descending colon.

2. **Paired Visceral Branches** i) Renal arteries are cranial and caudal artery. They arise from level of 2nd and 3rd lumbar vertebrae and supply cranial and caudal parts of kidney respectively. ii) Testicular artery is present in males in pair and they courses ventrolaterally and supply blood to male gonads. In females, these arteries are called ovarian arteries. iii) Parietal or Lumbar arteries are 6 pairs; arise from aorta extending between testicular or ovarian artery and point of bifurcation of aorta. They supply blood to lumbar vertebrae and body muscles adjoining these vertebrae. iv) Abdominal aorta at level of last lumbar vertebra bifurcates into 2 and gives rise to common iliac arteries which diverges caudolaterally from midline and then divides

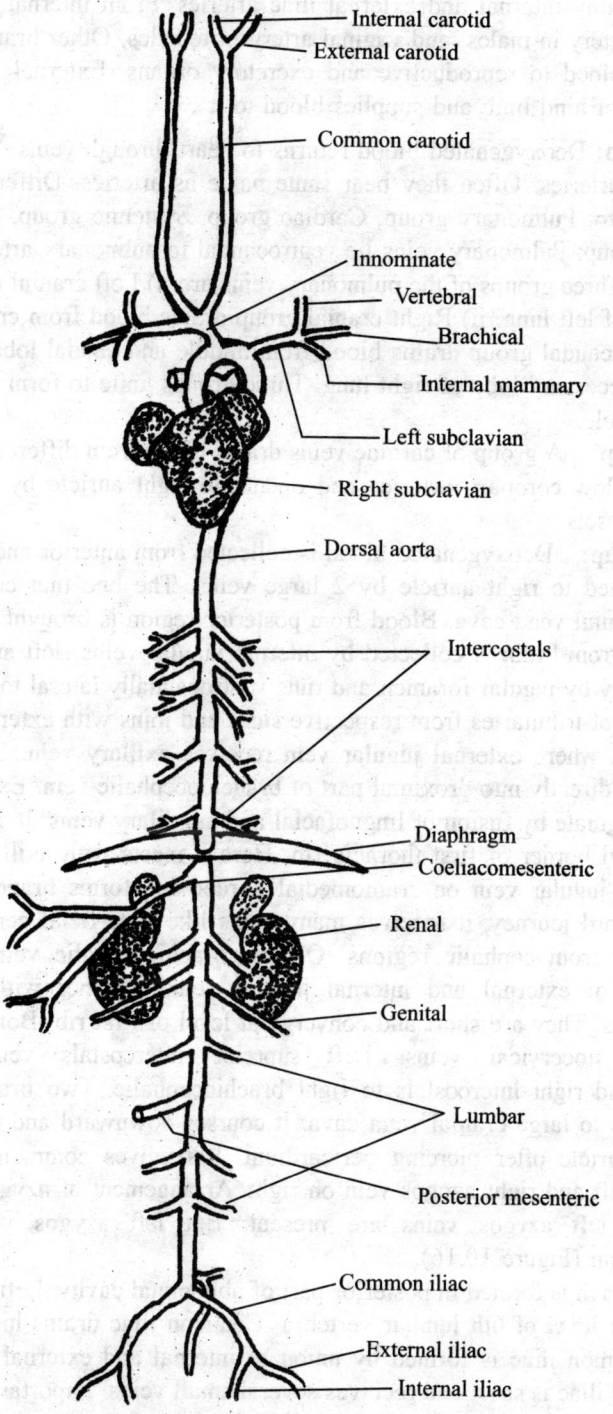

Internal carotid

External carotid

Common carotid

Innominate

Vertebral

Brachical

Internal mammary

Left subclavian

Right subclavian

Dorsal aorta

Intercostals

Diaphragm

Coeliacomesenteric

Renal

Genital

Lumbar

Posterior mesenteric

Common iliac

External iliac

Internal iliac

Figure 10.15(b) Arterial System of Guineapig.

into 2 forming internal and external iliac arteries. From internal iliac artery arises prostatic artery in males, and vaginal artery in females. Other branches arising from it supply blood to reproductive and excretory organs. External iliac artery enters thigh part of hind limb and supplies blood to it.

Venous System: Deoxygenated blood returns to heart through veins. Veins follow same general course as arteries. Often they bear same name as arteries. Different veins of body may be grouped into: Pulmonary group, Cardiac group, Systemic group, and Portal group.

Pulmonary Group: Pulmonary veins lie ventrocaudal to pulmonary arteries and dorsal to caudal vena cava. Three groups of the pulmonary veins are: i) Left cranial group drains blood from cranial lobe of left lung. ii) Right cranial group drains blood from cranial lobe of right lung. iii) Common caudal group drains blood from middle and caudal lobes of left lung and from middle and accessory lobe of right lung. Three groups unite to form a single trunk that opens into left auricle.

Cardiac Group: A group of cardiac veins drains blood from different regions of heart. They generally follow coronary arteries and open into right auricle by coronary sinus or directly by tiny vessels.

Systemic Group: Deoxygenated blood is collected from anterior and posterior parts of body and is returned to right auricle by 2 large veins. The one that collects blood from anterior part is cranial vena cava. Blood from posterior region is brought to heart by caudal vena cava .Blood from brain is collected by internal jugular veins (left and right) .It comes out of cranial cavity by jugular foramen and runs ventrocaudally lateral to trachea. Each one receives a number of tributaries from respective sides and joins with external jugular vein of same side at point where external jugular vein receives axillary vein. Sometimes internal jugular vein opens directly into proximal part of branchiocephalic vein. External jugular vein (left and right) originate by fusion of linguofacial and maxillary veins. It courses downwards and towards medial border of first thoracic rib. Here it meets with axillary vein on lateral sides and internal jugular vein on craniomedial border and forms branchiocephalic veins. During its downward journey, it receives many veins like superficial cervical, prescapular, and veins coming from cephalic regions. Of two brachiocephalic veins, each has been formed by union of external and internal jugular veins together with axillary vein of corresponding sides. They are short and converge at level of first rib. Both branches receive vertebral and costocervical veins. Left supreme intercostals vein opens to left brachiocephalic and right intercostals to right brachiocephalic. Two brachiocephalic veins unite and give rise to large cranial vena cava. It courses downward and enters craniodorsal aspect of right auricle after piercing pericardium. It receives common trunk of internal thoracic vein on left and right azygos vein on right. Arrangement of azygos vein is unusual. Often right and left azygos veins are present. But left azygos vein opens in left brachiocephalic vein (Figure 10.16).

Caudal vena cava is formed in posterior part of abdominal cavity. Left and right common iliac trunk unite at level of 6th lumbar vertebra. Common iliac drains hind limb and pelvic organs. Each common iliac is formed by union of internal and external iliac coming from hind limb. Internal iliac is short and receives several small veins. Important ones among them are prostatic vein in males and vaginal vein in females, visceral and parietal veins, and renal vein. External iliac vein extends between its point of meeting with femoral veins within hind

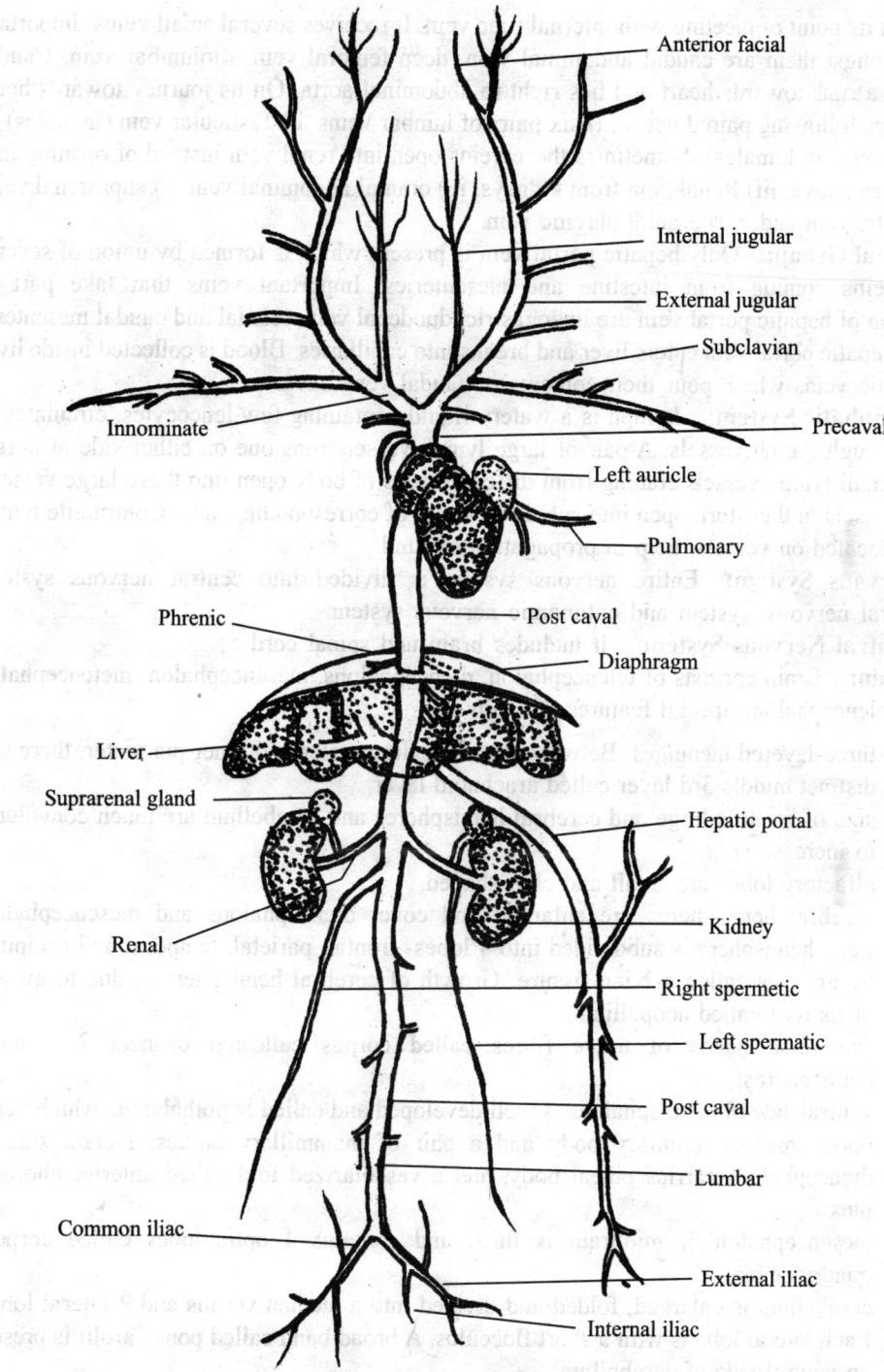

Figure 10.16 Principal viens.

limb and its point of meeting with internal iliac vein. It receives several small veins. Important ones amongst them are caudal abdominal vein, deep femoral vein, iliolumbar vein. Caudal vena cava runs towards heart and lies right to abdominal aorta. On its journey towards heart it receives following paired veins : i) six pairs of lumbar veins. ii) Testicular vein (in males) or ovarian vein (in females). Sometimes these veins open into renal vein instead of opening into caudal vena cava. iii) Renal vein from kidneys, iv) cranial abdominal vein. v) suprarenal vein, vi) hepatic vein and, vii) cranial phrenic vein.

Portal Group: Only hepatic portal vein is present which is formed by union of several small veins coming from intestine and mesenteries. Important veins that take part in formation of hepatic portal vein are lineogastric, duodenal vein, cranial and caudal mesenteric veins. Hepatic portal vein enters liver and breaks into capillaries. Blood is collected inside liver by hepatic veins which pour their contents in caudal vena cava.

Lymphatic System: Lymph is a watery liquid containing few leucocytes, circulates in body through lymph vessels. A pair of large lymph vessels runs one on either side of dorsal aorta. Small lymph vessels coming from different parts of body open into these large vessels. Large vessels in their turn open into subclavian vein of corresponding side. Contractile lymph hearts, located on vessels, help in propagation of fluid.

Nervous System: Entire nervous system is divided into central nervous system, peripheral nervous, system and autonomic nervous system.

Central Nervous System: It includes brain and spinal cord.

Brain: Brain consists of telencephalon, diencephalons, mesencephalon, metencephalon and myelencephalon. Special features of brain are

(i) three-layered meninges. Between the outer dura mater and inner pia mater, there is a distinct middle 3rd layer called arachnoid layer,

(ii) size of brain is large and cerebral hemispheres and cerebellum are much convoluted to increase area,

(iii) olfactory lobes are small and club-shaped,

(iv) cerebral hemispheres are enlarged and cover diencephalons and mesencephalon. Each hemisphere is subdivided into 4 lobes- frontal, parietal, temporal and occipital, by grooves called sylvian fissure. Growth of cerebral hemisphere is due to growth of its roof called neopallium,

(v) transverse bands of nerve fibres called corpus callosum connect 2 cerebral hemispheres,

(vi) ventral side of diencephalons is well developed and called hypothalamus which bears optic chiasma, pituitary body and a pair of mammillary bodies. Dorsal side of diencephalons carries pineal body, and a vascularized fold called anterior choroids plexus,

(vii) mesencephalon or midbrain is thick and contains 4 optic lobes called corpora quadrigemina,

(viii) cerebellum is enlarged, folded and divided into a median vermis and 2 lateral lobes. Each lateral lobe is with a short flocculus. A broad band called pons varolii is present on ventral side of cerebellum,

(ix) medulla oblongata is prominent and carries a vascularized posterior choroids plexus on its non nervous roof (Figure 10.17).

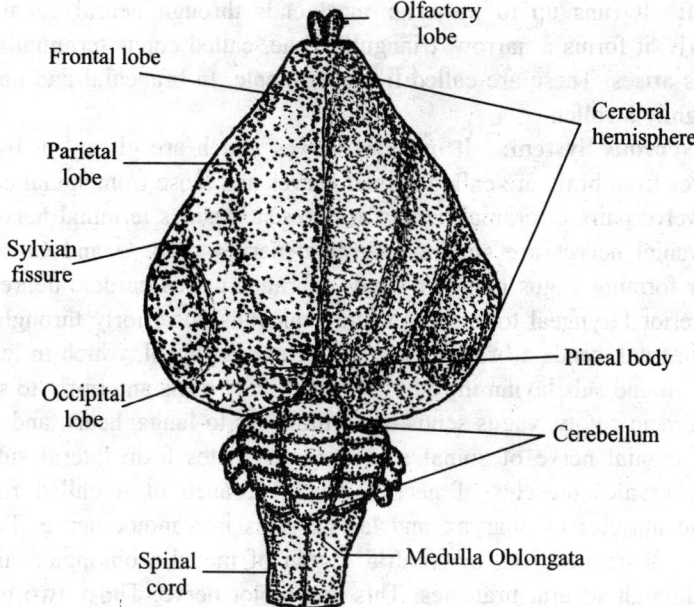

Figure 10.17(a)   Brain of Guineapig (Dorsal view).

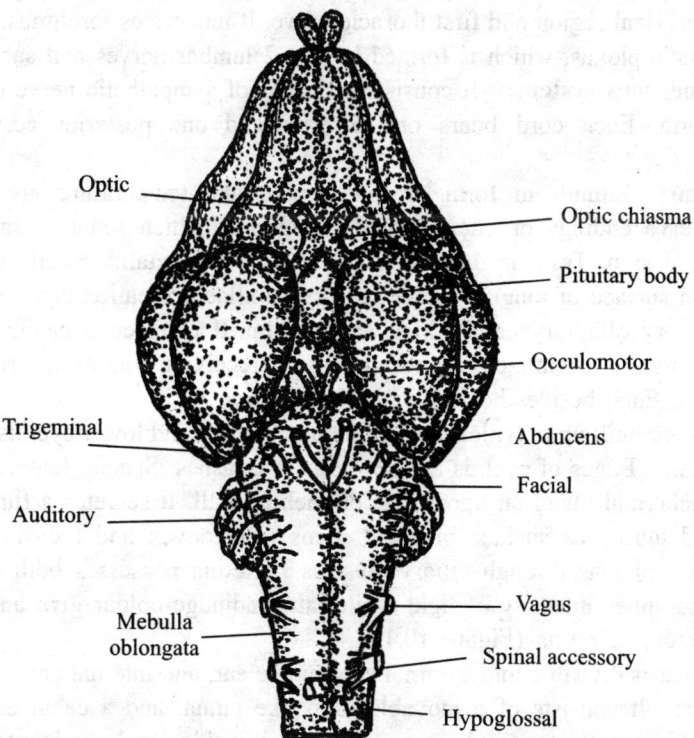

Figure 10.17(b)   Brain of Guinea-pig (ventral view).

Spinal Cord: It runs up to posterior most ends through neural canal of vertebral column. Posteriorly, it forms a narrow, triangular cone, called conus terminalis, from which a bunch of nerves arises. These are called filum terminale. In branchial and lumbar regions, spinal cord is slightly swollen.

Peripheral Nervous System: It includes nerves which are given out from brain and spinal cord. Nerves from brain are called cranial nerves and those from spinal cord are called spinal nerves. Twelve pairs of cranial nerves are present besides terminal nerve. Origin and distribution of cranial nerves are similar to cranial nerves of *Bufo* and *Calotes*. Tenth or vagus nerve after forming vagus ganglion sends a branch called cardiac depressor to heart, and a branch anterior laryngeal to larynx. Main trunk runs posteriorly through neck region. Near thorax, main trunk sends a branch called recurrent laryngeal, which in left side curves around aorta and around subclavian in right side and finally turns anteriorly to supply larynx. After entering thoracic cavity, vagus sends usual branches to lungs, heart, and other visceral organs. Eleventh cranial nerve of spinal accessory originates form lateral side of medulla oblongata and innervates muscles of neck region. A branch of it called ramus internus supplies nerves to muscles of pharynx and larynx. This is a motor nerve. Twelfth cranial nerve of hypoglossal begins from midventral region of medulla oblongata and innervates tongue muscles through several branches. This is a motor nerve. Thirty-two pairs of spinal nerves are present. On each side, 4th and 5th spinal nerves of cervical region unite as phrenic nerve to supply muscles of diaphragm. Brachial plexus is formed by participation of first 4 nerves in cervical region and first thoracic nerve. It innervates forelimbs. Hind limb is innervated by sciatic plexus, which is formed by last 2 lumbar nerves and sacral nerves.

Autonomic nervous system: It consists of a pair of sympathetic nerve cords, one on each side of aorta. Each cord bears one anterior and one posterior cervical ganglia (Figure 10.18).

Sense Organs: Stimuli in form of touch, pain and temperature are received by numerous free nerve endings or encapsulated corpuscles, which remain scattered within superficial layer of skin. Taste is determined by group of specialized cells which remain within papillae on surface of tongue. These sensory papillae are called taste buds. Smell is perceived by sensory olfactory cells which are distributed in mucous membrane of nasal cavity. Eyes and ears are specialized receptor organs for receiving stimuli in form of light and sound respectively. Ears, besides hearing maintains balance.

Eyes: Eyes are built up in typical vertebrate plan. Upper and lower eyelids are provided with small fine hairs. Edges of eyelids are devoid of eye lashes. Special features in eyes are 1. Presence of lachrymal gland on upper side of each eyeball. It secretes a fluid called tear which cleans and lubricates surface of eye. 2. Lens is biconvex and focusing is done by changing curvature of lens through ciliary muscles. 3. Retina possesses both rod and cone cells. Former determines intensity of light while latter adjudge colour give and idea of the microscopic structure of retina (Figure 10.19).

Ear: Each ear is divisible into external ear, middle ear, and internal ear.

External ear: It consists of a movable, flap like pinna, and a canal called external auditory meatus. Pinna collects sound waves and sends it inside ear through external auditory meatus.

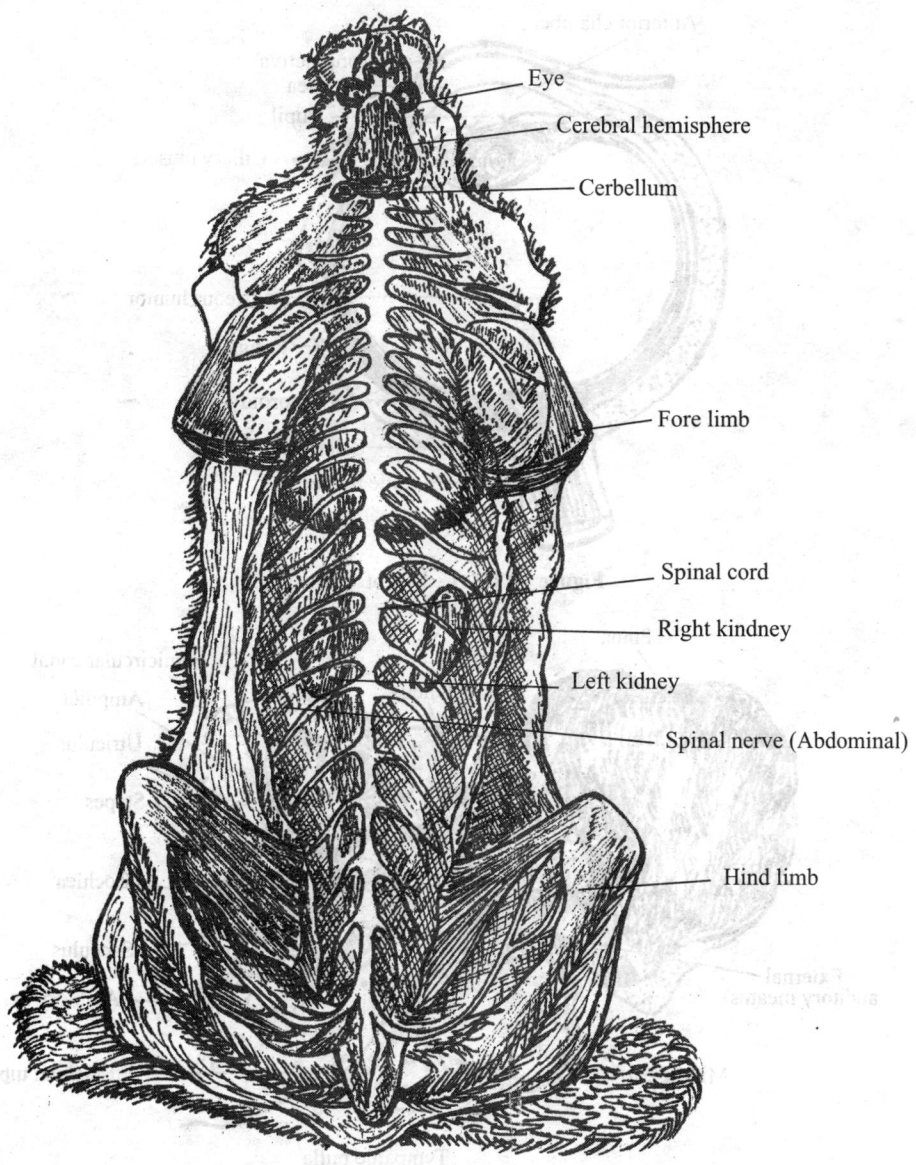

Eye

Cerebral hemisphere

Cerbellum

Fore limb

Spinal cord

Right kindney

Left kidney

Spinal nerve (Abdominal)

Hind limb

Figure 10.18 Nervous system (Dosal view).

Middle ear: External ear is separated from middle ear by a tightly stretched membrane called tympanum. Middle ear is in communication with buccal cavity by Eustachian tube. Tympanum is connected with internal ear by three ear ossicles- elongated malleus, slightly bent incus, and triangular, ring like stapes. Malleus is attached with tympanum and stapes is attached with opening in wall of internal ear, called fenestra ovalis. Incus is present between two. Ear ossicles carry sound waves to internal ear and Eustachian tube for regulating equilibrium of atmospheric pressure.

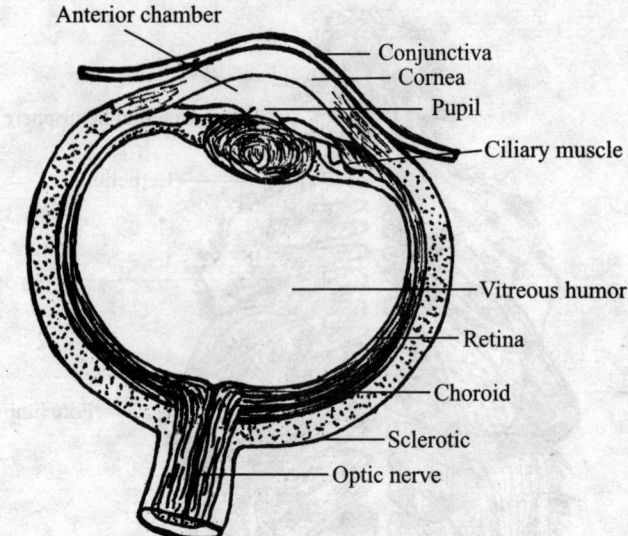

Figure 10.19(a)   Eye of Guineapig.

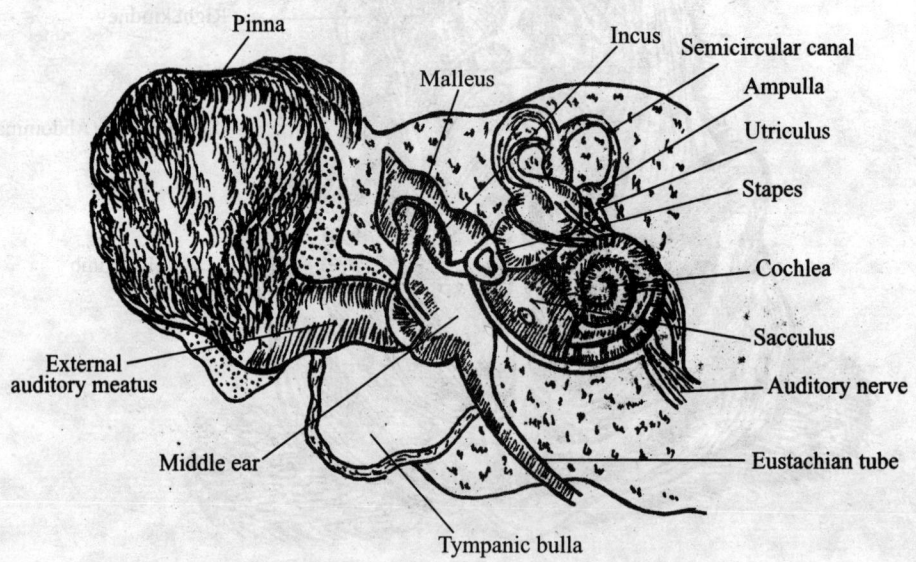

Figure 10.19(b)   Mammalian ear.

Internal ear: It consists of a bony labyrinth which is filled up with perilymph. Within perilymph, a membranous labyrinth is suspended which contains endolymph. Membranous labyrinth consists lower sacculus at the top of which lies utricles with which 3 semicircular canals are connected. Sacculus is drawn into spirally coiled cochlea which contains special receptor cells called organ of corti. Bony cochlear canal encloses 3 cavities. Scala vestibule is bounded by vestibular membrane and located above the cochlear duct. Scala tympani are situated below cochlear duct. Scala media is cavity of cochlear duct itself. Scala vestibule and scala tympani are perilymphatic spaces, while scala media is filled with endolymph. Reissner's membrane and basilar membrane are demarcating partitions between 3 cavities. Organ of Corti, receptor apparatus for hearing is supported on basilar membrane inside cochlear duct. Organ of Corti is composed in differentiated cells arranged in orderly rows. Semi circular canal's utriculus and sacculus are responsible for balancing.

Excretory System: Kidneys, together with several other structures, constitute excretory system. A pair of metancephric kidneys is located on dorsal posterior side of abdominal cavity. Right kidney is situated little up from level of left kidney. Bean shaped kidneys are notched on their inner surfaces. Notched region is called hilus. Renal artery and renal vein enter kidney through hilus. Internally, each kidney presents 2 regions. Cortex is outer and medulla is inner in position. Each kidney is made up of several nephron. Each nephron is made up of a Bowman's capsule having a tuft of capillaries called glomerulus's inside and long convoluted tube differentiated into proximal tubule, loop of Henle, and distal tubule. Cortex houses Bowman's capsules and medulla houses tubules. All tubules converge and open into collecting tubules which in their turn unite at point of hilus and form ureter which comes out through hilus. Ureter from each kidney runs posteriorly and opens into urinary bladder which is a small muscular elongated sac situated at posterior most region of abdominal cavity. Bladder continues posteriorly as urethra. Urethra in males receives male reproductive ducts and serves as common passage for urinary and reproductive substances. It opens outside through an opening located at tip of penis. In females, urethra is short, independent of the genital ducts and opens to outside through an aperture called urinary aperture.

Reproductive System: Sexes are separate but secondary sexual characteristics are not much elaborated. In both sexes, the reproductive system includes reproductive organs ducts and associated accessory glands. Male reproductive organs are testes, epididymis, spermatic cords, urethra, penis, accessory genital glands etc. In young male, testes remain inside abdomen. In adults, testes descend down and remain lodged in special fold of skin called scrotum. Scrotal cavity and abdominal cavity remain in communication with each other through inguinal canal. Paired testes lie in scrotal pouch and produce sperms. Each testis is oval in outline and its long axis remains oriented craniocaudally. Testis remains suspended within pouch by mesorchium dorsally and caudally by a fibrous ligament called gubernaculums. Each testis is divided internally into lobules by many thin septa. Each lobule contains tiny seminiferous tubules.

Epididymis: Each testis is associated with an epididymis which is a highly convoluted tube consisting of head, body and tail. Head is largest and most cranial portion of it. Body is narrow and runs along medial border of testis. Caudal part opens in ductus deferentus. Epididymis store sperms for ejaculation and helps in maturation of sperms.

Ductus Deferens: Paired deferents ducts are 40 to 60 mm long, divisible into a coiled epidiymal portion and an uncoiled portion and open separately in urethra. Openings are located ventral to seminal vesicles, dorsal to neck of bladder and medial toe ducts of coagulating and prostate glands. Two ducts then open into a median slit within urethra. Each deferents duct is supported by a mesoductus deferens which extends through spermatic cord as a separate fold of peritoneum.

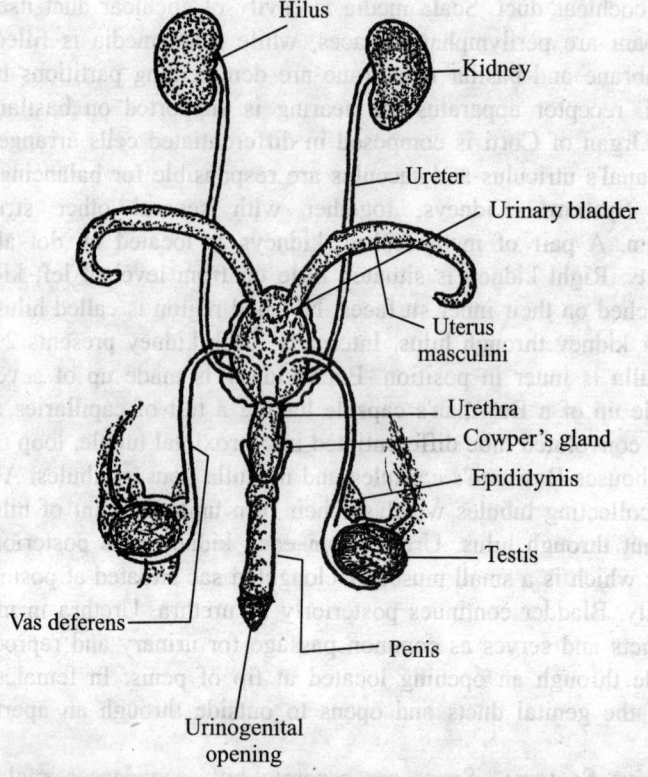

Hilus

Kidney

Ureter

Urinary bladder

Uterus masculini

Urethra

Cowper's gland

Epididymis

Testis

Vas deferens

Penis

Urinogenital opening

Figure 10.20   Urinogerital organs of male Guienapig.

Spermatic Cord: Two spermatic cords (one each for deferent ducts) contain deferent ducts and their mesenteries. Each cord begins at inguinal ring and ends within scrotal pouch.

Accessory Glands: Accessory genital glands consist of seminal vesicles, coagulating glands, ventral and dorsal lobes of prostate gland and bulb urethral gland. A common sheath envelops coagulating glands, prostate and urethral end of deferent ducts and seminal vesicle. Paired seminal vesicle is largest amongst accessory glands. Each seminal vesicle is cylindrical and elongated structure free end of which may by bifid. Converge medially and enter urethra by a pair of ducts that lie dorsocaudal to deferent ducts. Lumen of seminal vesicle remains filled up with a milky while fluid and its walls are granular. Coagulation gland are paired and pyramid shaped bodies. They lie later dorsal and in proximity to seminal vesicle. Each gland has single duct that opens in urethra, craniolateral to opening of deferents duct and seminal vesicles. Secretion of these glands coagulates and produces vaginal plug. Prostate gland consists of large dorsal and small ventral lobes which are joined by isthmus. It lies caudomedial to coagulation gland. A single pair of ducts come form ventral lobe and many pairs of ducts come from the dorsal lobe. These ducts open into urethra and openings are located craniolateral to opening of coagulating gland. Bulbourethral glands are small, lobulated, and lie ventrolateral to rectum. From each gland, a duct arises which open into urethra on dorsal surface. A peculiar structure called uterus masculinus present in male is

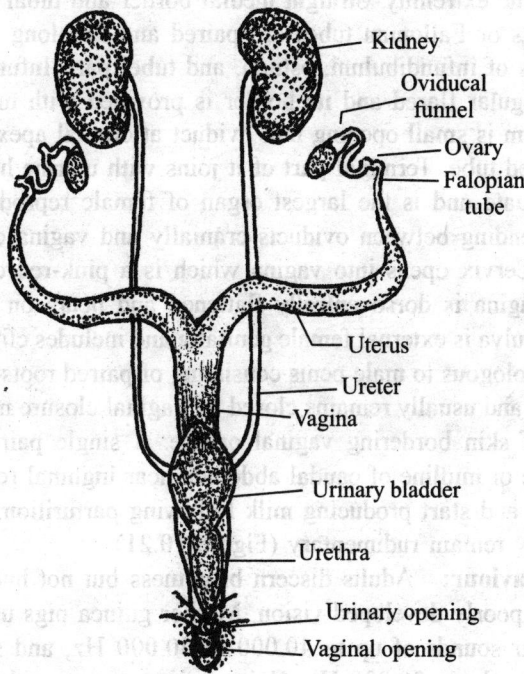

Figure 10.21   Urinogenital Organs of female.

considered homologous to uterus of female. This is a flat bilobed, hollow organ that lies between mesentery connecting deferent ducts and that of two seminal vesicles and its unpaired central body opens to urethra.

Urethra: Male urethra differs functionally from female urethra because it transports seminal fluids and urine. Urethra is divided into pelvic portion and spongy portion. Pelvic portion extends from neck of urinary bladder to penis and receives most of ducts of accessory glands. Spongy portion remains embedded within corpus spongiosum of penis. It opens externally through urethral orifice located at the tip of penis (Figure 10.20).

Penis: Penis is male copulatory organ, eversible but during sexual inactivity lies retraced within as sheath of skin called preputial sheath. Penis is made up of body and glans. Body is composed of 2 layers of corpora cavernosa on dorsal surface and its ventral midline is made up of corpus spongiosum, which houses spongy portion of urethra. Penis is enveloped by tunica albugenia. Glans is shorter than body, cylindrical and ends in a rounded tip. Tip bears uretharal orifice. The OS penis or baculum is present on dorsal surface of glans.

Female Reproductive System: It consists of a pair of ovaries which produced ova or eggs; two oviducts which carry ova to uterus; a uterus and a vagina which is a passage between uterus and external genitalia or vulva. Mammary glands may be considered under this system. Two ovaries are intraperitoneal oval and dorsoventrally flattened. Right ovary is caudolateral to right kidney, and left ovary is craniolateral to left kidney. Long axis of ovary lies parallel to axis of animals. Each ovary has a medial and a lateral surface and a cranial tube and caudal uterine extremity. Straight medial border and tubal extremity is invested by mesovarium. Oviducts or Fallopian tubes are paired and lie along lateral regions of ovary. Each oviduct consists of infundibulum, ostium, and tubel part. Infundibulum is most cranial part of oviduct, triangular flared and its border is provided with many papillary elevations called fimbrae. Ostium is small opening into oviduct at cranial apex of infundibulum. Tubal portion is a thin coiled tube. Terminal part of it joins with uterine horn.

Uterus is bicornuate and is the largest organ of female reproductive tract. It is a "Y" shaped structure extending between oviducts cranially and vagina caudally with 2 horns, a body and a cervix. Cervix opens into vagina which is a pink-red canal extending between cervix and vulva. Vagina is dorsoventrally flattened and bears on its internal side several longitudinal ridges. Vulva is external female genitalia and includes clitoris, vaginal orifice, and labia. Clitoris is homologous to male penis consisting of paired roots body and glans. Vaginal orifice is "U" shaped and usually remains closed by vaginal closure membrane. Labia are pair lateral thick folds of skin bordering vaginal orifice. A single pair of mammary glands is located on either side of midline of caudal abdomen near inguinal region. During pregnancy, glands become large and start producing milk following parturition. Glands are apocrine in nature. In males, they remain rudimentary (Figure 10.21).

Senses and Behaviour: Adults discern brightness but not hue. Youngs are born with their eyes open with poorly developed vision. Mother guinea pigs use scent to identify their young. They can hear sounds of up to 40,000 to 50,000 Hz, and some vocalizations have ultrasonic components above 20,000 Hz. Guinea pigs are active throughout the day. Under constant light and temperature, they do not show fixed sleeping or eating patterns. They seek contact with one another and tend to associate in groups of several to several dozen, and investigate unfamiliar guinea pigs by nose-to-nose contact and anogenital nuzzling and interact and socialize. In groups of guineapigs, males have a linear hierarchy. Alpha male has exclusive rights to mate with females in his group, and do not tolerate other male's attempt to mate with his females. Females are subordinate to male, and may have a loose social hierarchy among themselves. They cannot manufacture vitamin C and consume it in their diet. Sources vitamin C are fresh vegetable matter and supplements that caregiver add to food or water. They use coprophagy. They cannot digest cellulose themselves, but bacteria in their caecum can. They are unable to absorb nutrients from the caecumt. So, they eat special, soft fecal pellets.

Reproductive Cycle and Mating: Sexually mature female go into estrus every 13 to 20 days, except when pregnant. Vagina is closed by a membrane, which disappears at the start of estrus and reforms at the end. Male courts a female. After mating, the male grooms his genital region. In female, a plug made of sperm and vaginal cells forms in vagina, and falls out a few hours later. Pregnancy lasts about 68 days. They give birth to 2 to 3 young per litter. Each newborn pup weighs about 85 g to 95 g at birth. Mother eats the placenta and fetal membranes. During birth, the mother's pubic bones separate as pups are large. If a female does not breed before 6 months of age, her pubic bones may fuse. Pups are precocial and are born fully-furred, with well-developed sensory and locomotor abilities, consume solid food, can be weaned after 5 days, but they normally nurse for 3 weeks. Milk consumption decreases as solid food consumption increases. Mother grooms her young very little. Females reach sexual maturity at 2 months of age. Males reach it at 3 months. Pet guineapigs usually live up to 7 years if given proper care.

Fertilization is internal. During coitus, males insert numerous sperms in female genital tract. A few minutes after copulation during oestrous period, a vaginal copulation plug forms to prevent outflow of sperms after ejaculation. Plug is formed by secretion from coagulating and prostate glands. Plug is expelled a few days after copulation. Mature eggs liberated from Graffian follicles enter oviduct through oviducal funnel and wait in Fallopian tube for some days. One egg is fertilized by one sperm. Fertilized ovum travels along posterior part of oviduct and enters uterus. It then implants itself on uterine wall. Gradually, a placenta grows between uterine wall and developing embryo. Placenta is a joint structure formed by contribution of both uterine wall of mother and embryonic tissue. Development within uterus continues for weeks. This period is called gestation. A female usually produces 3 to 5 litters per year and each litter may contain 1 to 6 individuals. At the end of this period, parturition or expulsion of embryo to outside occurs. Youngs are born alive well-developed, fully hairy and with teeth. Guineapigs are viviparous. Immature young suck milk secreted by the mother and remains with her 12 to 14 days after birth. Animals attain sexual maturity early. Males become sexually mature in 70-75 days and the females become mature in 40-60 days.

Uses: Guineapigs are important part in the Andean region of South America for many years. Andean traditional medicine uses it for healing .Incans used them in religious ceremonies, and mummified them. People in Peru use guinea pigs in celebration of patron saint days. In South America; they are eaten on special occasions. They are source of meat in Nigeria, parts of Africa, and in Philippines. They are kept as family pets and bred for showing and have been used in scientific research since mid-1800s. They are used for studying collagen biosynthesis, as they can not produce vitamin C. Hearing studies use guineapigs since their inner ear is readily accesible. They are ideal for germ-free research, since precocial young have good chance at surviving from birth on solid food.

10.9 Human

As a member of class Mammalia, various biological aspects of human are described below:

Integumentary System: The system includes the skin which protects internal structures, prevents entry of pathogens, reduces water loss, produces vitamin D and sense touch, pain and temperature. Skin consists of 2 major tissues, dermis and epidermis. Dermis having fibroblasts, fat cells and macrophages provides the structural strength to skin and harbour nerve endings, hair follicles; smooth muscles and glands. It is divided into the outer papillary layer and inner reticular layer. The outer layer has projections; the papillae. The inner layer is dense and continuous with hypodermis.

Epidermis: This is created by stratified squamous and is separated from dermis by basement membrane. In it, the melanocytes and keratocytes produce the keratin. The deepest layer of epidermis produces nerve cells. New cells are formed and old cells are pushed to surface. The surface cells defend new cells and step by step the shape and nature of surface cells. They get filled by keratin through keratinization. During the process, the epidermis divides into 5 distinct regions namely, stratum basale, stratum spinosum, stratum granulosum, stratum lucidum and stratum corneum. Stratum basale, the deeper region consists of single layer of columnar cells. Keratinization starts here. Above this layer stratum spinosum has 8-10 layers of polygonal cells. Stratum granulosum is the next upper layer with 3-5 layers of flattened cells. On this layer, stratum lucidum occurs with several layers of dead cells. The top most layers, the stratum corneum consists of 20 layers of dead cells with keratin. Skin may be thick or thin. The 5 epithelial layers are found in thick skin which forms the soles of feet, palms of hands, and finger tips. Body surface bears thin skin with epithelial layer. Only 1 or 2 layers of cells are present in stratum granulosum. The regions of skin exposed to constant friction are thickened to form the callus with several cell layers cell in stratum corneum.

Skin Colour: Melanin produced by melanocytes cause colour of skin, hair and eye, protects the body from ultraviolet rays. Melanin production is genetically determined. Hormones and exposure to light alter skin colour. Thickness of stratum corneum and blood circulation influence skin colour.

Skin Derivatives: Hairs are integumentary structures with a root and a shaft. Shaft projects above the skin, while the root remains well below the surface. Base of root has a bulb with an expanded region. Shaft and most hair root are formed by dead keratinized epithelial cells which are arranged in 3 layers – the medulla, cortex and cuticle. The central axis of hair is formed by medulla. Major part of hair is formed by a single cell layer. During old age, melanin decreases in amount, and causes white hair. Grey hair consists of a mixture of faded, unfaded and white hairs. Hair grows due to addition of cells at the base of hair root. The growth stops at specific stage. After a period, new hair replaces old hair. Hairs on head grow for 3 years and rest for 1-2 years. Muscle cells associated with hair follicles are arrector pili. Contraction of such muscles makes the hair to stand on end. Skin has sebaceous and sweat glands. Sebaceous glands of dermis produce sebum and are connected to upper part of hair follicles. Mammary glands are modified sweat glands. The common sweat gland is the merocrine gland which is simple, coiled, tubular, and open directly through sweat pores. The gland has the deep coiled portion and duct which passes to skin surface. Number of sweat glands is more in palms of hands and soles of feet. Nail consists of nail root and nail body. Nail body, the visible part is covered by skin. Nail fold cover the proximal and lateral edges of nail .Stratum corneum of nail fold grows into nail body. Free edge of nail body is hyponchium. Small white region at the base of nail is lunula, which contains the nail matrix. Nails grow on an average rate of 0.5-1.2 mm per day.

Skeletal System: The system made by bones, cartilages and ligaments provides body shape and operates locomotion by holding weight. Bones attached with muscles serve as reservoirs of fat and minerals. Skull protects the brain. Bone marrow produces erythrocytes. Hands and legs have long bones. Short bones are broad. Carpals and tarsals are shorter. Flat bones are thin and flattened. Skull bones, ribs, sternum and scapula are flat. Verterbral and facial bones are irregular.

Typical Long Bone: Periosteum, an outer double layered sheath of dense collagenous layer with blood vessels and nerves covers bone. Growing long bone has 3 regions. The long bony part, the diaphysis or shaft is made by compact bone. The bone end consists of epiphysis. The outer surface of epiphysis is formed by compact bone. Between the epiphysis and diaphysis, epiphyseal or growth plate made up of hyaline cartilage lies. Growth in bone length occurs in this plate. The cavity inside diaphysis, the medullary cavity is lined by membrane called the endosteum. The cavity inside diaphysis in adults contains yellow marrow, mostly the adipose tissue. Medullary cavity of epiphysis contains red marrow which forms blood cell. Bones are named based on their position and are divided into axial - and appendicular skeleton. Axial skeleton consists of skull, hyoid bone, vertebral column, and thoracic cage. Appendicular skeleton consists of limbs and their girdles. Out of 206 bones, 80 remain in axial skeleton, 126 in appendicular skeleton. twenteight bones of axial skeleton remain in skull, 26 bones in vertebral column, 25 bones in thoracic cage and one remains as hyoid bone.

Axial skeleton forms the upright axis of body, protects brain, spinal cord and thoracic organs. The cranial capacity is about 1500 cm3.Skull supports the organs of vision, hearing, smell, and taste. Lower jaw or mandible remains attached to skull. Skull is covered by 8 bones like parietal, temporal, frontal, sphenoid, occipital, and ethmoid. Skull bones are joined as compact box like structure by sutures .In the front, of 14 facial bones maxilla, zygomatic, palatine, lacrymal, nasal and inferior nasal koncha remain as paired. Mandible or lower jaw and vomer are unpaired bones.

Parietal and occipital bones are major bones on posterior skull. Parietals are joined to occipital at back. Parietal and temporal bones form the side of head. The large hole in temporal bone; the external auditory meatus transmits sound waves to ear drum. On the lateral side just anterior to temporal, the sphenoid bone is seen, anterior to which zygomatic bone or cheek bone lies. Major bones in frontal view are frontal, zygomatic, maxillae and mandible. The marked openings in skull are orbits and nasal cavity. Two orbits accommodate the eyes. Bones of orbits protect eyes and are attachment sites for muscles that move the eyes. Bones forming the oribits are frontal, sphenoid, zygomatic, maxilla, lacrymal, ethmoid and palatine. Head region contains maleus, incus, and stapes. Medulla oblongata descends down as spinal cord through the foramen magnum.

Vertebrae: Vertebrae form the slightly S-shaped vertebral column of 26 bones which belongs to 5 types, cervical (7), thoracic (12), lumbar (5) sacral (1), and coccygeal (1) vertebrae. The load - bearing portion of a vertebra is the centrum. The centra of adjacent vertebrae are separated by intervertebral discs of cartilage. Dorsally, a vertebral arch projects from the centrum. Several bony projections are found on vertebral arch. On each side of centrum, 2 transverse processes occur. Dorsally, there is a neural spine. The first cervical vertebra, the atlas balances and supports the head. It lacks centrum. The second is the axis. The sacral vertebrae are fused and form the triangular sacrum. The coccygeal vertebra, a vestige, has no function. In the embryonic stage of about 34 vertebrae, 5 bones are fused to form a sacral bone. four or 5 coccygeal bones are fused to form a single coccyx. Ribs are 12 pairs, each articulate with a thoracic vertebra. The first 10 pairs are attached to sternum by costal cartilages. The first seven attached directly to sternum are called the true ribs. Cartilages of 8th, 9th and 10th ribs are fused and attached to 7th and are called the false ribs. Eleventh 12th pairs not attached to sternum are called floating ribs (Figure10.22).

Appendicular Skeleton: It consists of bones of upper and lower limbs and girdles by which they are attached to the body. Hands are attached to pectoral girdle consisting of 2 pairs of bones. Each pair has a scapula, or shoulder blade and a clavicle, or collar bone. The scapula is a flat, triangular bone. A glenoid fossa is located in superior lateral region of

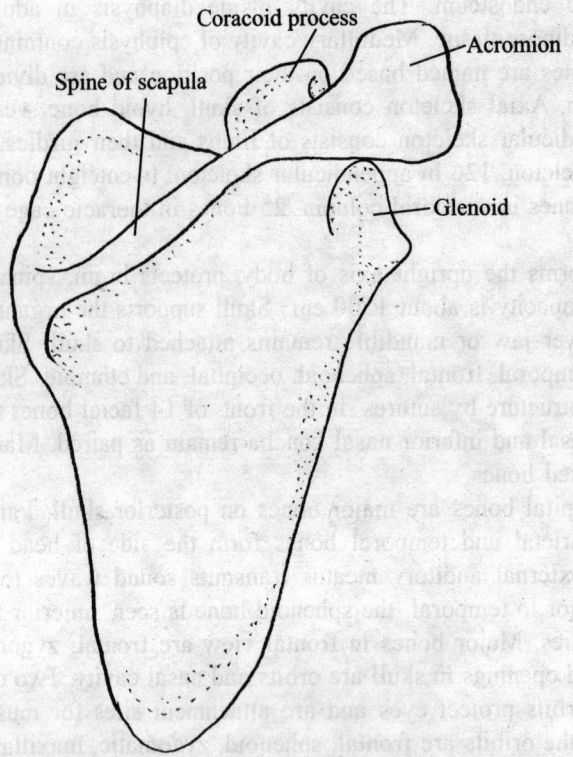

Figure 10.22(a)   Posterior aspect of right scapula in man.

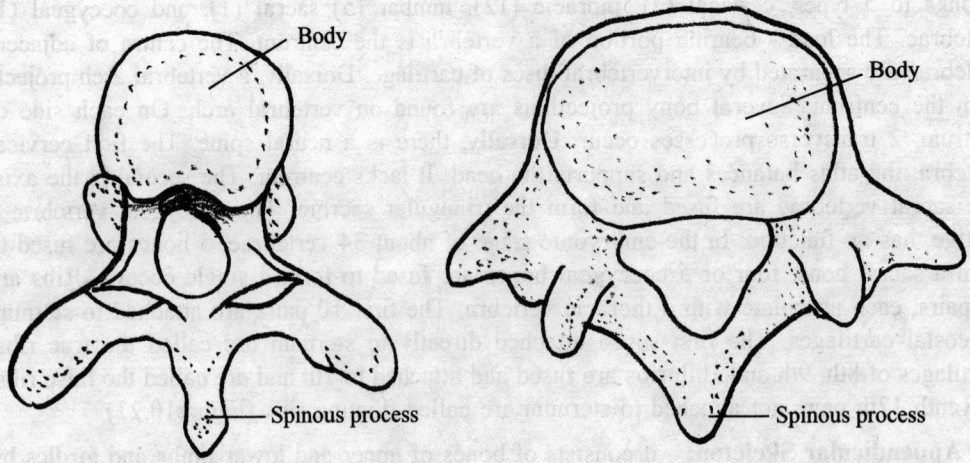

Figure 10.22(b)   Typical thoracic vertebra.   **Figure 10.22(c)**   Typical lumbar vertebra.

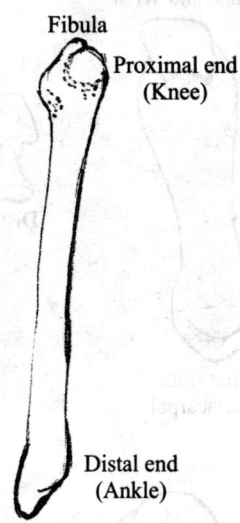

Figure 10.22(d)   Anterior view of right
                tibia (shin bone).

Figure 10.22(e)   Anterior view of right Fibula.

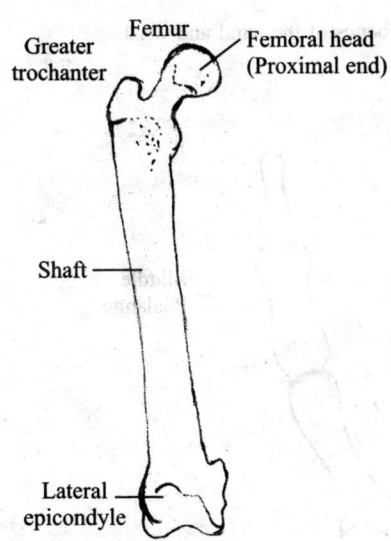

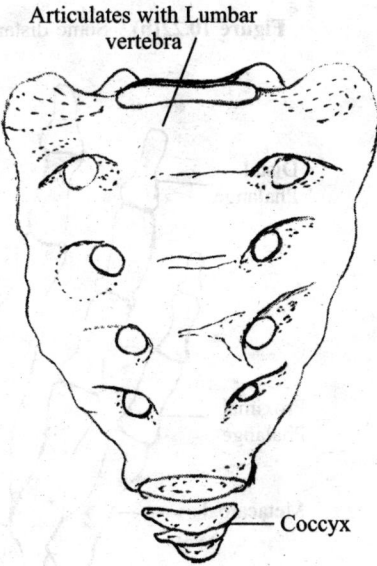

Figure 10.22(f)   Anterior view of femur
               (Thigh Bone).

Figure 10.22(g)   Sacrum and Coccyx
               (Tall Bone).

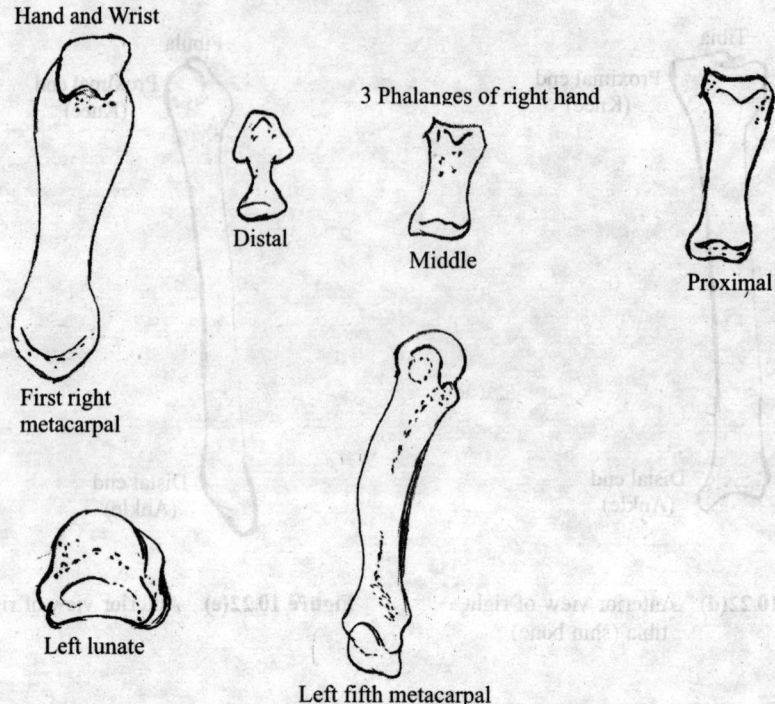

Hand and Wrist

3 Phalanges of right hand

Distal

Middle

Proximal

First right
metacarpal

Left lunate

Left fifth metacarpal

Figure 10.22(h)   Some distarticulated bones of the hand and wrist.

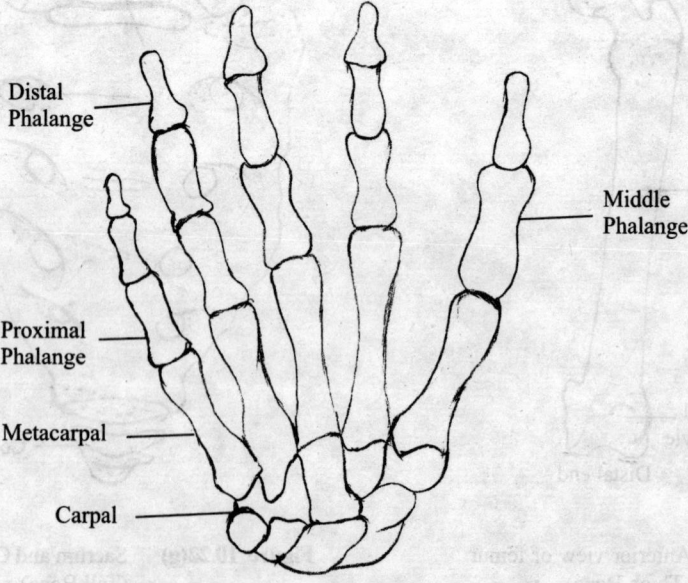

Distal
Phalange

Middle
Phalange

Proximal
Phalange

Metacarpal

Carpal

Figure 10.22(i)   Palm surface of right hand and wrist (articultated).

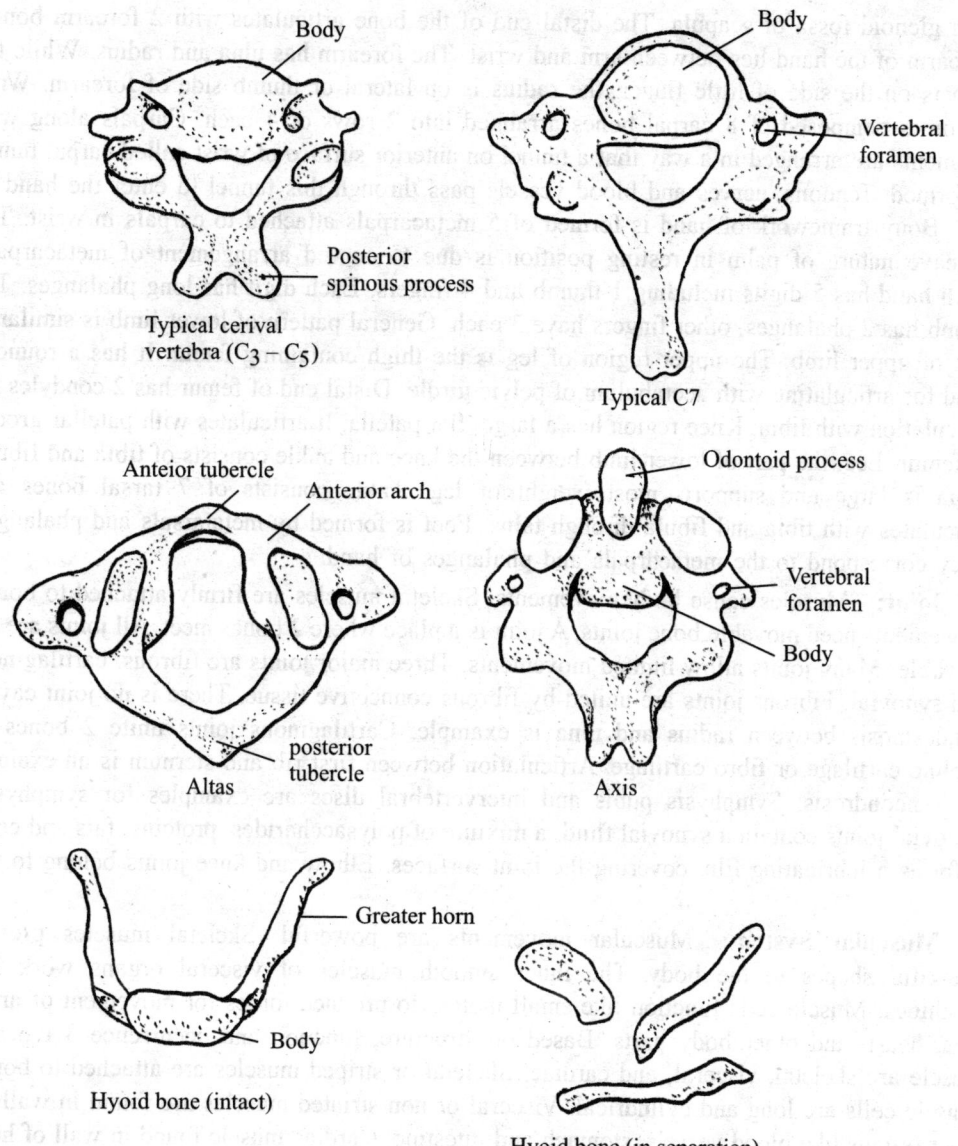

Typical cerival
vertebra ($C_3 - C_5$)

Typical C7

Altas

Axis

Hyoid bone (intact)

Hyoid bone (in separation)

Figure 10.22(j)

scapula. It articulates with head of humerus. The clavicle is a long bone with a slight S-shaped curve and holds the upper limb away from the body. Pelvic girdle is a ring of bones formed by sacrum, and paired bones called the coxae or hip bones. Each coxa is formed by fusion of ilium, ischium, and pubis. A fossa called the acetabulum is located on lateral surface of each coax and articulates to lower limbs. The part of upper limb from shoulder to the elbow is the arm containing one long bone called humerus. The head of humerus articulates with glenoid fossa of scapula. The distal end of the bone articulates with 2 forearm bones. Forearm of the hand lies between arm and wrist. The forearm has ulna and radius. While the ulna is on the side of little finger, the radius is on lateral or thumb side of forearm. Wrist region is composed of 8 carpal bones arranged into 2 rows of 4 each. Carpals along with ligaments are arranged in a way that a tunnel on anterior surface of wrist called carpal tunnel is formed. Tendons, nerves and blood vessels pass through this tunnel to enter the hand.

Bony framework of hand is formed of 5 metacarpals attached to carpals in wrist. The concave nature of palm in resting position is due to curved arrangement of metacarpals. Each hand has 5 digits including 1 thumb and 4 fingers. Each digit has long phalanges. The thumb has 2 phalanges, other fingers have 3 each. General pattern of lower limb is similar to that of upper limb. The upper region of leg is the thigh containing femur. It has a rounded head for articulating with acetabulum of pelvic girdle. Distal end of femur has 2 condyles for articulation with tibia. Knee region has a large, flat patella. It articulates with patellar groove of femur. Leg, the part of lower limb between the knee and ankle consists of tibia and fibula. Tibia is large and supports most weight of leg. Ankle consists of 7 tarsal bones and articulates with tibia and fibula through talus. Foot is formed by metatarsals and phalanges. They correspond to the metacarpals and phalanges of hand.

Joint: Muscles cause body movements. Skeletal muscles are firmly attached to bones. Movements need movable bone joints. A joint is a place where 2 bones meet. All joints are not movable. Many joints allow limited movements. Three major joints are fibrous, cartilaginous and synovial. Fibrous joints are united by fibrous connective tissue. There is no joint cavity. Syndesmosis between radius and ulna is example. Cartilaginous joints unite 2 bones by hyaline cartilage or fibro cartilage. Articulation between first rib and sternum is an example for syncondrosis. Symphysis pubis and intervertebral discs are examples for symphyses. Synovial joints contain a synovial fluid, a mixture of polysaccharides, proteins, fats and cells. It forms a lubricating film covering the joint surfaces. Elbow and knee joints belong to this type.

Muscular System: Muscular movements are powerful. Skeletal muscles provide beautiful shapes to the body. The inner smooth muscles of visceral organs work like machines. Muscle cells function like small motors to produce forces for movement of arms, legs, heart, and other body parts. Based on structure, function and occurrence 3 types of muscle are skeletal, visceral, and cardiac. Skeletal or striped muscles are attached to bones. Muscle cells are long and cylindrical. Visceral or non striated muscles are found in walls of inner organs like blood vessels, stomach and intestine. Cardiac muscle found in wall of heart is involuntary in nature

Skeletal Muscles: Skeletal muscles are attached to bones by tendons which help to transfer forces developed by skeletal muscles. The fascia, the sheets of connective tissue cover such muscles. The superficial fascia is found between skin and muscles. The deep fascia is collagen fibres found around musculature. They run between groups of muscles and connect with the bones. Tendons show little elasticity and are strongly attached to bones. Tensile strength of tendons is nearly half that of steel. Tendon with a diameter of 10 mm support 600 - 1000 kg. Based on general shape and the orientation, muscles can be grouped into 2 classes. Parallel muscle fibres are parallel to line of pull, may be flat, short, quadrilateral or long and strap like. Individual fibres run the whole length of muscle. Oblique muscle fibres are oblique to line of pull, may be triangular, or pennate. The pennate forms are divided into unipennate, bipennate, multipennate, or circumpennate. Some muscles show spiral, or twisted arrangement.

Naming of Muscles: The muscles are named according to their size, shape, position, and action.

Muscles of Head: Two groups of muscles are craniofacial and masticator muscles. Craniofacial muscles, also called muscles of facial expression, are related to eye orbital margins, eyelids, mouth, pinna, and the scalp. Facial expression is due to lip movement and positioning of lips. Such movements are caused by several muscles associated with lips and the skin around the mouth. Since orbicular and buccinator muscles provide lip movement for kissing, they are called "kissing muscles". Zygomasticus major and minor, levator anguli or sand risorius are associated with smile. Masticatory muscles called masseter temporalis and pterygoid move the mandible of lower jaw. Tongue movements are due to intrinsic and extrinsic muscles. Several muscles related to mouth, roof of pharynx, uvula and other regions are related with swallowing.

II. Muscles of Neck Region: Cervical, suprahyoid, infrahyoid, and vertebral muscles cause movements of the neck region.

III. Muscles of Trunk Region: Muscles of vertebral column help to bend and rotate the body. Strong back muscles help the trunk to maintain erect posture. The muscles of this region are the erector spine, longissimus, and spinalis. Four important thoracic muscle groups are associated with the breathing. Inspiration is due to scalene and external intercostals muscles. Expiration is performed by internal intercostals and transverse thoracic. Major breathing movement is due to diaphragm. Abdominal muscles aid in forced expiration, defecation, and urination. Inferior opening of pelvic bone is covered by pelvic diaphragm muscles. Below these muscles, perineum is present. Perineum and other "sub floor" muscles form the urinogenital diaphragm. Pelvic and urogenital diaphragm stretch in pregnancy due to weight of foetus.

Muscles of Upper Limb: Hands are attached to pectoral girdle and to vertebral column by muscles like trapezius, rhomboid major and minor; levator scapulae and lattissimus dorsi. The trapezius is flat, triangular and extends over the back of the neck and upper thorax. It maintains the level and poise of shoulder and helps to rotate the scapula forward and to bend the neck backwards and laterally. Latissimus dorsi is a large flat triangular muscle which stretches over the lumbar region and lower thorax. This muscle helps in adduction, extension, and medial rotation of humerus and in backward swinging of arm and in violent expiratory activities like coughing or sneezing. It helps in deep inspiration. Serratus anterior and pectoralis major connect the ribs to scapula. Pectoralis major, a fan shaped muscle extends from the upper thorax and abdomen to act on humerus and spreads between the clavicle and the 7th costal cartilage in front of chest. It helps to swing the extended arm forward and medially and in climbing, deep inspiration. Muscles of the upper arm are the coracobrachialis, biceps, triceps, and brachialis. Coracobrachialis arises from coracoid bone in shoulder and ends in humerus of upper arm. It helps to move the arm forward and medially. Biceps brachii, a large fusiform muscle is provided with 2 proximal heads for attachment. They are connected to coracoid and shoulder joint. The lower head ends in radius of lower arm and causes flexing of hand. The triceps arises by 3 heads from scapula and upper part of humerus on posterior side. Several extrinsic and intrinsic muscles cause the movements of the wrist, hand, and finger.

Muscles of Lower Limb: Thigh movements are caused by anterior, postereolateral, and deep muscles. Anterior muscles, the iliacs and psoas major help to flex the thigh. Gluteus maximus, form the mass of buttock. Leg movement is caused by anterior thigh muscles, quadriceps femoris and sartorius. Sartorius is the longest muscle that runs from the hip to knee. Muscle movement of ankle foot and toe are caused by groups of extrinsic and intrinsic muscles.

Digestive System: It includes alimentary tract and digestive glands which are described below

Tongue: It is a large muscular organ attached to floor of oral cavity. The anerior part of tongue is free. Thin fold of tissue called frenulum attaches the free end to mouth floor. Tongue is divided into 2 parts by a groove called terminal sulcus. Papillae cover two thirds of anterior surface, some of which contain taste buds.

Teeth: Four different types of permanent teeth are incisors (8), canines (4), premolars (8) and molars (12) with a total of 32 teeth in adult. This nature is called heterodontism. Each tooth consists of upper crown, middle neck and basal root regions. The crown region has one or more cusps. The tooth is made up of a calcified tissue called dentine which is covered by an extremely hard substance called enamel. The surface of dentine in root is covered with a bonelike substance called cementum, which anchors the tooth in jaw. In the centre of tooth, a pulp cavity called root canal contains blood vessels and nerves. The canal opens at base through apical foramen. The teeth are set in sockets along edges of upper and lower jaws. This region of jaw called gingiva is covered by dense fibrous connective tissue and stratified squamous epithelium. Salivary glands scattered throughout oral cavity are parotid, submandibular, and sublingual glands. Parotid glands are the largest and located just anterior to parotid gland. Submandibular glands are found on inferior borders of mandible. Sublingual glands are the smallest, lie immediately below mucous membrane in mouth floor. Other numerous small, coiled, tubular glands in mouth are lingual (tongue), palatine (palate), buccal, and labial (lips) glands (Figure 10.23).

Oesophagus: This part extends between the pharynx and stomach, about 25 cm long and lies in mediastinum of thorax, anterior to vertebra and posterior to trachea, passes through diaphragm and ends at stomach. This has thick walls; the inner wall is lined by a moist stratified squamous epithelium. The upper and lower ends have sphincters to regulate the movements of materials.

Stomach: It is a horizontally placed enlarged sac like structure found in upper part of abdomen and is divisible into cardiac and pyloric stomachs. Cardiac stomach lies at the left of abdomen. Oesophagus opens into cardiac stomach through gastroesophageal or cardiac opening. A part of stomach to left of cardiac region is the fundus. The largest part of stomach is the body which narrows to form pyloric region. Pyloric opening between the pylorus and intestine is surrounded by a ring of muscles called pyloric sphincter.

Intestine: Small intestine is about 6m long and consists of duodenum, jejunum and ileum. Duodenum is about 25 cm. long, curves within abdominal cavity and completes nearly

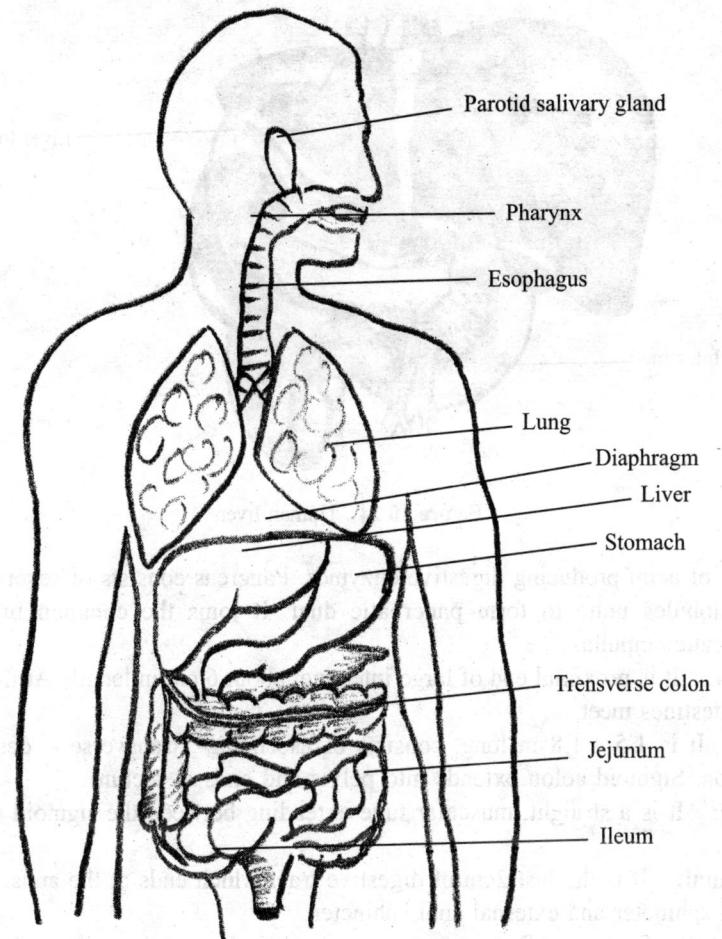

Figure 10.23 Digestive system in human.

180 degree arc. Liver and pancreas are associated with duodenum. Jejunum and ileum are 2.5m and 3.5m in length respectively. There is a gradual decrease in diameter of small intestine. The junction between ileum and large intestine is the ileocaecal junction which has ring of smooth muscles forming a sphincter, and a one way ileocaecal valve.

Liver: It is the largest visceral organ, weighs about 1.36 Kg, consists of 2 major left and right lobes, and 2 minor lobes caudate and quadrate. Bile secreted by liver is collected in gall bladder. Two hepatic ducts unite to form a single duct. The common hepatic duct is joined by cystic duct from gall bladder to form common bile duct. It empties into duodenum (Figure 10.24).

Gall Bladder: It measures 8 cm X 4 cm, sac like structure on the inferior surface of liver.

Pancreas: It is a complex organ composed of both endocrine and exocrine tissues. Endocrine part consists of pancreatic islets which produce insulin and glucagon. Exocrine

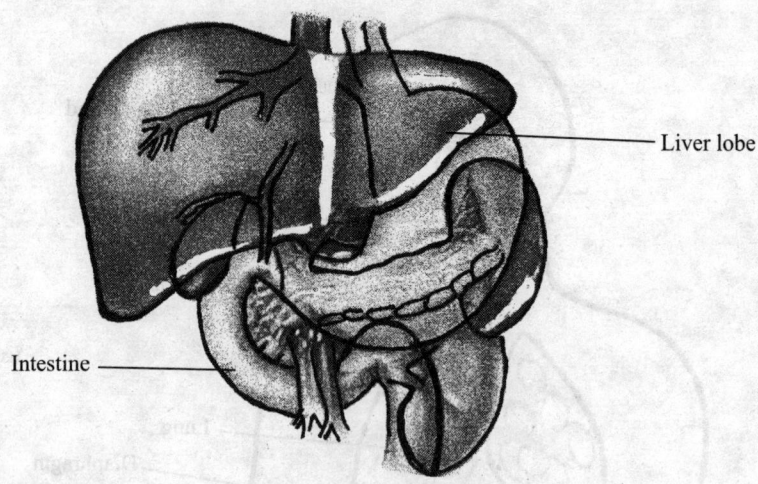

Figure 10.24   Human liver.

part consists of acini producing digestive enzymes. Pancreas consists of several lobules. The ducts from lobules unite to form pancreatic duct. It joins the common bile duct at the hepatopancreatic ampulla.

Caecum: It is proximal end of large intestine, about 6 cm in length. At this region large and small intestines meet.

Colon: It is 1.5 - 1.8 m long, consists of ascending-, transverse -, descending - and sigmoid colon. Sigmoid colon extends into pelvis and ends at rectum.

Rectum: It is a straight, muscular tube extending between the sigmoid colon and anal canal.

Anal canal: It is the last 2cm of digestive tract which ends at the anus. The canal has internal anal sphincter and external anal sphincter.

Respiratory System: Respiratory organs include nasal cavity, pharynx, larynx, trachea, bronchi, and lungs which are organized into upper and lower respiratory tracts.

Nasal Cavity: Nasal cavity follows the external nose. Internally it is supported by cartilage plates. Bridge of nose is formed by nasal bones and extension of skull bones. Respiratory passage is divided into 2 chambers by a median partition. Nasal passage opens to outside through external nostrils. It opens inside by internal nostrils at pharynx.

Pharynx: Buccal cavity and nasal passages open into pharynx which is divided into three parts. Nasopharynx extends from internal nostril to region of uvula. Uvula is a soft outgrowth hanging in between posterior part of oral cavity and pharynx. It prevents the entry of food into nasal cavity. Wall of nasopharynx is lined by ciliated columnar epithelium. Middle ear opens into nasopharynx through 2 auditory tubes for equalizing the air pressure between atmosphere and middle ear. Inner surface of nasopharynx contains pharyngeal tonsil for defence against infections. Enlargement of tonsil interfere with breathing. Oropharynx remains between uvula and epiglottis. Oral cavity opens into oropharynx. Near the opening of oral cavity, 2 sets of palatine tonsils and lingual tonsils are present. Laryngopharynx extends in between epiglottis and esophagus (Figure 10.25).

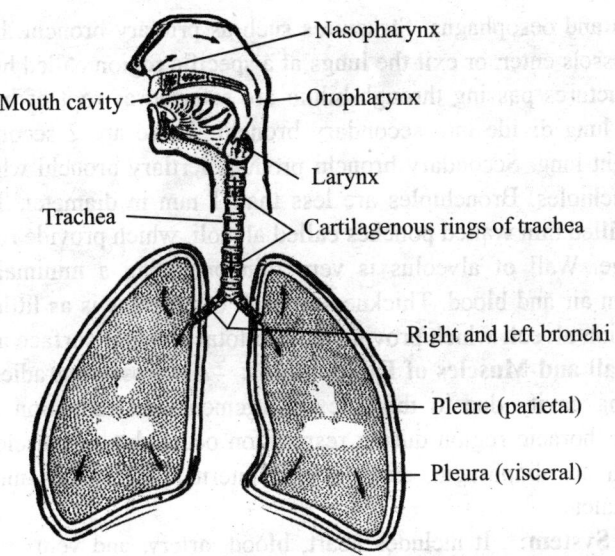

Nasopharynx
Mouth cavity
Oropharynx
Larynx
Trachea
Cartilagenous rings of trachea
Right and left bronchi
Pleure (parietal)
Pleura (visceral)

Figure 10.25 Respiratory tract.

Larynx: This lies just behind the pharynx and buccal cavity and is surrounded by cartilages (3 unpaired and 6 paired) which are interconnected by muscles and ligaments. The unpaired cartilages are thyroid, cricoid, and epiglottis. Thyroid cartilage is the largest called Adam's apple. Cricoid cartilage forms the larynx base. Other cartilages are placed above the cricoid. Epiglottis is attached to thyroid and projects as a free flap over opening of larynx. It prevents food particles from entering into tracheal tube. Ligaments inside larynx form vocal cords which are involved with sound production. Air moving past the vocal cords makes them to vibrate. Loud sounds are made by increasing the amplitude of vibrations. Frequency of vibrations can be altered by changing the length of vibrating segments of vocal cords. Length is altered by muscles attached to cartilage. Males usually have longer vocal cords than females. Sound made by vocal cords can be altered by tongue, lips, and teeth to form words.

Trachea: It is a membranous tube, 10-12 cm in length with an inner diameter of 12 mm. The wall made up of connective tissue and smooth muscles supported by 15-20 'C' shaped cartilage rings which protect trachea and keep it open all the time. The inner wall is lined by mucus membrane consists of ciliated columnar epithelium. Cilia of this epithelium help to propel mucus and foreign particles towards larynx. Trachea extends from larynx to level of 5th thoracic vertebra. Basal part of trachea divides to form 2 small tubes called primary bronchi. The cartilage ring found at basal region is called carina. Foreign objects reaching carina stimulate a powerful cough reflex.

Lungs: Lungs are conical in shape. Base of the lung rests on diaphragm. Right lung is larger than left and it weighs around 620g. Left lung weighs 560g. Right lung has 3 lobes and the left lung has 2. Lungs are placed in thoracic cavity. Separate pleural membrane surrounds each lung. The region inside pleural membrane is called pleural cavity, which is filled with pleural fluid. The region between 2 lungs is called mediastinum, a midline partition occupied by heart, trachea and oesophagus. Structures such as primary bronchi, blood vessels, nerves and lymphatic vessels enter, or exit the lungs at a specific region called hilum on inner margin of lungs. All structures passing through hilum are referred as root of lung. Primary bronchi on entering into lung divide into secondary bronchi. There are 2 secondary bronchi in left lung and 3 in right lung. Secondary bronchi produce tertiary bronchi which divide further to give rise to bronchioles. Bronchioles are less than 1 mm in diameter. Terminal bronchioles end in small air filled thin walled pouches called alveoli, which provide respiratory surface for gaseous exchange. Wall of alveolus is very thin providing a minimal barrier to gaseous exchange between air and blood. Thickness of wall of alveolus is as little as 0.05m in .There is about 300 million alveoli which provide a mean total alveolar surface area value of 143 m^2.

Thoracic Wall and Muscles of Respirations: Air pressure gradients between thoracic chamber and lung cavity due to thoracic enlargement and reduction cause ventilation of lungs. Change in thoracic region during respiration occur due to muscles of inspiration and expiration, which are diaphragm, external and internal intercostals muscles between ribs, pectorals, and scalene.

Circulatory System: It includes heart, blood, artery, and vein.

Heart: Heart is a hollow, fibro muscular organ, somewhat conical or pyramidal in form. An average heart measures 12 cm from base to the apex. Transverse diameter at its broadest region is 8-9 cm. It is 6 cm thick antero-posteriorly. Heart of adult male weighs 280-340 g, in female it weighs 230-280 g. Thoracic organs like heart, and trachea and esophagus form a midline partition called mediastinum. Heart lies obliquely in mediastinum and is surrounded by a double layered membrane called pericardium. The outer layer is called fibrous pericardium. The inner membrane is called serous pericardium. In between heart and pericardium, a pericardial space remains filled with pericardial fluid. The wall of heart is made up of epicardium, myocardium, and endocardium. The epicardium forms the smooth outer surface of heart. The middle myocardium is composed of cardiac muscle. This layer plays an important role in functioning of heart. The endocardium forms smooth inner surface. It is formed of squamous epithelium. Heart is a two-sided, 4-chambered structure with muscular walls. AV node called the pacemaker keeps heart beat regular. Heart beat is also controlled by nerve messages originating from autonomic nervous system.

Blood flows through heart from veins to atria to ventricles out by arteries. Heart valves limit flow to single direction. One heart beat includes atrial contraction and relaxation, ventricular relaxation and contraction, and a brief pause. Natural cardiac cycles (at rest) take 0.8 seconds. Blood flows into vena cava which empties into right atrium. At the same time, oxygenated blood from lungs flows from pulmonary vein into left atrium. Muscles of both atria contract, forcing blood downward through each AV valve into each ventricle. Diastole is the filling of ventricles with blood. Ventricular systole opens the SL valves, forcing blood out of ventricles through pulmonary aorta. Sound of heart contracting and valves opening and closing produces a characteristic "lub-dub" sound. Lub is related with closure of AV valves; dub is the closing of SL valves. Heart beats originate from the SA node near right atrium. Contraction of modified muscle cells send signal to other heart muscle cells to contract. The signal spreads to the AV node. Signals carried from AV node, slightly delayed, through bundle of His fibers and Purkinjie fibers as ventricles to contract simultaneously. Heartbeats are result of coordinated contractions of cardiac cells. When 2, or more cells are in proximity to each other their contractions synch up and they beat as one. Systemic circulation carries blood to lungs for oxygenation and returns it back to heart via the pulmonary circulation (Figure 10.26).

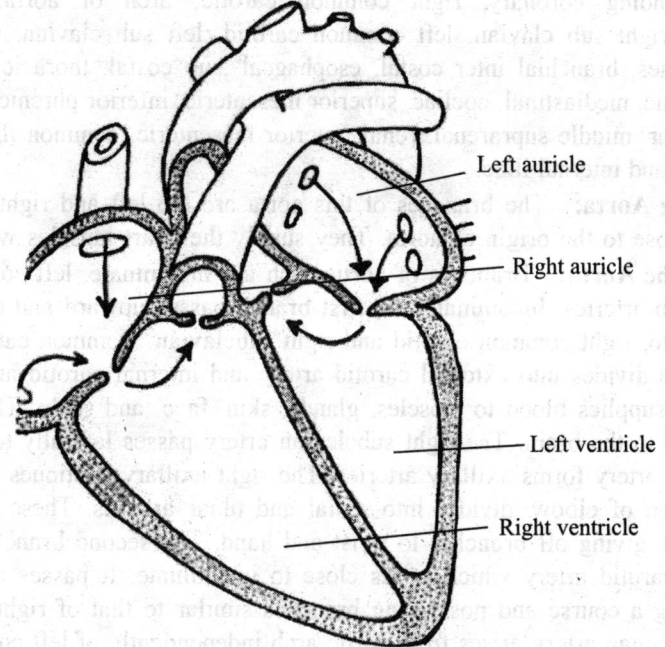

Figure 10.26 Human heart.

Systemic Circulation: Left atrium receives oxygenated blood from lungs through pulmonary vein. When the atria contract blood from left atrium is forced into left ventricle. Later ventricle contracts , blood leaves heart through aorta, the single systemic artery emerging from heart which gives rise to many arteries taking blood to all parts of body. Arteries divide into numerous arterioles. In the target organs, they produce 4 times as many capillaries. Similar number of venules unites into each other forming veins of increasingly larger size. The superior and inferior vena cavae return the blood to right atrium. Course of blood circulation takes place from left ventricles through body organs and back to atrium form systemic circulation.

Pulmonary Circulation: Venous blood from right atrium is conducted to right ventricle. Ventricle expels the blood via pulmonary trunk to lungs. Oxygenated blood later returns by pulmonary veins to left atrium.

Portal Circulation: In the systemic circulation, the venous blood passing through spleen, pancreas, stomach and intestine is not carried back directly to heart. It passes through hepatic portal vein to liver. This vein begins as capillaries from visceral organs and ends in liver as capillaries, which unite to form hepatic vein and joins the inferior vena cava, conveying blood to right atrium.

Arterial System: The main arterial trunk carrying blood from heart to tissues is the thoracic aorta. Its terminal portion, the abdominal aorta lies in abdomen. Thoracic aorta comprises the ascending aorta, arch of aorta and descending aorta, which passes downward through diaphragm and continues to level of 4th lumbar vertebra, where it terminates by dividing into 2 common iliac arteries. The main branches of various portions of aorta are as follows. Ascending coronary, right common carotid, arch of aorta, Brachiocephalic (innominate), right sub clavian, left common carotid, left sub clavian, visceral branches, parietal branches, branchial inter costal, esophageal ,sub costal, thoracic aorta ,pericardial ,superior phrenic, mediastinal, coeliac, superior mesenteric, inferior phrenic, abdominal aorta, gonadal, lumbar, middle suprarenal, renal, interior mesenteric, common iliac, external iliac, middle sacral, and internal iliac.

Ascending Aorta: The branches of this aorta are the left and right coronary arteries which arise close to the origin of aorta. They supply the heart muscles with blood.

Arch of the Aorta: Branches of aortic arch are innominate, left common carotid and left sub clavian arteries. Innominate, the first branch passes upward and diagonally to right and divides into, right common carotid and right subclavian. Common carotid artery passes up to neck and divides into external carotid artery and internal carotid artery. The external carotid artery supplies blood to muscles, glands, skin, face, and scalp. The internal carotid artery supplies to the brain. The right subclavian artery passes laterally to the arm. In arm, the subclavian artery forms axillary arteries. The right axillary continues as brachial artery, which in region of elbow, divides into radial and ulnar arteries. These continue in distal portion of arm, giving off branches to wrist and hand. The second branch of aortic arch is left common carotid artery which arises close to innominate. It passes up the left side of neck, following a course and possessing branches similar to that of right common carotid. The left sub clavian artery arises from aortic arch independently of left common carotid and supplies the left arum.

Thoracic Aorta: The branches of thoracic portion of descending aorta supply internal organs and body wall. The internal organs include bronchi, esophagus, pericardium, lungs; rib cage, and intercostals muscles. Superior phrenic arteries supply blood to upper surface of diaphragm. Abdominal aorta has many branches. It serves the entire lower body region. The following branches are given off from the abdominal aorta. The coeliac artery arises shortly behind the diaphragm and sends branches to liver, gall bladder, stomach, duodenum, and digestive glands like pancreas and liver. The superior mesenteric artery supplies the major portion of small intestine and a part of large intestine. The middle suprarenal arteries supply suprarenal glands. The renal arteries supply kidneys. The internal testicular arteries in male and ovarian arteries in female supply the testis and ovary, respectively. The inferior mesenteric artery supplies the large intestine and rectum. The common iliac arteries from the dorsal aorta enter into legs. Each iliac artery divides into sciatic and femoral arteries supplying blood to leg muscles.

Pulmonary Circulation: The pulmonary artery emerges from the superior surface of right ventricle, passes diagonally upward to left, and crosses the root of aorta. It divides into right and left pulmonary arteries, branches of which enter right and left lungs, respectively. Pulmonary veins are 4 in number. They carry oxygenated blood from the lungs to left atrium.

Venous System: Systemic veins collect blood from tissues. The principal systemic veins are the coronary sinus, inferior vena cava, superior vena cava and their branches, and portal vein which drains the abdominal viscera.

Coronary Sinus: This is a short vein lying on posterior side of heart and receives blood from heart tissues. Superior vena cava; a large venous trunk empties blood to heart from head, neck, upper extremities and thorax. It opens into right atrium. The principal veins draining the head and neck are the external and internal jugular veins. The right and left subclavian veins drain the upper extremities, each terminating at its junction with internal jugular vein to form innominate vein. Near their termination form each subclavian vein receives an external jugular vein. Each subclavian vein is formed by union of cephalic and axillary vein. Each innominate vein receives the deep cervical, vertebral, internal mammary, and inferior thyroid veins. The left innominate receives left superior intercostals vein and veins from thymus, trachea, esophagus, and pericardium. The subclavian veins also drain blood from veins of hand

Inferior Vena Cava and Its Branches: This is the venous trunk which receives most of the blood from regions of the body below the level of the diaphragm. It is the largest vein in the body. The inferior vena cava is formed by the union of the 2 common iliac veins. It extends forwards to the right of the aorta, passes through the diaphragm and opens into the right atrium. The inferior vena cava receives blood from following veins, inferiorphrenic, hepatic, right suprarenal, renal, right spermatic or ovarian, lumbar and common iliac, and the veins of the lower extremities.

Pulmonary Circulation: Venous blood from right atrium is conducted to the right ventricle. The ventricle expels the blood via pulmonary trunk to lungs. Oxygenated blood later return by the pulmonary vein to left atrium. This circulation from right ventricle to left atrium via lungs is termed the pulmonary circulation.

Portal Circulation: In systemic circulation, the venous blood passing through spleen, pancreas, stomach, and intestine are not carried back directly to heart. It passes through the hepatic portal vein to liver. This vein begins as capillaries from visceral organs and ends in liver again as capillaries. These capillaries converge to form the hepatic vein, which joins the inferior vena cava, conveying blood to right atrium. This route is the portal circulation.

Blood Vessels: Blood vessels carrying blood away from the heart are the arteries. The veins carry blood towards the heart. The arteries and veins are named and classified according to their anatomical position. They can also be classified according to their size and wall structure. Functionally, arteries are subdivided into conducting, distributing, and resistance vessels.

Conducting Vessels: These are large arteries from the heart and their main branches. The walls of such vessels are elastic in nature.

Structure of Blood Vessels: Blood vessel consists of a wall and a lumen. Walls are made up of 3 distinct layers– the tunica intima, tunica media, and tunica externa or tunica adventitia. Tunica intima is formed of an endothelium, a delicate connective tissue and elastic fibres. Tunica media contains smooth muscle cells and causes vasoconstriction and vasodilatation. Tunica externa is composed of connective tissue. Composition and thickness of layers varies with the diameter of blood vessels and type.

Types of Blood Vessels

1. **Large Elastic Arteries:** Walls of these arteries contain elastic fibres. The smooth wall measures about 1micron in thickness. It gets stretched under the effect of pulse and recoils elastically.

2. **Muscular Arteries:** There are larger and smaller muscular arteries. The larger muscular arteries are inelastic with thick walls. The wall is 30-40 microns in diameter in the layers of smooth muscles. Since they regulate blood supply, they are called distributing arteries. The small muscular arteries cause vasodilatation and vasoconstriction.

3. **Arterioles:** These are small vessels capable of vasodilation and vasoconstriction and conduct blood from the arteries to capillary bed.

4. **Capillaries:** These are fine vessels found between arterioles and venules, measure 5-8micron in diameter.

5. **Venules:** These are tubes of flat, oval or polygonal endothelial cells with diameters up to 30 microns and are formed by convergence of two, or more capillaries.

6. **Veins:** Veins run in between venules and large veins which transport blood to the heart. Veins with diameter above 2 mm have valves. They are of semilunar type. There are several valves in the medium veins.

Branching of Blood Vessels: When an artery divides into 2 equal branches, the original artery ceases to exist. Hence, the branches are called terminal branches. The smaller branching vessels formed on sides are called collateral branches. When arteries are joined to each other, it is called as anastomosis.

Blood Supply: Cells and tissue on the wall of blood vessel require nourishment. Some amount diffuses from blood in lumen. For vessels having diameter greater than 1 mm, diffusion of nutrients is difficult. Such vessels have minute vessels called vasa vasorum spread over them. They penetrate into wall of blood vessels. Walls of blood vessels are innervated by sympathetic nerve fibres. They regulate contraction of musculature and effect vasoconstriction.

Lmphatic System: Lymphatic and blood circulation helps to maintain fluid balance in tissues and absorbs fat from digestive tract, functions as body's defence system. The lymph, lymphocytes, lymph vessels, lymph nodule, lymph node, tonsil, spleen, and thymus gland constitute the system. Lymphatic organs contain lymphatic tissues which primarily consist of lymphocytes, macrophages, dendritic cells, and reticular cells. Lymphocytes a type of WBC originate from red bone marrow and are carried by blood to lymphatic organs and other tissues. B-lymphocytes synthesize antibodies for neutralizing alien macromolecules. T-lymphocytes selectively kill cells infected with viruses. B and T lymphocytes are produced from bone marrow stem cells. T lymphocytes mature after entering into thymus, a lymphoid organ.

Thymus: It is a roughly triangular, bilobed, located in mediastinum between sternum and pericardium. Its size varies with age. It is largest in early part of life. At birth, it weighs 10–15 g. After puberty, it decreases in size. Each thymus lobe is surrounded by a thin capsule. It has 2 layers. The inner layer is the medulla, the outer layer is cortex. Lymphocytes are found in cortex layer. Lymph nodes are small round structures. Their size ranges from 1-25 mm and are distributed throughout the lymphatic vessels. These nodes are found all over the body and as aggregations in 3 body regions namely the inguinal nodes in groin, axillary nodes in axillary region, and cervical nodes of neck. The lymph enters lymph nodes through afferent lymphatic vessels and exits through efferent vessels. Spleen is located on left side of abdominal cavity. It has a fibrous capsule. Spleen contains red pulp and white pulp. Tonsils are the largest lymph nodules that provide protection against pathogen. In adults, the tonsils decrease in size and may disappear. Of 3 groups of tonsils in pharyngeal walls, the palatine tonsils are called "the tonsils" which are large lymphoid masses on each side of junction between oral cavity and pharynx. Pharyngeal tonsil or adenoid is found near junction between nasal cavity and pharynx. Lingual tonsil is a loosely associated collection of lymph nodules on posterior tongue surface. Lymph from tissues is drained by lymphatic capillaries which present in many tissues but absent in epidermis, hairs, nails, cornea, cartilages, CNS and bone marrow. Lymphatic capillaries join into larger vessels which pass to local or remote lymph nodes. These vessels and associated lymph nodes are arranged in regional groups with its region of drainage. Nodes within a group are interconnected and are organized in head and neck, upper limbs, lower limbs, abdomen and pelvis, thorax. Regional vessels return to venous blood circulation via the right and left lymph venous portals. Nearly 8 lymphatic trunks converge at the site of vertebral column and open into venous portals nearer to neck.

Nervous System: Several billion cells constitute the nervous system. Basically the system is formed of neurons. Neurons transmit impulses, help in realizing, analyzing and storing messages, and stimulate muscles to work. The network of interconnected neurons in nerves, brain and spinal cord are complicated in methods of their functioning. A neuron has a basic cell structure called cyton, the projections of which are dendrites and dendrons. The inter communicating long projection is axon. Variations in shape of cyton, number of dendrons and nature of axon exist. A neuron is interconnected with dendrite of neighbouring neuron through end plate of axon. Such connections are called synapses. In terminal regions of effector nerves, the axon of nerve cells remains in contact with muscle tissue. These joints are called neuromuscular junctions (Figure 10.27).

Structure of Peripheral Nerve: A nerve is made up of several nerve fibres which are grouped in fasciculi. Each fibre contains axons with coverings called Schwann cells. Number and pattern of fasciculi vary in different nerves. Nerve trunk possesses many fasciculi and is surrounded by an epineuruium. A multilayered perineurium which surrounds the endoneurium or intra fascicular connective tissue surrounds the individual fasciculi. Epineurium constitutes 30–70% of total cross sectional area of nerve bundle. Thickness is more when there are more fasciculi. Layer of fat in epineurium provides a 'cushion' effect to nerve. Perineurium contains alternating layers of flattened polygonal cells. Endoneurium remains condensed around axons. Components of endoneurium remain bathed in endoneurial fluid. Fasciculi of nerve are supplied blood by vasa nervosum. These minute blood vessels radiate up to endoneurium.

The organs of nervous system are continuous in nature. It can be divided into the following.

Central Nervous System: This system includes brain and spinal cord. They are protected by surrounding bones. Brain is located in cranium; spinal cord is placed in vertebral canal of vertebrae. Through an opening called foramen magnum, the spinal cord descends down from brain.

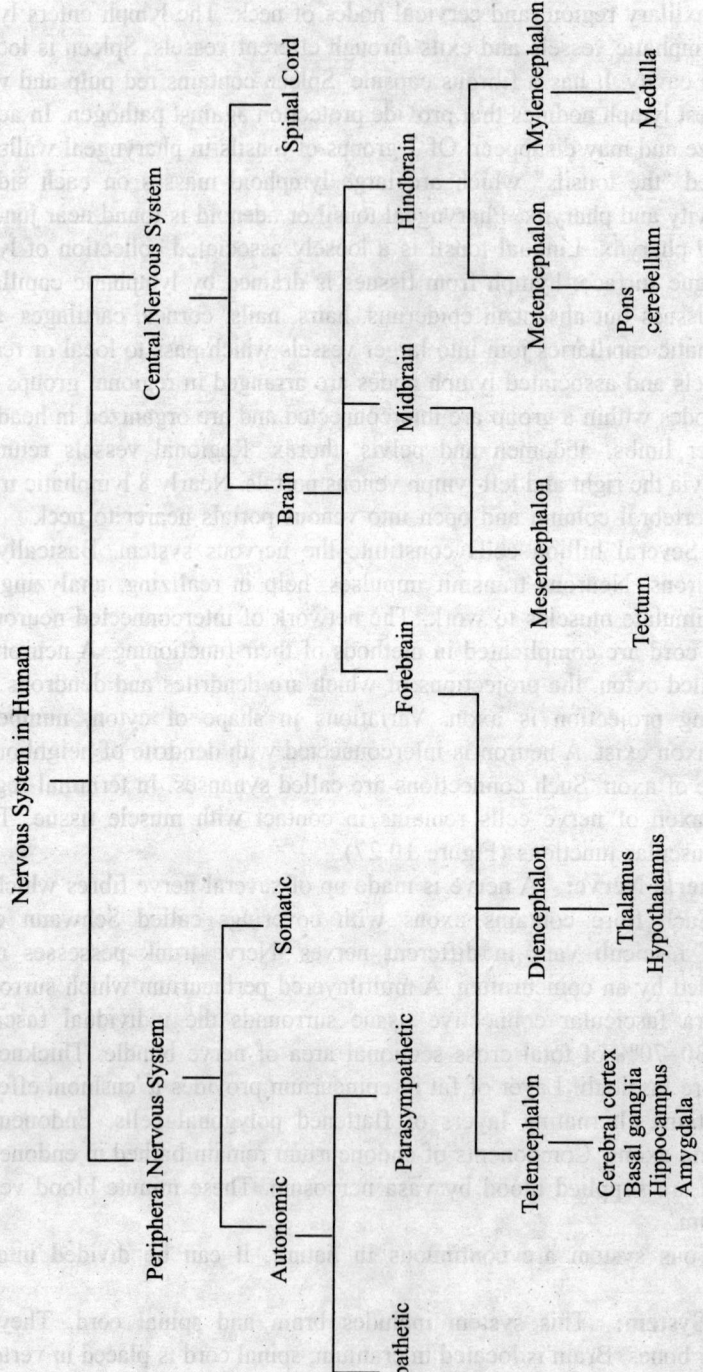

Figure 10.27(a)   Human nervous System.

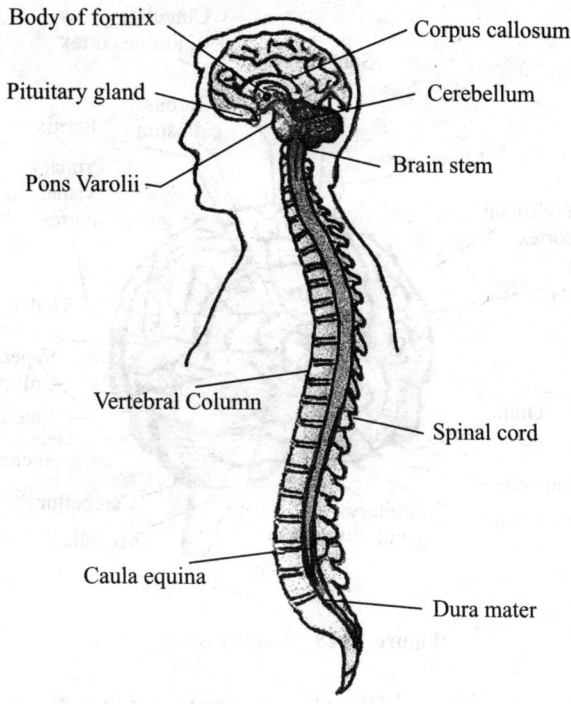

Body of formix

Corpus callosum

Pituitary gland

Cerebellum

Pons Varolii

Brain stem

Vertebral Column

Spinal cord

Caula equina

Dura mater

Figure 10.27(b) Human nervous system.

Peripheral Nervous System: It consists of nerves and ganglia. Nerves formed from brain are called cranial nerves. Cranial nerves are 12 pairs and spinal nerves are 31 pairs.

Autonomous Nervous System: Nerves transmit impulses from CNS to smooth muscles, cardiac muscles and glands. This system also called involuntary nervous system is subdivided into sympathetic and parasympathetic divisions.

Brain: Inside the skull, the brain is surrounded by 3 protective coverings which may be grouped under 2 divisions- Pachymenix which includes the duramater and leptomeninges which includes arachnoid mater and pia mater. Duramater is the outermost membrane, thick and inelastic in nature. Arachnoid mater is the middle covering over brain. In between arachnoid and piamater, a space called subarachnoid space contains cerebrospinal fluid and blood vessels. Piamater is a delicate membrane closely applied to brain contains blood capillaries supplying blood to brain cells. The brain weighs about 1.3 Kg. Based on embryological, development the brain can be divided as shown in Figure 10.28.

Prosencephalon: It consists of cerebrum and diencephalons. Cerebrum, the largest part is divided into right and left hemispheres by a longitudinal fissure. At the base, 2 hemispheres are connected by a sheet of nerve fibres called corpus callossum. The outer surface of cerebrum called cortex or grey mater is 2 to 4 mm thick. The inner content of cerebrum is the white mater. Surface of cerebrum has several folds called the gyri which greatly increase the surface area of cortex. Shallow grooves in between the gyri are called sulci. A central sulcus runs in lateral surface of cerebrum from superior to inferior region. Each cerebral hemisphere is divided into the frontal at the front, the parietal towards top of

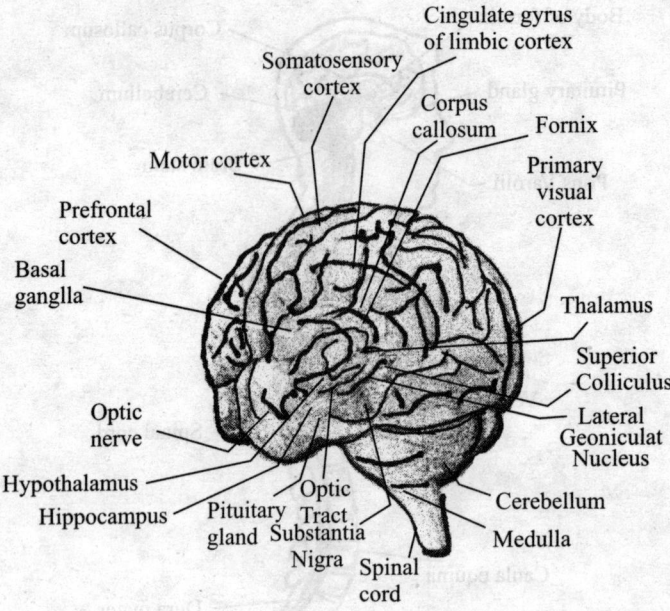

Figure 10.28 Human Brain.

head, the temporal on side and occipital at rear. Diencephalon containing thalamus and hypothalamus is found between cerebrum and brain stem. Thalamus has a cluster of nuclei which act as relays for particular sensory pathways. Just beneath the thalamus, the hypothalamus contains reflex centres linked to autonomic system. A funnel shaped stalk called infundibulum extends from its floor and is connected to neurohypophysis of pituitary gland.

Mesencephalon: It is the smallest region of brainstem. On its dorsal surface, there are 4 rounded corpora quadrigemina.

Rhombencephalon: Three main regions of it are medulla oblongata, pons varoli, and cerebellum. Cerebellum consists of 2 hemispheres. Its surface has many folia. Cerebellum consists of 3 parts which are small anterior flocconodular lobe, a narrow central vermis, and 2 large lateral hemispheres. Pons is located just superior to medulla oblongata. It contains ascending and descending nerve tracts. Medulla oblongata is 3 cm long, continuous with spinal cord, and remains as a bridge between brain and spinal cord. Medulla oblongata, pons and mid brain form brain stem to connect spinal cord with brain. Ten of the 12 cranial nerves enter, or exit the brain through brain stem. Spinal cord extends from foramen magnum to level of second lumbar vertebra and is short than vertebral column. There are cervical and lumbar enlargements. Below the lumbar enlargement, the spinal cord tapers to form a conus medullaris. A connective tissue filament, the filum terminale extends inferiorly from conus medullar to coccyx. Conus medullaris and nerves extend below resembling a horse's tail and are called cauda equina. Spinal cord has a central grey portion and a peripheral white portion. White matter consists of nerve tracts. Grey matter consists of neuron cell bodies and dendrites. Dorsal and ventral sides have long fissures. Thirty one pairs of spinal nerves arise from the spinal cord. Each nerve has dorsal and ventral root from spinal cord. Dorsal roots have dorsal root ganglia.

Ventricles: The entire CNS remains as a hollow tube inside adult brain forms ventricles. Each cerebral hemisphere contains large cavity called the lateral ventricle, which corresponds to hypothetical first and second ventricles. Two lateral ventricles communicate with 3rd ventricle located in centre of diencephalon. This connection is made through two interventricular foramina (foramen of Monro). Third ventricle opens into 4th ventricle found inside medulla oblongata through a narrow canal called cerebral aqueduct (aqueduct of sylvius). Fourth ventricle is continuous with central canal of spinal cord. Central canal extends nearly to full length of cord.

Cerebrospinal Fluid (CSF): This fluid fills ventricles of brain and central canal of spinal cord. About 80–90% of CSF is produced by epidermal cells within lateral ventricles. Remaining 10–12% is produced by similar cells in 3rd and 4th ventricles. These ependymal cells, their supportive tissue, and the related blood vessels together are called choroids plexuses which are formed by invagination of vascular piamater into ventricles.

Sensory Organs: Touch receptors in skin are the simplest single nerve cells responding directly to stimulus. Other receptors are complex sense organs, where stimulus is channeled into a receptive region of organ. Among the several organs, the most important are eyes and ears.

Eye: This is formed of 3 coats or tunics which are outer or fibrous sclera & cornea; middle or vascular choroid, ciliary body & iris; inner or nervous - retina. Sclera is the white outer layer which covers posterior five sixths of eye and provides shape and protects the internal structures. Small region of sclera is called the "white of the eye". In front; the outer layer forms a transparent cornea. It permits entry of light and is made up of a connective tissue having collagen, elastic fibres and proteoglycans.The middle tunic, vascular tunic of eyeball contains blood vessels and melanin containing pigment cells. It appears black in colour. Major part of vascular tunic is found in association with sclera called the choroid. Anteriorly, it forms the ciliary body and iris. The ciliary body consists of ciliary muscles. Contraction of ciliary muscles changes the shape of lens. The iris the coloured part of eye, may be black, brown or blue. A contractile structure surrounding an opening is called the pupil. Light enters eye through pupil. Iris regulates such entry by controlling the size of pupil. The inner most tunic is the retina. It consists of an outer pigmented retina and an inner sensory retina. Sensory retina is light sensitive. It contains nearly 120 million photoreceptor cells called rods and another 7 million cones (Figure 10.29).

Compartments of eye: Eye has a smaller compartment anterior to lens and a larger compartment behind the lens .The anterior compartment is divided into 2 chambers. Anterior chamber is found between cornea and iris. Smaller posterior chamber lies between iris and lens. The 2 chambers are filled with aqueous humor which helps to maintain intraocular pressure. The posterior compartment is larger and contains a transparent substance called vitreous humor. Eye lens is transparent and biconvex, made up of long columnar epithelial cells called lens fibres. Lens is placed between 2 eye compartments by suspensory ligaments. Functioning of eye is aided by structures like eye brows, conjunctiva, eyelids and lacrimal apparatus. Eyebrows prevent the sweat during perspiration from running down into eye. They help to shade eyes from direct sunlight. Eyelids and associated lashes protect eyes

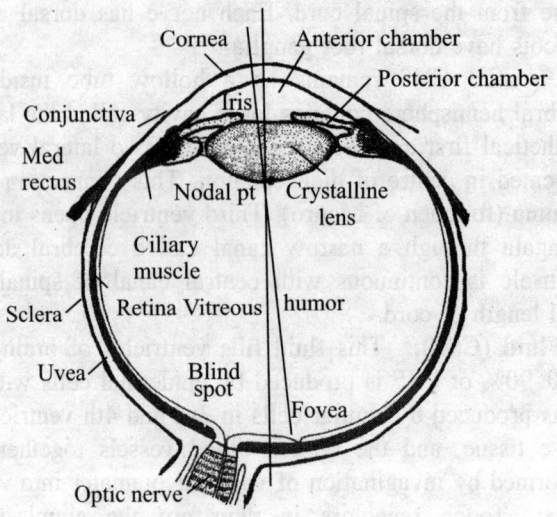

Figure 10.29   Structure of Human Eye.

from foreign objects. Medial region where the eyelids join has a small reddish-pink mound called caruncle containing modified sebaceous and sweat glands. There are 2 or 3 rows of hairs attached to free edges of eyelids. Modified sweat glands called ciliary glands open into follicles of eyelashes. It keeps them lubricated.

Melbomian Glands: These glands produce sebum for lubricating the eyelids. The inner surface of eyelids and anterior surface of eye are covered by a thin, transparent mucus membrane called conjunctiva. Lacrymal glands situated in supralateral corner of eye orbit produce tear at the rate of about 1 ml / day. It helps to moisten eye surface and wash away foreign substances. At the corners of eye, the small opening called the puncta opens into a lacrymal canaliculus that open into a lacrimal sac. This sac enters into a nasolacrimal duct which opens into inferior nasal concha. These ducts help to drain excess tear. The entire organization related to 'tear' is called the lacrymal apparatus.

Ears: Ears are the organs of hearing and balance with 3 parts, namely external, middle, and inner ears.

External Ear: Fleshy part outside the head called pinna is made up of elastic cartilage and skin, and is followed by external auditory meatus. This passage is lined with hairs and ceruminous glands. These glands produce cerumen or ear wax. Hair and wax prevent foreign objects from reaching the ear drum which is an oval, three layered structure. It separates outer and middle ears.

Middle Ear: It is an air filled cavity containing 3 auditory ossicles called the malleus, incus and stapes. Handle of malleus is in contact with inner surface of ear drum. Head of malleus is attached to incus. While stapes on one side is attached to incus, its other side fits into oval window. The oval window leads to inner ear.

Inner Ear: This region has tunnels and chambers inside temporal bone called bony labyrinth which contains 3 regions called cochlea, vestibule and semicircular canals. The oval window found in between middle and inner ears communicate with vestibule of inner ear. Organs of inner ear perceive the sound.

Urinary System: Urinary organs comprise two kidneys, ureters, urinary bladder, and urethra. Production of metabolic wastes causes a concentration gradient across the plasma membrane, diffuse wastes out of cells into extra cellular fluid.

Kidneys: Kidneys are bean shaped organs on posterior abdominal wall. The right kidney is slightly lower than left. It is because of the presence of liver superior to it. Each kidney measures 11 cm × 6cm in and 3cm in anteroposterior dimensions. In adult males, the average weight of kidney is about150g .The inner margin of kidney has a hilum. The renal artery and nerves enter and renal vein and the ureter exit at this region. Hilum opens into a cavity called renal sinus. Each kidney is enclosed by a fibrous connective tissue layer, called the renal capsule. Internally, the kidney is divided into cortex and medulla. Medulla consists of several cone-shaped renal pyramids. Extensions of pyramids called medullary rays, project from pyramids into cortex. Extension of cortex called renal columns, project between pyramids. Tips of pyramids are called renal papillae which are pointed toward renal sinus. Renal papillae are surrounded by minor calyces. Minor calyces of several pyramids join together to form larger funnel called major calyces. There are 8-20 minor calyces and 2 or 3 major calyces per kidney. Major calyces converge to form an enlarged channel called renal pelvis which then narrows to form ureter which leaves the kidney and gets connected to urinary bladder.

Nephron: Basic functional unit of kidney is the nephron. There are about 1.3 million nephrons in each kidney. Nephron consists of an enlarged terminal end, the renal corpuscle, a proximal tubule, a loop of Henle, and a distal tubule. Distal tubule opens into a collecting duct. Renal corpuscle, proximal tubule and distal tubules remain in renal cortex. Collecting tubules and parts of loops of Henle enter renal medulla. Most nephrons measure 50-55 mm in length. fifteen percent of nephrons are larger and they remain near medulla. These are called juxtamedullary nephrons which have larger loops of Henle. Renal corpuscle of nephron consists of a Bowman's capsule and capillaries called glomerulus. In the Bowman's capsule, the outer and inner layers are called parietal and visceral layers respectively. Outer parietal layer is composed of simple squamous epithelium. Inner visceral layer surrounds the glomerulus.It consists of podocytes. The walls of glomerular capillaries are lined with endothelial cells. There is a basement membrane between endothelial cells of glomerular capillaries and the podocytes of Bowman's capsule. Capillary endothelium, the basement membrane and podocytes of Bowman's capsule make up the filtration membrane. Glomerulus is supplied with blood by an afferent arteriole and is drained by an efferent arteriole. The cavity of Bowman's capsule opens into proximal tubule which is also called the proximal convoluted tubule. Posteriorly the proximal tubule continues as loop of Henle. Each loop has a descending limb and an ascending limb. The first part of desceding limb is similar in structure to proximal tubule. The loops of Henle that extend into medulla become thin near the end of loop. The first part of ascending limb is thin and consists of simple squamous epithelium, but it soon becomes thick. The distal tubules called the distal convoluted tubules are not as long as proximal tubules.

Ureters and Urinary bladder: Ureters extend inferiorly from the renal pelvis. They arise medially at the renal hilum to reach the urinary bladder. Bladder is a hollow muscular bag meant for temporarily storage of urine. It lies in the pelvic cavity. Size of bladder depends on presence, or absence of urine. Bladder capacity varies from 120-320ml. Filling upto500 ml is tolerated. Maturation will occur at 280ml. Ureters enter the bladder inferiorly on its posterolateral surface. Urethra exits the bladder inferiorly and anteriorly. At the junction of urethra with urinary bladder, smooth muscles of bladder form internal urinary sphincter. Around the urethra, there is another external urinary sphincter. Sphincters control the flow of urine through urethra. In male, the urethra extends to end of penis where it opens to the outside. In male, the urethra is 18-20cm long. In female, the urethra is shorter. It is about 4 cm long and 6 mm in diameter.

Reproductive System: Reproductive organs as internal and external genitalia are highly sophisticated yet simple in their functioning in accordance with psychological and endocrinological thresholds.

Male Reproductive System: It consists of testes, epididymes, ductus deferentia or vasa deferentia, urethra, seminal vesicles, bulbourethral glands, prostate gland, scrotum and penis. Testes are primary reproductive organs in male, remain suspended in scrotum by scrotal tissues. Sperms are temperature sensitive and do not develop normally at usual body temperatures. Hence, the testes and epididymes in which the sperm cells develop are located outside body cavity. Left testis usually is 1 cm lower than the right. A testis is 4-5 cm in length, 2-5cm in breadth. Its weight varies from 10.5-14g.The outer part of each testis is a thick, white capsule called tunica albuginea. Internally, the testis contains several incomplete septa which divide testis into 300-400 cone shaped lobules containing seminiferous tubules and interstitial cells or Leydig cells. Sperm cells develop within extensive seminiferous tubules. The combined length of tubules in both testes is nearly 800 metres. These tubules through straight tubules open into tubular network called rete testis. The rete testis opens into efferent ductules. Internally, the tubules and ductules are lined by ciliated columnar epithelium which helps to move sperm cells out of testis. Epididymis is formed of extremely convoluted ductules coming out of testis. It occurs on posterior side of testis. Maturation of sperms occurs within ductules of epididymis. Vas deferens or ductus deferens emerges from tail end of epididymis and ascends along posterior side of testis and becomes associated with blood vessels and nerves that supply the testis. These structures together constitute the spermatic cord which consists of vas deferens, testicular artery and venus plexus, lymph vessels, nerves and fibrous processes and muscles. This cord enters into the pelvic region. The end of vas deferens enlarges to form ampulla. At this region, the vas deferens is surrounded by smooth muscles capable of peristaltic contraction. They help to propel the sperm through ductus deferens.

Ejaculatory duct is situated near the ampulla of each vas deferens. There is a sac like seminal vesicle. It joins the ductus deferens to form ejaculatory duct which are about 2.5 cm long, project into prostate gland and end by opening into urethra. Male urethra extends from urinary bladder to distal end of penis is about 20 cm long. It is a passage way for both urine and reproductive fluids. Urethra is divided into prostatic urethra-, membranous -, and spongy urethra. Prostatic urethra is closest to bladder and passes through prostate gland. Membranous urethra is the shortest part of urethra and extends from prostatic urethra. Spongy urethra is the longest part of urethra. It extends from membranous urethra, through length of penis. Several minute mucus secreting urethral glands open into urethral passage.

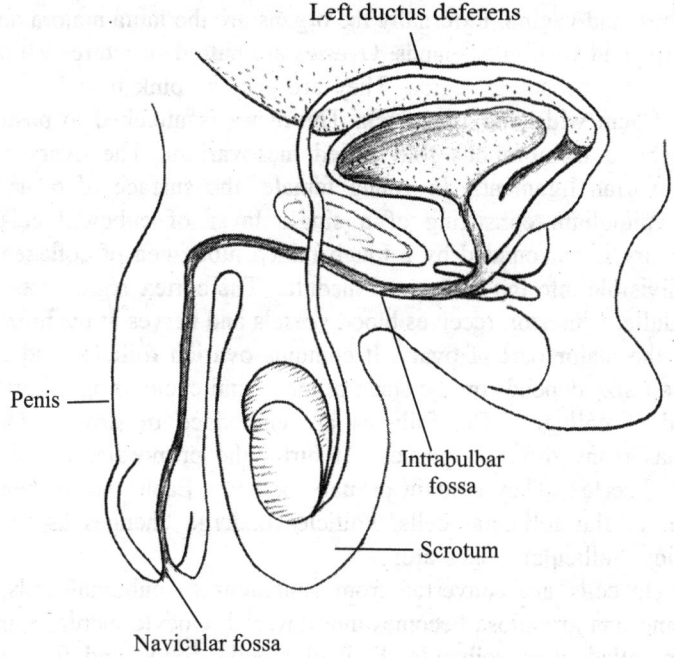

Figure 10.30 Male reproductive system in human.

Penis, the male copulatory organ consists of radix or root and the corpus or body. Radix attaches the penis to lower abdomen. Corpus is normally pendulous and is covered by a loose skin. Corpus consists of three masses of erectile tissue. Flooding these tissues with blood cause the penis to become firm. These tissues are the right and left corpora cavernosa and the median corpus spongiosum. Most of the corpus is formed of corpora cavernous. Corpus spongiosum surrounds urethra and near the end of penis it expands into a conical, glans penis. Its swollen base is corona glandis. The skin over the penis is thin and is loosely connected to tunica albuginea. At the tip of penis, it is folded to form foreskin. It overlaps the glans penis. Corona gland is and penile neck has numerous preputial glands.

Seminal vesicles are two sac-like structures located between bladder and rectum. Each vesicle is about 5 cm long. Their secretions contribute about 70% of seminal fluid. Prostate is a firm structure, partly glandular and partly fibro muscular. It is found around beginning of male urethra. It is about 3 cm in diameter and weighs about 8g. Muscular part of prostate help in dilating the urethra to hold seminal fluid during sexual excitement prior to ejaculation. After the middle age, the prostate often enlarges, may project into bladder and interrupt urination. Bulbourethral glands are small round masses about 1 cm in diameter, two in number. They lie lateral to membranous urethra. Its secretion may control genitourinary diseases. Scrotum is a fibro muscular sac containing the testes and their associated ducts. It is divided into right and left by cutaneous raphe. Its left side is usually lower. External appearance varies according to age and body temperature. Scrotal skin is thin and pigmented. It has numerous sweat glands and nerve endings (Figure 10.30).

Female Reproductive System: The internal reproductive organs are the ovaries, uterus, uterine tubes, and vagina. Externally the organs are the labia majora and labia minora, mons pubis, clitoris, and vestibular glands. Ovaries are paired structures which are placed on each side of uterus in the pelvic region. They are grayish pink in colour, almond shaped, about 3cm long, 1.5cm wide and 1cm thick. The ovary is attached to posterior surface of inner body wall by a membranous fold called mesovarium. The ovary is supported by suspensory and ovarian ligaments. In young female, the surface of ovary is covered by ovarian surface epithelium consisting of a single layer of cuboidal cells. Beneath the epithelium, the ovary is surrounded by a tough tunica albuginea of collagenous tissue. The ovary proper is divisible into the cortex and medulla. The cortex region contains the ovarian follicles. The medulla is interior, receives blood vessels and nerves at the hilum. After puberty the cortex forms the major part of ovary. It contains ovarian follicles and corpora lutea of various sizes. Their size depends on the stage of menstrual cycle or age. Cortex is filled with stroma composed of collagen. The follicles are embedded in stroma. The formation of female gamete has many different phases. At birth, the primordial follicles are found in superficial zone of cortex. They contain primary oocytes. Each one of them is surrounded by a single layer of flat follicular cells. Follicles undergo changes as the female attains puberty. The various follicular stages are:

Primary follicle cells are converted from squamous to cuboidal cells. The follicular membrane or membrana granulosa becomes multilayered. Oocyte increases in size. It has an outer thick layer called zona pellucida. Follicular cells divide and form granulosa cells. Secondary follicle is about 20μm thick. The granulosa cells surround the oocyte and form a mound of cells called cumulus ovaricus. The inner and outer theca becomes prominent. The theca interna is well established. Only one follicle reaches the tertiary stage. It increases in size. Now it is called the Graffian follicle. Oocyte and ring of cells surrounding the oocyte break away and float freely in follicular fluid. Finally, the wall of follicle ruptures and contents are released into peritoneum. Ovary of foetus at 5 months gestation has 7 million oocytes. At birth, the ovary of child contains about 1 million oocytes. Due to further degeneration at the time of puberty, only about 40,000 oocytes remain. Of the 40,000 oocytes only about 400 undergo ovulation during reproductive years. Corpus luteum is formed after ovulation. The walls of empty follicle collapse and fold extensively. The granulosa cells of theca externa get enlarged. They are now termed as luteal cells. They secrete hormones. In pregnancy, the corpus luteum persists. Otherwise, it degenerates after 10-12 days. The connective tissue cells get enlarged. It becomes white in colour and is now called as corpus albicans. In course of time, it shrinks and disappears.

Uterine tubes are two, one on each side of the uterus. Each one is associated with an ovary. Each tube is about 10 cm length. Terminal part of tube is enlarged to form the infundibulum. It opens into peritoneal cavity. Opening is called the ostium. The uterine tube consists of three parts. The part near to infundibulum is called ampulla. It is the longest part. That part of the tube near to uterus is called isthmus. It is narrow. The tubular part entering into the uterus is called uterine or intramural part. Uterus is a hollow, thick walled, muscular organ. It is pear shaped. It is about 7.5cm long and 5 cm wide. It weighs about 50g. During pregnancy, its weight may go up to 1kg. Its larger rounded part is called as the fundus. The narrower part is called as the cervix. The cervix is directed inferiorly. The middle part is the body. The uterus continues as the cervical canal and opens into the vagina through an opening called the ostium. The wall of the uterus is three layered. The outermost layer is the perimetrium or serous layer. The major part of the wall is made up of the next layer called the myometrium or muscular coat. The innermost layer is the endometrium or mucous membrane. The endometrium, a functional layer undergoes menstrual changes and sloughing during female sex cycle. Vagina, a fibromuscular tube is the female copulatory organ, about 10 cm long and extends from the uterus to outside. The vaginal passage is used during intercourse and it allows menstrual flow and child birth.

External Genitalia: The external genitalia, the vulva or pudendum consists of vestibule and its surrounding structures. The vestibular region remains in between two labia majora and contains vaginal opening and urethral opening and is surrounded by the mons pubis anteriorly and labia majora and labia minora on lateral sides. Mons pubis, a rounded eminence situated anteriorly is made up of subcutaneous adipose connective tissue and is covered by coarse hair at the time of puberty. It corresponds to similar structure in the male. Labia majora are two longitudinal folds of skin. They form the outer boundary for vestibule. Labia minora are two small skin folds lie between labia majora. They remain near the vaginal opening. Clitoris, an erectile structure is homologus with male penis and is found in anterior margin of vestibule. It is sensitive region having sensory receptors. Hymen vaginae, a thin mucous membrane is found within vaginal orifice. External urethral opening is found about 2.5 cm below the clitoris. It is anterior to the vaginal opening and remains as a small cleft.

10.10 Human Evolution

The picture of hominid evolution is a far cry from the "Australopithecus africanus begat *Homo erctus begat Homo sapiens*" scenario that prevailed 40 years ago– based to a great extent on fossils record since that time. Humans are thought to be originated in an ancestral line of apes (*Dryopithecus*) present in early Miocene. These Miocene apes differed from tree-dwelling primates in many aspects. Larger body modified to allow a radically new form of locomotion called brachiation. Selection for brachiation changed the ape's arms shoulders and upper body to allow suspension in a semi erect posture for long periods of time. In the early 1980's, it was accepted that human family tree had two branches. Presently many anthropologists believe that human tree has three branches.

Two-branch Tree: This view was proposed in 1979 by the discoverer of ***Australopithecus afarensis***, D. C. Johanson and his colleague, Timothy White. They suggested that two bipedal hominid lines branched off from a common ancestor (**A. afarensis**) about 3million years ago. One line, *Australopithecus*, showed an increase in robustness, larger bodies, a shortening of face with massive Jaws, larger and flatter premolar and molar, with a corresponding degree in size of front teeth (canines and incisors), some increase in brain size. This robust line began in Africa with *A. africanus* (living roughly 2.8 to, 2 million years ago), which was about the same size of *A. afarensis* (3 to 3.S feet tall) but less ape like and more thin. *A. robustus* came next which lived in southern Africa from roughly 2.3 to 1.8 million years ago and showed marked increases m robustness over *A. africanus* in body, ace, Jaws and teeth. Finally, the most robust form; A. boisei lived in eastern Africa from roughly 1.8 to 1 million years ago.

Second branch of Johanson White model, Hofom line, show a shortening of face, decrease in size of cheek and front teeth, and marked increase in brain size. This line starts with a transition from *A. afarensis* to *H. habilis* (handy man). *H. habilis* made and use tools, lived in Africa from roughly 2 to 1.5 million years ago. They had human like small teeth and a much larger brain than any Australopithecine (600-800 cm^3). *H. erectus* (1.6 million years to 300,000 years ago) was much large skull than *H. habilis* in body (5 to 6 feet tall) and brain size-(800- 200 cm^3). They had a larger skull with a low forehead. They first used fire at least 1.4 million years ago and first leave Africa, and their fossilized remains are found throughout Europe and Asia. Roughly 300,000, years ago *H. sapiens* appeared. They had larger brains than *H. erectus* (1200-1400 cm^3) in a rounder and more vaulted skull. They had reduced human like teeth with characters like primitive jaws and skull that they are considered to be "archaic" *H sapiens*. One of the archaic forms, *H. sapiens neanderthalensis*, lived in Europe and near east from 100,000 to, 35,000 years ago and may have been either an ancestor to modern humans or a side branch to extinction. They were more robust than modern humans with larger limb bones, larger front teeth, heavy brow ridges and a low forehead. But they had slightly larger brains than modern humans (1300-1750 cm^3 versus 1200-1600 cm^3). Perhaps the fully modern humans (*H. sapiens*) first appeared as a single population in Africa some 200,000 years ago. They migrated outward to rest of world some 100,000 years ago. By 35,000 years ago these people, now known as Cro-Magnons had completely replaced Neanderthals and all other archaic *H sapiens*.

Three-branch Tree: Alan Walker discovered a completely new type of hominid skull (known by its museum number KNM-WT 17000) in North Kenya in 1985. This skull dated to about 2.5 million years ago, is the most robust form ever found. It has massive teeth and jaws of *A. boisei* combined with a very primitive ape like brain. This indicates that *A. boisei* form was not the last australopithecine to evolve as indicated in two branch model, but rather one of the first Walker's discovery has led many to conclude that human family tree requires a third branch, from A. afarensis directly to A. boisei.

Cultural and Biologic Evolution: Cultural evolution is divided in 3 major stages-hunter-gatherer stage; agricultural revolution, and industrial revolution. All humans had hunter-gatherer conditions until about 10,000 years ago. Isolated populations still exist this way today. Males hunt big game and females stay in a home location where they care for young and gather nuts, seeds, fruit and small animals. In this life style, tool use and cooperative hunting became the strong selection pressures on biologic evolution. There was a continuous feedback between biologic and cultural changes. remarkable increase in body and brain size from *H. habilis* (3 to 3.5 feet in height; 600-800 cm^3 brain) to *H. erectus* (5-6 feet in height; 800-1200 cm^3 brain) is thought to be due to an interaction between consumption of more and more meat, with selection favoring progressively larger bodies for killing and dragging home the meat and need for better tools for butchering and preparing this meat, with selection favoring larger and more intelligent brains.

Rapid feedback between cultural and biologic evolution continued to second major stage, agricultural revolution. This began about 10,000 years ago in Middle East, when hunter gatherer nomads learned to plant to raise crops to domesticate and herd animals. Agriculture led to permanent settlements and supported a great increase in human population size. Industrial revolution in late 18th century England, with its first was of power driven machinery for mass production to the present, human populations have been more protected from basic forces of natural selection as starvation, sickness and temperature extremes. Thus, while biologic evolution has continued during the past 10,000 years, it has been at a much slower pace than cultural evolution (Figure 10.31)

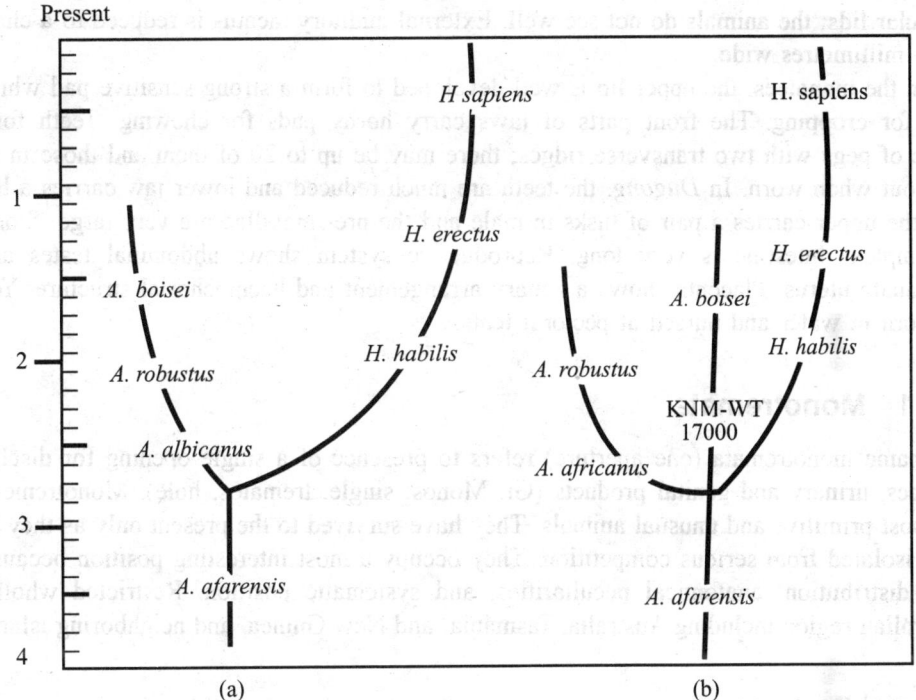

Figure 10.31   Two-branch (a) versus three branch, (b) model of the human family tree. KNM-WT is the new bosei fossil discovered in 1985.

Order Sirenia

Sea-cows are herbivorous, live along the coasts and rivers, highly adapted to aquatic life and have reached the present condition by modification of a basic ungulate type of organization. Two modern forms, the manatee of the Atlantic and *Dugong* of the Pacific and Indian oceans are different in many respects and show lines that were separate for a long time. *Manatus* has 3 species on the Atlantic coasts and in rivers of Africa and America. *Dugong* is a purely marine extending from the Red Sea throughout the Indian Ocean to Formosa and Australia. *Rhytina* was an Arctic form that became extinct in 18th century. Sea-cows have a streamlined body-form with few hairs and thick blubber. There are no hind-limbs and the pelvic girdle remains as small rods. The fore-limbs are large, the digits joined to form paddles with a full pentadactyl structure and no hyperphalangy or hyperdactyly. The caudal vertebrae are well developed and swimming is affected by body and tail. Vertebrae articulate with each other by flat surfaces, but there are zygapophyses. The characteristic structure of bone was probably produced by lack of stressing. The manatee has 6 cervical vertebrae. The ribs are round and diaphragm is oblique allowing the lungs to reach far back. The lungs contain large air-sacs. Sea-cows remain submerged only for10 minutes. The blood system shows rete mirabilia in the brain and elsewhere. The brain is small and ventricles are exceptionally large. The forebrain is rounded but the rhinencephalon is less reduced. Neopallium is smaller and less folded than in almost other mammal of comparable size. Eyes are small and protected by muscular lids; the animals do not see well. External auditory meatus is reduced to a channel a few millimetres wide.

In the manatees, the upper lip is well developed to form a strong sensitive pad which is used for cropping. The front parts of jaws carry horny pads for chewing. Teeth form a series of pegs with two transverse ridges; there may be up to 20 of them and those in front drop out when worn. In *Dugong*, the teeth are much reduced and lower jaw carries a horny pad; the upper carries a pair of tusks in male and the pre- maxillae are very large. Stomach is complex. Intestine is very long. Reproductive system shows abdominal testes and a bicornuate uterus. Placenta shows a zonary arrangement and haemochorial structure. Young are born in water and nursed at pectoral teats.

10.11 Monotremata

The name monotremata (one aperture) refers to presence of a single opening for discharge of feces, urinary and genital products (Gr. Monos, single, trematos, hole). Monotremes are the most primitive and unusual animals. They have survived to the present only as they have been isolated from serious competition. They occupy a most interesting position because of their distribution, anatomical peculiarities, and systematic position. Restricted wholly to Australian region including Australia, Tasmania, and New Guinea, and neighboring islands.

External Features

1. Body is covered with soft hairs. Hairs are coarse or spine-like on dorsal side.
2. Pinna is distinct but small.
3. Digits are provided with sharp claws and are webbed.
4. Tail is present, or absent
5. Male is provided with poison spurs on inner side of each hind leg.
6. Nictating membrane is present.
7. Mammary glands are devoid of teats.
8. A temporary mammary pouch develops during breeding season on abdomen of female.

Skull

1. Skull cavity is spacious and skull bones are smooth and thin. Sutures of skull are obliterated.
2. Auditory ring is incomplete.
3. Lacrymal and alisphenoid are absent.
4. Nasal and premaxilla are drawn out into a rostrum,

5. Pterygoid is present.
6. Jugal is reduced
7. Malleus is large, stapes is imperforated.

Jawbones

1. Angular and coracoid processes are ill developed
2. Mandibular symphysis is absent.
3. Dental formula in young stage is 0.1.2.3./5.1.2.3

Vertebral Columm

1. Epiphysis is ill developed.
2. Thoraco -lumbar vertebrate are 19, and sacral are 2-4 in number.
3. Caudal vertebrae are variable in number.
4. Zygapophyses in cervical vertebrae are developed.

Pectoral Girdle

1. Scapula is elongated and with out spines.
2. Acromian process is well developed.
3. Coracoid is large.

Pelvic Girdle

1. Acetabulum is perforated.
2. Ischio-pubic symphysis is present.
3. Humerus is provided with epicondylar foramen.
4. Sternum is T-shaped
5. Interclavicle is present in the sternum.

Limbs

1. Humerus is flat with a little developed olecranon process
2. Femur is flat with prominent trochanter.
3. Patella is large.

Digestive System

1. Tongue is long and sticky.
2. Saliva is of thick consistency.
3. Stomach is almost spherical.
4. Small and large intestines are differentiated.
5. Caecum of modest size is present.

6. Liver is large and prominent.
7. Spleen is spacious.

Respiratory System

1. Lungs serve as respiratory organ. Lungs of *Echidna* are small in proportion to total body size.

Circulatory System

1. Four-chambered heart is present.
2. Right auriculoventricular valve is incomplete and muscular.
3. Chordae tendineae is absent.
4. Single left aortic arch persists.
5. R.B.C. is small, circular, and non-nucleated.
6. Imperfectly worm blooded.
7. Body temperature ranges between 25-28 degree centigrade. (Figure 10.32, 10.33, 10.34)

Urinogenital System

1. Kidneys are metanephric.
2. Ureters open into urinogenital sinus, which does not traverse the penis
3. Testes are abdominal.
4. Penis consists of corpus spongiosum and corpus fibrosum, and bears a groove for transmitting spermatozoa.
5. Right ovary is reduced.
6. Oviducts open separately into cloaca.
7. Vagina and uterus are absent.

Nervous System

1. Brain is small, simple and without corpus callosum.
2. Cochlea is less coiled
3. Optic lobes are four

Development

1. Fertilization is internal; eggs are large, yolky.
2. Oviparous.
3. Segmentation is meroblastic.
4. Newly hatched young is immature and fed on milk in abdominal pouch till fully developed.

Example: *Echidna, Ornithorhynchus.*
Affinities of Monotremata

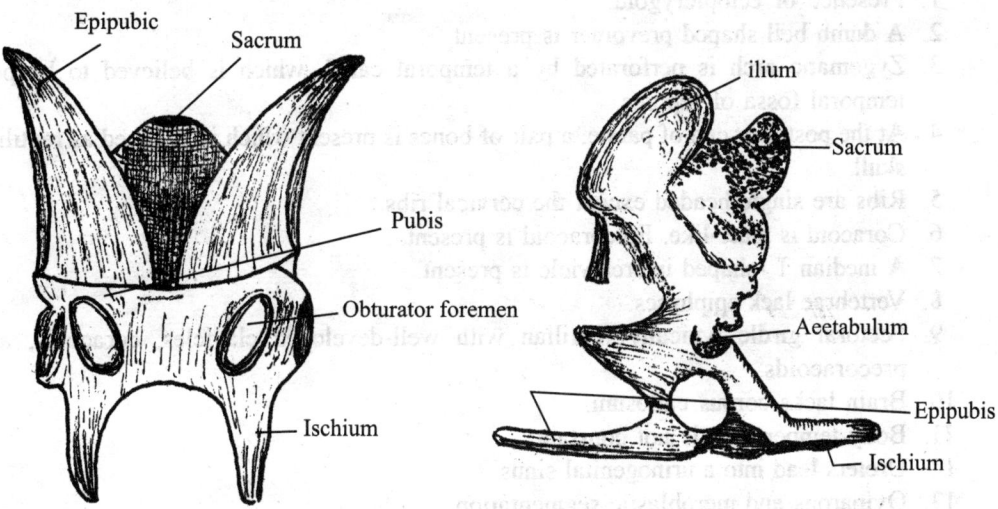

Figure 10.32 & 33   Pelvic girdles of monotreme showing epipubic bones.

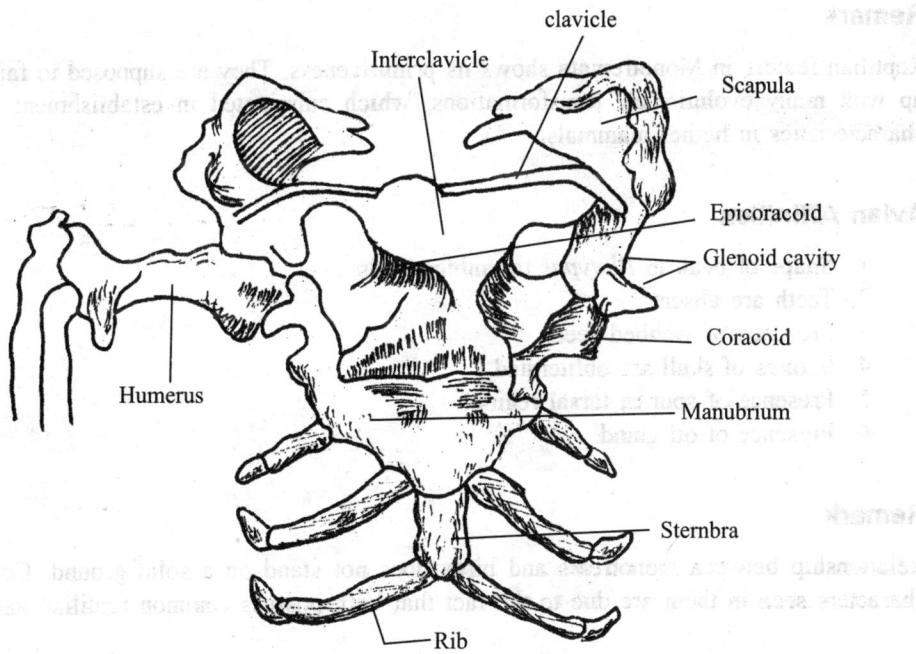

Figure 10.34   Pectoral girdle of monotreme showing T-shaped interclavicle.

Reptilian Affinities

1. Presence of ectopterygoid.
2. A dumb bell shaped prevomer is present.
3. Zygomatic arch is perforated by a temporal canal, which is believed to be post temporal fossa of reptiles.
4. At the posterior end of palate, a pair of bones is present which is believed as reptilian skull.
5. Ribs are single headed except the cervical ribs.
6. Coracoid is plate-like. Epicoracoid is present.
7. A median T -shaped interclavicle is present.
8. Vertebrae lack epiphyses.
9. Pectoral girdle typically reptilian with well-developed clavicles, coracoids, and precoracoids.
10. Brain lacks corpus callosum.
11. Body temperature is not constant.
12. Ureters lead into a urinogenital sinus.
13. Oviparous and meroblastic segmentation.
14. Oviducts separately open into cloaca, and without uterus and vagina.
15. Eggs are large with leathery shell.

Remark

Reptilian feature in Monotremata shows its primitiveness. They are supposed to fail to cope up with many evolutionary transformations, which culminated in establishment of better characteristics in higher mammals.

Avian Affinities

1. Shape of beak in *Platypus* resembles birds.
2. Teeth are absent.
3. Presence of webbed feet.
4. Sutures of skull are obliterated.
5. Presence of spur in tarsal region.
6. Presence of oil gland.

Remark

Relationship between monotrems and birds does not stand on a solid ground. Converging characters seen in them are due to the fact that both possess common reptilian ancestry.

Affinities with Marsupials

1. Structure of skull.

2. Presence of marsupial bone.
3. General contour of brain.
4. Bulbourethral gland.
5. Resemblance between foetal monotremes and marsupials.
6. Mode of milk secretion.

Remark

Considering similarities, Gregory proposed monotremes origin from some premarsupial stalk and their present features are due to specialization. He included both monotremes and marsupials in a subclass "Marsupiontia"

Mammalian Affinities

1. Presence of hair, mammary gland, oil gland, and sweat gland.
2. Pinna is distinct.
3. Skull is dicondylic.
4. Each ramus of lower jaw made of a dentary.
5. Sternum is segmented.
6. Cervical vertebrae are typically seven.
7. A typical mammalian diaphragm is present.
8. Lobes of liver are typically mammalian.
9. Heart is four chambered.
10. Only left aorotic arch is present.
11. RBC is small, circular, and non-nucleated.
12. Presence of four optic lobes.
13. Cerebellum is well developed;
14. Fertilization is internal.

Monotrems show affinity with other mammalian groups. The above characters unequivocally show close and firm affinity with mammals. In their brain, hair, warm blood, heart, and diaphragm, they are mammalian, but in skeleton and egg laying habit they resemble reptiles. They suggest an intermediate stage in many features between two groups. It is reasonable to conclude that monotrems originated as a side line from main line of mammalian evolution and have retained characters through which ancestors of higher mammals have passed.

10.12 Marsupials

The pouched mammals similar to placentals diverged from some early stage of the main mammalian stocks. They parallel the adaptive radiation accomplished by the placentals. Many features are specialized. Today some 230 species are found in Australasian region. In Eocene times, they occurred in Europe. The skull shows characters as found in Insectivora. Brain case is small with flat top. Orbit and temporal fossa remain fully confluent. Post-orbital bar is absent. The bony palate is incomplete posteriorly, there being large holes in palatine portion of it. The jugal reaches back to glenoid articulation of jaw. The lower jaw consists a single dentary with an inflected inner angle. The incisors are more numerous than in placentals, as many as 5 on each side in the upper and 3 in the lower jaw. Of the cheek teeth, only one, the third of series is replaced in modern forms. The cusps of teeth of many marsupials show a close approach to presumed primitive plan, but in addition to main triangle other cusps are present, especially on outside. Marsupials are often divided into two suborders, the polyprotodonts, found outside and within Australasia, and the diprotodonts which are more specialized and restricted. Distinction is based on presence of more than 3 pairs of incisors in each jaw in first group, while in other only 2 remain in lower jaw. In most diprotodonts, the second and third digits of hind- limb are fused to make a comb for cleaning hair which is absent in nearly all polyprotodonts.

10.13   Adaptive Radiation in Marsupials

Adaptation to environment is the most obvious and remarkable quality of living organisms which sums up nearly the whole result of evolution. Unrelated annual groups occupying the same habitat exhibit some common features or look alike and this phenomenon is called convergence. Convergence is especially manifest in features concern with locomotion, food getting, offense and defence. Animals of same or closely related groups exhibit divergence due to varied mode of life. Idea of divergence under new and strange conditions was recognized by Lamarck as "embranchement", while finally Osborn termed it as adaptive radiation. Adaptations describe characteristics of living form developed over a period of time. Certain structural and functional features enable them to reproduce and survive within limits of a particular environment. Once established, animals tend to invade different ecological niches and undergo diversification. Adaptive radiation leads to diversification to suit different modes of life and habits. Before a population occupying a new region, it shows some preadaptive characters. Preadaptation makes a wide range of tolerance for which it can encompass environmental conditions and allow individual to exist in a new habitat or perform new functions. The law of adaptive radiation as defined by Osborn is that each isolated region, if large sufficiently varied in its topography, soil climate and vegetation will give rise to a diversified fauna. The larger the region and the more diverse the conditions, the greater the variety of mammals will result. The impelling causes of adaptive radiation are the need of food or the need of safety. Over earth's surface where speed would become the great desideratium (cursorial) beneath the surface to subterranean realm (fossorial) resulting in degenerative specialization above the surface into trees (sansorial) into water (aquatic). Restricted differences arise within a group of closely related forms whose life habits are similar. Locally, habitats of black and white rhinoceros are distinct and have given rise to feeding and structural differentiation among them. Continental adaptive radiation embraces the entire fauna of a given class and have appeared more than once sometimes as repetitions of evolution within same area. Such type of adaptive radiation is recorded form Darwin gold finches of Galapagoes island. Contemporaneous adaptive radiation is found in three zoogeographical realms namely Arctogea, Neogea, and Notogea. Arctogea includes the entire northern hemisphere. Neogea includes South America while Notogea the Australia. These three realms have been the centers of three remarkable adaptive radiations of mammals during Tertiary time.

Origin and Distribution of Marsupial

Marsupials appear to have originated in early Cretaceous period in North America and Western Europe. They spread from there through South America, Africa and across Antarctica to Australia, as these lands were still broadly interconnected as Gondwana. The discovery in 1982 of marsupial fossils in Antarctica supports the notion of their spread into Australia by way of Antarctica. As Gondwana broke up, Antarctica moved south and became glaciated .Africa became connected to Europe and Asia, and South America became separated form North America. South America and Australia were effectively isolated as havens for marsupial evolution. Marsupial radiated widely and occupied most of ecological niches exploited elsewhere by parental mammals. Australia, Tasmania and New Guinea still have a tremendous adaptive diversity of marsupials like carnivorous Tasmanian wolf, ant eating species, arboreal phalangers and koala bears, kangaroos and rabbit like bandicoots.

Adaptive Radiation in Various Groups of Marsupials

Adaptive radiation is exhibited by different members of the group. They may be discussed as follows:

Superfamily Didelphoidea: Member of the group was once bound in North and Central America in addition to Australia. Majority are arboreal, only one species is aquatic. Opossums (*Didelphys*) are mainly omnivorous with a prehensile tail. They are small sized and are characterized by presence of an elongated muzzle, well developed nail and less opposable hallux .Marsupium is absent or incompletely formed in this group. Chironectes is a south and Central American Otter like form with webbed feet.

Superfamily Borhyaenoidea: Borhyaenoidea were successful carnivorous marsupials living in South America .They became extinct when placental carnivores arrived form North America in Pliocene. *Borhyaena* shows many similarities to *Thylacinus* suggesting that marsupials reached Australia form South America via Antarctica. Size of *Thylacosmilus* was that of a panther and its huge stabbing upper canine and other futures closely parallel the placental smilodon. Members of the group were short legged, large headed and carnivorous in habit (Figure 10.35).

Superfamily Dasyurordea: Members of this group are found mostly in Australia but about six stocks of them have reached New Guinea. Various types of adaptation are encountered among different members. They are:

(a) **Carnivorous;** Exemplified by Tasmanian devil *Sarcophilus* and *Thylacinus*. These animals have rudimentary pollex and small, clawless hallux. Most members are terrestrial with well developed four toed foot and marsupium. Tail is long and non-prehensile. In some nocturnal carnivorous polyprotodonts, teeth became modified for cutting flesh.

(b) **Semi arboreal:** Members are represented by phascogale. These are slender bodied rat -like creatures with bushy tail.

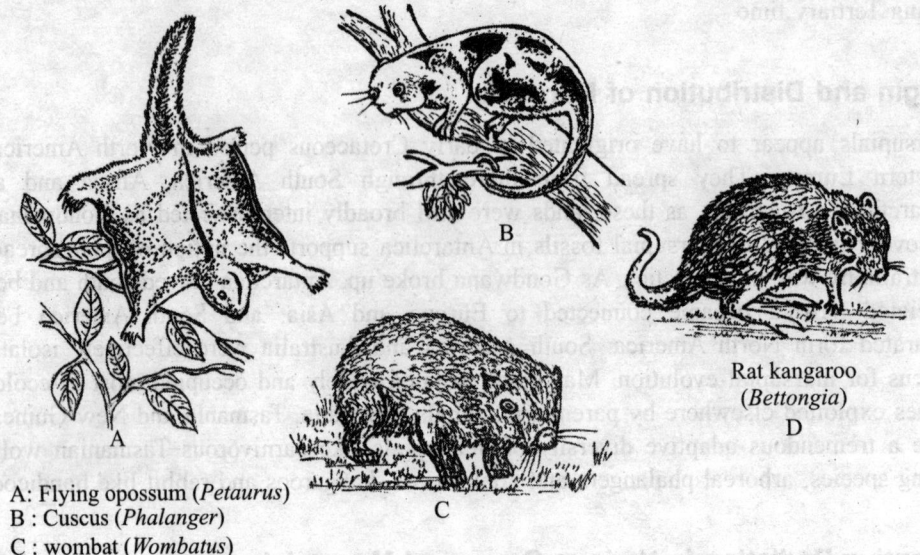

A: Flying opossum (*Petaurus*)
B : Cuscus (*Phalanger*)
C : wombat (*Wombatus*)

Rat kangaroo
(*Bettongia*)
D

Member of the superfamily
Phalangeriodea

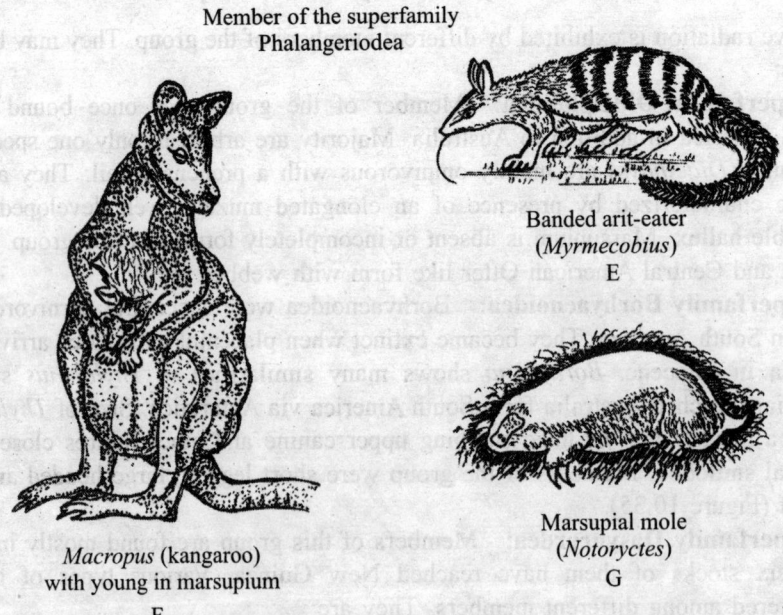

Banded arit-eater
(*Myrmecobius*)
E

Macropus (kangaroo)
with young in marsupium
F

Marsupial mole
(*Notoryctes*)
G

(c) **Ant-eater:** Members are represented by *Myrmecobius*. These are pouch less, small rat like animals characterized by presence of black bands across lumbar and sacral regions. *Myrmecobius* is an anteater, with long snout as many as 52 teeth.

(d) **Fossorial:** Exemplified by *Notoryctes* it is commonly known as marsupial mole found in South America and is characterized by reduced eyes, well developed forelimbs, fused cervical vertebrae and many features suitable for browning. They live in sandy deserts.

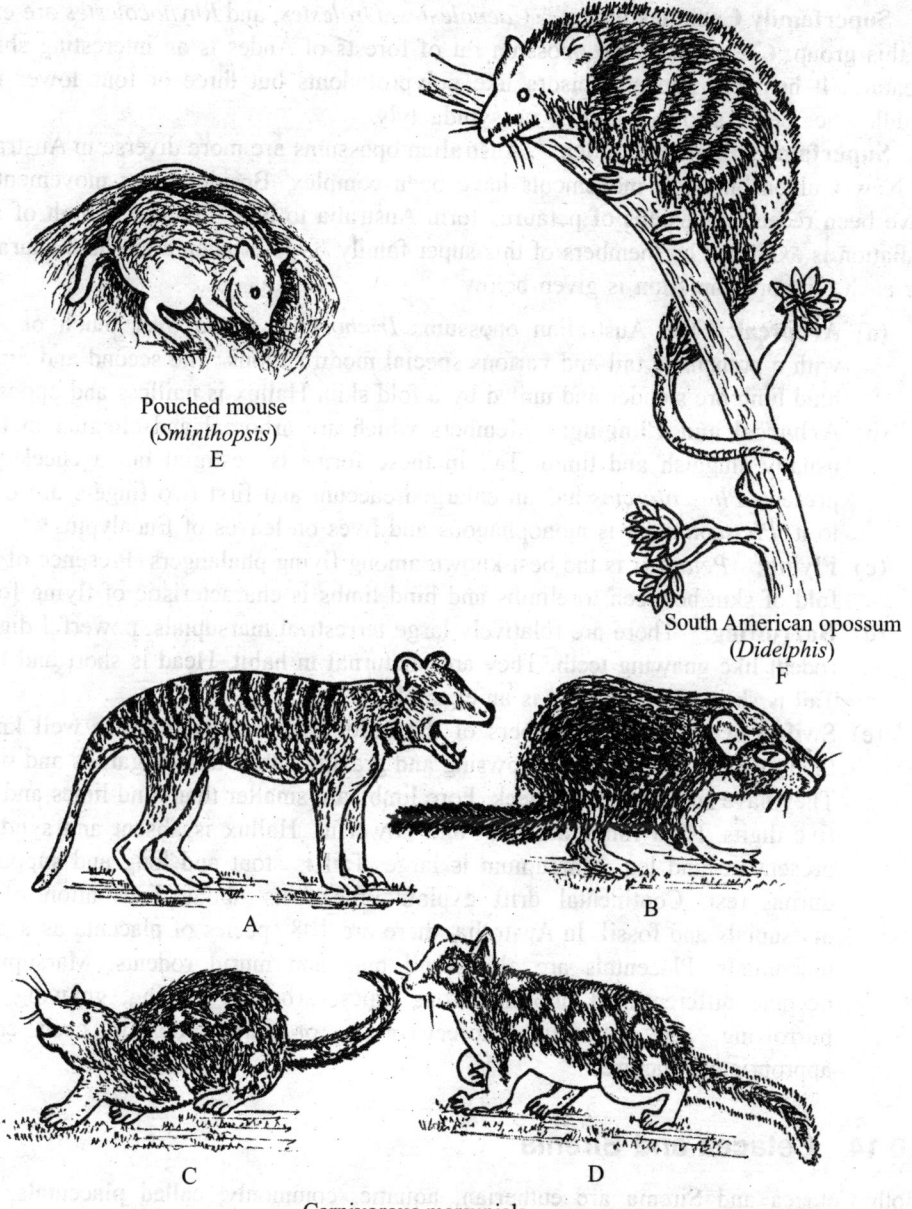

Pouched mouse
(Sminthopsis)
E

South American opossum
(Didelphis)
F

A

B

C

D

Carnivorous marsupials

Figure 10.35 (a) Tasmanian wolf (*Thylacinum*), (b) Tasmenian devil (*Sarcophilus*), (c) Tasmamian tiger cat (*Dasyurops*), (d) Eastern native cat (*Dasyurus*), (e) Pouched mouse, and (f) *Didelphis*.

Superfamily Perameloidea: Bandicoots (*Perameles*) with several species in Australia and New Guinea are burrowing animals, rabbit like but mainly insectivorous. They have a polyprotodont dentition, quadritubercular grinding molars, syndactyly of hind toes and an allantoic placenta.

Superfamily Caenolestoidea: *Caenolestes, Orolestes,* and *Rhynocolestes* are examples of this group. *Caenolestes,* the opossum rat of forests of Andes is an interesting shrew like creature. It has four supper incisors like polyprotodonts but three or four lower incisors, middle ones being strong. There is no symdactyly.

Superfamily Phalangeroidea: Australian opossums are more diverse in Australia than in New Guinea but their movements have been complex. Besides older movements, there have been recent extensions of petaurus form Australia to New Guinea. A high of adaptive radiation is exhibited by members of this super family. Type of adaptation land characteristic for each type or adaptation is given below

(a) **Arboreal:** The Australian opossum, *Trichosurus* found over much of Australia with a prehensile tail and various special modifications. The second and 3rd toes of hind limb are slender and united by a fold skin. Hallux is nailless and opposable.

(b) **Arboreal and Clinging:** Members which are arboreal and clinging in habit are usually sluggish and timid. Tail in these forms is vestigial but a cheek pouch is present. *Phascolarctos* has an enlarged caecum and first two fingers are opposable to it. *Phascolarctos* is monophagous and lives on leaves of Eucalyptus.

(c) **Flying:** *Petaurus* is the best known among flying phalangers. Presence of a lateral fold of skin between forelimbs and hind limbs is characteristic of flying forms.

(d) **Burrowing:** There are relatively large terrestrial marsupials, powerful digger with rodent like gnawing teeth. They are nocturnal in habit. Head is short and flattened. Tail is short. *Phascolomys* is an example,

(e) **Swift Locomotion:** Members of the order Marsupialia that are well known for their swift locomotion and browsing and grazing habits are kangaroos and wallabies. They have small head and neck. Fore limbs are smaller than hind limes and are with five digits. Hind limbs are long and powerful. Hallux is absent and syndactyly is present in hind leg. Marsupium is large. Tail is stout and long and supports body during rest. Continental drift explains the facts about distribution of modern marsupials and fossil. In Australia, there are 108 species of placenta as against 124 marsupials. Placentals are almost all bats and murid rodents. Marsupials have become differentiated into numerous types, arboreal grazing, gnawing, digging, burrowing, and ant eating, insectivorous, or carnivorous in each case with appropriate structure.

10.14 Cetacea and Sirenia

Both Cetacea and Sirenia are eutherian, aquatic, commonly, called placentals. Simpson suggested grouping of 26 eutherian orders in 4 main cohorts. Accordingly, Order cetacea is placed under cohort Mutica and order Sirenia is placed under cohort Ferungulata. Whales and sea cows are included in orders Cetacea and Sirenia respectively. Whales are successful set of aquatic, carnivorous mammals that have diverged from eutherian stock at a very remote date. Their specializations from typical mammalian plane for aquatic life are greater than those found in other order, and their organization shows in a remarkable-way the effects of habit of life and environment. Sea cows are highly adapted to aquatic life but have reached this condition by modification of a basic ungulate type of organization

Similarities

Both orders exhibit striking structural and physiological similarities due to their adaptive convergence for living in aquatic medium. Such similarities are as follows:

1. Streamlined body from.
2. Presence of thick layer of dermal flat which act as a heat insulator and food reservoir.
3. Teeth are peg-like structure.
4. Reduction, or absence of pinna.
5. Disappearance of hind limbs.
6. Fore limbs are paddle like.
7. Round ribs are chief agents of respiration.
8. Simplification of vertebrata
9. Modification of skull.
10. Abdominal testes.
11. Bicornuate uterus.
12. External auditory meatus is reduced to a channel.
13. Blood system shows rete mirabile in brain and elsewhere.

Dissimilarities

Disimilarities are discussed in the Table 10.1.

Cetacea	Sirenia
Some are colossal in size.	Comparatively small.
Large forms prefer open sea. Small forms live. near coast and ascend the rivers.	Found near coast, estuaries and river.
Carnivorous and predaceous.	Herbivorous.
Loss of all hair, except few sensory bristles round the shout in some.	Very sparse.
Caudal fin is horizontally expanded.	Horizontally flattened, either rounded or rhomboidal.
Dorsal fin is present	Absent.
Eyes are small. Third eyelid is not found.	Small and protected by third eyelid.
Nasal aperture is present on forehead as single or double blowhole is usually shifted to dorsal side of the head. Valvular arrangement is highly developed.	Separate from one another and placed on front of head. Valves and pads are less developed.
Mammae are inguinal in position.	Pectoral in position.
Teeth sometimes absent, if present are conical homodont, monphyodont and numerous.	Variables, Rhytina has no teeth, Halicare has rootless and enameless molars, manatus have numerous morals but no incisors.
Salivary Salivary glands are absent.	Present.
Stomach is divided into 3 chambers.	Not divided.
Trachea is very short with a 3rd bronchus.	Short with only 2 bronchus.
Intestine is moderate in length.	Very long and with large caecum.

Epiglottis & aryteniod cartilages are prolonged into nasal passage.	Not prolonged.
Brain is absolutely large in whales than in any other animals (up to 9.2 k.g.) and hemispheres are elaborately folded. Cerebellum is very large. Cerebral cortex shows many differences from that tomammals, especially the presence of a unique paralimbic lobe containing areas of both motor and granular type.	Brain is small and the ventricles exceptional large. Forebrain is rounded but the rhinencephalon is less reduced than might be excepted by comparison with whales. Neopallium is small and less folded than in almost any other mammal of comparable size.
Placenta is diffuse, epitheliochorial.	Zonary, heamochorial.
Bones are spongy and contain oils.	Heavy and compact.
In vertebral column epiphyses are distinct.	Epiphyses are absent.
Cervical vertebrae may 7 in number, more or less fused with one another and are disc like in appearance.	Cervical vertebrate are compressed, ankylosed, with incomplete neural arches.
Scapular spine is placed closeto anterior border.	Elongated and narrow.
Cranial cavity is large. Snout is formed of maxillae, premaxillae, and vomer and mesethmoid cartilage.	Cranial cavity is narrow. Snout is formed of maxillae and premaxillae.
Parietals are separate.	Parietals are united.
Movable joints are absent. In forelimbs number of digits is usually 5, may be 4 as in Balaena. Hyperphalangy and hyperdactyly is Present.	Movable joints present, number of digits is 5. Hyperphalangy and hyperdactyly are absent.
Sternum is short.	Elongated.
Adipose dorsal fin is present .	Absent.
Eeemplified by *Platanista, Balaenoptera*.	Exemplified by *Dugong, Manatus*.

10.15   Platyrrhina and Catarrihina

Modem primates are typically placed in one of two suborders: prosiroii and Anthropoidea. These complex mammals are end products of evolutionary processes that began some 6S million years ago. Much of primate evolution has taken plane in trees in response to arboreal selection pressures. Today there are two types of monkey New world (platyrrhine, or flat nosed) monkey found in central and south America and Old world (Catarrhine, or downward monkey from Africa and Asia.

Similarities

1. Fully developed prehensile hands and feet.
2. Opposable thumbs.
3. Big toes.
4. Delicate movement controls five digits of each hand and foot.
5. Brain, particularly the cortex, has greatly expanded in both forms, with a greater capacity for learning.
6. Reproduction is similar.
7. There is menstrual bleeding.
8. Haemochorial placenta.

Dissimilarities

Dissimilarities are described in the Table 10.2.

Table 10.2 Dissimlarities between Platyrrhina and Catarrhina

Points	Platyrrhina (new world)	Catarrhina (old world)
Distribution	Central and South America	Africa and Asia.
Disposition of Hairs	Usually directed downward	Variously disposed.
Nose	Flat nose	Drop nose
	(Fat, laterally directed)	(Pointed, directed downward)
Cheek pouch	Absent	Present
Nail	May be claw like	Flat
Tail	Long and prehensile	Not prehensile
Forelimbs	Almost equal to the hind limbs	May be longer than hind limbs
Skull	Narrow & Small	Rounded & large
Alisphenoid	Meets the parietal	Does not meet Parietal
Jugal	Comes in contact with the post-orbital process of parietal	Does not come in contact with the parietal
Tymapanic bulla	Large	Absent
Bony auditory Meatus	Ill developed	Well-developed
Auditory ring	Not fused with the skull	Fused with the skull
Internasal Septum	Broad	Narrow
Thoracolumbar Vertebtae	Usually 19-20	Usually 16-18
Caudal Vertebrae	Never less than 14	Variable
Sigmoid notch	Ill developed	Well-developed
Ischial colosities	Absent	May be present
Dentition	Second premolar is retained	Second premolar is lost
Color	Seldom brilliant	Highly colored skin
Signal system	Less complicated	Comparatively more complicated
Facial Musculature	Simple	Complicated
Sound production	Cooperation is ensured by a language of at least 9 distinct sounds	More than 30 sounds have been recorded
Postures	Relatively simple	Relatively complicated
Examples :	Aotus, Ateles, Cebus, Alouatta	Macaca, Homo, Simia.

New world monkeys do not share the derived features of old world monkeys, apes and humans, but have unique features of their own. The seventeen living genera of old world monkeys, apes and men are often classified together as Catarrhini because of the common characteristics in which they contrast with new world monkeys. Perhaps this union is justified and the catarrhini is a monophyletic group with a common ancestor in the late Eocene but the old world monkeys: the super family Cercopithecoidea, diverged very early from the apes and men, Hominoidea. Old world monkeys of Africa and Asia do not differ very strikingly in general habits and organization from monkeys of the new world, though they are mostly larger.

10.16   Dentition in Mammals

In symmetrodonts and pantotheres, the teeth were arranged to bite against each other. The lower cusps formed a triangle with a surface behind, the talonid or heel, with which the main cusp of upper molar made contact. The upper cusps also formed about a triangle. This tribosphenic condition of molars was the plan from which modern mammalian condition is derived. Apical cusp which lies on inner side of upper molars and outer side of lower molars was believed as the original reptilian cone and was called protocone in upper and protoconid in lower molars. Other 2 cusps are called paracone in front and metacone behind. Separation between these latter cones is not sharp in pantotheres, especially in upper jaw, and this condition again occurs in the earliest placental mammals. Various theories exist about how the triangular plan was reached. The original tritubercular theory supposed rotation into triangular position from 3 cusps in line of a triconodont. Even if this could be shown to have happened in a series of fossil teeth, change of morphogenetic process by which the rotation was produced is unknown. Attempts to explain the many cusped mammalian teeth due to fusion of several reptilian tooth germs, either those making one series on gum or teeth of successive series. Plausibly such changes in relative time and/or place of tooth development could occur in this way, leading to a partial fusion and production of many cusped structures. At present, there is little doubt about the tritubercular theory that shows the nature of the earliest mammalian cusp patterns. Indeed, nearly all the Eocene representatives of the various mammalian orders showed signs of a triangular cusp pattern. Whalebone whales, monotremes, Manis, and American Anteaters among the Edentata are devoid of teeth in adult. In some of them rudimentary teeth are found, which never cut the gums or else and lost easily in life. Latter is the case with *Ornithorhynchus* where teeth retain up to maturity. Teeth of a generalized mammal are heterodont. Homodont condition is found in cetaceans.

Types of Teeth

Four types of teeth in each jaw are generally present. Incisors have generally simple structure and single rooted. These have layer of enamel only on anterior face provided with sharp chisel-like edge on them due to the fact that harder enamel is worn away slowly than comparatively soft dentate. Incisors are totally absent in sloth. Number of incisor is usually two but up to five for marsupials. Canines are present in majority of mammals but absent in Rodentia. They are as rule simple conical teeth with a single root and are never more than one each half of the jaw. Premolars are anterior grinding teeth with cusps and may have two or more roots. Molars are posterior grinding teeth, similar to pre-molars in having cusps and have two or more roots. Molars are never shed like other teeth of milk set (Figure 10.36, 10.37).

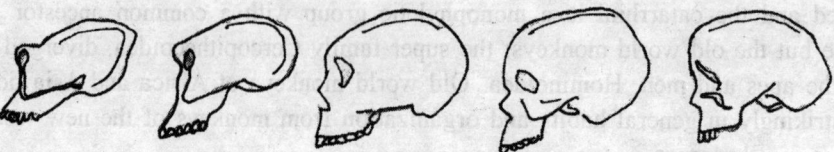

Figure 10.36   Changing structure of skull and teeth relation to evolution of man.

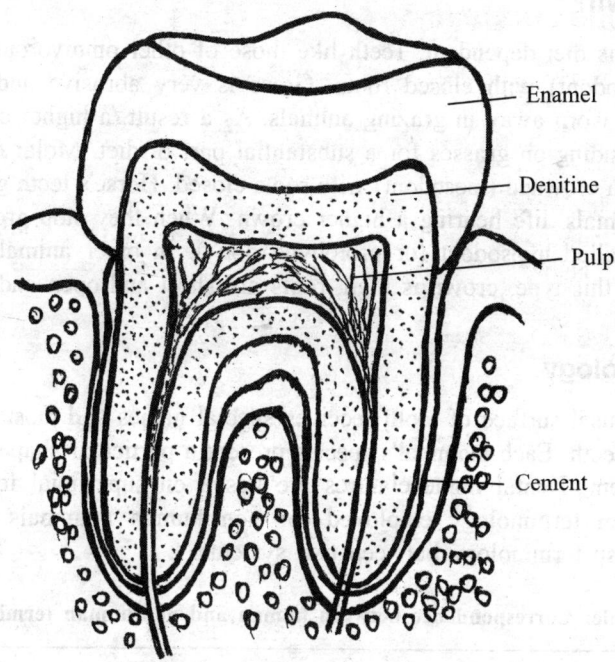

Figure 10.37 Structure of tooth.

Anatomical Structure

Teeth are white, hard, composed mainly of calcium phosphate in the form of crystalline hydroxyapatitic and collagen fiber in different percentages. It consists of a crown and a root, joined by a cervical region, or neck. Crown is coated by a layer of enamel, a hard crystalline tissue, while a bone like material called cement coats the root. Underlying these surface layers a very tough and resilient tissue of dentine forms main structure of tooth. Inside there is a pulp chamber. Dividing line between different parts are called enamel/dentine junction (EDJ), cement/ dentine junction (CDJ), and cement/ enamel junction (CEJ). Enamel is distinct from that of other tetrapods in having prisms at least in incipient form, where in flat, open prism sheaths and planar prism seams are present. Crown may be wider, taller or flatter; may include extra mounds, known as cusps, or folds in enamel layer. Cement may cover crown and root. The form and pattern of crowns and roots vary depending on kind of teeth and animal species. Carnivorous mammals bear narrower crowns with sharp and high cusps. In omnivores, crown is wider and cusps are numerous and blunt. In herbivores, in place of tubercles irregular crests delimit depressions. A tooth may possess one or mere roots that vary according to main group function, and location. Number of roots is more or less constant for each kind of tooth. They are fixed in sockets of mandible, pre-maxilla and maxilla by periodontal ligaments. Stretch receptors associated with these ligaments provide sensory feedback on tooth loading which is critical to precise tooth occlusion that characterizes mammals.

Size and Growth

Size and growth is diet dependent. Teeth like those of other omnivorous mammals are low crowned (brachyodont) with closed roots. Grass is very abrasive and brachyodont teeth would be rapidly worn away in grazing animals. As a result, a higher crowned cheek tooth has evolved depending on grasses for a substantial part of diet. Molar of most herbivorous are relatively high crowned (mesodont) with roots closed. Horse's teeth grow for a long time in relation to animals' life bearing a higher crown. When they stop growing, roots closed. Such teeth are called hypsodont (or protohypo don't). In other animals (sloths) teeth are ever-growing. In this type, crown is high, roots are short and open and there in no neck.

Crown Morphology

Structure on occlusal surface of tooth receives special names and most of them are present in all mammals' teeth. Each mammal's species present a particular shape of each accident or lack some of them. Dental nomenclatures are based on superficial features which differ substantially from terminology employed for non human mammals. Table I present a correlation of cusp terminology between two systems.

Table: Correspondence between human and non-human terminology

Teeth	UPPER		LOWER	
	Human	Non Human	Human	Non-Human
	Buccal	Paracone	Buccal	Protoconid
Premolar	Lingual	Protocone	Mesiolingual	Metaconid?
	Xxxxxx	xxxxxxxxx	Distolingual	Hipoconid?
First (I	Mesiobuccal	Paracone	Mesiobuccal	Protoconid
Third	Distobuccal	Metacone	Distobuccal	Hipoconid
Molars	Mesiolingual	Protocone	Mesiolingual	Metaconid
	Distolingual	Hypocone	Distolingual	Entoconid
	Xxxxxxxxx	Xxxxxxxxx	Distal	Hipoconulid
Second	Mesiobuccal	Paracone	Mesiolingual	Protoconid
Molar	Distobuccal	Metacone	Distobuccal	Hipoconid
	Mesiolingual	Protocone	Mesiolingual	Metaconid
	Distolingual	Hypocone	Distolingual	Entoconid

First upper molars sometimes bear a supplemented cusp united to and somewhat lingual to mesiolingual cusp, called cup or tubercle of Carabelli. This morphological character can take the form of well developed fifth cusp and is used to distinguish populations. Non-human mammals do not have any cusp in the same place as tubercle of Carabelli (mesiolingual to protocone). The closed feature is a style in mesial cingulum, a head of protocone (protostyle), but it can be present either in the first, or in the second upper molar and it is not the case in humans

Pattern and Adaptation to Diet: Dentition patterns among mammals involve differences in number, size, shape, and arrangement of cusps of cheek teeth. Size and shape of teeth have long been recognized as indicators of diet and dental function. Variety of dental features allows distinction of different type of crown morphology and each type presents a functional adaptation related to kind of food processing. Main types are bunodont, lophodont, Lambdadont, and secodont. Cheek tooth morphology possessed by omnivorous, frugivorous or generalist carnivorous mammals are termed bunodont. Bunodont cheek teeth are low crowned and rounded cusped and typical of pigs, bears, and humans. More folivorous mammals such as horses, rhinos and deers, for instance have lophed teeth. In these cheek teeth the low isolated cusps of bundonont form of tooth run together to form high cutting ridges, or lophs that act to shear and shred food as leaves. When lophs are crescentic and longitudinally oriented the tooth is called solenodont, whereas when they are transversely oriented, the tooth is termed lophodont. Secondont or carnassial is specialized type in carnivores in which cusps are mesiodistally elongated and straight. Upper and lower carnassials work together to provide a scissors action for shearing flesh. Insectivorous mammals present "lambdadont' tooth pattern which is characterized by one (zalambdadont) or 2 (dilambdodont) buccal ridges of inverted lambda (L) shape. These sharp bladed teeth cut and crush hard chiitinous exoskeletons of insects. Trilobite pattern of tooth is present in primitive relatives of armadillos. In this peculiar tooth, prominent ridges of osteodentine in middle of modified orthodentine exists which is little resistant to wear. These ridges are highly resistant to abrasion and very appropriate to folivorous (grazers) mammals in estimating age and body size in fossil mammals.

Age of mammals can be estimated by degree of tooth wear. Rate of wear depends on diet and environmental factors, overall morphology of crown. Area of occlusal surface thickens and microstructure of enamel and in some cases, hypoplasia. It is possible to assign absolute ages to fossil individuals by fitting comparative data on attrition rate from living population, but this procedure involves assumption that living animals of today are comparable in foregoing respect to those represented by fossils. In hominids proximate age of death can be determined, beside degree of wear on teeth by state of dental eruption. In Lucy's mandible example, third molar was erupted and just beginning to show wear, leading to conclusion that Lucy was fully adult when she died. Studies of mammalian paleoecology may be most useful predictor of species' adaptation to a habitat and also has an important role to play in studies of tempo and made of evolution. Estimates of body size base on limb measurement are more reliable than those based on dental measurement. Because of their identifiably, preservability and frequency in fossil assemblage teeth will continue to play a pre-eminent role in body mass estimation. The M1 and M2 have become widely used for prediction of body mass as in most extant species these are teeth with least size variation. In some groups, tooth length is a better estimator of body size than tooth width and area, which are more related to diet. Depending upon feeding habit and type of food taken, premolars and molars have under gone changes in their shape.

(a) **Bunodont:** When cusps in cheek teeth remain separate and rounded, the teeth are called bunodont type. In man and in some mammals, cheek teeth are bunodont type and used in grinding.

(b) **Lophodont:** If the cusps are joined to form ridges the teeth are called lophodont as cheek teeth of elephant

(c) **Secodont:** When cheek teeth are with sharp cutting crowns, they are secodont as found in terrestrial carnivores.

(d) **Solenodont:** A teeth with crescentic cusps as found in ruminants and horses.

(e) **Brachydont:** A teeth with a low crown is called brachydont.

(f) **Hypsodont:** When the crown is high.

Origin of Cusps in Molar Teeth: Basal type of mammalian molar teeth is called tritubercular and trituberculosuctorial with respect to lower jaw. Cope and Osborn proposed a theory of trituberculi and tried to show that all mammalian cusped grinding teeth have been derived from trituberculate tooth. Cope assumed that molar teeth of the earliest eutherian mammals commonly possessed a triangular crown with three main cusps. Later forms tended to have molars with four to six cusps. These he regarded as derived from earlier triangular type. Osborn developed this hypothesis, on the assumption that principal cusps are homologous throughout mammalia, gave them names that are universal in use ever since. Osborn assumed that a simple reptilian cone elongated in an antero posterior direction and two subsidiary cones to have arisen one on front and one, on hind border, thus producing a triconodont tooth. These two cusps were then supposed to have rotated in opposite directions in upper and lower jaws to form triangular and tritubercular teeth. Apex of triangle or trigon, points inwards in upper tooth and outwards in lower. Cusp at the apex of triangle, being supposed to represent the original reptilian cone, was named as protocone in upper and protoconid in lower teeth. Dentitions of Mesozoic mammals have led to modifications of Osborn's theory. It is widely held that, whereas in lower molar the protoconid does represent tip of originally simple reptilian tooth, this is not true of protocone of upper molar. Original upper molar cusp was paracone. Others think it was an amphicone, which split to form paracone and metacone. However the names, which Osborn gave to cusps, have become so widely applied to Cenozoic mammals that it is generally agreed to continue to call the lingual cusp of tritubercular tooth, protocone, even though it may not have arisen as Osborn supposed.

Type of upper molar, which is retained by living Opossums, the paracone and metacone were far removed from buccal edge of crown, which was occupied by a wide shelf bearing a number of styles, may be called pretritubercular. It is not very different from dilamblodent teeth of Tupaia. Probably, more typical tritubercular tooth has evolved from pretritubercular condition by reduction of buccal shelf to a cingulum, so that paracone and metacone come to stand near to buccal edge of tooth. Cope noticed that molars of earliest eutherian mammals posses three tubercles which have later become 4 to 6.

Number of Teeth and Dental Formula: Number of teeth in any particular species is constant. Maximum number in heterodont condition is 44. There are mammals with teeth fewer in number than this due to reduction in number of owner more types. Constancy of number of teeth is a tool for classification. Number of various sets of teeth is expressed in front of dental formula in which incisors, canines, premolars and molars are indicated by their initial letters. Teeth of upper jaw are placed as numerators, and those of lower jaw as denominators. Total when multiplied by two gives the total number of teeth on two jaws. Typical mammalian dentition consists of 44 teeth, which may be written as follows:

i3/3, clll,pm4/4, m3/3 = 22×2=44.

Above-mentioned typical dentition is found in horse and pig. In others, there has been a reduction of teeth. Dental formulas of some mammals are represented as follows:

Dog- i3/3, cl/l, pm4/4, m2/3 =42.
Man- i2/2, cl/l, pm2/2, m3/3 =32.
Sheep- i0/3, c0/1, pm3/3, m3/3 =32.
Rabbit- i2/1, cO/O, pm3/2, m3/3 =28
Rat- il/l, cO/O, pmOIO, m3/3 =16.
Cat- i3/3, Cl/1, pm3/2, ml/1 =30

10.17 Ruminating stomach

Many separate evolutions of fermentative digestion among vertebrates have resulted in distinctly different solutions to problems posed by plants as food. Monogastric and digastric or ruminant digestive systems are found among ungulates. Most derived stomach anatomy is ruminant digestive system, and animals that posses it (camels, giraffes, antelope, cattle, sheep, goat and deer) are called ruminants or cud chewing mammals. Fully developed ruminant stomach has four chambers viz rumen, reticulum, omasum, and abomasum. First three are lined by a stratified epithelium of oesophageal type, folded in muscular ridges.

Physiology of Digestion: Rumen is enormously large with a capacity of up to approximately 200 liters. Rumen acts as fermentation chamber where food is moistened, mixed with saliva and undergoes fermentation by mutalistic (symbiotic) microorganisms like bacteria, protozoan and fungi. Many of these produce celluloses, for digesting cellulose. Their presence is essential to ruminant, as they are unable to synthesize cellulose itself. End products of fermentation are carboxylic acids (ethanoic, propanoic and butanoic acids) CO_2 and methane. Acids are absorbed by host which uses them as a major source of energy in respiration In return, microorganisms obtain their energy requirements through chemical reactions of fermentations as microorganisms multiply, and they synthesize amino acids and proteins. Remarkably, nitrogen source includes some of cow's urea, which diffuses into rumen. Partially digested food; the cud is passed to second chamber reticulum, where it forms pellets. It is then regurgitated and thoroughly rechewed. This is called rumination or chewing the cud. Process of rumination depends upon oesophageal groove running from cardia to opening of omasum. When lips of these are brought together, food does not enter reticulum and is returned from rumen to mouth. After chewing, bolus is again swallowed, the groove opens, and food passes to reticulum. Food then passes to omasum. Here, water is pressed out and absorbed and remainder proceeds to abomasum, the true stomach with peptic glands. Elaborate digestive mechanism has no doubt contributed largely to success of artiodactyls. Efficient cellulose-splitting system enables them to make use of hard grasses and other unpromising sources of nutrient.

Modification of Ruminant Stomach: In camel, omasum is lacking. Pouch like diverticula called water cells arise from both rumen and reticulum. Their openings are guarded by sphincter muscles. Water cells or pouches do contain pure water but this, generally metabolic water drawn from other parts of body and is used to moisten food during digestion. Some water is stored in muscle and connective tissues but breakdown of glycogen, stored in muscles and of fat stored in hump results in metabolic water.

Advantages and Disadvantages with Ruminant Stomach: Foregut fermentation is extremely efficient because microorganisms attack plant material before it reaches small intestine where most absorption takes place. Rumen inhabiting microorganisms play role in detoxifying chemical compounds that would be harmful to a vertebrate. In foregut system food moves slowly and food can not pass out of rumen until it has been ground into very fine particles. Ruminants do not do well on diets that contain high levels of tannins or resins because these compounds suppress microbial function in rumen, and plants with high silica contents break down so slowly that they impede the' movement of food out of rumen (Figure 10.38).

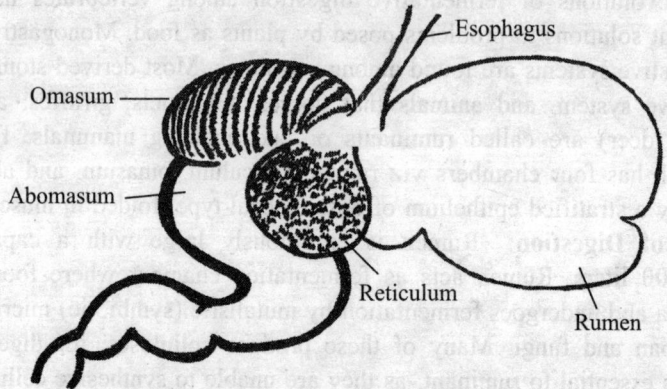

Figure 10.38 Ruminant stomach.

10.18 Echolocation in Bats

Some animals hunting in darkness or murky water have evolved a type of sonar called echolocation. Thoroughly studied in microchiropteran bats and toothed cetaceans, echolocation shows the way how an ancestral sensory capacity can be elaborated. As echolocation is used for hunting, let us discuss the hunting pattern of bats.

Three Phase Hunting Pattern

Initial Search Phase: In little brown bat (*Myotis lucifgus*), this phase is characterized by fairly straight flight and emission of pulsed sounds separated by silent periods of more than 50 milliseconds, each of the 10 pulses in a call is about milliseconds in duration and each pulse gradually decrease in frequency starting at 85 kilohertz and ending near 35 kilohertz. These bats call are therefore frequency modulated (FM). Other bats use different pulse length frequencies that vary from family to family.

Second Phase: It begins when a bat detects an insect. Fruit flies and mosquitoes are detected from a distance of about a meter. Interval between pulses shortens and silent intervals are less than 10 milliseconds for *M. lucifugus*. One hundred cries per second, each lasting only 0.5 to 1 millisecond are typical as bat alters its flight path to intercept its prey.

Terminal Phase: This phase is characterized by buzz-like emission of ultrasound; intervals between, pulses are less than 10 milliseconds; pulse duration is about 0.5 millisecond; frequencies drop to 25 to 30 kilohertz. When the bat is within a few millimeters of prey, it scoops with a wing or with membrane between its legs and pulls the insect towards mouth.

Sound Production: Situated just posterior to hyoid bone, larynx is a box of cartilages and muscles encircling esophageal end of trachea. Central opening of larynx may be closed by one or two paired folds of tissue that oppose each other across tracheal opening. Deepest of these folds are vocal cords, which have thickened edges and associated muscles. Sounds produced by larynx are emitted by open mouth, or nose depending on the family of bat. Forcing air produces sounds through slit between paired folds of tissue. Frequency modulation and loudness are by alteration of tension on vocal cords and entire larynx, amount of air expelled per unit time through structure, and sometimes by extra laryngeal resonating chambers.

Sound Dispersion and Echo Reflection: Calls travel through air in radially expanding waves about 34 cms per millisecond. The amount of sound energy striking a target decreases as square of distance traveled. A small object intercepts very little sound energy and thus can reflect very little. As the echo is reflected back towards the bat, its energy continues to diminish as square of distance. Thus, returned sound, despite its initial loudness is exceedingly faint. In addition, only that wavelength in emitted call that is about equal to or shorter than diameter of reflecting object will be returned. Time required for an echo to return is directly proportional to distance from bat to target. Change in return time between successive calls can show relative movement of bat and its target. As a bat approaches a target, call repetition rate increases, giving the bat more precise information about target's location.

Sensitivity of Hearing: Several features of morphology and neurology of auditory system of bats contribute to sensitivity of hearing. Tympanic membranes and ear ossicles are small and light, and are easily set into motion. Contraction of middle ear muscles briefly damps sensitivity of ear as each cry is emitted; thus bat does not deafen itself. A paddling of blood sinuses, fat and connective tissue isolates bony labyrinth of inner ear from the rest of skull and reduces direct bone conduction of sound into inner ear. Perception of direction of a returning echo is aided by large, complex pinnae and by a neural mechanism called contra lateral inhibition. Stimulation of cells sensitive to a particular frequency in inner ear on one side of head produces a transient desensitization of cells that respond to same frequency in ear on other side of head. Effect of that desensitization is to increase the contrast between the intensity of sound perceived by two ears, and thus to permit more precise determination of direction of an echo.

Using echolocation, bats can navigate and hunt insect prey in total darkness. Little brown bats can detect wires 1 millimeter in diameter from a distance of 2 meters and wires only 0.08 millimeter in diameter from shorter distances. Even irregularities on surfaces can be located. Fish-eating bats apparently locate fish by detecting ripples on water surface, and bats use echolocation to find cracks in rocks where they cling while they sleep. An object's size is showed by frequencies in echo; large objects reflect longer wavelengths than small ones. Extremely high frequencies of emitted calls are necessary to detect very small objects.

Character of reflecting surface is showed by character of echo. A smooth, hard surface such as the exoskeleton of a beetle returns a sharp echo, whereas a blurred echo shows a rough surface like body of a moth. One mechanism that probably contributes to a bat's ability to recognize echo of its own calls is a brief neural sensitization following emission of a call to sound of the same wavelengths that were emitted. Sensitive period begins about 2 milliseconds after end of call and lasts about 20 milliseconds. Because sound travels 34 cms per millisecond in air, this timing means that a bat is especially sensitive to echoes of its own call from objects at distances between 30 cms, and 4 meters. This same mechanism probably helps a bat on its final approach to prey.

10.19 Echolocation in Cetaceans

Some animals emit high-pitched sounds, often inaudible to humans, which get reflected back off the object and detected by ear or other sensory receptor of the animal concerned. Whales produce sounds of 20 to 200 kilohertz and detect objects around them through such sound. These mammals have no vocal chords. The sounds are made by release of air from lungs in sinuses of upper respiratory tract. A *Tursiops* in a tank recognized the sex of a new arrival in another tank, out of sight. It helps in establishing social relations for example, between mother & young. *Tursiops* can avoid obstacles in dark. Low frequency click or whistles provide them a general profile of environment whereas discrimination clicks help in identifying the details. To attract females in tropical banks and maintain contact, with a range of as much as 185 Km. Herd cohesion is certainly important / in this connection.

Nature of sound: Narrow-band pulses emitted by Blue whale and Mink whale (up to 30 KHz) are not directional like those of delphinoids. Humpback whales have a song lasting up to 35 minutes and repeated after surfacing. Fin whales (Balaenoptera) produce low-frequency sounds (20 Hz) of one- second duration several times per minute which are very loud for animal noises (75-80 dB).Whale could spread sound spherically for 800 km. or more. Baleen whales make good use of sound as they have sufficient ear separation to check direction. Songs consist of 15-20 syllable sung in a fixed sequence of six themes as low frequency moans and mups; yups; higher frequency modulated notes; whos or wos ; long low frequency moans, with ees and oos.; leven lower frequency sounds, and finally a rachet cry or surfacing. Other sounds in each phase repeat several times. Differences observed in consecutive years, suggest that dialects identify herd and individuals. Whales searching for food are silent which could bring them together in areas containing food .No one knows for certain, if fin whales can hear, 20 Hz, but if they cannot hear them why do they make the noises.

Difficulties in Cetacean Echolocations: Sound travels about five times as fast in water as in air. Wavelength of any sound frequency is five times longer in water than in air. As objects reflect only those wavelengths equal to, or shorter than their diameters, cetaceans must use exceedingly high-frequency sounds to produce wavelengths short enough for detection of small objects. Sound energy is not transmitted well across an air-water interface. About 99.9% of the energy that reaches such an interface is reflected and only 0.1% crosses boundary. Terrestrial vertebrates face the same problem because inner ear is fluid filled. Middle ear converts sound energy transmitted through air to mechanical movement of bones, and thence to displacement of fluid in inner ear. System does not work underwater, however, because a new water-air interface is created at the tympanic membrane and most of sound energy is reflected from tympanum back into water. High proportion of acoustic energy that impinges on a body surface submerged in water is absorbed and propagates through tissues, echoing back and forth to produce a diffuse buzz in inner ear that conveys no directional information. .

Features for Solving Difficulties: Middle ear serves as a sound receiving and transmitting organ. Inner ear is isolated from the sound propagated through body tissues, and sound is conducted to inner ear via a special fat body that extends from lower jaw to the auditory bulla. Bulla is composed of extremely dense bone, which does not transmit sound readily and is separated from rest of the skull by soft, sound absorbing tissues. Norris claimed that sound a cetacean perceives not receives via the middle ear but instead comes from lower jaw. Fat body is a characteristic feature of odontocete cetaceans and is responsible for external shape of the head. Sound passes from one medium to another as it leaves a cetacean's head. Because acoustic coupling between oil and water is good, little of the sound energy is lost by reflection as it would be in an air-to-water transmission, but the sound waves may be bent. It is likely that melon serves as flexible lens, changing its shape from moment to moment to beam the sound energy in different direction. Melon of some cetacean changes shape very conspicuously during echolocation.

10.20 Oestrous cycle in Mammals

The term estrus referred to existence of a period of strong sexual desire. It has become evident that estrus is close to time of ovulation. Actually, relatively brief periods within breeding season when female of most mammalian species (except higher primates) undergoes spontaneous ovulation and is most receptive of male is known as heat or estrus .Rhythmic or cyclic changes in development, release and maturation of egg shell, together with accompanying structural and physiological changes in reproductive system as well as behavioral changes, periodically prepare female for matins and are collectively termed estrus cycle.

Oestrous Cycle in Various Animals Groups: There is wide variation among animals in length of time occupied by this cycle. In some, it occurs only once in an entire year, estrus being so placed seasonally that when young are born, conditions are favorable for their rearing. Species having one breeding seasons in year are said to be monoestrous (dog, *Dasyurus*, Koala). Other animals exhibit several breeding periods in a year. They are said to be polyestrous (sheep, rat, and goat).

Physiology of Estrous Cycle: Various changes associated with estrus cycle are as follows. In beginning, thin endometrium of uterus thickens, its blood supply increases and its uterine glands deepen as follicular maturation proceeds. Graffian follicle ruptures, release ovum and collapses. Space created by collapse of follicle is invaded by proliferation of cells of granulose and theca interna and cells of ovarian stroma Structure thus formed is corpus luteum which secretes progesterone. Endometrium continues to thicken with appearance of corpus luteum. If fertilization occurs, zygote implants itself in endometrium. Oestrous cycle in, absence of pregnancy is followed through the following events. Egg cell destroys, corpus luteum begins to disappear and endometrium regresses returning to its original thin state. Endometrium does not build until beginning of next estrous cycle, Oestrous cycle is repeated regularly according to mating pattern of species.

Factors Associated with Estrus Cycle: Estrous cycle may be interrupted by many things other than pregnancy. Thus starvation, extreme exposure or severe sickness may cause suppression of the estrus. In mammals with short periods of gestation, such as rabbit, light is apparently a critical starting factor. Only when average daily amount of light gests above a certain threshold level hypophysis become active in production of enough follicle stimulating hormone (FSH) to set the whole reproductive cycle into operation.

10.21 Menstrual Cycle

Menstrual cycle refers to series of events those periodically modify female reproductive tract of humans and advanced primates. Uterus contracts squeezing out excess endometrial tissue and blood, as a result of erosion of blood vessels in endometrium is called menstruation (Latin, mensis; meaning "month"). Phases of menstrual cycle are follicular (proliferative), ovulatory, luteal (progestational), and menstrual. Follicular phase is characterized by progressive development and maturation of an ovarian follicle in one of the ovaries, redevelopment of uterine endometrium for reception of a fertilized egg. Ovulatory phase is characterized by rupture of follicle and release of ovum .This takes place 10 days after cessation of menstrual flow (or 14 days after start of menstrual flow). Luteal phase is characterized by development of uterine endometrium for implantation of embryo, inhibition of development of any other ovarian follicles, development of mammary glands to keep them in readiness for providing milk to new born infant. Menstrual phase is characterized by degeneration of corpus luteum, thickened endometrium begins to degenerate and sloughs off and there is bleeding.

Hormonal interactions: Neurosecretory cells of hypothalamus spontaneously release gonadotropin releasing hormone (GnRH), unless actively prevented from doing so by other hormones notably progesterone, GnRH stimulates anterior pituitary to release FSH and LH. Both FSH and LH circulate in blood stream and start development of several follicles within ovaries, and cells of these developing follicles secrete estrogen under combined influences of FSH LH and estrogen, follicles grow during next two weeks. Simultaneously, primary oocyte within each follicle enlarges, storing food and regulatory substances (mostly proteins and messenger RNA), which are needed for fertilized egg during early development. Only one or rarely two follicles complete development each month. As maturing follicle enlarges, it secretes ever-greater amounts of estrogen. Estrogen promotes continued development of follicle and of primary oocyte within .It stimulates growth of uterine endometrium. High levels of estrogen stimulates both hypothalamus and pituitary, resulting in surge of LH and FSH about 12th day of cycle. Surge of LH has three important consequences, triggering resumption of meiosis I in oocyte, resulting in formation of secondary oocyte and first polar body, causing final growth of follicle, culminating in ovulation and transforming remnants of follicle that remains in ovary into corpus luteum. Corpus luteum secretes estrogen and progesterone. Combination of these hormones inhibits hypothalamus and pituitary, preventing release of FSH and LH, thereby preventing development of more follicles. Simultaneously, estrogen and progesterone stimulates further growth of endometrium, which eventually becomes thick. In menstrual cycle in which pregnancy does not occur, corpus luteum starts to disintegrate about I week after ovulation.

This disintegration is precipitated by corpus luteum, through secreting progesterone that in turn shuts down LH secretion. Because corpus luteum can persist only while it is stimulated by LH, it induces its own destruction. With corpus luteum gone, estrogen anti progesterone levels plummets. Deprived of stimulation by estrogen and progesterone, endometrium of uterus also dies, and its blood and tissue are shed. This shedding forms menstrual flow that begins about the 27th -or 28th day of cycle. Reduced level of circulating .progesterone also means that it no longer inhibits hypothalamus and pituitary and spontaneous release of GnRH from hypothalamus resumes. Release of GnRH in turn stimulates release of FSH and LH, initiating the development of a new set of follicles and thereby restarting the cycle. Ovulation occurs about 14th day of menstrual cycle. Released egg can survive for 24-96 hours and therefore 14th to 15 the day is called as fertile period.

Hormonal Events During Pregnancy: During pregnancy, embryo itself prevents changes from occurring. Shortly after ball of cells formed by dividing fertilized egg embeds itself in endometrium, it starts secreting an LH-like hormone called chorionic gonadotrophin (CG). This hormone travels in blood stream to ovary, where it prevents breakdown of corpus luteum. Corpus luteum continues to secrete estrogen and progesterone, and uterine lining continues to grow, nourishing embryo. Embryo releases so much CG that hormone is excreted in mother's urine. In fact, most pregnancy tests use presence of CG in a woman's urine to determine pregnancy.

Remark: During first half of menstrual cycle, FSH and LH stimulate estrogen production by follicles. High levels of estrogen than stimulate mid Cycle surge of FSH and LH release (positive feedback). During second half of cycle, estrogen and progesterone together inhibit release of FSH and LH (negative feedback).Early positive feedback causes hormone concentrations to reach high levels, and later negative feedback shuts system down again unless pregnancy intervenes (Figure 10.39).

Summary

1. Mammals have 3 chief features in general viz .presence of 3 middle ear bones, hair and mammary gland .Malleus, incus, and stapes transmit vibrations from the tympanic membrane to inner ear. Differentiated teeth are replaced just once throughout a human life. Lower single jaw bone is called the dentary.

2. Eutherian young remains connected with mother through placenta. Female produces milk for young. Hair is an elongated rod of keratinized cells. Most mammals possess two kinds of hair. Dense hair provides insulation Keratin is a major component of epidermal structures. Hair cells area lives only in active growth site in root located at the base of a hair follicle.

3. Rodent lacks canines and a wide gap called diastema is found between the incisors and cheek teeth. Sebaceous glands associated with hair follicles, also present in hairless areas, produce sebum. Sebum maintains the hair and skin by keeping them

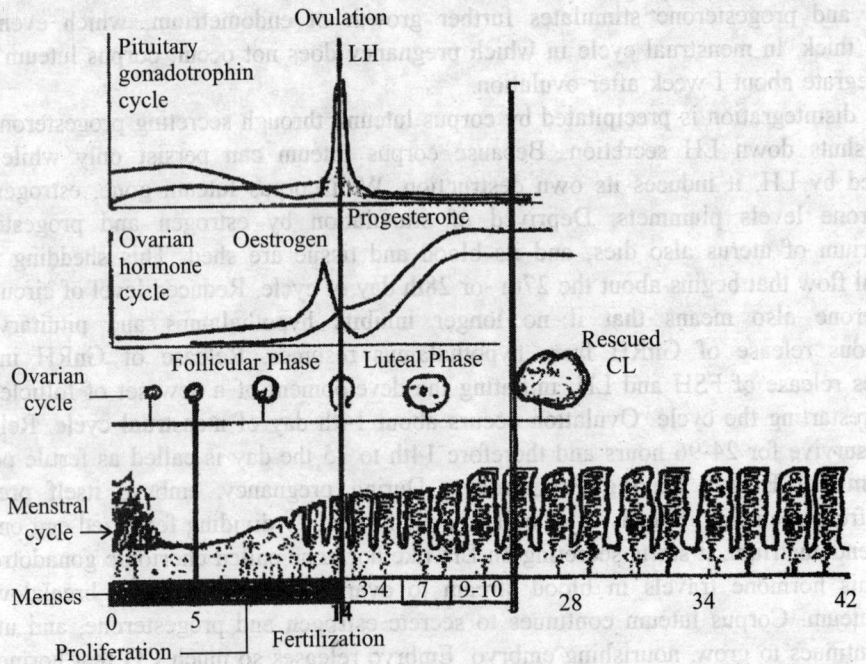

Figure 10.39 Menstrual cycle in human.

wet. Fossorial mammals lack pinnae completely. In semi-aquatic fur bearers, the pinnae are reduced.

4. The earliest eutherians of the Cretaceous were probably arboreal. Primates have retained this habit with features like five fingers, toes, clavicle and features which are strongly accentuated for arboreal life .Guinea pigs do not have tails, but have four toes on each front paw and three toes on each hind paw. They measure 20 to 40 cm from head to rump. Their incisors and molars grow continuously and wear down. As a member of class **Mammalia**, various biological aspects of human are described in this chapter.

5. The picture of hominid evolution is a far cry from the "Australopithecus africanus begat Homo erctus begat Homo sapiens" scenario that prevailed 40 years ago-- based to a great extent on fossils record since that time. Humans are thought to be originated in an ancestral line of apes (Dryopithecus) present in early Miocene. These Miocene apes differed from tree-dwelling primates in many aspects. Larger body modified to allow a radically new form of locomotion called brachiation. Selection for brachiation changed the ape's arms shoulders and upper body to allow suspension in a semi erect posture for long periods of time.

6. The name monotremata (one aperture) refers to presence of a single opening for discharge of feces, urinary and genital products (Monos, single, trematos, hole). Monotremes are the most primitive and unusual animals. They have survived to the present only as they have been isolated from serious competition. They occupy a

most interesting position because of their distribution, anatomical peculiarities and systematic position.

7. Sea-cows are herbivorous, live along the coasts and rivers, highly adapted to aquatic life and have reached the present condition by modification of a basic ungulate type of organization. Two modern forms, the manatee of the Atlantic and dugong of the Pacific and Indian oceans are different in many respects and show lines that were separate for a long time.

8. The pouched mammals similar to placentals diverged from some early stage of the main mammalian stocks. They parallel the adaptive radiation accomplished by the placentals. Many features are specialized. Today some 230 species are found in Australasian region. In Eocene times they occurred in Europe. The skull shows characters as found in Insectivora. Brain case is small with flat top. Orbit and temporal fossa remain fully confluent. Post-orbital bar is absent. The bony palate is incomplete posteriorly, there being large holes in palatine portion of it. The jugal reaches back to glenoid articulation of jaw.

9. Modem primates are typically placed in one of two suborders: prosiroii and Anthropoidea. These complex mammals are end products of evolutionary processes that began some 6S million years ago. Much of primate evolution has taken plane in trees in response to arboreal selection pressures. Today there are two types of monkey New world (platyrrhine, or flat nosed) monkey found in central and south America and Old world (Catarrhine, or downward) monkey from Africa and Asia.

10. Four types of teeth in each jaw are generally present in mammals. Incisors have generally simple structure and single rooted. These have layer of enamel only on anterior face provided with sharp chisel-like edge on them due to the fact that harder enamel is worn away slowly than comparatively soft dentine. Incisors are totally absent in sloths. Number of incisor is usually two but up to five for marsupials.

11. Many separate evolutions of fermentative digestion among vertebrates have resulted in distinctly different solutions to problems posed by plants as food. Monogastric and digastric or ruminant digestive systems are found among ungulates. Most derived stomach anatomy is ruminant digestive system, and animals that posses it is called ruminants or cud chewing mammals. Fully developed ruminant stomach has four chambers viz rumen, reticulum, omasum and abomasum.

12. Some animals hunting in darkness or murky water have evolved a type of sonar called echolocation. In micro chiropteran bats and toothed cetaceans, echolocation shows the way how an ancestral sensory capacity can be elaborated. As echolocation is used for hunting let us discuss the hunting pattern of bats.

13. The term estrus referred to existence of a period of strong sexual desire. It has become evident that estrus is close to time of ovulation. Actually, relatively brief periods within breeding season when female of most mammalian species undergoes spontaneous ovulation and is most receptive of male is known as heat or estrus .Rhythmic or cyclic changes in development, release and maturation of egg shell, together with accompanying structural and physiological changes in reproductive system as well as behavioral changes, periodically prepare female for matins and are collectively termed estrus cycle.

14. Menstrual cycle refers to series of events those periodically modify female reproductive tract of humans and advanced primates. Uterus contracts squeezing out excess endometrial tissue and blood, as a result of erosion of blood vessels in endometrium is called menstruation. Phases of menstrual cycle are follicular, ovulatory, luteal and menstrual. Follicular phase is characterized by progressive development and maturation of an ovarian follicle in one of the ovaries, redevelopment of uterine endometrium for reception of fertilized egg. Ovulatory phase is characterized by rupture of follicle and release of ovum.

Review Questions

Short Answer Questions

1. Give two features of mammals.
2. What the term "Monotremata " signifies?
3. Give an example of an oviparous mammal
4. Name the structure that allows mammals to breath while they chew.
5. What are antlers?
6. Give two characters of the order Insectivora.
7. Give two features of Primates.
8. Give two examples of Cetacea.
9. Give two examples of Sirenia.
10. What do you understand by adaptive radiation?

Long Answer Questions

1. Describe the main features of mammals.
2. Describe the evolution of mammals.
3. Distinguish among monotreme, marsupial, and placental mammals.
4. Describe the adaptive radiation of mammals during the Cretaceous and early Tertiary periods.
5. Compare and contrast the four main evolutionary clades of eutherian mammals.
6. Write a note on the evolution of Homo sapiens
7. Describe the general characteristics of primates. Note the particular features associated with an arboreal existence.
8. Distinguish between the two suborders of primates and describe their early evolutionary relationship.
9. Give an account of echolocation in bat.
10. Describe the physiology of menstrual cycle

11

General Notes

This chapter deals with some general account of vertebrates with emphasis on anatomy.

11.1 Skin

The ectodermal portion, on surface of embryo after differentiation and involution of CNS, forms epidermis which together with underlying derma makes up the skin. In earliest stage, the cuticle is single cell thick. Later alternative layers are fashioned on outside. In ganoids, teleosts, and Amphibia, the cuticle is 2- cells thick. In Amphibia, the outer layer is ciliated in young. Basal layer called the nervous layer produces the superficial layers, nerves and sense organs, and the outer one is cuticular layer. Derma is mesenchymatous in origin, and consists of fibrous animal tissue, smooth muscle cells, blood vessels, and nerves. The entire remains separated from deeper tissue by a connective tissue layer.

Structure of adult epidermis varies, becomes many layers thick, and is thicker in aquatic forms. In *Ichthyopsida*, the cells show less stratification than in higher teams, where various spherical glands remains loaded with slimy substance. On approaching the surface, these cells break, their contents adjoin the body manufacturing slimy condition.

In amniotes, the outer layer of cuticle hardened into stratum. The primary layer budding from the basal layer called epitrichium persists through major part of embryonic life on outside. Non-cornified cells rich in protoplasm are called Malpighian layer .In mammals; the skinny stratum lucidum between horny and Malpighian layers consists of flattened cells. Cells of basal layer are regularly divided to supply new cells that lie between basal layer and antecedently formed layers to extend the thickness of epidermis. Once these cells age, they pass into completely different layers,lucidum, corneum,and finally are cast off from the outer surface, either few cells at a time, or in larger sheets as in amphibians and reptiles. Strata stratum and lucidumare are protecting. In turtles, external layers of skin don't shed

Epiphysial Structure

Lachrymal glands absent in crocodiles are best developed in lizards on which the subsequent account is based. At associate early stage, a hollow outgrowth directed upwards and forwards arises from epithelial roof of this region. Its distal end keeps contact with epidermis of the top of head. The extremity of outgrowth expands into spherical vesicle, the pineal or membrane bone eye that lies on top of neural structure, sometimes on right side. Cells of pineal vesicle increase in range, cause a thickening of walls .Distal and proximal surfaces of vesicle differentiated, the previous into a clear lens-like body, the latter into sense cells supporting pigment cells, the entire creating up retina. Therefore, a camera eye is created. In lizards this lies on top head beneath the skin. One plate of dorsal surface bear a clear spot through that light reach the organ. Parietal eye differs from paired eyes.

An eye-like organ, the parapinealis from distal end of stalk often forms. Wherever the epiphysis is double, the pinealis arises from anterior, parapinealis from posterior out-growth. A 3rd outgrowth, the paraphysis, arises in front of epiphysis. It never develops sensory parts. Parietal eye in *Hatteria* retains its well developed elements throughout life and its nervous reference to brain persists which is useful to some extent. In other lizards, it lost such power, through deposit of pigment, or by degeneration of all equipment. In anura, the parietal eye lies between bones and skin. In several extinct vertebrates, pinealis was well developed and useful (Figure 11.1).

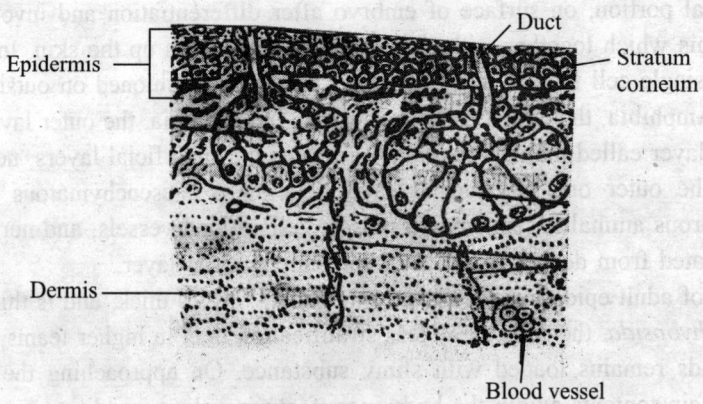

Figure 11.1   Section of skin of toad.

11.2   Exoskeleton

Hard structures are formed by accumulation of Keratin. Outer covering of epidermal scale prevents water loss through skin surface in reptiles. In lizards, scales are skinny, small, overlapping and moulted in little items. In sankes, scales are overlapping, enlarged on head called shields, and on ventral surface called scutes. Crocodilians and turtles have giant, thick rectangular scutes which are sloughed at intervals. Toothless horny beak of turtles, rattle at tail end of rattle snakes, and horns of horned toad are modification of stratum corneum. In birds, small dermal scales are found on lower leg, foot and base of beak. A bony projection of tarsometatarsus in males of some birds known as spur is roofed by epidermal sheath. Epidermal scales are found in feet and tails of rats and beavers. Giant scales of anteater shed individually. In armadillos, scales fuse into plates and bands.

In most primitive scales of elasmobranches, papillae of corium, organized in quincunx, push up into epidermis, carry basal layer of latter before them. External surface of every papilla and its base secretes the dentine with a central spine. Epidermis covering the papilla transforms into enamel. In ganoids, the first development is found in elasmobranchs, together with formation of plate, spines, and rudimentary enamel cap. Later spines and enamel cap disappear; outer facet of dentinal plate becomes coated by a tough, sleek layer called ganoin. In higher ganoids and in teleosts, dentinal papillae ensuing scales are dermal in origin without differentiation of layers. Scales might fuse into firm dermal armor enclosing body. In several fossils, ganoids, the external skeleton was extremely developed.

Existing amphibians except some caecilians lack scales. In several fossil amphibians, they were well developed. In caecilians, scales are dermal, and consist rings skirting the body. In stegocephalans, plates were confined to ventral surface. In class Reptilia, dermal skeleton of bony plates is well-developed. The plates in *Stegosaurus* are nearly 2 feet across. Similar dermal bones occur in alligators, and reach their extreme in turtles, wherever plates unite to create a bony box, composed of upper carapace and lower plastron .This shell firmly unites with true skeleton, and replaces it in some species. Vertebrae and elements of ribs are correspondingly reduced. In formation of reptile's scales, a papilla of derma forms. Scales arise from cornification of epidermal cells. Some tropical toads have bony plates beneath the skin of back. Claws of reptiles and horny beaks of turtles and birds are cornifications of cuticle. Scales and claws reappear in birds, however, dermal bones are never found.

Epidermal Body Covering

Both epidermis and dermis manufacture largely protecting and reconciling structures. Mucous, poison, sweat, sebaceous, ceruminous, mammary, and scent glands derive from epidermis. Horny structures derived from epidermis are dermal scales, feathers, hair, claws, nails, hoofs, spurs, true horns, beaks, and bills.

Feather: These are robust, light, elastic and waterproof, organized in skin in definite tracts of pterylae. Spaces lack feathers are known as apterae. There are 3 main varieties of feathers. Contour feather consists of a hollow, clear stalk, the quill or calamus, associated a distended distal portion, the vane or vexillum. Feather papilla fits into a tiny aperture, the inferior umbilicus at the bottom of quill. Another aperture, the superior umbilicus is found at the junction of quill with vane on central surface. A tiny tuft of down feathers called after shaft is found in neighborhood of superior umbilicus. Vane features a solid longitudinal axis, the rachis; on both sides of that distended vane thread-like barbs are found. Barbs extend obliquely outward from rachis, supplied with barbules on either side, by that barbs attach to each other.

Contour feathers could also be remiges, or rectrices, form covering of body and appendages. Remiges or wing quills are found on hind end of wing. Those connected to ulna are called cubitalis or secondaries. Rest is named primaries. Of primaries, those connected to metacarpal region are metacarpals. Those connected to phalanges are called digitals. Rictrices spring out from tail. Several rows of feathers filling the area between remiges and rectrices are called wing and tail coverts respectively.

Filoplume or hair feather consists solely of a long, slender shaft, and few barbs at its distal end. Shaft is embedded in skin and encircled by feather follicle at its base. Hair feathers are scattered over the body surface .Plumulae or down feathers are composed of a basal, short, hollow quill, embedded in skin. Various barbs arise from free end of quill. Barbs bear barbules on their edges. Down feathers of adult known as "Powder down" lie underneath giant contour feathers and form a heat insulating layer that helps in temperature management and assists in warming the eggs throughout incubation.

Feather tracts layer and epitrichium, becomes many cells thick. By continuous growth, the papilla becomes long and cylindrical, project from the body, the axial derma form the pulp of future quill, the epidermis surrounds the outgrowth. A circular depression round the base of papilla starts the formation of future feather follicle. In distal parts of those outgrowths, longitudinal ridges of pulp encroach cuticle step by step, dividing this layer into cylindrical rods, that finally remain in position by epitrichium. The derma retracts into feather follicle, carrying with it the basal layer of epidermis. A hollow epidermal out-growth, the quill, bears at its extremity dermal rods. Cells of those parts dried and cornfied. Epitrichium break away, and the rods separate as down-feather.

Later the contour feathers, develop from backward pulp, grows out once more. These develop like their predecessors. Rods of pulp are oblique to axis of outgrowth. As a result, the cornified rods proceed from associate undivided portion on dorsal facet of outgrowth. Once the epitrichium breaks away, these expand to create vane. On the facet wherever the shaft is created there are 2 longitudinal thickenings; with growth these become larger and bend inwards with formation of a solid rod close to tip. However, farther down in growth of cavity includes an area to make the proximal portion of shaft hollow. Dorsal and ventral sides of feather correspond to outer and inner surfaces of epidermis of feather papilla. At regular intervals the bird sheds or molts its feathers.

Development of Feather: Faeather develops from dermal papilla. It is outwardly coated by epidermis. Inner epidermal layer or stratum Malpighi consists of cylindrical cells that apace proliferate and ends in growth of papillae. Base of papilla sinks and forms a depression called feather follicle. Developing feather at this stage is named feather germ. Dermal mass comes into developing papilla to form pulp. Stratum corneum covering feather germ forms a covering sheath, the periderm. Cells of Malpighian layer in distal part of feather germ divide rapidly and form cornified longitudinal diverging ridges, the base. These cells become horny to from hollow quill.

Periderm ruptures and longitudinal ridges ultimately form barbs that develop barbules. Down feather differs from contour feather in absence of shaft, the barbs arising directly from the top of quill. These barbs never interlock, however stay soft and free from one another. In pin feathers, a hair-like shaft lacks barbs .Except in penguins and a few flightless bird birds, feathers don't seem to be uniformly distributed over whole body surface, occur in well marked pterylae. The apteria being sparsely coated with down or pin feathers. Feather tracts have importance in classification of birds. (Figure 11.2)

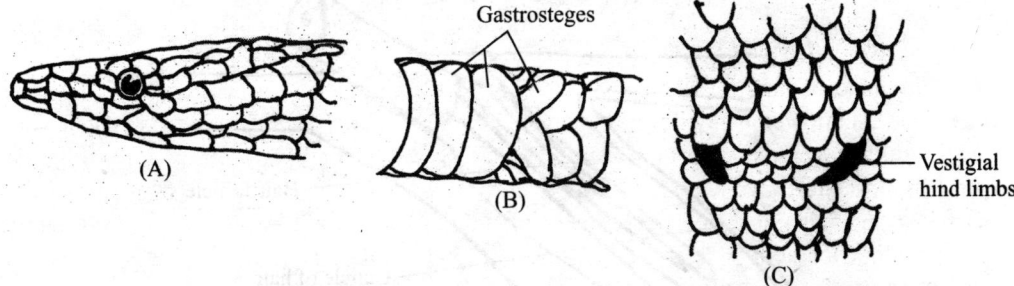

Figure 11.2(a)   Exoskeleton structure of snake (A) Head of a snake showing the arrangement of head shield, (B) Arrangement of scales on the ventral side of a snake, (c) Showing the vestigial hind limb of python.

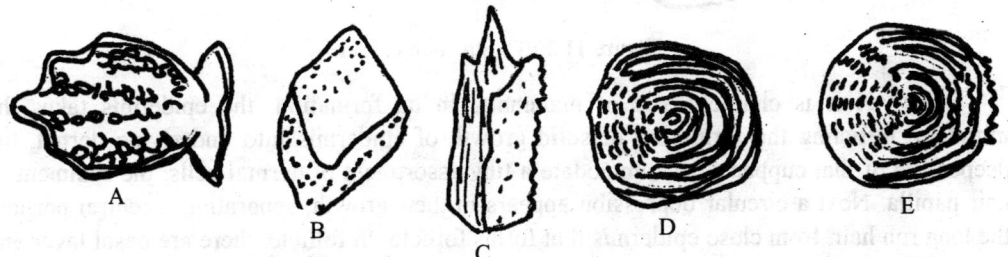

Figure 11.2(b)   Scales; A. Placiod, B. Cosmoid, C. Ganoid, D. Cycliod, and E. Ctenoid

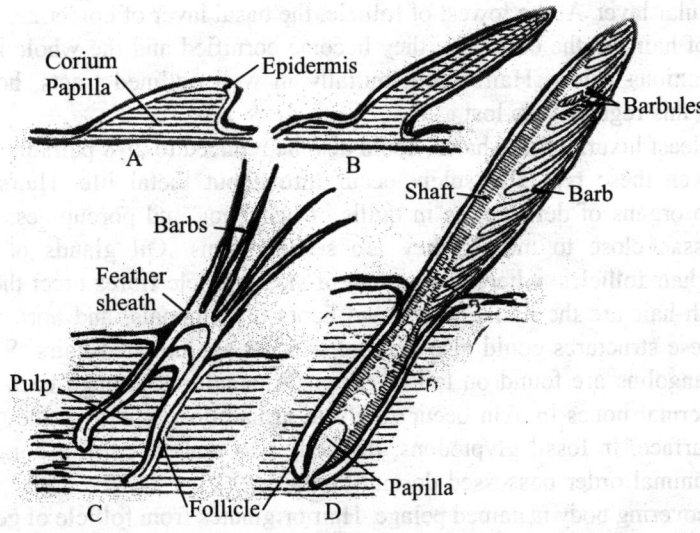

Figure 11.2(c)   Stages in the development of feather A, B, C : Development of down feather, D : Contour feather in feather sheath.

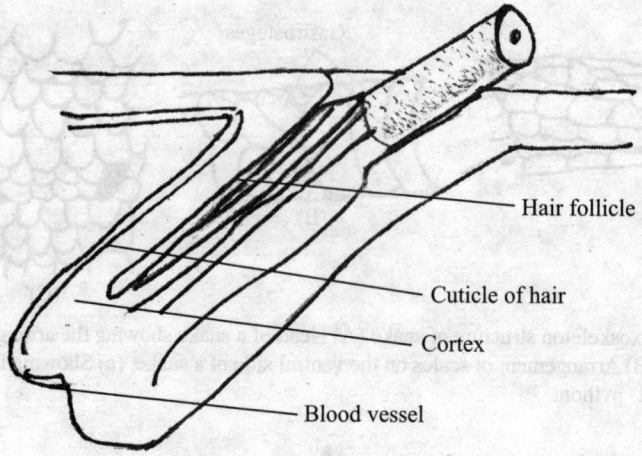

Figure 11.2(d)   Structure of hair

Hair: Hair is characteristic of mammals. In its formation, the epidermis takes the initiative, inflicting the formation of solid growth of epidermis into underlying derma, the deeper end of that cupped to accommodate a tiny assortment of dermal cells, the rudiment of hair papilla. Next a circular depression appears in their growth, separating a central portion, the long run hair, from close epidermis that forms follicle. In follicle, there are basal layer and superficial layers of epidermis without clear differentiation of stratum corneum and lucidum.

At the bottom of follicle, these pass directly into hair, on out facet of that is the inner root sheath composed of 2 layers. Inner root sheath doesn't reach the external surface. Hair consists of a central core or medulla, around that many layers of cells form cortex, and on outside a cuticular layer. At the lowest of follicle, the basal layer of epidermis, by cell division adds to base of hair. As the cells age, they become cornified and the whole is pushed out of follicle by additions below. Hairs seem initially in well outlined tracts, however later by multiplication, this regularity is lost.

Hairs are least luxuriant in whales, could also be reduced to 2 -8 pairs in neighborhood of mouth, and even these typically solely occur throughout foetal life. Hairs could also be developed into organs of defence, as in quills of hedgehog and porcupines, whereas just in case of vibrissae close to mouth, they are sense organs .Oil glands of racemose type connects with hair follicles, whereas a system of sleek muscle fibres erect the hairs. Closely associated with hair are the nails, claws, and hoofs of mammals, and horn of sheep, goats, and cattle. These structures could also be composed of agglutinated hairs. Scales that cover the body in pangolins are found on tail of rodent, Anomalnrns, though each is epidermal in origin. True dermal bones in skin occur solely in armadillos, wherever they form armor on dorsal body surface. In fossil glyptodons, the body was embedded in a bony case, whereas some extinct animal order possessed dermal bones.

All hairs covering body is named pelage. Hair originates from follicle of germinative layer of epidermis into dermis. A hair papilla contains blood vessels and nerves, nourishes swollen root or bulb, adding new cells forming shaft of hair. Cells of shaft become keratinized, hardened and shortly die. Hair protrudes on top of skin could be a dead structure and is lubricated by secretions of sebaceous gland. Hair shaft consists of associate external cuticle, middle cortex and inner medulla containing air areas in large hairs. Hairs have many modifications.

Beak and Bills: In turtle, tortoises and in modern birds, teeth are absent. Every jaw bone is roofed by a changed dermal scale that forms beak or bill. Seed eating birds have short, blunt beaks. Insect eaters possess long, slender beaks, that don't seem to be robust as those of seed eaters. Long, strong, hooked beaks of birds of prey are well fitted for getting food. Bill of duckbill is soft and isn't coated with a changed dermal scale.

Digital Cornifications: All digital cornifications are modifications of stratum corneum at the tips of digits and grow parallel to skin.

Claws: Claws of reptiles, birds and mammals are identical structure created by hard, pointed, narrow, dorsal plate called unguis, and a less hard ventral plate called subanguis. Unguis and sub unguis cover last tapering phalanx.

Nails: Claws are changed into nails in primates. Unguis is broad and flat whereas sub anguis is soft and reduced. Tip of digit forms a sensitive and extremely vascular pad over that cuticle invaginates to create a nail groove containing a nail root.

Hoofs: Hoofs are found in ungulates. Horny unguis is neither pointed nor flat however U or V shaped. Subanguis is U shaped, much thickened. Horse's shoe may be nailed into it. Sub unguis surrounds soft horny cuneus. Tip of digit forms a pad and contain a blunt phalanx. Other modifications of stratum corneum embrace whale bone plates of toothless whales, horny covering of horns of sheep and bovine and prong horns of antelopes (Figure 11.3, 11.4).

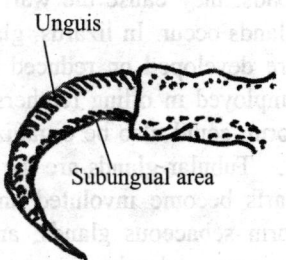

Figure 11.3   Claw of Eagle.

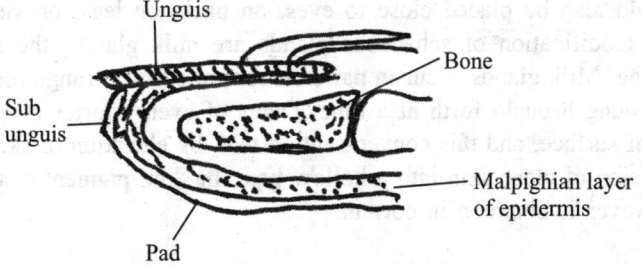

Figure 11.4   Section of finger tip of man.

Horns: These are found in artiodactyle and perissodactyle solely on head and are organs of offence and defense. True horns sometimes occur in each sex in goats, cattle. They are branchless, cylindrical, tapering permanent structures, grow through out life and never shed. They are products of a hollow dermal bony core arising from frontal bone and coated by an epidermal cap. Prong horns of antlered bovid (*Antilocarpa*) are formed by a tiny hollow central permanent bony core arising from frontal bone and coated by a skinny, hollow epidermal horn. Horny sheath of a prong horn bears 1 to 3 prongs, and is shed annually. Permanent bony core becomes the bottom around that new horn develops. Antlers are found solely on males of deer family, on each sex in reindeer and caribu.

Antlers are annual solid outgrowth of dense tissue connected basally to frontal bone and not true horns. Deposition of calcium salts makes antlers onerous. Throughout growth, it remains covered on surface by bushy and vascular skin or 'velvet'. Once growth is completed, the velvet wears off, exposing naked, branched horn develop. Giraffe horns are stunted, unbranched, permanent antlers found in each sex consisting of a brief bony dermal core, protrusive from frontal bone and remains coated with velvet that is rarely shed. Hair horns found in perissodactyl of each sex are perched on the chapped space of nasal bone. Indian rhinoceros features a single horn. African species has 2, one behind the opposite. These permanent structures if broken once more grow out and are known as fiber horns that are created by keratinized dermal fibers.

Dermal Glands: Epidermis produces various glands. In fish, mucous glands, and poison glands are noted. The poison glands in axilla of some catfishes are best noted. In toadfish, a gland in similar position is found. In Amphibia, glands in skin secrete acid juice. In toads, they cause the wart like skin. In Sauropsida, glands are few. In some snakes, stink glands occur. In lizards, glands are found within limb region of hind limbs. In birds, glands are developed on reduced tail, best developed in water birds, the oily secretion of that is employed in oiling feathers. In mammals, glands are well developed, and show a range of form, could also be organized in 2 classes, the hollow and indeterminate.

Tubular glands are the sweat glands that extend deep into the derma and in their deeper parts become involuted and convoluted. The racemose glands in their simplest condition form sebaceous glands, and are placed closely to roots of hair. In some mammals, the sebaceous glands of some body regions converts into scent glands, secretions of that serve for offence or defence; and will have price at rutting season as attractions for alternative sex. Defensive glands are found in skunk and polecat. The peculiar glands of the beaver, civet cat, and deer could also be placed close to eyes, on back, on legs, on ventral surface, or close to vent .A modification of sebaceous glands are milk glands, the secretion of that nourishes the young. Milk glands occur in pairs on ventral surface, range roughly correlating with number of young brought forth at a time. Ducts of every cluster of glands open on a restricted extent of surface, and this converts into a teat, by elevation of skin in which ducts open, or by extension of close skin into a hollow type .In skin, pigment could also be found in epidermis, however is common in corium.

11.3 Skeleton

Skeletal structures could also be membranous, cartilaginous, or bony; and in development some parts withstand all of those phases to achieve adult condition; or cartilage stage could also be skipped, membrane developing directly into bone; or cartilaginous condition could also be the ultimate stage of skeleton. Membranous skeleton consists of connective tissue, and in its highest development forms sheets. From it, cartilage is developed by increase in range of cells. The tissue called the procartilage consists of closely compacted polygonal cells with giant nuclei. These cells secrete intercellular substance, and tissue converts into cartilage (Figure 11.5).

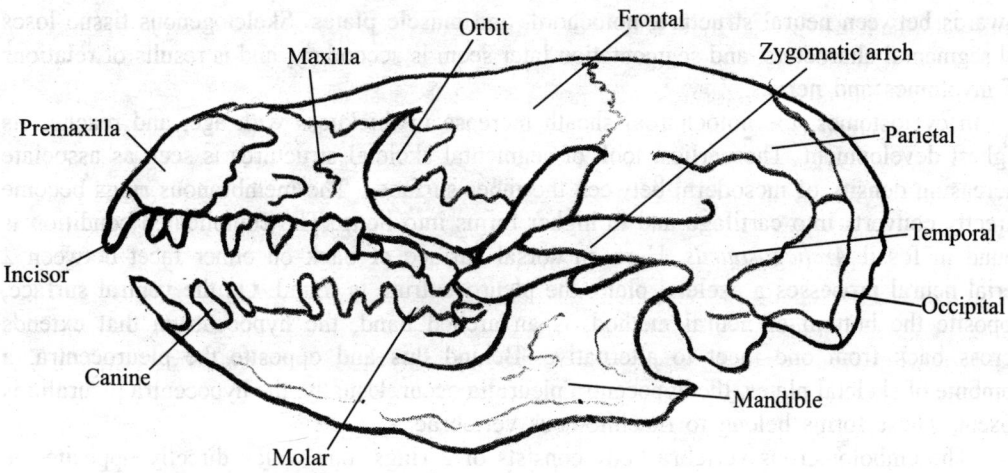

Figure 11.5   Typical skull of vertebrate.

In conversion of cartilage into bone the bone matrix is dissolved; and round the margins of cavities bone forming cells arranges them, secretes lime salts around themselves, and step by step build up the bone. In lower vertebrates, this method starts outside the cartilages and return toward interior. However, in higher forms, besides perichondrial ossification, centres of ossification appear within cartilage and from these ossification extends peripherally. In conversion of membrane into bone, there are an equivalent look of osteoblasts in and on the tissue, and these manufactures the bony substance in same manner. The end in either case is same.

Distinction between 2 is incredibly necessary. Increase in size of membranes is accomplished by additions of exterior substances and by increase in number of centre of ossification. From these centres ossification extends in all directions, except for a time there remains a cartilaginous region between ends and main portion during which increase in length is feasible. Later these epiphyses united or ankylosed to main portion and line of division can't be traced.

Skeleton could also be divided into internal and dermal parts. Internal skeleton consists of axial portion together with vertebral column, skull, ribs, associated breastbone; and an appendicular portion consists of skeleton of appendages and girdles supporting them. Vertebral column develops around back could be a rod-like structure of entodermal origin that lies between alimentary canal and central system nervous system, extend from infundibulum to posterior end. Its cells become gelatinous; migrate to fringe, wherever they organize in manner recalling epithelium. Notochordal mass consists of reticulum in meshes of solid jelly. The cellular envelope forms varied divisions of vertebrates. In cyclostomes, it continues to extend in size throughout life, and constitutes the most important portion of skeletal axis.

In other vertebrates, the development of vertebrae relegates it to subordinate terribly position in adult, wherever it's going to persist as obscure remnant. Vertebrae proper arise from mesenchymatous cells that bud off as sclerotomes from developing mesothelial tissues. Several such cells organize as never-ending envelope around back, whereas others wander inwards between neural structure, notochord, and muscle plates. Skeletogenous tissue loses all segmental characters, and segmentation later seem is secondary, and is results of relations of myotomes and nerves.

In cyclostomes, the notochordal sheath increase in thickness with age, and reaches its highest development. The earliest look of segmental skeletal structures is seen as associate increasing density of mesoderm between the inner surfaces. The membranous rings become directly converts into cartilage and in higher forms into bone. The complicated condition is found in fossil *Archegosaurus*. Here on dorsal surface of back on either facet between 2 serial neural processes a skeletal plate, the pleurocentrum is found. On the ventral surface, opposite the bottom of neural method, is an arched band, the hypocentrum that extends across back from one facet to alternative. Behind this and opposite the pleurocentra, a combine of skeletal plates, the hypocentra pleuralia occur. Usually, the hypocentra pleuralia is absent. These forms belong to rhachitomous vertebrae.

The embolomerous vertebra body consists of 2 rings, one which directly opposite the bottom of intercentrum is usually the hypocentrum arcale. The neural process is the other ring. In development, embolomerous condition springs from rhachitomous type by fusion of hypocentra pleuralia with pleurocentra to create one ring, whereas the opposite develops by dorsal extension of hypocentrum arcale. In others, no hypocentra pleuralia occur. The bone arises by ventral extension of pleurocentra. In birds and mammals; the bone arises initially by vertebral bone, passing beneath notochordal sheath and obliquely upwards and backwards to posterior limits of segment. In Amphibia and teleosts, the vertebral body arises from intercentrum and from hypocentrum arcale. The vertebrae don't seem to be precisely homologous throughout vertebrate phylum. Several authorities claim that centrum arises from hypocentrum arcale, and pleurocentra either contribute or produce the anterior zygapophyses.

A third style of vertebra is that the phyllospondylous in which vertebral body consists of right and left halves. This kind is found in fossil Branchiosauridae. Within the embryo, a ligament runs the body length simply dorsal to spinal cord. Wherever this passes between dorsal ends of neural processes, it converts into cartilage producing an extra component that together with 2 neural processes forms neural arch and protect the spinal cord. In caudal region of ichthyopsida and a few higher forms, the vertebra is completed below by similar arch which encloses caudal artery and vein. This arch consists of a combine of haemal processes and haemal spine.

Various elements of vertebrae arise one by one with a bent to fuse in adult, the fusion being most complete within the higher groups. The vertebrae are set down at an early development, and their number is not increased afterwards. Increase length of body is the result of longitudinal growth of centra of vertebrae. In fish additions are made, as it were, in layers, on circumference of centrum first formed every new layer being slightly longer than its forerunner. As a result, the centrum becomes concave on either end is amphicoelous. The elements of centrum first formed prevent farther increase of notochord in intravertebral regions; however intervertebrally it expands, filling up cavities between the serial vertebrae, and assuming the form of a string of beads. In higher forms, an intervertebral growth of cartilage produces a secondary series of constriction in notochord. Deposition of intervertebral cartilage forms a cup at one end of vertebra, and at alternative of a rounded extremity which inserts the cup at end of next vertebra. In extreme case, the intervertebral cartilage has been cut fully in 2 leading to formation of a ball and socket joint between serial vertebrae; whereas ossification extends that just about entire bone and a part of bone intervertebral cartilage converts into bone.

When this method ends in a centrum rounded before hollow behind, opisthocelous vertebrae result. Once rounded behind and hollow before, it is procoelous. Amphicoelous vertebrae occur in most fish, most perennibranch urodeles, some salamanders, some stegocephali, Gymnophiona, several dinosaurs, plesiosaurs, ichthyosaurs, precretaceous crocodiles, geckos, Rhynchocephalia, and fossil birds *Archaeopteryx* and *Ichthyornis*.

Opisthocoelous vertebrae occur in *Lepidosteus*, most salamanders, *Pipa, Discoglossus*, most dinosaurs, some vertebrae in penguins and auks, and neck vertebrae of most ungulates. Procoelous vertebrae occur in most members belonging to order Salientia, reptiles, and birds. In most mammal, vertebrae are flat on each end of centrum, amphiplatyan. In forms with amphiccelous vertebrae, true articulation of separate parts of vertebral column is lacking. The anterior or prezygapophyses have their 2-dimensional surface turned dorsally to articulate with ventral surfaces of posterior process of vertebra in front.

In snakes and in few lizards, these are strengthened by articular surfaces developed from neural spine. On the anterior surface of base of spine, a wedge-shaped method comes forward, its articulary surfaces directed obliquely outward and down. This fits corresponding cavity on posterior surface of neural spine of vertebra in front. In all abovementioned fishes, transverse processes occur. A diapophysis connects with neural process, and a parapophysis connect with vertebral column. One or alternative of those might stand out in development, and infrequently either could also be rudimentary.

In addition, names anapophysis and metapophysis are given to projections on the neural processes that have no great morphological significance. The vertebral column is constructed of these vertebrae, and in this structure 2 or a lot of regions is distinguished. In fish, there is a whole haemal arch, whereas within the trunk the haemal processes diverge and transform into ribs .In Amphibia, 2 other bone regions cervical and sacral occur. Sacrum intervenes between trunk and caudal vertebrae, and support hip that are supported by hind limbs. Trunk vertebrae bear true ribs, whereas vertebrae in neck lack ribs and transverse processes, or present in an exceedingly rudimentary condition. The line between cervical and trunk vertebrae is loosely drawn by girdle of forelimb. In sauropsida, an equivalent region may be derived as in amphibia; however each sacral and cervical regions are increased in extent, there being 2 or 3 sacral and a bigger range of cervical vertebrae.

In mammals, these regions increase by a division of trunk into a thoracic region, vertebrae of which bear ribs, and a lumbar region in which ribs want. In some regions a tendency towards the fusion of vertebrae occurs. Most often those sacrums unite into one piece, whereas fusions in caudal region are numerous, and are related to partial or entire disappearance of tail. In modern birds, there is a pygostyle, whereas in Salientia, the caudal vertebrae are consolidated into rod-like urostyle.In other regions this union is rare.

Of the anterior 2 vertebrae in amniotes, the primary one joins the bone is named atlas, the second as axis or epistropheus. Atlas bears on its anterior face articular surfaces for articulation with bone. Its neural arch is well developed, however centrum is absent, instead a thin bony arch, intercentrum is present. Axis bears, a cylindrical outgrowth, the odontoid process. This is often like bone of atlas, that has lost its reference to its neural arch, and has secondarily united with centrum of second vertebra, forming a pivot concerning that atlas turns. In crocodiles and *Hatteria*, a combine of plates, or one plate occur on dorsal anterior portion of neural arch of atlas. This is often called proatlas.

The name rib indicates 2 structures, one showing in ganoids, teleosts, and dipnoi, the opposite in Amphibia, amniotes, and in selachii. Ribs of fish are haemal processes of trunk vertebrae, which, in body cavity, extend from vertebral centres towards ventral surface between muscles and coelomic walls. Transitions from ribs into haemal arches may be derived in fish. In caudal region of urodeles, haemal arches are present, and besides these caudal vertebrae bear transverse processes that extend outward between the epi- and hypaxial muscles. Ribs are homologous throughout vertebrates. Outgrowth of sacral vertebra is enlarged, supports pelvic arch, whereas in presacral vertebrae transverse processes bear short articulated parts, the ribs.

So amphibian ribs don't seem to be resembling haemal processes in these animals, and that they are structures completely different from ribs of fishes. Ribs of amniotes are homologous with those of Amphibia.They are intersegmental in position, arise by condrification and complete ossification of part of myoicommatous tissue. In fish, the ribs are slender, and are firmly united to vertebral centra; or they are movably articulated to short basal stumps. In several physostomous fishes, some anterior ribs produce a sequence of bones connecting air bladder with ear. Some slender bones in fleshy parts in fish might represent ribs of upper forms. These epimerals, epicentrals, and epipleurals, are explicit to without cartilage stage.

Ribs of elasmobranchs are small and cartilaginous, and are intimately united with vertebral centra. In their relationships to muscles, they tally ribs of Amphibia, and are in no manner differentiations of haemal arches. From the Amphibia upwards, the ribs are articulated with vertebrae by 2 heads, a dorsal or tubercular head articulating with diapophysis, a ventral or capitular head resting on parapophysis. There's therefore produced a skeletal arch between rib and vertebra, through that passes vertebral artery.

In Amphibia, the 2 heads arise one by one and unite later. Varied modifications might occur. Head might disappear. The parapophysis could also be reduced to an articular surface. In anura, the ribs might fuse to diapophysis, or as in neck of mammals, to each di- and parapophysis. In crocodiles, each tubercular and capitular heads articulate with transverse process in most thoracic ribs. In Amphibia, the ribs are short, and are confined to region close to backbone. Pelvis doesn't articulate directly with outgrowth of sacral vertebra; however that association is established by intervention of a sacral rib. In caecilians, ribs occur on each bone except the primary and last. In amniotes, the ribs in trunk region acquire a great development, and extend around body cavity. They may be ossified throughout their extent. Ventrally they terminate freely; or they connect with a sternum.

In majority of birds, and in some reptiles, every rib bears a backwardly directed uncinate process and in cervical region, the ribs are shorter. They freely articulate to vertebrae, sometimes they're consolidated to transverse processes and centra. Sometimes caudal ribs are poorly developed. In some reptiles, abdominal ribs occur. These are chondrifications or ossifications in ventral wall of abdomen, behind ribs, and external to rectus muscles. These don't seem to be homologous with true ribs; the name gastralia has been given them.

In Amphibia, the sternum arises as a pair of longitudinal rods in connective tissue on ventral surface of body. These rods unite and form an odd plate in median line between origins of fore limbs. In urodeles, the sternum remains as small plate simply behind ventral portion of pectoral girdle;in Salientia, it extends farther forward. In urodeles, the sternum is cartilaginous; within the Salientia parts of omosternum, and the posterior portion, become ossified. The sternum is lacking in apodans.In amniotes, the sternum arises from ventral ends of ribs. The distal ends of those become separated from the remainder, and unite to form the unpaired structure. The sternum is lacking in snakes and turtles.

In birds, few ribs contribute to the sternum that could be a broadplate, and in birds bears a robust keel or carina on its ventral surface. In wingless birds, the keel is absent, and presence or absence of keel was once used as a method of dividing birds into Ratitae and Carinatae. In mammals, the sternum is elongated, and ribs contribute to its formation. Connected with sternum in several groups could be a structure the episternum. This first appears in stegocephali, however reaches its highest development in reptiles. It's distended before, and sometimes become T shaped. Arms support the clavicles, whereas shaft connects with, or may even be amalgamate with sternum. No episternum has been represented in birds; however in mammal one frequently exists.

11.4 Suspensorium or Jaw Suspension

Suspensorium is the manner of articulation or suspension of higher and lower jaws with chondrocranium. There are 5 principal variants or varieties of suspensoria as follows.

Amphistylic: This is often a primitive arrangement found in Crossopterygii and primitive sharks (*Heptanchus, Hexanchus*). The jaw has double articulation with cranium. Quadrate or basal and otic processes of mandibular arch are connected ligaments to chondrocranium. Equally upper end of hyomandibular of hyoid arch is connected to chondrocranium, whereas 2 jaws are suspended from its alternative end. Therefore, in amphistylic (amphi = both, style = bracing) jaw suspension, each articulator and hyoid arches are connected to cranium.

Hyostylic: It is found in most elasmobranchs, dogfish and every bony fish. Palatoquodrate is loosely connected by anterior ethmopalatine ligament and posterior spiracular ligament to cranium. Each jaw is braced against hyomandibular; upper of which inserts into sensory region of skull. Since hyoid arch braces or binds 2 jaws against cranium, this jaw suspension is termed hyostylic. It provides the jaws a wider movement and helps in swallowing larger preys.

Autodiastylic: Jaws are connected to cranium by anterior and posterior ligaments. Hyoid arch remains fully free and doesn't support jaws. Gill cleft in front of hyoid arch bears a whole gill and doesn't produce any opening. This condition is found in some earliest gnathostomes like acanthodians.

Autostylic: This condition is found in extinct placoderms, chimaeras, lung fishes and, most tetra pods. Hyomandibular doesn't participate, however becomes changed into columella or stapes for transmission of sound waves. Upper jaw rather than being connected by ligaments becomes fully amalgamate with cranium. Articular of mandible articulates with quadrate of upper jaw.Autostylic suspensorium is widespread and has a minimum of three variations or subtypes.

Holostylic: In subclass Holocephali (Chimaeras), jaw is firmly amalgamated with bone and mandible is suspended from it. Hyoid arch is complete, independent and not connected to bone.

Monimostylic: In several tetra pods, hyomandibular forms columella and articular articulates with quadrate. Quadrate remains immovably connected with skull.

Streptostylic: In some reptiles (lizards, snakes) and birds, quadrate is loosely connected and is movable at each ends, a condition called streptostylism.

Craniostylic: This kind is characteristic of mammals and may be a modification of autostylic suspension. Upper jaw fuses throughout its length with cranium, and hyomandibular forms stapes. Articular and quadrate become changed into malleus and incus. Consequently, dentary of mandible and squamosal of lower jaw offer articulation between jaws (Figure 11.6).

Chandrocranium or Neurocranium: Skull is derived from 3 major embryonic components, neurocranium or chondrocranium , dermatocranium , and splanchrocranium. Neurocraniurn includes brain box and 3 sense capsules - olfactory, optic and otic capsule.

Second visceral skeletal arch behind mandibular arch support the mouth floor. In cartilaginous fish it consists of a basihyal cartilage, 2 ceratohyals and 2 hyomandibular. Hyomandibular is dorsally braced against otic region of brain case and ventrally is tightly bound by ligaments to region of jaw joint.

11.5 Vertebrae

Vertebrae are composite structures of mineralized cartilages or of each endochondral and intramembranous bony parts. Six major elements are sometimes recognizable in bone. Centrum is a solid cylindrical spool surrounds and sometimes fully replaces or incorporates notochord and makes up body of vertebra. Neural arch grows dorsally to hide neural cord like a rigid tent. Haemal arch grows ventrally and encloses caudal blood vessels. Neural spine is bony blades that project into dorsal skeletogenous septa and supply sites for attachment of muscles and ligaments. Hemal spine is bony blade that project into ventral skeletogenous septa and supply sites for attachment of muscles and ligaments. Anapophyses are bilaterally symmetrical structures that project from vertebrae and fasten via muscles and ligaments to other skeletal parts. Among anapophypes characteristic of upper vertebrates are the following:

Zygapophyses: A combine of prezygapophyses arises from base of neural arch and project cephalod. A pair of postzygapophyses arises from neural arch and project caudad. Postzygapophyses articulate with prezygapophyses of next caudal vertebra. Zygapophyses are uncommon among fishes.

Diapophyses: These are transverse processes connected originally to base of neural arch, typically to centrum and lengthening lateral. Each articulates with dorsal head (tuberculum) of the 2 headed (bicipital) ribs common in tetra pods.

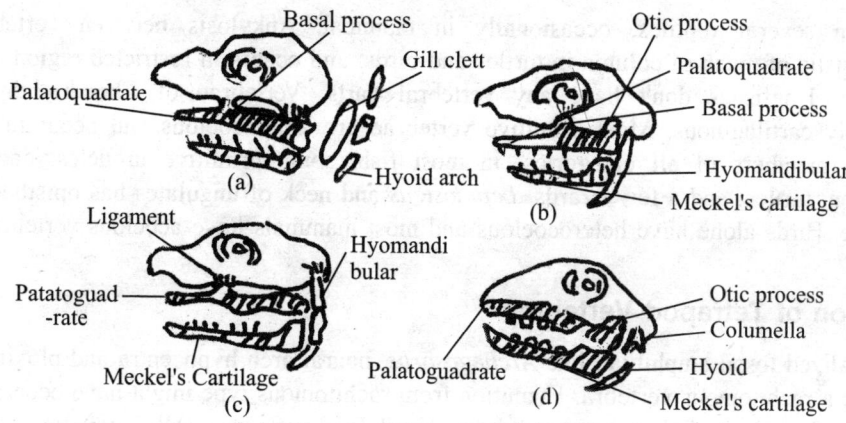

Figure 11.6 Jaw suspension in vertebrates (a) Autodiastylic, (b) Amphistylic, (c) Hyostylic, and (d) Autostylic.

Parapophyses: These are transverse processes of centrum articulating with ventral head (capitulum) of bicipital ribs. Though tetra pods ordinarily have ribs with 2 heads, either head could also be reduced, during which diapophyses or anapophyses could also be absent.

Basapophyses: These are ventrolateral processes of centrum, articulate with haemal arch once latter are present, and sometimes, meet ventrally to become haemal arch.

Hypapophyses: These are distinguished mid- ventral projections of centrum common on some vertebrae of reptile, birds, and mammals.

Pleurapophyses: Theses are transverse processes incorporating at their tips short ribs, that are amalgamate with them. Articulating cephalic and caudal ends of centre could also be convex, concave, flat or a mixture of those, and vertebrae could also be classified on basis of such variations.

Procoelous (pro = front + coelous = hollow): Anterior face of centrum is concave and posterior face convex. e.g. typical vertebrae of frog and most reptiles.

Opisthocoelous (Opistho = at the back): Centrum is concave posteriorly and convex anteriorly e.g. cervical vertebrae of some giant ungulates.

Amphicoelous (amphi=-both): Centrum is concave at each ends e.g. vertebrae of most fishes and caudate amphibians, eighth vertebra of frog.

Acoelous (a = absent): Centrum is flat at each ends without concavity or convexity. e.g. vertebrae of mammals.

Bi-convex (bi = two): Centum is convex at each ends e.g. sacral or 9th vertebra of frog.

Heterocoelous (hetero= asymmetrical): Ends of centra are saddle shaped .e.g. vertebrae of recent birds. Vertebrae from completely different regions of vertebral column of fishes exhibit abdominal and caudal vertebrae. In tetrapods, trunk vertebrae are divided into cervical, dorsal, and sacral. Dorsal are divided into thoracic and lumbar vertebrae when long curved ribs are restricted to anterior dorsal. Caudals sometimes numerous, are reduced to arch less centra at near end of series. They are few in birds and in some mammals and absent in anurans. First 2 cervical are specialized as atlas and axis in amniotes. A proatlas occurs in several reptiles, occasionally in mammal. Ankylosis between vertebrae is characteristic of much of column in turtles and birds, and occurs in restricted region in some mammals. Hagfishes don't have any vertebral parts. Vertebrae of Chondrichthyes are completely cartilaginous. Most primitive vertebrae are amphicoelous and occur in extinct primitive members of all categories, in most fish, some primitive urodeles, caecilians, sphenodon, turtles, and a few lizards. *Lepidosteus* and neck of ungulates has opisthocoelous vertebrae. Birds alone have heterocoelous and most mammals have acoelous vertebrae.

Evolution of Tetrapod Vertebrae

In generalized fossil amphibian like *Archegosarus*, neural arch hypocentra and ploxirocentra comprise a rachitomous vertebra. Evolution from rachitomous type might have occurred in 2 directions forming embolomerous and stereospondylous vertebrae. All 3 varieties are found among labyrinthodonts. Embolomerous type was transmitted from labyrinthodonts with modifications in amniotes. Stereorpondylous sort are transmitted to modern amphibians. In embolomerous vertebrae of extinct amphibians, hypocentra and pleurocentra form 2 discs like adult centra close, the notochord. In changed embolomerous vertebrae, hypocentrum (intercentrum) is reduced, whereas pleurocentra enlarge forming body of vertebra (*Seymouria*). Trend continues in modern amniotes. In stereorpondylous type, hypocentra enlarges to create body, whereas pleurocentra is indistinguishable and even disappear as separate parts. Centrum of modern amphibians might represent enlarged hypocentrum (Figure 11.7)

11.6 Muscular System

After separation from parts of primitive mesothelial tissue, myotomes form cuboid hollow bodies on either facet of notochord and gave rise to vertebrae. They manufacture voluntary muscular structure of body. The conversion of epithelial walls into muscle should be given importance. In most vertebrate, cells lose their original form and arrangement as a cylindrical epithelium, and form elongated cylinders, axes of that are parallel to longitudinal axis of body. Each of those primitive muscle cells initially contains single nucleus; however by division many arise. The peripheral protoplasm of cell becomes differentiated into numbers of longitudinal fibrillae.

The epithelial cell therefore becomes a somatic cell. The lateral or outer wall of myotome doesn't participate during this muscle formation, however produced deeper layer of skin. Histogenesis of muscle in cyclostomes differs from that given above. Myotomes increase rapidly in their dorsoventral dimensions, and step by step push between lateral plate and ectoderm within the ventral half of body, therefore manufacturing the muscular structure of this region, whereas dorsally their extension is less marked. During this process, the myocommata conjointly participate. This primitive condition could be recognized in trunk region of a fish; however it becomes greatly changed in birds and mammals. In fish, the ensuing muscles of trunk and tail become divided into dorsal and ventral or epiaxial and hypaxial systems, the line of division between the 2 following closely the lateral line organ, and being marked by a partition of connective tissue.

In Amphibia, this epi- and hypaxial parts are visible in tail, however farther forward the hypaxial system is reduced. This reduction is carried to amniotes, wherever virtually the only real traces of hypaxial system are to be found, greatly changed, in girdle and neck regions. In

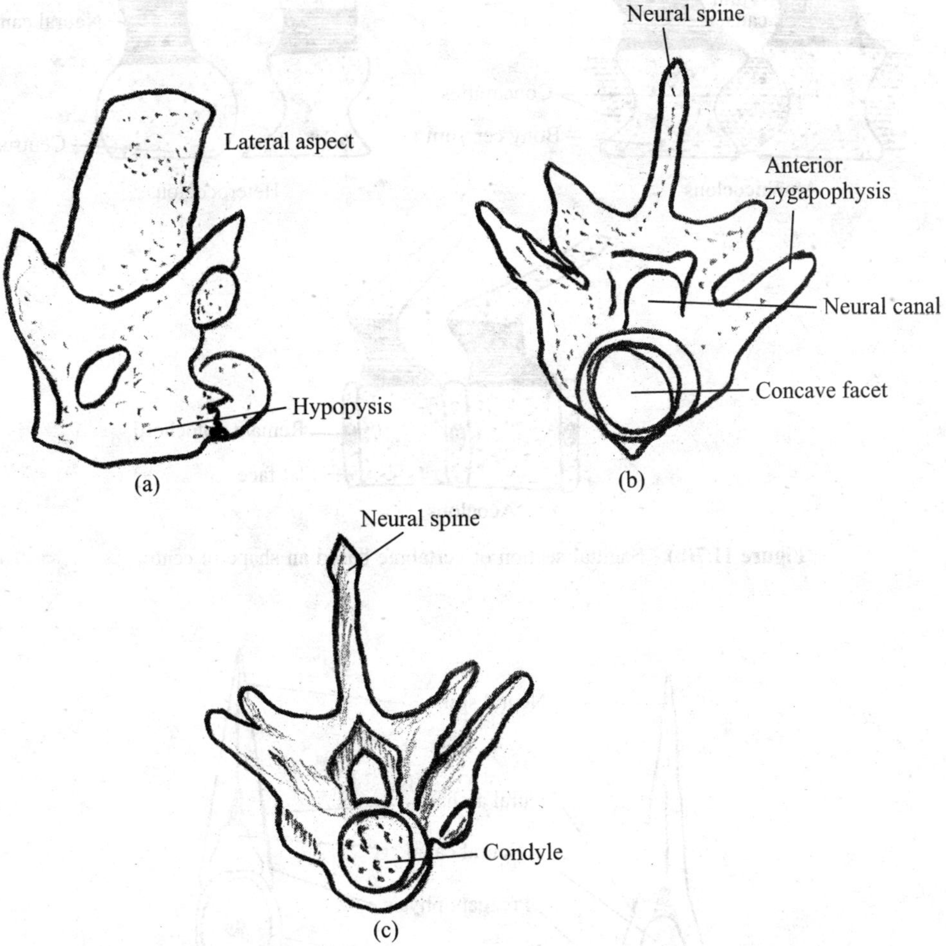

Lateral aspect

Hypopysis

(a)

Neural spine

Anterior zygapophysis

Neural canal

Concave facet

(b)

Neural spine

Condyle

(c)

Figure 11.7(a) Typical vertebrae of vertebrates.

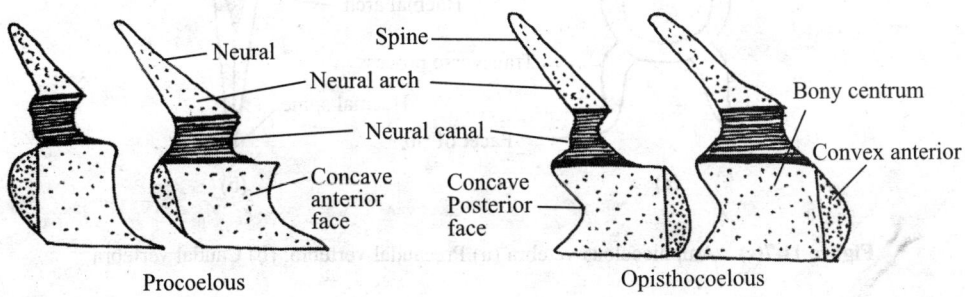

Neural

Spine

Neural arch

Neural canal

Concave anterior face

Concave Posterior face

Bony centrum

Convex anterior

Procoelous

Opisthocoelous

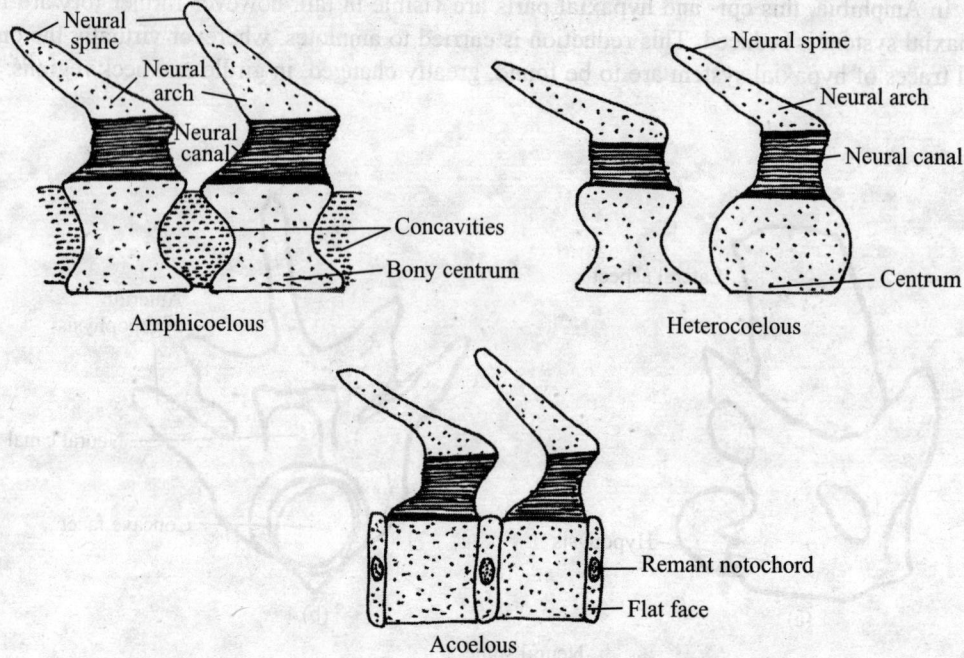

Figure 11.7(b)   Sagittal section of vertabrae based an shape of centra.

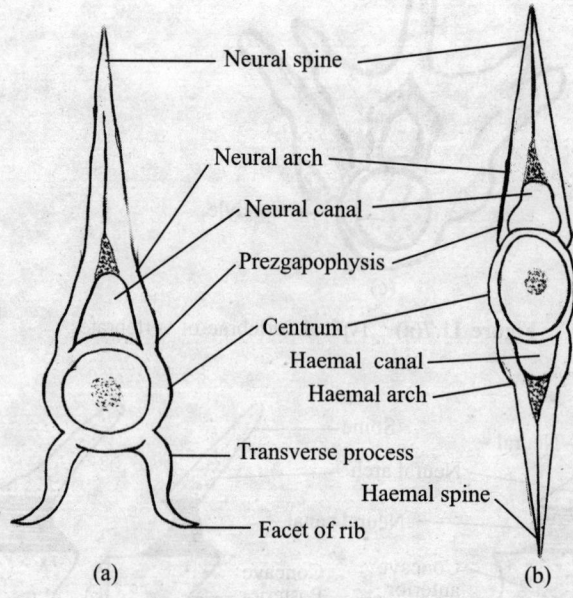

Figure 11.7(c)   Amphicoelous vetebra (a) Precaudal vertebra, (b) Caudal vertebra

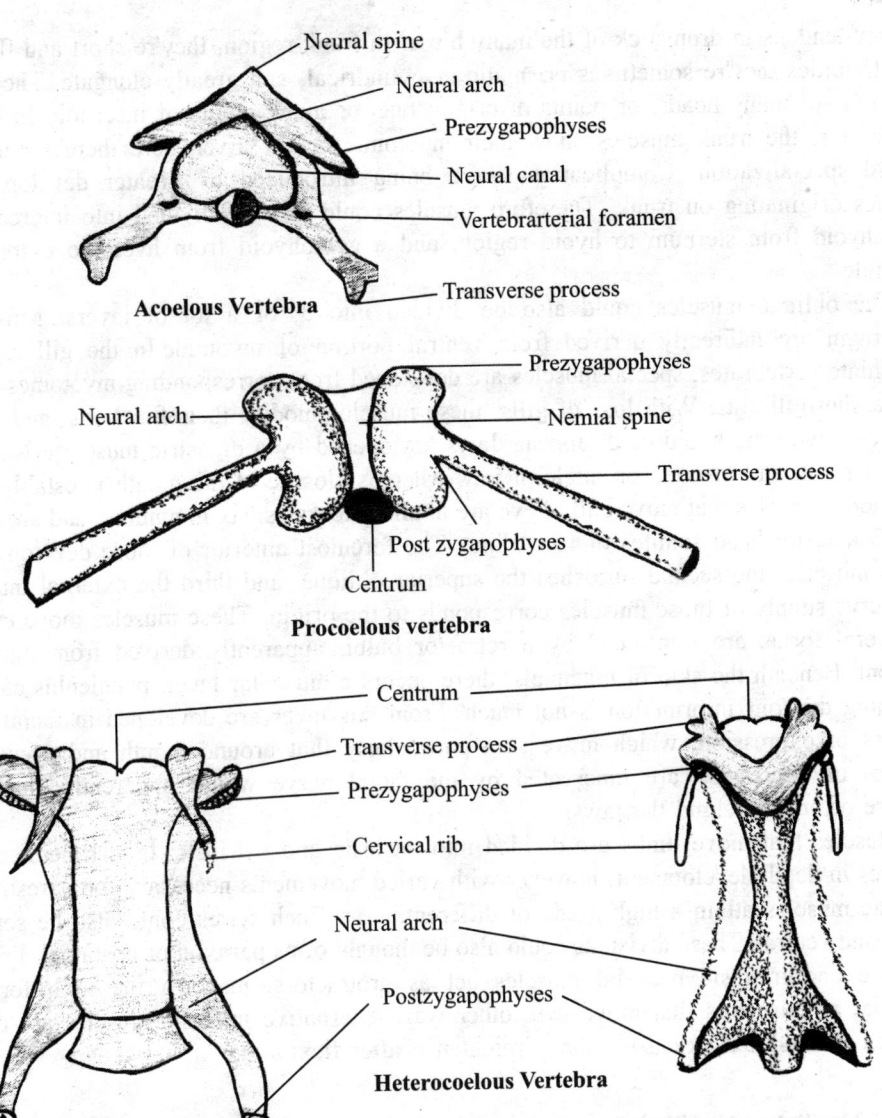

Figure 11.7(d)   Acoelous, Procoelous and heterocoelous.

fish-like forms, many somites shortly behind head bud from their lower surfaces cords of cells that extend out laterally, lose their distinctness, and form a common matrix out of which definitive muscles are later developed. In amniotes, somites intervene between the systems.

Perimysium which extends beyond contractile or true muscular portion forms the means for attachment of muscles to elements that are to be affected. Tendons might occur not solely at the ends, however middle of muscular tracts. Once forming broad, flat sheets, tendons are known as facia or aponeuroses. Often times ossification happens in tendons as in bony tendons in drumstick of the many birds. In trunk region, they're short and flattened; in extremities they're sometimes prismatic or cylindrical, and greatly elongate. They would have one, or many heads, or points of origin; one, or many points of insertion. In fish-like vertebrates, the trunk muscles show their myotomic origin. Even here there's a tendency toward specialization. Complications were being introduced by greater development of muscles originating on trunk. Therefore muscles could also be divided into intercostals, a sternohyoid from sternum to hyoid region, and a geniohyoid from hyoid to extremity of mandible.

The oblique muscles could also be divided into 3, or a lot of layers. Muscles of diaphragm are indirectly derived from ventral portion of myotome.In the gill region of branchiate vertebrates, special muscles are developed from corresponding myotomes to open and to shut gill slits. With loss of gills, these muscles modify their functions, and become connected with the hyoid or disappear. Jaws are opened by a digastric muscle arising from base of skull, and inserted on angle of jaw, whereas closure of the mouth is established by adductors. Muscles that move ball of eye are in all vertebrates, six in number, and are derived from 3 anterior head somites of van Wijhe. The foremost anterior of those develops into 3 rectus muscles; the second furnishes the superior oblique, and third the external musculus. The nerve supply of those muscles corresponds to the origin. These muscles move eye, and in several forms are reinforced by a retractor bulbi, apparently derived from third head segment. Beneath the skin of mammals, there occurs a muscular layer, panniculus carnosus, regarding that our information is not much. From this layer are developed in facial region muscles of expression, which move skin particularly that around mouth and eyes. These muscles of expression are innervated by the facial nerve would apparently show their purpose of origin behind the jaws.

Muscles that move limbs are divided into intrinsic and extrinsic .In fish neither series acquires in depth development; however with varied movements necessary for terrestrial life Intrinsic muscles attain a high grade of differentiation. Each series could also be sorted as dorsal and ventral. These divisions could also be thought of as paraxial or postaxial. Proximal extrinsic and intrinsic preaxial muscles act as protractors, maneuvering limb forwards, postaxial as retractors that move it in other way. Alternative intrinsic muscles are divided between flexors, and extensors, that straighten it after flexion.

11.7 Vertebrate Flight

Flight evolved 3 times in the 500 million years of vertebrate history. In invertebrates, flight has evolved once. Insects were the first animals to evolve flight. When an animal flaps its wings, then it is flying. Gliders and parachuters don't move wings, instead passively descend in the air. True flyers have control over their descent. They flap wings to produce thrust for added speed and lift for added height. True flyers move horizontally unlike gliders and parachuters. There are 4 levels of motion through air in vertebrate. These levels are as follows:

Parachuting: Parachuting is descent with an angle between the directions of falling, when the horizontal axis is more than 450. Better parachuters fall lowly, as they have more surface area relative to their weight and drag forces are higher, which slows down the object's descent. Parachuting focuses on maximizing drag forces to slow the object's fall. Plants use parachuting to disperse seeds. Humans use parachutes when jumping out of planes. Few vertebrates simply parachute as aerial locomotion. Most parachute in airborne condition, and get around using other type of locomotion.

Gliding: Gliding is an elaborate form of parachuting and occurs when angle of descent is less than 45 degrees and works by having a gliding airfoil design that generates lift forces, keeping the animal in air longer. Streamlining is important to gliders to reduce drag forces. Gliding is common in several reptiles, mammals, and even ray-finned fish. The earlier birds and their theropod dinosaur relatives show no evidence of gliding. Many lineages of gliding animals today include lizards, squirrels.

Soaring: Soaring animals appear to be only gliding because they don't flap their wings. Soaring requires specific physiological and morphological adaptations. Soaring uses the energy of surrounding air to keep animal at constant altitude and is analogous to "falling down an up escalator." Soaring animal do not flap wings except to take off, or to make adjustments in soaring. Only particularly large animals are efficient soarers. This form of flight has been achieved by animals in the course of evolution. The larger pterosaurs and some large birds are the examples of soaring animals.

Drag, Lift, and Thrust: Drag is a force exerted on an object moving through a fluid. It always orients in direction of relative fluid flow. Drag occurs when the fluid and object exchange momentum, creating force opposing the motion of object. Drag is high when surface area of object exposed to fluid flow is high, the object is moving faster, and fluid has more momentum, or inertia - this is low for air relative to other fluids. A dropped weight falls faster through air than through honey largely because of drag forces. Lift is another force exerted on an object moving through a fluid. It is generally directed upwards, opposing weight of animal that is pulling it down to earth.

Animals generating significant lift forces, the angle of wings against the flow of air creates the resistance that has net effect of moving the wing upward. Majority of lift in gliders and flyers is produced at proximal part (base) of wing, where wing area is largest. Lift is higher when the area of bottom of wing is larger. The animal is moving faster, and again, fluid viscosity and density are higher. Thrust is the third force present only in true fliers; produced by powered flight especially at the distal of wing. Thrust induced in direction of animal's flight, opposing the drag. To fly at a steady speed in a completely horizontal direction, a bird generates enough thrust to equal the drag on it. Thrust is produced by flapping wings, which creates a vortex wake that has the net effect of pushing the bird forward. Different kinds of wakes are formed in slow flight, fast flight, and bounding flight.

If the thrust force is greater than drag force, the animal accelerates; likewise the animal decelerates if the drag is greater than thrust, and when thrust force equals drag force, the bird moves at a constant speed. Thrust force depends on the power output of flight muscles. Drag forces should be minimized for fast flight; streamlining is a good way for this. Since drag increases rapidly with flight speed, drag always is limiting. Drag is very helpful, when a flying animal is trying to slow down or land; so in that case, animals spread out their wings. Lift is important to a flyer; to remain airborne; it should have forces holding it up. Body weight must be minimized. Flapping the wings quickly generates lift well, as does have a large wing area. Thrust is also a vital force for flying animals. Lacking the drag forces would slow the animal down enough to reduce lift forces. Large flight muscles are important to produce thrust.

11.8 Digestive System

In cyclostomes, alimentary tract shows slight differentiation into regions. The point of entrance of liver duct divides it into pre- and post-hepatic parts. In latter division, a small fold of internal surface forms a rudimentary spiral valve. In holacephali, some teleosts, and lower urodeles scarcely a lot of differentiation of digestive canal occurs. In other forms, digestive tract is split into regions. The pre-hepatic portion is differentiated into anterior slender tube, esophagus, and posterior widened portion with walls, the stomach. Length of esophagous correlates with neck. In birds, it becomes widened close to its middle into a glandular sac, the crop that is reservoir of food and in pigeons furnishes food for young.

Stomach shows various modifications. An enclosed fold along side well developed sphincter in its walls separates anterior portion of abdomen and forms pylorus. Opposite end of stomach lies near heart is called cardiac region. Stomach might lies parallel with body axis. In most fish, it is loop-like, or lie at right angles to that axis. In absence of teeth, stomach of bird becomes divided into 2 chambers, anterior glandular portion, proventriculus, and posterior muscular portion, the gizzard. In grain-eating birds, muscles of gizzard develop a disc on either facet, and inner surface is lined with a firm horny coat for grinding food. In mammals, line of division between stomach and oesophagous is sharp. Stomach, a straightforward sac or divided sac, lies parallel to body axis in seal. Elsewhere it's twisted into a transverse position.

In simple forms, the cardiac and pyloric regions are distinguished, and between them fundus region is characterized by distinction in glands lining the wall. Chambers correspond closely to those glandular regions. In ruminants, rumen, reticulum, omasum and abomsum are recognized. In Cetacea, diverticula in the pyloric region are found. Rumen and reticulum are not truly gastric but esophageal in nature, and function as digestive organ, and for storage of food. In lower forms, the liver duct opens close behind the pylorus. In higher forms, the pre-hepatic small intestine, might intervene between the 2. The post-hepatic portion is split into 2 regions, anterior midgut, the small intestine, and a posterior hind gut. In lower forms, this distinction is not sharp, being showed by character of internal walls, or by development of a cecal tube at the boundary between the 2.

From the Amphibia up wards, the line of division is sharp, an enclosed constriction, the ileocolic valve, forming the line of demarcation. Middle gut is the chief seat of intestinal absorption, and varied means are found for increasing the intestinal surface. In cyclostomes, invagination of inner wall follows a spiral course. In cartilaginous fish, this spiral valve is nicely developed, either growing out so the inside internal organ resembles a spiral stair. Spiral valve reappears in ganoids .In the higher fish, it is replaced by cecal tubes. Range of those varies from one in certain ganoids to over 150 in mackerel. In Amphibia and reptiles, mid-gut is straight and elongate. In birds, the middle gut bears a blind tube. In birds, intestinal surface is increased by lengthening of intestine and by development of minute finger-like projections.

Hind gut is hardly distinct in fish. From the Amphibia forwards, it acquires high individuality. It may consist simply of a straight tube, rectum, or might have terminal rectum connected with mid gut by a convoluted tube, the colon. Simply behind the ileo-colic valve in forms from turtles upwards a blind tube, the intestinal caecum, is clearly connected with increase of intestinal surface. In some birds, 2 caeca are found. In mammals, the cavity shows nice variations. It is lacking in some. In herbivores, it may equal the body length. In man, some apes and rodents all elements of caecum don't seem to be equally developed, the terminal portion known as the appendix vermiform is remain small than the remainder. In elasmobranchs, dipnoans, amphibians, sauropsida, and monotremes, the rectum doesn't open onto exterior, however into a terminal enlargement, the cloaca, into which the urinary and reproductive ducts empty; and from the cloaca contents pass to exterior through vent. In other vertebrates, cloaca is absent.

Liver: Liver develops as a diverticulum from ventral facet of primitive alimentary canal, branches repeatedly leading to a branched hollow gland, the proximal portion of tubes being specialized as ducts leading to intestine. In Amphibia and reptiles, this hollow condition is preserved throughout life, the minute lumen of glandular parts known as gall capillaries. In birds and mammals, the hollow condition shortly disappears, the gall capillaries running, without regularity between the cells. By the growth of connective tissue, the liver divides into lobules, the liver islands. In this connective tissue run the larger gall ducts, and branches of hepatic artery and portal vein. In centre of every island, there is a branch of hepatic vein , whereas capillaries extend through lobules from interlobular to interlobular blood vessels. There is a duct voidance from liver into intestine, and this connects with it by a lateral branch (cystic duct), a thin-walled gall-bladder. Once these conditions occur, the duct leading from liver up to the mouth of cystic duct is named the hepatic duct; from that point to intestines, the ductus choledochus. A separate hepatoenteric duct might lead directly from liver to intestine.

Liver could be a giant compact organ, largest in lower vertebrates, and bigger in flesh eating forms than in herbivores. In several fish, it forms one, undivided mass, in majority of vertebrates 2 lobes are present, and these successively could also be lobulated. Blood-vessels enter the liver in close relation to gall ducts, whereas veins departure it is widely separate. Liver is supported by a mesentery that connects it to ventral wall of alimentary canal, and that is commonly continued below as suspensory ligament of liver.

Pancreas: It develops as outgrowth from entodermal walls simply behind the liver outgrowth. Pancreas has recently been found to occur in many vertebrates wherever its existence was once denied. In some teleosts, it as a fragile tube lies within the peritoneum and its position in dipnoi simply outside the muscular walls of alimentary canal, caused it to be unnoticed for a long time. In elasmobranchs and other teleosts, it is a marked gland. In other forms, it is complicated. In ganoids, it arises by 2 dorsal and 2 ventral outgrowths; in Amphibia and in all higher forms, from 1 dorsal and 2 ventral outpushings, these later uniting into 1 organ mass. Ducts show varied modifications, all persisting, or either dorsal or ventral disappearing; or finally the ducts might acquire connection to those leading from liver. Two other structures, the spleen and bladder are better represented in connection to circulatory and excretory structures (Figure 11.8).

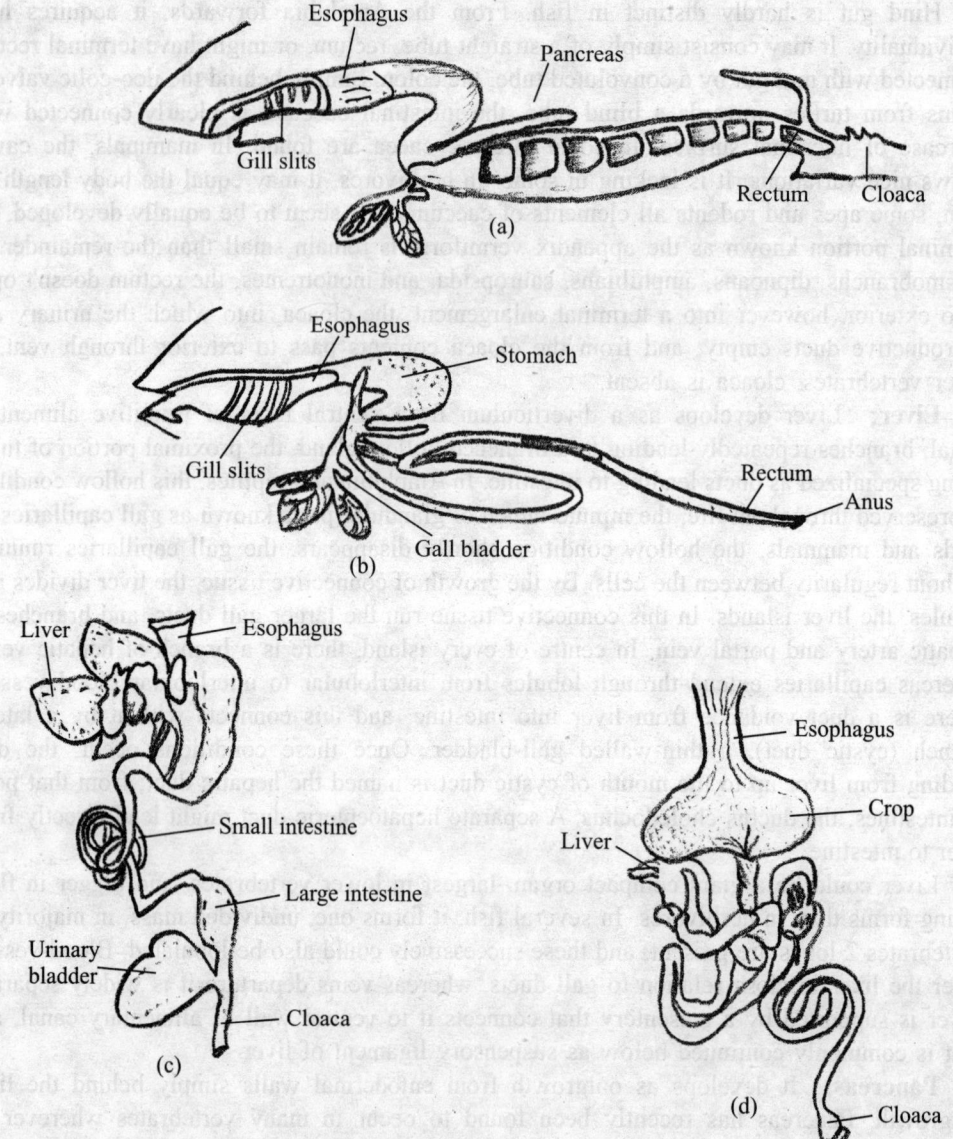

Figure 11.8 Diagestive System in (a) Elasmobranch, (b) Teleost, (c) Amphibia, and (d) Bird.

11.9 Evolution of Heart

Heart shows gradual complexity and specialization in structrure and function from lower chordates to mammals. Such complexity lies in increase in number of heart chambers and evolution of double circuit heart from single circuit one.

Single Circuit Heart: In primitive chordates, a true heart is absent. A part of ventral aorta below pharynx dilates and become muscular, single-chambered heart. Modifications which occur in this heart to become 2 -chambered heart of lower vertebrates are constrictions appear in cardiac tube to form chamber, each chamber tends to divide into separate chambers due to appearance of partitions, and heart shifts to thoracic cavity.

Two Chambered Heart: In cyclostomes and fishes (except dipnoi), heart is 2 chambered and S- shaped. Two chambers are an auricle and a vertricle. Two accessory chambers are a sinus venosus, and a conus arteriorsus. Deoxygenated blood enters heart through sinus venosus, then in auricle, ventricle and exit through conus arteriosus. Blood is carried to gills by ventral aorta for oxygenation. As blood goes through heart once, so it is called as single circulation and the heart as a single circuit venous heart. Due to shift from aquatic or branchial respiation to terrestrial or pulmonary respiration, heart gets modified to pump one stream of oxygenated blood and another independent stream of deoxygenated changes.

Transitional 3-Chambered Heart: Hearts of Dipnoi and amphibia are similar. Atrium is partly divided (dipnoans and urodels), or completely divided (anurans). Pulmonary oxygenated blood from lungs enter left chamber, deoxygenated blood from body enters sinus venosus which joins right chamber. Ventricle is partly divided in dipnoan, but single sac in amphibian. Conus is large and partly divided by spiral valve to prevent mixing of blood. The urodels, being less dependent in pulmonary circulation, has less effective doubt circuit pump. Reptiles have nearly complete double circulation. Sinus is vestigial except in turtles and attached to right atrium. Atrium is completely divided. Ventricle is partly divided due to incomplete interventricular septum whereas in crocodiles, ventricles are completely divided. Conus has been divided into 3 arches, a pulmonary arch leading to lungs and right and left systemic arches leading to whole body. Some amount of blood is mixed through foramen of panizzae.

Four Chambered Heart: In crocodiles, the foramen of Panizzae is absent. Birds and mammals have 4 - chambered heart and 2 complete separate circuits, one is a low pressure pulmonary circuit form right side of heart and a high pressure systemic circuit from left side of heart. Sinus venosus is vestigial in birds and absent is mammals. Atrium is divided and relatively small. Ventricle is divided by an oblique septum and is very strong as left side. Embryonic conus is divided into a pulmonary arch which joins the right ventricle and a systemic trunk joining the left. Systemic arch is single, right in brids and left in mammals. Despite similarities hearts of mammals and birds evolved double circulation independently (Figure 11.9).

11.10 Kidney

Vertebrate kidney is a pair of compact organs lying dorsal to coelom in trunk region, one on either side of dorsal aorta. Two sets of anatomical terms have been applied to kidneys. Typical mammalian metanephros is retroperitoneal, compact, bean shaped organ attached to dorsal body wall. Ureter leaves medial side at a depression called hilum, from where a renal vein leaves and a renal artery and nerves enter. Metanephros is surrounded by capsule of connective tissue under which lies cortex. Renal corpuscles and convoluted portions of secretory tubules are confined to cortical region. Immediately beneath cortex is medulla, with which it is immediately connected. It is made up of large areas called renal pyramids,

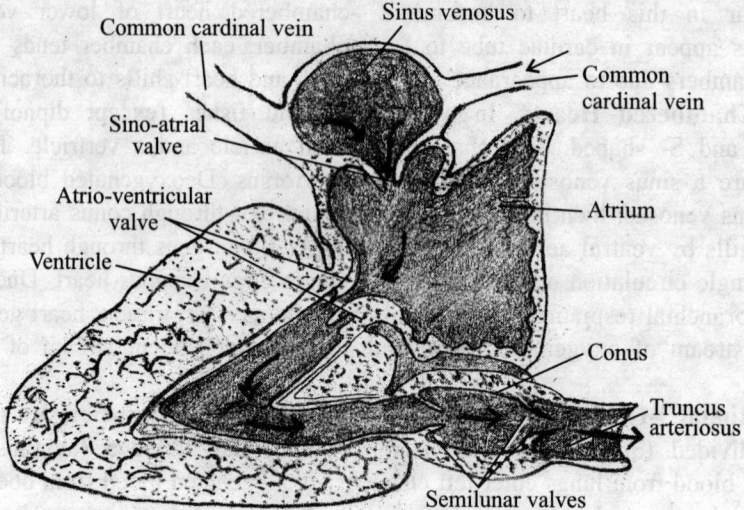

Figure 11.9(a)   Median longitudinal section of primitive fish heart.

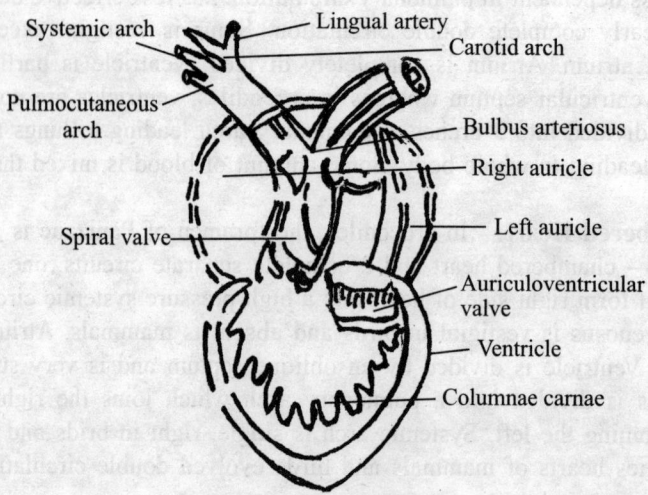

Figure 11.9(b)   Internal structure of heart of an aruran

between which extend renal columns of Bestin made up of cortical tissue. Each pyramid of medullary tissue is covered with a cap of cortical substance, the 2 constituting a lobe of kidney. Lobes are divided into lobules which are most evident at outer borders of kidney. Kidneys of rabbits, rats and others are called unilobar kidney since only a single pyramid with cortical cap is present. Multilobular kidneys of man are composed of from 6 to 18 lobes, each actually equivalent to an unilobar kidney. In center of medullary ray is a branched collecting tubule with which distal convoluted portions of numerous secretary tubules connect and into which they drain. Renal pyramid is striated in appearance with straight collecting tubules and loops of Henle in center of metanephros. It contains blood vessels,

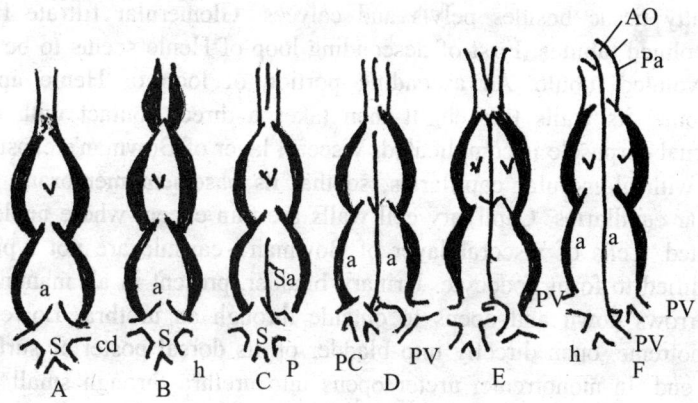

Figure 11.9(c) Different stages in the differentiation of heart. A : Elasmobrarch B : Telosts, C : Amphibia,
D : Lower reptius, E : Alligator, F : Birds and mammals, a : atrium, ao : aorta, b : conus,
pa : pilmanary artery, pc : pre and post caval vein, Pv : Pulmanary vein, S : sinus venarm,
SE : inter-auriculum septum, V : ventricles

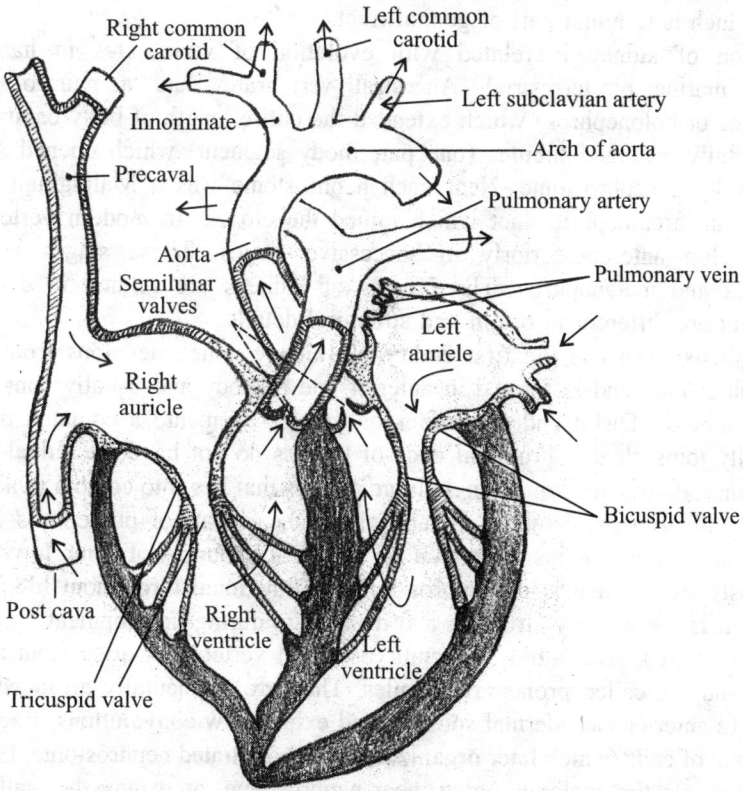

Figure 11.9(d)   Mammalian heart along with the associated valves and vessels.

nerves and fatty tissue besides pelvis and calyces. Glomerular filtrate first passes into proximal convoluted tubules. First of descending loop of Henle seems to be an extension of proximal convoluted tubule. As ascending portion of loop of Henle approaches distal convoluted tubule, its walls thicken. It then takes a direct contact with renal corpuscle. Structure of renal corpuscle is complicated. Visceral layer of Bowman's capsule is folded and interdigitated with glomerular capillaries, so that its basement membrane remains contact with glomerular capillaries. Capillary cell walls are thin except where nuclei of endothelial cells are located. Cells of visceral layer of Bowman's capsule are not typically squamous cells, but modified to form podocyte. Urinary bladder, present in all mammals, is muscular sac which narrows down and opens to outside through an urethra. Lower end of ureter, except in monotreme, open directly into bladder on its dorsal posterior surface and usually near urethral end. In monotreme, ureter opens into urethra through small papilla opposite base of the addere. Bladder musculature, as bundles of muscle fibers, or fascicles, continues down into urethra. Urine flow ceases when urethra is lengthened and diameter of its lumen is reduced. The entire urethra in female and prostatic portion of urethra of male control the maturation. In males, urethra passes through penis and opens at tip through external urethral orifice or meatus. In females condition varies; in some, as rat and mouse, urethra opens independently to outside, passing through clitoris; in others it enters urogenital sinus or vestibule, which is terminal part of genital tract.

Evolution of kidney is related with evolution of vertebrates in habitats such as freshwater, marine or terrestrial. Ancestral vertebrates had a pair of long kidneys (archinephros or holonephros) which extended the entire length of body cavity. Each kidney had segmentally arranged tubules (one pair /body segment) which opened separately into body cavity by a nephrostome. Near each nephrostome was a Malpighian body. Tubules opened into an archinephric duct which joined the cloaca. In modern vertebrates, kidney tubules develop anteroposteriorly in successive stages. These stages are pronephros, mesonephros and metanephros. Three types of kidneys are related in development and structure, but are different in origin and structural details.

Pronephros: This is the first embryonic kidney which develops from anterior most part of nephrostome and is formed in anterior end of body and usually consists of 1 to 12 uriniferous tubules. Distal end of uriniferous tubules open into a common pronephric duct which finally joins cloaca. Proximal ends of tubules do not have individual glomeruli, but several glomeruli unite to form a single large golmus that lies into coelom in neighborhood of nephrostome of pronephric tubule. In all vertebrates, a pair of pronephros appears during embryonic stages but remains functional for a time in embryos of some lower vertebrae. In many teleosts and hagfishes, pronephros remains functional throughout life, while in other vertebrates it is a transitory structure and disappears during development.

The very first kidney tubules in embryos of all vertebrates arise from anterior end of mesoderm and are called pronephric tubules. They are segmentally arranged, one opposite each of more anterior mesodermal somites, and exhibit few convolutions. Each tubule arises as a solid bud of cells, which later organize lumen and ciliated nephrostome. Glomerulus may be suspended into the coelomic cavity near nephrostome, or it may the wall of the tubule. Sometimes, 2 glomeruli develop for each tubule, one being suspended in coelom, ether lying in wall of tubule. Glomeruli are small knots of interstitial capillaries, each surrounded by a double walled structure called Bowman's capsule, 2 together being known as renal or Malphigian capsule .The outer walls of Bowman's capsule is parietal layer, the inner wall is visceral layer. Blood is brought to internal glomerulus by an afferent arteriole and exit through an efferent arteriole .Number of pronephric tubules is not large (13 in lampreys, 4 in sharks, 7 in 3mm human embryo at 3rd week of embryonic life). In few anamniote vertebrates, pronephros persists in adult stage and is called head kidney. Rest of the kidney, posterior to pronephric region is called the opisthonephros. Pronephric kidneys function in adults only among hagfishes. They function in immature fish and in larval amphibians until the functioning of caudal tubules. Then pronephric portion of kidney starts to disappear. Pronephric tubules develop in amniotes at an early stage but disappear before attaining a functional state (Figure 11.10).

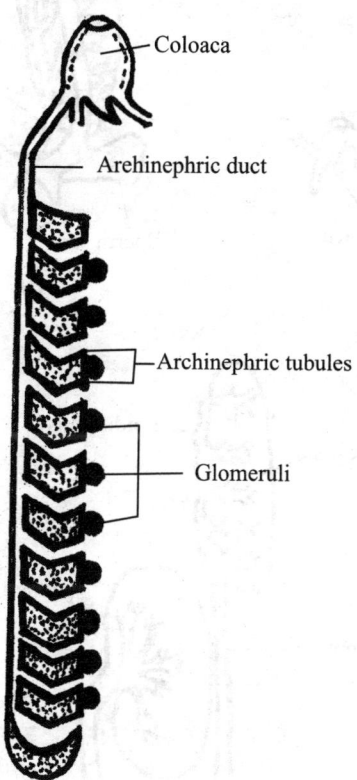

Figure 11.10 Structure of Archinephros.

Mesonphros: It arises from that part of nephrostome which lies behind pronephros. In beginning, it consists of segmentally arranged uriniferous tubules but later segmental arrangement is lost due to branching of tubules in each segment. Newly formed secondary tubules unite with those of primary tubules and form collecting tubules. Secondary tubules acquire their own Malpighian body. Unlike pronephric tubules, mesonephric tubules acquire independent glomeruli. Mesonephric tubules open in original pronephric duct, called the

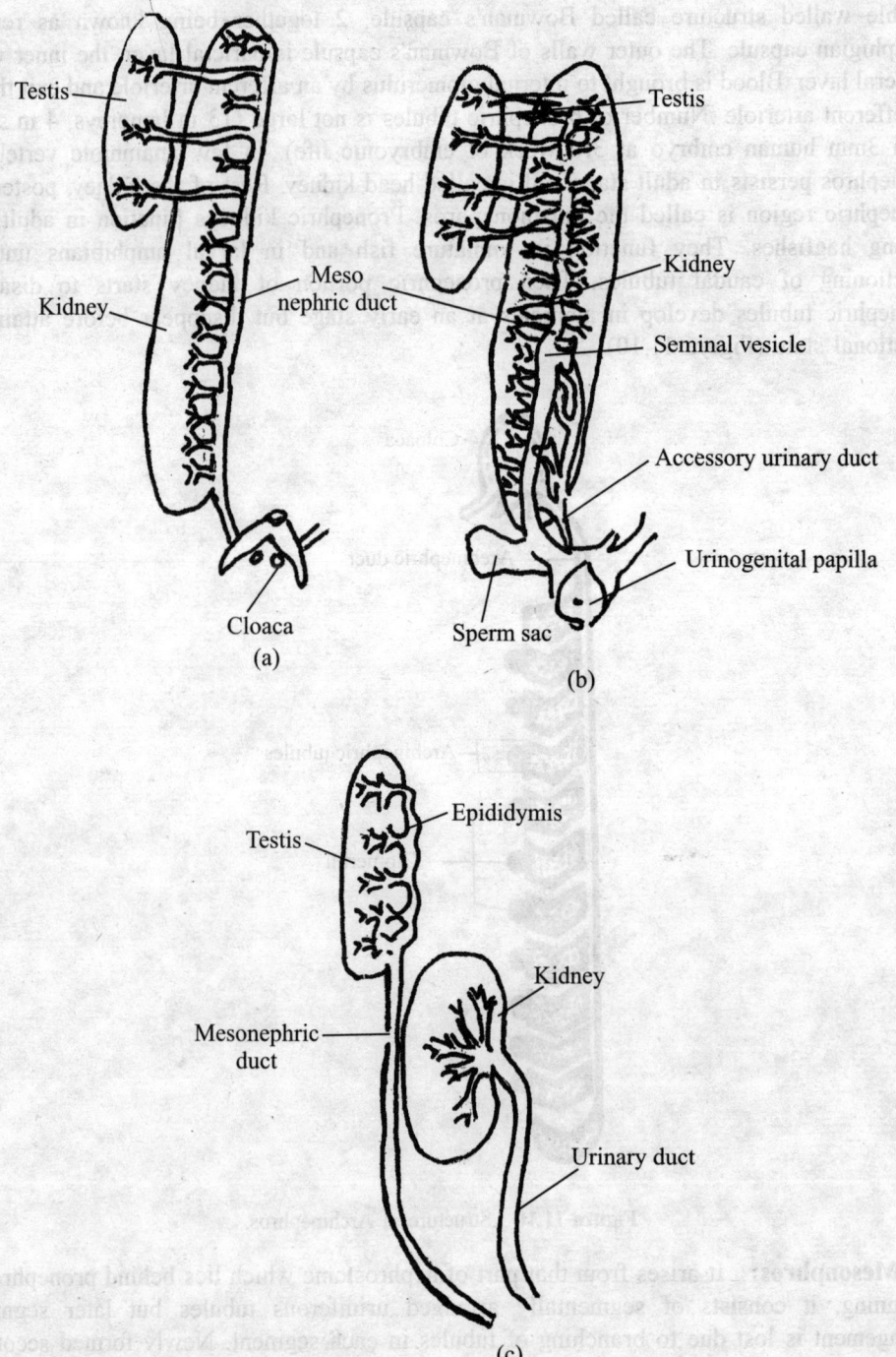

Figure 11.11 (a) Basic plan, (b) Dog fish, and (c) Mesonephric duct in amniotes

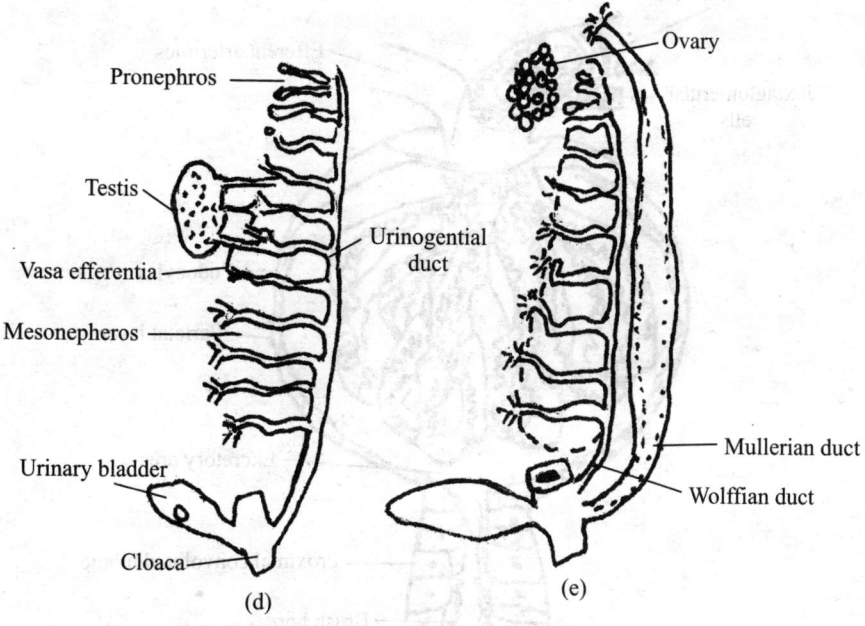

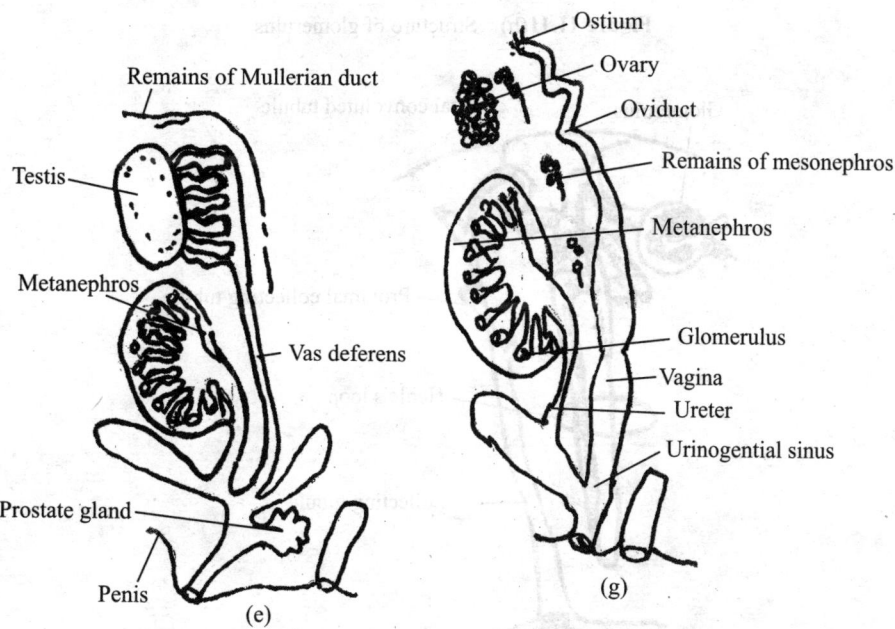

Figure 11.11 Urinogerital orgarn in mammal (d) Male, and (e) Female.

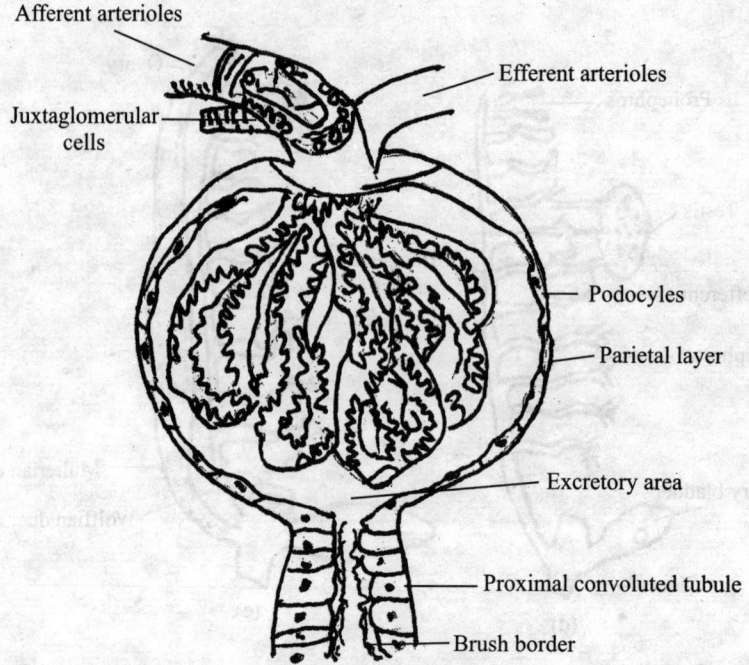

Afferent arterioles

Efferent arterioles

Juxtaglomerular cells

Podocyles

Parietal layer

Excretory area

Proximal convoluted tubule

Brush border

Figure 11.11(h) Structure of glomerulus.

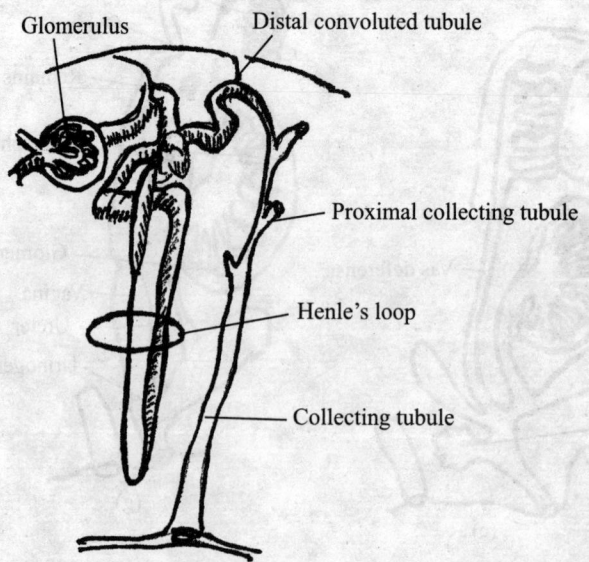

Glomerulus

Distal convoluted tubule

Proximal collecting tubule

Henle's loop

Collecting tubule

Figure 11.11(i) Structure of Henle's loop.

mesonephric duct. In elasmobranches and in some amphibians mesonephros develops. Pronephric duct divides longitudinally from its hinder end as far forward as anterior end of mesonephros. Of these 2 ducts, one remains connected with tubules of mesonephros and forms its excretory canal called Wolffian duct; other is called Mullerian duct which in females forms the oviduct. In amniotes, pronephric duct does not divide and remains solely as Wolffian duct. Oviducts in such case arise differently. Mesonephros forms the adult functional kidney of some cyclostomes, fishes and amphibians. It is also functional embryonic kidney of reptiles, birds and mammal .Mesonephros of adult is also sometimes called opishonephros. Mesonephros organizes caudal to pronephros and becomes a bulky elongated organ lying against dorsal body wall just behind peritoneum and often bulging into coelom. In shark and many fish, it extends a considerable distance along trunk and may even pass into base of tail since some of more anterior mesonephric tubules in males become modified to connect with testis and transport sperm to mesonephric duct. Cephalic end of mesonephros in male fish and urodeles may extend further forward than females .Mesonephric tubules are numerous than pronephric ones and more complex. Nephrostomes may persist throughout life in fishes and amphibians, especially anteriorly. They are prominent in embryonic mesonephric kidney in reptiles. In birds and mammals, they appear transitorily or not at all. External glomeruli are of no use to tubules lacking a nephrostome. An internal glomerulus with its associated Bowman's capsule is called Malphigian corpuscle. Mesonephric tubule emerges from corpuscle exhibits characteristic convolutions, which vary with the species. Distally the tubule opens into mesonephric duct or unites with other tubules to form accessory mesonephric ducts.

Metanephros: They develop only in amniotes. Embryonic mesonephros is replaced in adults by metanephros. Each metanephros develops from posterior part of nephrotome behind mesonephros near cloaca and from wolffian ducts. A tubular outgrowth arises which grows forward parallel to parent duct into nephrotome. In nephrostome, ureter gives off varying number of branches which from collecting tubules and calyces. Nephrogenic tissue produces numerous metanephrphic tubules. Each uriniferous tubule develops Bowman's capsules into which penetrate a glomerulus tubules which lack peritoneal funnels so that all connection with the coelom is lost. In mesonephros segmental arrangement of uriniferous tubules is lost. Metanephros is the adult functional kidney of amniotes. In some reptiles (lizards) and *Echidna*, mesonephros is functional for a short period after birth but is replaced by mesonephros. The part of mesoderm that differentiates to form kidney and gonad tissue is called the nephrotome (Figure 11.11).

Pro, Meso-, and Metnephric duct: Each pronephric tubule elongates as development progresses and growing distal tip turns caudal to connect with tip of next tubule. Longitudinal duct is thus formed which finally establishes a connection with cloaca. When mesonephric tubules organize caudal to pronephros, they also establish a connection with pronephric duct. Although pronephric tubules soon disappear, pronephric duct continues to serve mesonephric kidney. Tubules of metanephros develop from mesoderm immediately caudal to mesonephric portion. Tubules are homologous with pronephric and mesonephric tubules, but are more complex with convolution.

Table 11.1　Types of kidneys of vertebrates

Ontogenic classification	Phylogenetic classification
Pronephros	*Holonephros*
Develops in anterior most part of nephrogenic tissue, drained by archinephric duct	Hypothetical structure. Single tubule in each segment in the whole length of nephrogenic tissue.
Mesonephros	*Opisthonephros*
Develops in middle port of the nephrogenic tissue, drained by archinephric duct.	Develops at start in the middle and later in the posterior part of nephrogenic tissue.

Metanephros

Develops in the posterior part of nephrogenic tissue and drained by the uretes.

Tubular Portion

Proximal Convoluted Tubule: This is the longest and most convoluted segment which forms a major portion of the cortical substance. It has a length of about 14 mm and a diameter of 57 to 60 M. The tubule is lined by single layer of low columnar or pyramidal cells with round nuclei and granular cytoplasm. Brush border is present at the apical surface of cell. The border is composed of microvilli about 1.2 M lengths and about 0.03 M in width, and has a high concentration of alkaline phoshaptase and oxidative enzymes. The border also contains mucopolysaccharides. Invagination is often seen between the bases of microvilli and vesicles are present in apical cytoplasm. The sub nuclear portion of cell appears striated as a result of involution of plasmalemma and arrangement of mitochondria. The plasmalemma infoldings partially subdivide cell into compartments. The mitochondria are numerous, rod-shaped and oriented within the compartments parallel to long axis of cell.

Straight or Medullary Portion of the Proximal Tubule: It is lined by cells that appear structurally similar with those of convoluted portion. This portion forms the proximal portion of descending arm of Henle's loop. This loop is composed of a proximal thick segment, a thin segment and a distal thick segment.

Thin Segment of Henle's Loop: This is about 15 M in diameter and is lined by a single layer of flattened epithelial cells with nuclei that bulge into lumen. The extent of thin segment and the length of entire loop vary in different tubules. The tubules whose malpighian corpuscles lie near the junction of cortex and medulla have long loops and their thin segments entered nearly to apex of medullary pyramid. In them, the thin segment forms the entire crest of loop and continues a considerable distance up the ascending limb. Tubules associated with glomeruli of outer part of cortex have relatively short loops, and their thin segments are limited to small part of descending arm. There are many intermediate types between the 2 extremes described

Thick Distal Ascending Segment of Henle's Loop: This is about 9mm long and 30 M in diameter. It ascends to cortex and closely approaches the vascular pole of glomerulus form, which the nephron began. It is lined by narrower cuboidal cells. No brush border is observable, although few short of microvilli can be seen in E.M. E.M shows numerous infoldings of cell membrane and complex interdigitations along the lower portions of lateral borders of cells.

Distal Convoluted Tubule: It is less convoluted than proximal tubule with only 5 mm in length. It has an irregular outline with diameter between 22 and 50 M. It is lined by cuboidal cells containing a granular cytoplasm. No brush border is observable. E.M. shows fewer mitochondria and less basal infoldings in cells. In the region where the distal ascending segment becomes continuous with distal convoluted portion, the cells on side of tubule and adjacent to afferent arteriole a portion of efferent arteriole are taller and slender then elsewhere in distal tubules The region appears darker under light microscope. Hence it is named the macula densa. A thin basement membrane is only structure separating macula densa cells from the Juxtaglomerular cells of afferent arteriole.

Collecting Tubules: The straight, tubules receive several arched tubules (7 to 10) as they pass down through medullary rays of cortex, but receive no branches in outer zone of medulla. In inner zone of medulla, they unite with straight tubules and after several fusions ducts of Bellini are formed. The diameter of collecting tubule while it is still in medullary ray is about 40 M. The lining of collecting tubules consists of a single layer of cuboidal cells with round, darkly staining nuclei and clear faintly staining cytoplasm (Figure 11.12).

Counter-Current Mechanism: The unique feature of mammalian nephron is the loop of Henle. The descending and ascending limbs of loop of Henle lie parallel to each other, so that direction of flow of fluid in one is opposite to that in other. Active transport of chloride ions out of ascending limbs into interstial fluid, and their passive diffusion back into descending limb result in counter-current multiplying mechanism. Sodium and chloride ions move out of ascending limb go right back into descending limb. Recycling of these ions in loop of Henle, together with additional sodium and chloride ions being continually brought to loop in glomerular filtrate, results in an accumulation of sodium and chloride in loop of Henle and surrounding interstitial fluid of medulla. Degree of accumulation depends on the length of loop. Capillary loops associated with juxatamedullary nephrons provide a counter-current mechanism that permits sodium and chloride to diffuse back into blood entering medulla. This combined with sluggish rate of blood flow through these vascular loops means the little sodium and chloride is carried away from medulla by this route.

The osmotic gradient in such mechanisms makes extra water to be reabsorbed passively and for reduction of hyper osmotic urine. Water simply follows the osmotic gradient, moving from an area of low osmotic pressure to one of high osmotic pressure. The glomerular filtrate, which was isosmotic to the blood in proximal tubule, loses water as it passes down into loop of Henle. It does not regain this water as it ascends loop of Henle, for cells of ascending limb of loop have a low permeability to water. The filtrate becomes more dilute because of large amount of sodium and chloride pumped out. By the time the filtrate reaches the distal tubule it is again isosmotic, or sometimes hypoosmotic, to the blood. The filtrate now descends through medulla again, this time in the collecting tubule. It passes as a counter-current to the adjacent gradient of sodium chloride in filtrate of ascending loop of Henle and loses water to surrounding interstitial spaces. The filtrate now becomes very hype osmotic.

11.11 Nervous System

Central nervous system develops as a distinct structure from rest of ectoderm by formation of a neural or medullary plate on dorsal surface of embryo by infolding of a part of

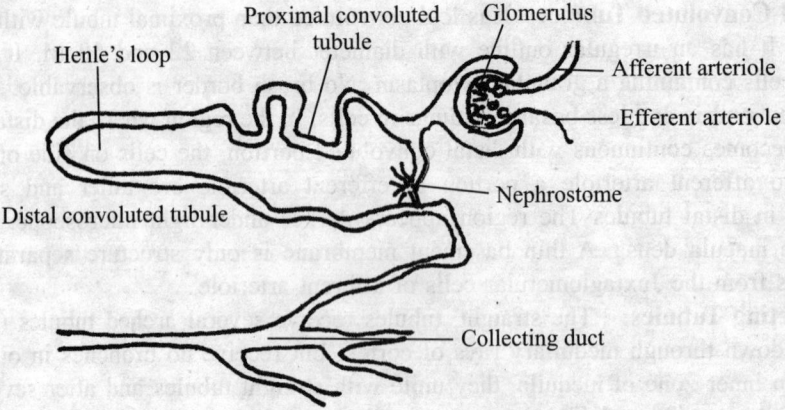

Figure 11.12(a)   Structure of an excretory tubule.

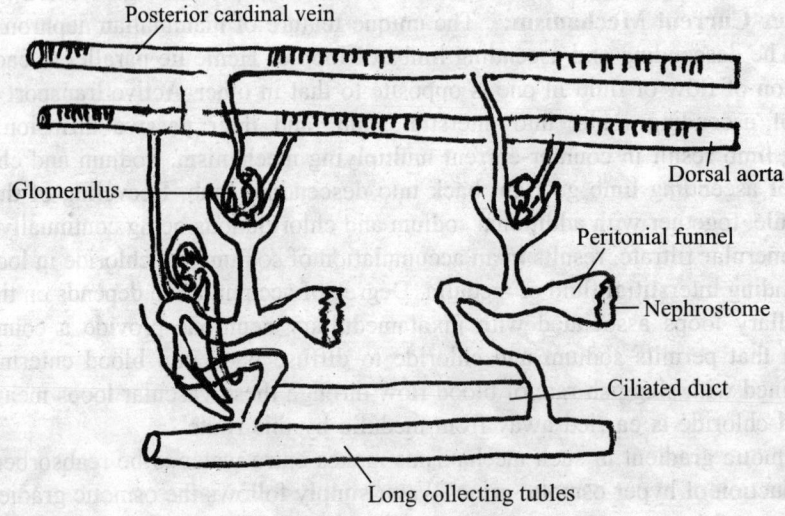

Figure 11.12(b)   Mesonephric tubules.

primitively external body surface. The inner surface of neural tube is morphologically external in origin. Before completion of infolding of neural tube, its anterior end enlarges for differentiation into brain and spinal cord. Spinal cord often retains its tubular character throughout life, although central canal does not increase its primitive diameter. In the earlier stages the cord is oval in section with thickened sides; in median line, above and below, it is thinner. These halves rapidly increase in size while central median portion lags behind, which result in a longitudinal groove along its ventral surface. Later a cleft appears on dorsal surface. These are the anterior and posterior fissures of human anatomy.

In sections of adult cord, an outer white matter and an inner gray substance is found; the latter takes the shape of the letter H, the ends of uprights being called the horns or cornua, the cross-bar is produced by fibres running from one half to other between bottoms of fissures and central canal. Horns extend towards the surface above and below dividing the white matter of each half of cord into three columns, dorsal, lateral, and ventral. The lateral column remains between 2 horns, the dorsal and ventral between the horns. In some forms, the development pattern varies, but the result is same. During the increase in size of cord, the cells produce protoplasmic outgrowths, some cells run forwards and backwards in ventral and lateral columns, while others pass outwards from the cord.

Dorsal Root: The white matter is composed of nerve fibres, the gray matter of nerve cells. Later the dendrites are formed. With increase in size, blood vessels and supporting tissue press into the cord. In its early stages, the spinal cord shows segmentation which causes alternate expansions and contractions of cord and its canal. Early segmentation disappears with growth. The same segments are later showed by roots of spinal nerves which are paired structures passing off from either side of the cord, each nerve arising by 2 roots, one dorsal, and the other ventral. Roots differ in structure and function. Dorsal roots are connected with dorsal horn of gray matter, and soon after leaving the cord each enlarges into a ganglion composed of cells to produce fibres.

Ventral roots are not ganglionated, but their fibres are connected with ganglion cells of gray matter of ventral horns of cord, from which they pass out into root. Beyond the ganglion of dorsal root, 2 roots of a spinal nerve unite, and fibres of each follow a common course. Dorsal roots are sensory and carry impulses from terminal sensory structures to CNS. Ventral roots are motor in function and the nervous impulses they transmit come from CNS and are carried to peripheral portions. Dorsal roots are often called afferent roots, while ventral roots are termed efferent. These roots differ in their mode of development. During closure of neural tube a thin sheet of cells is visible on both side of line of closure between the epidermis and tube. By unequal growth, this sheet of cells, or neural crest, convert into segments. Each segment develops into a ganglion of a dorsal root.

Apparently fibres grow out from this ganglion to enter cord, while others grow peripherally to connect with sense organs. No such crest is formed for ventral root; but fibres forming this are connected with ganglion cells of cord, and they lengthen with growth of animal to connect with muscles, and glands .Each spinal nerve divides into 2 branches, a ramus dorsalis supplying the dorsal region, and a ramus ventralis distributed on sides and ventral surface. This latter also gives off a ramus intestinalis to viscera. These latter connect with sympathetic system, a pair of longitudinal nerve cords with ganglia lying near junction of mesentery with dorsal wall of coelom. This system supplies digestive tract, vascular system, and glands, and in certain ichthyopsida it extend into head.

Typically the spinal nerves follow the septa between the muscle plates. Visceral and somatic motor fibres; visceral and somatic sensory tracts are recognized. Spinal cord enlarges where spinal nerves forming these plexuses are given off. In early stages spinal cord is since the region of body supplied by it, but with increase in size other tissues grow faster than cord. Brain is an enlarged and complicated anterior end of CNS. Soon after closure of neural tube the region which is to form the brain becomes differentiated into 3 hollow enlargements, which receive the names according to position.

A brachial plexus is formed in snakes and footless lizards. None is found in caecilians. Siren lacks sacral plexus and hind brains. These 3 regions are not exactly comparable to segments of spinal cord. Each of these vesicles contains an enlarged portion called primary ventricle, which in cord is called central canal. Soon the 3 vesicles differentiate by unequal growth into 5 regions; fore and hind brain each giving rise to 2, the mid brain remaining unaltered. From the fore brain arise prosencephalon and thalamencephalon. The extreme tip of fore brain in median plane remains stationary and forms a thin membrane, later called lamina terminalis. On both side of this fore brain grows outwards, and especially forwards, thus producing two lobes, right and left. These are cerebral hemispheres, while rest of primitive fore brain is thalamencephalon. Ventricle of primitive fore brain participates in this outgrowth, producing a cavity in each lobe; so that 3 ventricles in fore-brain region, the 1st and 2nd forming a pair, while the 3rd unpaired ventricle remains in thalamencephalon are produced. The paired ventricles remain in connection with 3rd by small opening, the foramina of Monro.

While this differentiation occurs in fore brain the mid brain remains almost stationary. The main change is the thickening of its walls so that primitive ventricle becomes a narrow tube. The iter or aqueduct connects the 3rd ventricle with ventricle in hind brain. In hind brains differentiation is largely confined to dorsal surface. It consists in outgrowth from anterior dorsal wall of a lobe which extends backward over the rest, and forms the cerebellum or metencephalon, the rest of hind brain forming medulla oblongata or myelencephalon, which passes into spinal cord behind. Different regions of these 5 divisions variously develops, the walls being thickened in parts. In cerebral hemispheres the ventral surface develops a large ganglionic mass, the corpus striatum, in either hemisphere. The rest of cerebral wall called the pallium undergoes modifications in different groups. In some fishes it is epithelial in nature. In other vertebrates it is nervous in nature, its outer surface being composed of ganglion cells. In all lower vertebrates the surface of cerebrum is smooth.

Even in mammals the septum pellucidum is epithelial in character. Fissures appear in its surface, separate convolutions or gyri which produces an increase in surface and cortical substance in correlation with intelligence. From its anterior ventral region each hemisphere gives off an olfactory lobe into which a ventricle may extend. Each olfactory lobe is connected with an olfactory ganglion which may be placed either in cerebrum itself, or may be carried out towards the end of olfactory lobe. In diencephalon the lateral walls become thickened into large tracts, the thalami, while dorsal wall retains its epithelial character, becoming variously folded to form anterior choroid plexus, which carries blood vessels into 3 anterior ventricles.

From the dorsal surface 3 structures, pinealis, epiphysis, and paraphysis are developed. From backwards and downwards, hypophysis or pituitary body develops from tissue derived directly from ectoderm. This ectodermal portion retains its connection with the parent layer by a cord of cells for a time, the hypophysial duct, which later disappears. Infundibulum may represent the invertebrate mouth. The ectodermal portion of hypophysis is a modified pair of sense organs. Mesencephalon has thickened walls to contain ventricle. In the higher groups, this is reduced to narrow aqueduct. Its dorsal surface is divided by a longitudinal groove into right and left lobes. These in turn may be divided transversely into 2. Leading ventrally and forwards from these lobes in all except cyclostomes are the optic tracts connecting with optic nerves. Floor of mid brain is formed by a pair of fibre tracts crura cerebri separated by a longitudinal fissure.

Cerebellum or metencephalon is a thickening of nervous matter on dorsal anterior end of hind brain. It may exist as a small transverse fold, or may be enlarged, extending forwards over part of mid brain, and backwards over anterior end of medulla. In older works, the anterior of these lobes were called nates; the posterior, testes. unpaired in appearance, or it may consist of a pair of lateral lobes or hemispheres separated by a median portion or vermis, ending in a small lobe, the valve of Vieussens, which roofs in the 4th ventricle in front.

Myelencephalon or medulla oblongata is the cranial extension of spinal cord, presenting behind but slight differences from that structure. In front it widens, while its roof thins out and becomes epithelial and folded, to form posterior choroid plexus for underlying fourth ventricle. This region, the fossa rhomboidalis is bounded in front by valve of Vieussens, and on either side by diverging dorsal columns of cord, which are divided into a median fasciculus gracilis and lateral fasciculus cuneatus. Each dorsal column receives in front fibres from the lateral column, the whole forming an enlargement, corpus restiforme on both sides, from which the posterior peduncles of cerebellum pass forward or upward into metencephalon. On ventral surface are the anteriorends of ventral columns, called pyramids. These can be followed forward until they pass into crura cerebri (Figure 11.13).

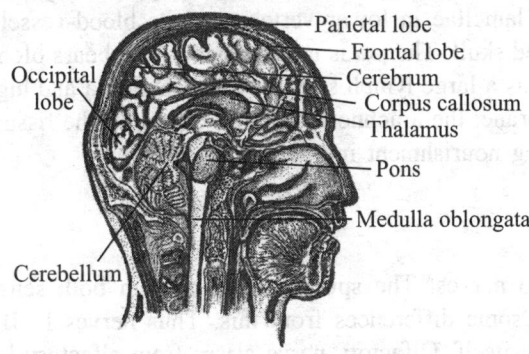

Figure 11.13 Structure of human brain.

In higher vertebrates, the anterior ends of pyramids are crossed by transverse bundles, forming pons Varolii, which act as commissures connecting 2 halves of cerebellum. Medulla oblongata is noticeable as it gives rise to greater part of cranial nerves. Various parts of brain are connected by longitudinal fibre tracts and by transverse fibres or commissures.

The chief longitudinal tracts are those of the pyramids, which may be followed through the crura cerebri to corpus striatum. Some fibres of lateral column and part of dorsal column enter the cerebellum through posterior peduncles of cerebellum, while majority from these columns end in medulla. From the cerebellum, fibres extend forward into mid brain through 2 bands of tissue called anterior peduncles of cerebellum, which enter posterior portion of optic lobes. The habenulae are longitudinal tracts; while the fornix, a part of which lies ventral to corpus callosum, is to be placed in same category, although this fibres seen in places to run transversely.

The pons Varolii, passing beneath the anterior pyramids of the cord, connects the cerebellar hemispheres in higher vertebrates. In its earlier stages, the brain lies in the same horizontal plane with spinal cord. Soon by unequal growth of its dorsal and ventral surfaces, bends or flexures appear. Most constant of these is the cephalic flexure between fore and mid brains, by which the axis of fore brain is bent ventrally at nearly right angles to the rest. Two other flexures may appear which are marked in mammals. The frontal flexure, in the region of pons Varolii, is in opposite direction; the nuchal flexure, in medulla, is ventral again. In *Ichthyopsida* these flexures largely disappear with growth.

In amniotes, they persist throughout life. In lower groups, the 5 divisions of brain are subequal in size, but in higher vertebrates, a great increase in size of cerebellar, and of cerebral regions occur, so that these completely cover over twixt and mid brains. Backward extension of cerebrum is marked in mammals. Connected with this overgrow formation of 5th ventricle, or pseudo-ventricle, a cavity in no way connected with true ventricles occurs, but lying morphologically outside the brain, between septa pellucida, fornix, and corpus callosum.

Brain and spinal cord are enclosed in envelopes of mesenchymatous origin, which hold them in position, and serve as bearers of nutrient vessels. These membranes from outside to inside are called dura mater and pia mater respectively. The dura is a denser connective tissue, consisting of 2 lamellae in lower vertebrates; its blood-vessels being distributed to walls of spinal canal and skull. The pia is more delicate, and bears blood vessels of brain and cord. Between 2 layers is a large lymph space, and in amphibia and higher vertebrates this is divided by a 3rd membrane, the arachnoid. The pia enters all the fissures and depressions in brain and cord, carrying nourishment into nervous mass.

Cranial Nerves

The brain gives rise to nerves. The spinal nerves contain both sensory and motor roots. Cranial nerves present some differences from this. Thus nerves I., II, and VIII are purely low outgrowths of brain itself. Olfactory nerve arises from olfactory lobe, and is distributed to sensory epithelium of nose. It is provided with its own ganglion, which may be included in brain, or may be carried out into close proximity with olfactory organ. The connection between olfactory ganglion and brain is made by olfactory tract. Optic nerves which arise primitively from ventral sides of diencephalon have their ganglia lying on superficial portion of retina.

They retain their connection with thalamencephalon throughout life. These optic tracts are so formed that nerves cross beneath the thalami that from the right eye going to leave optic lobe, and vice versa. There may be a simple crossing organ interlacing of fibres, or a complete union of trunks .Nerves III, IV, and VI are purely motor nerves, supplying the muscles which move the eye. Oculomotor arises from the crura cerebri, and supplies the muscles rectus superior, internus, inferior, and obliquus inferior. Trochlear arises from the posterior dorsal portion of mid brain, although its centre inside the brain lies ventrally. It supplies the superior oblique muscle. Abducens arises from anterior pyramids and is distributed to externus rectus muscle and to retractorbulbi, when this muscle is present. Oculomotor is always distinct, others may be fused with the 5th, and in some animals their existence is yet to be proved. Trigeminal nerve arises from anterior end of sides of medulla. It is always large, and in higher vertebrates at least has 2 distinct roots, the dorsal root bearing a ganglion. It has 3 branches, ophthalmicus profundus, distributed chiefly to nose and lachrymal region; maxillaries superior supplying the region of upper jaw ; and mandibularis or maxillaries inferior going to lower jaw, and in amniotes to tongue. Often the last 2 are united for a distance as a maxillary nerve.

In many *Ichthyopsida*, the 7th nerve is connected with fifth, and the roots of 2 cannot be distinguished. In higher vertebrates, 2 nerves are distinct throughout. Facialis is complicated than trigeminal, and may contain 4 components. In lower vertebrates, it is a mixed nerve, but in higher it is purely motor, and is connected largely with muscles of expression. In its extreme development, it produces 4 branches, ophthalmicus superficialis; hyomandibularis; buccalis; and palatinus. The first has its own ganglion and is purely sensory, supplying lateral line organs on top of head. It is found only in aquatic *Ichthyopsida*, the frog, for instance, losing it. Fibres from the 5th accompany the ophthalmicus superficialis. Hyomandibularis soon divides into an anterior or mandibular branch and a posterior division, which supplies the muscles of gill cover, and some of those of jaw.

When the first visceral cleft or spiracle is present, division takes place just above it, so that one branch is pre-trematic in front of opening, the other being post-trematic. Mandibularis goes to lower jaw; and its one branch, which unite with mandibularis branch of 5th nerve which is known in higher vertebrates as chorda tympani. Palatine branch supplies palate and roof of mouth. In lower forms, it is a mixed nerve. In mammals, it innervates only the muscles of soft palate. Auditory nerve is closely connected with seventh, and is often regarded as its dorsal root.

It goes directly to ear, dividing into 2 branches, which may leave skull through separate foramina. Vagus complex is composed of ninth, tenth, and eleventh nerves, which are closely connected, and show similarities to each other. They resemble closely the spinal nerves, especially in presence of distinct dorsal and ventral roots. Ear intervenes between these and nerves in front. The complex arises from side of medulla by four to eight, or more roots, the anterior pair being considered as those of glossopharyngeal. Usually in aquatic vertebrates its ganglion is fused with that of vagus. The glossopharyngeal nerve splits into 2 branches, the anterior going to pharyngeal region, other to muscles and mucous membrane of gill in fishes, and to sense organs of tongue in mammals. Pharyngeal branch also produces a nerve which unites with hyomandibularis of facial. Vagus or pneumogastric has a wide distribution. In branchiate vertebrates, the division occurs above the first true gill slit, and there are pre- and post-trematic branches. In aquatic vertebrates, it divides into 2 main trunks, a ramuslateralis, which is lacking in terrestrial forms, and a ramusintestinalis.

Lateralis branch runs the body length, either close beneath the skin, or deeper in muscles near the vertebral column. It is purely sensory, and is distributed to lateral line organs of trunk; and absence of these structures in amniote vertebrates explains disappearance of nerve. Ramus intestinalis is pneumogastric nerve of human anatomy. It is largely motor in functions. It is distributed to pharynx, stomach, respiratory apparatus, gills and lungs. Of the branches to gills, there are as many as there are gill clefts behind the one supplied by 9th nerve. Each branch divides above the gill cleft into pre- and post-trematicbranches.

The accessory of Willis is apparently a spinal nerve which in amniotes enters close association with vagus. Its distribution is chiefly to muscles connected with neck and shoulder girdle, e.g., sternocleidomastoid and trapezius. Hypoglossal nerve is, in adult vertebrate, purely motor, its branches being distributed to muscles of tongue and to those of hyoid region. In amniotes only this can be considered as cranial nerve. In *Ichthyopsida*, it does not enter skull. In larval stages of some forms, this nerve has a dorsal ganglionated root, while in some species 2 such roots are found. In spinal nerve, a distinction between the nerves of the body and those of viscera is clear. Each is made up of sensory and motor parts, so that 4 components are recognized: somatic sensory; somatic motor; visceral sensory; and visceral motor (Figure 11.14).

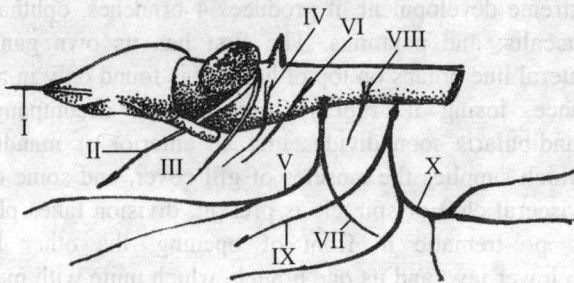

Figure 11.14   Origin of cranial nerves : I – Olfactory, II – Optic, III– Oculomotor, IV–Trochlear, V– Trigeminal, VI–Abdueens, VII–Facial, VIII–Auditory, IX–Glossopharyngeal, X–Vagus, XI– Accessory and XII–Hypoglossal.

The ganglion cells of first are situated in spinal ganglia, and nerves terminate in dorsal horn. Ganglion cells of somatic motor nerves lie in ventral horn, and nerves leave by ventral roots. The internal relations of visceral system are not clear; but both are possibly related to lateral horn region, the visceral sensory nerves, whose centres in trunk region are in sympathetic ganglia, entering by dorsal roots, while visceral motor nerves leave by both dorsal and ventral roots. In cranial region, the matter is complicated by a lateralis system, nerves of which are distributed to ear and to lateral line system, and to mother organs. In terrestrial vertebrates, where lateral line system is lost, lateralis nerves, with exception of eighth are lacking. Fibres of lateralis components terminate in tuber acusticum. The somatic motor nerves of the head include the eye muscle nerves (III, IV, VI). Visceral motor fibres are found in the fifth, seventh, ninth, and tenth nerves.

Sense Organs

All sensory organs arise from ectoderm. Some remain throughout life connected with body surface, epidermis, while others sink into special structures for their protection, the sense capsules. With few exceptions, sense organs are formed of specialized cells, sense cells, each of which is connected by afferent nerve fibres with central nervous system. Between the sense cells, ectodermal cells of supporting function are found. Sense organs situated in epidermis are generalized and are involved with sensations of touch, pressure, and temperature.

Lateral Line Organs

Some of these organs are irregularly distributed; others are grouped into regular series, and form the lateral line organs. In early stages, lateral line organs are present on surface. In amphibian, later they sink into pits, in Pisces into longitudinal grooves which maybe closed into tubes, with openings at regular intervals. With increase in size of the animal, the number of openings increases. The openings often perforate scales, while canals between them may become enclosed in bone, especially on head. By the presence of grooves and canals in skulls of many fossil forms, we infer that they possessed lateral line organs. Variation in distribution of lines of these organs exists, but following are the most constant series: lateral line of trunk which extends the body length between the dorsal and ventral musculature; this series gives the name to whole system; occipital series, crossing the back of head and connecting the systems of two sides; supraorbital, and infraorbital series, running respectively above and below the eye, mandibular series, on the lower jaw.

In these grooves, there are groups of sensory cells, the groups on head being innervated by ophthalmicus superficialis, bticcalis and mandibularis externus branches of the 7th nerve, those on the trunk by lateralis branch of 10th nerve. These organs occur in aquatic stages of amphibian; but in salamanders, frogs, the organs are lost and their nerves disappear. In selachians and ganoids ampullae and Savi's vesicles may sense pressure. Sense organs known as end buds consist of several sensory cells, each bearing sensory hairs, compacted into a bud-like mass, and surrounded by supporting cells. In cyclostomes and fishes, they are scattered over surface, but from dipnoi upwards they are confined to cavities of mouth and nose, and in higher forms to oral cavity. In mammals, they function as organs of taste.

Sense Corpuscles

In terrestrial vertebrates, the epidermal sense organs have various shapes due to modifications of accessory structures. These structures are buried in the deeper layer of epidermis. The simplest are oval cells, the deeper face of which seated in a cup-like expansion of a nerve termination. In the compound tactile cells of birds, 2 or more biscuit-shaped cells are found in a connective tissue, while connecting nerve becomes flattened out into disks between 2 cells. Complicated cell is found in corpuscles of Vater in the centre of which projects a sensory nerve. Tactile corpuscles include club-shaped aggregations of cells, around which the terminal fibrillae of a nerve are coiled. The long facial hair of mammals, the base of which is surrounded by a nerve network is tactile. Ears in all vertebrates are paired structures on either side of head between 7th and 9th nerves. In highly developed ears, 3 portions are distinguished, inner, middle, and outer, the first of which only is sensory and essential, and is the only part occurring in fishes; the other 2 are accessory in character.

Sensory portion of inner ear arises from ectoderm. At first it is a cup-like depression on either side of head. Then it sinks deeper and its edges unite, converting the cup into a closed sac, the primitive otic vesicle. In all forms, the closure lags at one point, and in this way, by the sinking of rest, a slender tube, ductus endolymphaticus or aqueducts vestibuli, is formed. In elasmobranchs, this tube opens to exterior by a small opening near middle line of top of head. Otic vesicle is at first spherical, or oval, but it soon divides by constriction into an upper portion, the utriculus, and a lower, sacculus which are connected by narrow utriculosaccular canal.

Flattened outgrowths arise from the utricles wall. The walls pinched together so that each outgrowth becomes converted into a semicircular canal, opening at either end into utriculus. In myxinoids, there is one such canal. The lampreys have 2 and all other vertebrates have 3. Two are in vertical planes at nearly right angles to each other, and from their position are called anterior and posterior canals. The third is horizontal in position and is called the external canal. Each canal bears an ampulla, at one end. In some forms, only the deeper layer of ectoderm participates in the formation of otic vesicle.

Single semicircular canal of myxinoids bears an ampulla at both ends. Sacculus is always connected with ductus endolymphaticus, and it gives off behind an out pushing known in lower vertebrates as the lagena. In mammals, this lagena becomes developed, and forms scala media of cochlea. As long as the otic vesicle remains a simple sac, it bears on surface a patch of sensory epithelium; but with differentiation of parts, the epithelium becomes correspondingly divided into several maculae and cristae. In lampreys, sacculus is absent; there are three patches, a crista in each ampulla and a macula in the vesicle. In other forms, there are 3 cristae, and at least one macula in utriculus, 2 in sacculus, and one in lagena. These ectodermal parts form the membranous labyrinth. It is filled with endolymph containing otoliths or particles of calcium carbonate, sometimes of microscopic size, but in teleosts forming large ear stones.

Otic capsules protect the membranous labyrinths . These are laid down in cartilage, but in all except lower vertebrates, the cartilage is finally replaced by bone. The inner walls of such capsules follow closely the contour of membranous labyrinth, thus constituting skeletal labyrinth, between which membranous portions is a space filled with perkily fluid. Walls of the capsules are perforated internally for passage of nerves, while on their lateral surfaces, in all groups above amphibia, are 2 openings, the fenestra ovalis and fenestra rotunda, through which sound waves pass.

In mammals, the skeletal labyrinth follows closely the membranous portion. The part called the lagena in lower vertebrates is well developed in mammal, and is drawn out and coiled in a spiral accompanied above and below by similar outgrowths of perilymphatic space. From the resemblance of these structures present to a spiral stairway these are called scalae, that part connected with membranous labyrinth being the scala media, the upper of perilymphatic spaces being the scala vestibuli, the lower the scala tympani. This whole structure, similar to a snail shell is called the cochlea.

In scala media, the macula lagenae of lower vertebrates becomes developed into a specialized organ of Corti. Besides hair and other cells, the organ consists of hard rods right angles to axis of scala. As spiral diminishes in size, from apex to base, these axises also diminish in size. The middle ear or tympanum first appears in anura. It is formed by expanded end of the first visceral cleft, which is closed externally by a thin tympanic membrane. Internally tympanic cavity connects with pharynx by proximal portion of cleft, here called the Eustachian tube.

Sound waves are conducted across the tympanic membrane. In urodeles and caecilians, the tympanic cavity is lacking, but single auditory ossicle, the stapes articulates with quadrate. Cavity of auditory ossicles extends from tympanic membrane to fenestra ovalis. In anura and sauropsida 2 ear bones, the stapes situated in fenestra ovalis, and columella extending from stapes to tympanic membrane are found. In mammals the columella is replaced by 2 bones, the incus and malleus. Stapes arises as a chondrification, and, later, ossification of membrane closing the fenestra ovalis. The columella is post-spiracular and corresponds to hyomandibular. The incus is the quadrate of lower vertebrates while malleus is proximal end of Meckel's cartilage, which becomes cut off from rest. In all anura and in many reptiles, the tympanic membrane remains on outer body surface. In mammals, an external ear is supported by cartilages and evidence show that this external ear is a derivative of the fish operculum, or of external branchial structures of amphibia. In fishes, the ears lack auditory functions and are solely organs of equilibrium. In terrestrial vertebrates, they are organs for hearing and equilibrium.

Olfactory Organ

The organ of smell is a single sac in cyclostomes, paired in all other vertebrates. Its vital part is the sensory epithelium, in which sensory cells are interspersed with supporting or isolating cells. Its nerve supply is the olfactory nerve. The powers of smell are directly proportional to the extent of sensory surface, and accordingly this may be increase the folded surface in longitudinal direction. In the primitive forms, the sensory surface is not uniformly distributed, but is gathered in patches separated by large masses of isolation cells. In some ganoids and amphibians, the nasal epithelium has a radiate appearance. From the amphibia upwards, outgrowths of cartilage or bone, either from the ethmoid or lateral walls tend to divide the cavity.

In Petromyzons and Pisces only external nostrils occur. In the forms with paired cavities there is a single nostril to each olfactory sac, but in the selachians and ganoids a fold of skin divides each nostril into two. In many teleosts, two distinct nostrils may occur on either side. These modifications clearly are to permit a current of water over the olfactory epithelium.

In all vertebrates above fishes, both external and internal nares open into the oral cavity. This condition is foreshadowed in selachians, where nasal groove leads back from the external nares of either side to the angles of mouth. In higher vertebrates, this groove converts during growth into a tube by the union of its edges. Thus a respiratory tract is formed on a side of olfactory surface, the posterior end of which opens inside mouth cavity. Similarly a nasolachrymal duct is formed leading from each eye into nasal passage. In terrestrial vertebrates, nasal glands are often present in connection with nose, the secretion of which moistens the olfactory epithelium.

The process is modified in some groups, where a solid cord of cells, instead of a groove is formed, the respiratory passage appearing later in cord. Connected with nose in all vertebrates above the fishes are the organs of Jacobson. They are outpushing of the wall of olfactory surface, supplied by branches of first and fifth nerves. In lower amphibia, these organs are placed on medial side of nasal cavities. In the highly developed amphibia, they rotate to lateral side of olfactory organ. In amniotes, they are either medial or ventral in position. In lower forms, these sacs connect only with nasal cavities. In mammals, a duct sometimes leads from them into mouth through foramina incisiva, between premaxillary and palatine processes of maxillary bones. In many mammals, these foramina are closed by membrane, and are vestigial. In mammals for the first time appears an external nose supported by cartilage. In some, like tapirs and elephants, this organ becomes enormously developed, and forms in the latter the well-known trunk.

Visual Organs

Sensory portion of eyes arises from brain, and in embryos optic areas can be recognized in medullary plate before its involution. Accessory portions are furnished by ectoderm, mesothelium, and mesenchyme.A hollow outgrowth arises on either side of primitive fore brain and extends towards skin. Distal portion expands into a globular optic vesicle, while proximal portion retaining its smaller size is called optic stalk. Thus, the cavity of the vesicle connects with ventricle of thalamencephalon by hollow stalk. Distal surface of each optic vesicle contacts with ectoderm of side of head at the place where lens is to form.

With formation of lens, the distal half of optic vesicle becomes invaginated into proximal part, thus partially obliterating cavity of the vesicle, and converting whole into a two-layered cup. Distal invaginated part of cup becomes retina, and outer layer forms pigmented epithelium of eye. This invagination is not confined to distal portion of optic vesicle, but extends along its lower surface and continues on optic stalk, resulting in a gap, the choroid fissure in ventral wall of optic cup which is produced as a groove along lower side of optic stalk. Through choroid fissure mesenchyme cells and later blood vessels enter the optic cup. Later when fissure closes, walls of stalk unite around blood vessels, which hence enter optic cup through centre of optic stalk (Figure 11.15).

The vitreous humor develops from immigrant mesenchyme cells, and fills optic cup. Blood vessels nourish the retina. At first the retinal layer is thin; but later increases in thickness by cell division and finally consists of several cell layers, which differentiate to form several strata .Those nearest the lens transform into ganglion cells, those farthest away the rod- and cone-cells, and between these granular layer being separated from other 2 by an inner and outer molecular layer. From some of the rod- and cone-cells, slender rods and cones grow out towards into pigmented epithelium, while others of this layer develop into isolating cells. From other side of each rod and cone cell, nerve fibre grows out towards front of eye, and breaks up into dendrites which interlace with other dendrites coming from cells of granular layer. Their fibrillations produce inner molecular layer. Outer layer is an interlacing of dendrites from granular and ganglion cells, the minute granulations which occasioned the name molecular layer being the sections of nerve fibrillations.

Ganglions produce from outer surfaces axons grow from cells to choroids fissure, and then through groove in optic stalk to brain. Axons form optic nerve appears after closure of choroid fissure, as if it left the eye through centre of retina. Sense cells are lacking at the exit point of optic nerve and this region forms blind spot. Optic nerve fibres grow from brain to eye: but some fibres arise in this way. Optic nerve resembles the dorsal root of a spinal nerve. In formation of eye by involution from skin, the layer of rods and cones is homologous with superficial layer of skin, while ganglionic layer corresponds to deeper surface of epidermis. Hence, light passing into eye transverses transparent deeper layer to reach superficial sense structure, the rod and cone.

At the place where optic vesicle reaches ectoderm of the side of head, the latter thickens, and then a portion of it becomes invaginated, and is at last cut off as an epithelial sac, the vesicle of lens. This body lies in aperture of optic cup with an anterior wall of cubical cells, while those of posterior surface are so strongly columnar that cavity is nearly obliterated. After the lens is cut off from ectoderm, the latter becomes a smooth, transparent sheet over front of eye, forming the epithelium called conjunctiva which is continuous with superficial layer of skin. In man, there are between 250,000 to 1,000,000 rods and cones to a square millimetre of retinal surface. The eye proper is about spherical. In fishes, it may be flattened, somewhat conical in birds in front.

In ichthyopsida, it lacks well developed external accessories for protection. In amniotes, movable lids can close over the organ. Some salamanders have feebly developed eyelids. In man, the nictitating membrane is reduced to a vestigial fold, the plica semilunaris which is visible at inner angle of eye. Free surface of eye is covered by conjunctiva and beneath this is a thicker, the cornea, composed of connective tissue fibres produced from mesenchyme cells. Laterally, the cornea is continuous with the sclerotic coat, which envelops the whole eyeball. This sclerotic forms a sense capsule, comparable to those enclosing the ears and olfactory organs. The choroid extends forward nearly to edge of optic cup; but beyond this point it is muscular. A portion of it forms a circular curtain, which extends from edge of optic cup into anterior chamber. This curtain called iris is opaque and is usually colored by pigment derived from edge of optic cup.

The opening in centre of iris, the pupil enlarges or contracts by muscles, and amount of light admitted to retina is regulated. Just inside the iris, the inner wall of optic cup develop into a strong ridge, the ciliary process, which extends inwards towards lens to which it is attached by fenestrated, suspensory ligament, thus partially separating the anterior from posterior chambers. Close to this region, the choroid develops a layer of ciliary muscles, which can move the lens nearer to or farther from retina, can alter its shape. This forms the apparatus of accommodation for viewing objects at different distances. In lower vertebrates, accessory glands connected with eyes are slightly developed. Lachrymal glands lubricate the surface.

Eletrical Organs

In *Torpedo*, *Malapterus*, and in some skates, some muscles change into electrical organ. This organ lies in *Torpedo* on either facet of head; in others in trunk or tail close to backbone. All organs consists of a series of capsules of connective tissue full of a gelatin like substance in which are electrical plates, nerve endings, and which are apparently changed motor end plates of muscle. Discharge of organ is in restraint of can, and varies in strength consistent with size of organ and its condition of fatigue. In Torpedo and electrical eel it is adequate to knock a person down, however in the others it is less in quantity.

11.12 Aquatic Adaptation

Two types of aquatic vertebrates are encountered, primary and secondary aquatic vertebrates. Secondary aquatic vertebrates are fishes that have evolved from aquatic progenitors. Fishes are perfectly adapted to aquatic environment. Besides fishes, many lung-breathers have gone back to primal aquatic home for food and safety and exhibit extreme modifications for aquatic life. Requisites of primary aquatic adaptation are

Body Contour: Body is compressed into a stream lined form, entire surface of which is accurately rounded with no protuberances which would retard swift passage of animal through water. Head is sub-conical edges of jaws and gill covers fit precisely, and even eyes conform accurately to curvature of head.

Locomotor Devices: Locomotion is primarily affected by lateral undulations of body. Fins help in process and are regarded as accessory locomotor organs. Tail fin is most important propelling organ. Besides unpaired fins, paired fins, comparable to tetrapod limbs, are present and serve normal function as stabilizer. Increase of resistant surface is obtained in fish through development of unpaired fin folds of skin, stiffened by fin rays of elastic bone or cartilage. These unpaired fins may be continuous from head along mid line of back, around tail and forward along under side as for as vent.

Hydrostatic Device: Swim bladder is largely hydrostatic. It maintains the fish at a certain depth of floatation. If the creature wishes to sink, body is slightly compressed through muscular contraction and with it the swim bladder. This lessens the bulk, and weight remaining constant, species gravity is increased, with a resultant loss of buoyancy. Relaxation of muscles has contrary effect; compressed gas in bladder expands, increasing fish's size and thereby decreasing specific gravity and creature rises.

Adaptation in Whale

Whales spend their whole lives in water, being conceived and born there. Large whales and dolphins sleep in the sea in many respects. Whales have reverted to characteristics of a fish form of life. Important morphological adaptations are as follows

1. Elongated head.
2. No neck.
3. Tapering and streamlined body.
4. Body surface is completely smoothed off by the loss of all hair, except for a few sensory bristles round the snout.
5. Horizontally placed tail flukes.
6. Paddle-like forelimbs and there is often a large dorsal fin.
7. Disappearance of hind limb.
8. Number of fingers is often reduced to four.
9. Skin glands are absent.
10. Forelimbs are modified to form flippers.
11. Dorsal fin and tail flukes are "neomorphs" folds of skin and hard connective
12. Tissue but no skeletal support.

Physiological Adaptation

Blubber: Thick layer of dermal fat (blubber) which besides acting as a heat insulator may provide a reservoir of food and perhaps when metabolized, of water. Fat reduces specific gravity of animal and provides an elastic covering to allow for changes in volume during deep diving.

Feeding Habits: Many toothed whales like porpoises and dolphins eat fish and teeth form a row of numerous (65/68) similar peg-like structures, usually in both jaws. With elongation of jaws, masticatory function of teeth has been reduced and they hold prey. *Orcinus*, killer whale has large powerful jaws and teeth and its diet includes dolphins, birds, seals and flesh of large whales. In whalebone whales, there are teeth only in fetus. Plankton as food is collected by fringed baleen which consists of rows of transverse plates of keratin. Stomach of whales has 3 or 4 parts. Fore-stomach is a muscular crop, often containing stones, lined by squamous epithelium and without glands. Main stomach has a folded mucosa and gastric glands. Thirdly, a smooth pyloric stomach with few glands leading to duodenum intestine may be 16 times as long as body.

Osmoregulation: They are hypoosmotic to their surrounding environment so that body fluids tend to lose water by osmosis and conserve salts by diffusion water is conserved in body through concentration of urine.

Lung Ventilation and Deep Diving: Whales can ventilate lungs more completely than terrestrial mammals and possible up to 90% of inspired air is used at each breath. Volume of air contained in lungs of a large whale is estimated to be 2,000 lbs. Whale which dive in deep water can not store much O_2 and have smallest lung capacity proportionate to their body size because lungs collapse completely in about 100 m depth due to water pressure . Muscle of whales contains large quantities of myoglobin, a protein that attracts O_2 from blood. This helps in long sub emergence of animals by storing O_2.

Bradycardia: Bradycardia means slowing of heart beat rate and occurs during submergence rate is usually reduced to 1/10 or 1/15 in normal condition and decreases in a deep diving. It is the oxygen saving mechanism.

Ratea Mirabile: Besides air in lung, there may also some provision for storage of extra oxygen in large blood volume of retea mirabile, networks of blood vessels which abound thought body especially in thorax. Retea mirabile is connected with accommodation of animal to varying hydrostatic pressures and temperatures.

Reproduction: Testes do not descend with sacs, but remain close to kidneys. Penis is long and curled when not erect. Uterus is bicornuate, but usually only one young is carried and is retained for a long time, Placenta is epitheliochorial, and there is large allantois. Birth takes place under water. A pair of teats in inguinal region and mammary glands is provided with many myoepithelial cells so that milk is pumped into mouth of young. Milk contains 50 percent of fat and is concentrated, economizing water and allowing rapid growth. Larger whales probably reproduce about 12 years of age and may live up to 60.

Cerebrum and Cerebellum: Brain functions as a controlling centre of all receptor and effector organs. It consists of hindbrain (rhombencephalon), midbrain (mesencephalon) and forebrain (prosencephalon). Hindbrain includes medulla oblongata, pons and cerebellum. Midbrain includes a sensory tectum and motor tegmentum. Forebrain includes telencephalon or cerebrum and diencephalons

A dome shaped folded extension of hind brain, often with highly convoluted surface this is divided into a medial corpus and paired lateral auricles. In fishes, cerebellum is relatively large due to extensive input from lateral line sensory system relating to water currents and electrical stimuli. In bottom - dwelling fishes like flounders and in inactive swimmers like lampreys, cerebellum is relatively small and performs a reduced role. Corpus sides of cerebellum expand into cerebral hemisphere in birds and mammals. Flocculus or flocculonodular lobe of tetrapod is homologous to dorsal half of fish auricle. As proprioceptive information and refinement of muscle action become important and place increased demands on cerebellum, it is large and prominent in terrestrial vertebrates.

Functions: To process information relating to touch, vision, hearing, proprioception and motor input from higher centers. Function to maintenance muscle tone and balance, run, jump, fly or swim. Cerebellum maintains positional equilibrium. It compares input and sends modified signals to motor centers. After removal of cerebellum, an organism can move but its movement is uncoordinated, insufficient and uneven. Cerebellum thus monitors and modifies rather than initiate action.

Cerebrum: Cerebrum possesses 2 regions, a dorsal pallium and a ventral subpallium. Pallium possesses medial, dorsal and lateral divisions. Medial pallium receives a small primary olfactory input along with substantial auditory, lateral line, somatosensory and visual inputs. Dorsal and lateral pallia receive ascending input including visual information relayed from thalamus. Agnathans possesses a pallium and subpallium. In lampreys, cerebral hemispheres consist of only lateral pallium and septum. Rest of pallium and subpallium are located just posterior to this in caudal telencephalon.In elasmobranch, lateral pallium receives main olfactory input via lateral olfactory tract. Parts of dorsal pallium receive visual, lateral line, thalamic and possibly auditory stimuli. Exchange of information between hemispheres occurs through medial pallium.In *Polypterus*, pallial and subpallial regions are recognized in a basal group.

Summary

1. Adult epidermis varies in structure, becomes many layers thick, and is thicker in aquatic forms. Derma is mesenchymatous in origin, consists of fibrous tissue, smooth muscle cells, blood vessels, and nerves. Lachrymal glands absent in crocodiles are best developed in lizards.

2. Exoskeletal hard structures are formed by accumulation of keratin. Outer covering of epidermal scale prevents water loss through skin surface in reptiles. In lizards, scales are skinny, small, overlapping and moulted in little items. In sankes, scales are overlapping enlarged on head known as shields, and on ventral surface known as scutes.

3. Horny structures derived from epidermis are dermal scales, feathers, hair, claws, nails, hoofs, spurs, true horns, beaks and bills. Hair is characteristic of mammals is formed by the initiative of epidermis. Epidermis produces various glands. In fish, mucus glands, poison glands are noted. Skeletal structures could also be membranous, cartilaginous, or bony.

4. Membranous skeleton consists of connective tissue, and in its highest development forms sheets. From it, cartilage is developed by increase in range of cells. Procoelous vertebrae occur in most members belonging to the order Salientia, reptiles, and birds. In most mammal, vertebrae are flat on each end of centrum,

amphiplatyan. In forms with amphiccelous vertebrae, true articulation of separate parts of vertebral column is lacking.

5. From the Amphibia upwards, the ribs are articulated with vertebrae by 2 heads, a dorsal or tubercular head articulating with diapophysis, a ventral or capitular head resting on parapophysis. In generalized fossil amphibian like *Archegosarus*, neural arch hypocentra and ploxirocentra comprise a rachitomous vertebra. Evolution from rachitomous type might have occurred in 2 directions forming embolomerous and stereospondylous vertebrae.

6. Flight evolved 3 times in the 500 million years of vertebrate history. When an animal flaps its wings, then it is flying. Gliders and parachuters don't move wings instead passively descend in the air. True flyers have control over their descent. They flap wings to produce thrust for added speed and lift for added height. True flyers move horizontally unlike gliders and parachuters.

7. Heart shows gradual complexity and specialization in structrure and function from lower chordates to mammals. Such complexity lies in increase in number of heart chambers and evolution of double circuit heart from single circuit one. Vertebrate kidney is a pair of compact organs lying dorsal to coelom in trunk region, one on either side of dorsal aorta. Two sets of anatomical terms have been applied to kidneys. Typical mammalian metanephros is retroperitoneal, compact, bean shaped organ attached to dorsal body wall.

8. Central nervous system develops from rest of ectoderm by formation of a neural or medullary plate on dorsal surface of embryo by infolding of a part of primitively external body surface. The brain gives rise to nerves. The spinal nerves contain both sensory and motor roots. Cranial nerves present some differences from this. Thus nerves I, II, and VIII are purely low outgrowths of brain. Olfactory nerve arises from olfactory lobe, and is distributed to sensory epithelium of nose.

9. All sensory organs arise from ectoderm. Some remain throughout life connected with body surface, epidermis, while others sink into special structures for their protection, the sense capsules. With few exceptions sense organs are formed of specialized cells, sense cells, each of which is connected by afferent nerve fibres with central nervous system. Between the sense cells, ectodermal cells of supporting function are found. Sense organs situated in epidermis are generalized and are involved with sensations of touch, pressure, and temperature.

10. Lateral line organs are irregularly distributed; others are grouped into regular series, and form the lateral line organs. In early stages, lateral line organs are present on surface. In terrestrial vertebrates, the epidermal sense organs have various shapes due to modifications of accessory structures. These structures are buried in the deeper layer of epidermis. The simplest are oval cells, the deeper face of which seated in a cup-like expansion of a nerve termination. In the compound tactile cells, of birds, 2 or more biscuit-shaped cells are found in a connective tissue, while connecting nerve becomes flattened out into disks between 2 cells. Connected with nose in all vertebrates above the fishes are the organs of Jacobson. They are outpushing of the wall of olfactory surface, supplied by branches of first and fifth nerves.

11. Two types of aquatic vertebrates are encountered, primary and secondary aquatic vertebrates. Secondary aquatic vertebrates are fishes that have evolved from aquatic progenitors. Fishes are perfectly adapted to aquatic environment. Besides fishes, many lung-breathers have gone back to primal aquatic home for food and safety and exhibit extreme modifications for aquatic life.

Review Questions

Short Answer Questions

1. What is hair feather?
2. What are plumulae?
3. What are down feathers?
4. Mention the difference between unguis and subunguis.
5. Name an animal where prong horns are found.
6. What do you understand by jaw suspension?
7. What do you understand by neurocranium?
8. Define amphicoelous vertebrae
9. Define procoelous vertebrae.
10. Comment on parachuting.

Long Answer Questions

1. Describe the evolution of kidney in vertebrates.
2. Describe the evolution of heart in vertebrates with suitable diagrams.
3. Describe various types of jaw suspension in vertebrates with suitable diagrams.
4. Describe the cerebellum in vertebrates.
5. Give a comparative account of digestive system in vertebrates

12

Threatened Wildlife and Conservation

There is no unanimous definition of wildlife, which includes plants, animals, and other organisms in their wild state. Wildlife occurs in all ecosystems, Deserts, rain forests, wetlands, mangroves, mountains and sea. The wildlife encompassing diverse range of species is preserved in protected areas. India is home to several rare and threatened species. Various ecosystems provide shelter to many endemic species. As a whole, 33% of plant species are endemic in India.

Forest cover extends through tropical rainforest, Western Ghats, and coniferous forest of the Himalaya. Sal-dominated moist deciduous forest, teak-dominated dry deciduous forest and babul-dominated thorn forest are home to various wildlife. India contains 2.9% of IUCN-designated threatened species like Asiatic lion, Bengal tiger, and Indian white-rumped vulture.

India has a rich heritage of wildlife and their conservation. Kautilya's Arthshastra contains the concept of Forest Reserves to protect elephants. Now wildlife is threatened due to various factors and activities. Most conservation practices are being adopted and enforced at the Government level. Some efforts are taken up by NGO's. The native animals in India include Bengal Tiger, Asiatic Lion, Leopard, Sloth Bear, Indian Rhinoceros, black panthers, cheetahs, wolves, foxes, bears, crocodiles, rhinoceroses, camels, dogs, monkeys, snakes, antelope, deer, bison, striped hyena, Macaque, langur and mongoose (Figures 12.1-12.4). Large mammals are the Asian Elephant, Wild Asian Water buffalo, Domestic Asian Water buffalo, Nilgai, Gaur, deer and antelope. Indian Wolf, Bengal fox and Golden Jackals are widely distributed. The dhole is the most endangered carnivore. The Himalayan Wolf is critically endangered and endemic.

A study shows that India lying within the Indomalaya ecozone provides shelter to about 60-70% of the world's biodiversity including 7.6% mammals, 12.6% birds, and 6.2% reptiles. In India, 12.6% mammals, 4.5% birds, 45.8% reptiles and 55.8% amphibians are endemic like Nilgiri leaf monkey and brown and carmine Beddome's toad of the Western Ghats. Ecoregions like the shola forests exhibit high endemism. India's forest cover ranges from the tropical rainforest of the Andaman Islands, Western Ghats, and Northeast India to coniferous forest of the Himalaya.

Figure 12.1   Tiger

Figure 12.2   Deer

Figure 12.3   Giraffe

Figure 12.4   Zebra

12.1   Threats to Wildlife

Human encroachment is a serious threat to wildlife. Exploitation of forest resources through hunting and trapping for food and sport has caused extinction of wild species. Birds like *Rhodonessa caryophyllacea* and *Ophrysia superciliosa* have gone extinct recently. Pink-headed Duck was thinly distributed through the wetlands, swamps and wilderness areas. White-winged Wood Duck is in danger of extinction due to deforestation, wetland drainage and hunting. Other ducks under threat are the Andaman Teal, the Fulvous Whistling Duck, the Comb Duck and the Cotton Pygmy Goose.

Loss of habitat, particularly the grasslands, is a major cause of decline of birds. Smaller game birds like francolins, partridges and quails are losing their breeding territories. Smaller grassland birds like bush chats and stonechats too are declining. About 8% of India's birds are facing serious threat. The plight of the vulture has officially banned the diclofenac. The loss of tall trees has affected the breeding of birds like hornbills, the Greater and Lesser Adjutants and the Black-necked Stork. The draining of wetlands has endangered Swamp Francolin. The numbers of the House Sparrow is declining rapidly, at least in the urban areas. Illegal netting of game birds is also a serious problem. The entire tribal belt in the north-east treats hunting as their right.

Anthropogenic extinction is recorded for at least 50,000 years. Humans armed with Stone Age weapons and fire was perhaps the effective killers. Mega fauna extinction on North American continent about 10,000 years ago has been correlated with activity of human armed with spears. During last past 500 years, anthropogenic extinction rate has increased exponentially. Sailors seeking spices, wood, and whale blubber released goats, pigs, sheep, and rats on remote islands and on Australia, which had never been home to those species. Such species drive many endemic species to extinction. Near the end of 20th century and beginning of 21st century, our activities have created "Crisis of extinction".

Marine animals are also threatened and endangered. Populations of Pacific leatherback sea turtles have plunged 95% in last 22 years due to inappropriate fishing and may extinct within 30 years. Unsustainable human population growth creates expanded demands for resources to serve vital need and increasing per capita consumption. In many regions, demands exist on limited supplies of fresh water, pasture for domesticated animals, and arable farm lands.

12.2   Man-Animal Conflict

Hunter-gatherer society had the knowledge about plant and animals of local environment as reflected in their beliefs and attitudes. People view themselves as inseparable from the nature and wildlife. The society was perhaps the best in their coexistence with wildlife. Humans lived in tribal, hunter-gatherer societies for most of the evolutionary history and collected food from natural ecosystems. Hunter-gatherer people with low population density were mostly migratory for food. There was little division of labour within the sexes with egalitarian social structure.

Hunter-gathers had small impacts on ecosystems in absence of technologies .In later period, extensive use of fire by the hunter gathers of Asia, Australia, and North America caused extinction of large mammal and bird. Early agrarian societies called horticultural societies plant species for food and the society which focus on livestock as primary food source are called herding, or pastoral society. Most early agrarian societies domesticated animals. Many wet tropical areas still support horticultural societies that practice small-scale slash and burn agriculture. The nutrient-poor soil of such regions cannot support large-scale, plowed farming. Permanent villages and small cities first occur in agricultural society. The greater agricultural productivity perhaps resulted in a complex social structure with more division of labour and in development of ruling class, religious class, and artisans. Few societies where family lineages are passed on by the mother and women have high status are usually horticultural.

Animals were associated with gods and goddesses, and were symbolic of a deity's power. Animals play important roles in mythology of such cultures. Dense human settlements started over exploitation of wildlife. Increased natality was common during transition from hunter gatherer to early agrarian societies. Growing human populations placed more demands on wildlife and natural ecosystems. Hunters and traders killed animals, and threatened many others. The growing human population and emerging technologies now allow manipulating the environment. Human-caused alterations of the biosphere are continuing to be serious threat to wildlife.

Biodiversity and natural resource conservation are now matter of policy issues. Differences between our lives and that of our hunter-gatherer fore bears are linked to technology. Technological change has caused the present day human condition as the size of world population, life expectancy, and anthropogenic activities on natural environment .Our first "human-like ancestors" evolved some 2 million years ago. The agricultural revolution started in the Fertile Crescent of the Middle East 10,000 years ago. The human population was about 5 million in hunter-gatherer society some 10,000 years ago, today reached over 6.6 billion. Many societal collapses have involved deforestation and habitat destruction, soil and water management, over hunting and over fishing, introduction of exotics, and human population growth.

Man-animal conflict is the most important issue that threatens wildlife in India. India is home for 60% of the world's tigers, 65% of its elephants, 80% of the Asian rhinos and 100%

of Asian lions and 131 of the world's mega-fauna. Large animals need space to live, breed and feed. Shrinking in wildlife habitat obviously creates conflict with human. Eleven elephants poisoned in 2001 in the reserve forests around Nameri Wildlife Sanctuary in Assam due to illegal encroachment of forest are notable in this regard.

Conflict reduction requires land use planning and implementation. Short term solutions include putting up barriers between man and animal, shifting animals, or encroachers out of conflict areas. Three major species specific conflicts that occur in India involve elephants, carnivores, and ungulates. Monkeys and bears also cause high levels of conflict in urban and Himalayan belts, respectively.

Elephants are large, migratory and require large habitats with well established movement paths. Man-elephant conflict can be addressed by making trenches, fences, or keeping repellants such as crackers, watcher squads between the elephant and man, change in cropping patterns including non-palatable crops, securing elephant corridors, and capture of rogues and problem herds.

Conflict between large carnivores like leopards, tigers, and wolves also require consideration. Such conflict is due to lack of the correct sized prey, and a lack of understanding of wildlife biology. Local subsistence hunting which depletes such prey-base should be stopped. Problem animals must be eliminated. Translocation of animals may be done under strictly scientifically monitored manner only.

Recurring and severe conflict between crop grazing ungulates such as nilgai, blackbuck and man is continuing in many parts. This drives local people into poisoning indiscriminately leading to casualties livestock and endangered animals. To develop alternatives to current cropping pattern, involving agriculture experts/institutions and taking into consideration the animal/bird species would be an appropriate measure. Poaching and trade seriously affect a wide range of wild species. Measures for field control of poaching vary in quality in protected areas. As a signatory to CITES, India is committed to enforce regulations arising out of it and the Committee on Prevention of Illegal Trade in Wildlife and Wildlife Products or the Subramaniam Committee in its report of 1994 had recommended measures for the control of poaching and trade.

12.3 Wildlife Conservation

Each species is a unique natural entity that, once lost, can never be revived. Biologists think about the intrinsic appeal of species. However, in the world dominated by politics and economics, it is necessary to measure the utility of species and habitats. Conservation decisions should base on historical, evolutionary, community, and species approaches blended with reality. A publication of World Resources Institute recommended the 3 rules of thumb to evaluate the trade-offs and value judgments made in setting priorities for protecting biodiversity. Distinctiveness rule emphasizes on preserving an entire species as more important than saving populations of with numerous representatives. Utility rule states that evaluation of conservation priority should evaluate utility. Tropical rainforests are important as they provide shelter to variety of living organisms and influence the global climate. Rule of threat emphasizes for saving the most beleaguered species and ecosystems first. Conservation priorities should focus on those areas that are at risk.

Central America's tropical rain forests are less threatened than the remaining fragments of tropical dry forest in the region, and conservation efforts should emphasize the latter. Conservation efforts attempt to protect relatively undisturbed areas, or to restore damaged areas. The amount of land and number of species likely to be protected in such way is likely to be small. Michael Rosenzweig has developed reconciliation ecology, which states to protect biodiversity incorporating the wildlife. The main use of bypass lands has been for agriculture, although it is important for fish and wildlife. When it floods, ducks and geese forage in its productive waters. Juvenile salmon that enter the flooded areas grow faster; when they leave the bypass, they are larger than river-reared salmon. Sacramento split tail, a large native fish, depends on the Yolo bypass for spawning and rearing of its young. Recognition of these values has resulted in the enhancement of parts of bypass as wildlife refuges, while other parts are maintained in productive agriculture.

Importance of wild life related recreational activities is growing. About 61% of Americans participated in non-consumptive wildlife related activities with 16% participated in trips to observe, photograph and feed wildlife as recorded elsewhere. Report says that 31 % of the U.S. populations aged 16 years and older participate in wildlife watching and 30 % expressed an interest in wildlife around their homes. Through interactions with animals, we may connect with spiritual sense of wonder of vast interconnected network. Natural world gives us a point of reference and support. In the past, humans were closer both physically and psychologically to natural world than they are now. Perhaps the industrial revolution provided the final coffin nails for human nature unity. Attitudes towards animals changed radically. However, an intrinsic value of animals is now becoming more recognized.

There are 2 main types of conservation viz in-situ- and ex-situ conservation. Protected areas are most important core "units" for in situ conservation. There are now over 100, 000 protected areas worldwide, covering over 12% of the earth's land surface. Importance of protected areas is reflected in their widely accepted role as an indicator for global targets and environmental assessments. Measuring the number and extent of protected areas provides one-dimensional indicator of political commitment to conservation. Appropriate management of animals for maintaining genetic diversity of the species is needed. All the genetic variations present in wild populations are potentially useful for improvement of domestic animals and plants and therefore should be preserved.

For preserving wildlife richness, knowledge about endemism, type of habitat, keystone species, effective population size and their behavior requires consideration. Geographic Information system (GIS) has been emerged as an effective tool for conservation of species and their habitat. In most cases, environmental problems are addressed in isolation but practically such problems are interrelated and originate from root cause of human population explosion and unsustainable development.

12.4 Conservation Biology

This branch of science attempts to make humanity more compatible with wildlife and wild ecosystems. It is a crisis discipline that is under-funded and under-appreciated. This is because its practitioners are mainly advisors and the real solutions are political. Conservation Biology developed when it becomes apparent that we are facing a global extinction crisis.

Conservation Biology has multiple origins with focus on applied, habitat, and ecosystem oriented solutions from fields like wildlife management and forestry. It is based largely on theory from biology, especially ecology and population biology. It has been strengthened with energy from the environmental movement, from people concerned about the direct effects of environmental change. It has a philosophical and spiritual foundation stemming from the world's religions and from ethical thinking of environmental philosophers.

Conservation Biology also includes components from social sciences, recognizing the extinction trends. Conservation Biology is thus an integrative discipline that is focused on understanding how humans are changing the world and on finding practical solutions to saving biodiversity. It is based on assumption that conservation of biodiversity will come from recognizing that humans are part of nature. It is in our own best interests to protect keep ecosystems, habitats, and species from extinction. Biologists have long realized that the smaller, the population, and the more susceptible it is to extinction.

12.5 Minimum Viable Population

Since the National Forest Management Act of 1976, the term minimum viable population size has come into wide use. This act required the U.S. Forest Service to maintain "viable populations" of all native vertebrate species in each National Forest. A definition proposed by Schaffer is that, "a minimum viable population for any given species in any given habitat is the smallest isolated population having a 99% chance of remaining extant for 1000 years despite the foreseeable effects of demographic, environmental, and genetic stochasticity and natural catastrophes."

In other words, it is the smallest number of a species that can maintain a population for indefinite future. Ideally, when determining a minimum viable population, a biologist must weigh the requirements of the individual species and the external factors that test the ability of a species to adapt. Factors that test the capabilities of a population to adapt include natural factors, including variation in demographic, environmental, and genetic factors, and natural catastrophes. The more frequent a population is exposed to these events, the more likely it is to become extinct.

Human factors, ranging from urbanization to noise created by vehicles, are additional stress on species .Natural factors included long-term droughts and major floods, and changes in ocean conditions. Human factors included activities that degraded freshwater habitats in a massive fashion and fishing in the ocean which reduced adult populations. The role chance events play in the survival and reproductive success of a population is termed demographic stochasticity. What it means is that the smaller the population, the greater the probability that all, or most offspring of the remaining females will not survive, or will be of one sex or will not be able to find one another to mate.

Variation of physical factors such as rainfall and biological factors like predation, competition, parasites, and diseases play an important role in the smaller population. Genetic factors also play an important role in the survival of species. In smaller population, the risk of loss of genetic material is high because not all individuals reproduce. Smaller populations have fewer genes on which evolution act. With less genetic material, populations may lose their vigor, have reduced fertility, and become more susceptible to genetical problems. For example, in zoos a strain of white tigers bred from few individuals is known for being cross-eyed and having abnormal hip joints. How small a population must be before such problems become irreversible, varies among species.

Sometimes populations go through a natural crash in numbers and then recover, but without the genetic variability, they once had. This is called a genetic bottleneck. Small populations are rarely complete representative of the original genetic composition of a population.

Certain traits that are desirable in evolutionary sense may be lost. When exposed to selective factors, populations that have gone through a bottleneck are less likely respond to selective pressure and more likely become extinct. The cheetah is the classic example. The cheetah occurs naturally in the African savannah in low densities, one per every 40 or 50 square miles. Cheetahs have been found to have extremely low genetic variability, showing that all cheetahs today are most likely descended from a very small population that experienced a genetic bottleneck. They are nevertheless successful predators, although there is concern that their low genetic diversity may make them exceptionally vulnerable to epidemic diseases or make it incapable to adapt to major climatic changes. Because the factors affecting minimum population size are often intertwined and inherently difficult to quantify, assessing the relative importance of each is at best guesswork.

The direct determination of a MVP size based on multiple factors has rarely been attempted. Experiments on extinction are for obvious reasons impossible to perform. Other direct attempts require large data sets and complicated computer models. Shaffer used a simulation approach for the grizzly bear in Yellowstone National Park. He found that grizzly bears survival was most affected by demographic and environmental affects. Mortality rate, cub sex ratio, and age at first reproduction mostly affected survival. His results showed that population of less than 30-70 bears occupying less than 2500-7400 sq km have less than a 95% chance of surviving for even 100 years. The system of national parks and protected areas, first established in 1935, was substantially expanded.

12.6 Organizations and Laws for Conservation

In 1972, India enacted the Wildlife Protection Act and Project Tiger to safeguard crucial habitat. The federal protections were promulgated in the 1980s.In 1963; the World Conservation Union passed a resolution calling for an international convention on regulations on export, transit and import of rare or threatened wildlife species, their skins and trophies. Twenty one countries signed the Convention on International Trade in Endangered Species of Wild Fauna and Flora (CITES). The main aim is to check the over exploitation through International Trade. In 1972, the endangering of various species due to Trade in skins of Lizards, Monitors, Snakes along with those of the Tiger, Rhino horns, Bear accelerated India to enact the Wildlife (Protection) Act 1972.

India joined CITES in 1976 by Ratification. The Wildlife (Protection) Act 1972 had some flaws and loopholes which were abused by unscrupulous traders which led to an amendment in 1986. There was a great influence on the Wildlife (Protection) Act 1972 by CITES, which led to an amendment in 1991 and caused the act to be more stringent. The salient features of the amendment were: That all hunting of Wildlife under Appendix I of Cites was prohibited. Collection and Trade in Specified plants was prohibited. Verification and Marking with identification of stock of wild life licensed dealers was required. Transportation of wildlife and wildlife products required a permit from an authorized officer that the product had been legally acquired. Trade in Ivory and its products were completely banned. Issue of firearms License within 10 kms of a sanctuary without the concurrence of the wildlife warden was prohibited. Vehicles, Arms, Vessels and Weapons used for committing offences under the Act were to be seized. Commercial felling and exploitation of Flora was banned. Individuals and NGOs were allowed to take instances of violations directly to courts. A Central Zoo Authority was setup to ensure sound management of Zoos.

CITES has played a very remarkable role in development of wildlife in India by working in coordination with Organizations like WWF -India and TRAFFIC-India by improving the enforcement of CITES by controlling trade in wildlife and wildlife products, organizing training courses, INTERPOL officials, Parliamentarians, and members of the Judiciary . CITES management Centers have been setup in Amritsar, Dehradun, Delhi, Guhati, Kolkata, Mumbai, Chennai, Cochin and Tuticorin. India is also responsible for submitting an annual report to CITES based on its developments.

Environmentalists and Nature Conservationists, NGOs like the World Wide Fund for Nature (WWF), Centre for Science and Environment (CSE), IUCN often play role in policy and management of forests. Historically, conservationists took up causes of protecting large animals like tigers, lions, elephants, and rhinoceros.

The PAs are constituted and governed under the Wild Life (Protection) Act, 1972. This Act is implemented and complemented by Indian Forest Act, 1927, Forest (Conservation) Act, 1980, Environment (Protection) Act, 1986 and Biological Diversity Act, 2002, Scheduled Tribes and Other Traditional Forest Dwellers (Recognition of Forest Rights) Act, 2006. The Wildlife Crime Control Bureau help the efforts of provincial governments in wildlife crime control through CITES and control of wildlife crimes having cross-border, interstate and international ramifications. India is a party to Convention on International Trade in Endangered Species of wild fauna and flora, International Union for Conservation of Nature (IUCN), International Convention for the Regulation of Whaling, UNESCO-World Heritage Committee and Convention on Migratory Species (CMS).

Wildlife conservation and management in India is facing both ecological and social challenges. Habitat loss and fragmentation, overuse of resources, increasing human-wildlife conflicts, livelihood on forests and wildlife resources, poaching and illegal trade exemplify the contemporary wildlife conservation scenario. The Government and the civil society are taking several measures to address these issues. Improved coordination among the stakeholders is needed to meet the challenges.

Sunderlal Bahuguna and Pandey in the Uttrakhand Himalayas started the 'Chipko' movement for protecting trees.The Bishnoi community in Rajasthan treat the conservation of forest and wildlife as a religious duty.Amrita Devi Bishnoi, sacrificed her life along with others for protecting 'khejri' trees. The Government of India has recently established Amrita Devi Bishnoi National Award .Damages caused to forest is not solely caused by communities staying close to and dependant on the forests. Ecotourism around forests and wildlife sanctuaries damage forest habitat. Hopefully, the government has realized that enforcement of forest laws alone, while ignoring the requirements of local people in the protected areas management cannot be successful.

The Government of India through a Scheme "Development of National Parks and Sanctuaries" provides the financial support to national parks and sanctuaries managed by the State Governments. The scheme provides 100% Central assistance on items of works of non-recurring nature. Under the scheme, an assistance of Rs 72.28crores was provided to the States during the IX Five Year Plan. The outlay for the X Five-Year Plan is Rs 350 crores.

12.7 Conservation Priority

The priority of conservation must be on in-situ conservation, where protection is accorded not only to species, but to habitats, ecosystems, biodiversity and wilderness. The guiding principle should be that no living species should be allowed to go extinct. Two methods of ensuring this as a safeguard against extinction in the wild are Propagation in captivity, and preservation of genetic material. India has a large number of zoos and safari parks, some owned by the government, others by municipal and other bodies.. Zoos must serve as centres for empathy for animate beings and love for and interest in the nation's fauna, and not just places for recreation.

In the Andaman and Nicobar Islands, the survival of the indigenous and ecologically adapted Andaman pig as a wild taxa is underthreat.Wild mangoes perhaps exist nowhere in India except in the Satpura National Park in Madhya Pradesh. Wild citrus, wild rice and others are also threatened by genetic infusion. Domesticated chickens and pigs should not be allowed within National Parks and efforts need to be made to segregate the wild and domestic stocks of these 2 animals. Genetic material will have to be kept of the pure wild buffalo. The same is true of many gravely endangered faunal and floral life forms.

12.8 Conservation Measures

Along with over 500 wildlife sanctuaries, India now hosts 15 biosphere reserves, 4 of which are part of the World Network of Biosphere Reserves; 25 wetlands registered under the Ramsar Convention. Article 48 of the Constitution of India specifies that, "The state shall endeavour to protect and improve the environment and to safeguard the forests and wildlife of the country" and Article 51-A states that "it shall be the duty of every citizen of India to protect and improve the natural environment including forests, lakes, rivers, and wildlife and to have compassion for living creatures." Charismatic mammals are important for wildlife tourism. Project Tiger, started in 1972, aims to conserve the tiger along with their habitats. Launched in1973, Project Tiger is one of the most successful conservation ventures. Thirty nine Project Tiger reserves in India covers an area of 37,761 km².Project Elephant, started in 1992, works for elephant protection (Table 12.1, 12.2).

Most of India's rhinos survive in Kaziranga National Park. National Board for Wildlife (NBWL) formulate policy framework for wildlife conservation in the country. The National Wildlife Action Plan (2002-2016) emphasizing the people's participation and their support for

Table 12.1 NPs and WLSs in Various Biogeographic Zones of India (Area in Km²)

Zone No.	Zone	% of India's Geographic Area	No. of NPs	Area of NPs	No. of WLSs	Area of WLSs	No. of NPs& WLs	Area of NPs &WLs	% of Biozone area of Nps & WLs
1.	Trans Himalaya	5.62	03	5809.00	04	10438.56	07	16247.56	8.79
2.	Himalaya	6.41	12	7366.92	65	16065.85	77	23432.77	11.12
3.	Deserts	6.51	01	3162.00	05	12914.09	08	16076.09	7.51
4.	Semi/Ard	16.41	10	1505.78	81	12410.66	91	13916.44	2.58
5.	Western Ghats	4.02	16	3673.52	47	10018.86	63	13692.38	10.16
6.	Deccan Peninsula	41.99	24	9712.24	127	44329.08	151	54041.32	3.92
7.	Gangetic Plain	10.79	06	2363.62	32	5473.24	38	7836.86	2.21
8.	Coasts	2.78	05	1731.18	20	2959.45	25	4690.61	5.14
9.	North East	5.21	13	2674.00	36	3418.62	49	6092.62	3.56
10.	Islagds	0.25	09	1156.91	96	389.39	105	1546.30	18.75
	Total	**100**	**99**	**39155**	**513**	**118417**	**612**	**15757.02**	**4.79**

Adapted from: National Wildlife Database, wildlife institute of India, 2009

Table 12.2 Biosphere Reserves in India

S.No.	Name	Total Geographic Area (Km) and name of Biogeographic Province	States
1.	Nigin	5.520 (6E Deccan Peninsula E)	Temil Nadu Kenata and Kamata
2.	Nanda Devi	5.860.69 (2B West Himalaya)	Uttarakhand
3.	Nokrek	820 (9B North East)	Meghalaya
4.	Manas	2837 (9A Brahampulra Valley)	Assam
5.	Sunderban	9.630 (8B East Coasts)	West Bengal
6.	Gulf of Mannar	10.500 (8B East Coasts)	Tamil Nadu
7.	Great Nicobar	885 (10A & 10B islands)	Andaman and Nicobar Islands
8.	Simlipal	4.374 (6B Chotta Nagpur)	Orissa
9.	Dibru-Sakhowa	765 (9A Brahampura Valiey)	Assam
10.	Dehang-Debang	5111.5 (2D East Himalaya)	Andhra Pradesh
11.	Kangchendzonga	2619.92 (4C Central Himalaya)	Sikkan
12.	Pachman	4926.28 (4B Gujrat Raiputana)	Madhya Pradesh
13.	Agasthymatal	3500.36 (5A Western Ghats)	Tamil Nadu and Karata
14.	Achanakmar Amarkantak	38351 (6A Deccan Peninsula)	Madhya Pradesh and Chhattisgarh
15.	Kachchh	12454 (3B-Kachchh)	Gujarat

Source: http:\\www.envfor.nic.in

wildlife conservation. India's conservation planning emphasizes to protect representative wild habitats. The Indian Constitution entails the subject of forests and wildlife in the concurrent list.

A network of 668 Protected Areas (PAs) has been established, extending over 1,61,221.57 sq. kms (4.9% of total geographic area), comprising 102 National Parks, 515 Wildlife

Sanctuaries, 47 Conservation Reserves and 4 Community Reserves, 39 Tiger Reserves, and 28 Elephant Reserves. UNESCO has declared 5 World Heritage Sites. Four categories of the Protected Areas are National Parks, Sanctuaries, Conservation Reserves and wildlife tourism has averaged 15% growth in India with increase in visitors to protect areas. A variety of efforts is going on at the international levels to conserve & preserve wildlife. Recently, corridors connecting wildlife sanctuaries have been made.

An all India estimation of tigers is done once in every 4 years, apart from Protected Areas and Tiger Reserves where it is done every 2 years, and in case of some reserves each year. The pug marks of tigers recorded through paper tracings, plaster casts and digital photographs are preserved in the concerned Forests Divisions. Other factors of the habitat are taken into account in the Geographical Information System (GIS) domain.

Recovery of Endangered Species

Several species of fauna and flora, listed under Schedule I of the Wildlife (Protection) Act, 1972, are critically endangered and need special recovery plans to prevent extinction. Under these individual plans, which need to be revised every 5 years or so, the threats would be assessed through the prevalent distribution of species, its coverage under the PA system. The concerned States, assisted and motivated by MoEF, would be responsible for implementation. Species covered under special projects like Project Elephant need not have such recovery plans. Species/taxon is included in different schedules of the Wildlife Protection Act depending on threat levels.

Relocation and Rehabilitation of Species

Relocation and rehabilitation is done mainly for 3 reasons. Firstly, to translocate excess or troublesome individuals and groups of species, secondly, to reintroduce species locally made extinct or to augment populations rendered critically low and thirdly to rescue temporarily displaced individual wild animals. For the first category, destroying individual animals may be necessary. The option of translocation and rehabilitation should be explored in case of harmful animals. Re-introduction of the lion in Kunu-Palpur has been delayed greatly. The techniques of mass capture, translocation and rehabilitation of herds of animals as social units, especially of "bothersome" species like nilgai, have not been developed yet in India.

Genetic Degeneration

The red jungle fowl with the grey one is the progenitor of all domestic fowl. However, studies have revealed that inter-breeding between the domestic fowl and wild red jungle fowl has occurred to greater extent with the loss of wild genetic resource. The ubiquitous domestic pig breeds with wild species, with the same result. In the Andaman and Nicobar Islands, they have their own indigenous races. Genetic "swamping" may cause the extinction of these isolated indigenous races on which the local tribes depend for their protein intake. Intrinsic value of these wild genetic resources as counterparts of country's most common domesticated animals and birds is incalculable.

Restoration Ecology

During the last 100 odd years, massive plantations of exotic trees have taken place. Sometimes prime forest was cut down to plant fast-growing, commercial timber and fuel wood trees. During the last 10 years, the Forest Department has stopped or curtailed growing such exotics in protected areas. In many protected areas, these exotics have matured but due to the national park status, the State Forest Departments have not harvested them. Trees growing outside forests, including farm forestry, play more important role in meeting national timber requirements than government forests. Present level of availability of timber is more from TOFs than government forests. Productivity is higher and cost of timber production is lower under farm forestry, as compared to forests.

12.9 Protected Area

The area of PAs should be at least 5% of the geographical area of the country. The first national park declared in 1935 is now famous as Corbett National Park. Steady rise in the number of PAs after the enactment of the Wildlife Protection Act in 1972 is noticeable. There are about 597 national parks and sanctuaries in India, encompassing 1, 54, 572sq km or 4.74% of the country's geographical area which should be ideally 870, totaling 1, 88,764 sq km or 5.74% of the geographical area. The Bombay Natural History Society has given leadership in identifying 463 important bird areas (IBAs) of which 199 are not protected.

The Wildlife Trust of India along with the Asian Elephant Research and the Conservation Centre has identified 88 elephant corridors. There are numerous sacred groves, some represent disappeared forest types of the area. There are small community conserved areas which serve as excellent habitats for waterfowl. The tribal reserves of Andaman and Nicobar are perhaps the best-protected forests. Knowledge about species, ecosystem and ecological processes is essential for management of PAs .Survey is required to know the carrying capacity of PAs and to reduce man-animal conflict, the impacts of long-term overgrazing, collection of minor forest products, fire, floods, tourism . Fragmentation of habitat/ecosystem creates small isolated populations.

However, long-term study on genetic deterioration of small populations appears to lacking. Research is essential for planning conservation management. There is need of knowledge about introduction, reintroduction and rehabilitation. In some well-managed PAs data on major vertebrate fauna have been collected for many decades. Corridors between protected habitats must be considered for conservation and conflict reduction. Such corridors must be planned keeping in mind animal migrations or movements, ecological gradients between habitats, needs of local communities.

The Wildlife Trust of India and the Asian Elephant Research and Conservation Centre have recently brought out a publication identifying all the important Elephant Corridors of India. The BNHS has coordinated the Important Bird Area programme. The Bio-geographic report also recommends the conservation of numerous identified corridors. Some of the important wildlife sanctuaries in India are Bandhavgarh NP in Madhya Pradesh, Corbett NP in Uttar Pradesh Gir NP & Sanctuary in Gujarat, Kanha NP in Madhya Pradesh, Kaziranga NP in

Assam, Periyar WS in Kerala. Sariska WS in Rajasthan, Sunderbans NP in West Bengal, Dachigam NP in Jammu & Kashmir.

National Park associate space of ecological, faunal, floral and geophysics importance is protected and preserved by law for preservation of its boundary. Grazing of cows, removal of wild animal is strictly prohibited in it. All rights of area are reserved with the Govt. Corbett NP, established in 1935, is India's initial NP. Yellow Stone park, USA, established in 1872 with a space of 8983sq km the first NP in world. There area eighty NPs in India. A number of such NPs of India are introduced here in brief.

Corbett National Park: Spread along the bank of the river Ramganga in Uttaranchal, this was constituted in 1936 with an area of 52,082 hectares. Important wildlife of this park is tiger, elephant, deer, wild boars, otters and several species of birds.

Hazaribagh National Park: Situated in Hazaribagh district of Jharkhand state, this park is the habitat of boar, sambhar, nilgai, tiger, leopard, sloth beer, hyena, and gaur in an area of 184 km of thick tropical forest. The park was notified in the year 1976 and its total space is more than 18 000 hectares.

Kanha National Park: This Park is a Tiger Reserve, located in Madhya Pradesh. Leopards, langurs, mongoose, cats, hyena, and porcupine are characteristics of this park. Sal and bamboos are principal trees. This park was notified in 1955 with an area of 94, 000 hectares.

Bandhav Garh National Park: This Park is located in Madhya Pradesh. White Tiger is remarkable. Notified in the year 1968, this spreads in 44,884 hectares.

Kajiranga National Park: It is located on the bank of Brahmputra in Assam and famous for One Horned Rhinoceros. Swamp cervid, bison, tiger, leopard, hoolock gibbon, wild buffaloes, pythons, monitor lizards, elephants are also mentionable. Tall elephant grass, Sal trees and bushes are important plants.. This park was notified in 1974 in 42, 996 hectares area.

Dudwa National Park: This Park was notified in 1977 in 49, 029 hectares area. Located in Lakhimpur Khiri district of Uttar Pradesh, it supports wild animals like crocodiles, leopards, jackals, sambhars and sloth beers including one- horned rhinos and swamp deer. Principal plants are grass, Sal.

Pench National Park: Located on the southern edge of Madhya Pradesh, this is named after the river Pench. It is the 19th Project Tiger Reserve in India. This park was notified in 1977 in 29, 286 hectares.

Dachigam National Park: This Park was notified in 1981 in Jammu and Kashmir with an area of 14, 100 hectares. Kashmiri Stag and Hangul are characteristics.

Gir Forests: Situated in Kathiawar district of Gujarat , this was notified in 1965 with an area of 115, 342 hectares. It is famous for Gir Lions.

Ranthambor Park: This National Park is found in Rajasthan .Constituted in 1980, this park is spread in 39, 200 hectares. The principal wildlife is crocodile, nilgai, gazelle, sambhar etc.

Palamau National Park: Located in Dalton Ganj District of Palamau area of Jharkhand state, this Park was notified in 1986 with an area of 21, 300 hectares. For its tropical forests it has been selected for the Project Tiger. The fauna comprises tiger, elephant, deer, panther, sloth beer, chital, gaur, nilgai, chinKara, and mouse deer.

Simlipal National Park: Located in the Mayurbhanj district of Orissa, this Park comprises Sal forest and has been chosen for Tiger. The fauna includes tiger, elephant, deer, pea foul, talking mainas, chital, sambhar, panther, gaur, hyena, and sloth beer. Notified in 1978, this park covers 135,500 hectares.

Tadoba National Park: This Park was notified in 1955 in ChandraPur district of Maharashtra in 11, 655 hectares and supports tiger, sambhar, sloth beer, lion, chital, chin Kara, barking deer, blue bull, four horned deer, langur, pea foul and crocodile.

Biosphere Reserve

A specified area, in which multiple use of land is allowed by dividing it into different zones and each zone remains specified for a particular activity, is called Biosphere Reserve. Each biosphere reserve has 3 zones- core zone, where human interference is banned completely; buffer zones, where human interference is allowed up to limited extent; manipulated or transition zone, where humans are free to perform their activities. The BRs are planned, managed and protected through joint efforts of the government, non- governmental organizations and the local people. India has declared 14 areas as BRs. These areas are aimed at biodiversity conservation, development of economic and human infrastructures, promotion of education, information - exchange and research concerning conservation and development. Biosphere reserves are helpful in conservation of ecosystems, species and other resources, in the promotion of economic development, in promoting scientific research and education.

Community Reserves

The Sanctuary is designated for protecting and developing wildlife. Certain rights of people living inside the Sanctuary are allowed. A Sanctuary is a protected area where wild animals and birds are kept and encouraged to increase their population. Presently, there are more than 490 sanctuaries in India covering a total area of 1, 48,848 sq km. National Park ,an area with adequate ecological, faunal, floral, geomorphological, natural significance is declared for protecting, propagating or developing wildlife, like that of a Sanctuary. The difference between Sanctuary and National Park lies in the vesting of rights of people living inside. In Sanctuary, certain rights can be allowed but in a National Park, no rights are allowed.

Olive Ridley Turtle Conservation Project

Olive Ridley turtle population nests at nesting sites along the eastern coast. These endangered sea turtles are a focus of attention of International community. The Sea Turtle Conservation Project started by the Ministry of Environment and Forests, in collaboration with UNDP in 1999, with a total allocation of Rs 1.29 crores has identified and made an inventory map of breeding sites of sea turtles, developed guidelines to minimize turtle mortality. Tourism guidelines for ecotourism in sea turtle areas have developed. The project has achieved success in the use of satellite telemetry to trace the migratory route of Olive Ridley turtles, and the sensitization of fishermen of Orissa in use of turtle excluder device.

Captive Breeding Programme

When a species reach its minimum viable population size, captive breeding programs is generally started for large vertebrates. Breeding of wild animals in zoos is practiced. Recently, the breeding of endangered species has been emphasized with the intent of reintroduction in wild. In 1972, Perry et al. found 162 rare or endangered mammal species in U.S. zoos, of which 73 had been bred and only about 30 had met with success. More recently zoos and game parks have drawn attention for success in Captive Propagation (CP). Captive breeding programme with crocodile is continuing with success at Bharatpur, West Bengal.

Conservation Reserves

These reserves are declared by the State Governments in area owned by the Government, particularly the areas adjacent to National Parks and Sanctuaries and those which link one Protected Area with another. Such declaration is made after consultations with local communities. CRs are declared for protecting landscapes, Seascapes, flora and fauna. The rights of people living inside a CR remain undisturbed.

National Institute of Coastal and Marine Biodiversity at Kanyakumari

A large part of India's population subsists on the resources available along the long coastal region, in 53 coastal districts of 10 maritime States and 6 Union Territories. The MoEF has started the process to establish a National Institute for Coastal and Marine Biodiversity. Proposal for allocation of Government land at Kanyakumari has been sent to the Government of Tamil Nadu.

UNESCO World Heritage Programme

The Government of India had received funding support for capacity building and awareness for world heritage sites, namely, Nanda Devi, Kaziranga, Manas and Keoladeo National Parks. The World Heritage Committee has encouraged India to prepare the list of outstanding naturalists. Recognition as World Heritage sites improves conservation values and status, socio-economic development, increases eco-tourism, and enhances the beauty of the site in the eyes of the people.

Indo-Russian Inter-Governmental Commission

The Working Group on Environment and Natural Resources is an important component of the Indo-Russian Inter-governmental Commission. In its meeting held at Moscow, several important decisions were taken which included the Siberian Crane Conservation Project.

12.10   Ecotourism

To develop tourism at the cost of wildlife interests must be resisted. Although tourism generates financial inputs to national parks, wildlife sanctuaries and other protected areas, this must never be at the cost of wildlife. Tourism prevents illegal activities such as illegal felling of trees, poaching, and encroachments. However, uncontrolled tourism disturbs wildlife and hinders their breeding. Properly regulated tourism can help in conservation, and create amongst the visitor's empathy for nature. Tourism zones should be clearly defined. The Tourism Plan must be updated. Development around the PA, particularly in the buffer zone, must protect the eco-system and exert a centrifugal pressure on human populations in the area.

Steps that serve to attract population to these sensitive areas do not serve the long-term interest of the PA. Tourism is also a feature of other forested areas, particularly those located in mountains near hill stations, along trekking routes and around water bodies. The authorities must take steps to educate the public about being eco-sensitive. Many PA's have temple associated with it and worshipers do want and need access. It is not practical to cutoff access to these sites. The forest authorities must ensure that the traffic is regulated and the safety of wildlife and pilgrims is ensured.

122.11   Bengal Tiger: *Panthera tigris tigris*

The Bengal Tiger is a beautiful regal animal and is the second largest cat alive today. Males weighing an average of 440 lbs and reach over 10' in length. Their population is the largest of the 5 remaining subspecies alive today. There are about 3000 to 4800 Bengal Tigers in wild. They are highly adaptable to live in wide range of habitats. Majority lives in mangrove of the Sunderbans. They range in high cold altitudes of the Himalayan forests. Sparse populations exist in Bhutan, Nepal, Bangladesh, and Myanmar. They hunt large prey like the sambar deer, wild cattle, pigs, monkeys, birds, and snakes.

Bengal tigers have been hunted for their fur .Tiger's body parts are sold as medicine. Tigers were protected under the Endangered Species Act of 1973.As per report, at least one tiger a day being killed in India. Extensive conservation program includes Corbett National Park, where there are over 100 Bengal Tigers. Tiger conservation program is running in Ranthambore National Park, and Kanha National Park.

At the turn of the 20th century, the tiger population in India was estimated to be 40,000. Census conducted in 1972 revealed the existence of only 1827 tigers. Various pressures in late 20th century led to decline of wilderness resulting in disturbance of tiger habitats.

At the IUCN General Assembly meeting in 1969, serious concern was voiced about the threat of several wildlife species. In 1970, national ban on tiger hunting was imposed. In 1972, the Wildlife Protection Act came into force. The framework was set up to formulate project for tiger conservation .Project Tiger launched on 1973, is one of the most successful conservation ventures. The project aims at tiger conservation in 'tiger reserves' which are representative of bio-geographical regions. It strives to maintain a viable tiger population.

12.12   Asian Elephants

As forest cover becomes fragmented, elephants raid plantations and crop fields for food or for moving. Poaching remains a threat to elephants. In 1989, the CITES banned the international trade in ivory. There are still some thriving but unmonitored domestic ivory markets in Asian and other countries. Ground challenges include securing habitat corridors, management of human-elephant conflict. There are 26,000 Asian elephants in India today, of which 3,500 are working. Their habitats are constantly being diminished by the expansion of human communities. Some 400 people are trampled to death every year in India by elephants and dozens of elephants are killed by villagers. Elephant habitats are under tremendous pressure.

Establishment of national elephant conservation authority, better management of elephant reserves and protection of almost 90 corridors appears to be urgent. Elephant being wide ranging animal requires large areas. Their requirement of food and water are very high and their population can be supported only by well managed forests. The status of elephant indicates the status of the forests. Asian elephants were widely distributed, from Tigris - Euphrates in West Asia eastward through Persia into Indian sub-continent, South and Southeast Asia including Sri Lanka , Java , Sumatra , Borneo and up to North China. Now, they are confined to Indian Subcontinent , South East Asia and some Asian Islands; Sri Lanka, Indonesia and Malaysia. About half of the Asian elephant population is in India.

Now distribution of wild elephant in India is confined to South India; North East including North West Bengal; Central Indian states of Orissa, South WB and Jharkhand; and North West India in Uttarakahnd and UP. Financial and technical support is being provided to major elephant bearing States in country. The Project is being mainly implemented in States / UTs, viz. Andhra pradesh, Arunachal Pradesh, Assam, Jharkhand, Karnataka, Kerala, Meghalaya, Nagaland, Orissa, Tamil Nadu, Uttranchal, Uttar Pradesh and West Bengal. Small support is also being given to Maharashtra and Chattisgarh .

Wild elephants in India are facing a variety of problems, mostly the habitat loss and human-elephant conflict. Human population explosion and the demands of economic development led to the clearing and cultivation of former elephant habitat. Concern for the threat to the elephant led to the formation in 1992 of the Government backed "Project Elephant". This scheme was intended to preserve habitat and establish elephant corridors, allowing for the traditional migration patterns of established elephant herds.

12.13   Indian Rhinoceros (*Rhinoceros unicornis*)

Rhinoceros unicornis literally means "a single nose horn". The Indian rhinoceros and Javan rhinoceros have one horn. The two African species and also the Sumatran rhinoceros each have 2 horns. The Indian rhinoceros has incisors. Indian rhinoceros is well recognizable by its large skin folds resembling a medieval knight's armor. Indian rhinoceros is that the largest Asian rhinoceros species, starting from 4,000-6,000lbs and standing 5-6ft tall at the shoulder. Males are larger than females. Found primarily in Northern Asian country and Southern Asian nation, they like floodplains and riverine grasslands which offer fruits, leaves, and ligneous plant branches. Tall grasses area most well-liked to shorter species. They use semi-prehensile upper lip to twist around long grasses and branches, putting the herb into its mouth. When feeding on shorter grasses, they tuck its upper lip in so it may forage closer to the ground. To avoid the mid-day heat, they often feed during the morning and evening hours.

Behavior

They are primarily solitary animals, except females with calves. Loosely formed groups are sometimes common near common wallowing and feeding areas. They spend vast time wallowing in mud and water, to avoid insects biting . To lessen the insects bite, they have developed relationship with birds that hunt the insects. Male rhinos have territories, but they do not strictly defend it.

They breed throughout the year. Males attain sexual maturity at 9 years. Females mature at 4 years, and bear calves at 6-8 years. For females males battle with one another using their tusk-like incisors. The winner mate with desired female. Gestation lasts for 16 months. Calves are precocial, weighing 160lbs at birth and walk shortly after. After about a year they start a solitary lifestyle.

The decline of Indian rhinoceros is mainly due to poaching for its horn and transformation of its riparian habitat for cultivation. The Indian rhino's former range extended from Pakistan to Burma, through Indian subcontinent. The demand for their horn in Oriental medicines and deforestation of natural habitat for farmland are the main causes of their decline. Rhino horn is used as an aphrodisiac.

12.14   Sunderban Mangroves

Located at the mouth of Ganga-Brahmaputra delta, Sunderban mangrove is the largest single mangrove chunk in the world and is listed in the UNESCO world heritage list. It covers an area of about 10,000 sq. km, of which Indian Sundarban shares 4482 sq. km of which 59% are found along the east coast (Bay of Bengal), 23% on the west coast (Arabian Sea) and remaining 18 %on the Bay Islands (A & N Islands in Bay of Bengal) and the rest is in Bangladesh. The Sunderban mudflats govern the food chain, and are found at the estuary and on the deltaic islands, where low velocity of river and tidal current occurs (Figure 12.5).

The sunderban has been recognized as a site of national and international importance for conserving biodiversity. The name Sundarban has 3 probable origins; one being the forest of Sundari tree; beautiful (sundar) forest (ban); and forests of the ocean (ocean, Samundra; forest, ban). Main estuaries from west to east in Indian Sundarbans are Hooghly, Saptamukhi, Thakuran, Bidya, Bidyadhari, Gosaba Kalindi and Raimongal.

Average tidal amplitude in the said estuaries varies between 3.5 to 5.0 m. Average annual maximum temperature is around 35°C. Average annual rainfall is 1,920mm. Average humidity is 82 % throughout the year. Three major mangrove coastal settings viz. deltaic, backwater-estuarine and insular categories exist in India. Ganga, Brahmaputra, Mahanadhi, Krishna, Godavari and Cauvery make the widespread deltaic mangroves along the east coast due to nutrient-rich alluvial soil. Backwater-estuarine mangroves characterized by typical funnel-shaped estuaries of Indus, Narmada, Tapti with delta formation exist in the west coast. Insular mangroves are present in the Bay Islands, where many tidal estuaries, small rivers, eritic islets, and lagoons are found.

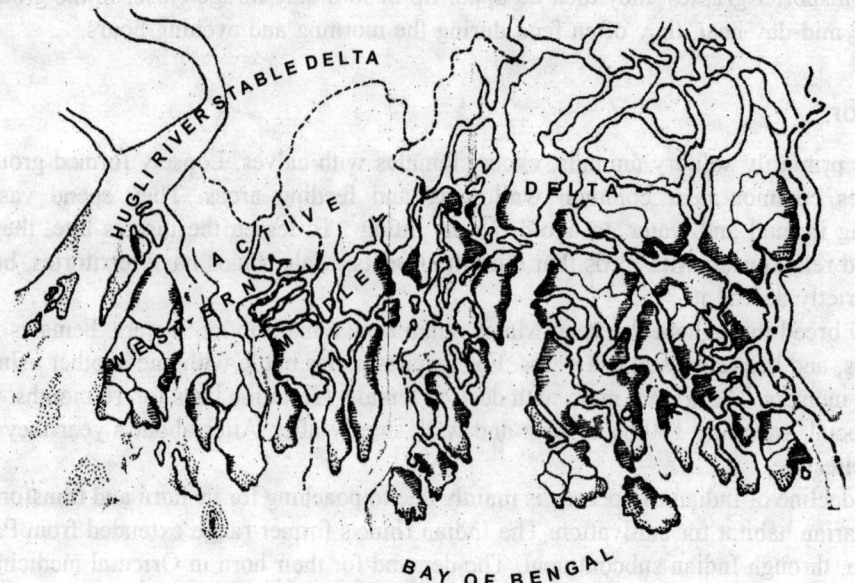

Figure 12.5 Sunderban delta in india
Modified after Paut et. al., 1998; Bose, 2004

Animals and Plants

Recorded number of species is 1,586 including 481 vertebrate, 1 hemichordate, 1,104 invertebrate. Total number of species included in Schedule I of WLP Act is known to be 40 that have 15 mammals, 8 birds and 17 reptiles. Total number species included in Appendix I of CITES Regulation are reported to be 14. Sundarban mangrove is the single largest home of the Bengal tiger, the flagship species, as its indigenous population. As per 2001 census, the number of tiger in Indian Sunderban is recorded as 271, out of which Sundarban Tiger Reserve had 245 tigers and South 24-Parganas forest division had 26 tigers. A total of 15 mammal - 8 bird - , and 17 reptile- species are included in the Schedules I and II of Wildlife (Protection) Act, 1972 (Table 12.3).

Mammals also include wild boars, spotted deer, porcupines, rhesus macaque, cetaceans including snubfin and Gangetic dolphin. Reptiles include king cobra, common cobra, common

Table 12.3 Floral and faunal diversity of Indian Sundarban mangrove

Species	Numbers
Plant	
Bacterial stain	24
Lichens	04
Algae	80
Fungi	22
Bryophytes	04
Pteridophytes	03
Hgher species	69
Animal	
Invertebrates	
Protozoa	104
Cnidaria	33
Ctenophora	02
Platyhelminthes	41
Nemathelminthes	68
Mollusca	142
Annelida	78
Arthropoda	476
Echinodermata	20
Vertebrates	
Chondricthyes	22
Fishes	154
Amphibia	08
Retilla	58
Ares	163
Mammalia	40

Source : Mandal, R.N. et. el. 2010
Science and culture 76 (4-8) : 275-282

krait, Russell's viper, python, chequered kil-back, dhaman, and green whip snake. Birds are herons, egrets, cormorants, storks, green pigeons, sand pipers, large and small spoonbills, darters, seagulls, teal, partridges and wild geese, ducks and migratory goliath heron. Fish, molluscs, crabs, and prawns inhabit the estuaries. *Peripthalmus* and *Baleopthalmus* are interesting. Crustaceans are *Uca* spp. and Tachepleursgygus and *Carcinoscropius rotundicauda*. Honey bee (*Apis dorsata*) has economic importance for the poor fringe people.

Threatened and extinct reptiles are *Crocodilus porosus, Varanus bengalensis, V. salvator, V. flavescens, Chelonia mydas*, Eretmochelys imbricata*, Lepidochalys olivacea, Caretta caretta*, Demochelys coriacea*, Lissemys punctata, Trionyx gangeticus, T. hurun* Batagur baska, *Python molurus*. Threatened and extinct birds are *Pelecanus philippinensis, Theskiornis* melanocephalus, *Leptoptilos javanicus*, Ardea goliath, Sarkiodornis melanotus*, Cairina scutulata**.

Threatened and extinct mammals are *Panthera tigris, Muntiacus muntjac*, Bubalis bubalis*, Rhinoceros sondaicus*, Cervus deruchea*, Axis porcinus*, Platanista gangetica* (*=Extinct species). The northern boundary and new depositions are inhabited by baen along with dhani grass. Baen is gradually replaced by genwa and goran. The southern and eastern associates include garjan, kankra and patches of sundari. Pure hental forest exists in relatively high land. Dhundul, passur and nipa palm swamps are extremely limited. A total of 26 true mangrove species, 29 mangrove associates, and 29 back mangrove species of 40 families and 60 genera are reported. Out of 60 varieties of mangrove species found in India, Sunderban accounts for 50, many of which are rare seemingly unlimited capacity to absorb pollutants from air and water.

Degradation of Sunderban Mangroves

Mangroves are over-exploited on large scale due to human pressure, conversion of mangrove areas for agriculture, horticulture, aquaculture especially for seeds of tiger prawns, and human settlement. About 50% of Indian Sunderban mangrove has been reported to be lost. Unscientific wood extraction destroys mangroves. National Remote Sensing Agency recorded a decline of 7000 ha of mangroves within a 6 year period between 1975 and 1981.

Sunderban is reported to loose 100 sq. km per year and would lose another 15% of its habitable land, displacing more than 30,000 people by 2020 as reported elsewhere. Large scale destruction of estuarine fish and prawn seed occurs in the Hooghly- Matlah estuarine system. Seed collectors harvest mostly tiger shrimp seeds as it has trade value. It is recorded that about 40,000 shrimp seed collectors annually harvest about 540 million seeds and destroy about 10.26 billion seeds of other fish and shrimp. Pollution, sedimentation and erosion, and embankment constructions also threaten the mangrove. Soils in many areas are rich in pyrite (FeS2), which transforms into sulphuric acid on exposure to air and makes the soil unfertile.

Effluents from factories also spoil the water. Sagar island, one of the first inhabited island in the world has been submerged by marine water. Vanishing of Bedford and Lohachara islands displaced climatic refugees. Inward migration of such refugees has caused mangrove degradation. Politics of gaining new lands for settlement exists here. Threats are also due to reduced flow of fresh water into mangrove, extension of non-forestry land use into mangrove, straying of tiger into villages, poaching, and uncontrolled fishing.

12.15 Globally Threatened Species

One-fifth vertebrate species of the world are in danger of extinction due to habitat destruction, habitat modification or fragmentation, pollution, pathogen, emergence and reemergence of diseases, and introduction of invasive species. Vertebrates are important for food, maintaining ecological balance, research and many other purposes. Almost 50% amphibians are going to be extinct. In birds, it is 13%. Most vertebrate losses occurred in the Southeast Asia, followed by Australia and Central and South America.

Occurrences of threatened species by country are also useful in providing a crosscheck to national Red Lists and vice versa with important implication for national conservation policy. Numbers of endemic threatened species in each country is necessary as they can guide a 'doctrine of ultimate responsibility' for each nation's contribution to global biological heritage .Besides 6 taxonomic groups viz. mammals, birds, amphibians, turtles, conifers and cycads for which coverage of species in the Red List is most complete; a 7th group the chondrichthyan fishes are included by IUCN as the largest marine group now on the Red List. Amphibians have very poor dispersal abilities over saltwater and do not occur naturally in many oceanic island nations important for threatened birds such as Mauritius or Vanuatu. Second countries with largest number of threatened and threatened endemic species lie in continental tropics. Those with highest proportion of threatened endemics are generally tropical island nations like Cuba with >50% of threatened species endemic for 5 of the 7 taxa. This is a combined result of low species richness of island and ecological naiveté of those species that do occur on island.

Countries or territories holding large number of threatened species include Colombia, India, New Caledonia, Peru, South Africa and Viet Nam while Colombia, India, Malaysia, Myanmar, New Caledonia, Papua New Guinea, the Philippines, South Africa and the United States are all among the top three countries for numbers of threatened endemics for at least one taxonomic group. Additional countries characterized by particularly high proportionate threat in multiple taxa include Madagascar as recorded in the Geography of the Red List compiled by IUCN.

Numbers and proportion of total and threatened mammals, birds, amphibians, turtles, chondrichthyan fishes, conifers and cycads occurring in terrestrial, freshwater and marine ecological systems are known. Absolute number of threatened species known from marine systems is low due to recording biases towards terrestrial and freshwater taxa. Only 187 threatened marine as opposed to 4,427 terrestrial species are recorded. A total of 1,388 freshwater species are listed as threatened.

A higher proportion of freshwater mammal species is threatened than of marine or terrestrial species. Among birds, turtles and chondrichthyan fishes, a higher proportion of marine species is threatened than of freshwater, or terrestrial species. A slightly higher proportion of terrestrial amphibians are threatened than of freshwater species. Across all seven taxa the proportion of threatened freshwater species is slightly higher (25%) than that for marine and terrestrial systems (22 and 21% respectively), as expected from intensity of threat to freshwater.

Biogeographic realms are 8 continents -scale terrestrial and freshwater regions distinguished by characteristic biota .Greatest number of threatened species for all taxa, occur in the tropical continents. In Australasia and Palearctic realms, less threatened species are many .Nearctic has more threatened amphibians than Australasia. Antarctic realm has almost no threatened species. Oceania with a low richness of threatened species has a high threat, due to vulnerability of oceanic island biodiversity. Proportionate threat is similar between biogeographic realms and taxa, although rather low for Nearctic mammals and Australasian amphibians and high for Indomalayan and Neotropical amphibians.

The FAO has defined 19 marine regions worldwide. Relative to seabirds, chondrichthyan fishes and seahorses marine mammals appear to contain largest numbers of threatened species occurring in northern Pacific Ocean regions. The greatest numbers of threatened sea birds, chondrichthyan fishes and seahorses are concentrated in "coral triangle" region of eastern Indian Ocean and southwest and western central Pacific. Arctic and Antarctic Oceans hold fewer threatened species across all taxa with exception of sea birds, several species of which occur in Antarctica.

Biomes represent global scale variation in structure, dynamics and complexity of terrestrial and freshwater communities and ecosystems that are driven key global scale patterns such as temperature and precipitations. Olson et.al identified 14 biomes world wide. Tropical/ Subtropical moist broadleaf forest is far and away the richest biome in terms of numbers of species and of threatened species for all three taxa, and is supposed to be only biome holding significant numbers endemics or of threatened endemics for any of 3 taxa. Large numbers of species and of threatened species for all 3 taxa with exception of latter are found in Tropical / Subtropical dry broadleaf forest. Tropical/ Subtropical Grassland, Savanna and Shrub land, Montane Grassland and Shrub land and Desert and Xeric shrub land. High latitude biomes of Boreal Forests/ Taiga and Tundra hold very few species. Forest habitats are most important for birds and amphibians.

Grassland and shrub land habitats hold high numbers of species. Inland wetland habitats are important particularly for amphibians, those have a larval stage. . Marine and desert habitats come out as having few species with exception of marine birds. Desert habitats are rich with reptiles. Species richness of all species of mammals, Western Hemisphere birds, freshwater turtles and amphibians is mapped. Species richness patterns are primarily driven by distributions of common, widespread species, but they do not provide context for threatened species distributions.

Tropics hold much higher species richness than do temperature boreal and Polar Regions. Variation explains some of this pattern in landmass across latitudinal bands. Species richness is much higher in tropics based on area alone, peaking around equator for all taxa. Species richness per grid cell is tightly correlated between mammals, freshwater turtles and amphibians. High correlation coefficients for each of these 3 taxa compared to birds are recorded in Western Hemisphere. Axon specific differences driven by particular biological traits exist. Birds have ability to disperse over water more than most taxa and occur in large numbers on islands while ectothermic reptiles flourish in desert.

High richness of mammal in East Africa and of turtles and amphibians in southeastern USA are recorded. Such differences increase with increasing evolutionary difference between taxa. More mobile species such as birds have wide distributions. Sedentary species like amphibians exhibit narrow distribution. Most species have small ranges and these narrowly distributed species tend to co occur in centers. Extreme concentration of centres of endemism in the tropics has also recorded, a manifestation of "Rapoports rule" which states that mean latitude of a species range correlates with species range size, although generalization of this rule has been questioned and it may be explicable by chance alone rather than by any underlying biological cause. Species are not evenly distributed across the planet and threats to species are not evenly distributed.

Species richness of threatened mammals, birds, freshwater turtles and amphibians show interesting similarities as well as differences between the groups. All four taxa exhibits marked concentrations of threatened species in southern Brazil, Madagascar, Western Ghats of India, eastern Himalayas, central china, mainland Southeast Asia, Sumatra, Borneo, and Philippines. Threatened mammals, birds and amphibians are also concentrated in Andes, West Africa, Cameroon, Albertine rift of Central Africa, Eastern arc mountains of Tanzania and Sri Lanka. These same regions are "biodiversity hotspots". Mammals map is not noteworthy in that there is at least one threatened mammal species in most parts of world.

Concentrations of threatened mammals also occur in eastern Amazon basin, southern Europe, Kenya, Sumatra, Java, Philippines, New Guinea and Australia. Mesoamerica and

Caribbean island are relatively less important for threatened mammals. In Caribbean, this is probably due to past extinction but on the other hand clearly stands out for amphibians. Areas of great importance for threatened birds include Caribbean island, cerrado woodlands of Brazil, highlands of south Africa, plains of northern India and Pakistan, Sumatra, Philippines, steepes of central Asia, Eastern Russia, Japan, Southeastern China and New Zealand .As for mammals, Mesoamerica and Australia are relatively less important. Amazon basin, Europe, Java and New Guinea are relatively less important for threatened birds. Threatened freshwater turtles exhibit different species richness pattern than other taxa. Their richness is very low in Atlantic Forest, Cerrado, Tropical Andes, Guinean, Forests of West Africa, Eastern Arc Mountains and Coastal forests and other hotspots holding so many threatened mammals, birds and amphibians. They also concentrate in Amazon ,eastern and southwestern United States and Asia Minor.

Most of the world is devoid of threatened amphibian species. Threatened amphibians occur densely in smaller areas than either mammals or birds. Majority of worlds known threatened amphibians occur from Mexico south to northern Peru, and on Caribbean island . Eastern Australia and southwestern cape region of South Africa are centres of amphibian threat. In mammals, birds, freshwater turtles and amphibians, the proportion of fauna in danger of global extinction is high in island ecosystems such as Caribbean, Madagascar, Sunderland, Philippines and New Zealand.

Map of relative threatened amphibian richness largely parallels that for all species. Relative distribution of threatened mammals and turtles is much more expansive. This covers threatened but species poor areas of temperate zone, such as California, fringes of Sahara and central China. Distribution patterns of threatened reptiles in particular lizards are likely to highlight the importance of some arid ecosystems. Some distribution patterns of threatened plants do not match those of most animal groups, notable example being the Cape Floral Region and Succulent Karoo of South Africa and deserts of Southwestern United States and Northern Mexico. There are also very different patterns of threat among some freshwater groups. For example, the Mississippi drainage system is probably the global centre for threatened freshwater mussels.

Threat Patterns in marine ecosystems are different and data on this regard are still largely unavailable. Ranges of globally threatened seabirds cover marine areas in Economic Exclusion Zones of many countries along with large parts of open oceans outside national sovereignty. Highest density of threatened birds at sea is found in international waters in southern oceans, with particular concentrations in Tasman Sea and southwestern Pacific around New Zealand .International cooperation is therefore required to conserve such species, many of which are threatened through incidental capture by commercial long line fisheries. As spatial resolution of data on geographic distributions of threatened species increases, so does the utility of these data for conservation but unfortunately, the effort required to compile the data does as well. Nevertheless the world's museums and herbaria represent a vast storehouse of such fine- scale geographic biodiversity data and several initiatives are underway that suggest that these data will become increasingly available in the future.

One important data set concerns the distribution of Critically Endangered (CR) and Endangered (EN) species restricted to a single locality. Maps of all sites hold the last remaining populations of a CR or EN mammal, bird, amphibian or conifer species and those reptiles assessed globally to date are available. Most of these sites lie in tropics, especially on island. The map shows stronger pattern in Latin America and Africa than it does in Southeast Asia, where sites are scattered liberally across the continent. In Africa all threatened birds species has been mapped at locality scale using Important Bird Areas approach of Birdlife International. Similar work has been completed in Middle East, Europe, Asia, Canada, Mexico and Andes and is on-going in pacific rest of Americas, Antarctica and marine areas.

To date, some 4,032 sites holding threatened birds have been identified world wide. Localities holding threatened birds species are highly clustered in Africa regions like Mediterranean coast. Upper the Cameroon highlands, the East and South African montane highlands, Madagascar and Indian Ocean islands, have particular concentrations of sites holding threatened birds. Miombo Mopane woodlands of south central Africa hold several sites, albeit spared fairly far apart and Sahara -Sahel, Congo forests and Kalahari have very few localities hosting threatened species.

The 2004 IUCN Red List covers 2, 140 assessments of infra-specific taxa or discrete subpopulations, of which 1,383are threatened. The 15,589 species threatened with extinction includes 12% of total bird species, 23% of total mammal species, 32% of total amphibian species and 34%of all gymnosperms. One in every 8 birds, one in every 4 mammals and one in every 3 amphibians and gymnosperms is facing a high to extremely high risk of extinction in near future. A total of 2.5% of world's described species have been evaluated for IUCN Red List with a strong bias towards terrestrial vertebrates and plants particularly found in well studied parts of world.

Number of mammal species has increased form 4,629 in second editions of Mammal Species of World to 5,416 in third editions. Number of reptile species has increased form 7,970 in 2000 to 8,163. Number of amiphibian species has increased from 4,950 in 2000 to 5, 7439. Number of fish species has increased from 25,000 in 2000 to 28,500. Number of bird species has remained fairly stable since 1996, with exception of albatrosses where species number has increased from 14 to 21. Number of invertebrate is highly provisional with as much as 20% uncertainty. Number of 259, 000 seed plant species is highly debated. Vertebrates are best evaluated with almost 40% species recorded on IUCN Red List. Birds and amphibians are fully evaluated. Number of mammals evaluated has declined form 100% in 1996 to almost 90%.The 2004 IUCN Red List includes 7,266 animals species threatened with extinction compared to 5,435 in 2000.

Comparing the numbers of threatened species in major taxonomic groups reported for 2000 and 2004 updates of the Red List, clearly overall number of threatened species has increased in all groups, with exception of mammals. Increase in numbers of threatened animals is mostly due to incorporation of assessment for all amphibian species for first time. Proportions of each taxonomic group threatened with exception of that for amphibians has remained same as in 2000. In birds for which since 2000 there has been a decrease in number of recognized species (9,946 to 9,917) and an increase in number of threatened (1,130 to 1,213). Often these apparent increases is the result of better knowledge or changes in taxonomy. Majority of Near Threatened animals species are mammals (587) and birds (773). If numbers in this category were combined with those listed as threatened then percentage of birds, mammals and amphibians that are threatened or near threatened would rise to 20%, 35% and 39% respectively.

There are still 111 animals species listed as LR/cd (lower-Risk, Conservation dependent) and time as these are all re -evaluated this category will persists as an artefact of previous classification system. The LR/cd category was rarely used for animals except for mammals which still have 64 species in this category, 39 of which are hoofed mammals or artiodactyls, and 14 are cetaceans. Rodents are the only mammalian with significantly fewer than expected threatened for extinct species despite having largest number of threatened mammal species on Red List. Significantly, more threatened species belong to five orders namely the Sirenia, Perissodactyla, Artiodactyla, Primates, and Carnivora than would be expected.

Families with higher numbers of threatened species include Ominidae, Tapiridae, Nesophontidae, Indridae, Equidae, Peramelidae, Lemuridae, Cpromyidae, Felidae Cercopithecidae, Bovidae and Pteropodidae. Highly threatened families are species poor. Bovidae, Cercopithecidae and Pteropodidae are relatively species rich. Major threats to Bovidae and Cercopithecidae include habitat loss and hunting, while for Pteropodidae, habitat loss due to extraction of timber, hunting for food, and general human disturbance. Apterygiformes, Sphenisciformes, Pelecaniformes, Procellariiformes, Ciconiiformes, Galliformes, Gruiformes, Columbiformes and Psittaciformes contained significantly more threatened species. Piciformes, Apodiformes, and Passeriformes have significantly fewer threatened species than average.

There are 15 extinction prone families viz. Mesitornithidae, Apterygidae, Gruidae Spheniscidae, Megapodiidae, Diomedeidae, Drepanididae, Phalcrocoracidae, Cracidae Procellariidae, Zosteropidae, Rallidae, Phasianidae, Columbidae and Psittacidae. Ten families contain significantly fewer than expected threatened species viz. Bucconidae, Dendrocolaptidae, Paridae, Capitonidae, Nectariniidae, Picidae, Trochilidae, Emberizidae and Muscicapidae.In amphibian, Gymnophiona are significantly less threatened than average, but this is misleading because 111 of the species are listed as Data Deficient. Caudata have significantly more threatened species than average, as species tend to have small ranges and are very sensitive to habitat loss. Average number of threatened species is determined by anurans by far the largest amphibians group with over 5,000 species. Families with significantly more threatened species than average include Astylosternidae, Hynobiidae, Rhacophoridae, Plethodontidae, Bufonidae, and Leptodactylidae. Astylosternidae are confined to west and Central Africa, with highest diversity being centred on Cameroon. Hynobiiade are very sensitive to habitat loss and have more threatened species than expected. High levels of threat in Rhacophoridae are mainly a reflection of large number of threatened species in genus *Philautus*. Many species in *Plethodontidae* also tend to have very small ranges.

Mexican and Central American members of this family are particularly threatened because of habitat loss. Bufonidae has the largest number of species that appear to be rapidly declining due to impacts of chytrid fungus. Most dramatically, 74 of the 77 species in genus *Atelopus* are threatened or extinct. Other high profile toad genera with high percentages of threatened species include Nectophrynoides and Nimbaphrynoides. Leptodactylidae is the largest amphibian family more than half of which are considered threatened. The family is dominated by the 700 members of Eleutherodactylus. Very small but phylogenetically significant families where all species are listed as threatened or extinct include Rheobatrachidae,

Nasikabatrachidae from India comprising a single evolutionally unique threatened species; Rhinodermatidae; Leiopelmatidae; and Sooglossidae.

Giant salamander are also worthy of mentioned, with one of three species being Critically Endangered, and other two being near Threatened. Families with significantly fewer threatened species than average include; Pipidae, Ichthyophiidae; Caeciliidae; Myobatrachidae; Mantellidae Hyperoliidae; Microhylidae; and Hyliade. Species in these families are less threatened than expected nearly all of them include several very seriously threatened species. Amphibians differ from mammals and birds in that many families with largest percentages of threatened species are species rich notably Leptodactylidae, Bufonidae, Rhacophoridae and Plethodontidae. Amphibian results are similar to mammals and birds in that there are several small highly threatened families that are phylogenetically unique.

Massive decline and increasing number of extinctions being observed in amphibians would lead to a disproportionate loss of evolutionary novelty. Two reptile orders have been completely evaluated namely Crocodylia and Rhynchocephalia. Crocodylia have 10 (43%) threatened species out of 23 described. *Alligator sinensis* is the most threatened crocodilian in world with a large population in captivity and an Action Plan has been drafted to reverse long trend of habitat loss and population decline for this alligator. Tuataras from New Zealand are only surviving members of their order are known only from the fossil record. One tuatara species is listed as threatened and other is considered to be Least Concern.

Testudines are well covered on the IUCN Red List with 205 (67%) of 305 described species, 128 (42%) of which are listed as threatened. *Amphisbaenia* have not been evaluated. Few snakes and lizards have been evaluated. Within lizards, the main focus has been on the iguanidae and other closely related families. Many snakes and lizards are cryptic, hard to find and poorly known.

Global Reptiles Assessment started in 2004 will greatly improve our knowledge of this group of vertebrates. The IUCN Red List includes 131 threatened marine fish species. Among the species included are seahorses and pipe fishes, groupers, wrasses, damselfishes, angelfishes and chondrichthyan fishes. While several species are restricted range coral reef fishes, some are widespread, commercially valuable species subject to fisheries. Slow life histories and low population growth rate of sharks, skates, rays and chimaeras limits their capacity to withstand over fishing and habitat destruction. To date, the IUCN/SSC Shark Specialist Group (SSG) has assessed one third (373 species) of world's chondricthyans and 17.7% are listed as threatened, 18.8% Near Threatened 37.5% Data Deficient and 25.7% Least Concern .Restricted- range species occupying heavily fished areas, such as sawfishes and deep sea dogfishes, typify some of more seriously threatened species. Least concern species share several features. They are abundant and /or widespread, occur in protected areas or areas with limited fishing, are not particularly susceptible to fisheries or are taken by well-managed fisheries.

All 7 species of sawfishes are listed as critically endangered or endangered. Sawfishes inhabit coastal tropical, subtropical and warm temperate regions, often in estuaries and freshwater. Their unique 'saw', a long rostrum studded with 'teeth'-makes them vulnerable to capture in nets and difficult to remove alive. With highly priced fins and 'saws' some fisheries targets these species but most mortality is other fisheries also, compounded by effects of extensive coastal development. Assessment for Tope Shark, *Galeorhinus galeus* lists it as Endangered, Vulnerable, and Near Threatened .

The species has valuable flesh but is having particularly low biological productivity and comparing the situation across regions demonstrates how inadequately managed species can be severely depleted. Four species of guitar fish ,fins of which are most valuable in world have been listed as threatened due to high levels of exploitation .Skates comprise a quarter of all chonndrichthyans and many have restricted geographic ranges and are potentially highly vulnerable to overexploitation. A total of 27% of the freshwater fish evaluated in East African region are listed as threatened. In North America where a recent analysis by Nature Serve of the status of 801 species of freshwater fish showed that 20% are threatened.

The International Council for Bird Preservation now called Birdlife International works on geographic ranges of birds; map them to identify Endemic Bird Areas (EBAs).

12.16 Biodiversity Conservation

Biological diversity consists of all plant and animal species, the genetic material they contain and the ecosystems which they inhabit. Genetic diversity means the variation of genes and genotypes between and within species. It is the sum total of genetic information found in living world. Diversity within species makes it tolerable to change in environment and climate. Ecosystems including major systems like grasslands, mangroves, coral reefs, wetlands and tropical forests, and agricultural ecosystems consist of interdependent communities of species and their interactions.

Loss of biodiversity is environmental, social, economic and political problem. Unsustainable consumption of resources by minority people, combined with impacts of poor people, has destroyed natural habitats. Several thousand plant species which were used for human food, now stands at about cultivated plants. Rice, maize and wheat supply about 60% of the calories. Genetic diversity enables crops to withstand drought and resist insect pests. Genetic material is the raw material for breeders and biotechnologists. Loss of biodiversity is a matter of serious concern.

Biodiversity provides the food, fibre for clothing, materials for shelter, fertilizer, fuel and medicines. Rural poor depend on bioresources for an estimated 90% of their needs. Access to bioresources is necessary for vast array of industrial products. Biodiversity maintains the ecological balance necessary for planetary and human survival. Human society depends on bioresources for food supply, medicines, clothing, fuel and building material and as an important part of mental and spiritual welfare. Reduced plant diversity impairs ecosystem functioning. Reduced biodiversity of grassland lowers the land productivity. Plant biodiversity is declining worldwide. Biodiversity loss may increase effects of climate change by reducing ecosystem ability to absorb carbon dioxide. Plants grow better in species-rich communities because each species has its own specialized way to gather resources for growth. Plants interact with each other in complementary and positive ways. Modern monoculture practices are probably less efficient in capturing and using resources. In "traditional" agriculture intercropping or companion planting were practiced. Grasslands that have lost some species are less resistant to environmental changes. Biodiversity buffers extreme climatic events like drought, flood, and fire. Extreme weather as seen during regular El Nino climate phenomenon is becoming more common as greenhouse gases accumulate in earth's atmosphere. Humankind lives on earth as a part of nature. Nature is a gift that provides us the food, water, and shelter.

Species diversity

Species are distinct units of diversity and their diversity or richness indicates the number of species within given area. Species richness is simply the relative abundance of species in a sample. About 1.75 million living species are described. Estimated number of total species ranges between 10 million to 50 million. Several levels of species richness patterns are known. Some taxa have more species than other. Number of beetle species is about 3 times than species number in any insect order. Two species-rich ecosystems are rainforests and coral reef ecosystem of tropical regions. Species richness is higher on larger islands. More species inhabit islands that are close to mainland. Local species richness is influenced by productivity, disturbances, like fires, and floods. In some communities, species richness is highest at intermediate disturbance. Frequent disturbance remove sensitive species. Infrequent disturbance allows time to superior competitors for removing less competitive species. New species arise through speciation. Species are lost by the process of extinction. Species richness varies through evolutionary time. Rapid speciation occurs during adaptive radiations.Rate of extinction vary among taxa and over time. At least 5 periods of mass extinction are known. Earth is now experiencing the 6th mass extinction. Some species or species groups that control energy flow and affect environment of other species to a major extent, are called ecological dominants. Their removal results in abrupt change in community. Removal of less or non-dominant species bring less change. Few dominants largely account for energy flow for trophic group. Number of rare species mostly determines the species diversity of whole community. Species diversity tends to be low in physically controlled ecosystems and high in biologically controlled ecosystems. Some species found in only one community are called exclusive species. Others live in many communities are called ubiquitous species. Diversity of species within habitat is called alpha diversity. Number of species in a region is called gamma diversity Ratio of gamma diversity to alpha diversity is called beta diversity

Ecosystem Diversity

Ecosystems can be divided into natural and anthropogenic. Marine ecosystems and sacred groves are examples of natural ecosystems. Mining areas and cities are largely anthropogenic. Human interference in natural ecosystems often reduces biodiversity. Some sustainable agro ecosystems which do not completely replace nature can actually increase local biodiversity which can serve as models as models for future sustainable agricultural development. In India, major ecosystems include forests, grasslands, deserts, wetlands, mangroves, coral reefs and marine ecosystems.

Ecosystems provide habitat for organisms. Healthy ecosystems deliver life sustaining service for free that humanity finds it practically impossible to substitute them. By affecting ecosystems functions through altering the food web, invasive species reduce the ability of ecosystems to deliver life sustaining services. Ecosystems recycle nutrients which include the elements of atmosphere and soil that are necessary for maintaining life. Biodiversity is essential in this process. Plants take up nutrients from soil and from air, and these nutrients form the basis of food chains. Soil's nutrient status is replenished by dead, or waste matter which is transformed by microorganisms; this may then feed other like earthworms which also mix and aerate the soil and make nutrients more readily available. Vegetation is integral for maintaining water and humidity levels and oxygen/carbon dioxide balance of the atmosphere.

Natural habitats afford sanctuary to breeding populations. Birds and nector loving insects roost and breed in natural habitats pollinate crops and native flora in surrounding areas. Populations of biota may end up with small genetic bases, which may lead to extinctions. Ecosystem degradation exacerbates the frequency and impact of natural hazards and can intensify competition and potential for conflict over access to shared resources such as food and water.

Forests canopies serves as particulate filters and reaction sites for regulating atmospheric composition and purifying air. Moist leaf surfaces provide sites for transforming polluting compounds. Clearing of forest for agriculture or other uses is of major concern, particularly in tropics. Deforestation can lead to change in regional and global climate. When a large forest is cleared, rainfall may decline and droughts may occur frequently. Deforestation contributes to global warming by releasing stored carbon into atmosphere and by eliminating a sink for atmospheric CO_2. Nitric oxide produced by soil microbes is a very reactive gas. NO combines with substances on leaf surfaces and does not reach the air above canopy.

Forests regulate water flows to downstream areas. Deforestation often alters natural flow pattern causing flood, or drought. Forest soils purify waters. In a healthy middle aged forest, rain falling enters with nitrogen load of about 8 pounds per acre each year while stream water leaving this forest often contain less than one tenth of nitrogen entering in rainfall as reported from New England.

Many weeds, insects, rodents, bacteria, and fungi compete with humans for food, shelter, and affect food production, or spread disease. Certain animals and microbes naturally control some pests. Scientists have developed biopesticides to replace traditional pesticides.

Forests and grassland provide natural protection to soils. Plant canopies reduce force with which rainwater hits soil surface. Roots prevent soil from washing down slopes. Following excessive rains, flood waters flow into floodplain, and wetland. When floodwaters recede, they leave nutrient for enhancing soil fertility. Natural flood plains provide habitat for plant and animals species. High cost of flood damage resulted in part, from drainage of floodplain wetlands, building of permanent structures on floodplain, and construction of levees. Salt marshes mangroves forest buffer the coastline against ocean storms. Mangrove plants stabilize submerged soil and prevent coastal erosion. Such ecosystems are also breeding grounds and nurseries for important fish and habitat for many species. Mangroves are effective in controlling raging floodwaters from tropical storms. Mangrove forests are under threat from coastal development, shrimp aquaculture and logging.

Land ecosystems are large storehouses of carbon both in plant and in soil organic matter. These ecosystems absorbing carbon and slow the accumulation of atmospheric CO_2. Burning fossil fuels are increasing atmospherics concentration of CO_2 and other gases.

Many flowering plants rely on animals to help them mate. About one third of world's food crops depend on natural pollinators. Toucans, monkeys and fruits bats consume tree fruits and scatter piles of seed rich dung across landscape which helps trees to populate their habitat.

Desertification affects 70% of world's dry lands due to unsustainable human activities. Desertification undermines food production. Stabilization of soil against water and wind erosion is diminished. Degraded land may cause downstream flooding reduced water quality sedimentation in rivers and lakes and the accumulation of silt in reservoirs and navigation channels. It can cause dust storms that exacerbate human health problems including eye infection, respiratory illnesses, allergies and cases of meningococcal meningitis. Critical habitat for plant and animals specie is lost as desertification proceeds leading to economic losses including those from declining tourism.

Over the 20th century, some 10 million square kilometers of wetlands have been drained globally. In the lower 48 states of the U.S., drainage has decreased wetlands. Pollution of air rain (and snow) surface waters and land diminishes ecosystem services. Air pollutant ozone reduces growth of agricultural crops and plant in natural ecosystems. Acid rain damages plants, impoverishes soils. Nitrogen pollution causes harmful algal bloom and depletes the waters of oxygen. Heavy metals form smelters accumulate in soils killing plant life and thus creating erosion problems. DDT and PCBs can alter food webs and thereby diminish ability of ecosystems to deliver services such as pest control. Microbes detoxify some human generated wastes. Oil spilled into ecosystems poses health risks to humans and other animals. Aesthetic values of our natural ecosystems and landscapes contribute to emotional and spiritual well-being of a highly urbanized population. Conservation of biodiversity also has ethical benefits. Presence of a wide range of living organisms reminds people that they are but one interdependent part of earth.

Genetic Diversity

Genetic diversity is the variety present at the level of genes. Genes are the building blocks that determine how an organism will develop and what its abilities will be. This diversity can differ by alleles, by entire genes, or by units larger than genes. Genetic diversity is measured at various levels. Genetic diversity is important because it provides the raw material for evolution and adaptation. More genetic diversity in species means greater ability to adapt to changes in environment. Less diversity leads to uniformity, which is a problem in long term, as it is unlikely that any individual in population would be able to adapt to changing conditions.

An increase in species diversity can also affect the genetic diversity. If there are many species, the genetic diversity at that level will be larger than when there are few species. Species that are closely related have similar genetic structures and makeup and therefore do not contribute much additional genetic diversity. Such species contribute to genetic diversity in the community less than more remotely-related species would. Genes are functional part of DNA that remains inside cells of living organisms. They carry variations of inherited biological information from parent to offspring. There are inherited genetic differences inside each living organism. Living organisms contain a "history" of specific genes from their parental line, which makes individual living organisms of the same species different from one another in some genetic ways. There are no exact duplicates of any living organism.

Conservation of genetic resources can be traced to 1910's. Origin of interest of agriculturalists in domesticated crops and in use of wild relatives of crops in breeding programmes goes back to the same year. By 1924, Russian botanist Nikolai I. Vavilov founded the All- Union Institute of Applied Botany and New Crops. Before the Second World War, the Institute sponsored 180 collecting trips in 65 countries. By 1940, it held about 200 thousand accessions of crop and vegetables. Number and size of crop gene banks has continued to grow dramatically ever since. Presently, we depend on a small proportion of existing plant diversity. Roughly 250,000 species of land plant are currently extant. Of those about 60,000 are believed to possess human food value. Over course of recorded human history, about 3,000 have been used by some cultures as sources of food stuffs and only about 150 have been commercially cultivated. In 1974, only seven plants including Wheat, Rice, Corn, Potato, Barley and Sweet potato, out of 30 major cultivated crops was harvested in excess of 100 million tons. Only three species i.e. Wheat, Rice, Corn accounts for over two-thirds of world's total grain crop. Genetic erosion, the reduction of diversity within and the main cause of extinction of a species, is a global threat to agriculture.

India's Biodiversity and Conservation

The two richest areas of India in terms of biodiversity are the Eastern Himalayas and the Western Ghats. Strategies for biodiversity conservation have comprised providing special protection to biodiversity rich areas by declaring them as protected sites. About 4.2% of the total geographical area of the country has been earmarked for extensive in-situ conservation of habitats and ecosystems. The country has 15 Biosphere reserves, 604 protected areas (97 National parks and 507 Wildlife Sanctuaries), covering more than 5% of the land surface. The Indian Council of Forestry Research and Education (ICFRE) have identified 309 forest preservation plots of representative forest types for conservation of representative areas of biodiversity. Six Indian Wetlands have been designated as wetlands of International importance under the "Ramsar Conventions". 'Assistance to Botanical Gardens' provides one-time assistance to botanical gardens to strengthen and institute measures for ex-situ conservation of threatened and endangered species in their respective regions. Programmes have been started for scientific management and proper use of wetlands, mangroves and coral reef ecosystems. A total of 21 wetlands, and mangrove areas and 4 coral reef areas have been identified for intensive conservation and management purposes. Mangroves conservation is the thrust areas of the Ministry of Environment and Forests.

Conservation of genes, species, and ecosystems is normally done in one of two ways. In situ conservation emphasizes protection of biodiversity in existing natural areas. Ex situ conservation involves the establishment of man-made areas where biodiversity is conserved outside of its natural habitat. In situ conservation is preferred because it allows living organisms to continually adapt, according to environmental conditions. This can be accomplished by establishing protected areas, and gene reserves. There is only one biosphere reserve in the Western Ghats, the Nilgiri Biosphere Reserve which helps conserve many endemic and endangered species, and their wild relatives.

Ex situ Conservation

A scheme 'Assistance to Botanic Gardens 'provides one time assistance to botanic gardens to strengthen the ex situ conservations measures. Another programme is the captive breeding programmes. Government of India started a crocodile breeding and management project in 1976 to save endangered salt water crocodile, fresh water Crocodile and the gharial. Thousands of crocodiles of 3 species have been reared at 16 centers. Eleven sanctuaries are designated for crocodile protection including National Chamber Sanctuary in Madhya Pradesh. Endangered white-winged wood duck was also bred in captivity and released into Protected Areas of the Northeast (MoEF, 2002).

The conservation of biodiversity is not limited to wild species. Farmers have selected crops and livestock, and maintaining many varieties and breeds which are being replaced by high-yielding strains. They are valuable sources of genes for livestock and plant breeders, and should be conserved. Farmers can help preserve them in situ by continuing their traditional farming practices. Universities and institutes maintain ex situ collections of seeds and live plants and animals to help maintain these valuable genetic resources.

In situ Conservation

Extensive system of protected areas (PA) encompasses at present 89 national parks and 496 sanctuaries (MoEF, 2001) covering an area of 1.83 lakhs sq km. Tura Range in Garo Hills of Meghalaya ,a gene sanctuary preserves rich native diversity of wild *Citrus* and *Musa* species. Sanctuaries for rhododendrons and orchids are located in Sikkim. Following the UNESCO/MAB criteria, 15 areas have been designated as Biosphere Reserves (BR). Kaziranga NP, Keoladeo Ghana NP, Manas WLS, Nanda Devi NP and Sunderbans NP have been declared as World Heritage Sites. Twenty seven tiger reserves in 14 states cover an area of 37 761 Km2 for providing protection to tiger and swamp deer, elephant, rhino and wild buffalo. Other special programmes/projects include the conservation of Indian elephant rhino, lion, certain primates, and aquatic mammals including river dolphins. A total of 309 plots of representatives' forest types have been identified for conservation of viable representative areas of biodiversity. Of these 187 plots are in natural forests and 112 in plantations, covering a total area of 8500 hectares. Fifteen mangroves and four coral reefs were identified for conservation in 1986/87. Another 15 mangrove areas have been added to the list. Coral reef research and monitoring training and capacity-building establishment of a database are carried out by Coral Reef Monitoring Networking. Conservation of medicinal plants is a priority issue. With collaboration of the State Forest Development of Kerala, Karnataka and Tamil Nadu, Foundations for Revitalization of Local Health Traditions have established 30 Medicinal Plant Conservation Areas and 15 parks for conserving the germplasm of these threatened, rare and endemic, medicinal plants. In situ conservation of agro-biodiversity including traditions of farming communities has been strengthened through steps taken by the Government and NGOs (MoEF, 2002)

EBAs

EBAs are areas with large number of endemic birds in restricted ranges. Locality records help identify bird species with breeding ranges. It is recorded that 27% of world's bird species have breeding ranges restricted to less than 50,000 km^2. Species of restricted range occur on islands or in isolated mountain. A total of 221 EBAs contains habitats of about 2,500 restricted range endemic birds. Most EBAs are found in tropics either on islands (46%), or in mountain in continental areas with few in northern temperate areas. Southern hemisphere with about 25% of world's land area contains more than half of EBAs (119).

The smallest (5 km^2)EBA is found in northwestern Hawaiian Islands and largest (174,000 km2) in northern South America in French Guiana, Surinam, Guyana, and Brazil as recorded by scientists. Regardless of their location, restricted range-endemic bird species are often listed as threatened (29 %), and most EBAs (85 %) contain threatened endemic species. Only 8 % of land area found in EBAs is under protection.

Nearly one third of individual EBAs have no protected areas coverage whatsoever and 35% have less than 5 % of their legally protected area .In Amazonia, there is a high overlap between centers of endemism for birds, some lizards, butterflies, and trees. Some general patterns of congruence in endemism between birds and other taxonomic groups appear. Overlap varies from region to region and distribution patterns for groups other than birds are poorly documented in most parts of world. The EBAs were classified into 3 groups to show increasing level of biological importance and classified separately into another 3 groups by degree of threat to area. These 2 classification systems are then combined into a conservation priority classification. Those EBAs with a combined score of 5 or 6 are classified as "critical," those with a combined score of 4 are "urgent," and those with a combined score of 2 or 3 are "high" conservation priority.

12.17   Major Wilderness Areas

Wilderness approach emphasizes identification of large relatively undisturbed natural areas with low human population densities and does not explicitly evaluate species diversity or endemism. Globally, relatively undisturbed habitats cover about one third of earth's land surface, much of it composed of desert, boreal, and arctic/Antarctic ecosystems. As relatively undisturbed habitats shrink in temperate, tropical and even boreal zones, such areas will become more important for biodiversity conservation. Wilderness areas in tropics, particularly the humid tropics are most important. Large wilderness areas in most of temperate world have disappeared and major wilderness areas in tropics are becoming increasingly rare. Few major tropical wilderness areas would contain large tracts of primary forest into next century.

Major wilderness areas should not be overlooked in assessment of conservation priorities. They can serve as "controls" against which the effects of human activities in managed ecosystems can be measured. They are biodiversity storehouses, where individuals of many species would continue to exist. They play a key role in maintaining local, regional, and sometimes because of their large size global climate patterns. They will be the last areas where aboriginal peoples can live their traditional lifestyles. Inventory of McCloskey and Spalding produced an area of about 48 million square kilometers of wilderness in 1,039 separate areas in 77 countries, covering 32.3% of the earth's land surface area. They showed that 41% of large wilderness areas are found in high arctic or Antarctic, 20% in warm desert areas, 20% in temperate regions, 11% in tropics, 45 in mixed mountain systems, 3% in cold desert regions and a small fragment in island regions. The region's rich and diverse wildlife is preserved in 89 national parks, 13 Bio reserves and 400+ wildlife sanctuaries across the country.

Several animals species are threatened in India. Wildlife management in the country aims to conserve such species. Wilderness inventory does not identify biodiversity conservation priorities per se. Conservation International has modified wilderness inventory approach to assess humid tropical forest conservation opportunities as part of its Rainforest Imperative program. Major tropical forest wilderness areas identified as conservation priorities include island of New Guinea, humid tropical forests of Zaire Basin and a major area of forest wilderness from southern Guyana and Suriname, across southern Venezuela, northern Brazilian Amazonia, and down through western Amazonian lowlands of Brazil, Colombia, Ecuador, Peru, and Bolivia.

Advantages of identifying major wilderness areas as priority sites for biodiversity conservation include lower management and maintenance requirements, large habitat areas for a range of species, especially large predators, encompassing most if not all species for ecosystems within wilderness and relatively few of socials and economic pressures that afflict most smaller habitat "islands" surrounded by intensive land uses. There may be little information on how much biodiversity is actually included in the wilderness area. Identifying wilderness areas may be a first step in defining more detailed priorities and a comprehensive approach to setting biodiversity priorities. Wilderness areas can help focus biodiversity assessments in areas where conservation prospects are less complicated by human activities.

12.18 Conclusion

India has about 63.5% million hectares of forests. It is extremely difficult to the forest department, to safeguard the forests along with the sustainable harvest of forest produce by its own effort. While the government, and conservation enthusiasts and organizations have a major role in the protection and conservation, the role of local players who have lived in harmony with natural resources is vital for achieving conservation success. This fact was exemplified in the West Bengal. In the Arabari forest range of Midnapore district, Mr. A K. Banerjeee involved the villagers in protection of degraded Sal forest covering 1,272 hectares. Villager were allowed fuel wood and fodder on payment of nominal fee and given employment in both silviculture and harvesting operations, 25% of the final harvest was given to the local community. The Sal forests showed a remarkable recovery by 1983.Such instances have now been accepted as a model in which decentralized economic growth and ecological conservation go parallel. (Box 12.1)

Box 12.1

environment

BIODIVERSITY LOSS:

A case for global concern

F.B. Mandal

Biodiversity is increasingly threatened due to habitat destruction, pollution and various anthropogenic activities. Joint action has to be taken by the government and NGOs to concerve it.

Biodiversity is the "variety and variability among living organisms and the ecolofical complexes in which they occur." Biodiversity loss is the most-serious and rapidly-increasing of all the global environment problems. Being an important component of biosphere, biodiversity embraces a number of levels of variation. Genetic diversity, i.e., genetic variation within species fules the engine of evolution. Species diversity is related to the number and relative abundance of species in a community. Diversity encompassing the varriety of habitats and biomes is termed ecosystem diversity. Species diversity at various geographical scales are expressed as (i) alpha diversity, the species number in a small area of relatively homogenous habital, (ii) gamma diversity, the number of species in a region, and (iii) betadiversity, the ratio of gamma diversity to alpha diversity. Total number of named species is currently about 15 million including 7,50,000 insects, 41,000 vertebrates and 2,50,000 plants. Earth is losing such species, the existence of which may never even be known. Estimates suggest that 99.9% of all species those have ever lived are now extinct during the past 570 million years of the earth. Since 1600, extinctions of over 200 vertebrate species, mostly birds and mammals are recorded with the highest extinction rate on islands. Scientists like E.O. Wilson have pointed out that as much as 20% of the world's existing diversity may be lost during the next 30 years.

Biodiversity loss is tragic for a number of reasons. Once a species is lost, the unique informations contained in its genes are lost which are unlikely to be recovered again and its scope for further evolution is lost. Published literature holds the view that the "human society depends on genetic resources for almost all the food supply, half of its medicines, most of its clothes and in some region virtually all of its fuel and building material and as well as of course an important part of its mental and spiritual welfare" Biodiversity is everybody's concern and particularly of our poorer section who depend entirely on biomass for sustenance. Instability and unproductivity including desetrification, water logging, mineralizatin and many other unwanted outcomes throughout the world are caused by high rate of biodiversity destruction. International Union for conservation of Nature and Natural Resources (IUCN) has established 5 conservation categories i.e. extinct, endangered, vulnerable, rare and insufficiently known species. Threatened species include those which are endangered, vulnerable and rare species. Threatened species include those which are endangered, vulnerable and rare species of IUCN categories. Of the three types of global extinction's, in the first type called the

background extinction, some species disappear while others appear due to charge in ecosystem. Such extinction is normal in the natural world. The second type, the extinction that results as a consequence of natural disasters such as volcanic eruption, hurricanse or droughts which occur either locally or globally causing disappearance of species is called mass extinction. And the third type, anthropogenic extinction is caused due to human activities.

Threats to about 60,000 plants and 2,000 animal species including species of fishes (343), amphibians (158), reptiles (170) are described by World Conservation Mointoring Centre (WCMC). In India, many species once present in abundance are now known as endangered. Indian cheetah, one hormed rhinoceros, Sikkim stag, mountain quali and Jerdon's courser are the examples of extinct species from India in recent times. It is also suggested that over one hundred species of wild animals need immediate protection in India alone.

During the last 150 years, human population has increaed from 1 billion in 1850, to 2 billion in 1930, to 5.3 billion in 1990 and probably will reach an estimated 6.5 billion by the end of the current year. Such incease in human population degrades natural habital in a number of ways. For example, an assessment of wildlife habital loss in tropical Asia in 1986 reported that India had already lost about 80% of its natural habital. Tropical rain forests, tropical dry forests, wetlands, mangroves and grasslands are the examples of threatened. Habitat fragmentation by roads, fields, canals, powerlines etc. limits the species potential for dispersal and colonization. Physical degradation of forest habital due to ground fire seriously affects the rich prennial wild plants along with entomofauna inhabiting the forest floor.

Pesticides, industrial chemicals and effluents, emissions form factories and automobiles, introducton of exotic species etc. cause undersirable change in the natural habitals. A total socioceconomic change of the human society has resultes as a consequence of agricultural and industrial development along with a decreased importance on biodiversity. It is difficult to provide all the niche requirement of a species

and hence the best method for conserving biodiversity is the preeservation of entire habita.

Human population growth needs immediate control for ameliorating the biodiversity problem. Another aspect of biodiversity conservation in cludes the DNA

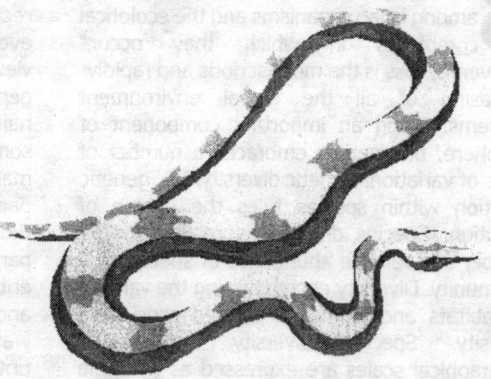

,libraries at molecular level (ex situ). Ex situ conservation is generally practised in the form of gene pool or gamete sotrage, sperm banks, germplasm banks in addition to captive breeding programme.

In India, conservation areas in the form of National parks (66), Sanctuaries (382), Tiger Reserves (15) and proposed Biosphere Reserves (13) covering an area of 1,32,000 sq. kms (15% of total endangered species. Regional strategies include the Protocol on Environmental Protection to Antarctic Treaty (1992) emphasis on monitoring the environmental damage and reporting on the progress of species protection measures. Lastly, it is important to mention that global priority must be given on biodiversity conservation with the implementation of effective conservation strategies.

Source : Environ, Vol/VIII 4 & 5, 2001

Summary

1. India has a rich heritage of wildlife and their conservation. Kautilya's Arthshastra contains the concept of Forest Reserves to protect elephants. Wildlife occurs in Deserts, rain forests, wetlands, mangroves, mountains and sea. The wildlife encompassing diverse range of species is preserved in protected areas. India's forest cover ranges from the tropical rainforest of the Andaman Islands, Western Ghats, and Northeast India to coniferous forest of the Himalaya. Ecoregions like the shola forests exhibit high endemism.

2. India home to several rare and threatened species provide shelter to many endemic species. As a whole 33% of plant species are endemic in India. India contains 2.9% of IUCN-designated threatened species like Asiatic lion, Bengal tiger, and Indian white-rumped vulture.

3. Now, wildlife is threatened due to various factors and activities. Human encroachment is a serious threat to wildlife. Exploitation of forest resources through hunting and trapping for food and sport has caused extinction of wild species. Growing human populations placed more demands on wildlife and natural ecosystems. Hunters and traders killed animals, and threatened many others.

4. Number and proportion of total and threatened mammals, birds, amphibians, turtles, chondrichthyan fishes, conifers and cycads occurring in terrestrial, freshwater and marine ecological systems are known. Absolute number of threatened species known from marine systems is low due to recording biases towards terrestrial and freshwater taxa. Only 187 threatened marine as opposed to 4,427 terrestrial species are recorded. A Total of 1,388 freshwater species are listed as threatened.

5. Man-animal conflict is the most important issue that threatens wildlife in India. India is home for 60% of the world's tigers, 65% of its elephants, 80% of the Asian rhinos and 100% of Asian lions and 131 of the world's mega-fauna.

6. Conservation decisions should base on historical, evolutionary, community, and species approaches blended with reality .Sometimes populations go through a natural crash in numbers and then recover, but without the genetic variability, they once had. This is called genetic bottleneck because small populations are rarely complete representative of the original genetic composition of population.

7. In 1972, India enacted the Wildlife Protection Act and Project Tiger to safeguard crucial habitat. The federal protections were promulgated in the 1980s.In 1963; the World Conservation Union passed a resolution calling for an international convention on regulations on export, transit, and import of rare, or threatened wildlife species, their skins and trophies. Twenty one countries signed the Convention on International Trade in Endangered Species of Wild Fauna and Flora (CITES).

8. The priority of conservation must be on in-situ conservation, where protection is accorded not only to species, but to habitats, ecosystems, biodiversity and wilderness A network of 668 Protected Areas (PAs) has been established, extending over 1, 61,221.57 sq. kms (4.9% of total geographic area), comprising 102 National Parks, 515 Wildlife Sanctuaries, 47 Conservation Reserves and 4 Community Reserves, 39 Tiger Reserves, and 28 Elephant Reserves. UNESCO has declared 5 World Heritage

Sites. Four categories of the Protected Areas are National Parks, Sanctuaries, Conservation Reserves and Wildlife tourism has averaged 15% growth in India with increase in visitors to protect areas.

9. Sunderban mangrove is the largest single mangrove chunk in the world and is listed in the UNESCO world heritage list. It covers an area of about 10,000 sq. km, of which Indian Sundarban shares 4482 sq. km of which 59% are found along the east coast (Bay of Bengal), 23 % on the west coast (Arabian Sea) and remaining 18 %on the Bay Islands (A & N Islands in Bay of Bengal) and the rest is in Bangladesh.

Review Questions

Short Answer Questions

1. Define wildlife.
2. Define conservation.
3. Name the ecozone to which India belongs.
4. What is the full form of NBWL?
5. Define sanctuary.
6. Define national park
7. Mention the difference between national park and sanctuary.
8. What do you understand by "crisis of extinction"?
9. What do you understand by in-situ conservation?
10. What do you understand by ex-situ conservation?
11. Define extinction.
12. What do you understand by "Minimum viable Population"?
13. Give the full form of CITES.
14. Give the full form of IUCN.
15. Define biodiversity.
16. Define endangered animal.
17. What do you understand by restoration ecology?
18. What do you understand by captive breeding programme?
19. What do you understand by ecotourism?
20. Define mangrove ecosystem.

Long Answer Questions

1. Discuss various issues for Asian elephant conservation.
2. Describe the conservation initiative of Indian rhinoceros.
3. State the importance of Suderban.
4. Write a note about the conservation of biodiversity.
5. Write a note on any conservation project.

Glossary

Abdomen: The hind portion of an animal's body, containing the viscera excluding the heart and lungs.

Abducens nerve: Cranial nerve VI which enervates the lateral rectus muscle of eye and rotates the eyeball laterally.

Abduction: In tetrapod locomotion, rotation of a limb upward in a vertical plane.

Abomasum: The fourth chamber of ruminant digestive tract.

Abscission: Shedding a part of an organism from the rest.

Accidental: Species that do not occur in a region normally.

Acclimation: Experiences of organism in response to changing environmental conditions to tolerate the new environmental conditions.

Acetabulum: Socket in pelvis for head of the femur, normally at the junction of pubis, ischium and ilium.

Acoelous: Centra flat on both ends, neither procoelous nor opisthocoelous.

Acoustic impedance: It is a measure of resistance of an interface to propagation of sound from one medium to next.

Acrocoracoid process: The distinctive hook-shaped process at the proximal end of coracoid that forms a part of the triosseal canal in neornithine birds. Actually, the process is not hook-shaped in ducks.

Acrodont: Teeth when attached to edge of jaw bone without sockets.

Acromion process: Outer end of spine of scapula that forms the outer angle of shoulder and articulates with clavicle.

Action potential: A reversal of the electrical potential in plasma membrane of a neuron that occurs when a nerve cell is stimulated.

Active transport: Movement of ions or other substances across a membrane towards increasing concentration.

Adaptation: Characteristic that enhances fitness by helping the organism to survive and reproduce.

Adaptive function: A mathematical expression that takes into account the fitnesses of a phenotype in each of several different environments to produce a measurement of the general fitness of the phenotype in a varied environment.

Adaptive radiation: Development of a variety of species from a single ancestral form which occurs when a new habitat becomes available to a population.

Adduction: In tetrapod locomotion, rotation of a limb downward in a vertical plane.

Adductor fossa: The opening in the palate which in life contained the adductor muscles.

Adductor muscles: All important jaw adductors are muscles that move the jaws together.

Aegithognathous: Type of avian palate with vomer broad and truncate anteriorly. Maxillopalatines do not join but touch basisphenoidal rostrum as seen in Passeriformes.

Aerial behavior: Behavior exhibited by dolphins and whales in which they come out above the surface of the water.

After-shaft: A double feather that grows from the shaft of a body feather.

Air sac: Thin-walled structures unique to the respiratory system of birds..

Airfoil: A structure that creates lift due to differential airflow over that occurs over its top and bottom surfaces.

Alisphenoid: Epipterygoid bone in mammals.

Allantois: One of the membranes of amniotic egg which provides a surface for gas exchange and waste removal.

Allele: One of multiple alternate gene forms at a chromosomal location.

Altricial: Born in an early developmental state requiring more parental care and protection.

Alula: A set of feathers on the leading edge of a bird's wing located close to the base of primary feathers. Small digit (thumb) which emerges from proximal base of carpometacarpus in birds.

Alveolus: A small cavity, sac, like the tiny cavities within the lungs, or the depression in which a tooth sits.

Amnion: Membrane that encloses an embryo of higher vertebrates.

Amniote egg: An egg with compartmentalized sacs that allowed vertebrates to reproduce on land.

Amniote: A group of tetrapod vertebrates that lay eggs that are specially adapted to survive in a terrestrial environment.

Amphibians: Class of vertebrates which lay their eggs in water but live on land as adults following a juvenile stage when they live in water and breathe through gills.

Amphicoelous: Vertebral centra concave at both ends.

Amphisbaenian: Reptile that has a short tail, and scales arranged in rings and well-adapted to burrowing.

Amphistylic: Jaw suspension in which jaw is suspended both by hyomandibula and by a direct connection between the jaw and braincase.

Amplexus: A mating position in which the male holds the female with its front legs and fertilization usually takes place outside the female's body.

Ampulla: Structure resembling a jug, such as ampulla of labyrinth in inner ear.

Anacanthous: Absence of dorsal fin spines in fish.

Anaerobic: Organism that is not dependent on oxygen for respiration.

Analogous structures: Body parts that serve same function in different organisms but differ in structure and embryological development.

Anastomosis: A branching and interconnecting network of tubes that branch and reconnect to form a plexus.

Anisodactyl: In birds, the basic digital configuration in which digit 1 (toe or hallux) points posteriorly.

Ankylosis: Ontogenetic fusion that occurs after formation of bones.

Annulus: Applied to any ring or ring-like structure.

Anterior trochanter: Same as the lesser trochanter in femur and same as illiofibularis tubercle in fibula.

Antimere: Opposite member of a paired structure.

Antitrochanter: A tuberosity contained in acetabulum.

Antler: Bony structure that grows on the head of a deer.

Antorbital fenestra: A hole in skull just in front of orbit.

Antorbital fossa: A depression in skull anterior to orbit.

Antrum: A cavity.

Aorta: Artery that carries blood from the left ventricle for distribution throughout tissues of body.

Apocrine glands: Sweat glands located primarily in armpits and groin area.

Apical: In mammalian dentition, toward the crown.

Aplesodic: Of a fin, the condition in which basals and radials do not reach to distal margin of fin.

Apocrine glands: Sweat glands associated with hair cells which secrete sweat.

Apomorph: Structural feature that arises in an evolving lineage that is dissimilar from ancestral line.

Aponeurosis: A sheet-like tendinous expansion, serve to connect a muscle with parts that it moves.

Aposematic: Characteristics that act as warnings to other animals and signal that an animal has defenses.

Appendicular skeleton: Bones of appendages and of pelvic and pectoral girdles that join appendages to the rest of skeleton.

Appendix: Blind sac at the end of large intestine; a vestigial organ in humans.

Apteria: Areas on skin of embryonic bird which do not develop feather primordia.

Arboreal: Pertaining to tree-dwelling animals.

Archenteron: Internal body cavity formed by gastrulation.

Arcocentrum: In elasmobranchs, the cartilaginous arch and its base in vertebrae.

Arterioles: The smallest arteries usually branch into a capillary bed.

Articulation: Condition in which bones are contacting each other.

Aspidospondyly: Condition in which all vertebral elements remain as separate units.

Aspondyly: Condition of having no vertebral centra.

Astragalar foramen: Opening of a canal on proximal surface of astragalus through which a nerve and vessels pass in primitive mammals.

Astragalus: One of the two proximal tarsals. It is the more medial of two and usually articulates with tibia.

Atlas: The first cervical vertebra of tetrapods, which articulates with skull via condyles which permit the skull to move dorsoventrally.

Atmosphere: Envelope of gases that surrounds the earth consisting largely of nitrogen (78%) and oxygen (21%).

Auditory bulla: The inflated-looking covering of middle ear and floor of skull in that region.

Auditory meatus external: A passage leading from the environment to tympanic membrane, often shaped to gather and concentrate sound from a particular direction.

Auditory meatus internal: In mammals, the common foramen for VIIIth and VIIth cranial nerves.An opening on posterior surface of petrous portion of temporal bone through which auditory and facial nerves pass.

Auditory ossicle: A small bone used to conduct sound energy. The term refers to the mammalian complement of malleus, incus and stapes. It applies equally to columella of many other tetrapods.

Autapomorphy: A character which is unique to a particular taxon.

Autodiastyly: A type of jaw suspension in which palatoquadrate is suspended from two articulations with the braincase.

Autonomic system: Portion of peripheral nervous system that stimulates smooth muscle, cardiac muscle, and glands and consists of parasympathetic and sympathetic systems.

Autopalatine: An endochondral bone consisting of anterior portion of palatoquadrate.

Autopodium: The manus or pes, including digits, metacarpals or metatarsals.

Autostylic: A form of jaw suspension in which upper jaw articulates or is fused with chondrocranium, lower jaw forms from mandibular cartilage, and jaw remains unsupported by hyomandibula.

Autotroph: An organism that acquires energy from their environment. Organisms that synthesize their nutrients and obtain their energy from inorganic materials.

Aves: Commonly known as 'birds', with features like feathers, endothermy, and amiotic eggs.

Axial: Toward an imaginary axis running antero-posteriorly through middle of organism or structure.

Axilla: Arm pit.

Axillary foramen: In some placoderms, an opening in anterior ventrolateral plate which allowed nerves and blood vessels to communicate with pectoral appendages.

Axis: The second cervical vertebra of terrestrial vertebrates.

Balanced Polymorphisim: Maintenance of more than one allele in a population .

Baleen: In mysticete whales, a keratinous substance derived from mucous membrane which forms sheets, plates.

Barbel: A long tubercle that serves as a sensory appendage and is attached to an animal's mouthparts.

Barbicels: Small hook-like structures on barbules that link adjoining barbules to form the rigid structure of feather vane.

Barbs: Structures that branch from main shaft of a feather and form feather's vanes.

Basal: Relating to the base.

Basement membrane: The basal layer of an epithelial wall.

Basibranchial: The most ventral of gill arch elements. In sharks, the basibranchial elements are fused in a triangular structure referred to as basibranchial copula.

Basihyal: The basibranchial of hyoid arch.

Basilar membrane: The outer membrane of cochlea facing the scala tympani. It vibrates in response to compression waves caused by middle ear ossicles. In mammals, the basilar membrane supports the Organ of Corti.

Basioccipital: One of the bones of occiput, the part of skull which articulates with spine. Basioccipital is located ventral to foramen magnum.

Basipterygoid process: Processes of the basisphenoid which act to join the braincase to palate.

Basisphenoid: It forms floor of braincase anterior to basioccipital. Ventrally it is covered by a dermal bone, the parasphenoid.

Beak: Narrow, protruding jaws that usually do not contain teeth.

Bilateral symmetry: Symmetry in which an organism's body possesses two equal halves that are symmetrical when compared on either side of a midline.

Bill: Bird's jaws which consist of bone and a hornlike outer layer of keratin.

Binocular: A type of vision that arises from the ability of an animal to view an object with both eyes at the same time.

Biogeography: Geographic distribution of organisms throughout the landscape.

Bioluminescent: Refers to organisms that emit light under certain conditions.

Biomass: The weight of living material in some unit such as an organism, population, or community, often cited as weight per unit area.

Biota: Living forms of an ecosystem or habitat.

Bladder: A hollow, distensible organ with muscular walls that stores urine.

Blastocoel: Fluid-filled cavity at the center of a blastula.

Blastocyst: Developmental stage of fertilized ovum formed from morula and consists of an inner cell mass, an internal cavity, and an outer layer of cells.

Blastopore: Entrance to archenteron where gastrulation begins.

Blastula : A ball of cells surrounding a fluid-filled cavity.

Blowhole: A hole on top of a cetacean's head through which air is inhaled and exhaled.

Bone: Rigid structural tissue primarily composed of apatite with varying amounts of structural protein.

Boss: A raised, thickened, normally round area of a dermal bone.

Boundary layer: A layer of slow-moving water or air that lies just above the surface of an object.

Bow riding: Behavior of cetaceans in which they swim the crests of ocean waves.

Brachiopatagium: The main flight membrane of pterosaurs.

Brachydont: Cheek teeth having low crowns.

Bradydont: Having slow tooth replacement.

Brain stem: Portion of brain that is continuous with spinal cord and consists of medulla oblongata and pons of hindbrain and midbrain.

Brain: The most highly developed portion of central nervous system.

Branchial Arch: Internal gills, the primitively, filter-feeding structures which secondarily developed as respiratory organs.

Branchiopercular: In fish the last branchiostegal ray, conceived as a continuation of opercular series.

Branchiosaur: Generic term for an aquatic, larval tetrapod, typically a temnospondyl.

Breeding System: Breeding behaviors characteristic of a population and the ways in which members of population adapt to these breeding behaviors.

Brevis shelf: A ridge running along the inside surface of ilium behind acetabulum.

Bristles: Long, stiff feathers often found near a bird's mouth or eyes.

Bronchi: Tubes that carry air from trachea to lungs.

Bronchioles: Small tubes in lungs that are formed by branching of bronchi.

Brood parasite: A bird that lays its eggs in the nest of another bird.

Brood parasitism: reproduction involving the laying of eggs in the nests of other birds. Eggs are left under the parantal care of host parents.

Brood patch: An area that develops on lower abdomen of birds in which feathers drop off and skin thickens.

Brood reduction: Strategy in which a female bird produces more eggs than she would be capable of raising.

Brooding: Behavior in birds in which the parents continue to warm nestlings that are unable to maintain their own body temperatures.

Buccal: Of dentition, the "outside" of the teeth, toward the cheeks.

Bucco-hypophyseal canal: A persistent remnant of Rathke's pouch, a short canal which links the sella turcica with the mouth.

Bulbourethral glands: Glands that secrete a mucus-like substance that is added to sperm for lubrication during intercourse.

Bulla: A lesion or structure which resembles a bubble.

Bunodont: Dentition consisting of low, rounded cusps characteristic of unspecialized omnivores.

Caching: Storage of food for later eating during times of limited food availability.

Caecilian: A group of amphibians that have long bodies, no limbs, and virtually no tail.

Calamus: Hollow, proximal portion of feather shaft that attaches the feather to skin.

Calcareous: Structures that contain calcium such as shells, exoskeletons, and bones and function to support or protect an animal.

Calcified cartilage: Cartilage containing deposited calcium salts; found in vertebrae of **cartilaginous fish.**

Camber: Curvature of a wing.

Call matching: A vocalization in birds characterized by the male and female of a pair duplicating the other's flight call.

Cambrian: Geologic period at the beginning of the Paleozoic Era and is marked by proliferation of animals with hard, preservable parts like brachiopods, trilobites.

Camouflage: Coloration that help an animal to appear to blend with its surroundings.

Campodactyly: When a muscle is improperly attached to bones in the little finger causing the finger to be permanently bent.

Canine flange: An expanded and/or reinforced section of maxilla in which the canines are rooted.

Canine tooth: A sharp tooth positioned near the front of the jaws present in mammals that has a single point for peircing and holding onto food. Of dentition, a single pair of elongate, pointed, recurved teeth in anterior jaw. Located between the incisors and premolars, most frequently found in carnivores, many omnivores.

Cannon bone: Bone created by fusion of two adjacent metapodials.

Capillaries: Small, thin-walled blood vessels through which gaseous exchange occurs.

Capillary bed: Network of capillaries supplied by arterioles and drained by venules.

Capitate: In mammal, one of distal carpals, essentially a proximal extension, also referred to as magnum.

Capitellum: Of the humerus, the rounded structure on which radius rotates.

Capitis lateralis: A superficial vein that drains the side of head.

Capitulum: Of ribs, the more ventral of articulations with vertebrae.

Carapace: A hard shell on the dorsal side of an animal's body.

Carbohydrates: Molecules composed of carbon, hydrogen, and oxygen that serve as energy sources and structural materials for cells.

Carboniferous Period: The penultimate period of Paleozoic Era. The Early Carboniferous is called the Mississippian. The North American Late Carboniferous is the Pennsylvanian.

Cardiac cycle: A heart beat consisting of atrial contraction and relaxation, ventricular contraction and relaxation and a short pause.

Cardiac muscle: Striated but branched muscle found in walls of heart.

Cardioid: Heart-shaped.

Cardiovascular system: The system consisting of heart and vessels that transport blood to and from the heart.

Carinate: Of birds, having a massively enlarged sternum to support flight muscles; generally any structure bearing a "keel" or sharp, longitudinal median ridge.

Carnasial tooth: A sharp, premolar tooth present in carnivores adapted for efficient tearning and slicing through the meat of their prey. In some carnivores, one or two of the cheek teeth may be carnassials, specialized blade-like cutting teeth.

Carnivore: Animal that consumes primarily the flesh of other animals.

Carotenoids: A class of organic pigments that absorb blue light and responsible for red, orange, and yellow hues.

Carotid artery: The major artery supplying the brain. Internal branch of carotid is of special significance in anatomy because it enters braincase. External carotid rapidly splits into numerous branches which supply mostly external structures.

Carpal: Relating to the carpus .Sometimes used to refer to all of the bones of the wrist and hand.

Carpometacarpus: Fused wrist plus hand unit

Carrying capacity: Maximum population size that can be supported by an environment.

Caruncle: A bright-colored area of skin on face, or neck of a bird.

Caste: A group of individuals belonging to same social group that share some specialized form or behavior.

Catabolic reactions: Reactions in which molecules are broken down for producing energy involving oxidation.

Caudad: Towards the tail, or posteriorly.

Caudofemoralis: A muscle which originates on tail and inserts on fourth trochanter of archosaurs and is active in retraction of limb during stance phase.

Cell body: The part of a neuron containing nucleus , most cytoplasm and organelles.

Cell cycle: Sequence of events from one division of a cell to next.

Cell theory: The theory states that all living things are composed of cell and cell is the fundamental unit of function

Cell: The smallest structural units of living organisms capable of functioning independently.

Cellular respiration: Transfer of energy from various sources to produce ATP in mitochondria of eukaryotes and cytoplasm of prokaryotes.

Cellulose: Carbohydrate composed of unbranched chains of glucose.

Cenomanian: The first age of Late Cretaceous.

Cenozoic Era: The present era. It has epochs: the Paleocene, Eocene, Oligocene, Miocene, Pliocene, Pleistocene and Holocene .

Central Nervous System: Part of nervous system that is made up of interneurons and exerts some control over the rest of nervous system.

Centrale: One of distal carpals or tarsals.

Centropostzygapophyseal lamina: Reinforcing ridge bone ridge in vertebrae (of sauropods) connecting the centrum with one of postzygapophyses.

Centrum: Central portion of vertebra, which in early vertebrates, was taken up by notochord. In derived forms, the centrum becomes ossified.

Cephalization: Concentration of sensory tissues in head.

Ceratal: Refer to more ventral of the two main elements of a gill arch, i.e. the ceratobranchials, the ceratohyal, and/or Meckel's cartilage.

Ceratohyal: In fishes, the ventral or ceratal main element of hyoid arch.

Ceratotrichia: Fin support structures primarily made of keratin, rather than bone. Ceratotrichia may replace fin radials, or simply act as a third level of support.

Cere: A raised and fleshy patch located at the base of upper mandible in birds.

Cerebellum: Part of brain concerned with motor coordination and body movement, posture and balance.

Cerebral cortex: Outer layer of gray matter in cerebrum consisting of neuronal cell bodies and dendrites in humans.

Cerebrum: Part of forebrain that includes cerebral cortex; the largest part of human brain.

Cervical: Relating to neck.

Cervix: Lower neck of uterus that opens into vagina.

Cetaceans: Marine mammals that includes toothed whales and toothless, filter-feeding whales.

Character displacement: Divergence of adaptations in two similar species in locations where the animals share habitat.

Chemotrophs: Organisms, usually bacteria that derive energy from inorganic reactions.

Chiasma: The site of exchange of chromosome segments between homologous chromosomes.

Chitin: A polysaccharide polymer that is found in invertebrate exoskeletons.

Chitin: A polysaccharide forming part of hard outer covering of insects.

Cholecystokinin: Hormone secreted in duodenum causing the gallbladder to release bile and pancreas to secrete lipase.

Chorion: Two-layered structure formed from trophoblast after implantation which secretes human chorionic gonadotropin.

Chronobiology: Biology that examines the rhythms of biological activity or phenomena. In birds, a non-vocal form of communication expressed by slapping together of upper and lower parts of the bill together.

Clitoris: A short shaft located where the labia minora meet.

Cloacal spur: A spur in boas and pythons that is a remnant of pelvic girdle used by male snake in courtship.

Clutch: A group of eggs produced by a female for a single breeding attempt.

Coelom: Body cavity derived from mesoderm between body wall and digestive system .

Coelomates: Animals that have a coelom.

Colostrum: Special milk consumed by young immediately after birth and for a short period thereafter.

Condyles: Rounded projections on a bone where two bones come together, one with condyle, other with a depression, to form a joint.

Competition: Competition arises when two or more individuals rely on the same limited resource.

Cones: Light receptors that operate in bright light, provide color vision and visual acuity.

Connective tissue: Animal tissue with cells embedded in a matrix, providing strength, storage, and flexibility.

Consumers: The higher trophic levels in a food chain consisting of primary consumers and secondary consumers.

Convergent evolution: Development of similar structures in distantly related organisms due to adaptation in similar environments.

Coronary arteries: Arteries supplying the heart's muscle fibers with nutrients and oxygen.

Corpus callosum: Tightly bundled nerve fibers connecting the right and left hemispheres of cerebrum.

Corpus luteum : Structure formed in the ovulated follicle in ovary; secretes progesterone and estrogen.

Courtship behavior: Behavioral sequences that precede mating.

Craniate: Craniates are a group of chordate animals that include hagfish, lampreys, and the jawed vertebrates

Cranium: The braincase composed of several bones.

Creche: A group of unrelated young birds gathered together for protection.

Cretaceous Period: The geologic period between Jurassic Period and Tertiary Period marked by a mass extinction with the reign of the nonavian dinosaurs.

Crop: An expandable pouch in esophagus of birds.

Crossopterygians: Type of lobe-finned fish with lungs ancestral to amphibians.

Cryptic: Pertaining to characteristics that serve to conceal an animal.

C-terminal: The region of a protein near the end with a free carboxyl group.

Dacriform: Shaped like a tear drop.

Dapingian age: The first age of Middle Ordovician .

Deciduous: Of mammalian dentition, relating to baby teeth.

Deme: A breeding population that occurs in nature and which consists of similar organisms that interbreeds more or less at random.

Dendrodont: Dentition marked by dentine with primary and secondary folds; secondary folds filled in with attachment bone.

Dental formula: A formula showing the number of upper and lower incisors, canines, premolars and molars, , the formula when written: 2/1, 0/0, 4/3-4, 3/3 indicate that the animal has, on each side, 2 upper incisors, 1 lower incisor, no canines, 4 upper premolars, 3 or 4 lower premolars, and 3 upper and 3 lower molars.

Dentary peduncle: In mammals, the stem which bears the dentary's articulation with squamosal. The process in which the dentary's contribution to the temporomandibular joint is built.

Denticle: A tooth-let, a small protuberance having the characteristic histology of teeth.

Denticulate: Characterized by having denticles.

Dentine: Dense, highly mineralized bone formed by mineralization of a collagen matrix by odontoblasts, mesenchyme-derived cells which form tubules in dentine as they travel through it.

Dermal bone: Bone which is formed directly, rather than pre-formed in cartilage.

Dermatocranium: This is composed of plates of dermal bone that cover the head and protect brain and gills. Six basic groups of dermal bones that make up the dermatocranium are the facial, orbital, temporal, vault, palatal and mandibular series.

Desmognathous: Of birds, a condition of palate characterized by vomers small or absent, maxillopalatines in contact in midline, pterygoids and palatines articulate with basisphenoidal rostrum.

Deuterostomes: All triploblasts characterized by development of blastopore into anus, rather than the mouth. Major Deuterostome taxa are Echinodermata and Chordata.

Developmental response: Physiological and morphological characteristics an organism develops in response to prolonged exposure to environmental conditions.

Devonian: Period of geologic time from 410 - 360 million years before the present.

Diapause: Temporary interruption in development of insect eggs or larvae associated with a dormant period.

Diaphragm: A muscular partition analogous to pulmonary fold, formed from a dorsal process of coelom which is invaded by migrating cervical mesenchymal cells. Mesenchyme differentiates into neuromuscular tissue which creates a muscular diaphragm bringing ventilation under direct muscular control.

Diaphysis: The middle, shaft region of long bones which is fully ossified and contains medullar cavity.

Diapophysis: Upper, articulating process of transverse process of neural arch which bears secondary articulation of ribs.

Diarthrodial: Freely moveable joints like knee, as opposed to slightly moveable or immoveable.

Diastataxy: Having a wing lacking a secondary feather associated with fifth secondary covert.

Diastema: Wide space or gap between incisors and molariform teeth, found especially in rabbits and rodents

Diastole: The filling of ventricle of heart with blood.

Diencephalon: Part of forebrain consisting of thalamus and hypothalamus.

Differentiation: Developmental change from an immature to a mature form especially in a cell.

Diffusion: Movement of particles of gas, or liquid from regions of high to low concentration. Spontaneous movement of particles from an area of higher concentration to an area of lower concentration.

Digestion: Hydrolysis of complex nutrient compounds into their building-block units.

Digitigrade: A style of locomotion in which main weight-bearing surfaces are the digits.

Dilambdodont: Having upper molars similar to zalambdodont teeth except that the ectoloph is W-shaped. The metacone and paracone are at the base of 'W.' Crests run from these cones to buccal stylar cusps and form the arms of the 'W.'

Dimorphism: Occurrence of two forms of individuals within a population.

Dioecious: Possessing either male, or female organs.

Diphycercal: Of fish tails, tail which has symmetrical top and bottom and usually comes to a point. The best example is probably the coelacanth, *Latimeria*.

Diphyodont: Having two sets of teeth during a lifetime as in human.

Diploblastic: Having body parts derived from two layers during embryonic development.

Disarticulated: Of a fossil, a condition in which bones are found separated or not contacting each other as they would in life.

Distal carpals: Small wrist bones between carpals and metacarpals.

Distal: Further away. Thus, the distal femur is the part that participates in knee.

Divergent evolution: Divergence of a single species into two or more descendant species.

Dominant: Occupying a high position in social hierarchy.

Dorsal: Pertaining to the back.

Dorsum sellae: The dorsum sellae forms the posterior bone margin of the pituitary fossa, and is also the most dorsal point of the sphenoid bone.

Double Pump: In fish respiration, mechanics of water movement across gills involving the combined pumping action of both the oral and opercular cavities.

Down feather: Feathers with a rachis shorter than the longest barb which makes excellent insulation but does not repel water.

Exocrine glands: Sweat glands under sympathetic nervous system control, primarily involved in thermoregulation.

Ectoderm: Outer layer of cells in embryonic development which gives rise to skin, brain, and nervous system.

Ectotherms: Animals with variable body temperature.

Ectothermy: Controlling body temperature through external means, like exposure to sunlight

Effector: In a closed system, the element that initiates an action in response to signal from a sensor.

Ejaculatory duct: In males, a short duct that connects the vas deferens from each testis to urethra.

Embolomerous: A condition in which intercentra and pleurocentra are of roughly the same size.

Embrasure: Space between two adjacent teeth.

Embryo: Term applied to the zygote after the beginning of mitosis that produces a multicellular structure.

Emigration: Movement of individuals out of a population.

Emminence: A raised section of a surface like a low ridge or tubercle.

Emphysema: Lung disease characterized by shortness of breath often associated with smoking.

Enameloid: A form of bone with greater density and mineralization than dentine, found as a superficial layer over a dentine structure; mesodermal derivative laid down at outer surface of mesodermal papillae.

Endemic: Confined to a certain region.

Endochondral bone: Bone which is pre-formed as cartilage.

Endocranium: More or less same as neurocranium, chondrocranium.

Endocrine: Pertaining to ductless glands.

Endoderm: Inner layer of cells that gives rise to organs and tissues associated with digestion and respiration.

Endometrium: Inner lining of uterus.

Endoparasite: A parasite that lives within the body of its host.

Endoskeleton: An internal supporting skeleton with muscles on outside; in vertebrates consists of skull, vertebral column, ribs, and appendages.

Endotherm: Organism that obtains its body heat from within by oxidative metabolism. Animals with ability to maintain a constant body temperature over a range of environmental conditions.

Endothermy: Ability to generate and maintain internal body heat.

Entoconid: In mammalian dentition, a major cusp on lingual side of talonid in lower molars.

Entocuneiform: One of the distal tarsals.

Entoplastron: one of the dermal bones in the plastron of turtles.

Entopterygoid: A key hinge bone in the suspensorium of certain actinopterygian fishes

Environment: Surroundings of an organism, including the plants, animals, and microbes .

Epididymis: A long, convoluted duct on testis where sperm are stored.

Epiglottis: A flap of tissue that closes off the trachea during swallowing.

Epinephrine: A hormone produced by adrenal medulla that contributes to "flight or fight" response.

Epithelial tissue: Closely packed cells in either single, or multiple layers which cover both internal and external surfaces.

Epoch: Subdivision of a geological period.

Eras: One of the major divisions of geologic time scale.

Esophagus: Muscular tube extending between and connecting the pharynx to stomach.

Estivation: A temporary state of inactivity during a time that the animal is usually active

Evolution: Change over time. Biological evolution is defined as descent with modification.

Excretion: Removal of waste products of cellular metabolism from the body.

Exoskeleton: A hard, jointed, external covering that encloses muscles and organs.

Extinction: Elimination of all individuals in a group by natural and human-induced means.

Extirpated: Local extinction

Extra cellular digestion: Digestion that takes place within lumen of digestive system and the resulting nutrient molecules are transferred into blood.

Facet: Any small relatively flat articular surface.

Facial nerve: Cranial nerve VII. In fishes, this is the motor nerve associated with hyoid gill arch. In tetrapods, it is associated with neck musculature. In mammals, it is the principal motor nerve controlling facial muscles.

Falcate process: A sort of faux paroccipital process derived from the basioccipital and apparent mostly on the basicranium.

Falciform: The falciform bone is an ossification of the falciform ligament at the base of the thumb (manus I).

Fang pair: Pairs of fangs on the inner row of bones in both upper and lower jaw. These are not pairs of fangs, but rather a single fang and its replacement pit.

Femoral: Relating to femur.

Femur: The upper leg bone.

Fenestra cochleae: Same as the fenestra rotunda.

Fenestra exonaria: The anterior nares in non-choanates are the fenestra exonaria anterior.

Fenestra ovalis: Oval window in inner ear which communicates with stapes or columella, and with operculum, if present.

Fenestra rotunda: Round window in inner ear which relieves pressure in inner ear.

Fenestra vestibulae: Same as fenestra ovalis.

Fertilization: Fusion of two gametes to produce a zygote that develops into a new individual.

Fibrocartilage: Cartilage containing collagen fibers found in intervertebral disks and pubic symphysis.

Fibula: The smaller and more lateral of the two lower leg bones.

Fibulare: The calcaneum of mammals.

Fishbase: Modified scales.

Flexor muscles: Muscles which flex the digits of fore-, or hindlimb in tetrapods.

Follicles (ovary): Structures in ovary consisting of a developing egg surrounded by a layer of follicle cells.

Fontanel: An opening in the skull.

Foramen magnum: The large hole in occiput through which the spinal nerves enter the brain.

Foramen pseudorotundum: The foramen in the anterior lamina of the petrosal by which maxillary branch of trigeminal nerve exits the braincase.

Foramen pseudovale: It is a foramen in anterior lamina of petrosal by which the mandibular branch of trigeminal nerve exits the braincase.

Fossa trochanterica: A gap on femoral head between median tuberosity and greater trochanter.

Fossa: Used to indicate the depressed areas.

Fossorial: Adapted for burrowing and spending much time underground

Fourth trochanter: A muscle attachment point, which lies on the inside surface of femur in archosaurs.

Fovea: Central depression in retina associated with particularly acute vision.

Fovea dentis: A depression on the inner, ventral surface of atlas for articulation of axis.

Frontal: One of the principal paired midline bones of vertebrate skull lying posterior to nasals and anterior to parietals, typically at the level of orbits.

Gametes: Haploid reproductive cells.

Ganoid: A heavy form of enamel characteristic of scales of various early fish and extant Polypteriformes.

Gastric: Relating to the stomach.

Gastrocentrous: A vertebral structure in which the neural spine rests on pleurocentrum and the intercentra are reduced to spacer elements appearing as arches over dorsal surface of notochord between pleurocentra.

Gastrocnemius: A large superficial muscle of leg which originates on femur and inserts on heel to extend the foot.

Gastrolith: A stone deliberately swallowed by an organism and kept in gut to cut and crush bulk food items.

Gastrosteges: Single row of wide ventral scales in snakes which are erected during locomotion to increase friction.

Gastrulation: Process by which blastula is invaginated to form a double-walled hollow sphere of cells in microlecithal development.

Gene flow: Exchange of genetic traits between populations by movement of individuals, gametes, or spores.

Gene frequency: Proportion of a particular allele of a gene in gene pool of a population.

Gene: Refers to the part of DNA molecule that encodes a single enzyme, or structural protein.

Genetic drift: Change in allele frequency in a population.

Genetic feedback: Evolutionary response of a population to adaptations of competitors, predators, or prey.

Genetic variance: Variation in a phenotypic value within a population.

Geniculate: Bent abruptly at an angle, normally a right angle.

Genome: Total complement of DNA of a cell.

Genotype: Genetic characteristics that determine the structure and function of an organism

Geographic range: Total area occupied by a population.

Geological time: Span of time that has passed since the formation of earth.

Germ cells: Collective term for cells in reproductive organs of multicellular organisms that divide to produce gametes.

Gestation: Period of time between fertilization and birth of an individual.

Ghost lineage: A phylogenetic lineage that is inferred to exist with out fossil record.

Gill rakers: One of a series of variously shaped bony or cartilaginous projections on inner side of branchial arch.

Gill slits: Opening or clefts between gill arches in fish. Water taken in by the mouth passes through gill slits and bathes the gills.

Gizzard: A chamber of digestive tract specialized for grinding food.

Glenoid cavity: Shallow cavity of upper part of scapula by which humerus articulates with pectoral girdle.

Glial cells: Nonconducting cells that serve as support cells in nervous system and help to protect neurons.

Glomerulus: A tangle of capillaries that makes up part of nephron.

Glove finger: In mysticete whales, a membrane forming the inner part of fibrous plugs occluding the external auditory meatus.

Gonads: The testes, or ovaries.

Gross Production: Total energy or nutrients assimilated by an organism, a population, or an entire community.

Guard hairs: The longer, outer hair that protects the under fur and contains pigment that gives mammals their typical coloration

Hair cell: A mechanosensory cell characterized by microvilli.

Hallux: Hind toe.

Hamular: Hook-shaped.

Hepatic diverticulum: An evagination of embryonic gut which invades surrounding coelom and, specifically, the transverse septum.

Hepatic: Relating to liver.

Herbivore: An organism that consumes living plants or plant parts.

Heterocercal: Same as epicercal, according to some sources, or referring to condition in which caudal fin is asymmetric as either epicercal, or hypocercal.

Heterodactyl: A specialized digital configuration in birds in which both digits 1 and 2 are reversed.

Heterodactyly: Arrangement of toes in which the inner front toe is turned backward such that two toes point forward and two backward.

Heterodont: Bearing teeth of more than one sort as seen in humans.

Heterogeneity: Variety of qualities found in an environment, or a population

Heterothermy: Metabolic temperature regulation, but regulation is variable.

Hibernation: State of winter dormancy associated with lowered body temperature and metabolism.

Histogenesis: Differentiation of tissue types during development.

Histology: Study of structure of tissues.

Holarctic: The boigeographic region including the northern parts of the Old and New Worlds, and that comprises the Nearctic and Palearctic regions.

Holocephalic: Of ribs, having one head with a single articulation with vertebrae.

Holospondyly: A condition in which all vertebral elements are fused.

Holostylic: Jaw suspension in which palatoquadrate is fused to braincase.

Holotype: A single specimen designated as name-bearing type of a species or subspecies when it was established, or single specimen on which such a taxon was based when no type was specified.

Home range: Area that an animal uses in course of its daily activities.

Homeostasis: Ability to maintain a relatively constant internal environment.

Homeothermy: Maintaining a constant body temperature in a range of ambient temperatures.

Hominid: Primate that includes humans and all fossil forms leading to man only.

Homocercal: Of caudal fins, having lower and upper fin lobes of approximately the same size and shape.

Homodont: All teeth in jaws are of same shape, although they may be of slightly different proportions and different sizes.

Homologous structures: Body parts in different organisms that have similar bones and similar arrangements of muscles, blood vessels, and nerves and undergo similar embryological development, but do not necessarily serve the same function.

Homology: Relationship between structures in different organisms which are united by modification of same structure.

Homoplasy: A character shared by two taxa, but not by their common ancestor which arose independently in the two lineages and represents a convergence.

Humerus: Nature's way of connecting the forearm to shoulder.

Hydrophilic: Water-loving.

Hydrophobic: Water-fearing.

Hydroxyapatite: A type of apatite in which monovalent ion is hydroxyl ion.

Hyodont: Teeth not ankylosed to jaw.

Hyoid apparatus: A collective term for the bones of the tongue and associated connective tissues.

Hyoid arch: The second hypothetical ancestral gill arch from which hyomandibula and other elements of splanchnocranium are derived.

Hyomandibula: The upper main element of hyoid arch.

Hyostylic: Jaw suspension in which upper jaw loses any major direct connection with braincase and the upper and lower jaws are supported solely by hyomandibula.

Hypapophysis: A ventral spine or keel of a vertebra.

Hyperphalangy: Condition of having numerous additional phalanges.

Hypobranchial: In fishes, small gill arch elements which forms joints, the ceratobranchials and basibranchials between the main ventral elements, which are usually fused to the ventral aspect of pharynx.

Hypocaudal lobe: Part of caudal fin below notochord.

Hypocercal: Caudal fin structure in which notochord, or vertebral centra extend only into lower fin lobe.

Hypocleidium: An enlarged, flattened ventral area of furcula at distal end, formed by fused clavicles. The pectoralis muscle attaches in part to hypocleidium.

Hypocone: The main cone on talon of a mammalian upper molar.

Hypoconid: On the lower molars, the buccal main cusp of talonid in mammal.

Hypoconule: In mammal, this is same as metaconule, a small cusp on an upper molar lying near line between the protocone and metacone.

Hypoconulid: In mammal a cuspule usually located near distal edge of a lower molar.

Hypoglossal foramen: The foramen for exit of hypoglossal nerve, located near base of occipital condyle.

Hypostyle: A stylar cusp near buccomesial corner of an upper tribosphenic molar.

Hypotarsus: In birds, a ridge or process located on posterior side of tarsometatarsus, near proximal end.

Iliac peduncle: Two peduncles of ilium come down on either side of acetabulum to form sides of acetabulum and meet pubis and ischium which are referred to as pubic peduncle and ischiac peduncle.

Iliac: Relating to ilium.

Iliofemoralis: A muscle which originates on the ilium and inserts on the lateral femur.

Iliotrochantericus: A deep, relatively small muscle of pelvis with separate cranial & caudal origins on lateral face of ilium.

Ilium: Dorsal bone of the three bones forming pelvis which supports the sacrum.

Imago: The fourth, or adult stage in the life of certain insects.

Imbricating: Overlapping, used to indicate a complex pattern made up of small parts.

Incisive foramen: A foramen in palatal process of premaxilla just posterior to incisors.

Incisor: Cutting teeth of anterior jaw.

Incudomalleal joint: Joint between the malleus and incus in middle ear of mammals.

Incumbent: Lying, or resting on something else.

Induced drag: Drag is created incident to production of lift. .

Infrahemal: Autogenous spines distal to hemal arches.

Infundibulum: Any of various funnel-shaped bodily passages, openings, structures, specially stalk of the pituitary gland, or calyx of a kidney.

Inhibition: Suppression of a colonizing population by another that is already established.

Insertion: The point of attachment of a muscle in the more movable of two structures which it joins.

Instinct: Inborn or innate behavior.

Integument system: Skin and its derivatives which protect against invading microorganisms and prevent the loss of internal fluids.

Intercalarium: An occipital bone of actinopterygian fishes and one of the Weberian ossicles in Otophysi.

Intercentrum: The intercentrum is formed in center by a single myomere.

Intergrading: merging of characteristics of two populations where their ranges come into contact.

Internal compass: Mechanism that allows organisms to orient themselves so as to proceed in the proper direction during long-distance movements.

Interneural arch: In the earliest vertebrates, each embryonic segment of the spine contained two dorsal arches. One arch developed in the middle of the segment which is referred to as interneural arches.

Intracranial joint: In Sarcopterygii, a complete transverse division of braincase into anterior and posterior halves which runs between basisphenoid & basioccipital ventrally and immediately anterior to otic capsule dorsally.

Interstitial fluid: Fluid surrounding the cells in body tissues.

Interstitial: Being situated within a particular organ, or tissue.

Intraprezygapophyseal lamina: Reinforcing ridge bone ridge in vertebrae connecting the prezygapophyses.

Invasion: Sudden large movement of individuals into an area where they are generally uncommon.

Involucrum: Any bone sheath or a sheath formed as an extension of the tympanic into middle ear.

Isolating mechanism: An obstacle to interbreeding, either extrinsic, or intrinsic.

Jacobson's organ: The vomeronasal organ; a chemosensory organ located in roof of mouth.

Jugal bar: In birds, the bony process which supports the upper beak on moveable quadrate.

Jugular canal: In many sarcopterygians, on the side of otoccipital portion of braincase, there is a sort of dorsoventrally oriented bone bridge bearing the articulations for hyomandibula. Under this bridge, a tunnel passes from otic capsule, posteriorly, and opens onto otic shelf anteriorly. The jugular vein passes through this tunnel, and it is called the jugular canal.

Jugular foramen: A small hole where jugular vein exits the braincase which marks the exit of the Xth and XIth nerves, as well, normally located along the articulation between the opisthotic and exoccipitals.

Jurassic Period: Middle period of Mesozoic era between 185-135 million years ago.

Keratin: A tough protein that is a major component to hair and nails

Kinesis: Ability of parts to move or flex relative to each other.

Kingdoms: Five broad taxonomic categories into which organisms are grouped.

Labial: Relating to the lips.

Labyrinthodont: Teeth characterized by folded sheets of dentine, common in sarcopterygians and basal tetrapods.

Lachrymal: An alternate spelling of lacrimal, a dermal bone of facial series.

Lacrimal foramen: Opening of tear duct.

Lagena: Region of inner ear related to normal hearing.

Lagenar crest: A crista in which walls off the lagena from vestibular elements of the recessus scala vestibuli.

Lambdoid: Relating to articulation between parietals and occipitals.

Laminar: Layered.

Laminate: In layers.

Lappet: A projecting, flap-like structure.

Lateral centrale: One of the carpals.

Lateral commisure: In braincase of sarcopterygians and basal tetrapods, a flange of bone on lateral surface of the otoccipital region that folds over jugular vein.

Lateral condyle: The condyle at proximal end of tibia which is further from the midline of body. It contacts the lateral condyle of femur and has some contact with proximal head of fibula.

Lateral line: A sensory structure in aquatic vertebrates which appears as a system of lines on surface of the animal made up of mechanoreceptive and/or electroreceptive cells.

Lateral septum: In fishes, a sheet of tough, fibrous tissue extending straight laterally from spine and dividing the body into dorsal and ventral halves.

Latissimus dorsi: A group of muscles which originate on laterodorsal body wall and insert on humerus.

Laurasia: Northern part of the supercontinent of Pangaea composed of present-day North America, Europe, and Asia.

Lenticular process: The long ventral process of the incus which articulates with the stapes.

Lepidotrichia: Fin rays. Scale-like structures that form segments of soft rays in bony fishes.

Lepospondyly: A condition in which elements of vertebra are fused into a single piece and centra have hollow core allowing for a continuous notochord to pass through centra.

Lesser trochanter: A trochanter located below greater trochanter on lateral, or anterior face of femoral shaft.

Levator arcus palatini: The muscle which always functions to expand the buccal cavity laterally. It may originate on the anterior neurocranium, the dermosphenotic, the parasphenoid, or any other similar location.

Levator bulbi: A thin muscle in floor of orbit innervated by the 5th cranial nerve that causes the eye to bulge outward and to enlarge the buccal cavity. This muscle is present in anurans and urodeles and in a modified form in caecilians.

Lift: Force opposing gravity in flight and is generated by well-behaved laminar airflows.

Lingual: Relating to tongue.

Loop of henle: A U-shaped loop between proximal and distal tubules in kidney.

Lophodont: Dentition characterized by lophs as in ruminants.

Loxodont: An extreme form of lophodont, or selenodont dentition.

Lumbar: The ribless region body between the thorax and sacrum.

Lumbricales: Intrinsic muscles of hand and foot originating in palm and inserting on radial side of metacarpals for flexing the digits at metacarpo-phalangeal joint and extending digits at the interphalangeal joints for digits II-V.

Lunate: In mammals, one of carpal bones, normally articulating distally with magnum and proximally with radius.

Lungfish: A type of lobe-finned fish that breathe by a modified swim bladder and by gills.

Lungs: Sac-like structures where blood and air exchange oxygen and carbon dioxide; connected to the outside by a series of tubes and a small opening.

Luteal phase: Second half of ovarian cycle when corpus luteum is formed; occurs after ovulation.

Luteinizing hormone : Hormone secreted by anterior pituitary gland that stimulates the secretion of testosterone in men, and estrogen in women.

Lymph hearts: Contractile enlargements of vessels that pump lymph back into veins; found in fish, amphibians, and reptiles.

Lymph: Interstitial fluid in lymphatic system.

Lymphatic circulation: A secondary circulatory system that collects fluids from between the cells and returns it to the main circulatory system.

Lymphatic system: A network of glands and vessels that drain interstitial fluid from body tissues and return it to circulatory system.

Lymphocytes: White blood cells that arise in bone marrow and mediate immune response including T cells and B cells.

Maastrichtian: Last age of Cretaceous and of the Mesozoic Era, about 71.3-65.0 Mya.

Macrolecithal: Having a large amount of yolk, a pattern of embryonic development characteristic of eggs with considerable yolk.

Macromolecules: Large molecules made up of many small organic molecules.

Macrophages: White Blood Cells derived from monocytes that engulf antigens, stimulates the production of antibodies against the antigen.

Macula: Sac-like areas of vestibular apparatus in inner ear involved in perception of linear acceleration, orientation in a gravity field, and low-frequency, or high-volume sounds.

Magnum: One of the distal carpals, also referred to as the capitate.

Malleus: One of the bones comprising the middle ear of mammals.

Malpighian tubules: A set of long tubules that open into gut of insects.

Mandible: The lower jaw.

Mandibular apron: A thin expansion of the posteroventral section of the jaw in certain mammals.

Mandibular arch: A hypothetical first gill arch.

Mandibular fenestra: A fenestra or hole in the dentary.

Mandibular symphysis: Area where two halves of lower jaw articulate.

Manubrium: Anterior 3 segments of sternum; generally the anterior portion of sternum.

Marsupials: Pouched mammals.

Mass extinction: When accelerated extinction rates cause more than 50% of all living species to become extinct.

Mast cells: Cells that synthesize and release histamine during an allergic response.

Maxilla: Upper jaw.

Maxillary fenestra: A fenestra or hole which pierces the maxilla anterior to antorbital fenestra.

Maxillary nerve: A sensory nerve which communicates sensations from teeth, nasal sinuses, and parts of the face lateral to nose.

Mechanical advantage: The ratio of output force to input force.

Mechanoreceptor: Sensory cell, or organ responsive to mechanical forces, such as water currents, touch, sound.

Meckelian bone: The Meckelian cartilage is normally ossified only at its ends. In sarcopterygians, and in a few other taxa, it may ossify completely.

Meckelian cartilage: The cartilage formed in mandibular process of branchial arch I; the primitive lower jaw, derived from ceratal element of hypothetical mandibular arch.

Medial centrale: One of the distal carpals.

Medial: Referring to the midline of body.

Median eustachian foramen: A foramen on ventral surface of basioccipital and/or basisphenoid which communicates with pharynx via eustachian tubes.

Median tuberosity: A bulge on proximal femur which is an attachment site for ligaments binding the femoral head in acetabulum.

Median: In the middle.

Medulla: Central portion of certain organs; e.g., the medulla oblongata of brain and adrenal medulla which synthesizes epinephrine and norepinephrine.

Medulla: Used for non-cortical part of some organs such as the kidney and adrenal.

Meiosis: Cell division in which each of resulting gametes (in animals) receive a haploid set of chromosome.

Melanin: Pigment that gives the skin color and protects the underlying layers against damage by UV light; produced by melanocytes in the inner layer of epidermis.

Melanocytes: Cells in inner layer of epidermis producing melanin.

Melon organ: In odontocete whales, a pocket of connective tissue and fats located on concave "forehead" formed by maxilla between and anterior to orbits. The melon organ focuses ultrasound clicks used for echolocation.

Menstrual cycle: Recurring secretion of hormones and associated uterine tissue changes typically 28 days in length.

Menstruation: Process in which uterine endometrium breaks down and sheds cells resulting in bleeding.

Mental foramen: A pit, hole, or channel in mandible leading to Meckelian cartilage, or bone.

Mental groove: In snakes, on the underside of head, an anterior mental scale generally is followed by large, paired chin shields and smaller gular scales.

Mentomeckellian: A bone developed by ossification of Meckel's cartilage in region where two halves of lower jaw meets.

Mesentery: Epithelial cells supporting the digestive organs.

Mesoderm: Middle layer of cells in embryonic development which gives rise to muscles, bones, and structures associated with reproduction.

Mesokinetic joint: Joint along articulation between frontals and parietals.

Mesolecithal: Having a moderate amount of yolk.

Mesomere: Median series of bones supporting the pectoral fin of sarcopterygian fish.

Mesoplastron: One of the dermal bones in plastron of turtles.

Mesopodium: Bones of ankle and wrist that is tarsals and carpals, variously known as astragalus, calcaneum, fibulare, intermedium, tibiale, ulnare, intermedium, and radiale, and all centralia.

Mesothelial: Relating to the tissue lining body cavities.

Mesozoic Era: The period of geologic time beginning 245 million years ago and ending 65 million years ago; the age of the dinosaurs.

Metacarpus: The wrist bones between the distal carpals and the finger bones.

Metacone: The most posterior cusp of the trigon on an upper tribosphenic molar.

Metaconid: The most linguodistal cusp of trigonid on a lower molar.

Metacromion process: In mammals, a lateral, posterolateral, or dorsolateral process of coracoid portion of the scapulocoracoid, appearing as an outwardly-produced extension of spinous process nears its ventral end.

Metakinetic joint: Joint occurs where parietal and supraoccipitals meet. A joint between the braincase and skull roof.

Metalophid: In mammalian dentition, a cutting edge running generally along the distal side of trigonid on a lower lophodont molar.

Metamere: A body segment.

Metameric: Characterized by serial repetition of segments.

Metaphysis: The sub-terminal, actively growing section of long bones.

Metapodials: The metatarsals and/or metacarpals; sometimes includes the phalanges as well.

Metapterygoid: Portion of palatoquadrate which actually articulates with braincase to form the dorsal and/or basal articulation.

Metastylid: A small, presumably stylar, cusp on lingual edge of a lower tribosphenic molar, typically located just distal to metaconid and more or less in break between trigonid and talonid.

Metatarsus: Metatarsals are the long bones inside the foot between the ankle bones and the toes.

Metautostylic: A type of jaw suspension in which the lower jaw is supported by quadrate; that is, the basic quadrate-articular jaw articulation.

Metotic foramen: A foramen by which the metotic fissure may open to outside and refers to an undivided opening which serves as lateral opening of recessus scala tympani at exit for various cranial nerves.

Microevolution: Small-scale evolutionary event like the formation of a species from a preexisting one.

Microfilaments: Rods composed of actin that are found in cytoskeleton and are involved in cell division and movement.

Microlecithal: Having little or no yolk.

Microsquamose: Having very small scales.

Molariform teeth: All the teeth posterior to the canines

Monoecious: Having one sexual form.

Monophyletic group: A group of organisms descended from a common ancestor.

Monospondyly: Condition of a spine having one vertebral centrum per segment.

Monotremes: Egg-laying mammals.

Morphological convergence: Evolution of basically dissimilar structures to serve a common function.

Morula: Solid-ball stage of pre-implantation embryo.

Mosaic evolution: An evolutionary pattern where all features of an organism do not evolve at the same rate.

Mosaic: Having a mixture of features characteristic of more than one lineage.

Motor neurons: Neurons that receive signals from interneurons and transfer the signals to effector cells to produce a response.

Motor output: A response to the stimuli received by the nervous system. A signal is transmitted to organs that can convert the signals into action like movement, or a change in heart rate.

Motor pathways: Portion of peripheral nervous system that carries signals from central nervous system to muscles and glands.

Motor units: Consist of a motor neuron with a group of muscle fibers, form the units into which skeletal muscles are organized enabling muscles to contract on a graded basis.

Mucus: A thick, lubricating fluid produced by the mucous membranes lining the respiratory, digestive, urinary, and reproductive tracts and serves as a barrier against infection .

Multicellular: Organisms composed of multiple cells exhibiting some division of labor and specialization of cell structure and function.

Multinucleate: Cells having more than one nucleus.

Muscle fibers: Long, multinucleated cells found in skeletal muscles made up of myofibrils.

Myofibrils: Striated contractile microfilaments in skeletal muscle cells.

Myomere: A unit of segmented muscle separated from contiguous units by a connective tissue, particularly the embryonic muscles blocks formed by mesodermal myotome .

Myosin: Thick protein filaments in the centre sections of sarcomeres.

Negative feedback: Inhibition of synthesis of enzyme by the accumulation of products of enzyme-mediated reaction.

Nektonic: Habitually swimming

Nephron: A tubular structure consisting of a glomerulus and renal tubule.

Nerve cord: A dorsal tubular cord of nervous tissue above the notochord of a chordate.

Nerves: Bundles of neuronal processes enclosed in connective tissue that carry signals to and from the central nervous system.

Neural arch: Any of the dorsal arches of spine.

Neural tube: A tube of ectoderm in embryo that form the spinal cord.

Neurapophysis: Structure forming either side of neural arch.

Neuromast organ: Unit of mechanoreception consisting of hair cells and support cells embedded in a gelatinous copula.

Neuromuscular junction: The point where a motor neuron attaches to a muscle cell.

Neurons: Highly specialized cells that generate and transmit bioelectric impulses from one part of body to another.

Neurotransmitters: Chemicals released from the tip of an axon into synaptic cleft when a nerve impulse arrives, may stimulate or inhibit the next neuron.

Nostrils: Openings of the nose through which air enters.

Notochord: A continuous dorsal rod of cartilaginous material running from head to tail.

Oocyte: A cell that undergoes development into a female gamete.

Notochordal pit: Refers to the cavity which receives convex anterior end of notochord in braincase.

Nuchal gap: Gap between cranial and thoracic armor of placoderms.

Oblique ridge: In mammalian lower molars, a ridge that proceeds from hypoconid, in direction of metaconid or middle of trigonid.

Obturator foramen: A depression or hole formed by pubis and ischium.

Obturator nerve: The nerve supplies the adductors of thigh and contributes to nerve supply to skin of medial side of thigh.

Obturator process: A blade-like process extending from ventral shaft of ischium.

Occipital condyle: A rounded projection(s) from the back of the skull usually formed by some combination of basioccipital and exoccipitals.

Occipital region: The most posterior region of the braincase.

Oculomotor nerve: Cranial nerve III, which controls the extrinsic eye muscles except for the superior oblique and lateral rectus.

Odontode: The basic unit of teeth and scales.

Odontoid process: A process of the axis (2nd cervical vertebra) which projects anteriorly into the atlas.

Olecranon: The protector of the elbow occurs as a proximal extension of ulna.

Omasum: The third chamber of ruminant stomach.

Oogenesis: The process of production of ova.

Operculum: Dermal bone covering the gill slit in actinopterygian fish.

Opisthocoelous: When the individual vertebrae are convex anteriorly and concave posteriorly.

Opisthopubic: A pelvic girdle with the pubic bones pointing back toward the tail.

Opisthotic: The posterior of the two bones making up the otic capsule and is usually fused to the prootic.

Opposable: Capability of being placed against the remaining digits of a hand or foot; the ability of thumb to touch the tips of fingers on that hand.

Opsins: Molecules in cone cells that bind to pigments creating a complex that is sensitive to light of a given wavelength.

Optic nerve: Cranial nerve II, which enervates the intrinsic eye muscles and the retina.

Orbitosphenoid: Portion of sphenoid that is visible in wall of orbit.

Orbitostylic: Jaw suspension in which upper jaw is attached to braincase near orbits.

Orders: Taxonomic subcategories of classes.

Ordovician Period: Geologic period of the Paleozoic Era after the Cambrian Period between 500 and 435 million years ago.

Organ of Corti: Hair cell in mammalian cochlea of inner ear which sorts sound by frequency and transduces it into nerve impulses. .

Organ systems: Groups of organs that perform related functions.

Organelles: Subcellular structures (usually membrane-bound and unique to eukaryotes) that perform some function, e.g. mitochondrion, nucleus.

Organism: A.. individual composed of organ systems.

Organs: Structures made of two or more tissues which function as an integrated unit.

Osmoconformers: Marine organisms that have no system of osmoregulation and change the composition of their body fluids as the composition of the water changes.

Osmoregulation: Regulation of movement of water by osmosis into and out of cells for the maintenance of water balance within the body.

Osmosis: Diffusion of water molecules across a membrane in response to differences in solute concentration.

Ossification: The process by which embryonic cartilage is replaced by bone.

Osteoderm: Dermal bone lying over epidermis as armor.

Otic articulation: One of the possible dorsal articulations of palatoquadrate in which palatoquadrate articulates with a process on otic capsule.

Otic capsule: Bone structures containing the organs of the inner ear, principally the labyrinth and the maculae.

Otic: Pertaining to the ear

Otolith: Minute calcium carbonate stones associated with neuromast organs in labyrinth.

Oviducts: Tubes that connect the ovaries and the uterus, transport sperm to the ova, transport the fertilized ova to the uterus, and serve as site of fertilization.

Ovoviviparous: Young are born from eggs, but eggs are incubated internally.

Ovulation: The release of oocyte onto surface of ovary; occurs at the midpoint of ovarian cycle.

Ovum: The female gamete, egg.

Oxidation: The loss of electrons from outer shell of an atom often accompanied by transfer of a proton and thus involves the loss of a hydrogen ion.

Pacinian corpuscle: A neurosensory structure which looks like an onion on a string.

Paedomorphosis: Retention of juvenile characters into adult life.

Palate: The bone of the roof of mouth.

Palatoquadrate: Primitive upper jaw formed from hypothetical mandibular arch.

Paleocene epoch: The first epoch of Cenozoic.

Paleontology: Study of ancient life by analysis of fossils.

Paleozoic era: The first era associated with the presence of abundant multicellular life.

Pancreas :A gland that secretes digestive enzymes into small intestine and also secretes hormones insulin and glucagon into blood.

Pancreatic islets: Clusters of endocrine cells in pancreas that secrete insulin and glucagon

Pangaea: A super continent that existed at the end of Paleozoic era and consisted of all the earth's landmasses.

Parabasal process: A posteriorly directed process of pterygoid near midline for articulation with basipterygoid processes of basisphenoid.

Parabasisphenoid: A term for conjoined basisphenoid and parasphenoid.

Paracone: In symmetrodont upper molars, the central cusp of trigon.

Parafibula: An unusual third bone of the lower leg normally restricted to area of knee.

Paraglossale: In avian hyoid apparatus, the arrow head-shaped bone which forms base of tongue.

Parallel evolution: Development of similar characteristics in organisms that are not closely related due to adaptation to similar environments.

Paraphyletic: A group which does not contain all of the descendants of its last common ancestor.

Parapophysis: The articulating surface for the lower, capitular branch of the rib.

Paraquadratic foramen: Either a foramen in the quadratojugal, or a foramen for quadratojugal located on the quadrate.

Parasagital: Parallel to the long axis; parallel to midline.

Parasitism: A form of symbiosis in which one species benefits at the expense of another species.

Parasympathetic system: Subdivision of autonomic nervous system that reverses the effects of sympathetic nervous system.

Parasphenoid: A dermal bone of palate.

Parastyle: In mammalian cheek teeth, a cusp on anterior stylar shelf nears the paracone.

Paraxonic: Foot type in which the plane of symmetry passes between two digits of roughly equal size.

Parietal foramen: Unpaired midline foramen located on parietals.

Parietal lobe: Lobe of the cerebral cortex that lies at the top of brain; processes information about touch, taste, pressure, pain, and heat and cold.

Paroccipital process: A usually long cross-bar across back of skull formed by exoccipitals and/or opisthotic.

Parotic plate: Denticulated palatal bones which covered the anterior end of notochord in some sarcopterygian groups.

Patagium: A flight membrane as in bats, pterosaurs, and gliding animals.

Pavement dentition: Teeth formed as a broad, crushing surface.

Pectinate: Adjective form of pecten.

Pectoral girdle: Bones or cartilaginous structures in trunk which articulate with forelimb, including any parts which articulate with spinal column or skull.

Pectoral girdle: Bony arch by which the arms are attached to the rest of skeleton and composed of the clavicle and scapula.

Pectoralis: The muscle which is responsible for flight stroke in birds.

Pedomorphosis: The retention of juvenile characters into adult life.

Peduncle: Two ventral "stalks" of the ilium which contact the pubis and ischium and forms the posterior and anterior sides of the acetabulum.

Pelvic girdle: Bones or cartilaginous structures in trunk which articulate with hind limb, including any parts which articulate with the spinal column.

Pelvic girdle: Bony arch by which the legs are attached to the rest of the skeleton and composed of two hip bones.

Pelvis: Hollow cavity formed by two hip bones.

Pelycosaur: A paraphyletic group comprising all Synapsida except therapsids.

Penultimate: Next to last.

Perichondral: A type of bone which grows as a layer over the surface of a bone from the fibrous perichondral covering.

Perilymphatic duct: The duct in inner ear that directs the perilymph along lagena and into metotic cavity.

Period: Units of geological time scale that are the major subdivisions of eras.

Permian Period: The last geologic time period of the Paleozoic Era, noted for mass extinction in earth history when nearly 96% of species died out.

Permian: The last period of Paleozoic Era between the Carboniferous and Triassic.

Pharyngobranchial: The most ventral element of a branchial arch.

Pharynx: Anterior portion of gut from which internal gills develop as a series of paired pouches.

Phenotype: The observed properties or outward appearance of a trait.

Phospholipids: Asymmetrical lipid molecules with a hydrophilic head and a hydrophobic tail.

Phylum: The broadest taxonomic category within Kingdoms.

Phytoplankton: A floating layer of photosynthetic organisms including algae that form the base of aquatic food chain.

Pineal gland: A small gland located between cerebral hemispheres of brain and secretes melatonin.

Pinna: External ear.

Piriform: Pear-shaped.

Pituitary gland: A gland located at the base of the brain consisting of an anterior and a posterior lobe and produces many hormones.

Placenta: An organ produced from interlocking maternal and embryonic tissue in placental mammals which supplies nutrients to embryo and fetus and removes wastes.

Placental mammals: One of three groups of mammals that carry their young in mother's body for long periods during which the foetus is nourished by placenta.

Placoid: Type of scale frequently found in chondrichthyans.

Plasma cells: Cells produced from B cells to synthesize and release antibodies.

Pleistocene: The first geologic epoch of Quaternary period of the Cenozoic era that ended 10,000 years ago.

Pleura: A thin sheet of epithelium that covers the inside of thoracic cavity and the outer surface of lungs.

Pleurocentrum: The vertebral centrum associated with the neural arches.

Pleurocoel: An internal cavity in bone.

Pleuroperitoneal cavity: One of the two primitive partitions of coelom.

Plicidentine: Infoldings of dentine at base of teeth, forming striations.

Pons : The region that with the medulla oblongata makes up the hindbrain which controls heart rate, constriction and dilation of blood vessels, respiration, and digestion.

Population: A group of individuals of the same species living in the same area at the same time and sharing a common gene pool.

Portal system: An arrangement in which capillaries drain into a vein that opens into another capillary network.

Precambrian: During this time the atmosphere and oceans formed, life originated, eukaryotes and simple animals evolved.

Precocial: Born in a later developmental state.

Predation: Biological interactions that limit population growth and occur when organisms kill and consume other living organisms.

Prehensile movement: Ability to seize or grasp.

Primates: Mammals that includes prosimians, monkeys, apes, and humans and characterized by large brain, stereoscopic vision, and grasping hand.

Progesterone: One of the two female reproductive hormones secreted by ovaries.

Prolactin: A hormone produced by anterior pituitary at the end of pregnancy to activate milk production by the mammary glands.

Pulmonary artery: Artery that carries blood from the right ventricle of heart to lungs.

Pulmonary circuit: Loop of the circulatory system that carries blood to and from the lungs.

Pulmonary vein: Vein that carries oxygenated blood from lungs to left atrium of heart.

Punctuated equilibrium: A model that states that the evolutionary process is characterized by long periods with little or no change interspersed with short periods of rapid speciation.

Pyloric sphincter: Ring of muscle at the junction of stomach and small intestine that regulates the movement of food into small intestine.

Quaternary Period: Most recent geologic period of Cenozoic era.

Rachis: The central stem of a feather.

Rachitomous: A condition in which the various elements of vertebrae are divided into a large, wedge-shaped intercentrum plus two pleurocentra.

Radial: Direction away from imaginary axis running antero-posteriorly through middle of the organism, or structure.

Radially symmetrical: Refers to organisms with their body parts arranged around a central axis.

Radiation: The evolution of several closely related species from a single ancestor to occupy many differnent habitats or ecological roles.

Radula: A rough, raspy tongue used to grate food.

Range: Geographic area, or spatial distribution in which a species is normally found.

Rathke's pouch: An embryonic invagination of the stomodeal ectoderm which migrates dorsally to contact with the diencephalon.

Rattle: Loosely interlocking remnants of shed skin that are vibrated to make a rattling sound.

Ray-finned: Fish like trout, tuna, and bass that have thin, bony supports holding the fins away from the body and an internal swim bladder that changes the buoyancy of body.

Reabsorption: Return of water, sodium, amino acids, and sugar to the blood that were removed during filtration mainly in the proximal tubule of nephron.

Receptor: Protein to which select chemicals can bind. Opiate receptor in brain cells allows both the natural chemical as well as foreign chemicals to bind.

Renal tubule: Portion of nephron where urine is produced.

Replacement pit: A pit paired with a tooth alveolus and joined to it by a channel.

Reptiles: Vertebrates characterized by scales and amniotic eggs.

Resident: Nonmigratory species that completes its annual cycle within a fixed area.

Reticulate: A network of raised lines creating a web-like pattern.

Reticulated: Arranged in a network pattern.

Reticulum: The second chamber of the ruminant stomach.

Retina: Inner, light-sensitive layer of eye that includes the rods and cones.

Reverse migration: Phenomenon in which migrating individuals orient in direction opposite the normal one for the species at that season.

Rhachitomous: Vertebrae with a large, dorsal, crescentic intercentrum and a small, dorsal, paired pleurocentrum.

Rhomboid: Diamond-shape.

Rhynchokinetic: Ability to flex the upper jaw independently

Ricochetal: Jumping locomotion in zigzag pattern of rabbits and some kangaroos.

Riparian: Associated with rivers and streams.

Rods: Light receptors in primates' eyes that provide vision in dim light.

Rostrum: In early vertebrates, an anterior medial projection of hard tissue from the head, containing the mouth. Structure that project from the head of an animal, such as a snout.

Rugose: Roughened. Rugose bone normally serves as a surface for muscle, or tendon attachment.

Rumen: The first chamber of ruminant digestive tract, used for storage of ingested food and digestion of proteins and simple carbohydrates.

Saprophagous: Feeding on dead and decaying matter.

Sarcomeres: Functional units of skeletal muscle; consist of myosin and actin.

Scale: Soft, usually overlapping body covering in fish, snakes, lizards.

Scapula: One of the two main bones of the pectoral girdle.

Scapular spine: A ridge running along the external length of scapula, characteristic of therian mammals.

Scapulocoracoid notch: A notch on anterior of shoulder girdle, where scapula & coracoid meet.

Scapulocoracoid: A fusion of the scapula and coracoid.

Schizognathal: Skull in which palatines do not meet at the midline.

Sclerotic ring: A ring of bone or small bones inside the sclera.

Scrotum: A pouch of skin located outside the body cavity into which the testes descend for providing proper temperature for the testes.

Scute: Large, well-defined scale.

Sebaceous gland: Any oil-producing gland.

Secodont: In mammals, cheek teeth with a cutting action adapted for a carnivorous diet.

Secondary feathers: Flight feathers originating on the ulna.

Secondary Palate: Structural separation of mouth and upper throat from nasal passages.

Sella turcica: A cavity in basisphenoid.

Semicircular canal: Organs closely associated with the brain which contains otoliths and hair cells.

Seminal vesicles: Glands that contribute fructose to sperm.

Seminiferous tubules: Tubules on interior of testes where sperm are produced.

Semiplumes: Feathers that lie beneath the countour feathers and that lack interlocking barbules and barbicels.

Semi-Precocial: Young with characteristics of precocial young at hatch.

Sensory (afferent) pathways: Portion of peripheral nervous system that carries information from organs and tissues of body to central nervous system.

Sensory cortex: A region of brain associated with parietal lobe.

Sensory input: Stimuli that the nervous system receives from the external, or internal environment.

Sensory neurons: Neurons that carry signals from receptors and transmit information about the environment to processing centers in the brain and spinal cord.

Shaft: A feather's stiff central structure, to which the vanes are attached.

Shell: Protective outer covering .

Sigmoid process: A wavy crest externally on tympanic bone of cetaceans.

Sigmoid sinus: Portion of the lateral venous sinus bulging prominently into the mastoid cavity.

Silurian period: The third period of the Paleozoic Era.

Sink population: A breeding group that does not produce enough offspring to maintain itself in coming years without immigrants from other populations.

Skeletal muscle: Muscle that is generally attached to skeleton and causes body parts to move; consists of muscle fibers.

Skeletal system: System that supports the body, protects internal organs, and with the muscular system allows movement and locomotion.

Skin: The outermost layer protecting multicellular animals from loss or exchange of internal fluids and from invasion by foreign microorganisms

Sliding filament model: Model of muscular contraction in which the actin filaments in sarcomere slide past the myosin filaments, and shortens the sarcomere and muscle.

Small intestine: A coiled tube in abdominal cavity, the major site of digestion and absorption of nutrients and composed of duodenum, jejunum, and ileum.

Smooth muscle: Muscle that lacks striations and is found around circulatory system vessels and in the walls of stomach, intestines, and bladder.

Song repertoire: Number of different individual songs produced by a single bird.

Source population: A breeding group that produces enough offspring to be self-sustaining.

Southern Ocean: The continuous expanse of ocean between Antarctica and the southern tips of the other continents.

Species: One or more populations of interbreeding or potentially interbreeding organisms that is reproductively isolated in nature from all other organisms.

Spermatogenesis: Development of sperm cells from spermatocytes to mature sperm including meiosis.

Sphenoid: Alternate name for the basisphenoid, especially when fused with the alisphenoid and pterygoids.

Spinal cord: A cylinder of nerve tissue extending from brain stem to receives sensory information and sends output motor signals and with the brain, forms the central nervous system.

Spiracle: A small respiratory opening.

Spleen: An organ that produces lymphocytes and stores erythrocytes.

Subelliptical: Egg shaped, rounded at both ends but elongated and tapering toward the rounded ends.

Subfamily: A subset of a family that contains one, or more genera.

Subspecies: A geographical subset of a species showing discrete differences in morphology compared to other members of species.

Subtropical: Habitats and climates that is tropical in nature but found north, or south of tropics.

Supercillium: Eye-brow.

Supplemental plumage: A generation of feathers, additional to the basic and alternate plumages, found in birds that have more than two molts per year.

Synapomorphy: A new trait that arises in an evolving lineage that is shared between two, or more sister groups.

Talon: In mammalian dentition, the distolingual extension of an upper molar.

Tarsal: Series of bones in the ankle.

Tarsometatarsus: In birds, the fused unit comprising the tarsus and metatarsus.

Tarsus: The ankle including proximal, distal tarsal and the metatarsals.

Tartarian: The last age of the Permian.

Taxon: Group of organisms comprising a given taxonomic category

Taxonomy: The science of classification of organisms.

Tectorial membrane: A membrane in the mammalian cochlea.

Tectum: A roof-like structure, particularly the dorsal part of midbrain.

Tegmen tympani: Thin plate of bone forming the roof of epitympanic recess in middle ear.

Temporal lobe: Lobe of cerebral cortex that is responsible for processing auditory signals.

Tendons: Bundles of connective tissue that link muscle to bone.

Testosterone: Male sex hormone that stimulates sperm formation, and is responsible for secondary sex characteristics.

Thermoregulation: Regulation of body temperature.

Thoracic cavity: Chest cavity in which the heart and lungs are located.

Trophoblast: Outer layer of cells of a blastocyst that adhere to endometrium during implantation.

Unciform: Anything hook-like.

Uncinate process: A roughly triangular posterior process on the dorsal ribs of birds which overlaps the shaft of next most posterior rib.

Underfur: Short, dense, insulating hair located beneath the guard hair

Ungual: Terminal phalanx of a digit which bears a claw.

Urethra: A narrow tube that transports urine from the bladder to outside of body. In males, it also conducts sperm and semen to outside.

Urine: Fluid containing various wastes that is produced in kidney and excreted from bladder.

Uropatagium: Skin membrane extending between the hind legs and often enclosing the tail.

Urostyle: The small upturned posterior tip of vertebral column generally formed of fused vertebrae.

Utriculus: Sac-like areas of vestibular apparatus in inner ear involved in perception of linear acceleration, orientation in a gravity field, and low-frequency or high-volume sounds.

Uterus: Organ that houses and nourishes the developing embryo and foetus.

Vacuity: Opening in a bony plate.

Vagus foramen: Located near the base of paroccipital process, between exoccipitals and opisthotic, and contains one, or more foramina for cranial nerves IX, X and/or XI.

Vagus nerve: The Xth cranial nerve.

Veins: Thin-walled vessels that carry blood to the heart.

Velum: A respiratory pump found in basal craniates.

Ventricle: Lower chamber of the heart through which blood leaves the heart.

Venules: Small veins that connect a vein with capillaries.

Vertebrate: Any animal having a segmented vertebral column.

Vestibular apparatus: Functional apparatus of inner ear.

Vestigial structures: Nonfunctional remains of organs that were functional in ancestral species and may be functional in related species.

Villi: Finger-like projections of lining of small intestine that increase surface area for absorption.

Volant: Flying.

Vomer: A median bone lying in the floor of nasal cavity.

Vomeronasal organ: A chemosensory organ located in roof of mouth.

White blood cell: Component of the blood that serves in immune function.

X-chromosome: One of the sex chromosomes.

Yolk sac: A membrane-bound compartment in amniotic egg which contains stored food for developing embryo.

Z lines: Dense areas in myofibrils at the beginning of sarcomere. Actin filaments are anchored in the Z lines.

Zygopodium: Lower part of limb.

Zygomatic arch: The arch running under orbit and temporal fenestra in synapsids.

Zygomatic plate: Zygomatic process of the maxillary bone in the form of a thin plate.

Zygomatic process: A process of either the maxillary or squamosal bone that contributes in formation of zygomatic arch.

Zygote: A fertilized egg.

Index